Advances in Intelligent Systems and Computing

Volume 1246

AF806121

The series "Advances in Intelligent Systems and Computing" contains publications on theory, applications, and design methods of Intelligent Systems and Intelligent Computing. Virtually all disciplines such as engineering, natural sciences, computer and information science, ICT, economics, business, e-commerce, environment, healthcare, life science are covered. The list of topics spans all the areas of modern intelligent systems and computing such as: computational intelligence, soft computing including neural networks, fuzzy systems, evolutionary computing and the fusion of these paradigms, social intelligence, ambient intelligence, computational neuroscience, artificial life, virtual worlds and society, cognitive science and systems, Perception and Vision, DNA and immune based systems, self-organizing and adaptive systems, e-Learning and teaching, human-centered and human-centric computing, recommender systems, intelligent control, robotics and mechatronics including human-machine teaming, knowledge-based paradigms, learning paradigms, machine ethics, intelligent data analysis, knowledge management, intelligent agents, intelligent decision making and support, intelligent network security, trust management, interactive entertainment, Web intelligence and multimedia.

The publications within "Advances in Intelligent Systems and Computing" are primarily proceedings of important conferences, symposia and congresses. They cover significant recent developments in the field, both of a foundational and applicable character. An important characteristic feature of the series is the short publication time and world-wide distribution. This permits a rapid and broad dissemination of research results.

**** Indexing: The books of this series are submitted to ISI Proceedings, EI-Compendex, DBLP, SCOPUS, Google Scholar and Springerlink ****

More information about this series at http://www.springer.com/series/11156

Sergii Babichev · Volodymyr Lytvynenko ·
Waldemar Wójcik · Svetlana Vyshemyrskaya
Editors

Lecture Notes in Computational Intelligence and Decision Making

2020 International Scientific Conference "Intellectual Systems of Decision-making and Problems of Computational Intelligence"

Springer

Editors
Sergii Babichev
Department of Informatics
Jan Evangelista Purkyně University
in Ústí nad Labem
Ústí nad Labem, Czech Republic

Waldemar Wójcik
Institute of Electronics and Information
Lublin University of Technology
Lublin, Poland

Volodymyr Lytvynenko
Department of Informatics
and Computer Science
Kherson National Technical University
Kherson, Ukraine

Svetlana Vyshemyrskaya
Department of Informatics
and Computer Science
Kherson National Technical University
Kherson, Ukraine

ISSN 2194-5357 ISSN 2194-5365 (electronic)
Advances in Intelligent Systems and Computing
ISBN 978-3-030-54214-6 ISBN 978-3-030-54215-3 (eBook)
https://doi.org/10.1007/978-3-030-54215-3

This Springer imprint is published by the registered company Springer Nature Switzerland AG
The registered company address is: Gewerbestrasse 11, 6330 Cham, Switzerland

Preface

Collecting, analysis and processing information are the current directions of modern computer science. Many areas of current existence generate a wealth of information which should be stored in a structured manner, analyzed and processed appropriately in order to gain the knowledge concerning investigated process or object. Creating new modern information and computer technologies for data analysis and processing in various fields of data mining and machine learning create the conditions for increasing effectiveness of the information processing by both the decrease of time and the increase of accuracy of the data processing.

The International Scientific Conference "Intellectual Decision-Making Systems and Problems of Computational Intelligence" is a series of conferences performed in East Europe. They are very important for this geographic region since the topics of the conference cover the modern directions in the field of artificial and computational intelligence, data mining, machine learning and decision making. The aim of the conference is the reflection of the most recent developments in the fields of artificial and computational intelligence used for solving problems in a variety of areas of scientific researches related to data mining, machine learning and decision making.

The current ISDMCI'2020 Conference held in Zalizny Port, Kherson region, Ukraine, from May 25 to 29, 2020, was a continuation of the highly successful ISDMCI conference series started in 2006. For many years, ISDMCI has been attracting hundreds or even thousands of researchers and professionals working in the field of artificial intelligence and decision making. This volume consists of 46 carefully selected papers that are assigned to three thematic sections:

Section 1. **Analysis and Modeling of Complex Systems and Processes**:

- Methods and tools of system modeling under uncertainty
- Problems of identification of complex system models and processes
- Modeling of operated complex systems
- Modeling of various nature dynamic objects
- Time series forecasting and modeling
- Information technology in education.

Section 2. **Theoretical and Applied Aspects of Decision-Making Systems**:

- Decision-making methods
- Multicriterial models of decision making under uncertainty
- Expert systems of decision making
- Methods of artificial intelligence in decision-making systems
- Software and tools for synthesis of decision-making systems
- Applied systems of decision-making support.

Section 3. **Computational Intelligence and Inductive Modeling**:

- Inductive methods of complex systems modeling
- Computational linguistics
- Data mining
- Multiagent systems
- Neural networks and fuzzy systems
- Evolutionary algorithm and artificial immune systems
- Bayesian networks
- Hybrid systems and models
- Fractals and problems of synergetics
- Image recognition and cluster analysis.

We hope that the broad scope of topics related to the fields of artificial intelligence and decision making covered in this proceedings volume will help the reader to understand that the methods of data mining and machine learning have become an important element of modern computer science.

June 2020

Oleh Mashkov
Volodymyr Stepashko
Yuri Krak
Yuriy Bardachov
Volodymyr Lytvynenko

Organization

ISDMCI'2020 is organized by the Department of Informatics and Computer Science, Kherson National Technical University, Ukraine, in cooperation with:

Black Sea Scientific Research Society, Ukraine
IT STEP University, Ukraine
Jan Evangelista Purkyně University in Ústí nad Labem, Úst nad Labem, Czech Republic
Lublin University of Technology, Poland
Taras Shevchenko National University of Kyiv, Ukraine
V.M. Glushkov Institute of Cybernetics NASU Ukraine
International Centre for Information Technologies and Systems of the National Academy of Sciences of Ukraine, Ukraine

Program Committee

Chairman

Oleh Mashkov	State Ecological Academy of Postgraduate Education and Natural Resources Management of Ukraine, Kyiv, Ukraine

Vice-chairmen

Volodymyr Stepashko	International Centre for Information Technologies and Systems of the National Academy of Sciences of Ukraine, Kyiv, Ukraine
Yuri Krak	Taras Shevchenko National University of Kyiv, Ukraine

Members

Natalia Axak	Kharkiv National University of Radio Electronics, Ukraine
Tetiana Aksenova	Grenoble University, France
Mikhail Alexandrov	Autonomous University of Barcelona, Spain
Svitlana Antoshchuk	Odessa National Polytechnic University, Ukraine
Sergii Babichev	Jan Evangelista Purkyně University in Ústí nad Labem, Úst nad Labem, Czech Republic
Olexandr Barmak	Khmelnitsky National University, Zhytomyr, Ukraine
Vitor Basto-Fernandes	University Institute of Lisbon, Portugal
Juri Belikov	Tallinn University of Technology, Estonia
Andrii Berko	Lviv Polytechnic National University, Ukraine
Oleg Berezkiy	Ternopil National Economic University, Ukraine
Oleg Bisikalo	Vinnytsia National Technical University, Ukraine
Peter Bidyuk	National Technical University of Ukraine “Igor Sikorsky Kyiv Polytechnic Institute”, Ukraine
Oksana Bihun	Mathematics Department of the University of Colorado Boulder, Colorado Springs, USA
Yevgeniy Bodyanskiy	Kharkiv National University of Radio Electronics, Ukraine
Oleksandra Bulgakova	V.O. Sukhomlynsky Mykolaiv National University, Ukraine
Yevhen Burov	Lviv Polytechnic National University, Ukraine
Volodymyr Buriachok	Borys Grinchenko Kyiv University, Ukraine
Yurii Biurher	Zoral Labs, Kyiv, Ukraine
Zoran Cekerevac	“Union – Nikola Tesla” University, Serbia
Sergiu Cataranciuc	Moldova State University, Kishinev, Moldova Republic
Klara Dimitrova	Varna University of Management, Bulgaria
Mykola Dyvak	Ternopil National Economic University, Ukraine
Michael Emmerich	Leiden Institute of Advanced Computer Science, Leiden University, the Netherlands
Oleg Garasym	Volvo IT, Poland
Fedir Geche	Uzhhorod National University, Ukraine
Larysa Globa	National Technical University of Ukraine “Igor Sikorsky Kyiv Polytechnic Institute”, Ukraine
Sergiy Gnatyuk	National Aviation University, Kyiv, Ukraine
Vladimir Golovko	Brest State Technical University, Belarus
Oleksii Gorokhovatskyi	Simon Kuznets Kharkiv National University of Economics, Ukraine
Aleksandr Gozhyj	Petro Mohyla Black Sea National University, Ukraine

Volodymyr Mashtalir	Kharkiv National University of Radio Electronics, Ukraine
Jiři Škvor	Jan Evangelista Purkyně University in Ústí nad Labem, Úst nad Labem, Czech Republic
Sergii Olszewski	Taras Shevchenko National University of Kyiv, Ukraine
Opeyemi Olakitan	Cornell University, UK
Volodymyr Osypenko	Kyiv National University of Technologies and Design, Ukraine
Nataliya Pankratova	National Technical University of Ukraine “Igor Sikorsky Kyiv Polytechnic Institute”, Ukraine
Anatolii Pashko	Taras Shevchenko National University of Kyiv, Ukraine
Dmytro Peleshko	GeoGuard, Lviv, Ukraine
Iryna Perova	Kharkiv National University of Radio Electronics, Ukraine
Eduard Petlenkov	Tallinn University of Technology, Estonia
Michael Pokojovy	Karlsruher Institut für Technologie (KIT), Universität Konstanz, Mannheim Area, Germany
Taras Rak	IT STEP University, Lviv, Ukraine
Hanna Rudakova	Kherson National Technical University, Ukraine
Yuriy Rashkevych	Lviv Polytechnic National University, Ukraine
Yuriy Romanyshyn	Lviv Polytechnic National University, Ukraine
Anatoliy Sachenko	Ternopil National Economic University, Ukraine
Yuri Samokhvalov	Taras Shevchenko National University of Kyiv, Ukraine
Silakari Sanjay	Rajiv Gandhi Technical University, Madhya Pradesh, India
Andrii Safonyk	National University of Water and Environmental Engineering, Rivne, Ukraine
Natalia Savina	National University of Water and Environmental Engineering, Rivne, Ukraine
Antonina Savka	Openet, Dublin, Ireland
Galina Setlak	Rzeszow University of Technology, Poland
Natalya Shakhovska	Lviv Polytechnic National University, Ukraine
Manik Sharma	DAV University, India
Ihor Shelevytsky	Kryvyi Rih Institute of Economics, Ukraine
Volodimir Sherstyuk	Kherson National Technical University, Ukraine
Galyna Shcherbakova	Odessa National Polytechnic University, Ukraine
Juergen Sieck	Humboldt-Universität zu Berlin, Germany
Miki Sirola	Institute for Energy Technology, Norway
Andrzej Smolarz	Lublin University of Technology, Poland

Volodymyr Stepashko	International Centre for Information Technologies and Systems of the National Academy of Sciences of Ukraine
Petro Stetsyuk	Glushkov Institute of Cybernetic of NAS of Ukraine, Ukraine
Sergey Subbotin	Zaporizhia National Technical University, Ukraine
Vasyl Teslyuk	Lviv Polytechnic National University, Ukraine
Roman Tkachenko	Lviv Polytechnic National University, Ukraine
Vasyl Trysnyuk	Institute of Telecommunications and Global Information Space, Kyiv, Ukraine
Ivan Tsmots	Lviv Polytechnic National University, Ukraine
Oleksii Tyshchenko	Institute for Research and Applications of Fuzzy Modeling, CEIT Innovations, University of Ostrava, Czech Republic
Kristina Vassiljeva	Tallinn University of Technology, Estonia
Viktor Voloshyn	IT STEP University, Lviv, Ukraine
Olena Vynokurova	GeoGuard, Kharkiv, Ukraine
Victoria Vysotska	Lviv Polytechnic National University, Ukraine
Waldemar Wojcik	Lublin University of Technology, Poland
Mykhaylo Yatsymirskyy	Institute of Information Technology, Lodz University of Technology, Poland
Yurii Yaremchuk	Vinnytsia National Technical University, Ukraine
Iryna Yevseyeva	Newcastle University, London, England
Danuta Zakrzewska	Institute of Information Technology, Lodz University of Technology, Poland
Elena Zaitseva	Zilinska Univerzita v Ziline, Slovakia
Yuriy Zaychenko	National Technical University of Ukraine "Igor Sikorsky Kyiv Polytechnic Institute", Ukraine
Polina Zhernova	Kharkiv National University of Radio Electronics, Ukraine
Jan Zizka	Mendel University in Brno, Czech Republic
Viacheslav Zosimov	V.O. Sukhomlynsky Mykolaiv National University, Ukraine

Organization Committee

Chairman

Yuriy Bardachov	Kherson National Technical University, Ukraine

Vice-chairmen

Volodymyr Lytvynenko	Kherson National Technical University, Ukraine
Yuriy Rozov	Kherson National Technical University, Ukraine

Contents

Analysis and Modeling of Complex Systems and Processes

Technique of System Operator Determination Based on Acoustic Emission Method

Volodymyr Marasanov[1], Dmitry Stepanchikov[1(✉)], Artem Sharko[1], and Alexander Sharko[2]

[1] Kherson National Technical University, Kherson, Ukraine
volodymyr.marasanov@gmail.com, {dmitro_step75,sharko_artem}@ukr.net
[2] Kherson State Maritime Academy, Kherson, Ukraine
avsharko@ukr.net

Abstract. The results of computer quantum-chemical modeling of the operator structure of the process of occurrence of acoustic emission signals under load are presented. The regularities of changes in the oscillatory properties of acoustic emission signals with force field parameters are established. The existence of a correspondence between changes in spectral characteristics and the moments of occurrence of acoustic emission signals is proved. Using the GaussView program with the Hartree-Fock method in the 3-21G basis and graph theory, the parameters of the loaded anharmonic oscillator were calculated. Based on the studies, an algorithm for a multifactor model of information diagnostics of the mechanical properties of materials is proposed. A meaningful interpretation of the operator structure of dynamic processes and numerical methods for recovering information about the strength properties of materials from the characteristics of sources of initiation of acoustic emission signals is given.

Keywords: Operator · Information recovery · Acoustic emission · Material structure · Damping · Anharmonic oscillator · Graph theory

1 Introduction

Interest in the creation of materials with specified physical and mechanical properties has increased significantly over the past two decades with the advent of nanotechnology, when it becames possible to control the structure of materials at the level of molecular clusters, individual molecules, and even atoms. The solution of such problems is impossible without a detailed study of the dependence of the properties of materials on their micro-structure. To do this, mathematical models of information diagnostics are needed, which are based on the generation of acoustic and vibration signals during load of structures.

S. Babichev et al. (Eds.): ISDMCI 2020, AISC 1246, pp. 3–22, 2021.
https://doi.org/10.1007/978-3-030-54215-3_1

Identification of acoustic signals and determination of mechanical properties of materials in the source of initiation of acoustic signals are the most difficult in studying the behavior of materials under load [11,19,21]. The close connection between destruction processes and presence of acoustic emission effect (AE) create the conditions to determine the degree of approximation of material state to destruction by means of registration of AE signals.

If the information processes in materials at the macro level can be described on the basis of stress-strain diagrams, then the evolution of a defect structure that initiates the formation of AE signals can be described only at the atomic level. Using information on the diagnosis of materials with the help of AE phenomenon, it is possible to determine the state of the material under changing external conditions [4,5,18], measure the level of stresses and strains [7,15], detect developing defects [16,17], and study crack growth [12,20], determine structural changes caused by the influence of aggressive environments [8], establish the coordinates of defects and determine the degree of their danger [2,13]. The most difficult moment of studying the behavior of materials under load is the identification of AE signals and the establishment of the state and mechanical properties of the material in the source of initiation of AE signals.

Since the destruction of the material is determined by the state space of the object, uncertainty arises in the diagnosis of the material, which is caused by limited information. When studying the characteristics of structural defects in order to predict failure, it is necessary to have the most complete information about the source of AE, which can be obtained not only through measurements, but also through the integrated use of theoretical models of the occurrence of AE and the calculation of their parameters with the use of changes in the structure of materials under load in the calculation of experimental data.

The hypothesis of the given work was as follows. If the force action and the equation of motion of the atoms of the structure under loading are known, then by setting the initial conditions it is possible to perform a computational experiment, comparing the results of which with experimentally determined quantities, to obtain information on the structure parameters that cannot be measured.

The relevance of studying the mechanisms of AE signals occurrence is due to the need to determine the initial stage of origin of changes in the structure of materials initiating possible destruction of structures.

The recovery of information about the AE source during deformation of the structure of materials can be performed based on the Schrödinger equation:

$$\varDelta\varPsi + \frac{2m}{\hbar^2}(W-U)\varPsi = 0, \tag{1}$$

where $\hbar = h/(2\pi)$ is Dirac constant, $\varPsi$ is the wave function, W is the total energy of a particle, U is the potential function, h is Plank's constant, m is the particle mass, $\varDelta$ is Laplace operator.

The wave function $\varPsi$ is an eigenfunction of the operator of the transition of an electronic structure from atoms to a solid.

The Schrödinger equation for atoms containing more than one electron cannot be solved analytically. Among the approximate methods for its solution, the most significant is the self-consistent field method. In the calculations it is assumed that each particle moves in some averaged self-consistent field created by all particles of the system. This method can be applied not only in atomic physics, but also for systems of interacting particles, which are the structure of materials. However, for complex atoms, self-consistent field calculations are very cumbersome and time-consuming.

An operator including spatial and temporal variables characterizing the occurrence and propagation of AE waves is postulated by analogy with the energy function W on the basis of experimental data obtained by the diagnostic system during loading of materials. A similar model of the operator of the dynamic process of AE signal generation was constructed by the authors during tests for four-point bend of samples made of St3sp steel [14].

At the same time, it has practical interest to construct a model of an operator of processes of occurrence and propagation of AE signals in a medium in the form of coatings from polymer composite materials on a metal substrate [1,6]. If for metals there is a model of operator of dynamic process of AE signals occurrence during deformation of material structure, for polymer structural materials such investigations were not performed before.

Prediction of the structural states of materials based on the establishment of links between structural changes at different stages of loading materials and the moments of AE signals is not only a scientific, but also a technical problem.

The aim of this work is creating a model of the operator of the process of AE signals appearance, which allows using information-computer modeling to establish the correspondence between the evolution of the defective structure and the kinetics of the AE signal appearance at different stages of loading of polymer composite materials.

2 Materials, Technology and Methods

As experimental data necessary for modeling the structure of the operator of AE signals occurrence, the results of acoustic emission and tensometric measurements were used to diagnose strain hardening of composite materials based on an ED-20 epoxy matrix with hardener polyethylene polyamine (PEPA) [3].

The most mobile tertiary amide groups, which are located at the ends of the hardener molecules, react with the epoxy matrix. The resulting fixed mesh structure becomes ductile and rigid. The more epoxy groups in the matrix, the more there will be branches that interact with each other using various chemical bonds and creating a plastic polymer [9].

The chemical formulas of the matrix and hardener are shown in Figs. 1 and 2.

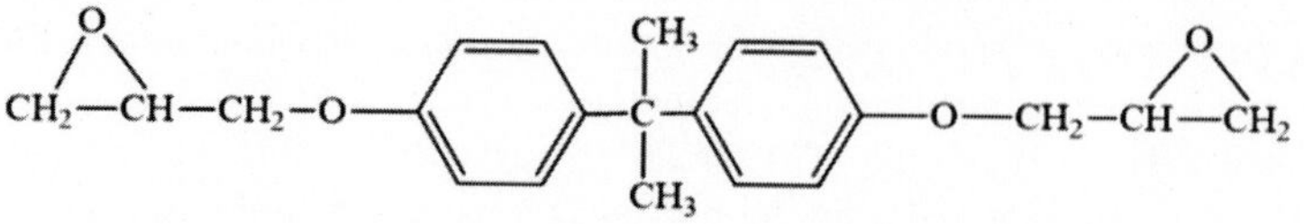

Fig. 1. The chemical formula of epoxy resin ED-20

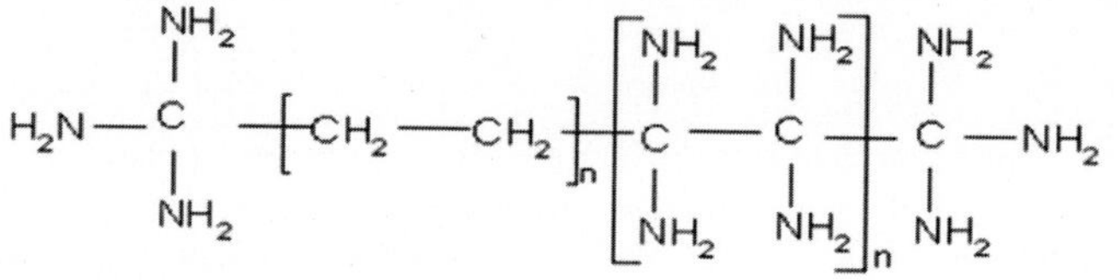

Fig. 2. The chemical formula of the hardener polyethylene polyamine (PEPA)

When ED-20 epoxy is reacted with a PEPA hardener at room temperature, the reaction shown in Fig. 3 takes place.

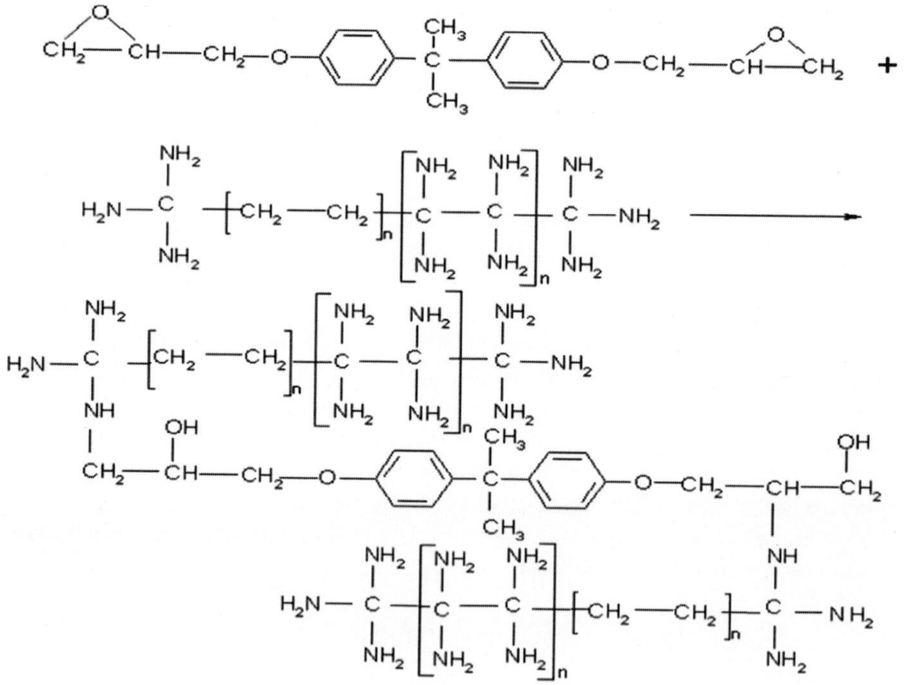

Fig. 3. Chemical reaction of the interaction of the epoxy resin ED-20 with the PEPA hardener

In polymer composite materials, the speed of propagation of acoustic waves depends not only on the density and elastic properties, but also on the viscosity, ductility and degree of structural heterogeneity.

Quantum chemical modeling of the molecular structure was performed using the Hartree-Fock method. This method is used to calculate equilibrium states and coupling sites and allows us to use information about the configuration of the electron shells of the atoms that make up the system in the in a dictionary-like structure of wave functions, to perform the calculation of the structure and properties of molecules. This makes it possible to predict the existence and properties of new materials which do not obtained experimentally yet.

3 Experiment, Results and Discussion

The analysis of the primary information on the number and type of structural units in the molecule of epoxy resin ED-20 with a PEPA hardener, molar mass, type, quantity and length of chemical bonds and the total binding energy for each type of bond is presented in Table 1. In this case, the amount n polymer branches in the hardener molecule was assumed to be one.

Table 1. Structural and energy characteristics of the epoxy resin molecule ED-20 with PEPA hardener

Number and type of atoms					
carbon C	nitrogen N	oxygen O	hydrogen H	Total	
33	20	2	76	131	
Total molar mass, g/mol					
396.35	280.13	31.99	76.60	785.07	
Number and type of chemical bonds					
C–H	C–N	C–C	C–O	O–H	N–H
36	22	30	6	2	38
Total bond energy, kJ/ mol					
12060	17314	18075	6422	849	13528
The average length of the chemical bond, Å					
1.08491	1.48440	1.53704	1.46816	0.96669	1.00353

Considering the elements of a dynamic system of an epoxy molecule together with a hardener (ED-20 + PEPA), using the Hartree-Fock method (Restricted Hartree-Fock method, RHF) in the 3-21G basis in the Gauss View program, the total initial energy of the W_0 oscillator was calculated.

The energy calculation of the molecules of epoxy resin ED-20 was performed using the Gaussian program for a fixed configuration of a polyatomic system (Single point Energy Calculation). The calculation of the energy of a molecule includes the kinetic energy of electron motion for a given internuclear state and all kinds of potential energy: the interaction of electrons with each other, the interaction of electrons with nuclei, and the interaction of nuclei with each other. Such calculations must be performed to select the method and basis of calculation in the case if there is experimental information about the characteristics of this configuration. The goal of optimization is finding the most stable molecular structures.

The structure of the epoxy resin molecule is optimized by searching for stationary points on the surface of potential energy. In the process of optimization, the magnitudes of length of bonds and valence angles of the investigated structure are changing. The search for an extremum ends with a stationary point closest to the launch configuration of the system, a calculation of the energy of zero vibrations and the thermodynamic properties of the molecules.

Using the GaussView program, a graphic image of the optimized molecular structures of the ED-20 epoxy resin (Fig. 4) and hardener (Fig. 5) was obtained, which enables to visualize the rotation of the molecule during analysis of the results.

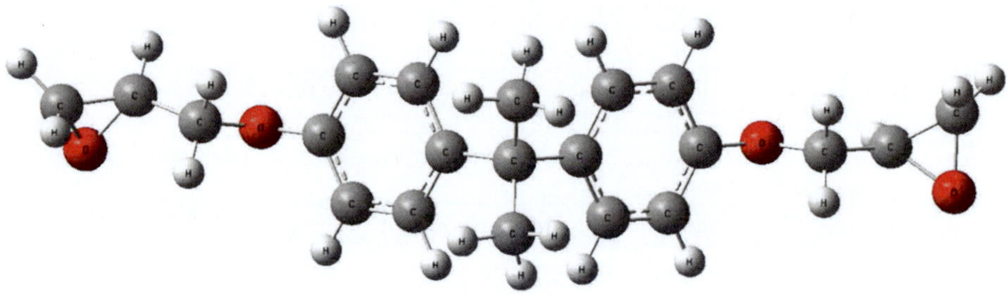

Fig. 4. The spatial structure of the epoxy resin ED-20

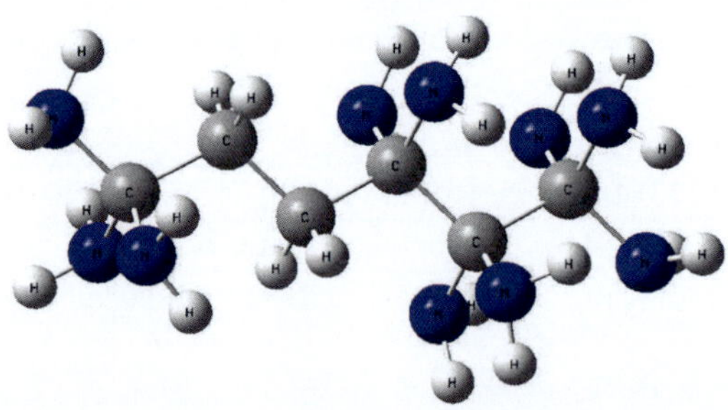

Fig. 5. The spatial structure of the hardener polyethylene polyamine (PEPA)

The main element of the operator of the process of AE signals occurrence is the anharmonic quantum oscillator, in which the vibrational energy is determined by the complex interaction of the dynamic component of the occurrence of AE signals and the static due to the linear deformation of the material.

The anharmonic oscillator model is a one-dimensional chain of atoms located in a sinusoidal potential relief (Fig. 6).

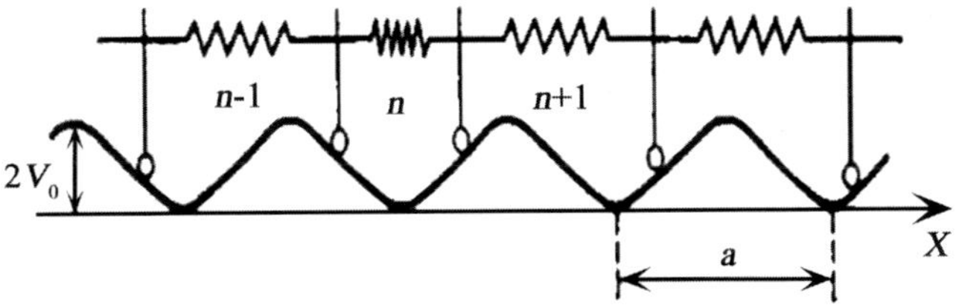

Fig. 6. Anharmonic oscillator model

The oscillatory dynamics of a solid can be considered on the basis of an ensemble of nonlinear anharmonic oscillators. The behavior of an individual anharmonic oscillator under the influence of an external force its generalization to the ensemble of such oscillators are analyzed.

It is assumed that elastic forces act between the atoms. The Hamiltonian of such a system has the form [10]:

$$\begin{aligned} H &= \sum_n \frac{p_n}{2m} \\ &+ \sum_n \left\{ \frac{f}{2} \left(x_{n+1} - x_n + u_{n+1} - u_n \right)^2 - \frac{g}{3} \left(x_{n+1} - x_n + u_{n+1} - u_n \right)^3 \right\} \\ &+ \frac{V_0}{2} \sum_n \left\{ 1 - \cos \left(Q \left(na + x_n + u_n \right) \right) \right\}, \end{aligned} \tag{2}$$

where m is the atomic masses, p_n is their impulses, f, g are the harmonic and anharmonic elastic constants, x_n is the mid positions of atoms, u_n is the dynamic motion of atoms, $Q = 2\pi/a$ is the wave vector.

To find f and g an explicit specification of the interaction potential is required. The potential well of a one-dimensional anharmonic oscillator near its bottom will be described by a cubic parabola of the form [10]

$$U(x) = \frac{f}{2} x^2 - \frac{g}{3} x^3. \tag{3}$$

The energy of a loaded oscillator can be represented as follows:

$$W = W_0 + A_0, \tag{4}$$

where W_0 is the initial energy of the oscillator, A_0 is the external force energy.

Under elastic loading, the work and increase in the oscillator energy are equal:

$$A_0 = \frac{A_{sp}}{n_0}; \; A_{sp} = \int_0^{\sigma} \sigma(\varepsilon)\, d\varepsilon \approx \frac{1}{2} \frac{\sigma^2}{E} = \frac{1}{2} E \varepsilon^2, \tag{5}$$

where A_{sp} is the work performed by loading force per unit volume, σ is the mechanical tension, ε is the relative strain, n_0 is the concentration of chemical bond breaks in the deformable volume of the sample, E is Young's modulus.

The total energy W of the loaded oscillator can be estimated from the experimentally detected AE signal during loading. As was shown in [14], AE signals in the polymer under study arise at the moments of a fairly rapid rearrangement of the material structure. The polymer molecules are rearranged in such a way that the "stoppers" that existed in the starting material disappear, the segments of the polymer macromolecules acquire a certain mobility, which results in the orientation of the macromolecules along the direction of deformation. Considering that the breaking of intermolecular bonds occurs primarily between atoms with the lowest binding energy, we believe that the energy of such a breaking is completely converted into the energy of a fixed acoustic signal W_Σ. Then we can write

$$\frac{W_\Sigma}{N} = A_0, \tag{6}$$

where N is the scission yield of bond, A_0 is the work expended by the external force to break a single bond.

Self-consistent methods are used to calculate the thermodynamic characteristics of the anharmonic oscillator. As an auxiliary system, a system of harmonic oscillators with a Hamiltonian is used

$$H_0 = \sum_n \frac{p_n}{2m} + \sum_n \frac{\phi_n}{2} u_n^2, \tag{7}$$

where ϕ_n is the set of variation force constants, along which the free energy is minimized.

The calculation of the thermodynamic properties of molecules consists in calculating the free energy by the formula:

$$\Phi \approx \Phi_0 + \langle H - H_0 \rangle, \tag{8}$$

where Φ_0 are the free energy of a system of harmonic oscillators and are determined by the expression:

$$\Phi_0 = kT \sum_n \ln\left(\frac{\hbar}{kT}\sqrt{\frac{\phi_n}{m}}\right), \tag{9}$$

where k is Boltzmann constant, T is temperature.

Then, the variational force constants are determined from the self-consistency equations in the following way:

$$\frac{\partial \Phi}{\partial \phi_n} = 0, \tag{10}$$

and the desired atomic positions from the equilibrium equations are:

$$\frac{\partial \Phi}{\partial x_n} = 0. \tag{11}$$

Replacing the discrete quantities x_n by continuous $u(x)$ and decomposing them up to the second order [10] we have

$$\frac{d^2u}{dx^2}\left(1-\frac{1}{\varepsilon^*}\frac{du}{dx}\right)-\frac{1}{\varepsilon^*}\frac{d}{dx}\lambda^2-AQ\sin\left(Qu\right)=0, \tag{12}$$

where the dimensionless constants ε^*, λ^2 correspond to the discontinuous deformation of the interatomic bond and the mean square of the amplitude of atomic vibrations and together with the constant A are determined by the dependencies:

$$\varepsilon^*=\frac{f}{2ga};\ \lambda^2=\frac{kT}{fa^2};\ A=\frac{V_0}{2fa^2}. \tag{13}$$

The solution of Eq. (12) has the form:

$$\varepsilon\left(x\right)=\varepsilon_0\left(x\right)\left(1+\frac{1}{\varepsilon^*}\left(\frac{1}{3}\varepsilon_0\left(x\right)+\frac{1}{2}\frac{\varepsilon_T}{1+\frac{C}{4A}-\pi^2\varepsilon_0\left(x\right)}\right)\right), \tag{14}$$

where $\varepsilon\left(x\right)=\frac{du}{dx}$ is deformation, $\varepsilon_T=\frac{gkT}{f^2a}$ is thermal deformation of a free atomic chain (without sinusoidal relief), $\varepsilon_0\left(x\right)=\sqrt{C-2A}\frac{1}{dn\left(x\right)}$, $dn\left(x\right)$ is Jacobi delta amplitude, C is integration constant, which is determined from the boundary conditions of the experiment.

Assuming that the n-th atom is at the bottom of the potential well, we have $\sin\left(Qu\right)=0$ and then the quantity $\varepsilon_0=\sqrt{C-2A}$ is determined by the positive radical of the quadratic equation:

$$\varepsilon_0^2+\varepsilon_0\left(3\varepsilon^*+\varepsilon_T\right)-2\varepsilon^*\varepsilon_T=0. \tag{15}$$

To determine the energy of a fixed acoustic signal W_Σ, an operator of the process of occurrence of AE signals is proposed. Finding the operator of a dynamic process is realized through the inverse problem of selecting a model that adequately describes the experimentally obtained signal from a developing defect.

The operator structure algorithm of the dynamic process of the occurrence of acoustic emission signals is presented in Fig. 7.

It was shown in [14] that the output signal $y(t)$ is described by an m-th order differential equation with constant coefficients:

$$\frac{d^my\left(t\right)}{dt^m}+\sum_{m=0}^{m-1}a_m\frac{d^my\left(t\right)}{dt^m}=0, \tag{16}$$

where m is order of the highest derivative of a differential operator.

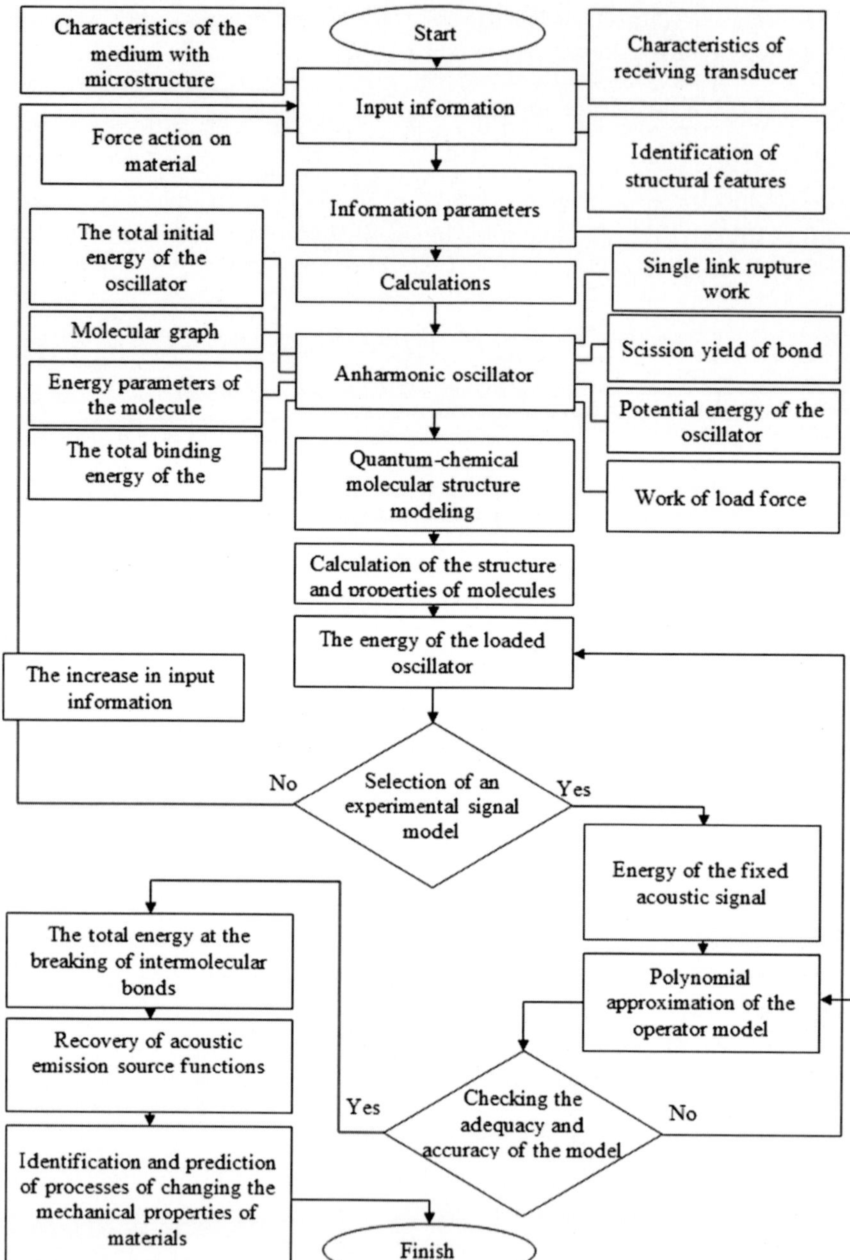

Fig. 7. The operator structure algorithm of the process of occurrence of acoustic emission signals

To obtain estimates of the characteristic polynomial of Eq. (16), the method of finite differences was used. This is because the experimental AE signal consists of two components: a trend and a random component. The trend (envelope of the AE signal) is quite smooth and it can be approximated by an m-th polynomial:

$$\hat{y} = a_0 + a_1 t + \ldots + a_m t^m, \tag{17}$$

where the coefficients of the characteristic polynomial $a_0, a_1, \ldots, a_m$ are determined by solving the system of normal equations by the least squares method.

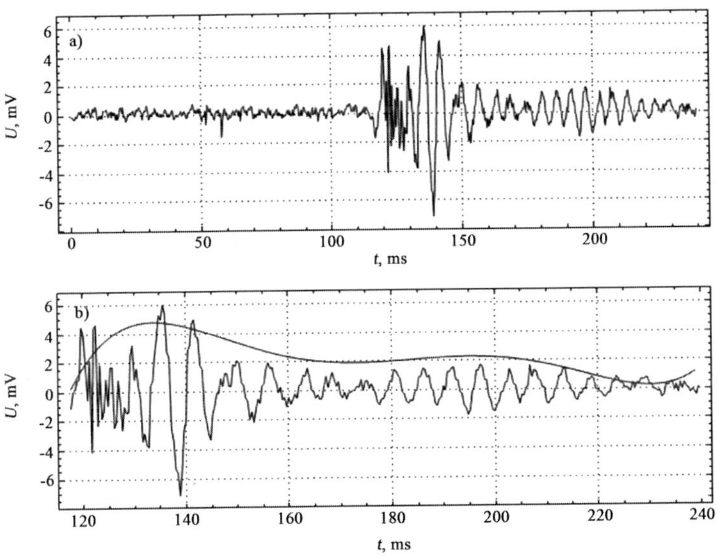

Fig. 8. Experimental AE signal arising from the deformation of the polymer sample (a), a fragment of the analyzed section of the AE signal and polynomial approximation of the envelope (b)

The isomorphism of models (16) and (17) is ensured by the fulfillment of four Gauss-Markov conditions [14], which make it possible to establish the adequacy and accuracy of the model under consideration for the operator of the process of occurrence of AE signals.

The order of polynomial (17) is determined using experimental information processed by the least squares method for finite differences. The polynomial equation is an approximation of the operator of the process of AE signals occurrence. The area under the envelope (17) determines the total energy during the breaking of intermolecular bonds of the aggregate of anharmonic oscillators of a solid in the deformation zone. Knowing the number of oscillators in the deformable volume, one can obtain the total energy in the Schrödinger Eq. (1).

The structural features of the damping mechanisms of samples of epoxy resin ED-20 were identified according to [3]. To find the energy of the operator of the process of changing the structure of the material under load, a fragment of the AE signal was selected that arose under a load of 70,26 kg on the studied polymer sample. The time of occurrence, duration, and termination of the AE signal (Fig. 8 (a)) and approximation of its envelope carried out through the local amplitudes U_{max} in the range of the signal existence (Fig. 8 (b)) were chosen as the analyzed parameter. The degree of the approximating polynomial (18) is chosen equal to five, the corresponding coefficients and approximation errors are given in Table 2.

Table 2. Numerical parameters of the polynomial approximant

Coefficients of polynomial approximation					
a_0	a_1	a_2	a_3	a_4	a_5
–3213	92.7410	–1.0540	$5.9070{\cdot}10^{-3}$	$-1.6340{\cdot}10^{-5}$	$1.7859{\cdot}10^{-8}$
Measure of inaccuracy of polynomial approximation					
Coefficient of determination			0.9594		
Adjusted coefficient of determination			0.9087		
Root Mean Square Error			0.3420		
Sum of Square Error			0.4678		

The integral of Eq. (17) allows us to estimate the total energy at break of the intermolecular bonds of the ensemble of anharmonic oscillators located in the area of deformation of the sample:

$$W_{\Sigma} = 2 \int_{t_1}^{t_2} \frac{\widehat{y}^2}{R} dt, \tag{18}$$

where t_1, t_2 are the times of both start and end of AE signal (see Fig. 8 (b)), $R \geq 60$ MOm is the value of the resistance of the piezotransducer of AE signal. The multiplier 2 in the formula (21) takes into account the symmetry of the AE signal with respect to the horizontal axis.

The results of calculating the integral (18) are given in Table 4. To find the number N of breaking bonds when an AE signal occurs, it is necessary to analyze the energy structure of the molecule of the polymer under study. The energy parameters of a molecule are represented as structural characteristics of a molecular graph. As follows from Table 1, the molecule of the studied polymer contains four types of atoms (carbon C, hydrogen H, oxygen O, nitrogen N) and six types of chemical bonds (C–N, C–H, C–C, C–O, N–H, O–H). The initial molecular graph G is shown in Fig. 9 (a). Each edge of the graph G corresponds to one type of chemical bond in the molecule of the studied polymer: a - N–H, b - C–N, c - C–O, d - O–H, e - C–H, f - C–C. Event S, which we will consider, is the simultaneous rupture of two types of chemical bonds.

A given event is determined by a model whose incidence matrix Ω has the following form:

$$\Omega = \begin{vmatrix} 0\,1\,1\,0\,0\,0 \\ 1\,0\,0\,1\,0\,0 \\ 1\,1\,0\,0\,0\,0 \\ 0\,0\,1\,1\,0\,0 \\ 0\,1\,0\,1\,0\,0 \\ 1\,0\,1\,0\,0\,0 \\ 0\,1\,0\,0\,1\,0 \\ 1\,0\,0\,0\,1\,0 \\ 0\,0\,0\,1\,1\,0 \\ 0\,0\,1\,0\,1\,0 \\ 0\,1\,0\,0\,0\,1 \\ 1\,0\,0\,0\,0\,1 \\ 0\,0\,0\,1\,0\,1 \\ 0\,0\,1\,0\,0\,1 \\ 0\,0\,0\,0\,1\,1 \end{vmatrix}; \; Z = \Omega^T \times \Omega = \begin{vmatrix} 5\,1\,1\,1\,1\,1 \\ 1\,5\,1\,1\,1\,1 \\ 1\,1\,5\,1\,1\,1 \\ 1\,1\,1\,5\,1\,1 \\ 1\,1\,1\,1\,5\,1 \\ 1\,1\,1\,1\,1\,5 \end{vmatrix}. \quad (19)$$

In the matrix Ω, each column corresponds to an edge of the graph G, and each row corresponds to a set of torn edges. The frequency relational matrix Z correspond to the matrix Ω. Elements of the matrix Z define the derivative $\frac{\partial G}{\partial S}$ of the graph G with respect to the event S, which is a graph whose support is $\{a, b, c, d, e, f\}$, and two vertices of this graph are adjacent if the value of the derivative on the arc formed by these vertices is different from zero or infinity. Calculating the values of the derivative $\frac{\partial G}{\partial S}$ on the edges of the graph, we have:

$$\frac{\partial G}{\partial S}(a, b) = \frac{z_a - 2z_{ab} + z_b}{z_{ab}} = \frac{5 - 2 \cdot 1 + 5}{1} = 8. \quad (20)$$

Formula (20) is written for a pair of edges (a, b). The same expressions for the derivative will be for the other combinations of pairs of edges due to the symmetry of the frequency matrix Z. Therefore, the graph $\frac{\partial G}{\partial S}$ will have six adjacent vertices (Fig. 9). Each vertex of such a graph corresponds to the type of chemical bond present in the studied polymer. Each vertex should be assigned a weight determined by the number of corresponding chemical bonds in the polymer molecule. The distances between pairs of adjacent vertices of the graph (Fig. 9 (c)) are determined by the total binding energy of chemical bonds. The number of such combinations is equal to:

$$C_6^2 = \frac{6!}{2!4!} = 15. \quad (21)$$

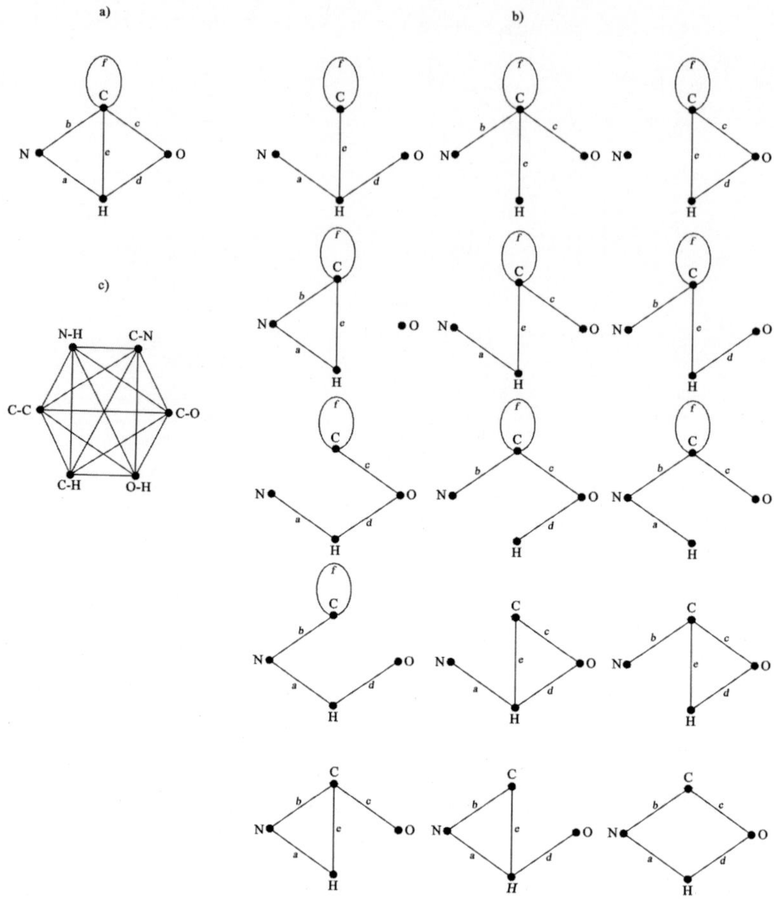

Fig. 9. Molecular graph G, reflecting the presence and types of chemical bonds in the studied polymer (a); combinations in which a given event S (breaking of two types of bonds) takes place (b); derivative graph $\frac{\partial G}{\partial S}$ (c)

Table 3 shows the results of calculations of energy distances between pairs of adjacent vertices of the graph $\frac{\partial G}{\partial S}$.

Analysis of Table 3 shows that the smallest energy distance corresponds to two vertices C–O; O–H and amounts to $W_{bin} = 7271$ kJ/mol. Assuming that at emergence of a signal of AE there is a rupture of these chemical bonds, it is possible to determine the number of moles ν and the number N of bond breaks when the AE signal occurs:

$$\nu = \frac{W_{\Sigma}}{W_{bin}}; \; N = \nu N_A, \tag{22}$$

where N_A – Avogadro constant.

Table 3. Energy distances between adjacent vertices of a graph $\frac{\partial G}{\partial S}$

Pair combination of bonds	Total binding energy, kJ/mol
C–N; C–O	23736
C–N; O–H	18163
C–N; C–H	29374
C–N; C–C	35389
C–N; N–H	30842
C–O; O–H	**7271**
C–O; C–H	18482
C–O; C–C	24497
C–O; N–H	19950
O–H; C–H	12909
O–H; N–H	14377
O–H; C–C	18924
C–H; C–C	30135
C–H; N–H	25588
C–C; N–H	31603

Further, based on the use of Eqs. (4, 6), it is possible to calculate the total energy of the oscillator W and the work A_0 of the external force when a single bond is broken.

As it is shown in [10], for the average energy of the loaded anharmonic oscillator approximate equality is achieved:

$$W \approx W_0 + \frac{1}{2}\frac{F_0}{f}, \tag{23}$$

where $F_0 = \frac{P}{N}$ is the external force acting along the bond, P is the load applied to the sample when the AE signal occurs, f is the harmonic constant of the oscillator from the Eqs. (2, 3, 12).

The external force F_0 can be represented through the gradient of the potential energy of the anharmonic oscillator in he following way:

$$F_0 = -\frac{dU}{dx} = -fx + gx^2, \tag{24}$$

where g is the anharmonic oscillator constant.

Solving Eqs. (23, 24) together, we can estimate the elastic harmonic f and anharmonic g constants of the oscillator and obtain the potential energy function (3). When calculating the value of the constant f the parameter x was taken equal to the width of the potential well a ($x = a$). The form of the potential energy function $U(x)$ of the anharmonic oscillator obtained from the results of the calculations is shown in Fig. 10.

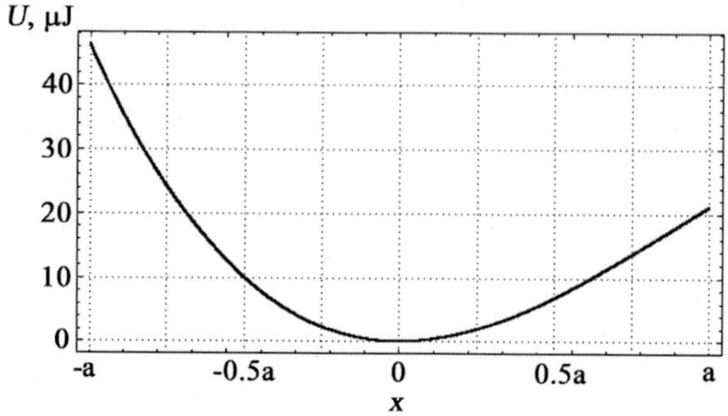

Fig. 10. The potential energy function of a one-dimensional anharmonic oscillator obtained from the results of the calculations

From Eq. (5), we can find Young's modulus:

$$E = \frac{2A_{sp}}{\varepsilon^2} = \frac{2A_0 n_0}{\varepsilon^2}, \tag{25}$$

where $n_0 = \dfrac{N}{V}$ is the concentration of molecular breaks in the deformable volume V of the sample, ε is the relative deformation, determined on the basis of the Eq. (14).

In the calculations, the width of the potential well of the anharmonic oscillator was taken equal to $a \approx 90\,\text{nm}$, which approximately corresponds to the total geometric dimensions of two atoms and the length of the chemical bond between them, the depth of the potential well was assumed to be equal to the work of the loading force when the bond was broken $2V_0 = A_0$ and the temperature was 300 K. The results of experimental measurements and theoretical calculations are shown in Table 4.

Finding the operator of the process of changing the structure of the material under load is part of the developed algorithm of the multifactor model of information diagnostics (Fig. 11).

Since it is impossible to establish a correspondence between the evolution of a defective structure and the kinetics of the occurrence of AE signals at different loading stages and to establish the state and mechanical properties of the material in the source of occurrence of AE signals using one algorithm, it is proposed to use the cross-structure algorithm of the operator structure of the dynamic process of AE signals occurrence and a multifactor model of information diagnostics. The general cross-algorithm takes into account the necessary operations and structural blocks for constructing a model of the operator of the process of AE signals occurrence under load. According to the generally accepted structuring of algorithms, they are divided into computational algorithms and control ones. A computational algorithm is a set of rules for converting source data into results, which is the energy of a loaded oscillator. The control algorithm is the rules and procedures for information diagnostics. In this case, a combination of algorithms and training of individual models on the same input data occurs, after which the results of both models are combined.

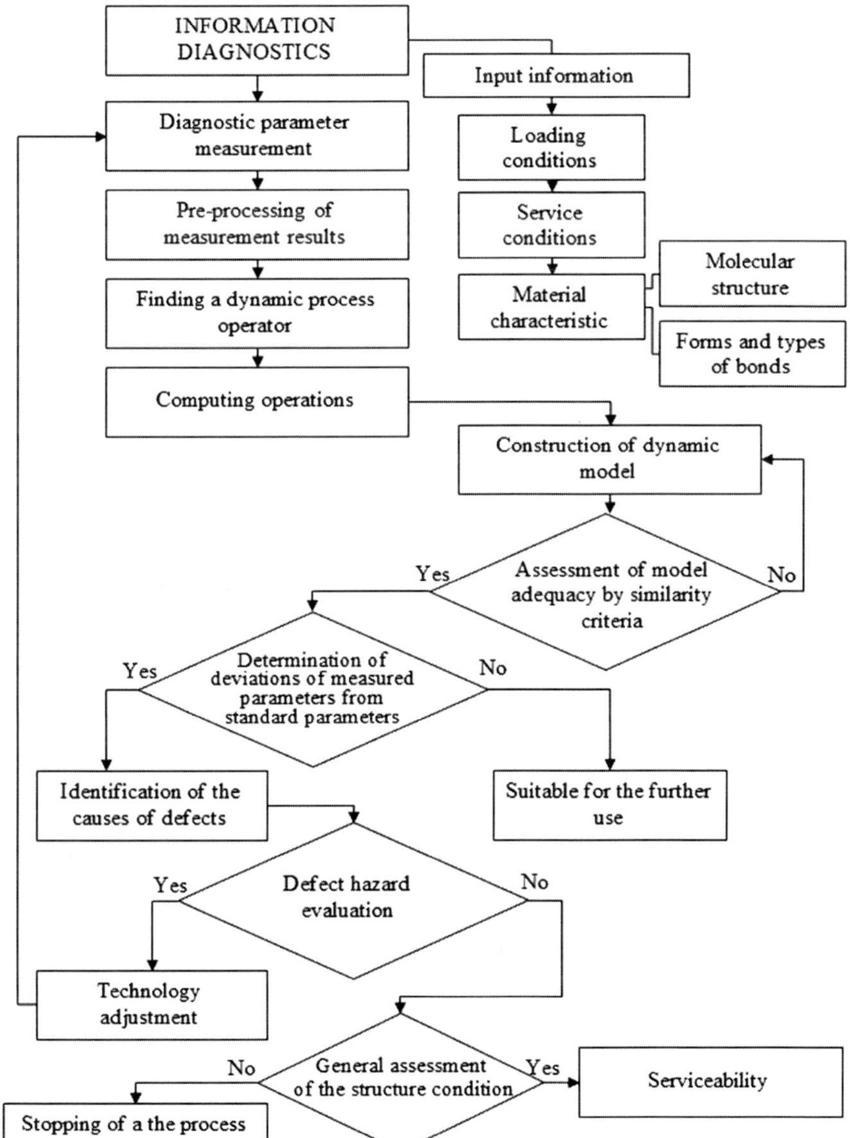

Fig. 11. Algorithm for multifactor model of information diagnostics

Table 4. The results of experimental measurements and theoretical calculations of the physicomechanical properties of ED-20 + PEPA molecules

Energy characteristics of the molecule	
The initial energy of the molecule ED20 + PEPA W_0, J	$8.56 \cdot 10^{-15}$
The smallest pair binding energy W_{bin}, kJ/mol	7271
Dipole moment of ED20 + PEPA molecule, C·m	$1.83 \cdot 10^{-29}$
Characteristics of the test sample	
The geometric dimensions of the deformable area of the sample, mm	$10 \times 15 \times 60$
Density ρ, g/sm^3	1.16
Young's modulus (theory) E, GPa	4.00
The strain at the AE signal σ, MPa	110.24
Load at occurrence of AE signal P, N	689.01
Scission yield of bonds when an AE signal occurs N	$3.42 \cdot 10^{9}$
The concentration of bond breaks in the deformable volume n_0, m^{-3}	$9.5 \cdot 10^{13}$
External force acting along the bond F_0, N	$2.01 \cdot 10^{-7}$
Work of load force when breaking one bond A_0, J	$1.21 \cdot 10^{-17}$
AE signal characteristics	
AE signal start time t_1, s	0.117
AE signal completion time t_2, s	0.239
AE signal energy W_Σ, J	$4.13 \cdot 10^{-8}$
Characteristics of the anharmonic oscillator	
Initial energy of the oscillator W_0, J	$8.56 \cdot 10^{-15}$
The width of the potential well a, nm	90
Oscillator total energy W, J	$8.57 \cdot 10^{-15}$
Amplitude of the sinusoidal potential V_0, J	$0.60 \cdot 10^{-17}$
Linear elasticity coefficient f, n/m	$8.31 \cdot 10^{9}$
Anharmonicity coefficient g, n/m^2	$5.19 \cdot 10^{16}$
Ruptural deformation of the interatomic bond ε^*	0.89
Young's modulus (calculation) E, GPa	7.74

The data obtained in the analysis of AE accompanying the process of destruction of materials allow us to conclude that for heterogeneous materials this process is a directed evolution, as a result of which a self-organized critical state is formed. Early detection of this condition is the main task of diagnosis.

4 Conclusions

Based on computer quantum-chemical modeling, a model of the operator of the process of occurrence of AE signals has been developed. The anharmonicity of the potential of intermolecular interaction in polymer samples based on ED-20 epoxy resin was evaluated. To study the energy structure of the molecule of the polymer under study, graph theory has been used. The influence of external mechanical loading on the internal vibrational dynamics in solids was analyzed, which may result in the breaking of intermolecular bonds and the appearance of acoustic emission signals. An analysis of the behavior of a single excited anharmonic oscillator under the action of an external force and the results are generalized to an ensemble of such oscillators has been performed. The form of the potential energy function of a one-dimensional anharmonic oscillator has been obtained. It has been shown that the found parameters of the anharmonic oscillator make it easy to calculate the macroscopic parameters of the material: Young's modulus, relative deformation, mechanical stress, which characterize the behavior of the material under load.

It has also been shown that the anharmonic solid and the anharmonic oscillator as an element of the internal dynamics of the body can act as systems or mechanisms in which under their mechanical elastic loading transitions of forms of internal energy occur.

An algorithm for the operator structure of the process of occurrence of acoustic emission signals and an algorithm for a multifactor model of information diagnostics has been developed.

Their combination has been proposed in the form of a cross algorithm for the computational and control components of the process of AE signal generation, where the computational component is a set of rules for converting source data into results, which is the energy of a loaded oscillator. The control algorithm is the rules and procedures for information diagnostics. In this case, a combination of algorithms and training of individual models on the same input data occurs, after which the results of both models has been combined.

References

1. Akimov, A., Buketov, A., Sapronov, O., et al.: Development of polymer composites with improved thermophysical properties for shipbuilding and ship repair. Compos.: Mech. Comput. Appl.: Int. J. **10**(2), 117–134 (2019). https://doi.org/10.1615/CompMechComputApplIntJ.2018026989
2. Aleksenko, V., Sharko, A., et al.: Identication by ae method of structural features of deformation mechanisms at bending. Tech. Diagn. Nondestr. Test. **1**, 32–39 (2019). https://doi.org/10.15407/tdnk2019.01.01
3. Aleksenko, V., Sharko, A., Smetankin, S., et al.: Application of acoustic emission and strain gaugez measurements to processes of diagnostics of strain hardening of epoxy matrix composites. Tech. Diagn. Nondestr. Test. **3**, 46–54 (2019). https://doi.org/10.15407/tdnk2019.03.07
4. Babichev, S., Škvor, J., Fišer, J., Lytvynenko, V.: Technology of gene expression profiles filtering based on wavelet analysis. Int. J. Intell. Sys. Appl. **10**(4), 1–7 (2018). https://doi.org/10.5815/ijisa.2018.04.01
5. Bohmann, T., Schlamp, M., Ehrlich, I.: Acoustic emission of material damages in glass fibre-reinforced plastics. Compos. B: Eng. **155**(15), 444–451 (2018). https://doi.org/10.1016/j.compositesb.2018.09.018

6. Buketov, A., Sapronov, O., Brailo, M., et al.: Dynamics of destruction of epoxy composites filled with ultra-dispersed diamond under impact conditions. Mech. Adv. Mater. Struct. **27**(9), 725–733 (2018). https://doi.org/10.1080/15376494.2018.1495788
7. Carpinteri, A., Lacidogna, G., Pugno, N.: Structural damage diagnosis and life-time assessment by acoustic emission monitoring. Eng. Fract. Mech. **74**(1–2), 273–289 (2007). https://doi.org/10.1016/j.engfracmech.2006.01.036
8. Cho, H., Shoji, N., Ito, H.: Acoustic emission generation behavior in a7075–t651 and a6061–t6 aluminum alloys with and without cathodic hydrogen charging under cyclic loading. J. Nondestr. Eval. **37**(83), 83 (2018). https://doi.org/10.1007/s10921-018-0536-7
9. Dickerson, R.E., Gray, H.B., Haight, G.P.: Chemical Principles, 3rd edn. The Benjamin/Cummings Publishing Company, Inc., Menlo Park (1979)
10. Gilyarov, V.: Angarmonicheskaya model frenkelya-kontorovoy dlya temperaturnoy soizmerimosti v kristallakh. FTT **29**(4), 1055–1060 (1987)
11. Kushchenko, V., Khomitskyi, D.: Vibration monitoring of steel shaft headgears. Lect. Notes Civil Eng. **47**, 227–234 (2020). https://doi.org/10.1016/j.measurement.2019.107320
12. Liu, S., Li, X., Li, Z., et al.: Energy distribution and fractal characterization of acoustic emission (AE) during coal deformation and fracturing. Meas J. Int Meas Confederation **136**, 122–131 (2019). https://doi.org/10.1016/j.measurement.2018.12.049
13. Madarshahian, R., Ziehl, P., Caocedo, J.: Acoustic emission Bayesian source location: onset time challenge. Mech. Sys. Sig. Process. **123**(15), 483–495 (2019). https://doi.org/10.1016/j.ymssp.2019.01.021
14. Marasanov, V., Sharko, A., Stepanchikov, D.: Model of the operator dynamic process of acoustic emission occurrence while of materials deforming. Adv. Intell. Sys. Comput. **1020**, 48–64 (2020). https://doi.org/10.1007/978-3-030-26474-1_4
15. Nedoseka, A., Nedoseka, S., Markashova, L., Kushnareva, O.: On identification of structural changes in materials at fracture by acoustic emission data. Tech. Diagn. Nondestr. Test. **4**, 9–13 (2016). https://doi.org/10.15407/tdnk2016.04.02
16. Ognezneva, L., Kalinichenko, A.: Akusticheskiye metody kontrolya i diagnostiki. Chast' II. Tomsk Polytechnic University, Tomsk (2009)
17. Pollock, A.: Acoustic emission inspection. Met. Handb. Ninth Ed. **17**, 278–294 (1989)
18. Salita, D., Polyakov, V.: Application of principal components analysis to the investigation of acoustic emission in plastic deformation of lead alloys. Izvestiya Altai State Univ. **102**(4), 26–30 (2018). https://doi.org/10.14258/izvasu(2018)4-04
19. Surace, C., Bovsunovsky, A.: The use of frequency ratios to diagnose structural damage in varying environmental conditions. Mech. Syst. Signal Process. **136** (2020). Article no. 106523. https://doi.org/10.1016/j.ymssp.2019.106523
20. Wang, K., Zhang, X., et al.: Application of improved leastsquare generative adversarial networks for rail crack detection by AE technique. Neurocomputing **332**, 236–248 (2019). https://doi.org/10.1016/j.neucom.2018.12.057
21. Zhao, X., Jia, M., Lin, M.: Deep Laplacian auto-encoder and its application into imbalanced fault diagnosis of rotating machinery measurement. J. Int. Meas. Confederation **152**, 107320 (2020). https://doi.org/10.1016/j.measurement.2019.107320

Technique of Gene Expression Profiles Selection Based on SOTA Clustering Algorithm Using Statistical Criteria and Shannon Entropy

Sergii Babichev[1,2](✉), Orest Khamula[2], Bohdan Durnyak[2], and Jiří Škvor[1]

[1] Jan Evangelista Purkyně University in Ústí nad Labem, Ústí nad Labem, Czech Republic
{sergii.babichev,Jiri.Skvor}@ujep.cz

[2] Ukrainian Academy of Printing, Lviv, Ukraine
khamula@gmail.com, durnyak@uad.lviv.ua

Abstract. In this paper, we have presented a results of the research concerning selection of informative genes based on the complex use of both statistical criteria and Shannon entropy. The main objective of the research is development a technique of selection of the groups of gene expression profiles which allow dividing correctly the investigated samples into previously known classes using results of both DNA micro array experiments or RNA molecules sequencing method. The DNA microchips of the patients which were investigated on lung cancer disease were used as the experimental data. At the first step, we have selected the informative genes in term of both the statistical criteria and Shannon entropy. The number of gene expression profiles was reduced at this step from 54675 to 21431. Then, we have performed the step by step clustering process using SOTA algorithm. The number of the obtained clusters was varied from 2 at the first clustering level to 512 at the ninth level. Finally, we have calculated the internal clustering quality criterion for investigated samples which are in each of the clusters. The less value of this criterion corresponds to higher separate ability of genes in this cluster. The proposed technique creates the conditions for development of both the diagnostic method and forecasting technique based on gene regulatory networks using results of both DNA microchip experiments or RNA molecules sequencing method.

Keywords: Gene expression profiles · Genes expression vector · Clustering quality criteria · Statistical criteria · Shannon entropy · SOTA clustering algorithm

1 Introduction

Gene regulatory network reverse engineering is one of the current direction of modern bioinformatics [2,11,28,30]. The term "reverse engineering" in this case

S. Babichev et al. (Eds.): ISDMCI 2020, AISC 1246, pp. 23–38, 2021.
https://doi.org/10.1007/978-3-030-54215-3_2

means that gene regulatory network of the biology object is reconstructed based on gene expression profiles data. The main difficulty of this problem is the following one: the nature and particularities of molecular elements interconnection are very complexity or unknown. Moreover, the number of genes in the investigated object is very large (tens of thousand of genes) and it is necessary at the first step to reduce the non-informative genes in terms of appropriate criteria. Selection of the groups of gene expression profiles which allow dividing correctly the studied samples into classes considering the solving problem can allows us to increase the exactness of the disease diagnostic process on the one hand and increase the quality of gene regulatory networks reconstructed based on the selected genes in terms of better understanding the genes interconnection particularities for appropriate type of disease on the other hand. These facts indicate the actuality of the current problem.

1.1 Problem Statement

Figure 1 presents the structural block chart of step-by-step procedure of sets of informative gene expression profiles formation based on results of DNA microchip experiments, which was implemented within the framework of the research.

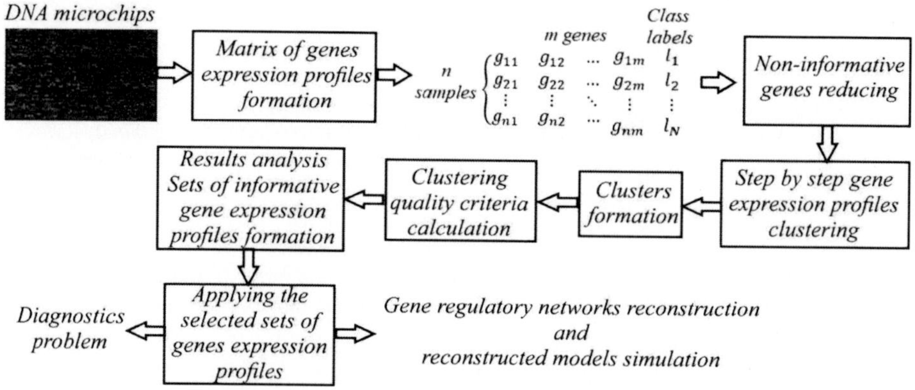

Fig. 1. A structural block chart of step-by-step procedure of sets of informative gene expression profiles formation

Initial data can be presented as array of light intensities in the case of DNA microchip experiments use or as matrix of genes counts, which correspond to appropriate type of protein, in the case of RNA molecules sequencing method apply. In any case, the first step involves formation of gene expression profiles matrix, there rows are the investigated samples and columns are genes, expression of which determines the level of the gene's activity. This value is proportional to quantity of genes which are able to generate appropriate type of protein. Gene expression profiles in this case is a vector of gene expressions, the values of which are determined for investigated samples. A vector of genes expression, values of which are determined for appropriate sample is named as genes expression vector. The main task within the framework of our research consists of selection of sets of gene expression profiles which allow dividing the investigated samples into previously known classes maximally adequately. It can allow us to create the diagnostic system of complex diseases based on the use of DNA analysis and reconstruct the gene regulatory networks which can help us to understand the particularities of genes interconnections for various diseases for further simulation process.

As it can be seen from Fig. 1, implementation of this procedure assumes:

- gene expression profiles reducing in terms of various quality criteria;
- step-by-step clustering process of gene expression profiles with following clusters formation at each step of this procedure implementation;
- calculation of the internal clustering quality criteria for vectors of genes expression in obtained clusters considering previously known nature of the samples distribution in the clusters;
- results analysis and selection of sets of the most informative gene expression profiles in terms of the internal clustering quality criteria extrema values.

1.2 Review of Literature

The issues concerning array of gene expression profiles formation using both the results of DNA microchip experiments or RNA molecules sequencing technique are presented in [20,37]. The authors describe in detail step-by-step procedure of the experimental data processing using various methods of this process implementation. In [26] the authors considered in detail the stages of DNA microarrays creation and the peculiarities of their processing. They have shown that there are four stages of this procedure implementation: background correction, normalization, PM correction and summarization. Classification and detail description of the background correction techniques are presented in [1,10,14,22]. The papers [3,10,13,18,34,36] present the techniques of DNA microchip data normalization. Necessity of this stage is determined by low correlation of the data which were determined when different conditions of the experiment performing. The aim of normalizing process is the reduction of the microchip empirical data to the same distribution. The results of the research concerning comparison analysis of various methods of PM corrections and summarization of the DNA microchip data are presented in [9,10,27,36]. PM correction stage is performed in order to reduce the nonspecific hybridization effect by correction of PM samples light intensities considering the light intensities of the appropriate MM samples. The summarization process assumes the calculation of gene expressions values from light intensities of the samples for investigated genes. Paper [5] presents the technique of DNA microarray data processing based on the complex use of Bioconductor tools and Shannon entropy for purpose of gene expression array formation.

The issues concerning gene expression reducing in terms of both the statistical criteria and Shannon entropy based on the use of fuzzy logic technique [39] are considered in [4,6,8]. The authors have assumed that if variance or average of absolute value of gene expression profiles is less and Shannon entropy is greater than appropriate boundary values, then these profiles are uninformative and they can be removed from data without significant loss of useful information. The technique to evaluate appropriate criteria boundary values was proposed by the authors in these works.

The technique which is focused to allocate mutually correlated genes and samples (bicluster analysis [23]) is described in [32,35]. The authors analysed various biclustering algorithms and specified their advantages and shortcomings. The paper [16] presents a comparative analysis of various biclustering algorithms effectiveness. The gene expression profiles were used as the experimental data in this case. In [25] the authors presented the results of the research concerning spectral biclustering algorithm use to allocate biclusters from gene expression profiles. In [29] the authors considered questions concerning bicluster analysis implementation for data which contain missing values. However, it should be noted, that in spite of the achieved results, an effective technique of biclustering algorithms optimal parameters determination based on qualitative criteria is absent nowadays. Moreover, implementation of biclustering technique lead to loss of useful information due to large quantity of little biclusters. This fact complicates the successful applying this technique. This task can be solved by using current techniques, models and information technologies which are used successfully in various fields of scientific research nowadays [24,31,33,38]. In the current paper we propose the solution of the problem of mutually correlated gene expression profiles grouping based on step-by-step clustering procedure using SOTA clustering algorithm.

The aim of the paper is development of a technique of step-by-step gene expression profiles processing in order to select the most informative genes in terms of statistical criteria and Shannon entropy.

2 Materials and Methods

Figure 2 shows the structural block chart of the algorithm to implement the stage of non-informative gene expression profiles reducing in terms of statistical criteria and Shannon entropy. Its implementation assumes the following stages:

Stage 1. Data preparation and setup of the algorithm initial parameters.

1. Formation a matrix of gene expression profiles, where rows are the studied samples and columns present itself the investigated genes. In the case of DNA microchips experiment use, initial data is presented as a matrix of light intensities and, to transform them into gene expression, it is necessary to perform background correction, normalization, PM correction and summarization. In the case of RNA molecules sequencing technique use the initial data is presented as quantity of genes for appropriate samples. In this case it is necessary to transform the data into suitable scale with

removing weakly expressed genes. In any case, the initial data contains tens of thousands of genes.

2. Calculation of variance, average of absolute values and Shannon entropy for each of the investigated gene expression profiles. The Shannon entropy criterion was calculated based on James-Stein shrinkage estimator [19]. This method is based on the complex use of two different models: a high-dimensional model with low bias and high variance, and a low-dimensional model with larger bias but smaller variance. Applying this technique assumes calculation of values distribution probability in the a priory determined sections as follows:

$$p_i^{JS} = \lambda p_i + (1-\lambda) p_i^{ML} \tag{1}$$

where p_i^{ML} is the probability in i-th section which is calculated using maximum likelihood method; $p_i = \dfrac{1}{n_i}$ is target probability in i-th section; n_i is the number of values in i-th section; λ is the intensity parameter which calculates in the following way:

$$\lambda = \frac{1 - \sum_{i=1}^{k} (p_i^{ML})^2}{(n-1) \sum_{i=1}^{k} (p_i - p_i^{ML})^2} \tag{2}$$

where n is the number of genes. Shannon entropy value is calculated by standard formula as follows:

$$H^{JS} = -\sum_{i=1}^{k} p_i^{JS} log_2 p_i^{JS} \tag{3}$$

In this case, minimum value of Shannon entropy corresponds to higher quality of the investigated vector.

3. Statistical analysis of the obtained results. Setup of both the ranges and steps of these parameters change.

Stage 2. Formation of both the clusters and function to calculate a clustering quality criterion.

4. Division of the investigated samples into clusters considering the experimental data annotation.
5. Formation of a function to calculate the clustering quality criterion. The internal clustering quality criterion which was used within the framework of this research considers both the objects distribution within clusters and the clusters distribution in the features space. The first component of this criterion is calculated as an average distance from the objects to the mass centers of the clusters, where these objects are allocated:

$$QCW = \frac{1}{N} \sum_{s=1}^{K} \sum_{i=1}^{N_s} d(x_i^s, C_s) \tag{4}$$

where K is the number of clusters; N is the number of genes expression vectors (number of samples); N_s is the number of samples in the cluster s; x_i^s is the i-th genes expression vector in the cluster s; C_i, C_j and C_s are the mass centers of the clusters i, j and s respectively; $d(\cdot)$ is the metric used to estimate the proximity level of the studied genes expression vectors. The second component of the internal clustering quality criterion was calculated as an average distance between the mass centers of the clusters:

$$QCB = \frac{2}{K(K-1)} \sum_{i=1}^{K-1} \sum_{j=i+1}^{K} d(C_i, C_j) \tag{5}$$

Considering the high dimension of features in the investigated vectors, we used the correlation distance as the proximity metric. In [7] the authors have proposed the complex internal clustering quality criterion which was calculated as multiplicative combination of Calinski Harabasz criterion [12] and WB-index [40]:

$$QC_{int} = \frac{QC_{WB}}{QC_{CH}} = \frac{K(K-1)QCW^2}{(N-K)QCB^2}; \tag{6}$$

The minimum value of this criterion corresponds to the best clustering.

Stage 3. Calculation of the internal clustering quality criterion within the range of the algorithm parameters change.

6. Initialization of initial boundary parameters of the system: variance, average of absolute values and Shannon entropy. Initialization of counter: m = 1.
7. Comparison of the values of variance, average of absolute values and Shannon entropy calculated for each of the investigated gene expression profiles with appropriate boundary values and removing non-informative genes in terms of the used criteria.
8. If counter value m less than maximal one m_{max}, then increment of the counter $m = m + 1$ and go to the step 7 of this procedure. Otherwise, results analysis and fixation of the boundary values of variance, average of absolute values and Shannon entropy, which correspond to minimal value of the internal clustering quality criterion calculated using formulas (4)–(6).

Stage 4. Final division of the gene expression profiles into informative and non-informative genes.

9. Formation a set of informative gene expression profiles considering the boundary values of the algorithm parameters determined at step 8 of this procedure.

The next stage of implementation of the procedure, shown in Fig. 1, is step-by-step gene expression profiles clustering with clusters of gene expression vectors

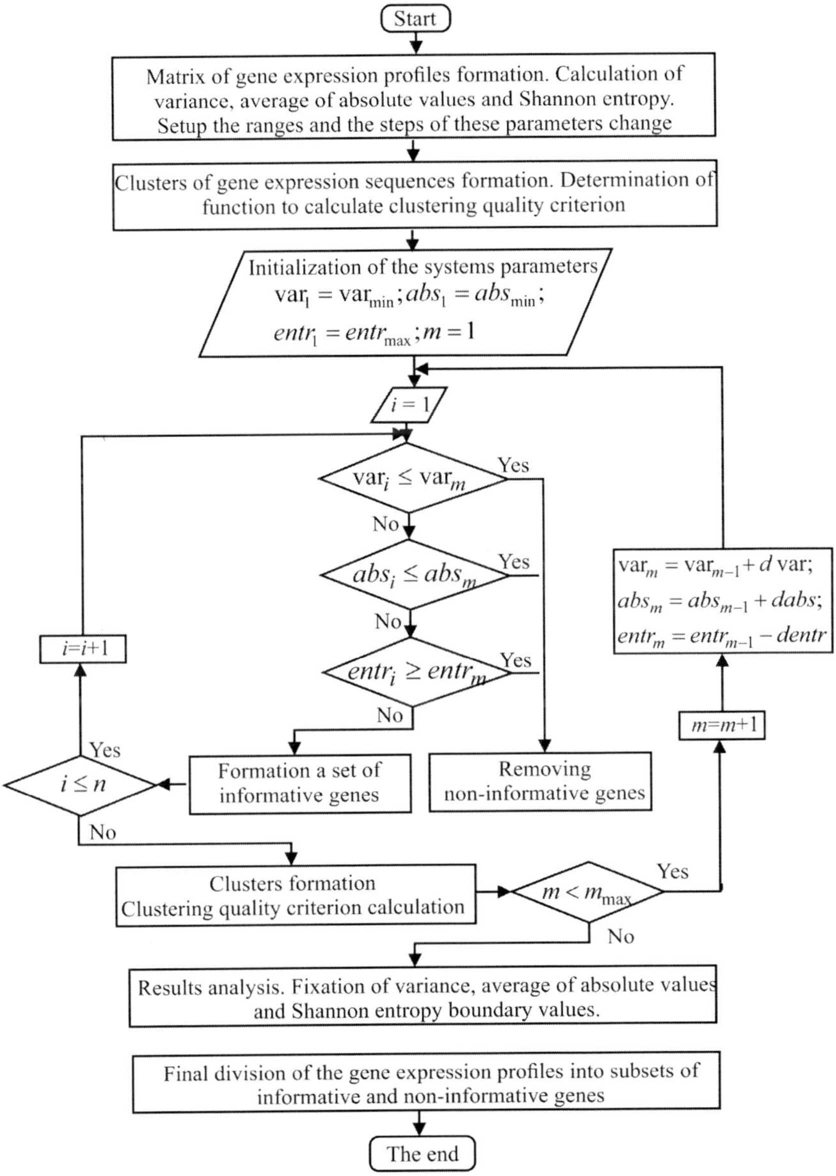

Fig. 2. A structural block chart of algorithm to implement the stage of non-informative gene expression profiles reducing

formation at each step of this procedure implementation. SOTA clustering algorithm (Self-Organizing Tree Algorithm) [15] was used as clustering algorithm within the framework of the proposed technique. This algorithm is a type of self-organizing neural networks based on the complex use of Kohonen maps and

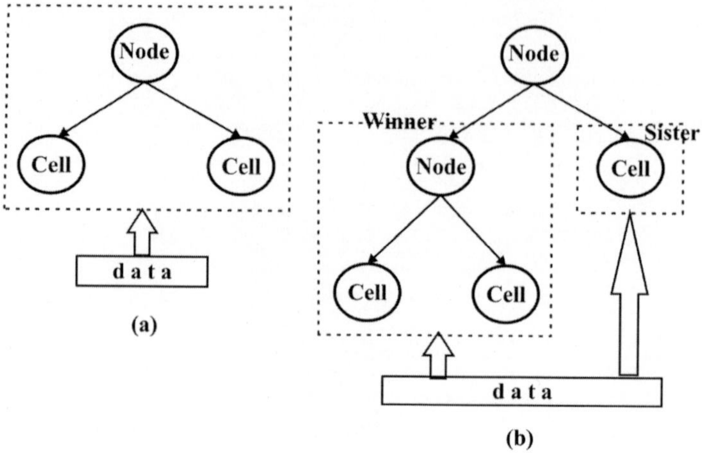

Fig. 3. Cells structure formation procedure: a) initial state of the system; b) state of the system after one cycle

Fritzke algorithm of spatial cell structure growing [17]. A binary topological tree is generated during this algorithm operation. Fritzke algorithm performs self-organization of network output nodes in such a way that quantity of the nodes increases in the field of higher density of objects concentration and decreases in the field of lower density respectively. Figure 3 shows the process of cell structure formation during SOTA clustering algorithm operation. Initially, the system consists of two cells that are connected through an external root node. Each of the cells or nodes is characterized by a feature vector, the number of elements in which is equal to the number of elements in the investigated vector. Implementation of SOTA algorithm assumes the following steps:

1. *Initialization.* Weights which are calculated as an average of the attributes of all of the investigated vectors assign to both the root node and cell vectors. Parameters for correction of the appropriate cells weight setup in accordance with the following conditions: $\alpha_w > \alpha_m > \alpha_s$, where α_w, α_m and α_s are parameters for weight correction of the winner cell, root (parent) node and adjacent cell respectively. Also, it is necessary to setup a boundary value of the variation coefficient E which determines the stopping condition of the algorithm operation.
2. *Adaptation.* The investigated vectors are applied to the input of all external cells during the algorithm operation The degree of proximity of the corresponding vector to the cell weight vector is calculated using the affinity function (in our case this function is the correlation distance). The winner cell, the vector of weights of which has the smallest distance from corresponding vector is allocated in accordance with the principle of "winner takes everything". The weights of the winner cell, its adjacent and root cells are adjusted

in accordance with the formula:

$$C_i(\tau + 1) = C_i(\tau) + \eta \cdot (P_j - C_i(\tau)),$$

where $C_i(\tau)$ and $C_i(\tau + 1)$ are weight vectors for i-cell at step τ and $\tau + 1$ respectively; P_j is j-th features vector which is entered to the system input; η is the parameter that determines the step of adjusting the weights of the winning cell. The alignment of the weights of the neighbouring relative to winning cell is carried out in accordance with the principle: if the cell adjacent to the winning cell has no offspring, the weights of the winning cell, the neighboring cell and the root node are adjusted. Otherwise, only the winner cells are adjusted. The parameter η at the t iteration is calculated as follows:

$$\eta_t = \alpha \cdot \frac{1 - t}{n} \cdot (1 - b\tau),$$

where t is the total number of vectors presented to the system input; n is the maximum number of the investigated vectors; τ is the number of operations in one cycle; b is the coefficient that determines the rate of the η parameter change; α is the parameter for the weights correction of the appropriate cells.

3. *Convergence of algorithm and network formation.* To determine the structure of a clustering tree, the coefficient of variation for each cell is calculated as an average distance from the cell weights to the feature vectors in a current cell:

$$R_i = \frac{\sum_{k=1}^{K} d(P_k, C_i)}{K},$$

where C_i is the weight vector for i-th cell; P_k is the feature vector in this cell; K is the maximum number of the investigated vectors in i-th cell. The total value of the variation coefficient is determined as the sum of the variation coefficients of all external cells:

$$\varepsilon_t = \sum_{i=1}^{S} R_i.$$

The criterion for estimating the algorithm convergence is the relative change in the total variation coefficient:

$$\left|\frac{\varepsilon_t - \varepsilon_{t-1}}{\varepsilon_t}\right| < E, \qquad (7)$$

where E is the boundary value of relative change of the variation coefficient. If the condition (7) is not fulfilled, the further growth of the tree after each cycle begins from the cell, which has the greatest value of relative change of the variation coefficient. This cell divides into two parts and becomes a root node (see Fig. 3b). The values of the weight coefficients of the daughter cells and root node are identical. The algorithm is stopped if condition (7) is satisfied. Thus, the adjusting the value of the boundary parameter allows us to achieve the desired network structure and, as a result, to get the desired structure of the objects distribution in the clusters.

The final stage of implementation of the procedure, shown in Fig. 1, is calculation of the clustering quality criteria for each of the cluster at each of the hierarchical level with following selection of the most informative subsets of gene expression profiles in terms of resolving ability of the investigated samples. This choice is done based on analysis of the minimal values of the internal clustering quality criteria.

3 Experiment

The expression dataset *GSE19188* for early stage NSCLC (lung cancer) [21] was used as the experimental data during the simulation process. This data contains a result of genome-wide gene expression analysis that was performed on a cohort of 91 patients. The authors have used 91 tumor- and 65 adjacent normal lung tissue samples. So, a total number of studied samples was 156. The *rma* method of data preprocessing was used to form the array of gene expression profiles. The initial quantity of genes was 54676. One gene was removed from the data at the first step as the result of statistical analysis. The variance value of this gene expression profile was significantly larger in comparison with variances of other genes. Thus, the initial dataset was presented as the matrix of size: (156×54675).

4 Results and Discussion

Figure 4 presents the box plots of variance, average of absolute values and Shannon entropy, which were calculated for gene expression profiles of the initial dataset. The analysis of these charts has allowed us to form the ranges of these parameters boundary values change. The maximal ranges were divided into 100 equal sections and were taken 85 the first section for variance and average of absolute values and the last 85 section for Shannon entropy. Then, the boundary values of both variance and average of absolute values were changed from minimum to maximum and Shannon entropy from maximum to minimum values in accordance with algorithm shown in Fig. 2. Figure 5 shows the charts of both the number of informative genes and the internal clustering quality criterion versus the step of the algorithm operation. The analysis of the obtained results allows us to conclude that internal clustering quality criteria achieves its minimum value at 31-*st* step. In this case the dataset contains 21431 of genes. Thus, the number of genes was reduced from 54675 to 21431 as a result of the first stage of the proposed technique implementation. Figure 6 shows the boxplots of distribution of variance, average of absolute values and Shannon entropy calculated for reduced dataset of gene expression profiles.

The next stage assumes step-by-step applying SOTA clustering algorithm to reduced dataset with calculation of the internal clustering quality criterion for the formed clusters of genes expression vectors at each hierarchical level of this procedure implementation. The result of the simulation with the use of high dimensional gene expression profiles has shown that applying SOTA clustering algorithm with correlation distance metric leads to division of the initial data

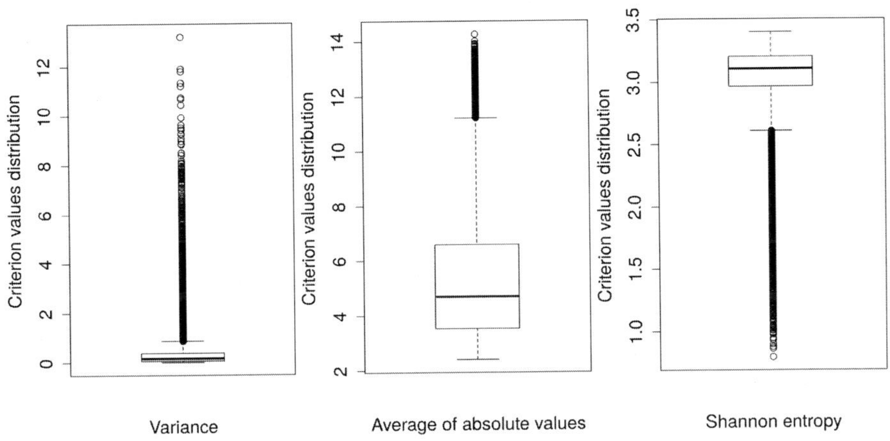

Fig. 4. Boxplots of variance, average of absolute values and Shannon entropy in the case of the initial dataset use

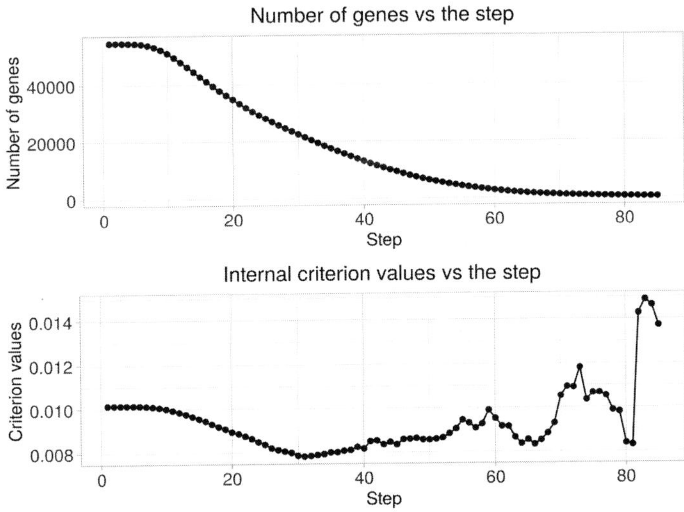

Fig. 5. Charts of the number of informative genes and the internal clustering quality criterion versus the step of the algorithm operation

into two subsets. Moreover, the algorithm stops at the second hierarchical level. This fact can be explained by repetition of the clustering results during the objects partitions. In this reason, we have implemented step-by-step process of SOTA clustering algorithm apply at hierarchical levels from 1 to 9. This process is illustrated in Fig. 7.

Concurrently, we calculate the internal clustering quality criteria for each of the obtained clusters at all hierarchical levels. The analysis of the obtained results has shown that the value of the internal clustering quality criterion was varied

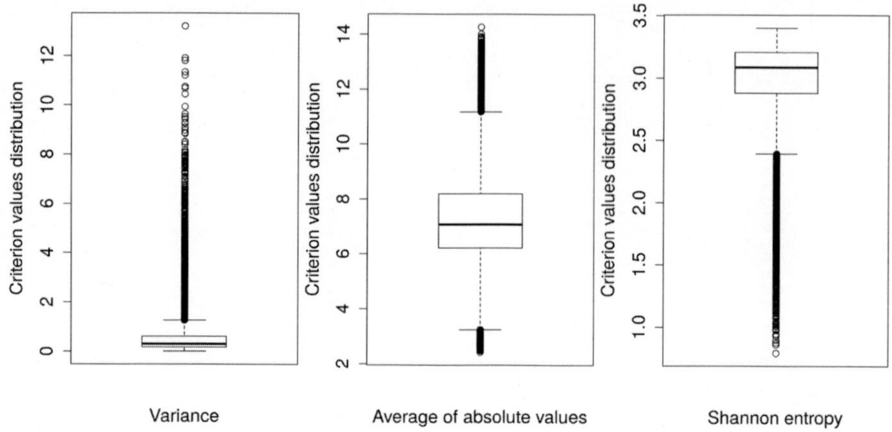

Fig. 6. Boxplots of variance, average of absolute values and Shannon entropy in the case of reduced dataset use

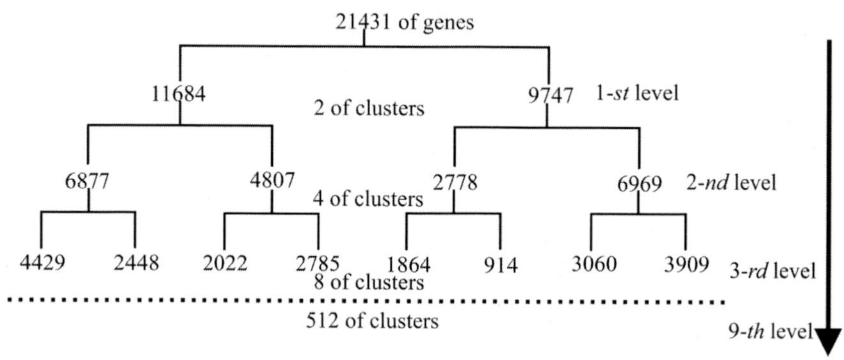

Fig. 7. Results of step-by-step procedure of SOTA clustering algorithm implementation

within the range from 0,001 to 36,27. The minimum value 0.001 of the internal clustering quality criterion is achieved at 9-*th* hierarchical level. In this case the cluster contains 116 of the most informative in terms of resolving ability gene expression profiles. The analysis of the simulation results has also shown that the nearest to minimum value of the internal clustering quality criterion (0,002) are achieved at 5-*th* and 8-*th* hierarchical levels. In these cases, the clusters contain 1126 and 237 of gene expression profiles respectively. These clusters also are interested for the following research in order to create both diagnostic systems or gene regulatory networks reconstruction. Figure 8 shows the heat map of genes expression which are allocated in the most informative cluster (0.001 value of the internal clustering quality criterion).

The analysis of the dendrogram for investigated samples confirms the assumption that these samples can be divided into two classes in dependence on state of the object (tumor or healthy) correctly.

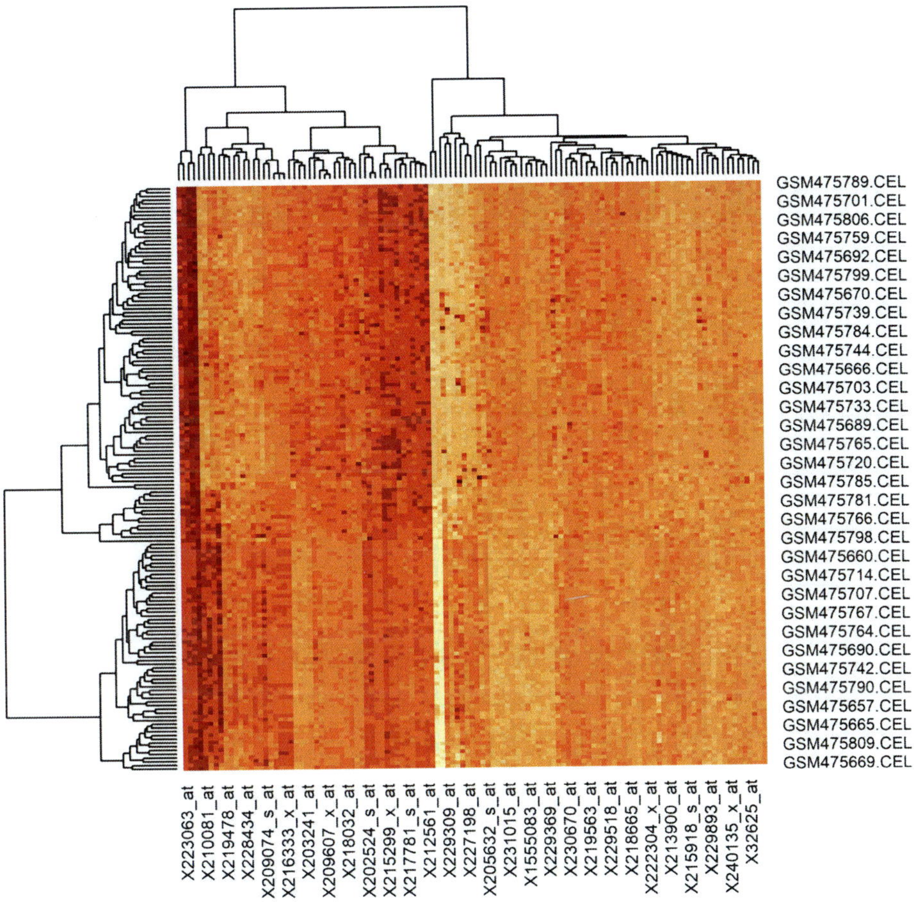

Fig. 8. Heat map of genes expression which are allocated in the most informative cluster

5 Conclusions

In this paper, we have presented the technique of gene expression profiles processing in order to select the most informative genes in terms of their resolving ability during the state of the investigated objects diagnostics. The expression dataset $GSE19188$ for early stage lung cancer disease was used as the experimental data during the simulation process. This data contained 156 total samples: 91 tumor- and 65 adjacent normal lung tissue samples. The rma method of data preprocessing has been used to form the array of gene expression profiles. The initial dataset has been presented as the matrix in size: (156×54675).

At the first step, we have removed non-informative genes in terms of statistical criteria (variance and average of absolute values) and Shannon entropy. It has been assumed that if variance and average of absolute values are less and Shannon entropy is larger than appropriate boundary values, then this gene removes from the data as non-informative. The internal clustering quality criterion has been used to determine the used parameters boundary values. The number of genes has been reduced from 54675 to 21431 as a result of this stage implementation.

The next stage has been assumed step-by-step applying SOTA clustering algorithm to reduced dataset with calculation of the internal clustering quality criterion for the formed clusters of genes expression vectors at each hierarchical level of this procedure implementation. The correlation distance has been used within the framework of the algorithm operation. The hierarchical level has been changed from 1 to 9. The clusters quantity was varied from 2 at the first level to 512 at the ninth hierarchical level. The internal clustering quality criteria for each of the obtained clusters at all hierarchical levels has been calculated during the simulation process. The analysis of the obtained results has shown that the value of the internal clustering quality criterion was varied within the range from 0.001 to 36,27. The minimum value 0.001 of the internal clustering quality criterion is achieved at 9-*th* hierarchical level. In this case the cluster contained 116 of the most informative in terms of resolving ability of gene expression profiles. The clusters, for which the value of the internal clustering quality criterion is minimal one are interested for the following research in order to create both the diagnostic systems or gene regulatory networks reconstruction.

References

1. Affymetrix: Statistical algorithms description document. Affymetrix, Inc., Santa Clara, CA (2002)
2. Alexiou, A., Chatzichronis, S., Perveen, A., Hafeez, A., Ashraf, G.M.: Algorithmic and stochastic representations of gene regulatory networks and protein-protein interactions. Curr. Topics Med. Chem. **19**(6), 413–425 (2019). https://doi.org/10.2174/1568026619666190311125256
3. Astrand, M.: Contrast normalization of oligonucleotide arrays. J. Comput. Biol. **10**(1), 95–102 (2003). https://doi.org/10.1089/106652703763255697
4. Babichev, S.: An evaluation of the information technology of gene expression profiles processing stability for different levels of noise components. Data **3**(4), art. no. 48 (2018). https://doi.org/10.3390/data3040048
5. Babichev, S., Durnyak, B., Zhydetskyy, V., Pikh, I., Senkivskyy, V.: Techniques of DNA microarray data pre-processing based on the complex use of bioconductor tools and shannon entropy. In: CEUR Workshop Proceedings, vol. 2353, pp. 365–377 (2019)
6. Babichev, S., Škvor, J., Fišer, J., Lytvynenko, V.: Technology of gene expression profiles filtering based on wavelet analysis. Int. J. Intell. Syst. Appl. **10**(4), 1–7 (2018). https://doi.org/10.5815/ijisa.2018.04.01
7. Babichev, S., Lytvynenko, V., Skvor, J., Fiser, J.: Model of the objective clustering inductive technology of gene expression profiles based on SOTA and DBSCAN clustering algorithms. Adv. Intell. Syst. Comput. **689**, 21–39 (2018). https://doi.org/10.1007/978-3-319-70581-1_2

8. Babichev, S., Barilla, J., Fišer, J., Škvor, J.: A hybrid model of gene expression profiles reducing based on the complex use of fuzzy inference system and clustering quality criteria. In: 2019 Conference of the International Fuzzy Systems Association and the European Society for Fuzzy Logic and Technology (EUSFLAT 2019). Atlantis Press (2019/08). https://doi.org/10.2991/eusflat-19.2019.20
9. Barbara, D., Wu, X.: An approximate median polish algorithm for large multi-dimensional data sets. Springer-Verlag London Ltd. Knowl. Inf. Syst. **5**, 416–438 (2003)
10. Bolstad, B.M., Irizarry, R.A., Åstrand, M., Speed, T.P.: A comparison of normalization methods for high density oligonucleotide array data based on variance and bias. Bioinformatics **19**(2), 185–193 (2003). https://doi.org/10.1093/bioinformatics/19.2.185
11. Byron, K., Wang, J.T.L.: A comparative review of recent bioinformatics tools for inferring gene regulatory networks using time-series expression data. Int. J. Data Mining Bioinform. **20**(4), 320–340 (2018). https://doi.org/10.1504/IJDMB.2018.094889
12. Calinski, T., Harabasz, J.: A dendrite method for cluster analysis. Commun. Stat. **3**, 1–27 (1974)
13. Chen, Y.J., Kodell, R., Sistare, F., Thompson, K.L., Morris, S., Chen, J.J.: Normalization methods for analysis of microarray gene-expression data. J. Biopharmaceutical Stat. **13**(1), 57–74 (2003). https://doi.org/10.1081/BIP-120017726
14. Chen, Z., McGee, M., Liu, Q., Kong, M., Deng, Y., Scheuermann, R.H.: A distribution-free convolution model for background correction of oligonucleotide microarray data. BMC Genom. **10**(1), 19 (2009). https://doi.org/10.1186/1471-2164-10-S1-S19
15. Dorazo, J., Carazo, J.M.: Phylogenetic reconstruction using an unsupervised growing neural network that adopts the topology of a phylogenetic tree. J. Mol. Evol. **44**(2), 226–260 (1997). https://doi.org/10.1007/PL00006139
16. Eren, K., Deveci, M., Kucuktunc, O., Catalyurek, U.V.: A comparative analysis of biclustering algorithms for gene expression data. Briefings Bioinform. **14**(3), 279–292 (2012)
17. Fritzke, B.: Growing cell structures a self-organizing network for unsupervised and supervised learning. Neural Netw. **7**(9), 1441–1461 (1994). https://doi.org/10.1016/0893-6080(94)90091-4
18. Gentleman, R., Carey, V., Huber, W., Irizarry, R., Dudoit, S.: Bioinformatics and Computational Biology Solutions Using R and Bioconductor. Springer, Heidelberg (2005)
19. Hausser, J., Strimmer, K.: Entropy inference and the James-Stein estimator, with application to nonlinear gene association networks. J. Mach. Learn. Res. **10**, 1469–1484 (2009)
20. Heather, J.M., Chain, B.: The sequence of sequencers: the history of sequencing DNA. Genomics **107**, 1–8 (2016)
21. Hou, J., Aerts, J., den Hamer, B., van Ijcken, W., et al.: Gene expression-based classification of non-small cell lung carcinomas and survival prediction. PLoS ONE **5**(4), art. no. e10312 (2010)
22. Irizarry, R.A., Hobbs, B., Collin, F., Beazer-Barclay, Y.D., Antonellis, K., Scherf, U., Speed, T.: Exploration, normalization, and summaries of high density oligonucleotide array probe level data. Selected Works of Terry Speed, pp. 601–616 (2012). https://doi.org/10.1007/978-1-4614-1347-9_15
23. Kaiser, S.: Biclustering: methods, software and application (2011)

24. Kanishcheva, O., Vysotska, V., Chyrun, L., Gozhyj, A.: Method of integration and content management of the information resources network. Adv. Intell. Syst. Comput. **689**, 204–216 (2018). https://doi.org/10.1007/978-3-319-70581-1_14
25. Kluger, Y., Basry, R., Chang, J.T., Gerstein, M.: Spectral biclustering of microarray data: coclustering genes and conditions. Genome Resourc. **12**(4), 703–716 (2003)
26. Kohane, I.S., Kho, A.T., Butte, A.J.: Microarrays for an Integrative Genomics, p. 236. A Bradford Book, The MIT Press, Cambridge (2003)
27. Lazaridis, E.N., Sinibaldi, D., Bloom, G., Mane, S., Jove, R.: A simple method to improve probe set estimates from oligonucleotide arrays. Math. Biosci. **176**(1), 53–58 (2002). https://doi.org/10.1016/S0025-5564(01)00100-6
28. Lesage, R., Kerkhofs, J., Geris, L.: Computational modeling and reverse engineering to reveal dominant regulatory interactions controlling osteochondral differentiation: potential for regenerative medicine. Front. Bioeng. Biotechnol. **6**, art. no. 165 (2018). https://doi.org/10.3389/fbioe.2018.00165
29. Li, J., Reisner, J., Pham, H., Olafsson, S., Vardeman, S.: Biclustering with missing data. Inf. Sci. **510**, 304–316 (2020). https://doi.org/10.1016/j.ins.2019.09.047
30. Liu, Z.P.: Towards precise reconstruction of gene regulatory networks by data integration. Quant. Biol. **6**(2), 113–128 (2018). https://doi.org/10.1007/s40484-018-0139-4
31. Mishchuk, O., Tkachenko, R., Izonin, I.: Missing data imputation through SGTM neural-like structure for environmental monitoring tasks. Adv. Intell. Syst. Comput. **938**, 142–151 (2020). https://doi.org/10.1007/978-3-030-16621-2_13
32. Mukhopadhyay, A., Maulik, U., Bandyopadhyay, S.: On biclustering of gene expression data. Curr. Bioinform. **5**, 204–216 (2010)
33. Naum, O., Chyrun, L., Vysotska, V., Kanishcheva, O.: Intellectual system design for content formation. In: Proceedings of the 12th International Scientific and Technical Conference on Computer Sciences and Information Technologies, CSIT 2017, vol. 1, pp. 131–138. Institute of Electrical and Electronics Engineers Inc. (2017). https://doi.org/10.1109/STC-CSIT.2017.8098753
34. Park, T., Yi, S.G., Kang, S.H., Lee, S.Y., Lee, Y.S., Simon, R.: Evaluation of normalization methods for microarray data. BMC Bioinform. **4**, 13 (2003). https://doi.org/10.1186/1471-2105-4-33
35. Pontes, B., Giráldez, R., Aguilar-Ruiz, J.S.: Biclustering on expression data: a review. J. Biomed. Inform. **57**, 163–180 (2015)
36. Raddatz, B.B., Spitzbarth, I., Matheis, K.A., Kalkuhl, A., Deschl, U., Baumgärtner, W., Ulrich, R.: Microarray-based gene expression analysis for veterinary pathologists: a review. Vet. Pathol. **54**(5), 734–755 (2017). https://doi.org/10.1177/0300985817709887
37. Schena, M., Davis, R.W.: Microarray Biochip Technology, pp. 1–18. Eaton Publishing (2000)
38. Tkachenko, R., Doroshenko, A., Izonin, I., Tsymbal, Y., Havrysh, B.: Imbalance data classification via neural-like structures of geometric transformations model: local and global approaches. Adv. Intell. Syst. Comput. **754**, 112–122 (2019). https://doi.org/10.1007/978-3-319-91008-6_12
39. Zadeh, L.: Fuzzy logic = computing with words. IEEE Trans. Fuzzy Syst. **4**(2), 103–111 (1996). https://doi.org/10.1109/91.493904
40. Zhao, Q., Xu, M., Fränti, P.: Sum-of-squares based cluster validity index and significance analysis. Lect. Notes Comput. Sci. (Lect. Notes Artif. Intell. Lect. Notes Bioinform.) **5495**, 313–322 (2009). https://doi.org/10.1007/978-3-642-04921-7_32

Comparison of Missing Values Handling Techniques Using MICE Package Tools of R Software and Logistic Regression Model

Vasiliy Nadraga[1(✉)], Volodymyr Smirnov[2], Oleksandra Boiko[2], and Vladyslav Dereko[2]

[1] Departments Personnel Management and Labor Economics, Ukrainian State Employment Service Training Institute, Kyiv, Ukraine
nadraga-vasil@ukr.net

[2] Military-Diplomatic Academy named after Eugene Bereznyak, Kyiv, Ukraine
{687120vd,vladyslavderek}@ukr.net, 2020Boikool@gmail.com

Abstract. The paper presents the result of the research concerning comparisson of various techniques of missing values multiple imputation by chained equations (MICE) with the use of logistic regression at the stage of the model verification. The presence of missing values in the data complicates the data processing and increases the risk factor in the process of solving various problems in various areas of data science techniques use. The simulation process was performed on the basis of the apply of both R and KNIME software tools. The Mammographic Mass dataset from Machine Learning Repository was used as the experimental data during the simulation process. Implementation of the step-by-step process of missing values handling involved the data analysis and missing values visualization at the first step. Then, we have performed the missing values handling with the use of various techniques which are available in MICE package of R software. The quality of the data processing at each step of this procedure implementation was estimated with the use of logistic regression model based on ROC analysis with calculation of the quantitative criteria: AUC (area under roc curve), Akaike and Bayesian ones. At the final step, we have compared various techniques of missing values handling for purpose of selection from them the best variants taking into account the used criteria.

Keywords: Missing values handling · Regression analysis · Akaike quality criterion · Bayesian quality criterion · Multiple imputation · Logistic regression · Roc analysis

1 Introduction

Missing values handling is one of the important steps of data processing. There are various reasons of missing values appearance in the studied dataset: defect

S. Babichev et al. (Eds.): ISDMCI 2020, AISC 1246, pp. 39–50, 2021.
https://doi.org/10.1007/978-3-030-54215-3_3

of both the equipment and data recording, data collecting and their forming can have malfunctions, the participants can fail response to appropriate questions, subjects can withdraw from studies before they will have complete results, etc. Of course, the easiest way of incomplete data processing is division them into complete and incomplete cases and removing samples containing the missing values at the next step of this procedure implementation. However, in this case we loss a large amount of useful information and in many cases this approach is not reasonable in terms of obtained results quality.

There are many techniques of missing values handling nowadays [3,9,10,20]. Mean or Median imputation are one of the easiest techniques of missing data processing. The use of these techniques assumes replace of missing values by the mean or median values of the available data respectively. The main disadvantage of these techniques is the following: they do not consider the statistical dependence of the investigated data. More advanced methods for missing values handling are based on models which take into account the degree of mutual influence of the studied data. These methods are based on K-Nearest Neighbours algorithm (KNN) [8], different types of regression models [30,31], Multiple Imputation by Chained Equations (MICE) technique [14,27], Random Forest method [27], Predictive Mean Matching (PMM) technique [19], etc.

However, it should be noted that each of the enumerated techniques has its advantages and shortcoming and is focused to appropriate type of data. Within the framework of our research, we compared various techniques of missing values handling which are available in MICE package of R software. The logistic regression with calculation the various quantitative quality criteria has been used for the obtained model verification at the final step of this procedure implementation.

1.1 Reviews of Literature

A lot of works are devoted to solve of the missing values handling problem nowadays. So, in [17,23] the authors considered the technique of missing data imputation using a high-speed neural-like structure of the Successive Geometric Transformations Model. The accuracy of the proposed technique is based on the apply of the evaluation criteria: errors of root mean squared. The authors empirically have shown that the proposed technique has a three times higher accuracy of the data handling in comparison with the arithmetic mean algorithm. However, the authors did not compare the proposed method with other well know and effective techniques of missing data handling. In [10] the authors compared the different methods missing data handling in the context of propensity score analysis. Four strategies of missing covariate values imputation considering both the propensity matching and propensity weighting were compared within the framework of their research. The authors have shown that the optimal way to imputate of missing values in covariates of propensity score models depends on the missing data structure and presence of the modification effect.

The results of the research concerning comparison analysis of various packages of R software which are focused to missing data handling are presented in [26]. The authors have compared the effectiveness of the following packages: missForest, Multivariate imputation by chained equations (MICE), and AMELIA-II. Two classifiers, support vector machine (SVM) and logistic regression (LR), were used for the quality evaluation. The authors have shown that effectiveness of the appropriate technique use depend on both the dataset size and the percentage of missing values in the data. In [28] the authors considered analysis of missing data using matrix-characterized approximations. The simulation process was performed on missing datasets with "do not care" conditions and missing datasets with lost values. Moreover, the experiments on datasets with different percent of missing data by using different thresholds were also carried out.

The issues concerning development of a new heuristic approach for treating missing values are considered in [12]. This paper presents an Artificial Bee Colony algorithm on the basis of the use of new approach for missing value imputation using four real-world discrete datasets. The authors have compared the proposed technique with other well-known six methods: Mean, Median, k Nearest Neighbour (k-NN), Multivariate Equation by Chained Equation (MICE), Singular Value Decomposition (SVD), and MissForest (MF). The authors have used the root mean square error and the classification error as the evaluation criteria for evaluate the imputation methods performance. They have also used Naive Bayes algorithm as the classifier within the framework of the research. The analysis of the obtained by authors' results has shown that the proposed technique is more effective in comparison with other used imputation methods. The paper [15] considers a complex model which combines a not missing at random (NAME) selection model with a generalized linear mixed model. The proposed technique is focused to longitudinal binary data. The proposed technique is based on the use of both the distribution of a missing values and the corresponding distribution of the outcome which observed via a fixed and known sensitivity parameter. The authors have shown that the proposed technique is completely transparent relative to assumed about missing data and it can be used as the basis method for sensitivity analysis.

In [11] the authors proposed a mechanism to use data with missing values for designing classifiers which is different from predicting missing values for classification. The proposed technique uses an auto-encoder neural network. The authors have used the training data without missing values to train the auto-encoder so that it is better equipped to predict missing values. In [22] the authors have proposed an extension of a multi-criteria majority-rule sorting model that allows the handling of problems where the decision alternatives contain imprecise or even missing evaluations. In [21], the improved KNN imputation algorithm is introduced to handle the missing values. The overall performance evaluation of the research method was performed using MATLAB simulation environment. The authors have shown that the proposed research method leads to provide the better performance in comparison with existing techniques. The issues concerning investigation of effects of replacing missing values of Volatile Organic Compounds (VOC) that represent gases emitted from human on the accuracy of classifying individual person are considered in [25]. The authors have shown that allowable range of missing gases number in human odor samples relative to human emitted gases are 4 and 3. Moreover, the best accuracy result when missing values are introduced in the odor dataset is achieved in the case of the ensemble Bagged Trees use.

However, we would like to note that despite of the achievements in this subject area, the problem of effective missing values handling has not unambiguous solution nowadays. It can be explained by variety of missing values types and reasons of their appearing. It is obvious that the effectiveness of appropriate machine learning system depends not only on missing values existence. There are other reasons which influence the effectiveness of the system operation [4,5,7,18]. The solution of this problem can be achieved using current techniques of data science which are successful implemented in various areas of scientific research nowadays [6,24].

The aim of the paper is comparison analysis of various techniques of missing values handling on the basis of complex use of MICE package tools of R software and logistic regression model.

2 Materials and Methods

Figure 1 presents the structure block chart of the missing values handling procedure which is implemented within the framework of our research. Implementation of this procedure involves the following stages:

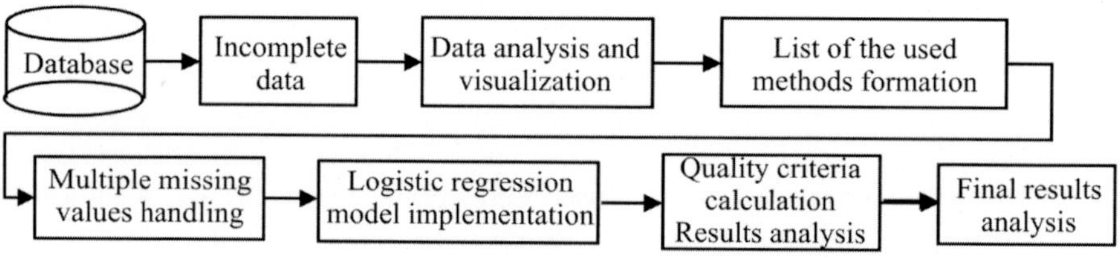

Fig. 1. A structural block chart of the missing values handling procedure

1. Formation of incomplete data. This data is formed as a data frame, where rows and columns present the examined objects and these objects attributes respectively.
2. Data analysis and visualization. Implementation of this step assumes the missing values analysis and visualization at the first step. Then, we perform the statistical analysis of the investigated data.
3. Forming the list of the used methods of the missing values imputation considering the type of the experimental data. The MICE (Multiple Imputation by Chained Equations) package [29] of R software was used at this step. Considering the numeric character of the used data, we have selected the following methods:

- *pmm* (Predictive mean matching);
- *midastouch* (Weighted predictive mean matching);
- *sample* (Random sample from observed values);
- *cart* (Classification and regression trees);
- *rf* (Random forest imputations);
- *mean* (Unconditional mean imputation);
- *norm* (Bayesian linear regression);
- *norm.nob* (Linear regression ignoring model error);
- *norm.boot* (Linear regression using bootstrap);
- *norm.predict* (Linear regression, predicted values);
- *quadratic* (Imputation of quadratic terms);

4. Missing values handling using selected model of missing values handling.
5. Logistic regression model creation at each step of this procedure implementation. In the easiest case, the model of dependence of output parameter versus the input variables can be presented using multiple linear regression equation:

$$y = b_0 + \sum_{i=1}^{n} b_i x_i \tag{1}$$

In this case, a logistic regression model is presented as follows:

$$p = \frac{1}{1 + exp(-(b_0 + \sum_{i=1}^{n} b_i x_i))} \tag{2}$$

where p is the probability of output parameter for the vector X of input variables. It should be noted that Eqs. (1) and (2) can be changed taking into account the character of dependence between appropriate variables.
6. Quality criteria calculation and result analysis. Three criteria were used to evaluate the quality of the used method:

- *Area under curve (AUC)*. This criterion is calculated based on ROC analysis (Receiver Operating Characteristics), implementation of which involves calculation of both one and two types of errors. ROC curve is presented as a plot of sensitivity versus the 1-specificity:

$$sensitivity = F(1 - specificity) \tag{3}$$

Higher value of area under the ROC curve corresponds to better quality of the regression model.
- *Akaike information criterion (AIC)*. This criterion is an estimator of the relative quality of statistical models for a given set of data. Let k be a number of estimated parameters in the model and $\hat{L}$ be the maximum value of the likelihood function for the model. Then, the AIC value of the model is calculated as follows:

$$AIC = 2k - 2ln(\hat{L}) \tag{4}$$

In this case the preferred model is the one with the minimum AIC value.

– *Bayesian information criterion (BIC)* also is a criterion for model selection among a finite set of models. The model with the lowest BIC is preferred. This criterion is formally defined as follows:

$$BIC = kln(n) - 2ln(\hat{L}) \tag{5}$$

where n is the number of observations.

7. Formation of the quality criteria table for the used missing values handling methods. Final results analysis. Decision making about choice of the missing values imputation technique.

3 Experiment

Within the framework of the simulation process, we used the Mammographic Mass dataset [13] from Machine Learning Repository [1] as the experimental data. This data contains the research results concerning breast cancer screening using mammography method. The data set can be used to forecasting the severity (benign or malignant) of a mammographic mass lesion from BI-RADS attributes and the patient's age. The set of the variables contains the following ones: BI-RADS assessment, the patient's age and three BI-RADS attributes together with the ground truth (the severity field) for 516 benign and 445 malignant masses which were identified on full field digital mammograms. The experiment has been performed at the Institute of Radiology of the University Erlangen-Nuremberg between 2003 and 2006. Six attributes in total are conteined in the data [13]:

1. BI-RADS assessment: 1 to 5 (ordinal, non-predictive!);
2. Age: patient's age in years (integer);
3. Shape: mass shape: round = 1 oval = 2 lobular = 3 irregular = 4 (nominal);
4. Margin: mass margin: circumscribed = 1 microlobulated = 2 obscured = 3 ill-defined = 4 spiculated = 5 (nominal);
5. Density: mass density high = 1 iso = 2 low = 3 fat-containing = 4 (ordinal);
6. Severity: benign = 0 or malignant = 1 (binominal, goal field!).

The simulation was carried out based on the KNIME software [2] using *R* tools [16]. A structure block chart of the model which were used during the simulation process is shown in Fig. 2.

4 Results and Discussion

Figure 3 shows the charts of the missing values visualisation. As it can be seen, there are total 162 of missing values in the data: BI-RADS assessment - 2; Age - 5; Shape - 31; Margin - 48; Density - 76. Output parameter Severity has not missing values. Moreover, in accordance with data description, the input variable BI-RADS is non-predictive. So, the regression model will contain only four of input variables.

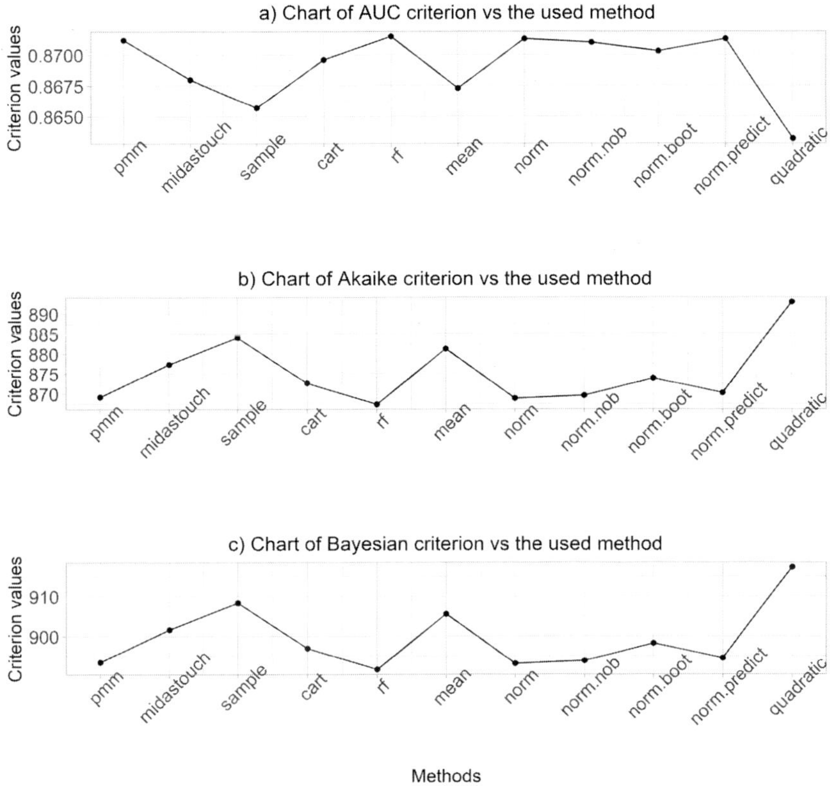

Fig. 5. Charts of the quality criteria vs the method of missing values handling: a) Area under ROC curve; b) Akaike criterion; c) Bayesian criterion

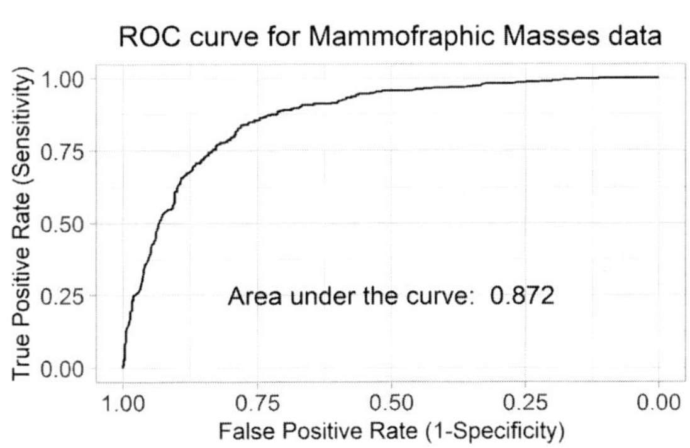

Fig. 6. Chart of ROC curve when the *rf* method of missing values handling has been used

The analysis of the obtained results allows us to conclude about reasonable of the proposed technique use for missing values handling. The optimal method can be determined automatically from list of available ones considering quantitative quality criteria of the data classification. Choice of the classification method and forming the list of methods of the missing value handling are done considering the type of the investigated data. Of course, the list of the quality criteria can be corrected too taking into account the used classification technique (Fig. 6).

5 Conclusions

In this paper, we have conducted the comparison of various methods of missing values handling which are available in package MICE of R software. The Mammographic Mass dataset from Machine Learning Repository has been used as the experimental data during the simulation process. This data contains the results of the research concerning breast cancer screening with the use of mammography method. The logistic regression model has been used as the classification technique to evaluate the quality of appropriate method of missing data handling. Three quantitative quality criteria have been calculated at each step of this procedure implementation: area under ROC curve, Akaike and Bayesian criteria. The list of 11 various methods of missing values handling has been formed considering the type of the used data. The analysis of character of the variables distribution based on correlation analysis has allowed us to form the equation of the logistic regression model which was used to evaluate the effectiveness of the appropriate method of missing data handling. The charts of the quality criteria versus the used method of missing data handling have been created as the simulation results. The analysis of the obtained results has shown that the best method of missing values handling in this case is random forest imputations (*rf*). The AUC criterion achieves its maximum value and AIC and BIC criteria achieve minima ones in the case of this method use. The presented research create the conditions for development of automated system of missing values handling with the use of optimal method of missing values imputation considering the type of the investigated data.

References

1. Center for machine learning and intelligent systems. Machine learning repository. http://archive.ics.uci.edu/ml/datasets/mammographic+mass
2. Knime. https://www.knime.com/
3. Allotey, P., Harel, O.: Multiple imputation for incomplete data in environmental epidemiology research. Curr. Environ. Health Rep. **6**(2), 62–71 (2020). https://doi.org/10.1007/s40572-019-00230-y
4. Babichev, S., Kornelyuk, A., Lytvynenko, V., Osypenko, V.: Computational analysis of microarray gene expression profiles of lung cancer. Biopolymers Cell **32**(1), 70–79 (2016). https://doi.org/10.7124/bc.00090F
5. Babichev, S., Škvor, J., Fišer, J., Lytvynenko, V.: Technology of gene expression profiles filtering based on wavelet analysis. Int. J. Intell. Sys. Appl. **10**(4), 1–7 (2018). https://doi.org/10.5815/ijisa.2018.04.01

6. Babichev, S., Lytvynenko, V., Škvor, J., Fišer, J.: Model of the objective clustering inductive technology of gene expression profiles based on SOTA and DBSCAN clustering algorithms. Adv. Intell. Sys. Comput. **689**, 21–39 (2018). https://doi.org/10.1007/978-3-319-70581-1_2
7. Babichev, S., Lytvynenko, V., Osypenko, V.: Implementation of the objective clustering inductive technology based on DBSCAN clustering algorithm. In: Proceedings of the 12th International Scientific and Technical Conference on Computer Sciences and Information Technologies, CSIT 2017, vol. **1**, pp. 479–484 (2017). https://doi.org/10.1109/STC-CSIT.2017.8098832
8. Barth, A., Wallerman, J., Stahl, G.: Spatially consistent nearest neighbor imputation of forest stand data. Remote Sens. Environ. **113**(3), 546–553 (2009). https://doi.org/10.1016/j.rse.2008.09.011
9. Chhabra, G., Vashisht, V., Ranjan, J.: A review on missing data value estimation using imputation algorithm. J. Adv. Res. Dyn. Control Sys. **11**(7), 312–318 (2019)
10. Choi, J., Dekkers, O., Cessie, S.: A comparison of different methods to handle missing data in the context of propensity score analysis. Eur. J. Epidemiol. **34**(1), 23–36 (2019). https://doi.org/10.1007/s10654-018-0447-z
11. Choudhury, S., Pal, N.: Imputation of missing data with neural networks for classification. Knowl. Based Syst. **182** (2019). Article no. 104838. https://doi.org/10.1016/j.knosys.2019.07.009
12. Cihan, P., Ozger, Z.: A new heuristic approach for treating missing value: ABCimp. Elektron. Elektrotech. **25**(6), 48–54 (2019). https://doi.org/10.5755/j01.eie.25.6.24826
13. Elter, M., Schulz-Wendtland, R., Wittenberg, T.: The prediction of breast cancer biopsy outcomes using two CAD approaches that both emphasize an intelligible decision process. Med. Phys. **34**(11), 4164–4172 (2007). https://doi.org/10.1118/1.2786864
14. Ferguson, K., Yu, Y., Cantonwine, D., McElrath, T., Meeker, J., Mukherjee, B.: Foetal ultrasound measurement imputations based on growth curves versus multiple imputation chained equation (MICE). Paediatr. Perinat. Epidemiol. **32**(5), 469–473 (2018). https://doi.org/10.1111/ppe.12486
15. Fitzmaurice, G., Lipsitz, S., Weiss, R.: Sensitivity analysis for non-monotone missing binary data in longitudinal studies: application to the NIDA collaborative cocaine treatment study. Stat. Methods Med. Res. **28**(10–11), 3057–3073 (2019). https://doi.org/10.1177/0962280218794725
16. Ihaka, R., Gentleman, R.: R: a language for data analysis and graphics. J. Comput. Graph. Stat. **5**(3), 299–314 (1996). https://doi.org/10.1080/10618600.1996.10474713
17. Izonin, I., Kryvinska, N., Vitynskyi, P., Tkachenko, R., Zub, K.: GRNN approach towards missing data recovery between IoT systems. Adv. Intell. Sys. Comput. **1035**, 445–453 (2020). https://doi.org/10.1007/978-3-030-29035-1_43
18. Kanishcheva, O., Vysotska, V., Chyrun, L., Gozhyj, A.: Method of integration and content management of the information resources network. Adv. Intell. Sys. Comput. **689**, 204–216 (2019). https://doi.org/10.1007/978-3-319-70581-1_14
19. Landerman, L., Land, K., Pieper, C.: An empirical evaluation of the predictive mean matching method for imputing missing values. Sociol. Methods Res. **26**(1), 3–33 (1997). https://doi.org/10.1177/0049124197026001001
20. Ma, S., Schreiner, P., at. al.: Multiple predictively equivalent risk models for handling missing data at time of prediction: with an application in severe hypoglycemia risk prediction for type 2 diabetes. J. Biomed. Inform. **103**, 103379 (2020). https://doi.org/10.1016/j.jbi.2020.103379

21. Meera, S., Rosiline Jeetha, B.: Missing value aware optimal feature selection method for efficient big data mining process. Int. J. Recent Technol. Eng. **8**(2), 354–360 (2019). https://doi.org/10.35940/ijrte.B1055.0982S1119
22. Meyer, P., Olteanu, A.L.: Handling imprecise and missing evaluations in multi-criteria majority-rule sorting. Comput. Oper. Res. **110**, 135–147 (2019). https://doi.org/10.1016/j.cor.2019.05.027
23. Mishchuk, O., Tkachenko, R., Izonin, I.: Missing data imputation through STGM neural-like structure for environmental monitoring tasks. Adv. Intell. Sys. Comput. **938**, 142–151 (2020). https://doi.org/10.1007/978-3-030-16621-2_13
24. Naum, O., Chyrun, L., Vysotska, V., Kanishcheva, O.: Intellectual system design for content formation. In: Proceedings of the 12th International Scientific and Technical Conference on Computer Sciences and Information Technologies, CSIT 2017, vol. 1, pp. 131–138. Institute of Electrical and Electronics Engineers Inc. (2017). https://doi.org/10.1109/STC-CSIT.2017.8098753
25. Sabri, A., Alfred, R.: Effects of handling missing values of VOCS gases emitted from human for human detection. Int. J. Recent Technol. Eng. **8**(2), 1405–1412 (2019). https://doi.org/10.35940/ijrte.B1075.0882S819
26. Sarkar, S., Pramanik, A., Khatedi, N., Maiti, J.: An investigation of the effects of missing data handling using 'R'-packages. Adv. Intell. Sys. Comput. **1079**, 275–284 (2020). https://doi.org/10.1007/978-981-15-1097-7_24
27. Shah, A., Bartlett, J., Carpenter, J., Nicholas, O., Hemingway, H.: Comparison of random forest and parametric imputation models for imputing missing data using mice: a caliber study. Am. J. Epidemiol. **179**(6), 764–774 (2014). https://doi.org/10.1093/aje/kwt312
28. Soe, T., Min, M.: Analysis of missing data using matrix-characterized approximations. Stud. Comput. Intell. **845**, 117–129 (2020). https://doi.org/10.1007/978-3-030-24344-9_7
29. van Buuren, S., Groothuis-Oudshoorn, K.: mice: multivariate imputation by chained equations in R. J. Stat. Softw. **45**(3), 1–67 (2011). https://www.jstatsoft.org/v45/i03/
30. Xiao, Y., Tian, Z., Guo, W.: Empirical likelihood for partially non linear models with missing response variables at random. Commun. Stat. Theor. Methods **44**(16), 3523–3540 (2015). https://doi.org/10.1080/03610926.2013.815211
31. Zhu, L.: Empirical likelihood for multidimensional linear model with missing responses. J. Probab. Stat. 473932 (2012). https://doi.org/10.1155/2012/473932

Investigation of Optimization Models in Decisions Making on Integration of Innovative Projects

Anna Kolomiiets(✉) and Victor Morozov

Taras Shevchenko National University of Kyiv, Kyiv, Ukraine
kolomietsa@fit.knu.ua, knumvv@gmail.com

Abstract. The issues of applying the knowledge system for managing innovative projects and programs to stimulate and improve innovation in Ukraine by analyzing the causes of weak innovation development and finding ways to improve this state are considered. The main directions of innovation activities of domestic enterprises, the reasons for failures to bring new products to the market are identified and analyzed. The types of innovations that are used for innovation activities in IT enterprises are highlighted. It is pointed out that the project approach to innovation is almost completely absent. At the same time, it is noted that traditional project management methodologies based on value systems do not fully allow us to effective management of innovative projects due to their significant complexity and risk. This leads to the need to find new approaches that would take into account the specifics of complex innovative projects and development programs, their elements, characteristics and parameters of connections. The issues of development and application of new models of innovative project management based on the value approach are considered. To do this, it is proposed to use a model of scientific cooperation based on franchising in terms of maximizing the values of all stakeholders of the project and the program, based on a comprehensive system approach to determining the state of activity in an innovative project. A structural scheme of interaction with the external environment in the implementation of innovative projects is proposed, which is a conceptual model of research. Based on this model and the scheme of innovative association, a mathematical model is proposed for determining the values of participants and stakeholders of an innovative project or program at all stages of functional interaction within the framework of the initiated project activity.

Keywords: Models · Optimization · Decision making · Innovative projects · Values · Stakeholders · Project management

1 Introduction

The current state of development of the Ukrainian economy indicates a high scientific and technical potential of Ukrainian developments. However, this applies mainly to the IT field, but in other industries, the offers are much more modest [29].

S. Babichev et al. (Eds.): ISDMCI 2020, AISC 1246, pp. 51–64, 2021.
https://doi.org/10.1007/978-3-030-54215-3_4

Systematically considering issues of increasing the competitiveness of domestic products, structural reforms, development of information infrastructure, etc. related to the further development of innovations in Ukraine, it can be concluded that officials pay less attention to the selection of priorities for innovative development, crediting innovations, and stimulating their creators. B.A. Malitskyi [13] focuses on aspects of the impact of the state of innovation on national security. He considers such threats to be chronic underfunding of research and development, reduction in the number of researchers, obsolescence of the experimental base of scientific research, etc. At the same time, the problems outlined above are not considered in sufficient detail in his work.

On the one hand, there is currently no state structure for the development of innovation activities, in contrast to the previous State Fund and the corresponding agency. In fact, there is no organization in Ukraine that would analyze innovation and investment activities in the business sector and support them. But entrepreneurs themselves do not always aim for innovation, they live in the present day, without thinking about the future. As a result, even today's successful machine-building enterprises may lose a market demand in the future.

On the other hand, there are also well-tested special methodologies for managing innovative programs based on the project approach [4]. In particular, this applies to methodologies for managing innovative projects and programs [21]. Their essence is to use a value-based approach to innovation management through project activities. However, modern innovation schemes for managing innovation finance are not considered there.

Thus, there is a scientific task of developing existing project management methodologies to create investment tools to support innovative projects and programs based on a value-based approach.

Because innovative projects cover various fields of expertise and use different technologies the monitoring process and analysis should include all project elements, product components and the emerging relationships between components and elements of the control system should be considered.

This setting of the task is also accelerated by the fact that the information environment for supporting the implementation of innovative projects is changing rapidly, which significantly affects the risk of successful completion of such projects. This makes the solution to this problem technically quite complicated.

2 Analysis of Recent Research and Publications

Questions of innovative activity improvement have been considered in the works of such scientists as: Slavianov [24], Morales [16], Maw-Shin Hsu, Yung-Lung Lai, Feng-Jyh Lin [14], Stephen D. Bryen and Bryen Shoshana [25]. However, project approach was completely absent there, which did not allow to use effective tools for certain results achieving.

Research related to value approach used for innovative projects carried out by such scientists as: Winter [27], Min Zheng, Hiroshi Tanaka [30], Carsten [28], Highsmith [11] and others. At the same time, support for project coordination was studied in the publications of Ukrainian and foreign scientists, such as

Kolesnikova [12], Morozov [17,18], Beloshchitsky [7] and foreign scientists - Nonaka , Takeuchi [20], Turner [26], Milosevic [15], Tom DeMarco [10] and others. However, the problems of innovative cooperation implementing with conditions of profit-maximizing for all interested parties increasing their value interest, are still insufficiently researched and require new development and improvements. These issues considered relevant and are the essence of this work research.

The purpose of the article is to analyze the state of innovation and find the ways for efficiency and effectiveness improving, implementation and applying it through a value approach. Research methods include the statements of social production efficiency theory, economic analysis, comparison, financing, project management, and innovation program management based on P2M methodology.

3 Research and Development of a Conceptual Framework

According to research conducted by Harvard Business Review [23], 72% of new products that have appeared on the market over the past five years have failed (failed to bring the expected profit). This research concerns both start-ups and large enterprises from each industry under consideration.

The main reasons for failure to bring new products to market include [9]: inadequate marketing analysis of the consumer and their needs - 24%, poor product quality - 16%, insufficient marketing strategy - 14%, costs higher than expected - 10%, competition - 9%, incorrect launch time - 8%, technological and production problems - 6%. It is worth noting that the main causes of ineffective launch innovations on the market are also mistakes in the monetization of the product, such as extra features in the product that future consumers are not willing to pay for; low price for innovation, which leads to an excess of demand over supply and the loss of manufacturer profits; the lack of use of existing ideas (patents) due to the company non-compliance pillar with the main type of activity; innovation that the consumers do not need [22].

Oslo Manual Innovation Data Collection and Analysis Guide [2] identifies four types of innovation: product, process, marketing, and organizational. Product innovation is the introduction of a product or service that is new or significantly improved in terms of its properties or uses. This includes significant improvements in technical characteristics, components and materials, in embedded software, in user-friendliness, or in other functional characteristics. Process innovation is associated with the introduction of a new or significantly improved method of production or delivery of a product. This includes significant changes in technology, production equipment and/or software, and new management technologies. Marketing innovation is the introduction of a new marketing method, including significant changes in the design (packaging) of a product, its promotion to the market, or the setting of the sale price. The company's first use of franchising is also considered a marketing innovation. Organizational innovation is the introduction of a new organizational method in a firm's business practice, workplace organization, or external relations.

Analysis of data on the introduction of innovations in the field of IT shows that only 22% of such enterprises implement innovations [3] (Fig. 1).

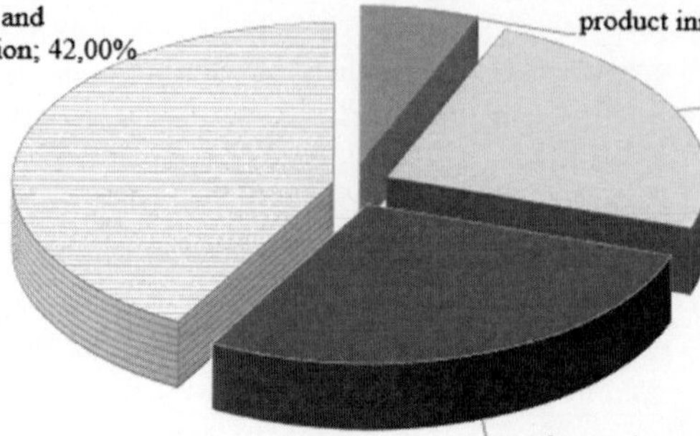

Fig. 1. Distribution of IT enterprises by type of innovation, 2017

The main source of financing (77%) of innovation activity is the enterprises' own funds. Only 12.8% of the companies collaborated with innovative partners from other countries. According to the results of the activities carried out in 2017, 11 patents for inventions and 75 patents for utility models were obtained. The distribution of these enterprises by types of innovation is shown in Fig. 1.

Thus, taking into account the above, it is relevant to search for and implement new approaches to implementing innovative projects at domestic enterprises. The authors consider it appropriate to use a value-based approach to the implementation of innovative IT projects, when the project is considered as a way to maximize value and is defined by P2M as "a commitment to creating value that is based on the project's mission, which must be completed within a certain period, within the agreed time, resources and operating conditions" [5].

Today, in a world where knowledge is spreading rapidly, companies cannot afford to rely only on their own research, they should consider buying licenses to use patents from other companies, and consider integrating with other companies to conduct and implement innovation. The paradigm of open innovation [8] is as follows: companies can and should use external ideas as well as internal ideas; internal inventions that are not used in business should be "extracted" from the company, for example, through licensing, joint ventures, business divisions; the boundaries between the firm and its environment have become more permeable and, accordingly, innovations can easily move in and out (Fig. 2).

4 Mathematical Model Development

Let's look at how you can evaluate whether an innovative association will create value for each of the stakeholders. From the point of view of management theory, a business implementing an innovative project can be represented as shown in Fig. 3) to evaluate the effectiveness of capital investment. At the entrance to the innovation cooperation, there is a multiple vector of management actions, which is transformed in an indeterminate internal system, and at the exit, it is already represented as a vector of the state or result of business functioning [6].

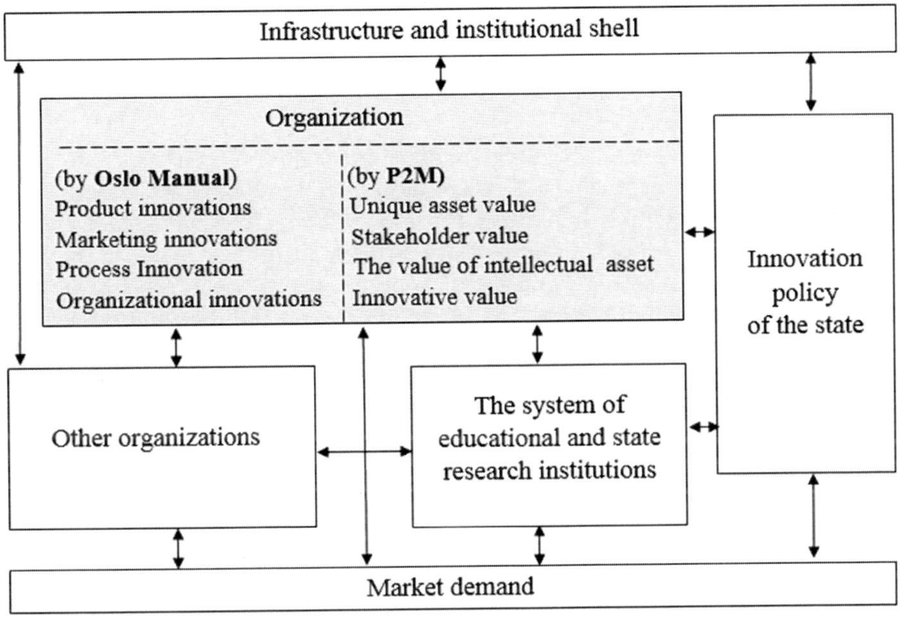

Fig. 2. Scheme of the interaction of the enterprise with the external environment in the implementation of innovative projects (based on [2,5,28])

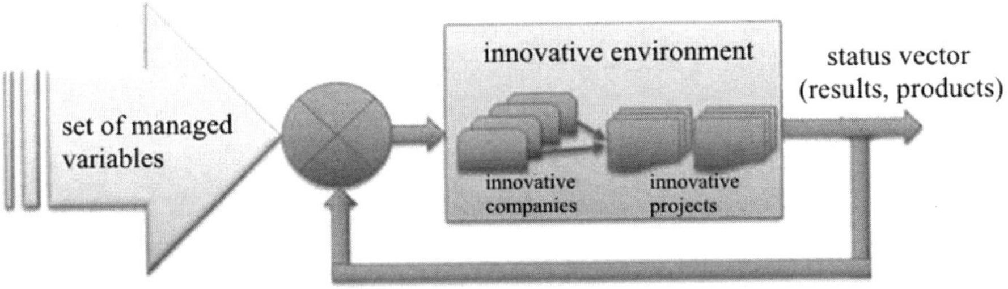

Fig. 3. Schematic representation of an innovative association

One of the forms of integration for creating innovative products is franchising. According to this form of cooperation, the franchisor (owner-value: patent for an invention, utility model, know-how) transfers to the franchisee (integration participants) the right to produce products (business organizations) under the license for intellectual property. Let's have a look, what parameters should be calculated to make a decision about implementing a cooperation.

When making a decision to conduct business on the basis of a franchise agreement, both parties proceed, first of all, from the principle of rationality. In this form of innovative integration, each stakeholder has its own values. For the owner, it is the development of the business concept, creating a mental space for the development of the innovative component of all involved parties, forming a knowledge base, increasing the profit that the system generates. For the integration participants, the value is the creation/implementation of innovative products, staff development, the introduction of a new methodology in the process of business implementation, getting more profit for alternative options.

Let's denote the income for a certain period of an independent enterprise and a potential integration participant by P and Pf, respectively. We will reflect the expenses of these companies for the same period through E and Ef (including royalties). Initial investment (or investment for production modernization) is denoted by I and If (including a lump-sum payment). The discount rate is denoted by d, and the project implementation period is denoted by T. Then, the value of creating for an independent enterprise and integration participant will be calculated using formulas (1) and (2) [19].

$$V = \sum_{t=1}^{T} \frac{P_t - E_t - I_t}{(1+d)^t} \tag{1}$$

$$V_f = \sum_{t=1}^{T} \frac{P_{ft} - E_{ft} - I_{ft}}{(1+d)^t} \tag{2}$$

After the calculations are made, there are three possible scenarios:

$V > V_f$, if the participation in innovative integration is not rational;

$V < V_f$, if there is a need to join integration;

$V = V_f$, if the decision must be made taking into account additional factors.

The principle of rational creation of the system for innovative integration of IT companies is as follows: the profits that will be provided to participants and owner of the values, together with the implementation of innovative cooperation must be greater than the profit which they would have received in individual work.

When making a decision to create a system of innovative integration, the owner must first assess the ratio of costs for its implementation to the potential level of income that it will bring, as well as analyze alternative ways of investing funds. When conducting this analysis, it is necessary to take into account the funds that the owner will invest to create and maintain the viability of the system, the entry fee of the integration system participants in the form of a franchise, as well as periodic payments (royalties).

The owner's income can be described using the formula (3).

$$M\left\{V_{fr}(D)\right\} = \\ -Io + Z + \sum_{j=1}^{n}\left[Fr_j + \sum_{t=1}^{T} \frac{\int_0^p R_0(r_t D - M_{1t}) f(D) dD - M_{0t}}{(1+d)^t}\right] \tag{3}$$

where I_0 is the investments of the owner to create integrating system; Z is the profit from other activities; n is the number of participants; F_r is the franchise cost; d is the rate of return (the discount rate); R_0 is the royalty; M_1 is the cost paid by the participant (to the insurance fund, advertising, training, etc); M_0 is the cost of expenditure on the maintenance of the insurance fund, advertising, training that pays the owner of the value; p is the maximum capacity; $f(D)$ is the distribution of demand for the products; T is the number of periods; r is the profit from sales per unit of output; D is the demand for products.

Investments made by the owner to create the system include: development of the project, documentation, training programs, brand book, communication, reporting, monitoring and control systems; creation of an expert commission, training and professional development department for franchisees, legal assistance department, intellectual property rights protection; launch of a pilot project; other financial expenses.

At the initial stage, in addition to expenses, it receives a profit in the form of a franchise, $Fr \geq \frac{I_o}{n}$ which is paid by the integration participants.

The format for paying royalties may differ depending on the agreement between the value owner and the integration participants. This may be the classic form of paying royalties for a period as a percentage of the profit received. In this case, the integration participant can resort to purposefully understating the profit in the reporting to reduce payments, which should be prevented by the owner by using an effective form of control and selection of potential participants in the system. When there is no royalty, it may mean that the owner's profit is already embedded in the product. However, the opposite situation with no payments is not always useful primarily for the participants themselves, since without making a profit after paying the deductible, the owner may be less interested in helping the participants.

Therefore, the cash flow of the corresponding period for the owner is the value consisting of monetary deductions (royalties), except for the costs of maintaining the viability of the system and paying taxes. Calculating the effect of using innovative integration for participants, you can use the formula (4).

$$M\left\{V_f(D)\right\} = -Fr + L + \sum_{t=1}^{T} \frac{\int_0^p (CP_t - R_o)\mathrm{f(D)dD} - M_{1t}}{(1+d)^t} \tag{4}$$

where CP is the net profit of the integration participant; L is the liquidation value of the equipment at the end of the agreement.

To apply the model for specific IT enterprises, it is necessary to determine the law of distribution of demand for innovative products of a particular participant in the integration system.

To analyze the regularities of stochastic demand for products, we will evaluate statistical characteristics using the normal law whose distribution density is written in the formula (5).

$$f(D) = \frac{1}{\sigma\sqrt{2\pi}} \exp\left[-\frac{(D-M)^2}{2\sigma^2}\right] \tag{5}$$

where M is the mathematical expectation; σ is the standard or root-mean-square deviation.

Thus, knowing the law of distribution of demand for innovative products, we can estimate the mathematical expectation of profit by substituting the formula (5) in (3) and (4) we get the formulas (6) and (7).

$$M\left\{V_{fr}(D)\right\} = -Io + Z$$
$$+\sum_{j=1}^{n}\left[Fr_j + \sum_{t=1}^{T}\frac{\int_0^p R_0\cdot(r_tD - M_{1t})\frac{1}{\sigma\sqrt{2\pi}}\exp\left[-\frac{(D-M)^2}{2\sigma^2}\right]dD - M_{0t}}{(1+d)^t}\right] \quad (6)$$

$$M\left\{V_f(D)\right\} =$$
$$-Fr + L + \sum_{t=1}^{T}\frac{\int_0^p (CP_t - R_0)\frac{1}{\sigma\sqrt{2\pi}}\exp\left[-\frac{(D-M)^2}{2\sigma^2}\right]dD - M_{1t}}{(1+d)^t} \quad (7)$$

The number of participants of integration depends on market size, competitive situation, the forecast of the market situation, monitoring of the system by the owner values and goals of the system.

Determining the optimal number of participants in the network is an important component of the successful implementation of the project. The value that the system creates does not depend directly on the number of integration participants. However, it should be taken into account that when the network partners are oversupplied in a certain territory, their profits decrease, and competition within the system increases, which can cause conflicts or even exit from it. Also, in this situation, the owner may lose control over the participants, which may lead to a deterioration in product quality and lead to increased costs for monitoring the system, which, accordingly, will reduce the level of its value. The number of participants will determine the profit of the owner and participants of the network, as well as the system as a whole; the cost of entrance fees; the possibility of maximum coverage of the target segment in a limited area; the cost of monitoring the system and advertising campaigns, and so on.

Let's consider in more detail the issue of determining their number, in terms of maximizing the value of the system for each of the parties to this form of cooperation.

When making a decision to optimize the number of participants in the system, the value owner should take into account the costs that it carries out by monitoring it, advertising campaigns, monitoring the quality of products and fulfilling its obligations, spending on research and development, and so on. With a significant increase in the number of participants, the growth of these costs is not directly proportional. At the point of saturation, a further set of partners ceases to be profitable for ensuring the value of the system.

However, given that this form of collaboration should be beneficial to all stakeholders, it is necessary to find the number of participants that maximizes the value that the system generates. The cash flow from participants' activities may grow, but even though payments increase with its growth, profits may grow more slowly than cash flow. Taking into account the need to find the optimal value for the entire system as a whole, rather than for an individual participant, it is necessary to find an optimal value for the number of participants that will maximize the value for both parties while meeting the requirement to minimize their costs. To solve this problem, you can use the modern apparatus of optimization theory and find the optimal Pareto set. Let's write the model in the following way:

$$\begin{gathered} F_1 = M\{V_{fr}(D), V_f(D)\} \rightarrow \max, \\ F_2 = \sum_{i=0}^{n} M_{0i} + \sum_{i=0}^{n} M_{1i} \rightarrow \min, \\ I_0 \leq I_{max} \\ \varphi(\bar{D}, \lambda_1, \lambda_2) = -\lambda_1 F_1 + \lambda_2 F_2 \rightarrow \max \end{gathered} \tag{8}$$

where $\varphi(\bar{D}, \lambda_1, \lambda_2)$ is the super goal function; λ_1, λ_2 are the weights of coefficients indicating the influence of participants on the system.

The first two conditions ensure maximization of participants' profit while minimizing their costs. The maximum profit of the system is limited by the I_{max} value. Using the super-goal method, we construct the convolution function (under the normal law of distribution of demand for products). Having solved this equation, we get the value of the optimal number of franchisees by finding a set of effective solutions-Pareto set.

5 Experiment, Results and Discussion

Today there are 504 franchise networks in Ukraine. They are most common in such areas as trade – 57.6%, restaurants – 19.2%, services – 18.4%, manufacturing – 2.5%, IT – 2%. Figure 4 shows a dynamics of growth in the number of franchisors in Ukraine. Examples of creating an innovative integration system based on franchising in IT are given in Table 1. Using the proposed mathematical model for these IT projects, we will perform calculations in Mathcad. Figure 5 shows the laws of distribution of demand for innovative products, the possible value that the system generates for the owner and integration participants based on the optimal number of integration participants for Fun pay, Fig. 6 – for 1 C and Fig. 7 for AppGlobal.

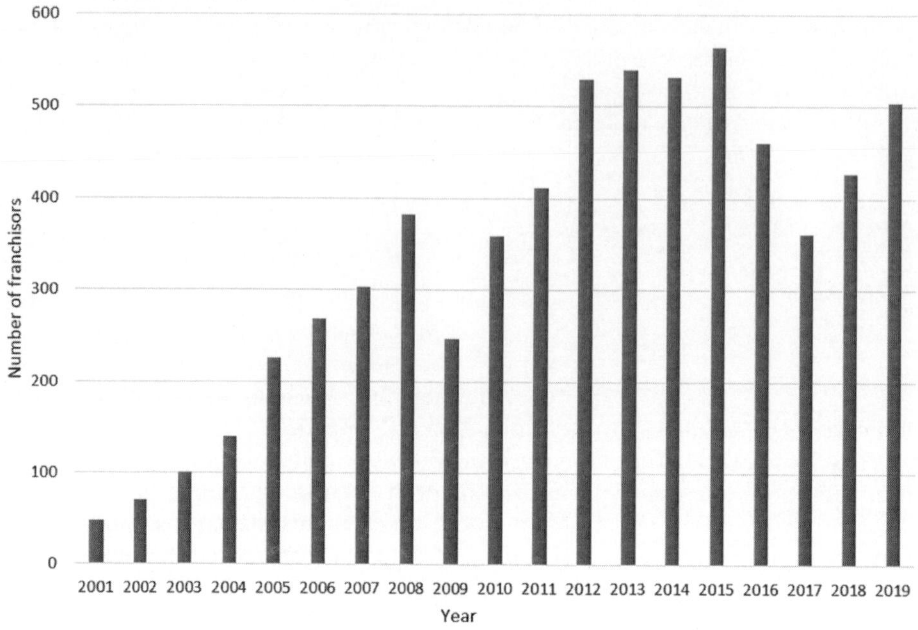

Fig. 4. Dynamics of growth in the number of franchisors in Ukraine [1]

Table 1. Examples of implementation of innovative projects in IT when creating an integration system, compiled on the basis of [1,19]

Title	Investment, $	Fee, $	Royalty	Description
FUNPAY	7500	2500	10% of profit	Payment for the service and goods of likes instead of real money
INFOVISION	2600–260000	–	It depends on the population	Online advertising sector
PROMBAZA	3000	1000–5000	120$ per month	Internet portal sectoral issues
AppGlobal	2000	2000	600$ per month	Mobile application developer
1C	3800	1200	40$ per month	Publishing and support of computer programs

Thus, it is established that the optimal number of participants (taking into account the law of distribution of demand for products) for Funpay is 157, for 1C – 281 franchisees, for AppGlobe – 263 franchisees. The number of system participants that maximize value for the owner and participants of the integration system is indicated.

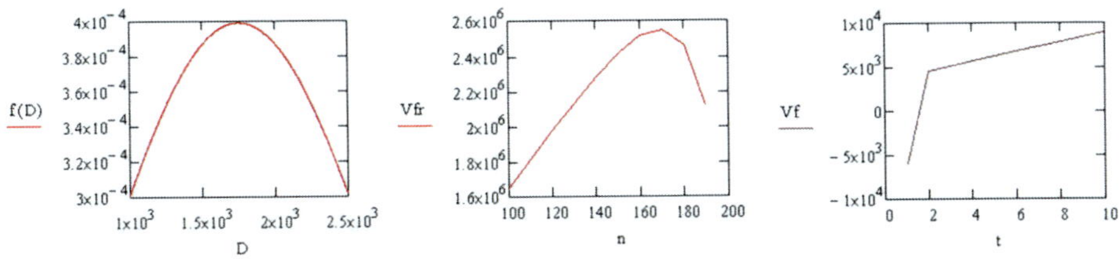

Fig. 5. Calculations for Funpay

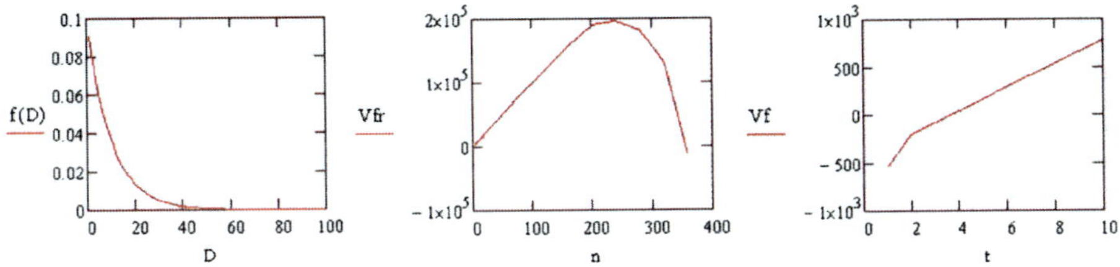

Fig. 6. Calculations for 1C

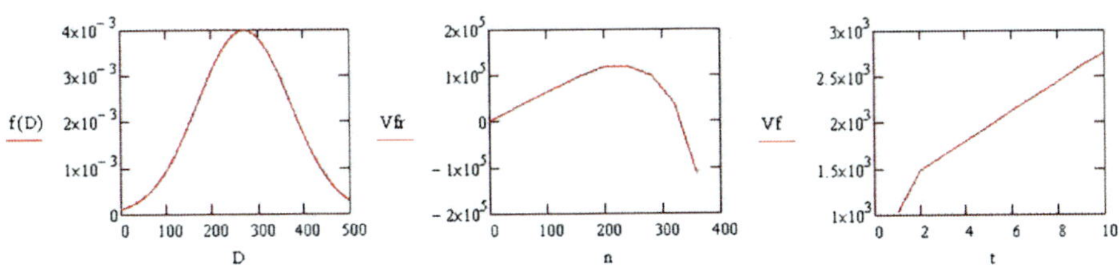

Fig. 7. Calculations for AppGlobal

6 Conclusions

The reasons for failure to bring new IT products to the market, as well as the state of innovation activity of domestic and international IT enterprises, are analyzed. The types of innovations for which innovative activity is carried out at such enterprises are highlighted. This analysis shows the relevance of introducing new management approaches, including value-based ones. Therefore, the issues of development and application of new models of innovative project management based on the value approach are considered. A structural scheme of interaction with the external environment in the implementation of innovative projects is proposed, which is a conceptual model of research. Based on this model and the scheme of innovative association, a mathematical model is proposed for determining the values of participants and stakeholders of an innovative project or program at all stages of functional interaction within the framework of the initiated project activity.

A comprehensive system approach to determine the state of innovation activity has been proposed. Moreover, we have constructed and analyzed the mathematical models. To implement this, we used a model of scientific cooperation based on franchising in terms of maximizing the values of all stakeholders of the project and the program, based on a comprehensive system approach to determining the state of activity in an innovative project. The use of the proposed mathematical apparatus helps to make a decision about the feasibility of using integration in the activities of IT companies and to calculate the optimal number of system participants, which maximizes the value for all stakeholders for such integration IT systems. For companies planning to join an integration association, this model allows us to calculate the effectiveness of such a solution.

Preliminary analysis has shown that the effectiveness of innovation activities is increasing when using a project-based and systematic approach. The authors consider it appropriate to conduct further research in this area, namely, to identify other forms of cooperation for the implementation of innovative activities of IT enterprises, to build mathematical models of such forms of interaction and conduct their analysis.

References

1. Six tips for taking your franchise global. https://www.entrepreneur.com/article/220386
2. Manual, O.: Guidelines for Collecting and Interpreting Innovation Data. The Measurement of Scientific Technological and Innovation Activities, 3rd edn. OECD Publishing, Paris (2005)
3. Science, Technology and Innovation. The organization for economic co-operation and development (2016). https://www.oecd.org/sti/STI-Stats-Brochure.pdf
4. A Guide to the Project Management Body of Knowledge (PMBOK®). Delaware, Pennsylvania, Newton Square 1907-3299, USA: Project Management Institute Four Campus Boulevard, 6th edn. (2017)
5. A Guidebook of Program and Project Management for Enterprise Innovation. Project management Association of Japan, 3rd edn. (2017)
6. Belenets, P., Embulaev, V., Latkyn, A.: The use of franchising in business as the basis for increasing the efficiency of sales of goods and services. VHUES, Vladyvostok (2016)
7. Biloshchytskyi, A., Kuchansky, T., Andrashko, Y., Biloshchytska, S.: Development of the Markovian model for the life cycle of a project's ben a method for the identification of scientists' research areas based on a cluster analysis of scientific publications. East.-Eur. J. Enterp. Technol. **89**(5/2), 4–10 (2017). https://doi.org/10.15587/1729-4061.2017.112323
8. Chesbrough, H.: Open Innovation: The New Imperative for Creating and Profiting from Technology. Harvard Business School Press, Boston (2003)
9. Cooper, R.: Winning at new products: creative value through innovation. Basic Books (2017). https://hbr.org/2014/09/the-silentkiller-of-new-products-lazy-pricing/
10. DeMarco, T., Lister, T.: Waltzing with Bears. Risk Management in Software Development Projects. Moscow: Izd. Company p.m.Office (2005)

11. Highsmith, J.: Agile Project Management: Creating Innovative Products, 1st edn. Addison-Wesley Professional, Boston (2004)
12. Kolesnikova, K., Olekh, T.: Development of the Markovian model for the life cycle of a project's benefits. East.-Eur. J. Enterp. Technol. **95**(5/4), 30–39 (2018). https://doi.org/10.15587/1729-4061.2018.145252
13. Malytskyj, B.A.: Nauka, tekhnolohyy, ynnovatsyy y natsyonal'naia bezopasnost': teoretycheskye y prykladnye voprosy [science, technology, innovation and national security: theoretical and applied issues]. KZhT, Sofyia, Makarov (2014)
14. Maw-Shin, H., Yung-Lung, L., Feng-Jyh, L.: Effects of industry clusters on company competitiveness: special economic zones in Taiwan. Rev. Pac. Basin Financ. Markets Policies **16**(03), 1350017 (2013). https://doi.org/10.1142/S0219091513500173
15. Milosevic, D.: A Set of Tools for Project Management. DMK Press, IT Co (2006)
16. Morales, J.: The most commonly applied methodologist in technology foresight. In: Proceedings of the UNIDO Technology Foresight Conference for Central and Eastern Europe and the Newly Independent States, pp. 170–178 (2001)
17. Morozov, V., Kalnichenko, O., Kolomiiets, A.: Research of the impact of changes based on external influences in complex It-projects. In: Proceedings of the 2019 IEEE International Conference on Advanced Trends in Information Theory. 18-20 December Kiyv, Ukraine, pp. 481–488 (2019)
18. Morozov, V., Kalnichenko, O., Mezentseva, O.: The method of interaction modeling on basis of deep learning of neural networks in complex It-projects. Int. J. Comput. **19**(1), 88–96 (2020)
19. Morozov, V., Kolomiiets, A., Steshenko, G.: Learning through practice in management projects master program implementation approach. In: Proceedings of the 2017 IEEE 9th International Conference on Intelligent Data Acquisition and Advanced Computing Systems: Technology and Applications, IDAACS, vol. 19(1), pp. 935–939 (2017)
20. Nonaka, I., Takeuchi, H.: The Knowledge-creating Company: How Japanese Companies Create the Dynamics of Innovation. Oxford University Press (2011)
21. Peter, M., Jeffrey, K.: The Wiley Guide to Project Organization and Project Management Competencies, 1st edn. Wiley, New York (2017)
22. Ramanujam, M., Tacke, G.: Monetizing Innovation: How Smart Companies Design the Product Around the Price. Wiley, New York (2016)
23. Sarah, G.: The Silent Killer of New Products: Lazy Pricing. Harvard Business Review (2014). https://hbr.org/2014/09/the-silent-killer-of-new-products-lazy-pricing
24. Slavianov, A.: Balance model of resource support of innovative development programs. Int. J. Prof. Sci., **11**, 6–9 (2019). http://scipro.ru/article/01-11-2019
25. Stephen, D., Bryen, S.: Take advantage of the reserves for innovation (2016). https://www.defensenews.com/opinion/commentary/2016/05/19/take-advantage-of-the-reserves-for-innovation/
26. Turner, R.: Guide to Project-based Management. Grebennikov Publishing House, Moskow (2007)
27. Winter, W., Smith, C., Morris, P., Cicmil, S.: Directions for future research in project management: the main findings of a UK government-funded research network. Int. J. Proj. Manag. **24**, 638–649 (2006). https://doi.org/10.1016/j.ijproman.2006.08.009

28. Wolff, C., Hussein, B.: Evaluating the impact of involving students in producing learning aids in project management. The animation project. In: Proceedings of the 2019 IEEE 10th International Conference on Intelligent Data Acquisition and Advanced Computing Systems: Technology and Applications, IDAACS, pp. 933–937 (2019). https://doi.org/10.1109/IDAACS.2019.8924296
29. Yakovlev, A.: Analiz stanu innovatsijnoi diial'nosti v ukraini ta shliakhy joho polipshennia. Nauka ta naukoznavstvo **2**, 29–44 (2018)
30. Zheng, M., Tanaka, H.: Impact of chinese confucian cultural value on management and employee performance in China's service industry: a case study of service innovation under program management environment. In: Proceedings of the 2019 IEEE 12th International Scientific and Technical Conference on Computer Sciences and Information Technologies (CSIT), pp. 787–791 (2019). https://doi.org/10.1109/STC-CSIT.2017.8099424

Model of the Facial Emotions Expressions Based on Grouping Classes of Feature Vectors

Olexander Barmak[1], Oleg Kalyta[1], Iurii Krak[2,3](✉), Eduard Manziuk[1], and Vladyslav Kuznetsov[3]

[1] Khmelnytskyi National University, Khmelnytskyi, Ukraine
barmakov@khnu.km.ua, oleg.kalyta@gmail.com, eduard.em.km@gmail.com
[2] Taras Shevchenko National University of Kyiv, Kyiv, Ukraine
krak@univ.kiev.ua
[3] Glushkov Cybernetics Institute, Kyiv, Ukraine
kuznetsowwlad@gmail.com

Abstract. The characteristic forms of facial expressions of the emotional states of a human are typical of a rather large degree of generalization on the basis of common physiological structures and the location of the muscles that form the human face. This circumstance is one of the main reasons for the commonality of human display of emotions that are reflected in the face. By the nature and form of facial expressions on the face it is possible, with high probability, to determine the emotional state of a human with some correction on the part of cultural characteristics and traditions of certain groups. In accordance with the existence of common mimic forms of emotional displays, an approach is proposed to create a model of recognition of emotional displays on a human's face with relatively low requirements for the means of photo, video-fixation. The creation of the model is based on the implementation of the hyperplane classification of mimic displays of major emotional states. One of the main advantages of the proposed approach is the low computational complexity, which makes it possible to implement a system of recognition of changes in the emotional state of a human by mimic displays without the use of specialized equipment. In addition, the model formed on the basis of the proposed approach allows obtaining proper recognition accuracy with low requirements for quality image characteristics, which allows extending the scope of practical application to a great extent. One example of practical application is the control of the driver while driving a transport, complex production operator and other automatic visual surveillance systems. The set of identified emotional states is formed in accordance with the assigned tasks and gives the opportunity to focus on the recognition of facial expressions and to group characteristic structural displays based on the set of distinguished characteristic features.

Keywords: Emotions · Facial expressions · Recognition · Hyperplane classification · Simplified classification model

S. Babichev et al. (Eds.): ISDMCI 2020, AISC 1246, pp. 65–76, 2021.
https://doi.org/10.1007/978-3-030-54215-3_5

1 Introduction and Literature Review

The natural display of a human's emotional state is the facial expression. There is also an inverse dependence, according to which, by the expression of a human's face, it is possible with a high probability to describe his/her internal emotional state. In addition, the facial expression is an integral part of "emotional communication". Thus people communication is accompanied by a certain facial expression and demonstrates the relationship of the person to the interlocutor and forms the emotional coloring of the conversation. It should be noted that ethnic-cultural differences at the comparison of the human emotional state and facial expression differ in a much smaller way than, for example, linguistic ones. Characteristic external facial expressions of emotions are practically common and exhibit some minor differences that are found only in the aspect of emotional coloring. There is also a difference in the extent of emotions display on the face, that is manifested in the arts, as an example - theatrical activity, and the cultural traditions of a particular group, as an example - the traditions indicate the undesirability of expressing emotions and promote emotional restraint, and therefore "stone" faces. Thus, the mimic displays of a human emotional state are practically the same for the whole population of people, especially in the main characteristic forms and are an integral part of communication and one of the main forms of visual expressions of human emotions. Based on this, the Swedish scientist Carl-Herman Hjortsjö (1969) developed a system of classification of emotions according to facial expression [8].

This is one of the first works in establishing the characteristic mimic forms. These studies were further developed by Ekman and Friesen (1978), who expanded the classification system [5] and, after conducting a series of in-depth studies, greatly expanded the range of emotional displays [6] (2002). Emotional display is characterized not only by the static shape of the face. This way, the most characteristic form of expression of emotion is shown and in its most expressed form. It is natural that the human emotional state is dynamic and is characterized by changes in the phases of extreme displays within one emotion and a general change in successive emotional states. Therefore, Ekman and Friesen suggested fixing the display of emotional state by encoding the movements of the facial muscles. This approach is the closest to the natural display of a human's emotional state in a dynamic form and has become widespread in a variety of fields, in particular, psychology and animation. Taking as a basis the systematic approach of the external displays of emotional state, a Facial Action Coding System (FACS) has been developed, which identifies faces on video, finds necessary signs and classifies emotions [2,7].

Based on the FACS system, almost any emotion can be encoded and described using so-called action units (AU), which means certain structural elements of emotion formation. These structural elements can be matched with such a dynamic display of the physiology of the parts of the human face as the tension or relaxation of one or more facial muscles [7].

By grouping compatible movements of muscular structures, the main areas of the face are distinguished, the changes of which form facial expressions (mimic), that is, characteristic displays that directly reflect the emotional state [3,13,17]. At a certain level of aggregation, the most influential on mimic expressions facial areas with eyebrows, eyes, and mouth can be distinguished. By grouping the structural components of mimic displays, a set of qualitative characteristics of displacements of points or groups of points can be formed, which are given in Table 1. Points are determined by face (by photo or video) and correspond to the areas of finding the muscular structures of the face, which most significantly affect the formation of facial expressions of emotions within the study.

Table 1. Qualitative characteristics of the facial areas

Emotions	Eyebrows	Eyes	Lips
Anger	Lowered	Without changes	Compressed
Grief	Without changes	Lowered upper eyelid	Lowered corners
Delight	Without changes	Skightly lowered upper eyelids	Without changes
Fear	Raised	Raised upper eyelids	Raised lips corners
Joy	Without changes	Raised outer corners of the eyes	Slightly raised lips corners

The purpose of the article is to conduct research in the direction of developing an approach for forming a model of representation of mimic displays in the form of separate sets of vectors of characteristic features that form the external representation of emotional states, that is, by identifying the sets of external mimic displays, implement a system of emotional states recognition [9–11,16]. Model development is based on hyperplane classification [1,15]. The main advantage of the proposed approach is the low computational complexity, which allows realizing the recognition of changes in the human emotional state by mimic displays without the use of specialized equipment.

2 Method of Model Synthesis

Based on the need for identification of facial expressions by means of conventional cameras with low resolution and according to the results of Table 1 the following gradation for features that are located on the facial areas is introduced:

1. Eyes: open, narrowed, normal;
2. Lips: stretched, compressed, normal;
3. Eyebrows: raised, lowered, normal.

According to the above graduation, the mimic displays of emotions obtained in the study are presented as follows (see Table 2).

The representation of mimic display in the context of emotional states shown in Table 2 will serve as a basis for the subsequent synthesis of the model by which detection will be carried out.

Empirically defined features are formally submitted as follows:

Table 2. Presenting mimic expressions of emotion by qualitative characteristics

	Joy	Grief	Fear	Anger	Delight
Eyes	Normal	Normal	Open	Narrowed	Normal
Lips	Stretched	Compressed	Normal	Normal	Stretched
Brows	Raised	Lowered	Raised	Lowered	Lowered

- x_1 is the sign of facial expressions of the eyes area;
- x_2 is the sign of facial expressions of the lips area;
- x_3 is the sign of facial expressions of the eyebrows area.

where $x_1, x_2, x_3 \in [0, 1]$, while $x_1 \in [0, 0.2]$ is for narrowed eyes; $x_1 \in [0.4, 0.6]$ is for normal eyes; $x_1 \in [0.8, 1]$ is for open eyes; $x_2 \in [0, 0.2]$ is for compressed lips; $x_2 \in [0.4, 0.6]$ is for normal lips; $x_2 \in [0.8, 1]$ is for stretched lips; $x_3 \in [0, 0.2]$ is for lowered eyebrows; $x_3 \in [0.4, 0.6]$ is for normal eyebrows; $x_1 \in [0.8, 1]$ is for raised eyebrows.

The gaps in the proposed synthetic model that are not used $((0.2, 0.4), (0.6, 0.8))$ serve to simulate good resolution between different emotional states at their classification.

3 Method of Hyperplane Classification

Thereby, the sets of characteristic features to identify the corresponding emotional states are formed. Mimic displays naturally have boundary states and at the same time are characterized by a typical set of features of the external displays of these states. Accordingly, the limits of displays of signs were empirically determined. However, it should be noted that there is a natural distribution and mentioned boundaries correspond to the most typical displays. To determine the forms of grouping states as a reflection of a set of features, we apply methods of data visualization to display characteristic forms of data grouping [1,15]. These grouped data determine the classes of emotional displays. An important aspect is the ability to set the boundaries of class division, in our case, emotional states. The location of borders can be adjusted depending on the quality and quantity of data. This form of adaptation allows a more flexible application of the technology of recognition of mimic expressions of emotions and adaptation to the data body. Although mimic displays due to physiological particularities are typical displays of emotions, the boundaries of data separation need to be adjusted for accurate distinguishment of classes of emotions. This allows the use of hyperplane classification. We adapt the use of hyperspace separation technology to the recognition of emotional displays.

The input data, and more specifically their features, must be transformed as follows:

1. reducing multidimensional space of features to two-dimensional (with a small number of features, the loss of information content is being minimized while reducing the dimension);

2. the visual representation of data in a two-dimensional space with the usage of a software tool;
3. construction the boundaries of classes of grouped data separation;
4. matching data groups to mimic emotional states;
5. display of the boundaries of emotional states classes in the original space of signs in the form of hyperplanes;
6. installation of spatial hyperbodies limited by hyperplanes with data belonging to these hypercubes;
7. construction of the rules for determining the belonging of a new data element to a class as finding its spatial position relative to hyperplanes.

Multidimensional scaling (multidimensional scaling - MDS) is used to reduce the dimensionality of the feature space [4,12,14,18,20]. The dimensional characteristic is taken to be the relative weighting of the objects relative to each other by a set of features. The distance measure of objects is used as a measure of difference.

The output data is presented in the form of a matrix "object-attribute" and we determine the distance between i and j objects as $\delta_{ij} = d(X_i, X_j)$. The position of the objects in the multidimensional space is defined as follows $X_i = \{x_{i1}, x_{i2}, ..., x_{in}\}$, $i = 1, ..., n$. The dimensionality of space is set according to the number of characteristics of emotional displays of mimicry that we choose. We apply the Euclidean metric to calculate distances between objects:

$$d(X_i, X_j) = \left(\sum_{k=1}^{n} (x_{ik} - x_{jk})^2\right)^{\frac{1}{2}} \tag{1}$$

When scaling space, especially of high dimensionality, the data may be distorted and lost. To minimize this effect, we use a measure of reflection quality as stress, and write it as a least squares function:

$$\sigma = \sum_{i<j} w_{ij} \left(d(Y_i, Y_j) - \delta_{ij}\right)^2 \tag{2}$$

Such a record allows minimizing the difference between the distances between the data in the reduced and the original space. Since it is quite difficult to minimize this function in general, we use the approximate method of replacing the original function with a more spacious and controllable surrogate function. This is based on the iterative use of the SMACOF algorithm [12]. Accordingly, the minimization of the space loss function is minimized.

To enable optimization and efficiency of space reduction, we use the following steps. First, we form a matrix of pairwise distances of the input data. Next, it is needed to form a matrix of the total distances between the objects. For the resulting matrix, double centering is required to optimally display relative position data. In this case, the visual display will be optimally centered in regards to the data. The eigenvalues and eigenvectors of the matrix are determined. The next step is to optimize the position of the data position using the surrogate function.

The goal is to obtain coordinate data that can be displayed on a plane, that is, in two-dimensional space. The data obtained will have two generalized coordinates. Since we use data to get a model for recognizing emotional displays, we use a training sample. That means the data has been pre-marked. In this way, we can observe the grouping of data on the visual plane and determine by which emotional displays this data is grouped. The most important aspect is the ability to split these data groups. The better they are separated and the greater the distance is between these groups, the greater the tolerance we can give to the dispersion of the values of the features. Thus, the separation of data by a set of features is important. The advantage of visualization is the ability to perform effective data analysis for resolution, emission, and boundary classes with effective dispersion tolerances for generalized features. In accordance with the grouping of data, we establish classes and graphically draw the boundaries of these classes.

Class boundaries are determined using piecewise linear separators [1,15]. The use of a discriminant linear function is due to the need to solve the projection problem in the initial space of features: A linear classifier $d(\boldsymbol{x})$ is defined by $d(\boldsymbol{x}) = \boldsymbol{W}^T\boldsymbol{x} + w_n$, where $\boldsymbol{x} = (x_0, x_1, ..., x_{n-1})^T$ - the feature vector that defines the image of the classification object; $\boldsymbol{W} = (w_0, w_1, ..., w_{n-1})^T$ - vector of classifier weight coefficient; w_n- the threshold value.

Class membership is determined by the classifier rule. It means the object location is determined according to the class line. A piecewise linear approach to class boundary formation allows you to form any shape by combining linear elements. If the spatial configuration of an object group is complex, the number of linear boundary elements increases. However, this does not violate the general approach of using a linear divider. Thus, it is possible to form a general nonlinearity of a class boundary by using local linear resolution of localization limit itself. Localization of linear delimitation is determined by visual analysis methods, which is an advantage of the proposed approach.

Further, the linear elements of the boundaries are projected into the original space and form hyperplanes therein. In hyperspace, hyperplanes intersect and form hypercubes. These hypercubes are the spatial volumes of classes. There is no need to find intersections of planes in order to determine the belonging of a new object to a class. It is important to create boundaries in two-dimensional visual space and the corresponding elements of the hyperplanes in the original space. A two-dimensional space defines the boundary of a class, that is, a defined contour. The new object in relation to this contour is located just like all other elements of this class. Thereby, we can form rules for the class belongings, rules for positioning the objects of the class relative to the piecewise contour of the boundary of the class. The location of the class objects in relation to the boundaries of the class happens in accordance with the rules of the decision tree. The application of the decision tree allows to establish a quick classification process at the obtained model.

The use of visual analysis of the boundaries of grouped objects is a key element in the formation of the classification model. The use of piecewise linear contours of class boundaries makes it possible to form discriminant separators of a rather complex nonlinear dependence. The visual definition of boundaries also makes it possible to optimize the calculations when forming a model by minimizing the number of linear elements and depends on the experience and qualifications of the analyst. If necessary, both the boundaries of classes and the number of linear elements can be changed. The most optimal is to provide the minimum required boundaries of classes.

A key element in the use of the proposed approach is the analyst's involvement in the learning of the classification model. Using the necessary tools, the analyst determines the boundaries of the classes, usually based on the visualized data. In the future, the learned emotion recognition system works independently and is fully automatic.

The learned data classification model is the decision tree for the relative location of data objects in relation to class boundaries. These rules determine the position of the new lens in the feature space relative to the boundaries of the classes and, accordingly, classify the new object.

4 Experiment, Results and Discussions

4.1 Model Analysis by Reference Images of Facial Expressions of Emotional States

The validity of the proposed model is verified by the data obtained in the study [19]. According to the images of relevant emotions from the above study, according to Table 2, the features x_1, x_2, x_3 are formed at appropriate intervals. The generated input is visualized in two-dimensional space by the proposed approach (see Fig. 1).

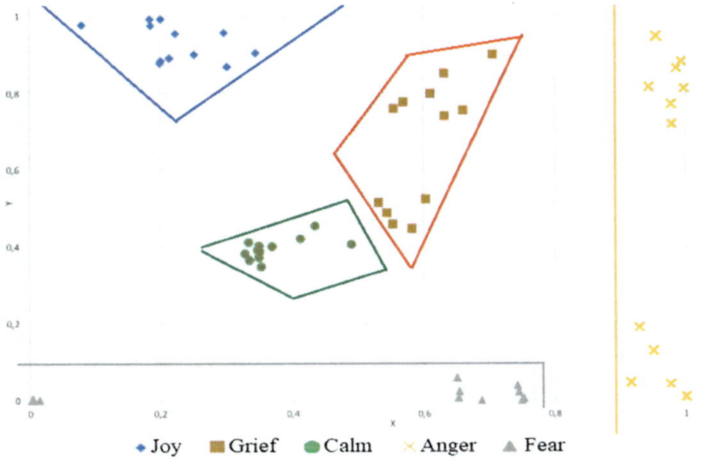

Fig. 1. Visualization of emotional states and piecewise-linear class dividers

Figure 1 shows that the synthesized data are grouped by emotion, which confirms the ability of the proposed model to be used to classify emotional states.

Further, the following steps of the proposed approach, the indicated piecewise linear dividers for the classes are transformed into the hyperplane parameters. Using the obtained parameters of the hyperplanes, a solution tree was constructed for the hyperplane classification of mimic expressions of emotional states.

As can be seen in Fig. 1, the classes of emotions "anger" and "grief" can be visually divided into two groups each. For the emotion of "grief", this is explained by the fact that some of the respondents in the photo have eyelids twisted ($x_1 \in [0, 0.2]$), and the rest of respondents have the eyelids in normal condition ($x_1 \in [0.4, 0.6]$) (see Fig. 2). For the emotion "Anger", some of the respondents in the photo have narrowed eyelids $x_1 \in [0, 0.2]$, and others have expanded ones ($x_1 \in [0.8, 1]$) (Fig. 2).

4.2 Comparison of FACS and the Proposed Model

The FACS system [2] has 100 Action-Unit and characterizes the severity of muscle changes in five levels: A. Minimal, B. Insignificant, C. Explicit, D. Strong or extreme, E. Maximum. For example, emotion "Happiness" is characterized by the 6 and 12 action-unit, emotion "Sadness" by 1, 4 and 15 action-unit, emotion "Fear" is 1, 2, 4, 5, 7, 20, and 26 action-unit.

In the proposed model, emotions are more easily defined than in Ekman's. For comparison, the following emotions are taken:

1. The mimic expression of emotion "Anger" by FACS is defined as (AU - Action-Unit):

(a) brow Lowerer (AU4);
(b) upper lid raised (AU5);
(c) lid Tightener (AU7);
(d) lip Tightener (AU23).

In the proposed by the authors model, "Anger" is characterized as follows:

(a) narrowed eyes;
(b) lowered eyebrows;
(c) compressed lips.

2. The mimic expression of emotion "Grief" by FACS is defined as follows:

(a) Inner Brow Raiser (AU1);
(b) Brow Lowerer (AU4);
(c) Lip Corner Depressor (AU15).

Grief				*Anger*			
	Eyes	*Brows*	*Lips*		*Eyes*	*Brows*	*Lips*
	0.2	0.1	0.5		0.8	0.1	0.1
	0.5	0.2	0.5		0.9	0	0.2
	0.2	0	0.5		0.9	0	0
	0.2	0.1	0.5		0.9	0	0.1
	0.5	0.2	0.5		0.2	0	0.1
	0.5	0.1	0.5		0.2	0	0.2
	0.5	0.2	0.5		0.1	0	0.1
	0.1	0	0.5		0.2	0	0
	0.2	0.1	0.5		0.1	0	0.2
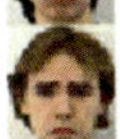	0.5	0.2	0.5		0.1	0	0.1
	0.5	0.2	0.5		0	0	0.1
	0.2	0.1	0.5		0.8	0.1	0.1

Fig. 2. The value of the features for emotions "Grief" and "Anger"

In the proposed by the authors model, "Grief" is characterized by:

(a) normal upper and lower lids;
(b) lowered eyebrows;
(c) compressed lips.

3. The mimic expression of emotion "Fear" by FACS is defined as follows:

(a) Inner Brow Raiser (AU1);
(b) Outer Brow Raiser (AU2);
(c) Brow Lowerer (AU4);
(d) Upper Lid Raiser (AU5);
(e) Lid Tightener (AU7);
(f) Lip Stretcher (AU20);

(g) Jaw Drop (AU26).

In the proposed by the authors model, "Fear" is characterized in the following way:

(a) Lowered upper lids and raised lower lids;
(b) Raised eyebrows;
(c) Normal lips.

4. The mimic expression of emotion "Joy" by FACS is defined as follows:

(a) Cheek Raiser (AU6);
(b) Lip Corner Puller (AU12).

In the proposed by the authors model, "Joy" is characterized in the following way:

(a) Lowered upper lids and raised lower lids;
(b) Normal eyebrows;
(c) Stretched.

The proposed model does not track the lowering of the jaw and corners of the lips. FACS examines the sequence of facial expressions. Sometimes these expressions are opposite, which indicates the excess of features and leads to manifestations of ambiguity. As with the emotion of fear (raising eyebrows and then lowering them). The proposed model tracks only one mimic expression (raised eyebrows), which most typically allows separating emotional states.

5 Conclusions

The proposed approach to the construction of a recognition model of emotional expressions by characteristic mimic displays has allowed us to determine a number of advantages, which are confirmed by the designed and implemented software on the basis of which the test researches have been conducted. Comparative analysis of the Fascial Action Coding System (FACS) and the model have proposed by the authors on the basis of the definition and selection of influential features makes it possible to realize the recognition of emotional expression on the human's face without reducing the qualitative performance indicators. The proposed approach allows us to determine the set of necessary emotional states and to generate the necessary set of characteristic features that create emotional display and are characterized by the proper parameters of the emotions separation by mimic forms.

The limitation of the number of structural features of mimic forms has shown its effectiveness due to the fact that significant features are identified, which together create the conditions for the good separation of these groups and, thus, the separation of emotional states. Reducing the number of structural features enables improving the separation of groups of emotional states by eliminating features that are not characteristic displays, and their presence increases the ambiguity of the detection of states, thereby worsening the expected result. Improving the delineation of groups of characteristic states of features which allows increasing the distances of their relative location, allows to reduce the requirements for qualitative characteristics of data for analysis and to allow some ambiguities in determining the objects of analysis in the photo and video data.

Testing the developed model according to the quality criteria of the results has indicated the possibility and allowed the use of data obtained by conventional means of fixation (camera phones, web-cameras), which significantly expands the scope of practical application due to low requirements for equipment. The developed approach of model creation is based on the application of the hyperplane classification, according to which the hyperplanes of the delimitation of groups of features are determined by the magnitudes of their displays directly on the set of these features. This fact allows us to greatly reduce the computational complexity during the model usage phase and the use of appropriate computing equipment.

The proposed approach allows formating the set of mimic emotional expressions that require detection based on the use of only a characteristic set of well-resolved features, together with low requirements for image quality and computational requirements for the practical use of the model with the proper qualitative parameters of the results of the calculations allows to greatly expand the scope of practical application.

Further authors' research will be aimed to identifying a face on the image obtained from low resolution or long distance camcorders, model parameters in the proper format, and creating appropriate information systems.

References

1. Barmak, A.V., Krak, Y.V., Manziuk, E.A., Kasianiuk, V.S.: Information technology of separating hyperplanes synthesis for linear classifiers. J. Autom. Inf. Sci. **51**(5), 54–64 (2019). https://doi.org/10.1615/JAutomatInfScien.v51.i5.50
2. Duchenne de Boulogne, G.B., Cuthbertson, A.: The Mechanism of Human Facial Expression. Cambridge University Press, Cambridge (1990)
3. Brunelli, R., Poggio, T.: Face recognition: features versus templates. IEEE Trans. Pattern Anal. Mach. Intell. **15**(10), 1042–1052 (1993). https://doi.org/10.1109/34.254061
4. Cox, T.F., Cox, M.A.A.: Multidimensional Scaling, 2nd edn. Chapman and Hall/CRC, Boca Raton (2001)
5. Ekman, P., Friesen, W.V.: Manual for the Facial Action Code. Consulting Psychologist Press, Palo Alto, CA (1978)
6. Ekman, P., Friesen, W.V., Hager, J.C.: The Facial Action Coding System. Salt Lake City, UT Research Nexus eBook (2002)
7. Hamm, J., Kohler, C.G., Gur, R.C., Verma, R.: Automated facial action coding system for dynamic analysis of facial expressions in neuropsychiatric disorders. J. Neurosci. Methods **200**(2), 237–256 (2011). https://doi.org/10.1016/j.jneumeth.2011.06.023

8. Hjortsjo, C.H.: Man's Face and Mimic Language. Studentlitteratur/Lund, Sweden (1970)
9. Krak, Y.V., Barmak, A.V., Baraban, E.M.: Usage of NURBS-approximation for construction of spatial model of human face. J. Autom. Inf. Sci. **43**(2), 71–81 (2011). https://doi.org/10.1615/JAutomatInfScien.v43.i2.70
10. Kryvonos, I.G., Krak, I.V.: Modeling human hand movements, facial expressions, and articulation to synthesize and visualize gesture information. Cybern. Syst. Anal. **47**(4), 501–505 (2011). https://doi.org/10.1007/s10559-011-9332-4
11. Kryvonos, I.G., Krak, I.V., Barmak, O.V., Ternov, A., Kuznetsov, V.A.: Information technology for the analysis of mimic expressions of human emotional states. Cybern. Syst. Anal. **51**(1), 25–33 (2015). https://doi.org/10.1007/s10559-015-9693-1
12. de Leeuw, J., Mair, P.: Multidimensional scaling using majorization: SMACOF in R. J. Stat. Softw. **31**(3), 1–30 (2009). https://doi.org/10.18637/jss.v031.i03
13. Li, S.Z., Anil, K.L.: Handbook of Face Recognition. Springer, New York (2005)
14. van der Maaten, L., Postma, E., van den Herik, H.: Dimensionality reduction: a comparative review. Technical report TiCC-TR 2009-005. Tilburg University (2009)
15. Manziuk, E.A., Barmak, A.V., Krak, Y.V., Kasianiuk, V.S.: Definition of information core for documents classification. J. Autom. Inf. Sci. **50**(4), 25–34 (2018). https://doi.org/10.1615/JAutomatInfScien.v50.i4.30
16. Martinez, A., Du, S.: A model of perception of facial expressions of emotion by human: research overview and perspectives. J. Mach. Learn. Res. **13**(1), 1589–1608 (2012)
17. Romanyuk, O., Vyatkin, S., Pavlov, S., et al.: Face recognition techniques. Informatyka, Automatyka, Pomiary W Gospodarce I Ochronie Srodowiska **10**(1), 52–57 (2020). https://doi.org/10.35784/iapgos.922
18. Vorontsov, K.: Lectures on algorithms of clustering and multidimensional scaling (2020). http://www.ccas.ru/voron/download/Clustering.pdf
19. Wingenbach, T.S.H., Ashwin, C., Brosnan, M.: Validation of the Amsterdam dynamic facial expression set - bath intensity variations (ADFES-BIV): a set of videos expressing low, intermediate, and high intensity emotions. PLoS ONE **11**(1), 1–28 (2016). https://doi.org/10.1371/journal.pone.0147112
20. Zhambyu, M.: Hierarchical cluster analysis and matching. Finance Stat. (1988)

Scientific and Methodological Approach to Assessing the Level of Banks' Financial Security

Ihor Kulyniak(✉), Solomiya Ohinok, and Halyna Rachynska

Lviv Polytechnic National University, Lviv, Ukraine
{ihor.y.kulyniak,solomiia.v.ohinok,halyna.v.rachynska}@lpnu.ua

Abstract. The purpose of the study is to develop theoretical provisions and applied recommendations for assessing and improving the financial security of banks. To achieve this objective, the following tasks are fulfilled in the research: a) it is proposed to distinguish such components of the bank's financial security by functional approach as: "security of deposit transactions", "security of credit transactions", "security of cash transactions", "security of investment transactions", "security of currency transactions"; b) a scientific and methodological approach is proposed for estimating the level of the bank's financial security by functional components, which is based on the calculation of an integral index with further ranking and grouping of banks by their financial security level; c) a set of measures to increase the level of banks' financial security by each functional component is formed. The study results are based on the use of the following methods: economic and statistical (calculation of the integral index of the bank's financial security level), the method of hierarchy analysis and an expert survey (determination of the importance of financial security indicators), the normalization principle (bringing indicators that are different by type to information unidirectionality and dimension), the method of ranking and grouping of banks by the level of financial security, logical generalization and a systematic approach (formation of measures to increase the level of financial security). The proposed scientific and methodological approach to assessing the level of the bank's financial security is tested on the example of Ukrainian banks; the calculations are made according to the official statistics of the National Bank of Ukraine as of 01.09.2019.

Keywords: Financial security · Financial security components · Security of deposit transactions · Security of credit transactions · Security of cash transactions · Security of investment transactions · Security of currency transactions · A financial security indicator · Level of financial security

S. Babichev et al. (Eds.): ISDMCI 2020, AISC 1246, pp. 77–97, 2021.
https://doi.org/10.1007/978-3-030-54215-3_6

1 Introduction

For the overall development of the country and the welfare of its population, it is necessary to take care of all spheres of its activity. The financial system is by far the most important in shaping the stability of the country's economic development. The financial sector is a driving mechanism in the overall structure of the country's economic development, as all the country's cash flows go through this industry. On the one hand, it is a stimulating factor for the rapid and productive development of the industry; on the other hand, it provides financial security and independence of the country and its economy in case of crises, both external and internal.

Banking institutions are central links to ensure the country's financial system functioning. Now it is difficult to imagine any financial transactions conducted without the involvement of commercial banks. Undoubtedly, commercial banks participate in all socio-economic processes, accumulating financial resources of society and ensuring their efficient and rational distribution. Any country is actively involved in global globalization processes, including the financial sphere. However, given the political instability, economic problems and social tension in society, as well as the significant influence of external factors, including military confrontation, the financial system is riddled with a large number of threats that are difficult to detect and prevent in the initial stages of their emergence. This should be facilitated by ongoing and systematic research, timely identification of potential factors and threats to financial security in the banking sector. Not all threats can be prevented, but their timely detection will reduce their negative impact on the efficiency of the bank's functioning.

In today's conditions of an interbank competition strengthening, not quite stable both political and economic state of the country, the factor of timely detection of threats to financial security is very important for the development and implementation of protection measures. Ensuring the financial security of commercial banks is also very important because the betrayal of confidence in credit institutions is detrimental to the financial security of the entire country. And, finally, finance is today the main channel of external control and possible impact on the financial and banking systems, and therefore the national economy as a whole.

The assessment of the financial security level of the country's banking institutions is necessary to ensure the efficiency of functioning of both the individual bank and the country's banking system as a whole, in particular for timely formation of measures to protect against external and internal destabilizing factors and prevent their negative impact. It is worth noting that assessing the level of the commercial bank's financial security, as well as identifying and preventing threats to financial security is important not only for the state but also for investors, as a high-security level gives greater guarantees for the preservation and multiplication of investments; this issue is especially relevant in the context of the risk and instability of the Ukrainian economy's financial sector.

The range of problems of assessing the financial security of business entities, in particular banks, is new and not well studied by the scientific community. Most researches are devoted to the study of theoretical aspects of financial security, among them the works [9,14,23,26,27] have great importance. The analysis of the studies shows the availability of some approaches to assessing the financial security level of the banking sector; in particular, the approaches of most scholars are based on the use of coefficients and indicators to assess the level of the bank's financial security. Thus, the paper [18] substantiates the feasibility of using economic standards to assess the financial and economic security of the Ukrainian banking system. In [10,11,13] the authors offer a system of indicators and ratios. The paper [20] proposes to evaluate the financial stability of the banking system based on two groups of indicators: the first one characterizes the financial stability of the central bank and the second one determines the level of financial stability for state banks, banks with private and foreign capital.

The Ministry of Economic Development and Trade of Ukraine, in its "Guidelines for Calculating the Level of Economic Security of Ukraine" [4], also proposes to determine the level of banking security using of indicators based on statistical data.

Another area of the assessment of banks' financial security is based on the evaluation of banking risks. In [17] the author proposes that the bank's financial security indicators be grouped according to the degree of risk-taking into account banking economic standards. In [21] carried out the security risk analysis in their paper research to evaluate the banking company management through the use of a questionnaire to determine the level of risk that the customer of a financial institution is likely to encounter.

Another category of scientists [12] recommend the use of economic and mathematical modeling methods to assess banks' financial security.

Despite the considerable achievements of scientists, the methodological tool for assessing the level of the commercial bank's financial security requires further improvement, which is a comprehensive manner, taking into account the previously accumulated experience of assessment and experience of the Ukrainian and international scientists in the field of financial security, would allow the government, research institutions, and other stakeholders to promptly assess the state of financial security of the banking sector to make managerial decisions on the analysis, prevention, and minimization of the impact of real and potential threats in the banking sector.

This research is a continuation of the authors' work cycle in the field of poly-parametric evaluation of banks' activities and the development of methods for their rating. In particular, in the article [24] the theoretical and applied provisions for estimating the level of the bank's marketing activity, based on the definition of an integral index taking into account all elements of the marketing complex and their partial indicators are developed. The banks are ranked by the level of their marketing activity in the studied period and a system of concrete measures to increase the level of banks' marketing activity is formed in the article.

Given the important role of the bank's financial security in ensuring the security of both the banking sector and national financial security in general, this study will focus on the scientific and methodological assessment and improvement of banks' financial security, especially in the context of financial and economic instability and risk in this area.

The aim of the research is development of theoretical provisions and applied recommendations for assessing and improving the financial security of banks.

2 Materials and Methods

2.1 Separation of the Bank's Financial Security Components

To comprehensively assess the bank's financial security, it is necessary to distinguish its main components. In scientific sources, a versatile vision on the fact what components from the bank's financial security system can be found. So, in [16] the author proposes to investigate the structure of the bank's financial security in a combination of three components of its level assessment: subsystem of economic independence, a subsystem of sustainability and stability of banking activity and subsystem of the assessment of the ability to self-development and progress. The paper [29] defines the following constituents as the main components of the bank's financial security: risk management system, financial stability, level of the capitalization and equity adequacy to cover banking risks, quality of credit and investment portfolios and, as a result, banking assets, on the whole, level of profitability of banking activities, financial potential, level of corporate control, the competitiveness of the bank and banking services in the market, the share of foreign capital in the authorized capital, etc.

In scientific sources, the authors identify the components of financial security by functional approach, in particular, in [25] the author thinks that the financial and economic security system of a commercial bank should be presented as a set of subsystems related to credit, currency, cash, investment transactions, as well as transactions related to the collection, processing, storage, use and prevention of the spoilage, forgery, and theft of information, both with and without the use of technical and technological resources. Particular attention should be paid to ensuring the security of transactions carried out by banks with payment documents and the formation of a system of protection of the bank and the results of its business activities against fraudulent and criminal activity by third parties and employees of the bank. In [28] the author holds a somewhat similar opinion when considering the components of financial security of the national economy's banking sector, distinguishing the following components by resource and functional orientation: resource-forming component of the banking sector, deposit, credit, currency, inflationary, investment, debt.

Any banking service is exposed to the state of insecurity, which according to the Law of Ukraine "On Banks and Banking Activities" [7] includes:

1. Attraction of funds and precious metals from an unlimited number of legal entities and individuals into deposits.

2. Opening and maintaining current (correspondent) accounts of customers, including in precious metals, and conditional storage accounts (escrow).
3. Placement of funds and precious metals on their behalf, on their own terms, and at their risk, including on current accounts into deposits.

The bank's deposit (passive) transactions are a set of actions related to attracting funds of legal entities and individuals to deposits on fixed terms or call deposits. The bank's cash transactions are a complex of the bank's actions, which consists of ensuring the movement of funds in open accounts of the bank's customers at their instruction. Active transactions are divided into credit and investment ones. The bank's credit transactions are a set of actions related to providing loans to customers and repaying banking loans. The bank's investment transactions are a set of actions related to investing funds in securities of enterprises (public, collective or private) for a relatively long period.

As the bank's transactions can be carried out in both national and foreign currencies, it is appropriate to emphasize the importance of the bank's currency transactions in the system of ensuring the bank's financial security. Currency transactions are transactions that involve the shift or transfer of the ownership of currency values. The bank's currency security is the optimum ratio of claims (entries) and liabilities of a commercial bank in foreign currency (between the number of liabilities by the currency sold and the number of claims, between the volume of claims by purchased currency and the number of liabilities).

Therefore, taking into account the theoretical provisions of the research of scientific and legislative sources, the authors propose to distinguish the following functional components of the bank's financial security: security of deposit transactions, the security of credit transactions, the security of cash transactions, the security of investment transactions and the security of currency transactions.

2.2 Methodology for Assessing the Level of Financial Security of Ukrainian Banks by Functional Components

Based on the elaboration of the approaches of scientists and own experience, the authors propose to evaluate the level of financial security of Ukrainian banks by functional components through calculating the integral index in such a sequence.

2.2.1. Formation of a List of Indicators by Each of the Components of the Bank's Financial Security. The list of indicators of each of the components of the financial security of Ukrainian banks is formed taking into account the principles of representativeness (the most significant indicators that affect the level of the bank's financial security are selected), information availability (availability of the statistical data to calculate this or that indicator) and reliability (indicators adequately represent the bank's financial security).

The security of the bank's deposit transactions will be evaluated by the following indicators:

1.1 Share of deposits in the bank's assets (I_{da}):

$$I_{da} = \frac{D_b + D_c}{A} \cdot 100\%, \quad (1)$$

where D_b is the amount of banks' deposits (funds); D_c is the amount of customers' deposits (funds); A is the amount of the bank's assets.

1.2 Share of the amount of possible compensation by the Deposit Guarantee Fund in the total amount of individuals' deposits (I_{comp}):

$$I_{comp} = \frac{PC}{D_i} \cdot 100\%, \quad (2)$$

where PC is the amount of possible compensation by the Deposit Guarantee Fund; D_i is the amount of deposits of individuals.

1.3 Share of on-demand funds in customers' total assets (I_{ca}):

$$I_{ca} = \frac{F_{en} + F_i}{D_c} \cdot 100\%, \quad (3)$$

where F_{en} is funds on call of economic entities and non-banking financial institutions; F_i is funds on call of individuals; D_c is the total amount of customers' funds (economic entities and non-banking financial institutions and individuals).

1.4 Share of deposits in the total amount of funds attracted by the bank (I_{df}):

$$I_{df} = \frac{D_b + D_c}{F_b} \cdot 100\%, \quad (4)$$

where D_b is the amount of banks' deposits (funds); D_c is the amount of customers' deposits (funds); F_b is the total amount of funds attracted by the bank.

The security of the bank's credit transactions will be evaluated by the following indicators:

2.1 Maximum credit risk exposure per single counterparty / Regulatory capital (no more than 25%) ($R7$) [1,5]:

$$R7 = \frac{AI_a}{RC} \cdot 100\%, \quad (5)$$

where AI_a is aggregate indebtedness by deposits, loans, factoring, financial leasing, bills of exchange, securities, other active banking transactions and financial liabilities of the bank to one counterparty (or a group of related counter parties); RC is the bank's regulatory capital.
Indicator is established to limit the credit risk arising from the failure of individual counterparties to fulfill their obligations.

2.2 Total large credit risk exposure / Regulatory capital (no more than 800% of Regulatory capital) ($R8$) [1,5]:

$$R8 = \frac{AI_l}{RC} \cdot 100\%, \tag{6}$$

where AI_l is aggregate indebtedness by deposits, loans, factoring, financial leasing, bills of exchange, securities, accounts receivable, other active banking transactions, and financial liabilities of the bank by all large credit risks, provided by the bank to all counterparties (or a group of related counterparties), all individuals related to the bank; RC is the bank's regulatory capital.

Indicator is established to limit the concentration of credit risk by an individual counterparty or a group of related counterparties.

2.3 Maximum credit exposure to related parties (no more than 25%) ($R9$) [1,5]:

$$R9 = \frac{AI_{ind}}{FC + AC - C_1} \cdot 100\%, \tag{7}$$

where AI_{ind} is aggregate indebtedness by deposits, loans, factoring, financial leasing, bills of exchange, securities, other active banking transactions and financial liabilities of the bank to individuals related; FC is fixed capital (Tier 1 capital); AC is additional capital (Tier 2 capital); C_1 is the carrying amount of assets.

Indicator is established to limit the risk that arises when engaging in insider transactions, which may have a direct or indirect impact on the bank's activities.

2.4 Share of a loan portfolio in the bank's assets (I_{la}):

$$I_{la} = \frac{LD}{A} \cdot 100\%, \tag{8}$$

where LD is the amount of customers' loans and debts; A is the amount of the bank's assets.

2.5 Share of interbank loans in total liabilities (I_{ibl}):

$$I_{ibl} = \frac{IB}{GL} \cdot 100\%, \tag{9}$$

where IB is the amount of attracted interbank loans (funds received from the NBU and banks); GL is the bank's general liabilities.

The security of the bank's cash transactions will be evaluated by the following indicators:

3.1 Core capital adequacy ratio (no less than 7%) ($R3$) [1,5]:

$$R3 = \frac{FC}{A_r + GDP_b - UCR} \cdot 100\%, \tag{10}$$

where FC is the bank's fixed capital; A_r is total assets reduced by the amount of appropriate reserves/write-downs under financial assets, by the amount of coverage, and weighted by the appropriate risk coefficient depending on the risk group to which the asset is assigned; GDP_b is the aggregate amount of the bank's GDP by all foreign currencies and precious metals; UCR is uncovered credit risk.
Indicator determines the minimum level of the ratio of the bank's fixed capital to the total assets.

3.2 Short-term liquidity ratio (no less than 60%) ($R6$) [1,5]:

$$R6 = \frac{A_q}{L_l} \cdot 100\%, \tag{11}$$

where A_q is quick assets with a maturity date of up to one year; L_l is the bank's liabilities with a maturity date/a sale deadline of up to one year.
Indicator determines the minimum amount of assets required to meet their obligations within one year.

3.3 Profitability of cash transactions (P_{ct}):

$$P_{ct} = \frac{NCI}{A} \cdot 100\%, \tag{12}$$

where NCI is the bank's net commission income; A is the amount of the bank's assets.
Indicator shows how much net commission income the bank accounts for 1 UAH of total assets.

The security of the bank's investment transactions will be evaluated by the following indicators:

4.1 Investment in securities issued by single entity (no more than 15%) ($R11$) [1,5]:

$$R11 = \frac{BF_{in}}{C_{auth}} \cdot 100\%, \tag{13}$$

where BF_{in} is the bank's funds invested in the acquisition of stocks (stakes/shares) and investment certificates separately for each institution; C_{auth} is the bank's authorized capital.
Indicator is established to limit the risk associated with investing in stocks, units, shares and investment certificates of an individual legal entity.

4.2 Total investment in securities (no more than 60%) ($R12$) [1,5]:

$$R12 = \frac{BF_{en}}{C_{auth}} \cdot 100\%, \tag{14}$$

where BF_{en} is the bank's funds invested for the purpose of acquisition of stocks (stakes/shares) and investment certificates of any legal entities; C_{auth} is the bank's authorized capital.
Indicator is established to limit the risk associated with the bank's investment activity.

The security of the bank's currency transactions will be evaluated by the following indicators (Table 5):

5.1 Total long open FX position (no more than 10%) (L_{13_1}) [3]:

$$L_{13\_1} = \frac{VC - VL}{RC} \cdot 100\%, \tag{15}$$

where VC is the volume of claims by foreign currency and precious metals; VL is the volume of liabilities by foreign currency and precious metals; RC is the bank's regulatory capital.
Established by the NBU quantitative limit in the percentage of the daily value of long currency positions opened by the bank, namely: total in foreign currencies and precious metals and separately by groups of relevant foreign currencies and precious metals, as well as by a corresponding currency or precious metals, by separate active transactions with currency valuables to the bank's regulatory capital.

5.2 Total short open FX position (no more than 10%) (L_{13_2}) [3]:

$$L_{13\_2} = \frac{VL - VC}{RC} \cdot 100\%, \tag{16}$$

where VL is the volume of liabilities by foreign currency and precious metals; VC is the volume of claims by foreign currency and precious metals; RC is the bank's regulatory capital.
Established by the NBU quantitative limit in the percentage of the daily value of short currency positions opened by the bank, namely: total in foreign currencies and precious metals and separately by groups of relevant foreign currencies and precious metals, as well as by a corresponding currency or precious metals, by separate active transactions with currency valuables to the bank's regulatory capital.

5.3 Liquidity Coverage Ratio in foreign currency (no less than 100%) (LCR) [6]:

$$LCR = \frac{HLA}{NO} \cdot 100\%, \tag{17}$$

where HLA is high-quality foreign currency liquid assets; NO is a net cash outflow within 30 days in foreign currency.
Indicator sets a minimum required level of foreign currency liquidity to cover the net expected cash outflow within 30 calendar days, taking into account a stress scenario.

5.4 Ratio of the foreign currency volume to national currency volume (R_{fn}):

$$R_{fn} = \frac{F_{fc}}{F_{nc}} \cdot 100\%, \tag{18}$$

where F_{fc} is funds in foreign currency; F_{nc} is funds in national currency.

5.5 Ratio of granted loans to attracted funds in foreign currency (R_{lf}):

$$R_{lf} = \frac{F_{fc}}{F_{cfc}} \cdot 100\%, \tag{19}$$

where F_{fc} is loans in foreign currency; F_{cfc} is customers' attracted funds in foreign currency.

2.2.2 Determination of the Characteristic Values of Indicators. The characteristic values of indicators of the bank's financial security components will be determined according to the methodology developed by the Ministry of Economic Development and Trade, which is given in the "Methodological Recommendations for Calculating the Level of Economic Security of Ukraine" [4].

The characteristic values of indicators of the bank's financial security components are presented in Table 1.

Table 1. Characteristic values of indicators of the bank's financial security components, %

Indicator's name	Characteristic values of indicators, %									
	XL_{cr}	XL_{haz}	XL_{unsat}	XL_{sat}	XL_{opt}	XR_{opt}	XR_{sat}	XR_{unsat}	XR_{haz}	XR_{cr}
I_{da}	10	20	30	40	50	60	70	80	90	100
I_{comp}	50	70	80	90	100					
I_{ca}						20	30	40	50	60
I_{df}	85	90	94	97	100					
$R7$						0	10	15	20	25
$R8$						0	30	100	150	800
$R9$						0	10	15	20	25
I_{la}	20	30	35	40	45					
I_{ibl}						0	10	15	25	40
$R3$	1	4	7	10	20					
$R6$	30	40	60	80	100					
P_{ct}	0	0.5	1	1.5	2					
$R11$						0	4	8	12	15
$R12$						0	20	30	40	60
L_{13_1}						0	1	2	3	5
L_{13_2}						0	1	2	3	5
LCR	40	50	80	100	200					
R_{fn}						10	20	30	40	50
R_{lf}	50	70	80	85	90	110	130	140	160	180

Note. L (left side) is for indicators that are stimulants, rationing occurs by left side criteria.
R (right side) is for indicators that are de-stimulants, rationing occurs by right side criteria.
X_{cr} is critical value of the indicator.
X_{haz} is hazardous value of the indicator.
X_{unsat} is unsatisfactory value of the indicator.
X_{sat} is satisfactory value of the indicator.
X_{opt} is optimum value of the indicator.
Source: proposed by the authors

2.2.3 Collecting Statistical Data to Calculate Indicators. The basis for collecting statistical data for the calculation of indicators is the official website or official publications with statistics on banks' activities (for this study it is the official site of the National Bank of Ukraine [2], which provides statistical data on Ukrainian banks' activities as of 01.09.2019).

2.2.4 Indicator Rationing. Bringing indicators that are different by type (de-stimulants, stimulants, mixed type) to informational unidirectionality and dimension of the interpretations of characteristic values occurs through rationing. Indicator rationing is performed using a linear function so that indicators' characteristic values fall into intervals comparable in magnitude. The transition from absolute to normalized values of indicators allows measuring indicators on a scale from 0 to 1 or in a percent: 0 corresponds to 0%, 1 to 100%. Thus, the normalized indicator's value obtained characterizes the degree of approaching the optimum value 1 by its magnitude. The indicator rationing formulas are given in the methodology developed by the Ministry of Economic Development and Trade, which is given in the "Methodological Recommendations for Calculating the Level of Economic Security of Ukraine" [4].

2.2.5 Determining Indicators' Weight. To determine indicators' weight the authors propose to use the analytic hierarchy process (Saaty's method) [22,24], which provides for the calculation of vectors of alternative priorities concerning the selected criteria. Pairwise comparisons are defined as an advantage of one element over the other according to the relative importance scale (Table 2).

Table 2. The scale of the relative importance of the analytic hierarchy process of T. Saaty's

Point, k	Definition	Characteristic
1	Equal importance	Equal contribution of two elements to the overall estimate
3	Moderate advantage	Slight advantage of one element over the other
5	Significant advantage	Significant advantage of one element over the other
7	Considerable advantage	Practically considerable advantage of one element over the other
9	Fairly considerable advantage	Obvious advantage is the dominance of one element over the other
2, 4, 6, 8	Intermediate values	Applied in transitional cases
1/k	Values of symmetric elements	Used to evaluate non-dominant elements

Source: adapted according to [19,22]

The components of the vector of local priorities are calculated by the formulas:

$$w_i = \frac{\bar{w_i}}{\sum_{i=1}^{n} \bar{w_i}}; \; i = \overline{i, n}, \tag{20}$$

$$\bar{w}_i = \sqrt[n]{\prod_{j=1}^{n} a_{ij}};\ i = \overline{i, n}, \tag{21}$$

where a_{ij} is i-element of j-column of the matrix of pair wise element comparisons, n is the number of elements being compared.

2.2.6 Calculation of the Integral Index of Financial Security and Indices of the Bank's Financial Security Components. To calculate the integral index of the bank's financial security (I) the following formula will be used:

$$I = I_{dep} \cdot w'_{dep} + I_{cred} \cdot w'_{cred} + I_{cash} \cdot w'_{cash} + I_{inv} \cdot w'_{inv} + I_{cur} \cdot w'_{cur}, \tag{22}$$

where I is the integral index of the bank's financial security; I_{dep} is the index of security of the bank's deposit transactions; I_{cred} is the index of security of the bank's credit transactions; I_{cash} is the index of security of the bank's cash transactions; I_{inv} is the index of security of the bank's investment transactions; I_{cur} is the index of security of the bank's currency transactions; w'_{dep}, w'_{cred}, w'_{cash}, w'_{inv}, w'_{cur} are weighted coefficients of the indices of security of the bank's deposit, credit, cash, investment and currency transactions respectively.

The indexes of security of the bank's deposit transactions (I_{dep}), credit transactions (I_{cred}), cash transactions (I_{cash}), investment transactions (I_{inv}), and currency transactions (I_{cur}) will be determined by the formulas:

$$I_{dep} = \sum_{i=1}^{n} w_{depi} \cdot s_{depi}, \tag{23}$$

$$I_{cred} = \sum_{i=1}^{n} w_{credi} \cdot s_{credi}, \tag{24}$$

$$I_{cash} = \sum_{i=1}^{n} w_{cashi} \cdot s_{cashi}, \tag{25}$$

$$I_{inv} = \sum_{i=1}^{n} w_{invi} \cdot s_{invi}, \tag{26}$$

$$I_{cur} = \sum_{i=1}^{n} w_{curi} \cdot s_{curi}, \tag{27}$$

where w_{depi}, w_{credi}, w_{cashi}, w_{invi}, w_{curi} are the values of weight of i-indicator of security of the bank's deposit, credit, cash, investment and currency transactions respectively, determined using the method of hierarchy analysis; s_{depi}, s_{credi}, s_{cashi}, s_{invi}, s_{curi} are normalized values of i-indicator of security of the bank's deposit, credit, cash, investment and currency transactions respectively; n is the number of indicators in the bank's financial security component.

The value of the integral index of financial security and indices of the bank's financial security components ranges from 0 to 1. The higher the value of the integral index is, the higher the level of financial security of the bank and its components and vice versa is.

2.2.7 Determining the Level of the Bank's Financial Security. In the scientific literature, there is no unambiguous scale of bank grouping by the level of financial security, so the authors suggest using the universal scale of Harrington's [15]. The level of the bank's financial security will be determined according to this scale depending on the value of the integral index of the bank's financial security (I) (Table 3).

Table 3. Characteristic of levels of the bank's financial security according to the Harrington's scale [15]

Intervals of the Harrington's scale (value I)	Level of the bank's financial security
(0.8–1]	High level
(0.63–0.8]	Sufficient level
(0.37–0.63]	Average level
(0.2–0.37]	Low level
[0–0.2]	Critical level

3 Results and Discussion

To determine the indicators' weight, a matrix of pair wise comparisons with the definition of local priorities is formed. The criteria and their weight are determined based on the expert estimation. The experts were 26 employees of the relevant departments of such banks as: PJSC CB "PryvatBank", JSC "Ukreximbank", JSC "Raiffeisen Bank Aval", PJSC "Kredobank", JSC "Pivdennyi Bank". The experts were asked to compare in pairs the advantage of one element of the bank's marketing complex over the other according to T. Saaty's 10-point scale (see Table 2). The summary results of determining the indicators' weight based on the expert estimation using the Saaty's method are given in Table 4.

The results of ranking and grouping of the Ukrainian banks by the high value of the integral index of the bank's financial security in descending order as of 01.09.2019 are presented in Table 5; by the sufficient value of the integral index - in Table 6; by the average value of the integral index - in Table 7. The results of determining the level of financial security of the Ukrainian banks according to the Harrington scale are also presented in the table (see Table 3).

The banking sector is an extremely important element of the country's financial and credit system. Still, it is the area through which all financial transactions

Table 4. Summary results of the indicators' weight of the bank's financial security components

Bank's financial security components	Weight of the bank's financial security components, w_i'	Indicators of the bank's financial security components	Indicators' weight, w_i
Security of the bank's deposit transactions	0.329	I_{da}	0.511
		I_{comp}	0.084
		I_{ca}	0.279
		I_{df}	0.126
Security of the bank's credit transactions	0.365	$R7$	0.388
		$R8$	0.175
		$R9$	0.190
		I_{la}	0.175
		I_{ibl}	0.073
Security of the bank's cash transactions	0.098	$R3$	0.359
		$R6$	0.124
		P_{ct}	0.517
Security of the bank's investment transactions	0.145	$R11$	0.667
		$R12$	0.333
Security of the bank's currency transactions	0.063	L_{13_1}	0.325
		L_{13_2}	0.325
		LCR	0.142
		R_{fn}	0.078
		R_{lf}	0.131

Source: calculated by the authors.

go, which, in one way or another, are related to all the other spheres of the state's activity. That is why its protection is the first function. The detection of threats to the financial security of a commercial bank is accompanied by the search for measures to prevent, eliminate or neutralize them.

The essence of financial security in the banking system is to ensure the best use of its resources to prevent threats to banks and create conditions for stable, efficient functioning and profit maximization. The level of financial security of banking activity is determined by the fact of how effectively banking departments and services manage to prevent threats and eliminate losses from a negative impact on the banking system.

The proposed measures to increase the level of security of the banks' deposit transactions are: analysis of cash on customers' accounts; increase in long-term

Table 5. Results of ranking and grouping of the banks by high financial security level as of 01.09.2019

Name of bank	Integral index of the bank's financial security (I)	Bank rating	The values of the indices of the financial security components of the bank				
			I_{dep}	I_{cred}	I_{cash}	I_{inv}	I_{cur}
JSC"Idea Bank"	0.951	1	0.974	0.960	0.839	0.999	0.844
JSC "A - Bank"	0.893	2	0.934	0.822	0.913	1.000	0.823
JSC "Okci Bank"	0.883	3	0.910	0.836	0.821	1.000	0.846
JSC "Bank Forward"	0.882	4	0.994	0.796	0.746	1.000	0.736
JSC "Universal Bank"	0.865	5	0.765	0.906	0.946	1.000	0.716
JSC "SEB Corporate Bank"	0.836	6	0.681	0.963	0.667	1.000	0.797
JSC "Procredit Bank"	0.820	7	0.624	0.948	0.690	0.999	0.886
JSCB "Lviv"	0.818	8	0.722	0.833	0.790	1.000	0.850
JSC "AP Bank"	0.803	9	0.551	0.948	0.923	1.000	0.641
JSC CB "Zemelny Capital"	0.802	10	0.932	0.603	0.779	1.000	0.850

Source: calculated by the authors.

deposits; calculate the cost and cost-effectiveness of services for each customer in advance; introduction of an information-analytical system of decision-making support during the formation of a deposit portfolio; creation of a deposit guarantee system; taking into account the ratio of expensive and low-paid deposit resources at the level of ensuring appropriate; profitability of the banking institution and balance liquidity; use of funds and ways of reducing interest expenses by deposit resources; mutual consistency and optimization of credit and deposit transactions by amounts and terms of the placement and repayment of liabilities; implementation of effective regulation and control of the bank's deposit base; conducting analytical observations on an ongoing basis regarding limits and sizes of risks.

The proposed measures to increase the level of security of the banks' credit transactions are: formation of credit limits; limiting a loan amount per one person; analysis of customers' financial activities, their solvency; setting clear limits on the timing of loan repayment; control over the presence and condition of the collateral, behavior and state of guarantors (warrantors) and insurers; ensuring the rapid response of banks to changes in the economic situation in the country; formation of reserves for credit operations; loan portfolio diversification.

The proposed measures to increase the level of security of the banks' cash transactions are: audit of cash, valuables, documents in the bank's cash depository and cash departments; study of the results of previous revisions, their frequency; control over timely display of cash transactions in accounting; periodic inspection of the rules for storing keys (and their duplicates) from repositories, safes; systematic internal control of cash transactions at each stage of their implementation; implementation of the system of electronic cash reporting.

The proposed measures to increase the level of security of the banks' investment transactions are: improving the system of investment risk insurance to encourage banks to extend long-term loans; creating a system of diagnostics and

Table 6. Results of ranking and grouping of the banks by sufficient financial security level as of 01.09.2019

Name of bank	Integral index (I)	Bank rating	The values of the indices of the financial security components of the bank				
			I_{dep}	I_{cred}	I_{cash}	I_{inv}	I_{cur}
JSC “ING Bank Ukraine”	0.795	11	0.744	0.828	0.602	1.000	0.693
JSC “Kredobank”	0.792	12	0.542	0.950	0.852	1.000	0.601
JSC “Raiffeisen Bank Aval”	0.791	13	0.553	0.902	0.912	0.971	0.791
JSC CB “PrivatBank”	0.781	14	0.700	0.792	0.861	0.998	0.511
JSC “Sky Bank”	0.780	15	0.844	0.648	0.828	1.000	0.626
JSC “RadaBank”	0.773	16	0.749	0.679	0.970	1.000	0.613
JSC “CB ”Globus”	0.769	17	0.680	0.741	0.861	1.000	0.720
JSC “Rws Bank”	0.769	18	0.714	0.727	0.927	1.000	0.518
JSC “Piraeus Bank ICB”	0.761	19	0.588	0.749	1.000	1.000	0.812
JSC “UkrconstinvestBank”	0.760	20	0.683	0.677	0.976	1.000	0.752
JSC “Creditwest Bank”	0.759	21	0.770	0.621	0.786	1.000	0.898
JSC “Poltava-bank”	0.755	22	0.781	0.558	0.991	0.999	0.833
JSC “MetaBank”	0.752	23	0.714	0.613	1.000	1.000	0.805
JSC “Bank ”Grant”	0.749	24	0.778	0.596	0.878	0.986	0.733
JSC “CIB”	0.744	25	0.719	0.595	0.972	0.999	0.796
JSC “Cleaning house”	0.738	26	0.701	0.643	0.873	1.000	0.663
JSC “Tascombank”	0.735	27	0.717	0.650	0.776	1.000	0.649
JSC “Cominvestbank”	0.735	28	0.723	0.588	0.956	1.000	0.697
JSC “Unex Bank”	0.734	29	0.674	0.635	1.000	1.000	0.601
JSC “EIB”	0.732	30	0.739	0.634	0.826	1.000	0.506
JSC “ALTBank”	0.731	31	0.630	0.645	1.000	1.000	0.723
JSC “Polikombank”	0.727	32	0.863	0.441	1.000	0.971	0.677
JSC “Ibox Bank”	0.718	33	0.705	0.534	0.970	0.999	0.815
JSC “Alpari Bank”	0.718	34	0.387	0.808	1.000	1.000	0.833
JSC “Alfa-Bank”	0.716	35	0.651	0.605	0.898	1.000	0.761
JSC “Pumb”	0.716	36	0.644	0.640	0.830	0.993	0.719
JSC “Asvio Bank”	0.713	37	0.630	0.608	0.990	1.000	0.658
JSC “Bank Sich”	0.711	38	0.633	0.617	0.944	1.000	0.636
PJSC “Settlement center”	0.710	39	0.466	0.813	0.618	1.000	0.867
JSC “Credi Agricole Bank”	0.709	40	0.524	0.715	0.815	0.997	0.817
JSCB “IndustrialBank”	0.706	41	0.796	0.492	0.906	0.965	0.571
PJSC “Bank ”Ukrainian Capital”	0.706	42	0.738	0.506	0.947	1.000	0.642
PJSC “MTB Bank”	0.705	43	0.516	0.698	0.937	0.986	0.720
JSC “JSCB ”Conkord”	0.703	44	0.627	0.598	0.933	1.000	0.661
JSCB “Arcada”	0.702	45	0.706	0.629	0.981	0.633	0.833
JSC “Bank 3/4”	0.701	46	0.643	0.596	0.990	1.000	0.479
JSC “Mega Bank”	0.697	47	0.807	0.408	0.866	0.999	0.841
JSC “Oschadbank”	0.692	48	0.816	0.489	0.680	0.998	0.538
JSC “Credit Europe Bank”	0.692	49	0.695	0.576	0.559	1.000	0.852
JSC “First Investment Bank”	0.686	50	0.772	0.495	0.803	0.936	0.594

(continued)

Table 6. (*continued*)

Name of bank	Integral index (I)	Bank rating	The values of the indices of the financial security components of the bank				
			I_{dep}	I_{cred}	I_{cash}	I_{inv}	I_{cur}
PJSC "Bank Familny"	0.685	51	0.298	0.815	1.000	1.000	0.734
JSC "Pravex Bank"	0.680	52	0.695	0.489	0.826	1.000	0.753
JSC "Sberbank"	0.675	53	0.609	0.592	0.642	1.000	0.802
JSC "CrystalBank"	0.674	54	0.530	0.594	0.992	1.000	0.648
JSC "Bank Credyt Dnepr"	0.673	55	0.752	0.477	0.723	1.000	0.561
JSC "Bank Alliance"	0.670	56	0.673	0.479	0.858	1.000	0.716
JSC "Motor-Bank"	0.670	57	0.666	0.523	0.984	0.844	0.649
JSC "Bank Avangard"	0.664	58	0.555	0.634	0.647	1.000	0.654
JSC "Pivdennyi Bank"	0.663	59	0.524	0.644	0.773	0.987	0.581
JSC "OTP Bank"	0.659	60	0.600	0.535	0.949	0.917	0.642
JSC "CityBank"	0.653	61	0.449	0.654	0.744	1.000	0.775
JSC "Misto Bank"	0.645	62	0.891	0.304	0.515	1.000	0.716
PJSC "Bank Vostok"	0.644	63	0.435	0.635	0.881	1.000	0.598
JSC "BTA Band"	0.642	64	0.431	0.702	0.651	0.997	0.570

Table 7. Results of ranking and grouping of the banks by average financial security level as of 01.09.2019

Name of bank	Integral index of the bank's financial security (I)	Bank rating	The values of the indices of the financial security components of the bank				
			I_{dep}	I_{cred}	I_{cash}	I_{inv}	I_{cur}
JSC "Deutsche Bank DBU"	0.637	65	0.481	0.574	0.712	1.000	0.869
JSB "UkrgasBank"	0.632	66	0.504	0.598	0.629	0.983	0.691
JSC "First Investment Bank"	0.627	67	0.549	0.437	0.981	0.999	0.726
JSC "Ukrsibbank"	0.625	68	0.545	0.498	0.944	0.964	0.502
JSC "Bank of Trust-Capital"	0.598	69	0.346	0.604	0.756	1.000	0.705
PJS "CB "AccordBank"	0.592	70	0.525	0.365	0.990	0.999	0.702
JSC "UkrsotsBank"	0.587	71	0.484	0.463	0.709	0.998	0.710
JSC "Bank "Portal"	0.582	72	0.353	0.595	0.699	1.000	0.562
JSC "Ukreximbank"	0.577	73	0.545	0.468	0.560	0.998	0.434
PJSC "Prominvestbank"	0.445	74	0.262	0.368	0.651	1.000	0.240

Source: calculated by the authors.

risk prevention; control throughout the period of investment placement; controlling current risks and preventing future ones; analyzing information reliability on investment objects; protection of confidential information concerning the conduct of major investment transactions; personnel training on investment security issues.

The proposed measures to increase the level of security of the banks' currency transactions are: checking monetary notes; implementation of the system of rapid response to changes in the exchange rate; control of open currency position limits; study of the behavior of exchange rates, exchange rate forecasting, determination of exchange rate volatility; netting, hedging and reservation; currency portfolio diversification; transfer pricing.

Thus, the bank's financial security consists of stable functioning of the institution, eliminating the possibility of harming the bank, ensuring its stable and maximally effective functioning, the balance of its state and resistance to internal and external threats.

In the authors' opinion, it is necessary to start with monitoring, find reliable information on the bank's activities and factors that affect them and understand what threats are inherent in a particular situation and prevent them from occurring. As for external threats, it is necessary to monitor the market (competitors, the state's political and financial and economic conditions, etc.).

Besides, it is necessary to follow the bank's normalized performance in order not to reach the critical risk limit and, as a result, not to be declared insolvent.

It is very important to assess the impact of a particular threat so that the bank's management can focus on more significant problems and quickly resolve them. Unfortunately, today, the banks of Ukraine think of financial security only if there are already some problems.

The main measures to prevent external and internal threats are to strengthen the state regulation of the banking sector, to introduce new methods and technologies of information processing and analysis to evaluate and ensure the financial security of an individual bank, as well as an adequate level of financial control over both financial resources and transactions performed and bank staff's actions. Such measures should be developed using minimum resources and time spent to achieve the desired result.

4 Conclusions

The study results show that as of 01.09.2019, the ten banks (10) of Ukraine are characterized by a high level of financial security, which indicates a significant performance of managers, compliance with mandatory standards, high efficiency of managing financial resources and financial transactions, a high-quality system of financial risk management and counteracting external and internal threats. The fifty-four (54) Ukrainian banks are characterized by a sufficient level of financial security, which indicates the banks' inability to fully anticipate the onset and withstand financial threats. This group includes four state-owned banks in Ukraine. The group with the average level of financial security includes 10 banks, which indicates negative financial results, the existence of deviations in compliance with the regulations, the inability of banks to effectively counteract financial threats. The lowest level of financial security (0.445) was found in PJSC "Prominvestbank". Also, the state-owned JSC "Ukreximbank" belongs to the group of banks with average financial security, which was also potentially problematic according to the results of stress tests [8] carried out by the NBU.

The practical significance of the study is that the proposed scientific and methodological approach to assessing the level of the bank's financial security by functional components and a set of measures to improve the level of the bank's financial security components can be applied in practice to both Ukraine's banks and authorities to formulate plans and strategies of the development of banking institutions to enhance the efficiency of the financial market banking sector development.

The assessment of the level of Ukrainian banks' financial security is tested according to the NBU's data only as of 01.09.2019, so the generated rating reflects the situation for the period for which the data were collected and cannot be applied for other periods. In this case, it is necessary to recalculate the level of the bank's financial security according to new statistics or data for another study period.

As an advantage of this study, it should be noted that in assessing the level of banks' financial security the indicators are used in combination, which allows getting results that will be obtained taking into account many factors that may affect the level of the bank's financial security. By applying the hierarchy method, the recommendations of the Ministry of Economic Development and Trade and the National Bank of Ukraine, subjective factors for assessing indicators and the level of the bank's financial security are minimized.

As a limitation, two main points should be noted: 1) in the formation of a system of indicators, it is proposed to use those indicators and standards that are accepted in the practice of banking regulation in Ukraine and can be specific only for Ukraine; 2) this methodology cannot be applied to other banking institutions that are not commercial banks and cannot be applied to the National Bank of Ukraine or the Central Banks of other countries.

The prospect of further research raises the need to develop strategic actions to balance the state's regulatory influence and monitoring of the level of financial security of the banking system and market-oriented methods of managing and controlling the risks of banks' activities. The problem of forming a complex of measures for preventing threats in the system of increasing the level of banks' financial security should be considered as part of the overall strategy of the banking sector development in the system of ensuring the national security of Ukraine.

References

1. Economic ratios calculation method of banks activity regulation in Ukraine no. 803-rsh. https://bank.gov.ua/file/download?file=Method_15122017_803rsh_economic_ratios.pdf
2. The main indicators of banks activity/supervisory statistics/national bank of Ukraine. https://bank.gov.ua/statistic/supervision-statist/data-supervision#1
3. On amendments to some legal acts of the national bank of Ukraine and setting limits on the bank's open currency position: NBU resolution no. 290. https://zakon.rada.gov.ua/laws/show/z0946-05

4. On approval of methodological recommendations for calculating the economic security level of Ukraine: Order of the ministry of economic development and trade of Ukraine no. 1277. https://zakon.rada.gov.ua/rada/show/v1277731-13
5. On approval of the instruction on the procedure for regulating the activity of banks in Ukraine: NBU resolution no. 368. https://zakon.rada.gov.ua/laws/show/z0841-01#n25
6. On approval of the methodology for calculating the liquidity coverage ratio (LCR): NBU resolution no. 101-rsh. https://zakon.rada.gov.ua/laws/show/vr101500-18
7. On banks and banking activity: law of Ukraine no. 2121-iii. https://zakon.rada.gov.ua/laws/show/2121-14
8. Ukreximbank and oschadbank are potentially problematic - the results of NBU stress tests. https://www.5.ua/ekonomika/
9. Barilyuk, M.M.: Theoretical aspects of defining the essence of the term "financial security of a commercial bank". Black Sea Econ. Stud. **15**, 176–181 (2017)
10. Burkhanov, A., Akhmedov, F.: Indicators to assess financial security of the banks. Int. J. Innov. Technol. Exploring Eng. Regul. Issue (IJITEE) **9**(1), 4465–4468 (2019). https://doi.org/10.35940/ijitee.a3894.119119
11. Epifanov, A., Plastun, O., Dombrodskyi, V.: Financial security of enterprises and banking institutions. UABS NBU, Sumy (2009)
12. Franchuk, V., Melnyk, S.: Scientific and methodical approaches to commercial bank economic security level assessment. Sci. Bull. Lviv State Univ. Intern. Aff.: Econ. Series **1**, 48–58 (2012)
13. Halkiv, L., Karyy, O., Demchyshyn, M.: Bank crediting within the structure of national economy of Ukraine: parametric evaluation and problems with activation. Actual Prob. Econ. **6**(180), 314–320 (2016)
14. Han, R.: Financial internationalization and financial security issues. Open Access Libr. J. **5**, e4874 (2018). https://doi.org/10.4236/oalib.1104874
15. Harrington, E.: The desirability function. Ind. Qual. Contol **22**(10), 494–498 (1965)
16. Hubareva, I., Shtayer, O.: Determination of indicators set for assessing the economic security of a bank. Prob. Econ. **3**, 66–75 (2010)
17. Kallaur, P.: System of providing sustainability of banking sector of Belarus Republic. Money Credit **6**, 23–29 (2018)
18. Krupka, I.: Financial and economic security of Ukraine's banking system and perspectives of national economy development. Bus. Inform **6**, 168–175 (2012)
19. Kuzmin, O., Ovcharuk, V., Zhezhukha, V., Mehta, D., Gregus, J.: Diagnosing the administration systems as a prerequisite for enterprises business processes reengineering. In: International Conference on Intelligent Networking and Collaborative Systems, pp. 513–524 (2019)
20. Kuznyetsova, A., Pogorelenko, N.: Assessment of the banking system financial stability based on the differential approach. Banks Bank Syst. **13**(3), 120–133 (2018). https://doi.org/10.21511/bbs.13(3).2018.12
21. Noah, G., Shafi'I, A., Joseph, O.: Security risk analysis and management in banking sector: a case study of a selected commercial bank in Nigeria. Int. J. Inf. Eng. Electron. Bus. (IJIEEB) **11**(2), 35–43 (2019). https://doi.org/10.5815/ijieeb.2019.02.05
22. Saati, T.: Decision Making. Radio and Communication. Analytic Hierarchy Process, Moscow (1993)
23. Sak, T.: Financial and economic security of Ukrainian commercial banks. Econ. J. Lesya Ukrainka East. Eur. National Univ. **2**, 66–71 (2015)

24. Shpak, N., Kulyniak, I., Gvozd, M., Malynovska, Y., Sroka, W.: Estimation of the marketing activity of banking structures. Acta Universitatis Agriculturae et Silviculturae Mendelianae Brunensis **68**(1), 229–242 (2020). https://doi.org/10.11118/actaun202068010229
25. Shpyliovyi, B.: The structure and components of banking institutions financial and economic security system. Econ. Soc. **2**, 632–637 (2016)
26. Vasilchyshyn, O.: Financial security of the banking system of Ukraine: philosophical determinants. Ternopil (2017)
27. Vernimmen, P., Le Fur, Y., Dallochio, M., Salvi, A., Quiry, P.: Financial securities. In: Corporate Finance: Theory and Practice, 5th edn. (2017). https://doi.org/10.1002/9781119424444.part5
28. Vovchenko, R.: Financial security of the banking sector of the Ukrainian economy. Thesis. University of Banking, Kyiv (2016).
29. Vovk, V.: Theoretical principles of ensuring the financial bank security. Prob. Econ. **4**, 200–204 (2012)

Methods of Classifying Foreign Language Communicative Competence Using the Example of Intending Primary School Teachers

Anatolii Pashko[1](✉) and Iryna Pinchuk[2]

[1] Taras Shevchenko National University of Kyiv, Kyiv, Ukraine
aap2011@ukr.net

[2] Oleksandr Dovzhenko Hlukhiv National Pedagogical University, Hlukhiv, Ukraine
pinchukiryna6@gmail.com

Abstract. This article deals with the methods of classifying communicative competence. Finding effective ways to evaluate the level of intending primary school teachers' foreign language communicative competence is becoming increasingly relevant today, as objectifying the measurement process through feedback provides an opportunity to coordinate the developing that competence components (motivational, cognitive, communicatively active, reflexive) and further improving the students' levels of foreign language communicative competence.

In order to effectively evaluate developing the foreign language communicative competence, it is necessary to focus not only on explicitly expressed, but also on implicitly expressed traits and qualities of an intending professional. It means that the assessment, in addition to traditional parameters, should determine those personal qualities that are hidden for direct observation and which are only indirectly manifested in the professionally oriented communicative activity of the individual.

All above mentioned information necessitate the development of a special assessment scale, the creation of which is objectively difficult due to the complexity of identifying and describing implicit parameters of professionally relevant components of communicative competence. During this process it is important to use the more sophisticated mathematical methods and information technologies.

The minimax classifications, method of k-means, Bayesian classification are proposed with the aim to assess the levels of students' (intending primary school teachers') foreign language communicative competence. Students' ratings were determined by the hierarchy analysis method. Decision-making procedures are used to select one of two or four possible options.

Keywords: Foreign language communicative competence · The hierarchy method · The Bayesian classification · Intelligent method · Competence assessment

S. Babichev et al. (Eds.): ISDMCI 2020, AISC 1246, pp. 98–113, 2021.
https://doi.org/10.1007/978-3-030-54215-3_7

1 Introduction

Finding effective ways of assessing the level of foreign language communication competence of intending primary school teachers is becoming increasingly relevant today, as the objectification of the measurement process by providing feedback enables us to coordinate intending specialists' professional developing.

In today's world, assessment is being transformed as the only means for measuring ability levels. Response or feedback, mutual evaluation, self-assessment, dynamic evaluations are modern effective means of interpreting university students' knowledge, skills and competences [6]. Control and assessment of the level of competence development is an integral structural element of the educational process, and is the final component of the content block [15].

Assessment makes the necessary adjustments in the educational process to improve the content, methods and forms of teaching. Assessment is the process of determining the level of students' academic achievement in mastering the content of the subject compared to the requirements of the State standard and programs [18]. The evaluation process is a special (diagnostic) component of the control, and the rating (mark) is its result. Rating (mark), being a quantitative indicator of the learning outcomes quality, indirectly characterizes the development of intending professionals, their abilities, attitudes to the profession, which can be expressed in the form of qualitative verbal characteristics [18].

The context of evaluating intending primary school teachers' communicative competence is also presented by several considerations for assessment. Firstly, the used assessment means are not to be focused only on knowledge about foreign language competence, but on attitudes and skills too. Assessment is supposed to show university students, what potential issues they themselves could include in their own learning and teaching in the form of loop-input [20]. It is important intending teachers' language proficiency requirement has to be taken into account (sufficient linguistic proficiency, approximately on the B2 - C1 level) and show knowing the related terminology.

In connection to what has been said above, there is one more condition, concerning the need to diversify (in comparison with the existing traditional set) types and forms of assessment that acquires special significance. It is necessary to abandon adherence to purely traditional forms of assessment (tests, quizzes, surveys, etc.), they should be supplemented with innovative forms that are aimed specifically at identifying the level of the formation of the professional and communicative competence of the future specialist. Such forms can be, for example, referring to the solution (by means of a foreign language as a means of communication) of problematic professional tasks, holding scientific conferences on professional issues, etc. [19].

This condition necessitates the development of a special criteria scale of assessment, the creation of which is objectively difficult due to the complexity of identifying and describing implicit parameters of professionally significant competencies.

It means that the assessment, in addition to traditional objects, must identify those personality parameters that are hidden for direct observation and which only indirectly manifest themselves in the professionally oriented communicative activity of an individual. (These objects of assessment were almost completely and totally ignored as immeasurable). Among such parameters, in particular, we can mention such personality traits as the speed of decision making, energy, self-discipline, the wish to study English, motives of future profession etc.

The solution of this problem can be regarded as another condition that ensures the effectiveness of assessment of the development of students' foreign language communicative competence.

The aim of the article is to analyse the approaches and peculiarities of assessing the future primary school teachers' foreign language communicative competence, the selection and justification of the criteria and methods for assessing future primary school teachers' communicative competence. To justify the criteria and methods for assessing communicative competence, in addition to analysing the problem of assessing communicative competence, statistical classification methods are used: the K-means method, minimax method, Bayesian classification. A hierarchy analysis method is used to rank grades. For the proposed criteria for assessing the future primary school teachers' foreign language communicative competence, statistical methods of analysis and classification were first applied.

2 Review of Literature

A lot of scientists from different countries research the problem of assessing the process of foreign language communicative competence. In [12], the scientist analysed the used assessment means, explored their advantages and disadvantages for the particular teaching situation, the profile of undergraduates and the purposes of the course, highlighted the issues that are relevant for assessment developers in teacher education, researched the process of assessing intercultural communicative competence in pre-service English language teachers in the process of considering all the aspects that make the assessment in the course complex and challenging, after implementation.

We analysed scientific literature according the issue of evaluating students' communicative competence and got to know that at the present time there are several attempts to solve the problem of assessing foreign language communicative competence of intending specialists. In particular, assessment criteria of a lesson for creativity (ACLC) was developed in [13] to promote teaching for creativity in Vietnam. It was used in guiding and assessing the teachers' preparation and implementation of lessons for creativity.

In [19], the authors prove that the first and significant condition in the context of teaching a foreign language is assessment procedures of the features and characteristics of professional activities of intending specialists. This means that assessment and evaluation in all their forms and techniques should be aimed at revealing the degree of development of those students' qualities, properties, characteristics, knowledge, and skills that are necessary for professional communicative activity in a specific sphere of employment. The parameters of assessment and evaluation will be determined by the core spheres of the specialist's professional activity. Only such approach can verify the quality of professional training. It was defined that it is necessary to indicate one more condition for the effectiveness of assessment of the development of professional and communicative competence. It is related to the need for the assessment and evaluation to be focused not only on the explicitly but also on the implicitly expressed features and qualities of a professional. Beside cross-sectional measurement (one-time testing, for example), they advise to provide longitudinal measurements of professional communicative competence. Only in this way we can check the sustainability of the acquired material, objectively determine the permanence and stability of the actualization of the professionally significant qualities.

The article [17] deals with understanding the contribution of using assessment criteria to the develop-ment of self-regulation. Self-regulation is essential for learning, but it is neither innate, nor spontaneous. It is imperative for the subject to clear up the reference in order to have an effective auto-regulation. The teacher has the responsibility to create op-portunities for students to construct this reference, in order to permit the appropriation of the learning objectives' meaning.

In [1], the author explored how assessment can be linked to instruction and the multimodal composing process by inviting students to generate the grading criteria for new media assignments; and showed how this approach influenced students' composing and understanding of multimodal texts. It was firstly helped the class learn to construct the evaluative criteria for a digital video project, describing the course curriculum, instructional approaches, and assignments. Then it was presented the perceptions on how the collaborative construction of grading standards affected the students' learning and comprehension of new media rhetoric. The author described with pedagogical recommendations for instructors who teach multimodal digital writing and who seek to integrate the collaborative construction of grading criteria into their classroom.

The study [3] aims to highlight the diachronic evolution of the word bilingualism and the emergence of neologisms pluriligualism, multilingualism and plurililingual intercultural competence and the development of meanings of these terms. In conclusion, the authors propose a sort of self-assessment plurilingual intercultural competence scale.

The other condition necessary for assessment procedures is specified by the prob-lem-based content of assignments [2]. As a consequence, it is necessary to revalue the ratio of productive and reproductive types of assessment tasks in favor of the former. This particular condition seems to provide the solution to the problem of optimization and modernization of assessment in connection with the need to develop students' professional and communicative competence. Competence in its essence is associated with a real professional context, which is often unforeseen, being sophisticated in nature. Therefore, it is necessary that in the process of assessment students should exhibit their ability to solve new professional tasks using a foreign language. In the course of such activities, students should be able to express their own opinion, reflecting their individual professional worldview and demonstrating individual professional and communicative profiles. We think that, traditionally, this parameter of training is almost completely ignored even in the course of final testing at University.

With a problem-intensive approach to the assessment of competences, many difficulties arise. According to the authors [2], one of the most significant difficulties is that students and teachers have different perceptions, leading to differing viewpoints in assessments. In other words, there is no correlation between university student grades (self and peer) and teacher assessment.

Moreover, the important factor in connection with the competence approach to higher professional pedagogical education is the condition connected with taking into account the individual traits of students. Each undergraduate manifests his/her professionally important feature and personal qualities in different ways, forming an individually distinctive profile of the professional activity, an individual style of professional and communicative activity. In this regard, it is necessary to provide the variability of testing assignments, through variation of tasks wording, scope, and quality of such assignments.

Taking into consideration the active position of students is another important condition for the effectiveness of assessment. The students should be acquainted with the evaluation criteria, with the requirements for the procedure and the result of the assessment [5]. Students should know the objects subjected to assessment, they should be aware of the requirements to the quality of their activities. The assessment should be recognized as ineffective if a student is not able to prepare himself for testing independently, evaluate himself before, during and after it, and outline ways to improve the achieved results. Thus, assessment is regarded as a procedure that gives the opportunity for students to prove their professional and communicative autonomy, their responsibility in the process of the implementation of foreign-language communicative activities.

Competence-based assessment requires self-, peer- and teacher assessments [16]. In order to obtain a holistic view of students' contribution, self-assessment and peer-assessment are often used in addition to teacher assessment to arrive at the overall course grade of a student.

Despite the efforts of the authors, the issue of assessment and developing criteria, aimed at identifying and assessing communicative competence, has not been finally solved and requires further attention.

3 Methods of Classification

The classification problem is one of the main tasks of intellectual data mining. The statement can be formulated in the following way. Let the classification consist of N objects, each of which is characterized by an n-dimensional feature vector W, that is, a set of vectors $\{W_i, i = 1, ..., N\}$ is given. Each object S must be assigned to one of L classes: $\Omega_1, \Omega_2, ..., \Omega_L$, where the number of classes L may be known in advance or may not be known. The main task of the classification problem is the definition of the class.

There are several approaches to the classification problem. In the research [14], the issue of ranking students' competencies by the method of hierarchy analysis was studied. Based on the ratings obtained, students could be classified into several classes. The possibility of using Bayesian classification was also considered there.

In order to determine clusters on a data set, it is necessary first of all to introduce a measure of images similarity, which can be the basis for the rule of classifying images characterized by some center of the cluster. The following measures are widely used as similarity measures:

- Euclidean metric $d_{ij} = \sqrt{(\sum_{k=1}^{p}(x_{ki} - x_{kj})^2)}$;
- Minkowski metric $d_{ij} = \left(\sum_{k=1}^{p}(x_{ki} - x_{kj})^r\right)^{\frac{1}{2}}$;
- Hamming metric $d_{ij} = \sum_{k=1}^{p} \|x_{ki} - x_{kj}\|$;
- Chebishev metric $d_{ij} = \sup_{1 \leq k \leq p} \|x_{ki} - x_{kj}\|$.

After choosing the similarity measure, it is necessary to determine the criterion of clustering. The clustering criterion, as a rule, is based on the minimization (or maximization) of a certain quality indicator. The clustering method, which involves the use of a quality indicator, is associated with the development of procedures that will minimize or maximize the selected quality indicator. Bayesian decision theory underlies a statistical approach to the classification tasks. The advantages of the Bayesian approach are pointed out in a large number of works [10], as well as the disadvantages. Among its shortcomings, which were called by researches, are statistical data independence [9], and impossible direct processing of continuous variables [4]. However, at present, the Bayesian classification is gain-ing popularity and is used in many applied problems [7,8]. A hierarchical Bayesian classifier was created in the work [11]. Consider the most popular classification methods.

3.1 Simple Algorithm

Suppose a set of N images $\{S_1, ..., S_N\}$ is given. Let us to consider the centre of the first class M_1 also coincides with any of the given images, for example S_1. Let an arbitrary non-negative threshold value T be determined. We calculate the distance $d(M_1, S_2)$ between the existing cluster centre M_1 and the image $S2$. If the distance is greater than the threshold value T, a new class with a centre $M_2 = S_2$ is created. Otherwise, the image S_2 is included in the class, centre of which is M_1. At the next step, the distances $D(M_1, S_3)$ and $d(M_2, S_3)$ to the image S_3 from the centres of the clusters M_1 and M_2 are calculated. If both distances are greater than threshold T, a new class with centre $M_3 = S_3$ is created. Otherwise, the image S_3 is included to the class l which centre M_l is closer to it. Similarly, the distances from each new image $S_j, j = 1, ..., N$ to each known cluster centre $M_l, l = 1, ..., L$ are calculated and compared with the threshold value. If all these distances exceed the value of the threshold T, then a new class with the centre $M_j = S_j$ is created. Otherwise, the image is included to the class with the centre closest to it. The results of the algorithm are determined by the choice of the first centre of the class, the order of viewing the images, the value of the threshold value T, and, of course, and the characteristics of the images.

3.2 Maximum Distance Algorithm

An algorithm based on the principle of maxi-min (mini-max) distance is another procedure using Euclidean distance or any other. This algorithm is similar to the previous one, except that it first identifies the most distant classes.

1. The centre of the first cluster M_1 is randomly selected. It is conveniently to choose as the cluster centre M_1 that features vector S, which has some extreme properties. For example, the removal of which from the average of all vectors from the selection is maximum

$$M_1 = \arg\max_S d\left(S, \frac{1}{N}\sum_{i=1}^{N} S_i\right)$$

2. The centre of the second cluster M_1 is selected. The vector S that is the farthest from the first cluster centre is used as the centre:

$$M_2 = \arg\max_S d\left(S, M_1\right)$$

3. The cluster centre M_L is selected. For this purpose, all distances between the remaining images (feature vectors) S and already existing cluster centres $M_1, M_2, ..., M_{L-1}$ are calculated. For each remaining image x, the cluster centre l is found out, the distance to which is minimal:

$$l = \arg\min_{i=1,2,...,L-2} d\left(S, M_i\right)$$

4. The vectors are distributed among clusters by the criterion of proximity to their center. The one pretending to be the new cluster center M_L is taken the feature vector S, which has its minimum distance is maximum:

$$M_L = \arg\max_S \min_{i=1,2,...,L-1} d\left(S, M_i\right)$$

5. The distance obtained for the selected vector

$$d_{min} = \min_{i=1,2,...,L-1} d\left(M_L, M_i\right)$$

from it to the nearest cluster is compared with some specified distance d_p between the clusters. If the obtained distance is greater than the specified one, then this vector becomes the new cluster center. Otherwise, the process of cluster allocation stops. The choice of a given distance can be carried out in various ways.

K-means algorithm minimizes the quality index specified as the sum of the squares of the distances of all points in the cluster area to the centre of the cluster.

3.3 K-means Algorithm

K-means algorithm minimizes the quality index specified as the sum of the squares of the distances of all points in the cluster area to the centre of the cluster.

1. K initial cluster centres are selected at the first iteration ($r = 1$): $M_1(r)$, $M_2(r), ..., M_k(r)$. This choice is made arbitrarily, and usually the first k images from the given set $\{S_1, ..., S_N\}$: $M_1(r) = S_1, ..., M_k(r) = S_k$ are used as the initial centres. The iteration number is increases: $r = r + 1$.
2. At the r-th iteration ($r = 2, 3...$) the initial set of images $\{S_1, ..., S_N\}$ is distributed among k clusters according to the proximity rule. That is, some image S_j is assigned to the cluster $\Omega_l(r), l = 1, 2, ..., k$ with the centre $M_l(r-1)$ calculated in the previous step, if this centre is the closest:
$$S \in \Omega_l(r), l = \arg \min_{i=1,2,...,k} d(S, M_i(r-1))$$
3. Thus, $\Omega_l(r)$ is the set of images included in the cluster with number l at the r-th iteration of the algorithm. If the distances from a certain image to several centers of the clusters are equal, the decision to assign this image to one of them is made arbitrarily.
4. Based on the results of (2), new cluster centers $M_1(r), M_2(r), ..., M_k(r)$ are determined at the r-th iteration of the algorithm. They are determined from the condition that the sum of the squares of the distances between all the images belonging to the cluster $\Omega_l(r)$ and the new center of the cluster $M_l(r)$ should be minimal. In other words, new cluster centers are chosen in such a way as to minimize indicators
$$F_l(r) = \sum_{S \in \Omega_l(r)} d(S, M_l(r))^2, l = 1, 2, ..., k$$
and, therefore, the integral indicator of the quality of clustering at the r-th step
$$F(r) = \sum_{l=1}^{k} F_l(r) = \sum_{l=1}^{k} \sum_{S \in \Omega_l(r)} d(S, M_l(r))^2$$
For the case of Euclidean distance, the new cluster center $M_l(r)$, which ensures minimization of the corresponding particular exponent $F_l(r)$, is, in principle, a selective average determined by the set of images in the cluster: $\Omega_l(r)$: $M_l(r) = \frac{1}{N_l} \sum_{S \in \Omega_l(r)} S, l = 1, 2, ..., k$, where N_l, is the number of images entering the cluster $\Omega_l(r)$ at the r-th step.
5. The equality of the centers of the clusters at neighboring steps $M_l(r-1) = M_l(r), l = 1, 2, ..., k$ is a condition for the convergence of the algorithm, and when it is reached, the implementation of the algorithm ends. Otherwise, the algorithm repeats from the 2-nd step with a new iteration number $r = r + 1$. The quality of the k-means algorithm depends on the number of selectable cluster centers k, the choice of cluster centers at the first iteration $M_1(1), M_2(1), ..., M_k(1)$, and the data parameters. The results do not depend on the sequence of data viewing.

3.4 Bayesian Classification

When using classification methods, the choice of one class from two (in the general case of m) is made by determining the class number to which the classified object belongs. Bayesian classification algorithm implementation assumes the following steps:

- let's assume that R_{ij} is losses from the erroneous classification of the object, the object belongs to the class Ω_i, and it is attributed to the class Ω_j;
- risk function $H_i(d) = \sum_{j=1}^{m} R_{ij} P(\Omega_i|\Omega_j)$.

A decisive rule follows from the condition of minimizing risk – a classified object d, which is described by a vector of features W, belongs to the class Ω_i, if

$$\sum_{k=1}^{m} P(\Omega_i) R_{ki} f_k(x) = \min_{1 \leq j \leq m} \left(\sum_{k=1}^{m} P(\Omega_j) R_{kj} f_k(x) \right)$$

The number of classes for classification is four, this Ω_1 is "unsatisfy", Ω_2 - "satisfy".

The Bayesian formula for the posterior probability of each class has the form:

$$P(\Omega_j|W) = \frac{P(W|\Omega_j)P(\Omega_j)}{\sum_{i=1}^{2} P(W|\Omega_i)P(\Omega_i)}, j = 1, 2,$$

where $P(W|\Omega_j)$ is the conditional probability of indicators for different classes, $P(\Omega_j)$ is the a priori probability of each class.

Probability values $P(W_i|\Omega_j)$ are determined either by expert judgment or by preliminary estimates. In this case, the results of the preliminary assessments are considered as a training sample.

4 Results and Discussion

In the study it was revealed the extent to which intending primary school teachers possess a foreign language communicative competence. The general research concept is presented in Fig. 1. In Fig. 2, the scheme of the hierarchy is shown [14].

According to the experts' advice, the criteria of the most accurately charactering the level of proficiency in foreign language communicative competence by intending primary school teachers were chosen. In order to determine the level of intending specialists' foreign language communicative competence formation in the field of primary education among the students of primary education faculties, a survey, testing, questionnaire and evaluation were conducted by experts, which provided for the assessment of students by the following criteria:

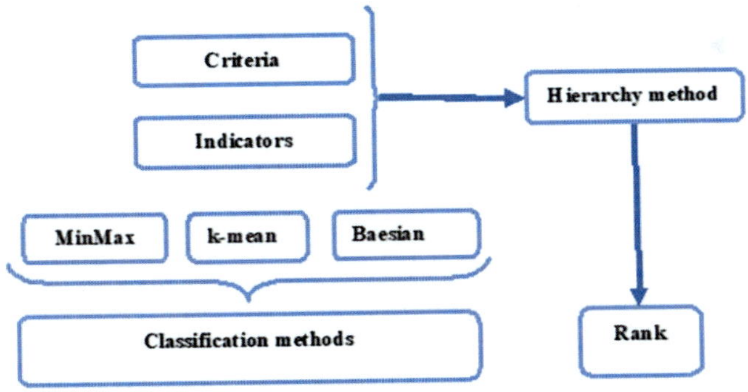

Fig. 1. General research concept

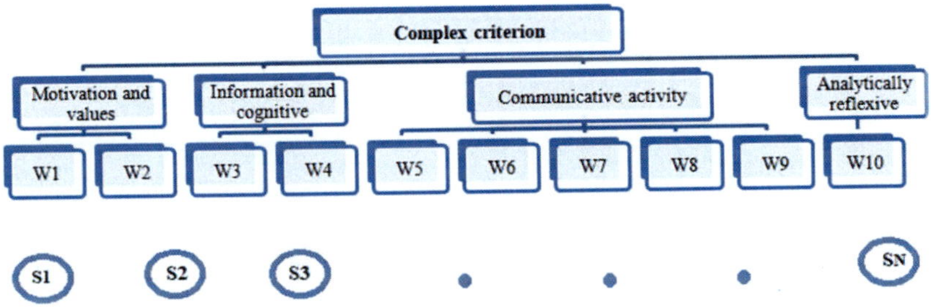

Fig. 2. Scheme of the hierarchy

1. professional orientation (the attitude to the profession of primary school teacher);
2. value orientations, attitudes and motives for learning a foreign language (English);
3. content understanding of communication;
4. relevance to the topic;
5. grammatical correctness of expressions;
6. lexical correctness of communication (quality and expediency of linguistic means);
7. logic and coherence of speaking;
8. composition of communication, correspondence of genre, style;
9. volume of expression;
10. activity and personal initiative expression in professional and self-development, adequate self-assessment

The value of each of the criteria for the levels of foreign language communication competence was calculated for each student with a 100-point scale, respectively, in the interval scale, the highest level of foreign language communicative competence corresponded to the score of 100 points, the lowest – 1.00.

During defining the model of the process of forming foreign language communicative competence, we relied on specific data on the levels of the students' investigated quality formation on the primary education faculties at higher education establishments. Further activity on forming the investigated quality of intending primary school teachers implies balancing the criteria indicators of its acquisition process. Therefore, there was a need of empirical studying the levels of foreign language communicative competence of intending primary school teachers (Table 1).

Table 1. Indicators and marking of intending primary school teachers' foreign language communicative competence

Indicators	Marking
1) professional orientation (the attitude to the profession of primary school teacher);	W1
2) value orientations, attitudes and motives for learning a foreign language (English);	W2
3) content understanding of communication;	W3
4) relevance to the topic;	W4
5) grammatical correctness of expressions;	W5
6) lexical correctness of communication (quality and expediency of linguistic means);	W6
7) logic and coherence of speaking;	W7
8) composition of communication, correspondence of genre, style;	W8
9) volume of expression;	W9
10) activity and personal initiative expression in professional and self-development, adequate self-assessment	W10

Diagnosis of formation of intending teachers' foreign language communicative competence has implemented not only an information function (gaining some knowledge about the object under study). It was intended that it should stimulate the growth of the foreign language communication skills. In this regard, the self-esteem of intending teachers and the degree of its correspondence with the experts' assessing became especially important. Adequate self-esteem is based on the ability to evaluate multipositionally the real interaction and to predict a constructive way out of a difficult communicative situation. Therefore, there is a connection between intending teachers' self-esteem of their own foreign language communicating competence and the ways of overcoming their difficulties in communicating with pupils (choosing a productive or non-productive way to solve communicative problems) (Table 2).

Table 2. Criteria and indicators of intending primary school teachers' foreign language communicative competence

Criteria	Indicators
Motivation and values	W1 – professional orientation (the attitude to the profession of primary school teacher); W2 – value orientations, attitudes and motives for learning a foreign language (English);
Information and cognitive	W3 – content understanding of communication; W4 – relevance to the topic;
Communicative activity	W5 – grammatical correctness of expressions; W6 – lexical correctness of communication (quality and expediency of linguistic means); W7 – logic and coherence of speaking; W8 – composition of communication, correspondence of genre, style; W9 – volume of expression;
Analytically reflexive	W10 – activity and personal initiative expression in professional and self-development, adequate self-assessment

The level of formation of intending primary school teachers' foreign language communicative competence during the second stage of the study was determined by analysing the results of externally expressed activity by means of observing method and expert evaluation. They made it possible to correlate intending teachers' self-esteem with experts' evaluation, thereby determining the degree of adequacy of students' self-esteem and, as a consequence, the presence or absence of their reflective position in their professional activities.

Having analyzed the data of the table, we can conclude about the levels of formation of foreign language communicative competence of primary education faculties' students at higher education establishments in Ukraine. We can deduce that according to the first criterion students' professional orientation is at a very low level, respectively, about 25% of students have a positive attitude towards their future profession. More students are motivated and interested in learning a foreign language (about 28–30%), but it can be noticed that they are not always the students whose attitude to the profession of primary school teacher are positive. According to the results of the 2–9-th criteria, we can see quite low levels of the different components indicators of the students' foreign language communicative competence. In particular, grammatical correctness of a foreign language communication is at the lowest level, while students understand the requirements for the content and volume of communication. In addition, we may notice a rather low level of activity and personal initiative in professional and self-development; moreover, intending teachers do not always have adequate self-esteem. If we analyze the dependence of foreign language communicative competence and adequate self-esteem, in most cases the higher the level of foreign-language communicative competence - the more adequate the student's self-esteem (Table 3).

Table 3. Indicators' rating for intending primary school teachers' foreign language communicative competence

	W1	W2	W3	W4	W5	W6	W7	W8	W9	W10
S1	22	24	28	42	26	36	26	38	56	28
S2	26	56	66	68	68	56	68	56	68	58
S3	42	38	38	64	32	44	34	38	38	42
S4	26	74	62	92	64	76	56	68	74	68
S5	46	16	38	44	16	26	58	42	52	32
S6	52	28	46	36	28	28	56	44	36	68
S7	42	58	58	52	58	48	52	56	58	66
S8	64	66	62	62	68	62	64	72	68	72
S9	22	48	52	58	56	54	42	36	44	58
S10	16	28	28	26	22	28	36	32	24	24
S11	88	56	54	64	82	76	62	56	78	82
S12	64	82	84	88	74	72	84	92	94	92
S13	82	78	72	86	80	82	70	84	92	88
S14	38	26	24	38	32	36	26	28	26	38
S15	16	18	36	24	36	38	24	26	32	28
S16	38	88	46	88	62	72	56	74	58	58
S17	96	42	22	36	38	44	34	58	28	68
S18	54	56	48	86	38	58	68	58	72	64
S19	68	76	82	84	76	80	86	74	92	80
S20	86	74	78	72	72	82	76	84	92	84

In the Table 4 the results of estimating the average value, complex assessment by the hierarchy method, classification results are shown: Class1 – splitting the Simple algorithm into two classes, Class2 – splitting the Simple algorithm into four classes, Class3 – the Bayesian classification into two classes.

As a result, two groups were received in Class1. The students in one of them have higher level of foreign language communicative competence while in the

Table 4. The classification results

	Average value	Hierarchy method	Classes 1	Classes 2	Classes 3
S1	32.6	27.5	1	3	1
S2	59	58.3	1	2	1
S3	41	40.3	1	3	1
S4	66	66.9	0	1	0
S5	37	30.8	1	3	1
S6	42.2	41.7	1	3	1
S7	54.8	56.6	1	2	1
S8	66	65.5	1	2	0
S9	47	48.9	1	2	1
S10	26.4	26.9	1	3	1
S11	69.8	64.3	1	2	0
S12	82.6	82.6	0	0	0
S13	81.4	78.8	0	0	0
S14	31.2	29.6	1	3	1
S15	27.8	25.4	1	3	1
S16	64	67.6	0	1	0
S17	46.6	44.5	1	3	1
S18	60.2	57.8	1	2	1
S19	79.8	78.7	0	0	0
S20	80	77.6	0	0	0

other the students are weaker. But we can't indicate that one group satisfies the required level, and the other does not. The same applies to the division into 4 classes. Each class has the advantages, but it is also impossible to say that in one class the mark is excellent, in the other it is good, in the third it is satisfactory, in the fourth it is unsatisfactory.

5 Conclusions

In the paper the approaches to the selection of criteria for assessing the future primary school teachers' foreign language communicative competence are suggested. Classical classification algorithms are considered. Classification and assessing criteria that consider foreign language communicative competence skills during professional qualities developing are offered in the paper, as well as personal parameters that are hidden for direct observation and which are only indirectly manifested in the professionally oriented communication activity of the individual.

Intending primary school teachers' communicative competence was assessed according to four criteria: motivation and values, information and cognitive, communicative activity and analytically reflexive. The results of studying groups are consistent. The work assumes that the parameters are independent.

The findings revealed some problems, namely that student groups cannot be divided into classes based on criteria. This leads to a classification by their level of knowledge. And it does not reflect their competences by the requirements that are imposed on them.

The studies showed the features of each classification method, confirmed the peculiarities of the proposed approach to the selection of criteria for assessing communicative competence.

Further research will be based on the assessment of foreign language communicative competence when the indicators are dependent; the analysis will be carried out using the principal component analysis. It is also planned to use the Bayesian classification, which takes into account the average value of the class.

References

1. Adsanatham, C.: Integrating assessment and instruction: using student generated grading criteria to evaluate multimodal digital projects. Comput. Compos. **29**(2), 152–174 (2012). https://doi.org/10.1016/j.compcom.2012.04.002
2. Alias, M., Masek, A., Md Salleh, H.: Self, peer and teacher assessment in problem based learning: are they in agreements? Procedia-Soc. Behav. Sci. **204**, 309–317 (2015). https://doi.org/10.1016/j.sbspro.2015.08.157
3. Birova, J., Barancova, M., Simkova, Z.: A note on terminology: bilingualism, plurilingualism and plurilingual intercultural competence or are we all plurilingual? XLinguae **9**, 114–127 (2016). https://doi.org/10.18355/XL.2016.09.02.114-127
4. Brand, E., Gerritsen, R.: Naive-bayes and nearest neighbor. DBMS **7** (1998)
5. Brantmeier, C., Vanderplank, R., Strube, M.: What about me?: individual self-assessment by skill and level of language instruction. System **40**(1), 144–160 (2012). https://doi.org/10.1016/j.system.2012.01.003
6. Budas, Y.O.: Stvorennya psykholohichno spryyatlyvoyi atmosfery dlya formuvannya profesiynoyi inshomovnoyi kompetentnosti [creating psychologically favorable atmosphere for the forming professional foreign language competence]. Visnyk Zhytomyrskoho derzhavnoho universytetu imeni Ivana Franka. Seriya Pedahohichni nauky **3**(81), 41–46 (2015)
7. Castillo, E.F., Gutierrez, J.M., Hadi, A.S.: Sensitivity analysis in discrete Bayesian networks. IEEE Trans. Syst. Man Cybern. Part A: Syst. Hum. **27**(4), 412–423 (1997). https://doi.org/10.1109/3468.594909
8. Chickering, D., Geiger, D., Heckerman, D.: Learning Bayesian networks: the combination of knowledge and statistical data. Mach. Learn. **20**, 197–243 (1995). https://doi.org/10.1007/BF00994016
9. Friedman, N., Geiger, D., Goldszmidt, M.: Bayesian network classifiers. Mach. Learn. **29**, 131–165 (1997). https://doi.org/10.1023/A:1007465528199
10. Heckerman, D.: Bayesian networks for data mining. Data Mining Knowl. Discov. **1**, 79–119 (1997). https://doi.org/10.1023/A:1009730122752
11. Langseth, H., Nielsen, T.: Classification using hierarchical Naive Bayes models. Mach. Learn. **63**(2), 135–159 (2006). https://doi.org/10.1007/s10994-006-6136-2

12. Lazarevic, N.: Assessment of intercultural communicative competence in pre-service English language teachers. Nastava i vaspitanie **67**(3), 471–487 (2018). https://doi.org/10.5937/nasvas1803471L
13. Lieu Tran, T.B., Ho, T.N., Mackenzies, V., Long, K.L.: Developing assessment criteria of a lesson for creativity to promote teaching for creativity. Thinking Skills Creativity **25**, 10–26 (2017). https://doi.org/10.1016/j.tsc.2017.05.006
14. Pashko, A., Pinchuk, I.: Intellectual methods of estimation of intending primary school teachers' foreign language communicative competence. In: IEEE International Conference on Advanced Trends in Information Theory, vol. 1, pp. 439–444. ATIT (2019). https://doi.org/10.1109/ATIT49449.2019.9030429
15. Petryk, L.V.: Klassifikatsyya mediasredstv v nauchno-pedagogicheskoy literature. [classification of media in scientific and pedagogical literature]. Inshomovna osvita pedahoha: vyklyky, problemy, perspektyvy **5**, 143–149 (2015)
16. Roffe Gomez, A.: Foreign language testing through competence assessment rubrics and oral interviews in the European higher education area. Procedia-Soc. Behav. Sci. **141**, 524–532 (2014). https://doi.org/10.1016/j.sbspro.2014.05.091
17. Santos, L., Pinto, J.: The development of self-regulation through assessment criteria. Procedia-Soc. Behav. Sci. **112**, 907–915 (2014). https://doi.org/10.1016/j.sbspro.2014.01.1248
18. Savchenko, O.Y.: Uminnya vchytysya yak klyuchova kompetentnist' neperervnoyi osvity. [skills to learn as a key competence in continuing education]. Mach. Learn. **2** (2010)
19. Tareva, E., Tarev, B.: The assessment of students' professional communicative competence: new challenges and possible solutions. XLinguae **11**, 758–767 (2018). https://doi.org/10.18355/XL.2018.11.02.59
20. Woodward, T.: Loop input. ELTJ. **57**(3), 301–304 (2003). https://doi.org/10.1093/elt/57.3.301

Analyzing Performance of Apache Spark MLlib with Multinode Clusters on Azure HDInsight: Spark-Perf Case Study

Sergii Minukhin[1](✉), Natalia Brynza[1], and Dmytro Sitnikov[2]

[1] Simon Kuznets Kharkiv National University of Economics, Kharkiv, Ukraine
minukhin.sv@gmail.com, natalia.brynza@hneu.net
[2] Kharkiv National University of Radio Electronics, Kharkiv, Ukraine
dmytro.sytnikov@nure.ua

Abstract. In this paper, we present an analysis and results of experimental research into determining the performance of solving machine learning problems via the library Apache Spark MLlib for the ecosystem Microsoft Azure HDInsight with the help of the test dataset Spark-Pref. In order to solve the defined problems, software and information support methodology have been developed based on the monitoring system SparkMeasure and Ambari. Metrics have been suggested for analyzing the performance of Apache Spark computations. These metrics use statistical characteristics of learning and testing processes when benchmark Spark-perf tests are carried out. There have been suggested formulas for determining settings for Apache Spark parameters. These formulas provide a time minimization as compared to the standard values of Spark parameter settings for executing sets of machine learning test tasks for heterogeneous and homogeneous cluster configurations of Apache Spark Azure HDInsight. In order to assess computing performance for machine learning methods in Spark-Pref a metric has been proposed, which is calculated as the ratio of the average testing time and the average training time. The results of the computational experiments have been demonstrated. They confirm the effectiveness of the proposed algorithms for Apache Spark settings relative to the standard values for heterogeneous and homogeneous clusters deployed on the platform Apache Spark Azure HDInsight, machine learning methods for a Spark-Pref test set being implemented.

Keywords: Azure Hindsight · Apache Spark · Resilient Distributed Dataset · Mlib · SparkMeasure · Ambari · Benchmark · Spark-Perf

1 Introduction

Spark is a fast, fault-tolerant, and general-purpose cluster computing platform for big data processing. Apache Spark is a popularized data analytic platform

S. Babichev et al. (Eds.): ISDMCI 2020, AISC 1246, pp. 114–134, 2021.
https://doi.org/10.1007/978-3-030-54215-3_8

that is being adopted by numerous organizations (Yahoo!, eBay, Baidu, Netflix [23]) due to its in-memory computation framework. The Apache Spark framework, being open source, offers a great amount of tuning capabilities and customization. Its configuration has a total of 100 parameters that can be changed easily by the user and customized according to the application and cluster he uses. The main types and description of the Apache Spark parameters are represented in [16,22].

Zaharia et al. [25,26] proposed Spark as a streaming engine which is capable of processing large scale data and reduces latency in processing the data. A Spark cluster consists of multiple distributed objects, Resilient Distributed Datasets (RDDs) [24] stored in memory, and it also executes operations in parallel. Spark supports one file system for a particular task and consists of several nodes for resource management. For example, if Spark is set in the Hadoop environment, it is necessary to allow it to access HDFS, whereas on empty machines it is much simpler to launch Spark. In order to improve performance of clusters deployed on the basis of Apache Spark, at present new methodologies and tools are developed, which are built based on: optimizing standard cluster parameter settings by using heuristic rules for modifying their values (user settings); as a basis of them 2 data processing technologies are used - RDDs and Data Frames as well as machine learning methods that allow obtaining more flexible AI settings, software and architectural solutions on cloud platforms, such as Microsoft Azure [9], AWS [1], Google Cloud Platform [8] etc. These solutions, which use relatively simple tools and methods for cluster deployment, allow providing the solution of lots of applied problems in different subject domains with high efficiency and performance. An actual problem of investigating the effectiveness of clusters for distributed and parallel calculations is the assessment of the performance of computational processes for Big Data. In the Big Data era, the increasing data volume, the requirement of quick data processing (data velocity) and the wide variety of data types, structures, and sources give rise to various new challenges at which traditional data management systems are inadequate. Benchmarking Big Data Systems: A Review. For monitoring performance standard test sets are used (benchmarks), which represent a universal set of different tests, which represent a universal set of different tests (data sets) that are related to different categories. For a benchmarking study, workloads capture operations commonly used to implement most applications, hence the generation of relevant, portable, and scalable workloads determines the usefulness of benchmarks [17] Benchmarking Big Data Systems: A Review. Most popular benchmarks are as follows: Yahoo! Cloud Serving Benchmark (YCSB) [5], BigBench [15], HiBench [18], Cloud-Suite [21], and BigDataBench [14]. Besides, one should consider specialized test sets that are used in popular frameworks – for example, for assessing the Apache Spark performance the following benchmarks are used: SparkBench, Spark-Perf (Databrick's Spark Perf suite - databricks/spark-perf), TeraSort benchmark for Spark, Spark Tpch & Tpcds Suites, Spark Benchmarks, for the cloud platform - 2016 CloudSort Benchmark, CloudRank-D. The given test sets are designated for testing performance for solving problems of sorting, Machine Learning, Graph Computation, SQL Queries, Streaming Application.

The most important investigation domain for improving performance on clusters is the application of cloud platforms. Lately. Such platforms mainly use Google Cloud Platform (GCP), Amazon Web Services (AWS), IBM Cloud, Microsoft Azure and others. For implementing frameworks that support high-performance cluster computations Amazon Elastic MapReduce (EMR), Azure Hindsight, Google Cloud DataProc (in GCP) are applied. They utilize the version Apache Spark 2.1.0 f and in this case, Apache Spark MLlib, Apache Spark ML, TensorFlow library are used. It should be noted that issues have not been sufficiently studied related to the application of ML methods under the condition of the scalability of cluster architectures on cloud platforms, determining framework settings values, which allow minimizing the task execution time as to the standard settings, and also task performance evaluation for machine learning when test data sets are used, which allow the maximum coverage of machine learning methods.

2 Problem Statement

The problem under consideration is related to the investigation of the distributed computations performance for machine learning methods from the library Apache Spark MLlib [10] on the platform Microsoft Azure HDInsight under the conditions of the scalability of cluster solutions for homogeneous and heterogeneous Apache Spark architectures and experimental definition of their optimal values when machine learning problems are solved based on using benchmark Spark-Perf [11].

The aim of this investigation is to increase the performance of distributed computations for different configurations of Apache Spark Azure HDInsight clusters when Apache Spark MLlib machine learning problems are solved based on benchmark Spark-Perf. Our contributions are as follows:

- a methodology for processing Apache Spark MLlib test results has been developed; it includes preprocessing and analyzing the test results for standard Apache Spark settings values for different cluster configurations, the cluster using different types of virtual machine configurations with memory optimization, and further determining optimal (from the time minimization viewpoint) minimizing performing the configuration for a selected set of tests;
- software and information manipulation tools have been developed for obtaining, processing and analyzing the results of passing ML tests in the framework benchmark Spark-Perf;
- metrics have been suggested for the analysis of Apache Spark computations performance, which use statistical characteristics for processes of learning and testing when benchmark Spark-Perf tests are passed;
- algorithms for tuning parameters Apache Spark for cluster homogeneous and heterogeneous configurations that provide increasing the computational performance for the library Apache Spark MLlib when benchmark Spark-Perf is used.

3 Overview and Related Works

3.1 Benchmarks

Normally benchmarks can be subdivided into any categories according their scope of benchmarking. Benchmark suites including HcBench and MRBS provide workloads for Hadoop-related systems, while HiBench, CloudSuite, and BigDataBench cover workloads of different big data systems. The major types of benchmarks are shown in Table 1 [17]. (Benchmarking Big Data Systems: A Review), the structures of separate tests – HiBench and SparkBench – are shown in Tables 2 and 3 correspondingly [19].

Table 1. Overview of the state-of-the-art open source big data benchmarks

Systems	Type	Micro benchmark	End-to-end benchmark	Benchmark suite
Hadoop-related systems	Hadoop MapReduce and HDFS	Hadoop built-in micro benchmarks, NNBench, TestDFSIO, DFSCIOTest, ThreadedMapBench, HiBD, TPCx-HS	GridMix, SWIM, MRBench	HiBench, PUMA, MRBS, CloudSuite, BigDataBench
	High-level languages on Hadoop		PigMix	HiBench, BigDataBench
	SQL on Hadoop	AMPLab benchmark	YCSB, YCSB++	HcBench, MRBS, BigDataBench
Data stores	DBMSs	CALDA	Six active TPC benchmarks, BG, BigBench	HiBench, BigDataBench
	NoSQL data stores	AMPLab Benchmark	YCSB, YCSB++, BG	CloudSuite, BigDataBench
Specialized systems	Graph data		LinkBench, WGB, Graphalytics, GraphBIG	CloudSuite, BigDataBench
	Stream data	Stream Bench	Streams BigDataBench	HiBench, CloudSuite
	Scientific data	DSIMBench	GenBase	BigDataBench

From the above tables one can conclude that machine learning tests are popular enough, which confirms the relevance of their usage for testing the existing machine learning libraries.

Table 2. HiBench workloads

Category	№	Workload
Micro Benchmarks	1	Sort
	2	WordCount
	3	TeraSort
	4	EnhancedDFSIO
Web Search	5	Nutch Indexing
	6	PageRank
Machine Learning	7	Bayesian Classification
	8	K-means Clustering
Analytical Query	9	Hive Join
	10	Hive Aggregation

Table 3. SparkBench workloads

Application Type	Workload
Machine Learning	Logistic Regression
	Support Vector Machine
	Matrix Factorization
Graph Computation	PageRank
	SVD++
	TriangleCount
SQL Queries	Hive
	RDDRelation
Streaming Application	Twitter
	PageView

3.2 Machine Learning Libraries in Distributed Computations

Apache Spark MLlib. Spark MLlib are components of Apache Spark machine learning. One of the main Spark peculiarities is the possibility of massive computa-tions scalability. The MLlib library consists of two program packages: spark.mllib includes an original API built on RDD; spark.ml is the implementation of a higher level API with using DataFrames for constructing machine learning conveyers, which are always called ML Pipelines. In Spark MLlib support for several data times, input data representation and output data has been implemented. Machine learning tasks that can be solved with in-built in Spark MLlib machine learning algorithms are as follows: linear regression; generalized linear regression; decision tree regression; random forest regression; gradient-boosted tree regression); survival regression; isotonic regression; basic statistics: summary statistics maximum, minimum, mean, standard deviation, number of filled values, and also general number of values; correlation between several rows with the help of Pearson and Spearman correlations; stratified sample including sampleBykey and sampleByKeyExact; hypotheses verification with the help of the Pearson criterion and chi-squared test relating to whether or not the result is statistically significant or rando; random data generation via the methods RandomRDD, Normal and Poisson; classification: logistic regression; binomial logistic regression; multinomial logistic regression; decision tree classifier; random forest classifier; gradient-boosted tree classifier; linear support vector machine); naive Bayes; k-means clustering.

Apache Mahout. Apache Software foundation has launched an open source free project Apache Mahout [3,13]. The main objective was to create scalable machine learning libraries. Mahout was supposed to become scalable to large enough data sets. The core algorithms for clustering, classification and batch based collaborative filtering are implemented on top of Apache Hadoop using the MapReduce paradigm. The framework of Mahout allows using a variety of tools for finding automatically meaningful and convenient patterns for generating Big Data [13]. Mahout has been created on top of Apache Hadoop using the MapReduce paradigm. At present, Mahout allows resolving problems associated with the following: Recommendation, Clustering, Classification, Frequent item set mining.

Amazon Machine Learning (AML). Amazon Machine Learning is designated to facilitate ML technologies. It provides developers with tools for creating scalable, robust, and intelligent applications. AML gives a possibility to avoid getting into the details of ML methods and algorithms to provide powerful Machine Learning technology [2].

MLbase [20]. It should be noted that MLbase has not yet been fully developed, but research into this field has been going on at Berkeley's AMP lab on the MLbase platform wraping Spark, MLlib, and other projects. The main purpose is to make machine learning on datasets of all sizes accessible to a broader range of users. Besides MLlib and Spark, one should mention MLI, an API for feature extraction and algorithm development, and ML Optimizer, which automates the hyperparameters tuning.

One of the most popular machine learning libraries is TensorFlow, for which distributed solutions have been developed. These solutions use Apache Spark – Tensor-Flow on Spark, Distributed TensorFlow, which allows significant increasing machine learning methods performance when the Apache Spark cluster has been deployed.

4 Methods

4.1 Apache SPARK Azure HDInsight

The investigation of Apache Spark cluster performance is carried out on the platform Azure HDInsight [4], version 3.6 (OS Ubuntu 16.0.4 LTS) for virtual machines with optimized memory. It includes the following components: Apache

Table 4. Characteristics of virtual machines, series Dv2

Size	vCPU	Memory, GB	Temp. storage (SSD), GB	Max temp. storage throughput: IOPS/Read MBps/Write MBps	Max data disks/throughput: IOPS
Standard_D11_v2	2	14	100	6000/93/46	8/8 × 500
Standard_D12_v2	4	28	200	12000/187/93	16/16 × 500
Standard_D13_v2	8	56	400	24000/375/187	32/32 × 500
Standard_D14_v2	16	112	800	48000/750/375	64/64 × 500
Standard_D15_v2	20	140	1000	60000/937/468	64/64 × 500

Table 5. Characteristics of virtual machines, series Esv3

Size	vCPU	Memory, GB	Temp. storage (SSD), GB	Max cached and temp. storage throughput: IOPS/MBps (cache size, GB)	Max uncached disk throughput: IOPS/MBps
Standard_E2s_v3	2	16	32	4000/32 (50)	3200/48
Standard_E4s_v3	4	32	64	8000/64 (100)	6400/96
Standard_E8s_v3	8	64	128	16000/128 (200)	12800/192

Hadoop and YARN(2.7.3), Apache Tez (0.7.0), Apache Pig (0.16.0), Apache Hive (2.1.0, 1.2.1), Apache Tez Hive2 (0.8.4), Apache Ranger (0.7.0), Apache HBase (1.1.2), Apache Sqoop (1.4.6), Apache Oozie (4.2.0), Apache Zookeeper (3.4.6), Apache Storm (1.1.0), Apache Mahout (0.9.0+), Apache Phoenix (4.7.0), Apache Spark (2.3.0, 2.2.0, 2.1.0), Apache Livy (0,4, 0,4, 0,3), Apache Kafka (1.1, 1.0), Apache Ambari(2.6.0), Apache Zeppelin (0.7.0), Mono (4.2.1).

Instances of virtual machines with optimized memory for HDInsight are determined by a high ratio of RAM to processor performance, which is appropriate for relational database servers, caches of middle and high levels, and in-memory analysis (Tables 1, 2). Dv2-series sizes [6] run on the Intel®Xeon®8171M 2.1 GHz (Skylake) or the Intel®Xeon®E5-2673 v4 2.3 GHz (Broadwell) or the Intel®Xeon®E5-2673 v3 2.4 GHz (Haswell) processors and support Intel Turbo Boost 2.0. Azure Compute suggests sizes of a virtual machine that are allocated for a particular hardware type and are given to a single client. These sizes of virtual machines are mostly useful for workloads that require a high level of isolation from other clients. In Table 4 the main characteristics for virtual machines of series Dv2 are shown. Esv3-series instances [7] are based on feature the Intel®Xeon®8171M 2.1 GHz (Skylake) or the Intel®Xeon®E5-2673 v4 2.3 GHz (Broadwell) processor, Intel Turbo Boost Technology 2.0, and use premium storage. In Table 5 the main characteristics for machines of series Esv3-series are represented.

4.2 Tests for Spark-Perf Machine Learning Benchmark

Spark-Perf is a framework for testing performance of Apache Spark, versions 1.0+. It has been written in the languages Python and Scala. Spark-Perf has been developed by the company DataBricks for checking the performance of the MLlib, library, which implements machine learning algorithms. Also it is possible to test streaming applications, SQL-queries and the core of Spark (Spark Core tests), which is now under development. The main advantages of Spark-Perf over other frameworks are as follows: the possibility of testing the effectiveness of Spark, PySpark, Spark Streaming and MLlib; test parametric configurations; easy configuring (the configuration file is written in Python); it has sets of parameters to test simultaneously several Spark versions and test configurations; automatic loading and creating Spark, which provides caching successful combinations for quick testing on several Spark versions.

In Table 6 a description of Spark-Perf machine learning methods tests that have been implemented in the library MLlib is represented.

4.3 Software and Information Tools for Collecting and Processing Test Results for the Benchmark Spark-Perf

To analyze the results obtained with the help of the benchmark Spark-perf, the soft-ware Spark Benchmarks and Performance Metrics has been developed. The test results for Spark-Perf have been represented in a log-file with the extension .out, which is processed for obtaining and analyzing results with the following modules.

Module Spark-PerfReport. For collecting and processing test results the program module Spark-PerfReport has been developed in Python3.7. In the module Spark-PerfReport a statistical analysis of the obtained results is carried out for each test. Namely, mean and mean squared value for the learning and testing procedures, and also the ratio between the mean testing time and the mean testing time for learning. The sequence diagram for module in UML for module SparkPerfReport is shown in Fig. 1.

Module SparkMeasureReport. It is built based on the monitoring system SparkMeasurr. Spark executor task metrics provide instrumentation for workload measurements. They are exposed by the Spark WebUI, Spark History server, Spark EventLog file and from the ListenerBus infrastructure. Apache Spark Task Metrics contains more than 30 parameters that allow analyzing in detail states of accomplished stages of the tasks processed by the cluster (executorRunTime, executorCpuTime, executorDeserializeTime, executorDeserializeCpuTime, shuffleWriteMetrics, shuffleReadMetrics etc.). In the module SparkMeasureReport a table with metrics for all conducted tests and graphs built on its basis are output. Sequence diagram of the implementation of the module SparkMeasureReport in UML is shown in Fig. 2.

Table 6. Characteristics of machine learning tests for Spark-Perf benchmark

Code	Name	Number of experiments	Number of test examples
Regression and classification algorithm			
glm-regression	Generalized Linear Regression Model	10	5000
glm-classification	Generalized Linear Classification Model	10	5000
naive-bayes	Naive Bayes	10	5000
Decision trees			
decision-tree	Decision Tree	16 × 10	5000
Recommendation algorithm			
als	Alternating Least Squares	10	5000
Clustering algorithm			
kmeans	K-Means clustering	10	5000
lda	Latent Dirichlet allocation	10	
pic	Power Iteration Clustering	10	
gmm	Gaussian Mixture Model	10	5000
Statistical algorithm			
summary-statistics	Summary Statistics	10	
pearson	Pearson's Correlation	10	50000
spearman	Spearman's Correlation	10	50000
chi-sq-feature/gof/mat	Chi-square Tests	10	100000
Linear algebra algorithms			
svd	Singular Value Decomposition	10	
pca	Principal Component Analysis	10	
block-matrix-mult	Matrix Multiplication	10	
Algorithms for association rules generation			
fp-growth	FP-growth frequent item sets	10	
prefix-span	PrefixSpan	10	
Feature search algorithms			
word2vec	Word2Vec distributed presentation of words	10	

Module for Monitoring Metics Ambari. Apache Ambari is a tool for administering with an open code. It is deployed on Hadoop clusters and used for tracking programs launched on the cluster. Apache Ambari is a Web application that controls, checks and provides a safe functioning of the Hadoop clusters.

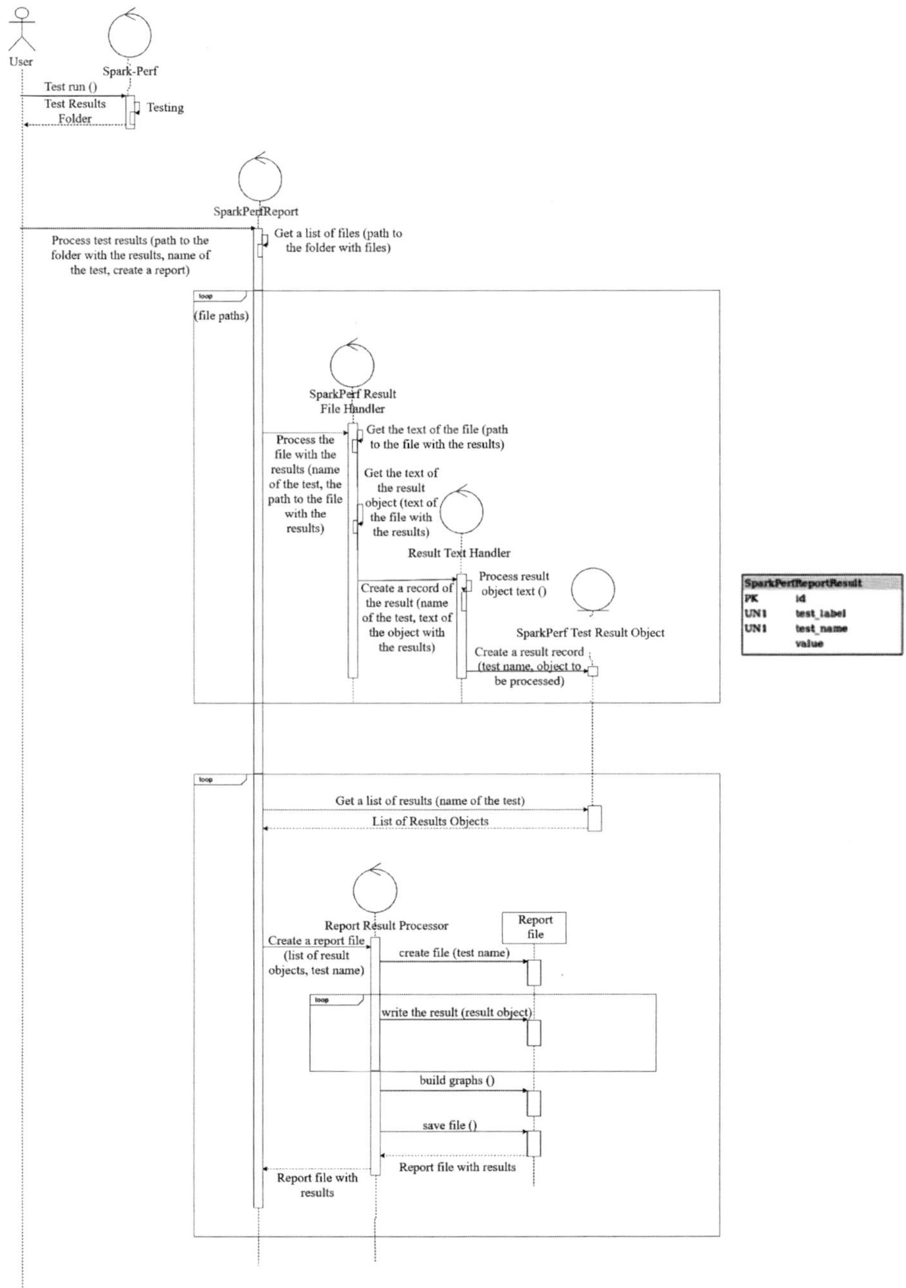

Fig. 1. Sequence diagram: module SparkPerfReport

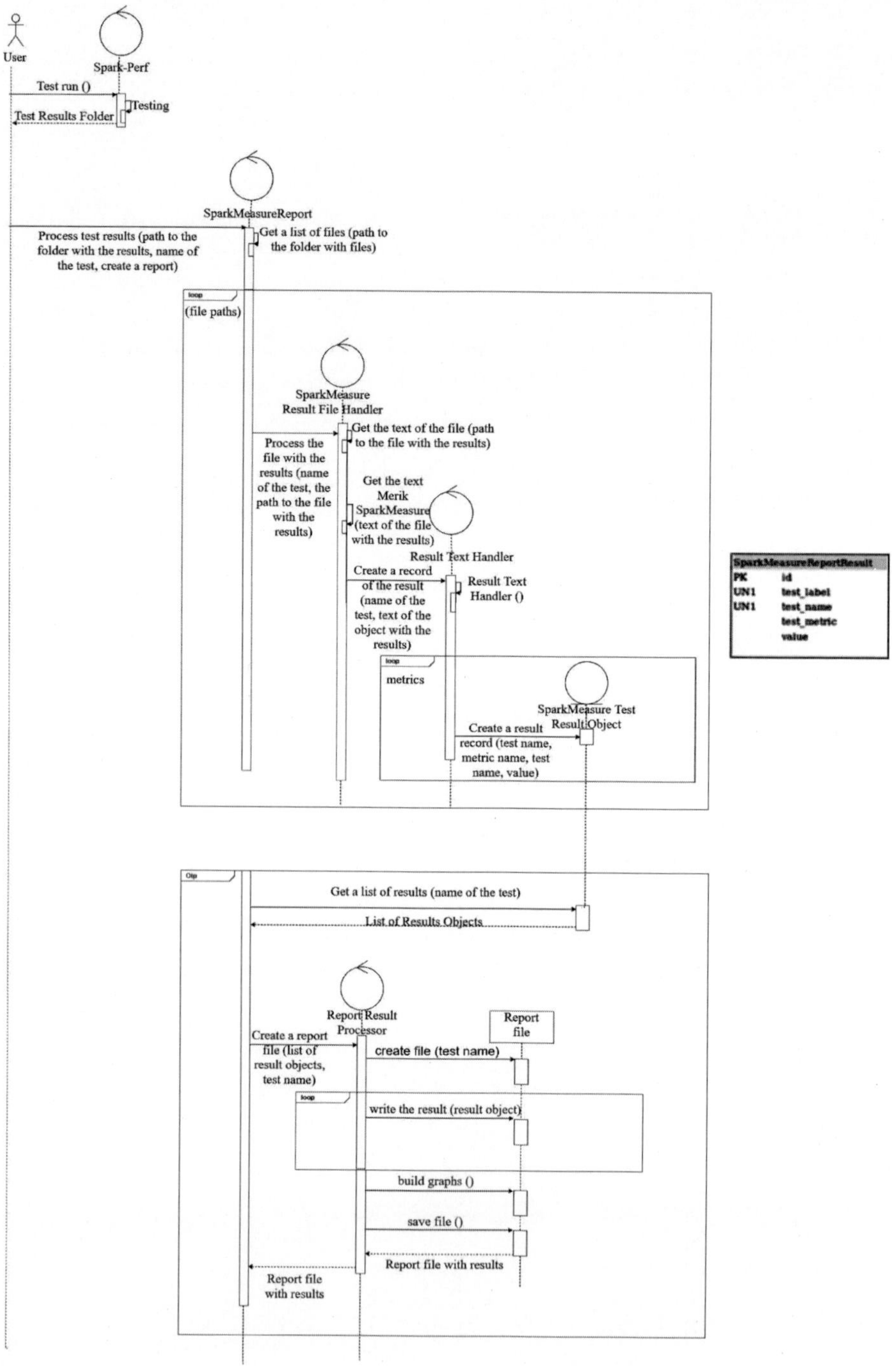

Fig. 2. Sequence diagram: module SparkMeasureReport

It provides highly interactive panel, which allows administrators to visualize the functioning and state of each program that is run on the Hadoop cluster. Its flexible and scalable user interface allows setting up on the cluster a row of tools such as Pig MapReduce, and Hive. For collecting and analyzing metrics for monitoring cluster resources a special soft-ware has been developed. It consists of the following components: Ambari, Grafana, Graphite; Ambari2Grafana. Apache Ambari is based on the architecture master-slave, where the main node issues instructions to subordinate nodes to accomplish certain actions and inform on the state of each action. Grafana is a free software that allows visualizing and formatting metric data. It is distributed in accordance ith the licence Apache 2.0.2. It allows creating information panels and graphs from different sources, including time serious databases, such as Graphite, InfluxDB and OpenTSDB. For collecting information from Ambari database the program Ambari2Graphite written in Python3.7 is used. The input data for the program is settings file having access to Ambari and Graphite.

Performance Metrics. Elapsed Time: it is the difference between beginning time and an ending time of task; SpeedUp: it is the ratio elapsed time of 2 and 3 workers.

5 Experimental Results and Evaluation

According to the suggested methodology of conducting experiments, they include the following stages.

Stage 1. Preparatory Stage (Stage of Preparatory Processing and Analysis). The aim of this stage is the determination of a Spark cluster for the values of its parameter setting used by default and providing minimum execution time for Spark-Perf test tasks (tests MLLIB_TESTS). Experiments have been carried out, which include executing tasks on Apache Spark clusters with different configurations (see Tables 4, 5) for their standard (default) settings. For processing and analyzing results, the software for retrieving data from the monitoring system SparkMeasure and the given results in log-files of Spark-Perf tests has been used. The results obtained for testing the parameter Executor Run Time of the system SparkMeasure are shown in Fig. 3.

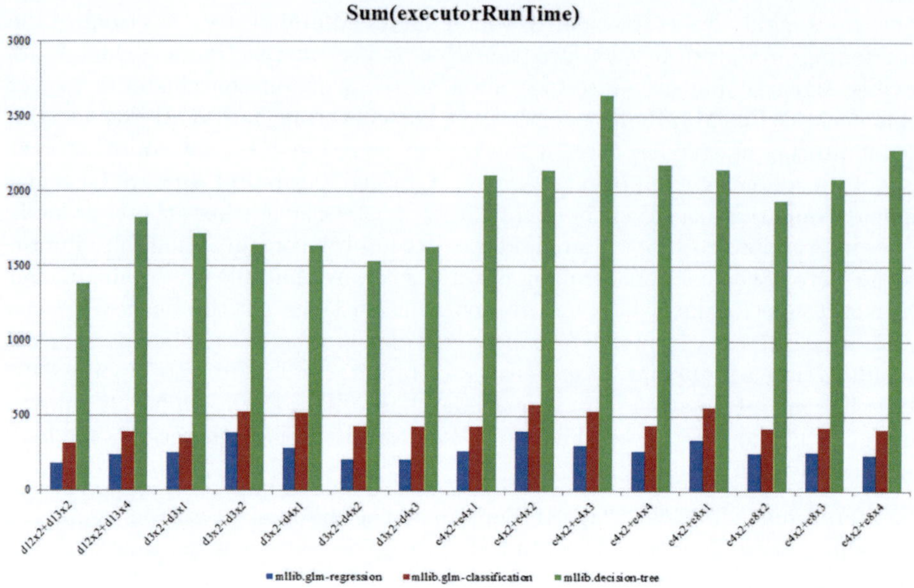

Fig. 3. The result of parameter calculation Executor Run Time for different configuralion Apache Spark

In Fig. 3 the following notation has been used: 2xD12v2 + 2xD12v2 - configurations: 2 masters D12v2 and 2work nodes D12v2 (**Env1**); 2xD12v2 + 3xD12v2 - configurations: 2 and 3 work nodes D12v2 (**Env2**) (**homogeneous cluster**); 2xD12v2 + 2xD13v2 – configuration: 2 masters D12v2 and 2 work nodes D13v2 (**Env3**)); 2xD12v2 + 3xD13v2 – configurations: 2 masters D12v2 and 3 work nodes D13v2 (**Env4**)) (**heterogeneous cluster**).

Thus, the above results let us conclude that when standard cluster settings are tuned, the most preferable is the configuration with the instances D12v2 and D13v2.

Stage 2. Selecting Spark cluster configuration for testing Spark-Perf. In accordance with the obtained results, when experiments have been conducted, the following configurations have been used (Table 7).

Stage 3. Deploying and tuning cluster parameters. The usage of 2 data sources is a peculiarity of deploying and tuning the cluster Apache Spark Azure HDInsight: Ambari and configuration file (config.py). In Table 8 one can see sources for some tunable cluster parameters.

Stage 4. Determining settings for Apache spark parameters that provide a minimal execution time. At this stage there have been suggested and investigated algorithms for computing cluster settings parameter values and carried out a comparative analysis with standard default values. Default Apache Spark values for different cluster configurations.

Table 7. Configuration of Spark clusters for testing

Cluster type	Master-node		Worker-node	
	Number of cores	Instance type	Number of cores	Instance type
Env1, Env2	2	Standard_D12_v2	2	Standard_D12_v2
	2	Standard_D12_v2	3	Standard_D12_v2
Env3, Env4	2	Standard_D12_v2	2	Standard_D13_v2
	2	Standard_D12_v2	3	Standard_D13_v2

Table 8. Data sources for tuning Apache cluster parameter tuning

Parameter	Source
spark.yarn.executor.memoryOverhead	Ambari
spark.executor.cores	Ambari
spark.executor.memory	Ambari
spark.yarn.driver.memoryOverhead	Ambari
yarn-scheduler-maximum-allocation-mb	Ambari
yarn-scheduler-maximum-allocation-vcores	Ambari
minimum/maximum container size	Ambari
spark.driver.memoryOverhead	config.py
spark.executor.cores (customized modified)	config.py
spark.executor.instances (customized modified)	config.py
spark.executor.memory (customized modified)	config.py
spark.executor.memoryOverhead (customized modified)	config.py

In this paper the following algorithms for computing Spark parameters have been suggested.

Basic Parameters: Number of cores and number of accessible cores for one core of the virtual machine (VM): $Cores_{total}$ and $Cores_{available}$.

Calculated Parameters: Number of Executor_Cores for work of 1 executor is determined by the value 0,7-0,8 ($Cores_{total}$ **OR** $Coresav_{ailable}$);

The ratio of the RAM memory RAMInstance to the number of virtual cores of the instance of the virtual core: $RAM_{Instance}/vCores_{Instance}$. Then the parameter spark.executor.memory is calculated according to the following formula:

$$spark.executor.memory = RAM_{Instance}/vCoresInstance* \\ spark.executor.instances;$$

$$spark.executor.memoryOverhead = 0,1 * RAM_{Instance}/vCoresInstance\cdot \\ spark.executor.instances.$$

Thus, only the value Executor_Cores is a **heuristic determined parameter**: 20–30% (not less than 10%) characterizes a deployed operation system, Hadoop (HDFS) and planner Yarn depending on the type of the instance used by the virtual machine.

Parameter values calculated according to the suggested algorithm for different configurations of the cluster Apache Spark are shown in Table 9.

Table 9. Values of Apache Spark parameters determined according to the suggested method for different cluster configurations

Instance configuration	Env1	Env2	Env3	Env4
spark.executor.instances	4	5	4	5
spark.yarn.executor.memory Overhead	2,1 GB	2,1 GB	4,2 GB	4,2 GB
spark.executor.memory	18,9 GB	18,9 GB	37,8 GB	37,8 GB
spark.yarn.driver.memory Overhead	2,1 GB	2,1 GB	4,2 GB	4,2 GB
spark.driver.memory	9,45 GB	9,45 GB	9,45 GB	9,45 GB
spark.executor.cores	3	3	6	6
spark.driver.cores	3	3	3	3
spark.default.parallelism	32	40	48	60

Stage 5. Comparative analysis.

Conducting the comparative analysis has been implemented based on the results obtained for the configuration 2 master-nodes "Standard_D12_v2" and 3 worker-nodes "Standard_D13_v2". As metrics there have been used scalability and the ratio of the average testing time to the average learning time $K_{perf} = T_{testing}/T_{training}$.

In Tables 10, 11 one can see test results for the test set Spark-Perf (MLIB_TESTS), in Tables 12, 13 one can see a comparative analysis of the scalability effect for the configuration Env4 for the stages of learning and testing correspondingly. A comparative analysis for performance of clusters with homogeneous and heterogeneous structures is shown in Figs. 4, 5.

Table 10. Results for the test set "Decision Tree" (DTR)

№	Id	Testing		Training		K_{perf}
		Average value, sec	StD	Average value, sec	StD	
1	DTR1	0,1061	$\pm 0,0343$	2,1784	$\pm 1,1105$	0,0487
2	DTR2	0,1317	$\pm 0,0655$	2,7923	$\pm 1,3532$	0,0472
3	DTR3	0,1004	$\pm 0,0389$	2,2272	$\pm 1,1368$	0,0451
4	DTR4	0,1114	$\pm 0,0365$	17,3074	$\pm 2,6651$	0,0064
5	DTR5	0,2121	$\pm 0,1167$	5,7753	$\pm 2,073$	0,0367
6	DTR6	0,4085	$\pm 0,0583$	16,4465	$\pm 1,7365$	0,0248
7	DTR7	0,1532	$\pm 0,0332$	5,6087	$\pm 1,6385$	0,0273
8	DTR8	0,4374	$\pm 0,1165$	55,6258	$\pm 4,6424$	0,0079
9	DTR9	0,106	$\pm 0,0521$	2,1714	$\pm 1,1393$	0,0488
10	DTR10	0,1318	$\pm 0,0446$	2,5651	$\pm 1,3227$	0,0514
11	DTR11	0,0999	$\pm 0,0371$	2,2829	$\pm 1,2607$	0,0438
12	DTR12	0,1129	$\pm 0,0358$	17,5106	$\pm 3,0733$	0,0064
13	DTR13	0,17	$\pm 0,0892$	4,4882	$\pm 1,7729$	0,0379
14	DTR14	0,2316	$\pm 0,0509$	5,9187	$\pm 1,6501$	0,0391
15	DTR15	0,175	$\pm 0,0807$	4,7227	$\pm 1,8071$	0,0371
16	DTR16	0,3543	$\pm 0,1$	49,345	$\pm 4,5763$	0,0072

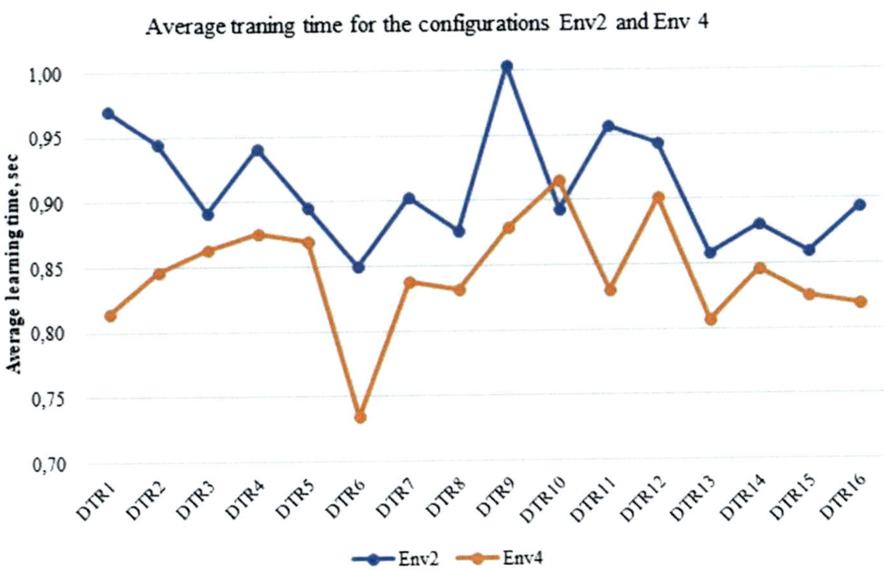

Fig. 4. Dependence for the average training time for the configurations Env2 and Env4

Table 11. Results for the test set MLlib

Test name	Id	Testing		Training		K_{perf}
		Average value, sec	StD	Average value, sec	StD	
als	ML1	0,5624	$\pm0,1194$	2,7009	$\pm0,8796$	0,2082
lda-1	ML2	15,5694	$\pm3,2071$	-	-	
lda-2	ML3	9,036	$\pm2,1171$	-	-	
svd	ML4	63,1744	$\pm5,7117$	-	-	
pca	ML5	11,4606	$\pm1,4769$	-	-	
summary-statistics	ML6	0,7669	$\pm0,845$	-	-	
block-matrix-mult	ML7	0,5783	$\pm0,3724$	-	-	
pearson	ML8	5,0412	$\pm0,6905$	-	-	
spearman	ML9	7,0741	$\pm1,8997$	-	-	
chi-sq-feature	ML10	5,1979	$\pm0,4478$	-	-	
word2vec	ML11	9,9814	$\pm1,2726$	-	-	
fp-growth	ML12	61,9484	$\pm1,3969$	-	-	
prefix-span	ML13	1,2658	$\pm0,6271$	-	-	
glm-regression	ML14	0,0948	$\pm0,034$	3,3957	$\pm1,997$	0,0279
glm-classification-1	ML15	0,1013	$\pm0,0337$	3,4001	$\pm1,3187$	0,0298
glm-classification-2	ML16	0,1006	$\pm0,0389$	8,1728	$\pm2,7251$	0,0123

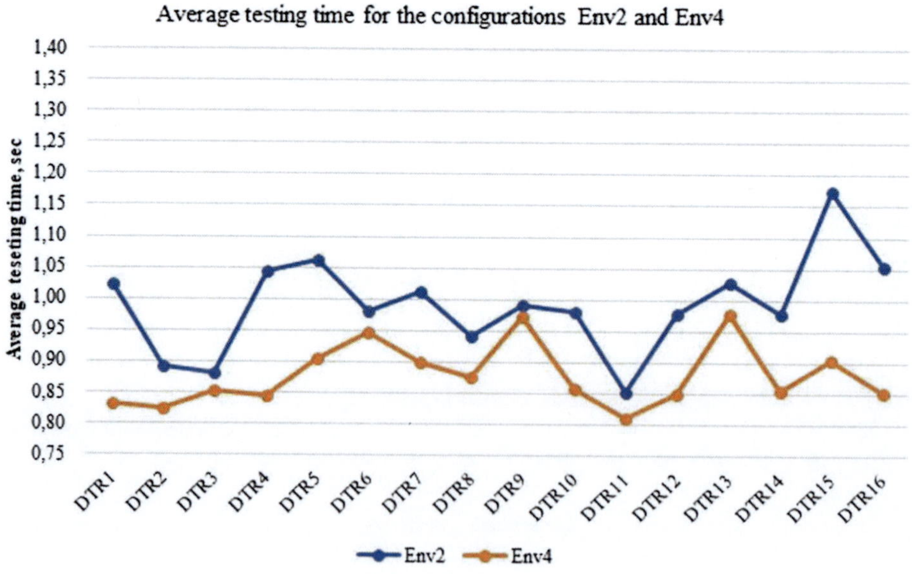

Fig. 5. Dependence of average testing time for the configurations Env2 and Env4

Table 12. A comparative analysis of training time for the configuration Env4

Test identifier	Average training time, sec		SpeedUp
	2 computing nodes	3 computing nodes	
DTR1	0,8309	0,8239	1,008448512
DTR2	0,8571	0,7927	1,081333749
DTR3	0,8807	0,8533	1,032072368
DTR4	0,8566	0,7828	1,094240698
DTR5	0,8095	0,7776	1,041028781
DTR6	0,9813	0,9478	1,035324308
DTR7	0,8757	0,8231	1,063874073
DTR8	0,9421	0,8767	1,074612522
DTR9	0,8084	0,7981	1,012891699
DTR10	0,9807	0,8138	1,110991238
DTR11	0,8531	0,7679	1,110991238
DTR12	0,9783	0,9111	1,073783658
DTR13	0,8487	0,8067	1,063829787
DTR14	0,9785	0,9390	1,04196309
DTR15	0,8435	0,8256	1,041666667
DTR16	0,8814	0,8733	1,009218224

Table 13. A comparative analysis of testing time for the configuration Env4

Test identifier	Average testing time, sec		SpeedUp
	2 computing nodes	3 computing nodes	
DTR1	1,0227	0,8309	1,230834035
DTR2	0,8927	0,8231	1,084558377
DTR3	0,8807	0,8533	1,032110629
DTR4	1,044	0,8453	1,235064474
DTR5	1,0609	0,9047	1,172653918
DTR6	0,9813	0,9478	1,035345009
DTR7	1,0132	0,9008	1,124777975
DTR8	0,9421	0,8767	1,074597924
DTR9	0,9907	0,9723	1,0189242
DTR10	0,9807	0,8568	1,144607843
DTR11	0,8531	0,8097	1,053600099
DTR12	0,9783	0,849711	1,15133263
DTR13	1,0291	0,9783	1,051926812
DTR14	0,9785	0,8539	1,145918726
DTR15	1,1753	0,9059	1,297383817
DTR16	1,0532	0,8533	1,234266964

6 Conclusions

For the analysis of the Apache Spark MLlib machine learning library performance, a set of specialized Spark-Perf tests has been used. For these tests there have been investigated scalability possibilities for the cluster of Apache Spark of the platform Azure HDInsight for homogeneous and heterogeneous cluster configurations. A methodology of assessing Apache Cluster performance has been developed. This methodology is based on a preliminary analysis of data received from data processing result log-files and the monitoring system SparkMeasure for tests from the Spark-Perf set. The tests provide statistical processing the obtained results for learning and testing stages when we have standard Apache Spark settings. Software and information support procedures have been developed for testing MLlib results based on the system SparkMeasure and Ambari. Formulas for determining the Apache Spark parameters for heterogeneous and homogeneous cluster solutions that provide a time minimization for task execution has been suggested, which allows minimizing execution time as compared to the standard Aparch Spark parameter settings. The experimental results obtained have shown that when optimally tuning the cluster parameters, it is possible to improve performance for a variety of configurations of Apache Spark in the context of Apache Spark MLlib machine learning methods by 15%–20%. Our further research will be directed towards the influence of Apache Spark settings on the TensorFlow [12] library to obtain scalable solutions for distributed computing based on the Apache Spark cluster.

References

1. Amazon AWS: Complete business guide to the world's largest provider of cloud services. https://www.zdnet.com/article/amazon-aws-everything-you-should-know-about-the-largest-cloud. Accessed 5 Mar 2020
2. Amazon Machine Learning. https://docs.aws.amazon.com/machine-learning/latest/dg/what-is-amazon-machine-learning.html. Accessed 5 Mar 2020
3. Apache Mahout For Creating Scalable Performant Machine Learning Applications. https://mahout.apache.org/. Accessed 5 Mar 2020
4. Azure HDInsight Azure HDInsight documentation. https://docs.microsoft.com/en-us/azure/hdinsight/. Accessed 5 Mar 2020
5. Cloud Serving Benchmark. https://research.yahoo.com/news/yahoo-cloud-serving-benchmark. Accessed 5 Mar 2020
6. Dv2 and DSv2-series. https://docs.microsoft.com/en-us/azure/virtual-machines/dv2-dsv2-series. Accessed 5 Mar 2020
7. Ev3 and Esv3-series. https://docs.microsoft.com/en-us/azure/virtual-machines/ev3-esv3-series. Accessed 5 Mar 2020
8. Get started with Google Cloud. https://cloud.google.com/docs. Accessed 5 Mar 2020
9. Microsoft®Azure Official Site — Create Your Free Account Today. https://azure.microsoft.com/en-us/free/search/. Accessed 5 Mar 2020
10. MLlib is Apache Spark's scalable machine learning library. https://spark.apache.org/mllib/. Accessed 5 Mar 2020

11. Spark-perf (homepage) Performance tests for Spark. https://spark-packages.org/package/databricks/spark-perf. Accessed 5 Mar 2020
12. TensorFlow on Spark. TensorFlow. https://docs.microsoft.com/en-us/azure/databricks/applications/deep-learning/single-node-training/tensorflow. Accessed 5 Mar 2020
13. Aziz, K., Zaidouni, D., Bellafkih, M.: Big data processing using machine learning algorithms: Mllib and mahout use case, pp. 1–6 (2018). https://doi.org/10.1145/3289402.3289525
14. Gao, W., Zhan, J., Wang, L., Luo, C., Zheng, D., Wen, X., Ren, R., Zheng, C., He, X., Ye, H., Tang, H., Cao, Z., Zhang, S., Daig, J.: Bigdatabench: A scalable and unified big data and ai benchmark suite (2018). https://arxiv.org/abs/1802.08254
15. Ghazal, A., Ivanov, T., Kostamaa, P., Crolotte, A., Voong, R., Al-Kateb, M., Ghazal, W., Zicari, R.: Bigbench v2: the new and improved bigbench, pp. 1225–1236 (2017). https://doi.org/10.1109/ICDE.2017.167
16. Gounaris, A., Torres, J.: A methodology for spark parameter tuning. Big Data Res. **11** (2017). https://doi.org/10.1016/j.bdr.2017.05.001
17. Han, R., John, L., Zhan, J.: Benchmarking big data systems: a review. IEEE Trans. Serv. Comput. **PP**, 1 (2017). https://doi.org/10.1109/TSC.2017.2730882
18. Huang, S., Huang, J., Dai, J., Xie, T., Huang, B.: The hibench benchmark suite: characterization of the mapreduce-based data analysis, pp. 41–51 (2010). https://doi.org/10.1109/ICDEW.2010.5452747
19. Ivanov, T., Rabl, T., Poess, M., Queralt, A., Poelman, J., Poggi, N., Buell, J.: Big data benchmark compendium. In: Nambiar, R., Poess, M. (eds.) Performance Evaluation and Benchmarking: Traditional to Big Data to Internet of Things - 7th TPC Technology Conference, TPCTC 2015, Kohala Coast, HI, USA, August 31 - September 4, 2015. Revised Selected Papers. Lecture Notes in Computer Science, vol. 9508, pp. 135–155. Springer, Heidelberg (2015). https://doi.org/10.1007/978-3-319-31409-9_9
20. Kraska, T., Talwalkar, A., Duchi, J., Griffith, R., Jordan, M., Franklin, M.: A distributed machine-learning system (2013)
21. Palit, T., Shen, Y., Ferdman, M.: Demystifying cloud benchmarking, pp. 122–132 (2016). https://doi.org/10.1109/ISPASS.2016.7482080
22. Petridis, P., Gounaris, A., Torres, J.: Spark parameter tuning via trial-and-error. In: Angelov, P., Manolopoulos, Y., Iliadis, L.S., Roy, A., Vellasco, M.M.B.R. (eds.) INNS Conference on Big Data. Advances in Intelligent Systems and Computing, vol. 529, pp. 226–237 (2016). http://dblp.uni-trier.de/db/conf/inns/inns2016.html#PetridisGT16
23. Wang, K., Maifi Hasan Khan, M., Nguyen, N., Gokhale, S.: A model driven approach towards improving the performance of apache spark applications. In: 2019 IEEE International Symposium on Performance Analysis of Systems and Software (ISPASS), pp. 233–242 (2019)
24. Zaharia, M., Chowdhury, M., Das, T., Dave, A., Ma, J., McCauly, M., Franklin, M., Shenker, S., Stoica, I.: Resilient distributed datasets: a fault-tolerant abstraction for in-memory cluster computing. In: NSDI, pp. 15–28 (2012)

25. Zaharia, M., Chowdhury, M., Franklin, M.J., Shenker, S., Stoica, I.: Spark: cluster computing with working sets. In: Proceedings of the 2nd USENIX Conference on Hot Topics in Cloud Computing, HotCloud 2010, p. 10. USENIX Association, USA (2010)
26. Zaharia, M., Xin, R.S., Wendell, P., Das, T., Armbrust, M., Dave, A., Meng, X.,Rosen, J., Venkataraman, S., Franklin, M.J., Ghodsi, A., Gonzalez, J.,Shenker, S., Stoica, I.: Apache spark: a unified engine for big dataprocessing. Commun. ACM **59**(11), 56–65 (2016). https://doi.org/10.1145/2934664

Management of Tourists' Enterprises Adaptation Strategies for Identifying and Predicting Multidimensional Non-stationary Data Flows in the Case of Uncertainties

Marharyta Sharko[1](✉), Ivan Lopushynskyi[1], Natalia Petrushenko[1], Olena Zaitseva[1], Volodymyr Kliutsevskyi[1], and Yuliia Yarchenko[2]

[1] Kherson National Technical University, Kherson, Ukraine
mvsharko@gmail.com, doctordumetaua@meta.ua, natalia.velikaya@gmail.com, lenzai@ukr.net, kliutsevskyi@gmail.com

[2] State Higher Educational Institution "Pryazovskyi State Technical University", Mariupol, Ukraine
july-248@ukr.net

Abstract. All the motivations of adaptive strategies in conditions of dynamic environmental changes come down to how effectively they describe the situation. If the market segment and the characteristics of consumer preferences remain unchanged for a long time, then the accumulated statistics can be the basis for forecasting and making appropriate management decisions. If the market segments and consumer preferences are constantly changing and under the influence of the environment, then synergetic helps to predict their behavior. The structural scheme of the model of tourist enterprises adaptivity to the dynamic changes of the environment is offered. Its main stages are qualitative analysis of input information about the dynamics of the production system, mathematical studies and their interpretation. An explanation of the nature of connections in a production system, which develops in the face of dynamic environmental changes based on synergistic representations in the form of entropy models of a set of production factors and information systems, is given. The tools, operations, and competitive advantages of adaptation to environmental changes are outlined. It has been proven that the greatest success in their development is achieved by those tourist enterprises that use short-term forecasting strategy. They move from one goal to another using the detection of anomalous economic values in the initial row by means of the Irwin criterion and smoothing of indicators by the moving average method.

Keywords: Management · Identification · Adaptation · Forecasting · External environment · Uncertainty · Strategies · Tools

S. Babichev et al. (Eds.): ISDMCI 2020, AISC 1246, pp. 135–150, 2021.
https://doi.org/10.1007/978-3-030-54215-3_9

1 Introduction

This research is caused by the fact that in conditions of unforeseen changes in the environment that affect the functioning of production, high competition and market segmentation of the enterprise are forced to develop their strategies for adaptation to the expected changes. When solving management problems under uncertainty, one of enterprise adaptability tasks is to obtain the required volume and quality of input information and to forecast economic indicators. Today, tourism market is shaped in terms of needs, the cost of decision-making is increasing, the requirements for product quality are increasing, the management techniques, which goal is to eliminate the imbalance between the existing and the real capabilities of the enterprise, are changing. The consumers who shape the image and attractiveness of the tourism product determine the influence of the environment on the functioning of tourism enterprises. The travel agency has no ability to influence the environment and needs to adapt to it, keep track of changes, forecast and respond in a timely manner.

Tourism management in the conditions of dynamic environmental influences does not have a developed methodological base. Management decisions are made on the base of stochastic recommendations, industry and territorial characteristics based on experience with extrapolation to future trends without taking into account the consequences, risks and limitations caused by changes in political, social and economic life. That in the current conditions of tourism enterprises functioning introduces great uncertainty and inadequate decision-making. Difficulties in forecasting are explained by the multicriteria of estimates, the multitude of dimensions of factors, and the multidirectionality of indicators. One of the problem areas for tourism development forecasting is the preparation of high-quality input information and its processing to build trends.

2 Literature Review

Background information on the identification of problems of unstable development of the tourism market is contained in the relation of internal and external destabilizing factors. The analysis of modern publications shows that the problem is common in many countries of the world: Great Britain, – Williams A., Balaz V. (2014) [20], India – Anupria K., Abhilasha C., Yajulu M. (2016) [1], Republic of Korea – Han Jinseok, Kim Namjo (2019) [5], Austrian – Reitsamer B.F. and Brunner-Sperdin A. (2015) [13], Japan – Elaine YTC, Siti A. (2014) [4], Ukraine – Shpak N., Dvulit Z., Maznyk L., Mykytiuk O., Sroka W. (2019) [18], Russia – Kozlov DA (2016) [9], Portugal – Hwang, Stewart (2016) [6], Sri Lanka – Mustafa A.M.M. (2019) [12].

A critical analysis of the information contained in the indicated literary sources, in addition to the particular problems of the territorial development of tourism, reveals the presence of general problems associated with decision-making risks at all levels of the tourism market and the tourism industry regulation. Incorrect decisions, as well as decisions made without proper statistical processing of incoming dynamic information, often lead to irreparable consequences.

The current state of tourism development forecasting methods uses the following methods: expert estimates [17], logical modeling [21], scenario approach [3], econometric methods [10], multiple regression methods [8], information entropy model [15], trends [7], strategic management decision [16], simulation modeling [2,19], neural networks [11], multivariate singular spectral analysis [14] and requires statistical processing of the time series when it is introduced. However, the quality of used input information is not checked or analyzed. At the same time, abnormal bursts and outliers of indicators in different periods of observations when building forecasts are either discarded or make distortions into the display of the real situation, which reduces the accuracy of forecasting. The exclusion of these negative aspects when building trends is the essence of the proposed research.

In accordance with Internet analysis technique, the citation of definitions of the conceptual categorical apparatus is estimated by determining the proportion of interpretations from the total amount of information available on the economic essence of adaptation concepts. The key aspects to which various authors respond and focus are presented in Fig. 1.

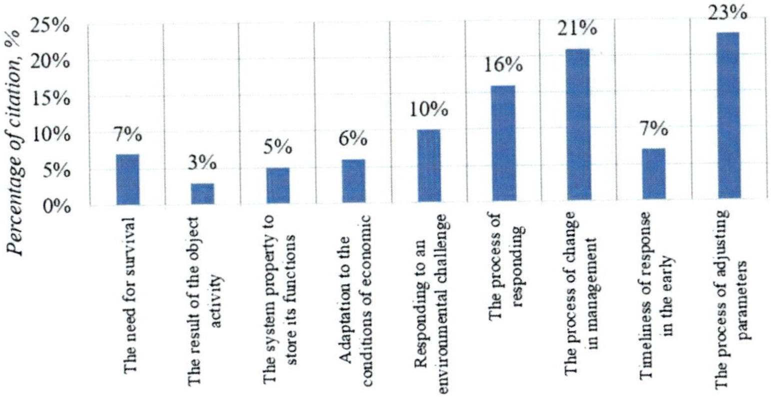

Fig. 1. Internet analysis of the conceptual apparatus

In accordance with the conceptual core in Fig. 1, all interpretations can be divided into result and process group. While the result group is a congealed substance, the adaptation process can be controlled, that is done in this paper. Based on the analysis of the above formulations, the author's interpretation of this concept is offered. The system can be called adapted if it is able to detect and respond on external challenges in time, to make early management changes quickly based on self-change and self-development and dynamic equilibrium with the environment. Concerning the development of an enterprise adaptation strategy to dynamic environmental changes, the synergistic model in its final implementation is to use the theory of control to modify complex structures to minimize the impact of external disturbances in situations that are adequate to the real situation.

In any production that develops, there are simultaneously features of order and uncertainty of future states. Multivariate and ambiguous behavior of multi-element production structures, that do not degrade but develop, are characterized by openness and inflow of financial, material and information flows, nonlinearity of internal processes, exacerbation and presence of stable states. Entropy is the quantitative measure of disorder. It characterizes the steadiness and stability of production functioning that is the spatial-energy interaction of the probability of finding the system in this state. All the motivations of adaptive strategies in the face of dynamic environmental changes come down to how effectively they describe the situation. If the market segment and the characteristics of consumer preferences remain unchanged for a long time, then the accumulated statistics can be the basis for forecasting and making appropriate management decisions. If the market segments and consumer preferences are constantly changing and being influenced by the environment, then synergetics helps predict their behavior.

A synergistic approach to the synthesis of control systems links the trajectories of development with the selection of possible controls from the set of possible directions on the base of invariant relations, which reflect the laws of production system sustainability. When exchanging with the external environment, there are processes of self-organization, that is, the birth from the chaos of some stable ordered structures with new properties. The main concept of synergetics is bifurcation, that is, the distribution of some dependence $x(\lambda)$ on several branches when changing any parameter of the system. A cascade of bifurcation is shown in Fig. 2.

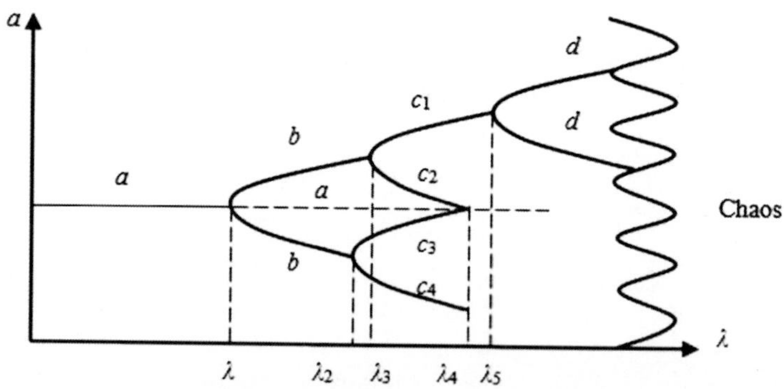

Fig. 2. Cascade of bifurcations in the synthesis of control systems

If the parameter λ passes through some fixed value of λ, the number of branches increases abruptly from one a_0 to three: two constant branches b_1 and b_2 and one unstable branch a_1. The choice of a constant branch b_1 and b_2 after the bifurcation point determines the fluctuations of the external perturbations, which determine the branches that will further develop the system. To manage processes in complex systems, the managed parameter λ must be changed. At bifurcations, the production system determines the following states by itself: - branches of development $b_1 \rightarrow c_1 \rightarrow d_1$ or stagnation $b_2 \rightarrow c_4$, etc. When constructing a synergistic model of enterprises adaptation to dynamic changes in the environment, its main stages are the qualitative analysis of input information, formulation of the task, determining the predicted values of the dynamics of the production system, performing mathematical studies and their interpretation.

The essence of synergetics lies in the fact that open economic systems, which exchange with the external environment, have processes of self-organization and birth from the chaos of some stable ordered structures with new properties. Competitive advantages based on the introduction of new technologies are possible only in a very short time. The competitive environment is design, guarantee, service, image, intelligence and immense assets. Regardless of the production system nature, there are a number of regularities the violations of which by external management leads to erroneous results. Stereotypes of management and marketing as part of strategic management require updating.

The purpose of this work is development of the methodological basis for adaptation of tourism enterprises to the dynamic changes of the environment.

3 Materials and Methods

One of the key theoretical and methodological problems of enterprises adaptation to dynamic changes of the environment and their influence on the functioning of production facilities is to substantiate its economic essence. The imperative of ensuring the sustainable functioning of tourism enterprises is anticipation of possible changes in the external environment in order to develop preventative measures and means of eliminating threats. There is some correlation between information about the current state of the object and the potential for environmental impact caused by different economic situations. This is motivated by forecasting trends in the implementation of additional operations related to probabilistic estimates of expected events. The detection of anomalous values in the time series using the Irwin criterion and the alignment of time series by the moving average method is considered as technological forecasting operations in the conditions of unpredictable changes in the influence of the external environment. Since the forecasting of tourism development in modern dynamic conditions is of particular relevance for many countries in the world, the statistics of world tourism is used as initial reference information.

4 Technology of Calculation

The tourism industry is one of the fast growing industries. According to the UNWTO, from 2015 to 2017, the international tourist flow increases by 4.2%

annually, which is significantly higher than the average annual economic development rate (2.2%). The global tourism industry in 2017 provided a vacation for 1326 million people. According to UNWTO forecasts, in 2030 this figure will increase to 1.8 billion people at the expense of tourists from emerging economies. The amount of tourism revenues increased by more than 4.5 times. According to forecasts, the volume of tourists in 2030 will amount to 1900 billion USA dollars. However, the forecast is variable in nature, due to the quality of the input information. Global trends bring new and often invisible challenges, threats and opportunities. The globalization and integration of the economic, political and information space has led to changes in the tourism system, its typology, including tourism risks. The statistics of time series of tourists and the volume of international tourism revenues were used as input (Table 1).

Table 1. Dynamics of tourism development in the world

Indexes	Years									
	2000	2005	2010	2011	2012	2013	2014	2015	2016	2017
Number of tourist arrivals in the world, million people	680	809	952	995	1035	1087	1136	1195	1240	1326
Amount of international revenues, billion USD	479,2	680	927	1042	1078	1159	1202	1221	1242	1340

Preliminary analysis of economic time series centers on the identification and elimination of anomalous values of these indicators. If a specific anomalous value remains as a link in the time series, then the trend model is not considered adequate. According to the Irwin method, it is necessary to compare the calculated values of the trend characteristics with the tabular ones. The calculated values of the characteristics λ in the Irwin criterion were determined by the formula:

$$\lambda_t = \frac{|y_t - y_{t-1}|}{\sigma_y} \qquad t = 2, 3, \ldots, n$$

where λ_t is the current value of the indicator at time t; $y_{(t-1)}$ is the previous value of this indicator; n is the total number of indicators; σ_y is average squared deviation.

The value of σ_y is calculated as follows:

$$\sigma_y = \sqrt{\frac{\Sigma_{i=1}^{n}(y_t - \bar{y})^2}{n-1}}$$

where $\bar{y}$ is the average of all analyzed indicators:

$$\bar{y} = \frac{\Sigma_{i=1}^{n} y_i}{n}$$

Table 2. Theoretical values of the Irwin criterion

n	2	3	4	5	6	8	9	10
λ_α	2,8	2,3	2,1	1,9	1,8	1,7	1,6	1,5

Tabular value of the Irwin criterion λ_α for the level of $\alpha = 0.05\%$ are given in Table 2.

The results of calculations of construction of functions of forecasting of the tourism market development processes provided by the dynamic changes in the number of tourists x_1 and the volumes of revenues x_2 are given in Tables 3, 4, 5. If the estimated values of the Irwin criterion are greater than theoretical, then the corresponding values of the economic indicators of the time series are considered anomalous.

Table 3. Calculations of deviations from the average in the trend indexes

Indexes	Average value $\bar{y}$	Deviation from the average $y_t - \bar{y}$								
		2000	2010	2011	2012	2013	2014	2015	2016	2017
x_1	1045,5	365,5	236,5	93,5	50,5	10,5	−41,5	−90,5	−49,5	−4,5
x_2	1036,02	566,82	356,02	109,02	94,02	−41,98	−22,98	−165,98	−195	−205,98

Table 4. Calculations of the average squared deviation in the trend indexes

Indexes	Average squared deviation σ_y	Deviation from the average $y_t - \bar{y}$								
		2000–2005	2005–2010	2010–2011	2011–2012	2012–2013	2013–2014	2014–2015	2015–2016	2016–2017
x_1	163,28	129	143	43	40	54	49	59	45	86
x_2	269,04	200,8	247	115	36	81	143	19	21	98

The causes of anomalies can be errors in accounting and information transfer when aggregating indicators and errors in calculations. These anomalies can be eliminated by repeatedly retrieving information. If such errors are not eliminated in the re-calculations of the statistical indicators, then the noted anomalies are fixed and used as regularities of the dynamic processes. Since the data correlation of the Table 2 with the results of calculations of the Irvine index given in Table 5 showed that for both investigated time series, in all cases $\lambda_\alpha > \alpha_t$, then no anomalies were detected in the trend indexes. Equalization or smoothing of time series must be performed to identify general trends and hidden random deviations in the fluctuations of economic dynamics, when constructing trends

Table 5. Calculations of the Irwin criterion in the trend indexes

Indexes	Irwin criterion λ_t							
	2000–2005	2005–2010	2010–2011	2012–2013	2013–2014	2014–2015	2015–2016	2016–2017
x_1	0,79	0,87	0,26	0,25	0,33	0,30	0,36	0,52
x_2	0,74	0,91	0,42	0,13	0,30	0,53	0,08	0,36

in the conditions of stochastic changes of the external environment. For this purpose, for the first three values of the time series, their arithmetic mean is calculated. This mean is the smoothed value of the series levels. The smoothing by means of sliding averages is based on the fact that the deviations in the mean values are mutually canceled due to the replacement of the initial levels of the arithmetic mean within the selected time interval. Then the smoothing interval is shifted into one period to the right and again the average mean is calculated, etc. Thus, in each case, the mean is centered as it is related to the midpoint of the smoothing interval. The smoothing interval m must be less than the length of the time series n. The aligned row is shorter than the initial $n-1$ observations.

The method is used in the short-term prediction of time series that have a linear trend:

$$y_{t+1} = m_{t-1} + \frac{1}{n}(y_t - y_{t-1}), \qquad n = 3$$

where $t-1$ is the forecasting period; $m_{(t-1)}$ is the moving average two periods before the forecast, $t+1$ is the forecast period.

Using this formula, we calculate a moving average for every three nearby periods of input information and build forecasts for the years ahead. The relative error was determined by calculating the difference between the indicators of the Table 1 y_{fact} with moving average values y_c divided by the values of the analyzed metrics y_{fact} followed by a calculation. Data for calculating the smoothing of time series of international tourism development are summarized in Table 6 for the number of tourist arrivals in the world and Table 7 for international arrivals.

Table 6. Smoothing of time series of the number of tourists arrivals in the world

Years	2010	2011	2012	2013	2014	2015	2016	2017
Moving average, m	–	994	1039	1086	1139,3	1190,3	1253,7	–
Calculations of the average relative error, %	–	0,10	0,39	0,09	0,29	0,39	1,1	–
Result, %	2,37							
Average relative forecast error, %	0,39							

Forecast errors were determined by the formula:

$$\varepsilon = \frac{1}{n} \cdot \sum_{i=1}^{n} \left[\frac{|y_f - y_c|}{y_f} \cdot 100 \right]$$

Table 7. Smoothing the time volume of international revenues

Years	2010	2011	2012	2013	2014	2015	2016	2017
Moving average, m	–	1015,7	1093	1146,3	1194	1221,7	1267,7	–
Calculations of the average relative error, %	–	2,53	1,39	1,09	0,67	0,05	2,07	–
Result, %	7,80							
Average relative forecast error, %	1,30							

Table 8. World tourism development forecast

Indexes	Years				
	2018	2019	2020	2021	2022
Number of tourist arrivals in the world, million people	1282,3	1268	1287	1285,7	1279,9
Amount of international revenues, billion USD	1300,3	1281	1301	1300,5	1293,9

5 Results and Discussion

The calculation data of the forecasts of world tourism development are presented in the Table 8.

The implementation of this method in the research is provided by its polynomial approximation, however, since $n = 3$ and $m = 10$, the smallest approximation is the figure three, that has a cubic approximation of the form:

$$x = x_0 + x_1 t + x_2 t^2 + x_3 t^3$$

where x_0 is indicator value; x is linear increase of this indicator.

The trend line of tourist arrivals in the world is presented in Fig. 3 and the volume of international revenues is shown in Fig. 4.

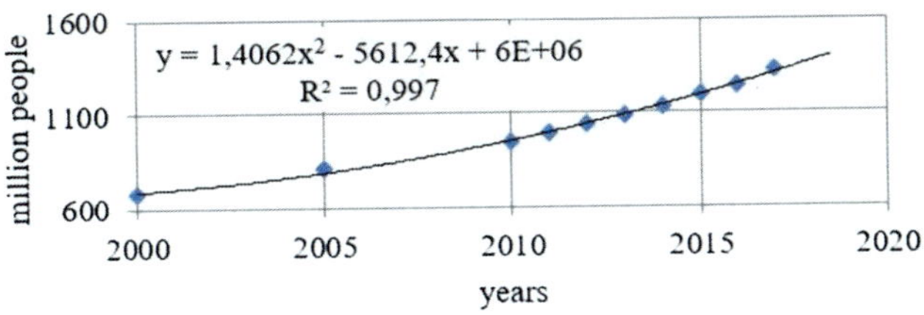

Fig. 3. Trend line of tourist arrivals worldwide, million people

Next to the polynomial approximation equation are the values of the determination coefficients R^2.

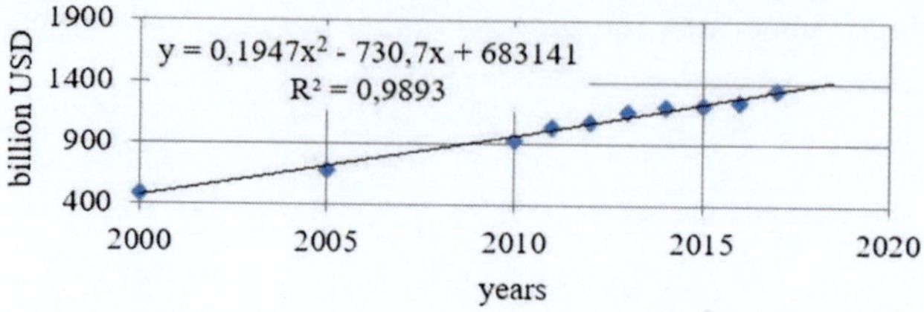

Fig. 4. Trend of international revenue volumes, billion USA

A discussion of the obtained results in the context of the cited previous literature [3–6, 14, 16, 18, 20, 21] shows the presence of problems of identification of factors of unstable tourism development factors, typical for many worldwide countries. The calculated accounting of world tourism development forecasting is in satisfactory agreement with the long-term forecasting presented in the UNWTO statistical reports [11]. However, the proposed forecasting techniques based on calculations of the Irwin criteria and smoothing of time-series methods make it possible to perform a more accurate and detailed and short-term forecast with a prediction interval of 1 ... 1.2 years. Such information is extremely useful for all world tourism organizations as a means of avoiding risks and ways to overcome them. The structural model of adaptation to the environmental conditions, compiled by the authors, is presented in Fig. 5.

The external influence of the market in the short-term interval looks chaotic, while in the long-term it is possible to distinguish the steady stages of rising stabilization and falling, alternating with states of uncertainty and fluctuation. The positive feedback with market transformations is achieved at these points. The market consumption system is self-regulating, becoming stable. At different stages of the evolution of its development, the production system has a different sensitivity to external influences. In general, coherence, i.e. the coordination of individual elements development can be manifested in the large-scale activity of the entire production system due to the different correlations of their interactions.

Formation of information flows to effective management and adaptation to changing market conditions requires the right tools that provide flexibility for managing production activities. This tool for adapting to changes in enterprise functioning and business sustainability should describe the main components of management, the set of which at each phase of the product life cycle is adequate to the state of the external and internal environment. An example of such tools is given in Fig. 6.

The transition from synergistic management principles to quantitative ratios of the tourism product attractiveness can be expressed through the entropy of the production system, which may underlie the adaptability of tourism enterprises to the dynamic changes of the environment. In order for adaptive processes to occur in the production system, they must have a constant influx of environmental information that leads to entropy change. The amount of information I depend

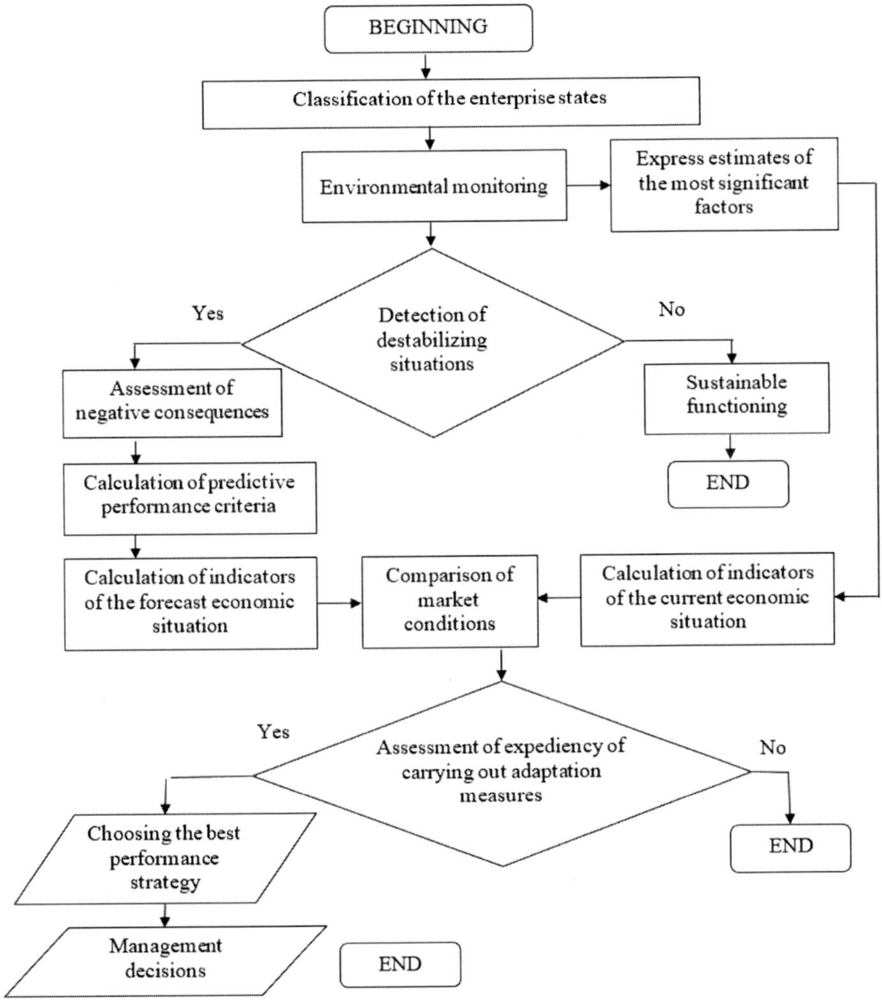

Fig. 5. Algorithm for managing enterprise adaptation strategies to environmental conditions

on the total number of elements of the production system n and the number of states m in which each element may reside

$$I = n \log_2 m$$

The more the entropy is, the more the economic uncertainty of the production object behavior. As new information arrives, uncertainty is reduced. Assessment of production state and its dynamic changes is a necessary element of diagnostics and forecasting of the state, as well as the adaptability of the enterprise to external disturbances. The minimum of entropy corresponds to the constant value of the enterprise state. Non-equilibrium processes in an isolated system

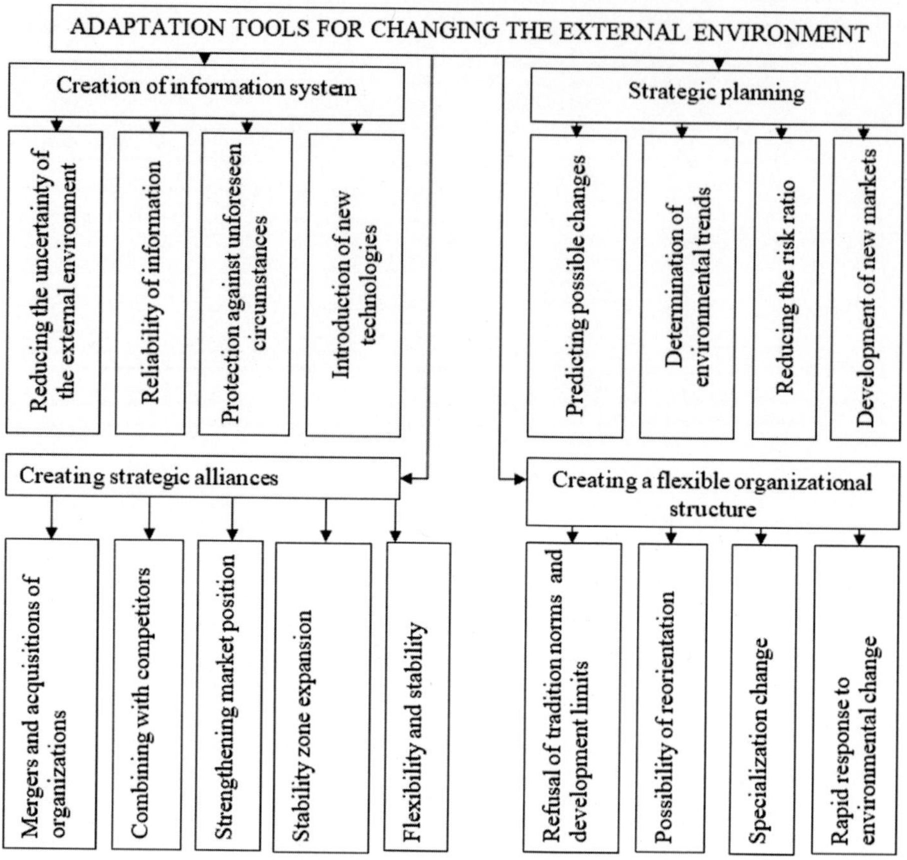

Fig. 6. Tools, operations and competitive advantages of adaptation to environmental changes

lead to an increase in entropy. Entropy E is related to the probability of the system being in this state W by the ratio

$$E = k \ln W$$

where k is Boltzmann's constant.

The physical meaning of entropy E, as defined by Clausius, is explained by the expression:

$$dE = \frac{dQ}{T}$$

where dQ is the amount of heat needed to move the system from one state to another; T is the temperature of this moving.

This value depends on the moving path and is defined as its change instead of a specific E. Entropy is a general indicator of the system state at this point of time. Its value can be calculated by the formula:

$$E = -\sum_{i=1}^{m} P_i \log_2 P_i$$

where P_i is the probability of finding the system in this state; m is the number of system states.

The entropy of the production system is a logarithmic means of the information source inversion. The theory of nonequilibrium processes and dissipative structures proves that disequilibrium is a source of development. The link between information and the entropy of the production system is traced on the base of the following considerations. The amount of information in a production system I is a measure of orderliness or organization of industrial relations, while system entropy is a measure of disorganization or disorderliness, so such correlation is fair:

$$I = E$$

According to the law of large numbers, with a large number of observations, their arithmetic mean is constant. With a small number of observations, it may be random. Analysis of information about the state of the production system is to determine its probability $P(t)$ as the ratio of the number of states of the production system $N(t)$ at a given time to their total possible number N.

$$P(t) = \frac{N(t)}{N}$$

From this point of view, the entropy of the production system can be represented as a mathematical expectation of information about the states of the production system $M[X]$. According to the tenets of probability theory, this value is given by the formula:

$$M[X] = \frac{\sum_{i=1}^{n} x_i P_i}{\sum_{i=1}^{n} P_i}$$

where x_i is the random variable; P_i is the probability of their occurrence in the general ensemble of observations.

The entropy of associations of statistically independent information sources about the state of an object is equal to the sum of entropies:

$$E = \sum_{i=1}^{n} E_i$$

where E_i is entropy of the information source; i is the number of states of these sources; n is the number of information sources.

Based on the above, we can conclude that an isolated system consisting of two interacting subsystems in their organization is subject to entropy compensation. It is that the entropy of one system can be reduced only if there is a compensating increase in the entropy of another system by the same amount, that is, an increase in order in one part leads to an increase in chaos in the other. Achieving a sustainable state of production systems depends on the intensity of exchange of financial and material flows of fixed and circulating assets. In a completely isolated system, such information is absent, production potential remains unchanged, entropy of the system either remains unchanged or absent at all. Open systems exchange information with the external environment, thereby increasing their orderliness, changing their structure, organization and reducing entropy. Practical interest is not the absolute value of entropy, but its change. If information is the process of acquiring knowledge about the state of the production system, then its quantitative measure at a given time is determined not by a function of the state of the system E, but by its increase ΔE. In an open system, the entropy increase ΔE consists of two components: ΔE_1, which determines the results of the exchange with the external environment, and ΔE_2, which determines the entropy change in the middle of the system, i.e.

$$\Delta E = \Delta E_1 + \Delta E_2$$

For irreversible processes $\Delta E_2 > 0$, whereas for inverse processes this value is zero. If the system is isolated and there is no exchange of information, then $\Delta E_1 = 0$. Under external conditions, the entropy of the equilibrium state is increased, and if there are no resistances, the entropy reaches a minimum. From the above considerations it follows that the efficiency of production potential is determined by the synergistic effects and entropy transformation within the system.

6 Conclusions

Adaptability of the enterprise to the consideration of irreversible processes dynamics to dynamic changes of the environment is based on the synergistic approach of system interactions and their transformations. The change of accents in the management system from the planning to the management in the conditions of uncertainty, which requires the involvement of new ways of forecasting and making appropriate management decisions is revealed. The novelty and the theoretical significance of the methodologies of the tourism enterprises adaptability strategies to changes in the external environment lies in the accounting and elimination of anomalies and smoothing of time series when collecting and analyzing data for trends, which significantly improves the accuracy of approximation and forecasting. The developed methodological approaches for forecasting and diagnostics allow monitoring efficiency, developing and implementing stabilization measures in the market of tourism services.

Development of forecasting in tourism provides an information base for making adequate managerial decisions, allows you to identify problems that impede business development, the reasons for their occurrence and ways to resolve it, identify future trends in the tourism market and reduce risks. The synergistic model of enterprise adaptability to environmental changes, which is based on entropy reduction in open structures far from thermodynamic equilibrium, should take into account possible cooperative and intersectoral relations, form entropy changes within the production system, and exchange network interactions, carry out network and transaction planning.

References

1. Anupria, K., Chauhan, A., Yajulu, M.: Destignation image of Indian tourism destinations. An international using correspondence analysis. Asia Pacific J. Market. Logistics **28**(3), 499–524 (2016). https://doi.org/10.1108/APJML-05-2015-0074
2. Babichev, S., Škvor, J., Fišer, J., Lytvynenko, V.: Technology of gene expression profiles filtering based on wavelet analysis. Int. J. Intell. Syst. Appl. **10**(4), 1–7 (2018). https://doi.org/10.5815/ijisa.2018.04.01
3. Chatkaewnapanon, Y., Kelly, J.: Community arts as an inclusive methodology for sustainable tourism development. J. Place Manag. Dev. **12**(3), 365–390 (2019). https://doi.org/10.1108/JPMD-09-2017-0094
4. Elaine, Y., Siti, A.: Destination image as a mediator between perceived risks and revisit intention: a case of post-disaster Japan. Tourism Manag. **40**, 382–393 (2014). https://doi.org/10.1016/j.tourman.2013.07.008
5. Han, J., Namjo, K.: Q-methodology analysis of perceived risks in tourists and local residents towards natural disaster. The 2016 gyengju earthquake in Republic of Korea. Travel Tourism Res. Assoc.: Adv. Tourism Res. Global. **46** (2019). https://doi.org/10.1111/0272-4332.213125
6. Hwang, D., Stewart, W.: Social capital and collective action in rural tourism. J. Travel Res. **56 (1)**, 80–91 (2016). https://doi.org/10.11772/0047287515625128
7. Jiao, E., Chen, J.: Tourism forecasting: a review of methodological developments over the last decade. Tourism Econ. **25(3)**, 469–492 (2019). https://doi.org/10.1177/1354816618812588
8. Kang, C., Gu, J.: Analysis of tourist flow forecasting model based on multiple additive regression tree. IOP Conf. Ser.: Mater. Sci. Eng. **490**(4), 42001 (2019). https://doi.org/10.1088/1757-899X/490/4/042001
9. Kozlov, D.: Forecasting of Russian outbound tourism. Int. J. Appl. Fundam. Res. **1**, 232–237 (2016). https://doi.org/10.14505//jemt.9.6(30).16
10. Liu, H., Ying, L., Wang, Y., Pan, C.: Hot topics and emerging trends in tourism forecasting research: a scientometric review. Tourism Econ. **25**(3), 448–468 (2019). https://doi.org/10.11772/1354816618810564
11. Livieris, I., Pintelas, E., Kotsilieris, T., Stavroyiannis, S., Pintelas, P.: Weight-constrained neural networks in forecasting tourist volumes: a case study. Electronics (Switzerland) **8**(9), 1005 (2019). https://doi.org/10.3390/electronics8091005
12. Mustafa, A.: Impact of tourism and foreign direct investment on gross domestic production: forecasts for the case of Sri lanka. Int. J. Recent Technol. Eng. **8**(2), 260–267 (2019). https://doi.org/10.35940/ijrte.B1044.0982S1019
13. Reitsamer, B., Brunner-Sperdin, A.: Tourist destination perception and well-being: what makes a destination attractive? J. Vacation Market. **23**(1) (2015). https://doi.org/10.1177/1356766715615914
14. Saayman, A., Klerk, J.: Forecasting tourist arrivals using multivariate singular spectrum analysis. Tourism Econ. **25**(3), 330–354 (2019). https://doi.org/10.1177/1354816618768318
15. Sharko, M., Gusarina, N., Petrushenko, N.: Information-entropy model of making management decisions in the economic development of the enterprises. Adv. Intell. Syst. Comput. 304–314 (2019). https://doi.org/10.1007/978-3-030-26474-1

16. Sharko, M., Doneva, N.: Methodological approaches to transforming assessments of the tourist attractiveness of regions into strategic managerial decisions. Actual Probl. Econ. **8**(158), 224–229 (2014)
17. Sharko, M., Sharko, A.: Innovative aspects of management of development of enterprises of regional tourism. Actual Probl. Econ. **7**(181), 206–213 (2016)
18. Shpak, N., Dvulit, Z., Maznyk, L., Mykytiuk, O., Sroka, W.: Validation of ecologists in enterprise management system: a case study analysis. Polish J. Manag. Stud. **19** (2019). https://doi.org/10.17512/pjms.2019.19.1.29
19. Shpak, N., Kyrylych, T., Greblikaitė, J.: Diversification models of sales activity for steady development of an enterprise. Sustainability **8**(4), 393 (2016). https://doi.org/10.3390/su8040393
20. Williams, A., Balaz, V.: Tourism risk and uncertainty. J. Travel Res. **54**(3), 271–287 (2014). https://doi.org/10.1177/0047287514523334
21. Zhang, B., Pu, Y., Wang, Y., Li, J.: Forecasting hotel accommodation demand based on LSTM model incorporating internet search index. Sustainability (Switzerland) **11**(17), 4708 (2019). https://doi.org/10.3390/su11174708

The Expanding Space Method in Sphere Packing Problem

Sergiy Yakovlev(✉)

National Aerospace University "Kharkiv Aviation Institute", Kharkiv, Ukraine
svsyak7@gmail.com

Abstract. A problem of packing unequal spheres, which radii are known, into containers of various shape are considered. An equivalent mathematical model to a standard sphere packing model is formed based on the assumption that the radii may be variable. In the new model formulated in a higher dimension space, a combinatorial structure is derived, and additional constraints are formed involving variable radii and providing its equivalence to the original packing problem. This approach of treating radii as variables allows improving local solutions of the original problem. Some practical tasks are indicated, which are reduced to packing unequal spheres.

Keywords: Packing problem · Sphere · Container · Optimization · Permutation

1 Introduction and Literature Review

Sphere packing problems (SPP) are the well-known family lying on ajoint of computational geometry and combinatorial optimization. SPP is known to be NP-hard, therefore, heuristic approaches are most commonly used, which allow to obtain approximate solutions. G. Wäscher et al. [34] proposed a typology of cutting and packaging problems (C&P). According to the typology of SPP can be divided into the following main types: Open Dimension Problems (ODP), Placement Problems (PP), Knapsack Problems (KP). The simplest and most researched class of SPPs is Circle Packing (CPP) [2, 5, 6, 9, 18, 25, 44, 45]. A more general case is the packing problem of unequal spheres into containers of arbitrary shape.

Hifi and R'Hallah [10] reviewed the most efficient models and methods for packing circles and spheres. We also indicate some modern approaches to solving SPP.

Burtseva and Salas [4] analyze different model and approaches for packing monosized spheres into cylindrical containers and their applications.

A global optimisation approach to solve unequal sphere packing problems is proposed in Sutou and Day [31]. The optimisation problem is formulated as a nonlinear optimisation problem with quadratic constraints and a linear objective. Paper [30] offers an optimisation method for packing unequal spheres into

S. Babichev et al. (Eds.): ISDMCI 2020, AISC 1246, pp. 151–163, 2021.
https://doi.org/10.1007/978-3-030-54215-3_10

a cuboid based on the decremental neighbourhood method and a local optimisation method. The second-order methods for 2D and 3D packing problems are presented in [3].

Kubach et al. [16] adapt the parallel greedy algorithms proposed in paper [14] for the 3D case. A hybrid algorithm ELPGD for packing unequal circles and spheres into a larger circular (or spherical) container is proposed in [17].

Stoyan et al. [27] modify the jump algorithm [25] to pack unequal spheres into various containers. A methodology to solve sphere packing problems in different dimensions is proposed in [43].

Several heuristic approaches for solving the unequal sphere packing by combining dichotomous search, tree search, hill-climbing and other ones are suggested in [1,11,12]. The authors make use of the truncated tree search and local search. Hifi and Yousef [13] develop an improved local search-based method.

Note that all existing methods for solving problems are approximate in general case. Therefore, approaches to improving solutions obtained by one method or another are of particular interest.

This article describes the idea of expanding the space of variables [36] in order to overcome the neighborhood of attraction of a local extremum. As a result, there is the possibility of directional enumeration of local solutions.

2 Materials and Methods

2.1 Problem Statement

Consider the following spheres packing problem in a container. Let there be given a set of spheres S_i, $i \in J_n$, the radii of which are equal $\hat{r}_i$, $i \in J_n$ respectively, and some container $S_0(\mu)$ with metric parameters (sizes) $\mu = (\mu_1, \mu_2, ..., \mu_l)$. The problem is to place the spheres S_i, $i \in J_n$ inside the container $S_0(\mu)$ under the conditions of their pairwise non-intersection.

In practice, containers that are in the shape of a sphere, a cube, a cylinder, an annular cylinder and a spherical layer are most often considered.

A sphere-shaped container $S_0^1(\mu)$ with metric parameter $\mu = R$ is described in the coordinate system $Oxyz$ by the inequality

$$x^2 + y^2 + z^2 - R^2 \leq 0.$$

A cuboid-shaped container $S_0^2(\mu)$ with metric parameters $\mu = (a, b, h)$ is described by the inequality

$$0 \leq x \leq a,\ 0 \leq y \leq b,\ 0 \leq z \leq h,$$

where $a \geq 2 \max\limits_{i \in J_n} \{\hat{r}_i\}$, $b \geq 2 \max\limits_{i \in J_n} \{\hat{r}_i\}$.

A cylinder-shaped container $S_0^3(\mu)$ with metric parameters $\mu = (R, h)$ is described by the inequality

$$x^2 + y^2 - R^2 \leq 0,\ 0 \leq z \leq h,$$

where $R \geq \max_{i \in J_n} \{\hat{r}_i\}$, $h \geq 2 \max_{i \in J_n} \{\hat{r}_i\}$.

Annular cylinder-shaped container $S_0^4(\mu)$ with metric parameters $\mu = (R, \rho, h)$ is described by the inequality

$$x^2 + y^2 - R^2 \leq 0, -x^2 - y^2 + \rho^2 \leq 0, R > \rho, 0 \leq z \leq h,$$

where $R - \rho \geq \max_{i \in J_n} \{\hat{r}_i\}$.

Container in spherical layer shape $S_0^5(\mu)$ with metric parameters $\mu = (R, \rho)$ is described by the inequality

$$x^2 + y^2 + z^2 - R^2 \leq 0, \ -x^2 - y^2 - z^2 + \rho^2 \leq 0, \ R > \rho.$$

In general, the metric parameters of containers or some of them are variable and are subject to optimization. Denote the coordinates of the centers of spheres S_i by $p^i = (u_i, v_i, w_i)$, $i \in J_n$, which are called the placement parameters.

Thus, the problem of packing spheres into container is an optimization one, and in the general case the objective function $F\left(\mu, p^1, p^2, ..., p^n\right)$ depends both on the metric parameters of the container $\mu = (\mu_1, \mu_2, ..., \mu_l)$ and on the placement parameters of the spheres $p^i = (u_i, v_i, w_i)$, $i \in J_n$. The metric parameters of the container and the placement parameters of the spheres must satisfy the restrictions formalizing the conditions of non-intersection and inclusion. It is clear that the constraints on non-intersection for spheres depend only on the placement parameters $p^i = (u_i, v_i, w_i)$, $i \in J_n$, while the constraints on inclusion also on the metric parameters μ of the container. Then the mathematical model of the problem has the form

$$F\left(\mu, p^1, p^2, ..., p^n\right) \rightarrow extr \tag{1}$$

subject to

$$\Phi_{ij}(p^i, p^j) \geq 0, \ i \in J_n, \ i \in J_n, \ i < j, \tag{2}$$

$$\Phi_{0i}(\mu, p^i) \geq 0, \ i \in J_n. \tag{3}$$

The functions in inequalities (2) describe the conditions non-intersection of spheres S_i and S_j have the form

$$\Phi_{ij}(p^i, p^j) = (u_i - u_j)^2 + (v_i - v_j)^2 + (w_i - w_j)^2 - (\hat{r}_i + \hat{r}_j)^2.$$

The functions in inequalities (3) describe the conditions for the inclusion of spheres S_i, $i \in J_n$ into a container $S_0(\mu)$ and depend on its shape. If the container shapes described above, we offer the following representation:

$$\Phi_{0i}\left(p^i, \mu\right) = -u_i^2 - v_i^2 - w_i^2 + (R - \hat{r}_i)^2$$

for sphere-shaped container $S_0^1(\mu)$;

$$\Phi_{0i}\left(p^i,\mu\right) = \min\{x_i - \hat{r}_i,\ y_i - \hat{r}_i,\ z_i - \hat{r}_i,\ a - x_i - \hat{r}_i,\ b - y_i - \hat{r}_i,\ h - z_i - \hat{r}_i\}$$

for cuboid-shaped container $S_0^2(\mu)$;

$$\Phi_{0i}\left(p^i,\mu\right) = \min\{-u_i^2 - v_i^2 + (R - \hat{r}_i)^2,\ w_i - \hat{r}_i,\ h - w_i - \hat{r}_i\}$$

for cylinder-shaped container $S_0^3(\mu)$;

$$\Phi_{0i}\left(p^i,\mu\right) = \min\{-u_i^2 - v_i^2 + (R - \hat{r}_i)^2, u_i^2 + v_i^2 - (\rho + \hat{r}_i)^2,\ w_i - \hat{r}_i,\ h - w_i - \hat{r}_i\}$$

for annular cylinder-shaped container $S_0^4(\mu)$;

$$\Phi_{0i}\left(p^i,\mu\right) = \min\{u_i^2 + v_i^2 + w_i^2 - (\rho - \hat{r}_i)^2,\ - u_i^2 - v_i^2 - w_i^2 + (R - \hat{r}_i)^2\}$$

for container in spherical layer shape $S_0^5(\mu)$.

Note that the above expressions define the Φ- functions [26] of spheres $S_i,\ i \in J_n$ and corresponding containers $S_0^j(\mu),\ j \in J_5$.

2.2 Lifting Model of the Unequal Sphere Packing Problem

The main idea of the Expanding Space Method proposed in the paper does not depend on the shape of the container. In this regard, we restrict ourselves to the problem of packing unequal spheres into a sphere of minimum radius. Based on the foregoing, the mathematical model of the problem (1)–(3) will take the form

$$R \to min \tag{4}$$

subject to

$$u_i^2 + v_i^2 + w_i^2 \leq (R - \hat{r}_i)^2, i \in J_n, \tag{5}$$

$$(u_i - u_j)^2 + (v_i - v_j)^2 + (w_i - w_j)^2 \geq (\hat{r}_i + \hat{r}_j)^2, i \in J_n, j \in J_n, i < j. \tag{6}$$

Thus, we have a nonlinear optimization problem with variables $R, u_i, v_i, w_i,\ i \in J_n$, which we call *Problem 1*.

Problem 1 is multiextremal due to the nonconvexity of constraints (6). The use of numerical methods for nonlinear optimization allows finding only its local solutions, depending on the choice of initial values $R, u_i, v_i, w_i,\ i \in J_n$. Traditionally, methods of directed search of local solutions are further used. At the same time, multi-start schemes are effective, in which starting points are generated randomly or according to some special rule. This paper proposes a new approach that provides an opportunity to improve local solutions. This approach is based on the idea of the method of artificial expansion of space dimension, first proposed in [36], and allowing to distinguish the combinatorial structure of placement problems. Note that the use of radii as independent variables in the implementation of the Jump Algorithm for solving some classes of packing problems confirmed the prospects of this direction [25,27].

Without loss of generality, suppose

$$\hat{r}_1 \leq \hat{r}_2 \leq ... \leq \hat{r}_n.$$

Despite the fact that in the original problem the radii of the spheres $\hat{r}_i$, $i \in J_n$ are fixed, we introduce variable radii $r_1, r_2, ..., r_n$ and form the system

$$\sum_{i=1}^{n} r_i = \sum_{i=1}^{n} \hat{r}_i, \tag{7}$$

$$\sum_{i \in W} r_i \geq \sum_{i=1}^{|W|} \hat{r}_i \ \forall W \subset J_n, \tag{8}$$

$$\sum_{i=1}^{n} (r_i - \tau)^2 = \sum_{i=1}^{n} (\hat{r}_i - \tau)^2, \tag{9}$$

where elements of subsets $W \subset J_n$ of capacity $|W|$ are ordered in increasing, and

$$\tau = \frac{1}{n} \sum_{i=1}^{n} \hat{r}_i. \tag{10}$$

System (7)–(9) is such that the set of its solutions coincides with the permutation set of real numbers $\{\hat{r}_1, \hat{r}_2, ..., \hat{r}_n\}$. This property is a partial case and follows from the general concept of continuous representations and functional extensions of combinatorial sets [20,21]. Representation (7)–(9) is called polyhedral-spherical [38,42], because it describes a set that is the intersection of the permutation polyhedron (7)–(8) and the hypersphere (9) with center at point (10). Optimization problems on polyhedral-spherical sets have interesting properties, which are investigated in the theory of convex extensions on vertex-located sets [40]. A review of modern Euclidean combinatorial optimization methods is given in [28].

Let us consider the function optimization problem (1) with restrictions

$$u_i^2 + v_i^2 + w_i^2 \leq (R - r_i)^2, i \in J_n, \tag{11}$$

$$(u_i - u_j)^2 + (v_i - v_j)^2 + (w_i - w_j)^2 \geq (r_i + r_j)^2, i \in J_n, j \in J_n, i < j, \tag{12}$$

where variables r_i, $i \in J_n$ satisfy conditions (7)–(9).

The problem of nonlinear optimization (4), (7)–(12) with variables R, u_i, v_i, w_i, r_i, $i \in J_n$ is called *Problem 2*.

Theorem 1. *Both Problem 1 and Problem 2 are equivalent in the sense of the coincidence of their global solutions.*

Problem 1 has a dimension $3n + 1$, and the dimension of *Problem 2* is equal $4n + 1$. Thus, we have implemented an approach of artificial expansion of the space of variables, based on the general scheme proposed in [36]. Applied to the problem of packing circles and spheres, this approach will be called the variable radius method.

2.3 Decomposition Approaches in the Method Implementation

The Expanding Space Method allows one to realize the idea of improving solutions of *Problem 1*. Indeed, we choose the local solution obtained as the starting point and consider the radii of the spheres as independent variables $r_1, r_2, ..., r_n$ satisfying constraint system (7)–(12). As a result, it is possible by variables to overcome the neighborhood of attraction of a local extremum and direct to a new, possibly better, local solution. The initial values of the variable radii can be chosen as equal to the initial values, or generated randomly in (r_1, r_n).

The main difficulties in implementing the proposed approach are associated with an increasing in the dimension of *Problem 2*. It is easy to see that the number of linear constraints in system (8) is equal to $2^n - 2$, which already at $n > 20$ leads to difficulties in applying modern numerical methods of nonlinear optimization. On the one hand, using the properties of linear and quadratic functions on combinatorial polyhedra allows one to partially overcome the arising difficulties [28,29,38–40,42]. However, this does not greatly expand the capabilities of methods in large-scale problems.

On the other hand, it is possible to select only $l < n$ circles with variables radii. Let $M' = \{m_1, m_{2,...,} m_l\} \subset J_n$ be the numbers of circles, whose radii $\hat{r}_{m_1} \leq \hat{r}_{m_2} \leq ... \leq \hat{r}_{m_l}$ are fixed, and the set $M'' = J_n \backslash M'$ be of ordered numbers of circles with variables radii r_i, $i \in M''$. Form a system of restrictions

$$u_i^2 + v_i^2 + w_i^2 \leq (R - \hat{r}_i)^2, i \in M' \tag{13}$$

$$u_i^2 + v_i^2 + w_i^2 \leq (R - r_i)^2, i \in M'' \tag{14}$$

$$(u_i - u_j)^2 + (v_i - v_j)^2 + (w_i - w_j)^2 \geq (\hat{r}_i + \hat{r}_j)^2, i \in M', j \in M', i < j \tag{15}$$

$$(u_i - u_j)^2 + (v_i - v_j)^2 + (w_i - w_j)^2 \geq (r_i + r_j)^2, i \in M'', j \in M'', i < j \tag{16}$$

$$(u_i - u_j)^2 + (v_i - v_j)^2 + (w_i - w_j)^2 \geq (\hat{r}_i + r_j)^2, i \in M', j \in M''. \tag{17}$$

$$\sum_{i \in M''} r_i = \sum_{i \in M''} \hat{r}_i \tag{18}$$

$$\sum_{i \in W} r_i \geq \sum_{i=1}^{|W|} \hat{r}_{m_i} \ \forall W \subset M'' \tag{19}$$

$$\sum_{i \in M''} (r_i - \hat{\tau})^2 = \sum_{i \in M''} (\hat{r}_i - \hat{\tau})^2, \tag{20}$$

where

$$\hat{\tau} = \frac{1}{l} \sum_{i=1}^{l} \hat{r}_{m_i}. \tag{21}$$

The problem of nonlinear optimization (4), (13)–(21) in the space of variables R, u_i, v_i, w_i, $i \in J_n$ and r_j, $j \in M''$ we call *Problem 3.*

Theorem 2. *The sets of global solutions of both Problem1 and Problem 3 are the same for any* $M' = \{m_1, m_2, ..., m_l\} \subset J_n$

To summarize the above reasoning consider a set $Q = \{r_1, r_2, ..., r_n\}$ and part it onto $q+1$ pairwise disjoint subsets as follows. Let be

$$\bigcup_{k=0}^{q} M^k = J_n, \tag{22}$$

where $M^k = \{m_1, m_2, ..., m_{l_k}\}$, $m_1 < m_2 < ... < m_{l_k}$,

$$k \in J_q^0 = \{0\} \bigcup J_q, \sum_{k=0}^{q} l_k = n, l_0 \geq 0$$

Let us consider the set

$$Q = \bigcup_{k=0}^{q} Q^k \tag{23}$$

where $Q^0 = \{\hat{r}_{m_1}, \hat{r}_{m_2}, ..., \hat{r}_{m_{l_0}}\}$, $Q^k = \left\{r_{m_1}, r_{m_2}, ..., r_{m_{l_k}}\right\}$, $k \in J_q$.

Form a system of restrictions

$$u_i^2 + v_i^2 + w_i^2 \leq (R - \hat{r}_i)^2, i \in M^0 \tag{24}$$

$$u_i^2 + v_i^2 + w_i^2 \leq (R - r_i)^2, i \in J_n \backslash M^0 \tag{25}$$

$$(u_i - u_j)^2 + (v_i - v_j)^2 + (w_i - w_j)^2 \geq (\hat{r}_i + \hat{r}_j)^2, i \in M^0, j \in M^0, \ i < j \tag{26}$$

$$(u_i - u_j)^2 + (v_i - v_j)^2 + (w_i - w_j)^2 \geq (r_i + r_j)^2, \tag{27}$$

$$i \in J_n \backslash M^0, j \in J_n \backslash M^0, i < j$$

$$(u_i - u_j)^2 + (v_i - v_j)^2 + (w_i - w_j)^2 \geq (\hat{r}_i + r_j)^2, i \in M^0, j \in J_n \backslash M^0. \tag{28}$$

For each set Q^k, $k \in J_q$, we have

$$\sum_{i \in M^k} r_i = \sum_{i \in M^k} \hat{r}_i \tag{29}$$

$$\sum_{i \in W} r_i \geq \sum_{i=1}^{|W|} \hat{r}_{m_i}, \ \forall W \subset M^k \tag{30}$$

$$\sum_{i \in M^k} (r_i - \tau_k)^2 = \sum_{i \in M^k} (\hat{r}_i - \tau_k)^2, \tag{31}$$

where

$$\tau_k = \frac{1}{l_k} \sum_{i=1}^{l_k} \hat{r}_{m_i}. \tag{32}$$

The problem of nonlinear optimization (4), (24)–(32) in the space of variables R, u_i, v_i, w_i, $i \in J_n$, and r_j, $j \in J_n \backslash M^0$ we call *Problem 4*.

The statements of **Theorems 1, 2** can be generalized by formulating the following one.

Theorem 3. *The sets of global solutions of both Problem 1 and Problem 4 are the same for any* $M^0 = \{m_1, m_2, ..., m_{l_0}\} \subset J_n$.

The choice of the method of partitioning $Q = \{r_1, r_2, ..., r_n\}$ onto a sets $Q^k = \left\{r_{m_1}, r_{m_2}, ..., r_{m_{l_k}}\right\}$, $k \in J_q^0$ with the subsequent formation of constraints (15)–(17), forms a family of modifications of the method of variable radius.

It is easy to see that the described method trivially extends to the 2D class of circle packing problems.

3 Experiment, Results and Discussion

The choice of various strategies for the implementation of the Expanding Space Method depends on the method of forming the starting points, the rules for changing the variable radii of the spheres and of the sequential search for local solutions in order to improve them. The process of improving decisions is iterative. Suppose that at the first stage (iteration) a local solution $R^{(1)}$,

$u_i^{(1)}, v_i^{(1)}, w_i^{(1)}, r_i^{(1)} = \hat{r}_i$, $i \in J_n$ was obtained. A change in this solution is associated both with the formation of the parameters of the placement of the spheres, and with the rule of choosing their radii, considered as independent variables.

As a rule, in the future, a new starting point from an extended neighborhood of the placement parameters $u_i^{(1)}, v_i^{(1)}, w_i^{(1)}$ with fixed radii $r_i^{(1)} = \hat{r}_i$ and a corresponding new local solution $R^{(2)}$, $u_i^{(2)}, v_i^{(2)}, w_i^{(2)}, r_i^{(2)} = \hat{r}_i$, $i \in J_n$ is searched for. At the k-th iteration, the initial values of the placement parameters corresponding to the best current value of the local solution for fixed values of the radii of the spheres are selected.

A distinctive feature of the approach proposed in present paper is that at each iteration, the placement parameters of the spheres are fixed and equal to the corresponding best current solution. New local solutions are formed taking into account the radii of the spheres as independent variables.

The following numerical experiments were carried out. To search for local extremes, the IPOPT software package (https://projects.coin-or.org/Ipopt) was used, which implements the internal point method for nonlinear programming problems. A PC with the following specifications of the i3/8G/SSD 256G was used.

Test problems with the number of spheres less than 20 were considered. 50 test problems were generated in which the radii of the spheres were randomly generated uniformly distributed on the interval (1, 20). In each test, the placement parameters (coordinates of the centers of the spheres) were similarly randomly selected from the cube $(-50, 50) \times (-50, 50) \times (-50, 50)$. The obtained values were chosen as initial values for the implementation of the local optimization method in *Problem 1*.

Then, with the same initial data, *Problem 2* was solved, in which the radii of the spheres were considered as independent variables. The initial values for the variable radii in *Problem 2* were randomly generated from the interval (1, 20). Thus, we compared the solutions of both *Problem 1* and *Problem 2* using the same local optimization method for the same starting point.

In 83% of test problems, local solutions to *Problem 2* turned out to be better than in *Problem 1*, which certainly confirms the effectiveness of the Expanding Space Method. The average runtime for *Problem 2* was only 1.5 times longer. Initially we intended to improve local solutions. Therefore, in the future, solutions to *Problem 1* were chosen as a starting point for the implementation of *Problem 2*. In all cases, improvements were obtained for local solutions. However, when we tried to improve the solution to *Problem 2*, we succeeded in 79% of cases. The attempt to improve the new best solution at the next iteration was successful only in 47% of cases. As a rule, after the 7th iteration, there were no improvements. This is naturally explained, because the better the initial local solution, the more difficult it is to improve it.

In the packing problem of a large number of spheres ($n \geq 20$), a numerical experiment was carried out in two ways. On the one hand, we examined two groups of spheres (*Strategy 1*). The radii of the spheres in the first group were fixed, and in the second group the variables with $l = 7$. Thus, the formulation of

Problem 3 was considered. On the other hand, *Problem 4* was posed, in which all the radii of the spheres were variable (*Strategy 2*).

A series of 20 test problems was formed in which the radii of the spheres were randomly generated in the interval (1, 30). For each test, *Problem 1* was solved, in which the initial values of the coordinates of the centers of the spheres were randomly selected from a cube $(-50, 50) \times (-50, 50) \times (-50, 50)$. Then, the obtained solutions were improved using *Strategies 1, 2*. Improvements were obtained in all the tests examined. The average runtime for *Strategy 2* was 1.7 times longer, and the quality of the solution was better by 4,4%. The choice of the method of decomposition of the set has a significant impact on the result. Experimentally, the best results corresponded to the groups obtained by partitioning the set $Q = \{\hat{r}_1, \hat{r}_2, ..., \hat{r}_n\}$, in order $\hat{r}_1 \leq \hat{r}_2 \leq ... \leq \hat{r}_n$ (*Strategy 3*).

Note that similar numerical experiments were previously performed for the problem of packing circles in a circle of minimum radius [35]. The results correlate with those obtained in present paper.

4 Conclusions

This article proposed a new approach for improving local solutions to the problem of packing unequal spheres into containers of various shapes. The main idea is to expand the space of variables in the optimization problem, assuming that the radii of the spheres are considered as independent variables. In this case, an additional constraint system for variable radii is constructed using the permutation structure of the problem. The described approach can be generalized to a wide class of packing and covering problems with fixed metric parameters. For this, we use the concept of configuration spaces of geometric objects [24,37,41], the generalized variables of which are their metric and placement parameters. In these cases, assuming that the metric parameters are independent variables, a rigid system of constraints is formed by analogy with spherical objects. The combinatorial structure of the problem will be the same, which will form a system of constraints for metric parameters.

Finally, we note the areas of practical application of sphere packing problems. SPP arise in chemistry, biology, in manufacturing automobiles, electric motors (cutting, cable bundling problems); topology optimization problems products for 3D-printing; coding and information security; in studying the structure of nanomaterials, crystals, concrete, granular materials; in casting techniques; packing catalysts in chemical catalytic reactors, columns for gas distillation and absorption, fuel elements in nuclear reactors and heat exchangers; in radio-surgical treatment of tumors and retinal laser coagulation and so on [7,8,15,19,22,23,32,33].

References

1. Akeb, H.: A two-stage look-ahead heuristic for packing spheres into a three-dimensional bin of minimum length. In: Fidanova, S. (ed.) Recent Advances in Computational Optimization: Results of the Workshop on Computational Optimization, WCO 2014, pp. 127–144. Studies in Computational Intelligence. Springer, Cham (2016). https://doi.org/10.1007/978-3-319-21133-6_8
2. Amirgaliyeva, Z., Mladenović, N., Todosijević, R., Urošević, D.: Solving the maximum min-sum dispersion by alternating formulations of two different problems. Eur. J. Oper. Res. **260**(2), 444–459 (2017). https://doi.org/10.1016/j.ejor.2016.12.039
3. Birgin, E.G., Sobral, F.N.C.: Minimizing the object dimensions in circle and sphere packing problems. Comput. Oper. Res. **35**(7), 2357–2375 (2008). https://doi.org/10.1016/j.cor.2006.11.002
4. Burtseva, L., Salas, B.V., Werner, F., Petranovskii, V.: Packing of monosized spheres in a cylindrical container: models and approaches. Rev. Mex. Fis. **61**(1), 20–27 (2015)
5. Castillo, I., Kampas, F.J., Pintér, J.D.: Solving circle packing problems by global optimization: numerical results and industrial applications. Eur. J. Oper. Res. **191**(3), 786–802 (2008). https://doi.org/10.1016/j.ejor.2007.01.054
6. Chen, D.Z.: Sphere packing problem. In: Science, M.Y.K.P.O.C. (ed.) Encyclopedia of Algorithms, pp. 871–874. Springer, Heidelberg (2008). https://doi.org/10.1007/978-0-387-30162-4_391
7. Conway, J., Sloane, N.J.A.: Sphere Packings, Lattices and Groups, 3rd edn. Springer, New York (1999)
8. Corwin, E.I., Clusel, M., Siemens, A.O.N., Brujic, J.: Model for random packing of polydisperse frictionless spheres. Soft Matter **6**(13), 2949–2959 (2010). https://doi.org/10.1039/C000984A
9. He, K., Huang, M., Yang, C.: An action-space-based global optimization algorithm for packing circles into a square container. Comput. Oper. Res. **58**, 67–74 (2015). https://doi.org/10.1016/j.cor.2014.12.010
10. Hifi, M., M'Hallah, R.: A literature review on circle and sphere packing problems: models and methodologies. Adv. Oper. Res. **7**, 1–22 (2009). https://doi.org/10.1155/2009/150624
11. Hifi, M., Yousef, L.: Handling lower bound and hill-climbing strategies for sphere packing problems. In: Fidanova, S. (ed.) Recent Advances in Computational Optimization: Results of the Workshop on Computational Optimization, WCO 2014. Studies in Computational Intelligence, pp. 145–164. Springer, Cham (2016). https://doi.org/10.1007/978-3-319-21133-6_9
12. Hifi, M., Yousef, L.: A global dichotomous search-based heuristic for the three-dimensional sphere packing problem. Int. J. Oper. Res. **33**(2), 139–160 (2018). https://doi.org/10.1504/IJOR.2018.095195
13. Hifi, M., Yousef, L.: A local search-based method for sphere packing problems. Eur. J. Oper. Res. **274**(2), 482–500 (2019). https://doi.org/10.1016/j.ejor.2018.10.016
14. Huang, W.Q., Li, Y., Akeb, H., Li, C.M.: Greedy algorithms for packing unequal circles into a rectangular container. J. Oper. Res. Soc. **56**(5), 539–548 (2005). https://doi.org/10.1057/palgrave.jors.2601836
15. Ke, C. (ed.): Recent Advances in Nanotechnology, 1 edn. Apple Academic Press (2011)

16. Kubach, T., Bortfeldt, A., Tilli, T., Gehring, H.: Greedy algorithms for packing unequal spheres into a cuboidal strip or a cuboid. Asia-Pacific J. Oper. Res. (APJOR) **28**(06), 739–753 (2011)
17. Liu, J., Yao, Y., Zheng, Y., Geng, H., Zhou, G.: An effective hybrid algorithm for the circles and spheres packing problems. In: Du, D.Z., Hu, X., Pardalos, P.M. (eds.) Combinatorial Optimization and Applications. Lecture Notes in Computer Science, pp. 135–144. Springer, Heidelberg (2009). https://doi.org/10.1007/978-3-642-02026-1_12
18. López, C.O., Beasley, J.E.: Packing unequal circles using formulation space search. Comput. Oper. Res. **40**(5), 1276–1288 (2013). https://doi.org/10.1016/j.cor.2012.11.022
19. Pichugina, O., Kartashov, O.: Signed permutation polytope packing in VLSI design. In: 2019 IEEE 15th International Conference on the Experience of Designing and Application of CAD Systems (CADSM) Conference Proceedings, pp. 4/50–4/55. Lviv (2019). https://doi.org/10.1109/CADSM.2019.8779353
20. Pichugina, O.S., Yakovlev, S.V.: Continuous representations and functional extensions in combinatorial optimization. Cybern. Syst. Anal. **52**(6), 921–930 (2016). https://doi.org/10.1007/s10559-016-9894-2
21. Pichugina, O.S., Yakovlev, S.V.: Functional and analytic representations of the general permutation. Eastern Eur. J. Enterprise Technol. **79**(1(4)), 27–38 (2016). https://doi.org/10.15587/1729-4061.2016.58550
22. Shekhovtsov, S.B., Yakovlev, S.V.: Formalization and solution of a class of covering problems in the design of monitoring and control systems. Akademiya Nauk SSSR. Avtomatika i Telemekhanika **5**, 160–168 (1989)
23. Shirokanev, A., Kirsh, D., Ilyasova, N., Kupriyanov, A.: Investigation of algorithms for coagulate arrangement in fundus images. Comput. Opt. **42**(4), 712–721 (2018). https://doi.org/10.18287/2412-6179-2018-42-4-712-721
24. Stoyan, Y.G., Yakovlev, S.V.: Configuration space of geometric objects. Cybern. Syst. Anal. 716–726 (2018). https://doi.org/10.1007/s10559-018-0073-5
25. Stoyan, Y., Yaskov, G.: Packing unequal circles into a strip of minimal length with a jump algorithm. Optim. Lett. **8**(3), 949–970 (2014). https://doi.org/10.1007/s11590-013-0646-1
26. Stoyan, Y.G., Chugay, A.M.: Mathematical modeling of the interaction of non-oriented convex polytopes. Cybern. Syst. Anal. **48**(6), 837–845 (2012). https://doi.org/10.1007/s10559-012-9463-2
27. Stoyan, Y.G., Scheithauer, G., Yaskov, G.N.: Packing unequal spheres into various containers. Cybern. Syst. Anal. **52**(3), 419–426 (2016). https://doi.org/10.1007/s10559-016-9842-1
28. Stoyan, Y.G., Yakovlev, S.V.: Theory and methods of Euclidean combinatorial optimization: current state and promising. Cybern. Syst. Anal. **56**(3), 513–525 (2020)
29. Stoyan, Y.G., Yakovlev, S.V., Emets, O.A., Valuĭskaya, O.A.: Construction of convex continuations for functions defined on a hypersphere. Cybern. Syst. Anal. **34**(2), 27–36 (1998). https://doi.org/10.1007/BF02742066
30. Stoyan, Y.Y., Yaskov, G., Scheithauer, G.: Packing of various radii solid spheres into a parallelepiped. Central Eur. J. Oper. Res. (CEJOR) **11**(4), 389–407 (2003)
31. Sutou, A., Dai, Y.: Global optimization approach to unequal global optimization approach to unequal sphere packing problems in 3D. J. Optim. Theory Appl. **114**(3), 671–694 (2002). https://doi.org/10.1023/A:1016083231326

32. Wang, J.: Packing of unequal spheres and automated radiosurgical treatment planning. J. Comb. Optim. **3**(4), 453–463 (1999). https://doi.org/10.1023/A:1009831621621
33. Wang, Y., Nagarajan, M., Uhler, C., Shivashankar, G.V.: Orientation and repositioning of chromosomes correlate with cell geometry-dependent gene expression. Mol. Biol. Cell **28**(14), 1997–2009 (2017). https://doi.org/10.1091/mbc.E16-12-0825
34. Wäscher, G., Haußner, H., Schumann, H.: An improved typology of cutting and packing problems. Eur. J. Oper. Res. **183**(3), 1109–1130 (2007). https://doi.org/10.1016/j.ejor.2005.12.047
35. Yakovlev, S., Kartashov, O., Korobchynskyi, K., Skripka, B.: Numerical results of variable radii method in the unequal circles packing problem. In: 2019 IEEE 15th International Conference on the Experience of Designing and Application of CAD Systems (CADSM), pp. 1–4 (2019). https://doi.org/10.1109/CADSM.2019.8779288
36. Yakovlev, S.V.: The method of artificial space dilation in problems of optimal packing of geometric objects. Cybern. Syst. Anal. **53**(5), 725–731 (2017). https://doi.org/10.1007/s10559-017-9974-y
37. Yakovlev, S.V.: Formalizing spatial configuration optimization problems with the use of a special function class. Cybern. Syst. Anal. **55**(4), 581–589 (2019). https://doi.org/10.1007/s10559-019-00167-y
38. Yakovlev, S.V., Pichugina, O.S.: Properties of combinatorial optimization problems over polyhedral-spherical sets. Cybern. Syst. Anal. **54**(1), 99–109 (2018)
39. Yakovlev, S.V., Valuiskaya, O.A.: Optimization of linear functions at the vertices of a permutation polyhedron with additional linear constraints. Ukrainian Math. J. **53**(9), 1535–1545 (2001). https://doi.org/10.1023/A:1014374926840
40. Yakovlev, S.: Convex extensions in combinatorial optimization and their applications. In: Optimization Methods and Applications. Springer Optimization and Its Applications, pp. 567–584. Springer, Cham (2017). https://doi.org/10.1007/978-3-319-68640-0_27
41. Yakovlev, S.V.: On some classes of spatial configurations of geometric objects and their formalization. J. Autom. Inf. Sci. **50**(9), 38–50 (2018). https://doi.org/10.1615/JAutomatInfScien.v50.i9.30
42. Yakovlev, S.V., Pichugina, O.S., Yarovaya, O.V.: Polyhedral-spherical configurations in discrete optimization problems. J. Autom. Inf. Sci. **51**(1), 26–40 (2019). https://doi.org/10.1615/JAutomatInfScien.v51.i1.30
43. Yaskov, G.: Methodology to solve multi-dimentional sphere packing problems. J. Mech. Eng. **22**(1), 67–75 (2019). https://doi.org/10.15407/pmach2019.01.067
44. Ying, S.: A novel method for packing unequal circles into a rectangular container. Appl. Mech. Mater. **513**, 3942–3945 (2014). https://doi.org/10.4028/www.scientific.net/AMM.513-517.3942
45. Zong, C.: Sphere Packings. Springer, New York (1999)

Information Resource Management Technology Based on Fuzzy Logic

Victoria Vysotska[1(✉)], Andriy Berko[1], Vasyl Lytvyn[1], Petro Kravets[1], Lyudmyla Dzyubyk[1], Yuriy Bardachov[2], and Svitlana Vyshemyrska[2]

[1] Lviv Polytechnic National University, Lviv, Ukraine
{Victoria.A.Vysotska,Andrii.Y.Berko,Vasyl.V.Lytvyn,
Petro.O.Kravets}@lpnu.ua
[2] Kherson National Technical University, Kherson, Ukraine
Y.Bardachov@kntu.net.ua, printvvs@gmail.com

Abstract. The content management method is proposed as a stage of its life cycle, which is based on the application of fuzzy logic. This method describes the processes of generating commercial information resources and simplifies content management technology. The article describes the methods and procedures for forming design decisions in the management of commercial web-projects and some characteristics of the project are incomplete and inaccurate. The analysis of the main factors of project decision making is carried out, the reasons and nature of occurrence of incompleteness and inaccuracy of design characteristics were determined. The proposed method makes it possible to create tools for processing web-resources and implement the content management subsystem. Procedures for reducing the level of incompleteness and inaccuracy of project characteristics based on fuzzy logic have developed. Content management tasks are formation and rotation of operational and retrospective databases; personalization of users' work, saving personal requests of users and sources, keeping statistics of work; providing search in databases; generation of initial forms; information interaction with other databases; formation and development of a web-resource. The content management subsystem is implemented by caching it or information blocks.

Keywords: Commercial content · Indistinct data · Data fuzzy · Web resource · Content analysis · Project management · Web project · Project decision making · Internet marketing · Indistinct logic

1 Introduction

The rapid development of the Internet contributes to the growing need for rapid acquisition of data of various nature and the new forms implementation for web services through modern IT [5,6,11,23]. An information technology product that is produced and distributed for profit is, called commercial content, and the Internet tools through which it is disseminated is e-commerce content systems. Today,

S. Babichev et al. (Eds.): ISDMCI 2020, AISC 1246, pp. 164–182, 2021.
https://doi.org/10.1007/978-3-030-54215-3_11

content e-commerce is an objective reality and a promising business process. The Internet is a business environment, and commercial content is the product with the highest demand and sales. Content can immediately ordered, issued, paid and received online as a product [2,3,8]. The Internet sells all spectrum of commercial content - scientific and non-fiction software, photos, movies, books, music, articles, etc. Well-known corporations that develop e-commerce content are Apple through Apple Store, I-Tunes, Google– the Play Market, Amazon – Amazon.com [1,4,6,10,15]. Electronic commerce (e-commerce) content systems (ECCS) are built on a closed principle as one-off projects. Most of the decisions and studies are made at the level of specific projects. The analysis of the above factors makes it possible to conclude that there is some contradiction between the active development and spread of IT and ECCS on the one hand, and the relatively small amount of scientific research on this topic and their locality on the other [2,3,6,9,12,13,16–22,24]. This contradiction raises the problem of holding back the innovative development of the e-commerce sector by creating and implementing the latest emerging IT, which negatively affects the growth rate of this part of the market [2,3,6,7,14].

2 Related Works

Initial information of the process of management of commercial content through processing of information resources of e-business systems is data on the purposes and conditions of work of the information system [2,3,6], which determine the main purpose of the simulation and allow us to formulate requirements for CMS [6]. The process for managing commercial content is presented as

$$Y = \langle X, Q, C, V, H, Z, T, \delta \rangle \tag{1}$$

where X is the plural of data from different sources, Q is the plural of information requests from users, C is the plural of commercial content, V is the plural of conditions for the maintenance of commercial content and external environmental influences on the system, H is the plural of conditions for the processing of commercial content, Z is the plural of components of the information resource, T is the time of content processing transactions, Y is a collection of initial characteristics of the system, δ is operator of formation of results of analysis of functioning statistics CMS [5–7,11,14,23]. The process of developing information resources from (1) is describe as

$$y_j \left(t_{p+1}\right) = \delta \left(x_i, q_d, c_r, \nu_l, h_k, t_p, z_w\right) \tag{2}$$

According to Google Analytics [23] the value y_j is a collection of data over a specified period of time, where y_1 is number of visits, y_2 is average time to visit the information resource, y_3 bounce rate (%), y_4 is search goal achieved, y_5 is content dynamics (%), y_6 is the total number of pages viewed, y_7 is the number of pages viewed per visit, y_8 is new visits (%), y_9 is unique visitors, y_{10} is source of traffic in y % (search, direct or referral traffic) etc. The effects of dimension

c_r, ν_l, h_k, z_w and y_j as a result of CMS are unknown. A formal CMS description does not disclose the links and dependencies between input, commercial content, output, and information processing processes. Analysing the dynamics of content flow and building stages of processing information resources is important and relevant [5,6,11,23]. To effectively implement the commercial content management process, multiple content is broken down c_r to relevant/irrelevant and issued/unissued subsets (Table 1), where the number of content plural is equal $C = a + b = d + g$ [5,6,11,23].

Table 1. Content search performance metrics [5,6,11,23]

Coefficient	Characterizes	Plurality of content	Formula
Completeness	Issued relevant a	Relevant	$p_1 = a/(a+g) = 1 - p_6$
Accuracy	Issued relevant a	Issued	$p_2 = a/(a+b) = 1 - p_3$
Noise	Issued irrelevant b	Issued	$p_3 = b/(a+b) = 1 - p_2$
Precipitate	Issued irrelevant b	Irrelevant	$p_4 = b/(b+d) = 1 - p_5$
Specificity	Issued irrelevant d	Irrelevant	$p_5 = d/(d+b) = 1 - p_4$
Remainder	Issued relevant g	Relevant	$p_6 = g/(a+g) = 1 - p_1$
Fuzzy	Issued relevant g	Unissued	$p_7 = g/(g+d) = 1 - p_8$
Ambiguities	Unissued irrelevant d	Unissued	$p_8 = d/(g+d) = 1 - p_7$

Table 2 presents comparisons of the features of ECS and ECCS on the implementation of stages of the content lifecycle. ECCS has more opportunities to implement all stages of the content lifecycle to enhance functionality and improve the performance of these systems. Existing ECCSs do not support the entire content lifecycle (its generation, management, and support) [1–24]. ECCS is characterized by a complex set of interrelated operations, methods, techniques, which form the technological process of its operation. The technological process of ECCS consists of six phases: familiarization with the system, search for goods; obtaining information about the product; selection of product category; calculation; delivery of goods. The technological process of functioning of the ECCS is simplified due to the intangibility of the goods, but is complicated by additional stages of the content life cycle in the process of processing information resources: formation, management and support of commercial content. E-commerce content reduces the number of e-commerce organization employees, but increases the number of e-commerce participants and provides greater opportunities to meet their needs and reach the target for both commercial content creators and ECCS users. Lack of design recommendations and general requirements for the ECCS structure leads to the problem of defining and forming methods for processing information resources of these systems. This explains the purpose, relevance and direction of the study.

First block: traditional sale, shop, showcase signboard, working stocks of products (newspapers, magazines books) in the hall, seller consultant, bas-

Table 2. Features comparison ECS and ECCS

The name of the characteristic	ECS	ECCS
Intangible products	–	+
The constant quantity of product	–	+
Product creation	–	+
Automatic product management	+/–	+
Increase in quantity of variety of product	+/–	+
Lack of composition	–	+
Storage of products in databases	–	+
Product promotion efficiency according to keywords	+/–	+
Keyword product search performance	+/–	+
Products duplication detection and elimination	–	+
Determination of the aging of the goods in content	–	+
Determination of relevance of goods	+/–	+
Audience analysis	+/–	+
Digests formation	–	+
Goods distribution between participants	+/–	+
Digests distribution among employees	–	+
Product formation	–	+
Product management	–	+
The impact of user experience on increasing sales	+/–	+
Automatic maintenance of goods	+/–	+

ket, cash desk, packaging of purchase. Second block: electronic content sales, e-commerce system, web showcase, web directory, working stocks of products (newspapers, magazines books) in the warehouse, dialog interface, moderator consultant, electronic shopping cart, registration, ordering, identification, payment, delivery system, packaging of purchase.

Familiarization, content search, getting product information, choice of content, calculation, receiving content. Third block: electronic content sales, e-commerce content system, content creation, web showcase, content management, content support, web directory, navigation system, dialog interface, electronic shopping cart, registration, ordering, identification, authorization, transaction, delivery system.

Analysing the dynamics of content flow and building stages of processing information resources is important and relevant. The development and implementation of ECCS is one of the strategic directions of e-business development. A characteristic of such systems is the ability to automatically process information resources to increase the volume of content sales to a regular user, actively engage potential users and expand the boundaries of the target audience. In particular, the principles and technologies of e-commerce are actively used in the creation of systems of on-line/off-line sales and analysis/exchange/storage of content, e-commerce, cloud storage/computing. The lack of a common standardized approach to the design of ECCS and the processes of processing information resources leads to a number of problems when developing the typical structure of the respective systems. The advanced classification of the ECS provided the opportunity to distinguish and analyse the specific features of the ECCS. Improving the classification of ECCS allows problems identification of their functioning due to the lack of methods of processing information resources as the formation and maintenance of commercial content, as well as improving the method of managing commercial content in these systems by introducing additional management parameters. This solves the problems of implementation of the relevant software and substantiates the purpose, relevance and directions of the research.

The article deals with the scientific and practical problem of developing a method of managing an information product by automating the processing of information resources in e-commerce systems. The work was performed within the framework of scientific researches of Information Systems and Networks Department in the Lviv Polytechnic National University on the topic "Research, development and implementation of intelligent distributed information technologies and systems based on database resources, data warehouses, data spaces and knowledge in order to accelerate the processes of modern the information society." The research results were carried out within the framework of the research work on the topic "Development of methods, algorithms and software tools for modelling, designing and optimization of intellectual information systems based on Web technologies WEB". Scientific research was also conducted within the framework of the initiative topics of the ISM department of Lviv Polytechnic National University on the topic "Development of intelligent distributed systems based on ontological approach for the integration of information resources".

3 Materials and Methods

The main process of functioning of the ECCS is the development of information resources (Fig. 1), which do the following circuits of the main components describe:

1. for the process of formation of information resource of the system the scheme is as follows: *content* → *content formation* → *content database* → *content management* → *information resource*,
2. for the process of forming a response to a user's request, the scheme is as follows: *user request* → *content management* → *information resource* → *content support* → *user database*,
3. for the process of forming the report of the system for the moderator the scheme is as follows: *moderator request* → *content support* → *user database* → *content management* → *report for moderator*,

4. for the process of moderation of the internal parameters of the system the scheme is as follows: *moderator request* → *content formation* → *rule base* → *content support* → *rule base* → *content management* → *result*.

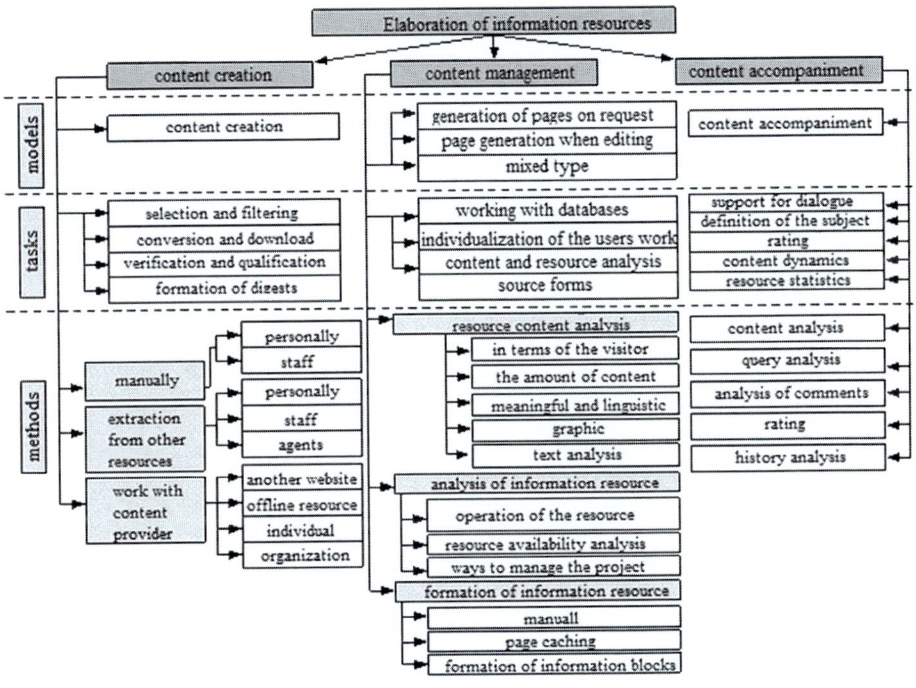

Fig. 1. Features of the process of processing information resources ECCS

The main subsystems for processing information resources in the ECCS are content creation, management and support (Fig. 1), whose connection scheme is as follows: *content creation* → *content management* → *content support*, The e-commerce content system is presented as

$$Y = \langle X, Q, H, C, V, Z, T, \alpha, \beta, \gamma \rangle \tag{3}$$

where α is commercial content creation operator, β is commercial content management operator, γ is commercial content support operator y. Than with (1–2) $\delta : X \to Y$ is superposition of functions $\delta = \gamma \circ \beta \circ \alpha$.

The commercial content formation operator α is submitted as

$$c_r = \left\{ \bigcup_{i}^{n_X} x_i \middle| \begin{matrix} \forall x_i \in X_{u_f},\ x_i \notin X_{\overline{u_f}},\ \exists u_f \in U_{x_i},\ u_f \notin U_{\overline{x_i}}, \\ X = X_{u_f} \cup X_{\overline{u_f}},\ U = U_{x_i} \cup U_{\overline{x_i}},\ f = \overline{1, n_U} \end{matrix} \right\} \tag{4}$$

where the plural of conditions u_f is defined as

$$u_f = \left\{ \bigcup_{j}^{k} u_{jf} \middle| \begin{array}{c} \forall u_{jf} \in U_{x_i},\, \exists x_i \in X_{u_f},\, u_{jf} \notin U_{\overline{x_i}}, \\ U = U_{x_i} \cup U_{\overline{x_i}},\, X_{u_f} \subseteq X,\, f = \overline{1, n_U},\, i = \overline{1, m} \end{array} \right\} \tag{5}$$

A commercial content management operator β is a mapping of content c_r to a new state c'_r as $\beta : (q_d, z_w, c_r, h_k, u_M, t_p) \rightarrow (c'_r, h'_k, z_{w+1}, t_{p+1})$, where $q_d \in Q$, $h_k \in H$, $h_k = \{h_{1k}(c_r, q_d), \ldots, h_{n_H k}(c_r, q_d)\}$ is plural of content management conditions c_r, at

$$z_w = \left\{ \bigcup_{r=1}^{n_C} c_r \middle| \begin{array}{c} \forall c_r \in C_{q_d},\, \exists q_d \in Q,\, \exists h_k \in H_{c_r},\, c_r \notin C_{\overline{q_d}},\, h_k \notin H_{\overline{c_r}}, \\ C = C_{q_d} \cup C_{\overline{q_d}},\, H = H_{c_r} \cup H_{\overline{c_r}},\, d = \overline{1, n_Q},\, k = \overline{1, n_H} \end{array} \right\} \tag{6}$$

where the plural of values of the determining parameters is formed as $h'_k = h_k + \Delta h$.

Commercial content operator γ is submitted as

$$\gamma : (c_r, q_d, \nu_l, h_k, z_w, u_S, t_p) \rightarrow y_i, \tag{7}$$

The output is implemented

$$y_j = \left\{ \bigcup_{l}^{n_V} \nu_l \middle| \begin{array}{c} \forall \nu_l \in V_{q_d} \cup V_{z_w},\, \exists q_d \in Q,\, \exists z_w \in Z,\, \exists h_k \in H_{c_r},\, \nu_l \notin V_{\overline{q_d}},\, \nu_l \notin V_{\overline{z_w}}, \\ V_{q_d} \subset V,\, V_{z_w} \subset V,\, d = \overline{1, n_Q},\, w = \overline{1, n_Z},\, r = \overline{1, n_C},\, k = \overline{1, n_H} \end{array} \right\} \tag{8}$$

The collection y_i $\{y_{1j},\, y_{2j},\, \ldots,\, y_{gj}\}$ describes the functioning of the ECCS with the main processes of processing information resources such as content creation, management and support. The ECCS statistics are analysed according to the analysis of reactions to this system of permanent / potential user (visits, queries, keyword searches, etc.). Content management operator $\beta : C \rightarrow Z$ submitted by superposition $\beta = \beta_4 \circ \beta_3 \circ \beta_2 \circ \beta_1$, where β_1 is operator of editing and modification of commercial content; β_2 is operator of determining the weight of the block of commercial content; β_3 is operator of formation of values of defining parameters of management of content; β_4 is operator of formation and submission of information resource page. Plural of operators $\beta = \{\beta_1, \beta_2, \beta_3, \beta_4\}$ are adequate in the commercial content management process described by the view operator $\beta = \langle C, Q, H, U, T, Z, \beta_1, \beta_2, \beta_3, \beta_4 \rangle$, where

1. The operator of editing and modification of content $\beta_1 : (c_r, h_k, u_l, t_p) \rightarrow c'_r$.
2. The operator of determining the weight of the block of commercial content $\beta_2 : (c'_r, y_j, u_l, t_p) \rightarrow c''_r$.
3. The operator of formation of values of defining parameters of management of commercial content $\beta_3 : (c''_r, h_k, u_l, t_p) \rightarrow h'_k$.
4. The operator of formation and submission of information resource $\beta_4 : (c''_r, h'_k, z_w, q_d, t_p) \rightarrow z_{w+1}$, where $h_k \in H$, $h_k = \{h_{1k}, h_{2k}, \ldots, h_{mk}\}$ is plural of parameters of process of management of commercial content (h_{1k} – topicality, h_{2k} – relevance, h_{3k} – completeness, h_{4k} – authenticity, h_{5k} – authenticity

of the content); $u_l \in U$, $u_l = \{u_{1l}, u_{2l}, \ldots, u_{nl}\}$ – a plural of criteria for the content management process (u_{1l} – the block placement factor in commercial content, u_{2l} – placement of keywords, u_{3l} – the statistical significance of keywords, u_{4l} – keyword query placement for a user's query). Formation of information resource pages is given as

$$z_w = \left\{ \bigcup_{r=1}^{n_C} c_r \,\middle|\, \begin{array}{l} \forall c_r \in C_{q_d},\ \exists q_d \in Q,\ \exists h_k \in H_{c_r},\ c_r \notin C_{\overline{q_d}},\ h_k \notin H_{\overline{c_r}}, \\ C = C_{q_d} \cup C_{\overline{q_d}},\ H = H_{c_r} \cup H_{\overline{c_r}},\ d = \overline{1, n_Q},\ k = \overline{1, n_H} \end{array} \right\} \tag{9}$$

Commercial content topicality h_{1k} is calculated from profitability I_p and content demand growth I_g for content. Content authentication is a subjective value within $(0.5; 1\}$ and is determined by the moderator ECCS. Content management method is a set of measures to support the maintenance of defining parameters of commercial content $h_k \to h'_k$ (topicality, completeness, certainty) in accordance with the set requirements for a set of content management criteria $z_{w+1}\left(h'_k, c''_r, t_{p+1}\right) = \beta\left(q_d, z_w, c_r, h_k, u_M, t_p\right)$.

Let A is the formation of a plurality of pages. N is be the dimension of the generated plurality of pages of the information resource. M is be the dimension of the processed array of relevant commercial content for forming the plurality of pages. U is be the dimension of the entire array of commercial content for forming the plurality of pages ($U = M + L$). L is the dimension of the array of irrelevant content. Let $Z = \beta_{A4}(C, M)$ us denote a function that gives the upper limit of the maximum number of basic search and filter operations that algorithm A, has to accomplish by solving the dimension M. The search or uniform search method is a method of finding the values $\bigcup_{j=1}^{m} c_j(q_i, t_r)$ of truly meaningful functions by a set of comparison criteria $\exists h_k \in H_c$ to form a plurality of pages Z of an information resource.

The design process used to determine the best parameter values $\exists h_k \in H_c$ management of commercial content C to optimize the algorithm for generating multiple pages Z by a complete search of relevant commercial content $\bigcup_{j=1}^{m} c_j(q_i, t_r)$. This optimization is related to optimal parameter values $\exists h_k \in H_c$ with a given structure of commercial content c_j and is parametric optimization. It is necessary to find among the elements $\forall c'_j \in C_q$, $c'_j \notin C_{\overline{q}}$, $C = C_q \cup C_{\overline{q}}$ that match the parameters $\exists h_k \in H_c, h_k \notin H_{\overline{c}}$, $H = H_c \cup H_{\overline{c}}$ and reduces the value $\beta_{A4}(C', M) = \beta_{A4}(C', (U - L))$ of the given function $\beta_{A4}(C, U) = \beta_{A4}(C, (M + L))$. Reducing the dimension $M < U$ of the processed array of relevant commercial content before forming multiple pages results in a linear decrease in the computational and temporal complexity of algorithm A. The time complexity of the algorithm is estimated using the operator $\beta_{A4}(C', M) = O(\beta_{A4}(C, U))$, that is $\beta_{A4}(C, U)$ decreasing as $\beta_{A4}(C', M)$ for $M < U$. Exact values $\beta_{A4}(C', M)$ depend on the particular implementation, whereas $O(\beta_{A4}(C, U))$ are a characteristic of algorithm A. One method of increasing the speed of search for algorithm A is to parallelize the formation of multiple pages with the construction of the conveyor or with the division into disjoint subsets.

1. Let the relation algorithm $Z = \beta_{A4}(C, M)$ be represented in the form of a chain of simple operations of search of commercial content according to criteria's $\exists h_k \in H_c$ for formation of plural of relevant content: $g_1, g_2, \ldots, g_N$. An element a_i in a queue N of time-aligned processes $a_1, a_2, \ldots, a_N$ performs three time-consuming operations:
   - receiving content from a_{i-1} process;
   - conducting a compliance g_i check operation h_k;
   - data transmission a_{i+1} process.

   Then a pipeline of N time-aligned processes $a_1, a_2, \ldots, a_N$ commercial content at speed $\nu/3$, where ν – the speed of performing one operation g_i one process a_i the formation of a plurality of pages from a plurality of relevant commercial content.
2. The plurality C is subdivided into intersecting $C_1, C_2, \ldots, C_N$ subsets according to their topic heading. The content management module sets the sets simultaneously $C_1, C_2, \ldots, C_N$. The subsets whose theme does not match the user's theme query are not sorted through q_i, at $C = C_q \cup C_{\overline{q}}$ part of subsets $C_1, C_2, \ldots, C_N$ ignored. The average number of steps of the algorithm for forming multiple pages is M/N.

The application of a marketing strategy in content management processes is intended to manage maximizing the scope of its implementation. This goal is achieved by monitoring, controlling and setting a set of regulatory indicators. H, which determine the main technological and consumer properties of the content. In practice, content management uses, in particular, such parameters as relevance, completeness, topicality, authenticity, certainty, popularity, profitability. For a web resource, additional parameters can defined that characterize its marketing value and ability to perform certain tasks.

4 Experiment

The analysis of the listed indicators shows that the determination of their exact values in practice is quite difficult, sometimes impossible task. The fact that the content of a web-resource in its properties meets the requirements of the situational management model [6] allows organizing the process of making management decisions based on incomplete, inaccurate or fuzzy values in without losing efficiency and quality of the result. This makes it possible to override the values of the content management parameters by some generic values that allow you to formulate management decisions and evaluate the result of making the appropriate changes. Another feature of parameters that determine content properties is the lack of formal methods and procedures for setting values for many of them. For example, the h_1 content rating is determined based on the h_6 profitability metrics and the h_7 popularity based on the content. The h_1 content rating is a subjective value that is determined by a moderator or administrator of the relevant web property through peer review. This approach creates the possibility of applying the principles of fuzzy logic [6], according to which the results of the assessment are presented in verbal linguistic form and management is carried out on the basis not of their own values, but of their fuzzy analogues. Therefore, in content management IT, instead of the values of the parameters h_1–h_7, their fuzzy correspondences are used h_1^*–h_7^*. The use of fuzzy logic in processes and systems of control implies the operation of the scheme: a clear value → determination of function → phasing (switching to fuzzy) → fuzzy calculations → defasification (moving to exact values). Content features and the use of a situational management model to create IT require another way of generating and applying fuzzy parameter values in management processes. The first step is to form a verbal review directly h_i^*, $i = \overline{1,n}$ the value of the i-th control parameter without setting its exact value and determining the membership function. Forms of presentation and interpretation of fuzzy estimates of different parameters are different. To share them in content management processes and tools, the next step is to standardize them - summarizing various fuzzy parameter values h_1^*–h_7^* to a single syntax and interpretation. Normalization is carried out by semantic differentiation [6] using a special scale. Such action involves the replacement of verbal assessment h_i^*, $i = \overline{1,n}$ numeric value d_i^*, $i = \overline{1,n}$ taking into account the content and the mutual correlation of linguistic meanings. In this case, numerical values do not define any quantitative concepts, but only formalize the corresponding fuzzy verbal estimates and the relationships between them. Consider the use of a scale of semantic differentiation in the range of numerical values [0; 1] (Table 3, 4, 5, 6, 7, 8 and 9, Fig. 2). For hierarchical presentation and further analysis of content parameters, a method of hierarchy analysis is used, which consists of decomposing the problem into simpler components. In addition, it is further processing the sequence of evaluation of the components of the problem by experts and decision makers, by paired. To make subjective paired comparisons of Web resource management parameters with respect to their effect (weight or intensity) on their overall characteristics (optimal Web resource management), the scale given in the Table 4. The result is the relative degree (intensity) of interaction of elements in the hierarchy. These estimates are quantitative. The method of hierarchy analysis consists of procedures for synthesis of multiple estimates, obtaining priority of parameters (criteria) and finding alternative solutions. The values obtained in this way are ratios on a scale and correspond to so-called rigid estimates. The solution to this problem is a systematic process of setting priorities (Table 5). The first stage identified the most important elements of the problem (Table 6), the second step is the best way to test the observations, experiments and evaluation of the items. The next step is to create ways to apply the solution and evaluate its quality. The proposed method systematizes the process of solving the multi-stage problem of managing a Web resource.

Table 5 shows a paired comparison of the parameters at the lower level. In Table 7, the + sign indicates a very strong advantage, the +/- parameter has a significant ad-vantage, and the + / - value is given a moderate parameter when managing a Web resource.

Table 3. Variants of linguistic and normalized content parameter estimates

h_i^*	Parameter	Contents of the parameter	Linguistic meanings	Value
h_1^*	Topicality	An indicator of the relevance of the data and information that contains the content to the value currently valid	Actual	1
			Partially actual	0.5
			Out of date	0
h_2^*	Completeness	A measure of the availability of content in the content of the web-resource that meets the needs of all categories of users	Full	1
			Quite full	0.75
			Not full enough	0.5
			Incomplete	0.25
			Completeness is missing	0
h_3^*	Relevance	Characteristic of the ratio of the total amount of content of a web-resource, with the amount of data required by users	High	1
			Sufficient	0.75
			Average	0.5
			Low	0.25
			Missing	0
h_4^*	Authenticity	A measure that assesses the authorship, propriety, and connection to the origin of information contained in the content of a web resource	Authentic	1
			Partially authentic	0.5
			Not authentic	0
h_5^*	Certainty	An indicator of the relevance of the content to real values and the reliability of the information source	High	1
			Sufficient	0.75
			Average	0.5
			Low	0.25
			Missing	0
h_6^*	Popularity	A demand characteristic that determines the number of effective visits to a web resource	High	1
			Sufficient	0.75
			Average	0.5
			Low	0.25
			Missing	0
h_7^*	Profitability	Characterizing the amount of financial revenue from the use of commercial content	Super-profitable	1
			Profitable	0.75
			Not unprofitable	0.5
			Unprofitable	0.25
			Critical	0
h_8^*	Uniqueness	A measure of the quality of the content that is present in the content of the web resource of the original copyright data	High	1
			Good	0.75
			Satisfactory	0.5
			Unsatisfactory	0.25
			Missing	0

With Table 5 yields eight 3×3 judgments matrices, as there are eight second-level criteria and three A situations (preference is given to parameters 1–3), B (important parameters 4–6), C (preference is given to parameters 7–8), which are compared in pairs by each parameter (Table 7).

From the group of matrices of paired comparisons, a set of local priorities is formed that expresses the relative influence of the set of elements on the element of the previous level. (Tables 8, Fig. 2). Found the relative strength, magnitude,

Table 4. Scale of the relative importance of Web resource management parameters

Weight	Definition
1	Equal importance/value of two parameters
3	Slight margin of preference for one over the other
5	Significant or strong superiority of one over the other
7	Significant advantage. One type of activity is given such a strong advantage that it becomes practically significant
9	Very strong advantage. The obvious advantage of one parameter over another is most strongly confirmed
2, 4, 6, 8	Intermediate solutions between two adjacent parameters. Apply in a compromise case
Reverse quantities	If one of the above numbers (for example, 3) is obtained when comparing one type of parameter with another, then we obtain a reciprocal (that is 1/3) when comparing the second type of parameter with the first

Table 5. Web Resource Management: pairwise comparison matrix for level 2

h_i^*	h_1^*	h_2^*	h_3^*	h_4^*	h_5^*	h_6^*	h_7^*	h_8^*	Vector of priorities	A	B	C
h_1^*	1	4	2	5	6	7	3	8	0.297472	+	+/-	+/-
h_2^*	1/4	1	1/5	3	4	6	1/2	7	0.108468	+	+/-	+/-
h_3^*	1/2	5	1	6	7	8	4	9	0.286965	+	+/-	+/-
h_4^*	1/5	1/3	1/6	1	2	4	1/9	7	0.056587	+/-	+	+/-
h_5^*	1/6	1/4	1/7	1/2	1	3	1/8	5	0.041313	+/-	+	+/-
h_6^*	1/7	1/6	1/8	1/4	1/3	1	1/5	9	0.030128	+/-	+	+/-
h_7^*	1/3	2	1/4	9	8	5	1	6	0.164928	+/-	+/-	+
h_8^*	1/8	1/7	1/9	1/7	1/5	1/9	1/6	1	0.014138	+/-	+/-	+

value, desirability and probability of each individual object by the solution of matrices, each of which has inverse symmetric properties. For this purpose, the set of eigenvectors for each matrix is calculated, and then the result is normalized to one, thus the vector of priorities is obtained. A useful side product of the theory is the so-called consistency index (CI), which gives information about the degree of violation of quantitative (cardinal) and transitive (ordinal) consistency. To improve consistency, it is recommended that you look for more information and view the data used to build the scale. All measurements, including those using parameters and control criteria, have measurement errors as well as errors due to inaccuracies in the measurement mechanism itself. These errors can lead to inconsistent conclusions. For example, parameter A is better than B, B is better than C, but C is better than A. Lack of coherence is a serious factor

Table 6. Web Resource Management: The matrix of priority vector calculations

N	1	2	3	4	5	6	7	8			x	Y
1	1.00	4.00	2.00	5.00	6.00	7.00	3.00	8.00	40320.0	3.76	0.2975	2.66
2	0.25	1,00	0.20	3.00	4.00	6.00	0.50	7.00	12.6	1.37	0.1085	0.94
3	0.50	5.00	1.00	6.00	7.00	8.00	4.00	9.00	30240.0	3.63	0.287	2.64
4	0.20	0.33	0.17	1.00	2.00	4.00	0.11	7.00	0.069136	0.72	0.0566	0.52
5	0.17	0.25	0.14	0.50	1,00	3.00	0.13	5.00	0,00558	0.52	0.0413	0.37
6	0,14	0.17	0.13	0.25	0.33	1.00	0.20	9.00	0.000446	0.38	0.0301	0.32
7	0,33	2.00	0.25	9.00	8.00	5.00	1.00	6.00	360.0	2.09	0.1649	1.63
8	0.13	0.14	0.11	0.14	0.20	0.11	0.17	1.00	0.000001	0.18	0.0141	0.15
	2.72	12.89	4,00	24.89	28.53	34.11	9.10	52.00	L	12.7	1	9.21
	0.81	1.40	1.15	1.41	1.18	1.03	0.13	0.74	IC	0.172169		
OC										0.122105		

limiting the study of some management issues. However, perfect consistency in measurements with even the most accurate instruments is difficult to achieve in practice.

Together with the matrix of paired comparisons, we have a measure of estimating the degree of deviation from consistency in Table 8. When such deviations exceed the set limits, it is necessary for the decision maker to check them in the matrix.

The consistency index in each matrix and for the entire hierarchy is approximately manually obtained. Firstly, each column of judgments is summed up, then we get the product of the sum of the first column and the value of the first component of the normalized vector of priorities, the sum of the second column and the second component etc. Then the resulting numbers are summed up. So we get the value L. For the index $IC = (L - n)/(n - 1)$, where n – number of items to compare. Then compare this value with what we would get by randomly selecting quantitative judgments from a scale $1/9, 1/8, 1/7, \ldots, 1, 2, \ldots, 9$ when forming an inverted symmetric matrix. If we divide by a number that corresponds to the random consistency of the matrix of the same order, we obtain the ratio of the consistency of the OS (Table 7). The OS value should be about 10% or less to be acceptable. In some cases, it is possible to assume 20%, but not more. If the OS goes beyond, then you need to investigate the problem and check your own judgment. The synthesis principle is then used (Table 8, Fig. 2). Priorities are synthesized starting from the second level down. Local priorities are multiplied by the priority of the corresponding higher-level criterion and added each element according to the criteria affected by that element.

Table 7. Content management: pairwise comparison matrix for level 2

Topicality	A	B	C	Priorities vector	Completeness	A	B	C	Priorities vector
A	1	5	9	0.751405	A	1	7	4	0.704936
B	1/5	1	3	0.178178	B	1/7	1	1/3	0.084144
C	1/9	1/3	1	0.070418	C	1/4	3	1	0.21092
L				3.03	L				3.032367
IC				0.014532	IC				0.016183
OC				0.025055	OC				0.027902
Relevance	A	B	C	Priorities vector	Authenticity	A	B	C	Priorities vector
A	1	6	8	0.753554	A	1	1/5	5	0.234411
B	1/6	1	4	0.181135	B	5	1	5	0.685422
C	1/8	1/4	1	0.065311	C	1/5	1/5	1	0.080167
L				3.135611	L				3.294779
IC				0.067805	IC				0.14739
OC				0,116906	OC				0.25412
Certainty	A	B	C	Priorities vector	Popularity	A	B	C	Priorities vector
A	1	1/2	1/2	0.189709	A	1	1/5	1/7	0.072354
B	2	1	3	0.547216	B	5	1	4	0.642435
C	2	1/3	1	0.263074	C	7	1/4	1	0.28521
L				3.135611	L				3.338932
IC				0.067805	IC				0.169466
OC				0.116906	OC				0.292183
Profitability	A	B	C	Priorities vector	Uniqueness	A	B	C	Priorities vector
A	1	1/4	1/8	0.065311	A	1	1/2	1/3	0.15146
B	4	1	1/6	0.181135	B	2	1	1/4	0.218443
C	8	6	1	0.753554	C	3	4	1	0.630098
L				3.135611	L				3.107847
IC				0.067805	IC				0.053924
OC				0.116906	OC				0,092972

Table 8. Content management: a matrix of defined global priorities

	1	2	3	4	5	6	7	8	Priorities
	0.2975	0.1085	0.2869	0.0566	0.0413	0.0301	0.1649	0.0141	
A	0.7514	0.7049	0.7536	0.2344	0.1897	0.0724	0.0653	0.1515	0.552423597
B	0.1782	0.0841	0.1811	0.6854	0.5472	0.6424	0.1811	0.2184	0.22782074
C	0.0704	0.2109	0.0653	0.0802	0.2631	0.2852	0.7536	0.6301	0.219755605

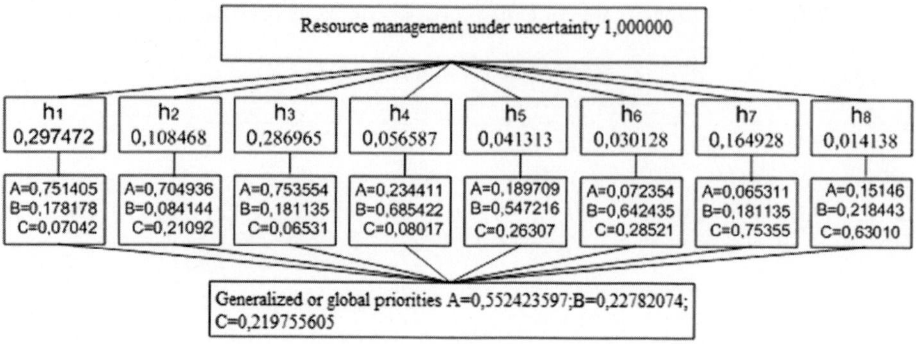

Fig. 2. Hierarchy tree in the case of three alternatives to managing Web resources

5 Results and Discussion

One, multiplies each element of the second level that is, by the weight of a single higher-level destination. This gives a global priority to this element, which is then used to explore the local priorities of the elements that are compared to it as a criterion. The procedure continues to the lowest level. The collection $y_j = \{y_{1j}, y_{2j}, \ldots, y_{gj}\}$ describes the functioning of ECCS with the main processes of processing Web resources as the formation, management and maintenance of content. The ECCS statistics are analysed according to the analysis of reactions to this system of permanent / potential users (rating, recording, keyword search). This helps to effectively analyse the response of the target and potential audience. For a detailed analysis of the functioning of the ECCS newspaper and electronic journal, eight different systems have been designed and implemented to provide different content lifecycle stages for each of them. That is, not all components have developed for the various systems implemented or no subsystems for processing information resources have been developed, such as content generation, management and maintenance. Table 9 presents a list of implemented ECCS indicating the existence of implemented subsystems for processing information resources with maintenance of the life cycle of commercial content. In the Tables 10–11 shows the performance of the developed systems according to Google Analytics over a period of time 2015–2020 years.

Analysis of the results of commercial content support allows determining the reasons for the formation of the target audience by the set of characteristics of the functioning of the ECCS. Figures 3–4 presents the results of the developed systems in the form of graphs, from which it follows that in the presence of all stages of the content lifecycle, the volume of visits and unique users is significantly increased. By adjusting the content set, its uniqueness, timely formation and adequate management according to the individual needs of the regular user, it is possible to model the boundaries of the target social audience and the number of unique search engine visitors. The practical value of the results is that commercial content creation, management and maintenance software is devel-

Table 9. Implementation of information resources processing in developed ECCS

Information resource	Type	Resource address	α	β	γ
Photo gallery	Journal	fotoghalereja-vysocjkykh.com	+/–	+	+/–
Students news	Journal	victana.lviv.ua	–	+/–	+
Poems	Journal	tatjana.in.ua	–	+/–	+/–
Lviv news	Newspaper	presstime.com.ua	+/–	+	+
Ukraine news	Newspaper	vgolos.com.ua	+	+	+
Lviv school	Newspaper	lviv3.in.ua	–	–	–
Ukraine business	Newspaper	dobryjranok.com	+/–		+/–
Currency rates	Newspaper	kursyvalyut.com	+	+	–
Business news	Newspaper	goodmorningua.com	+/–	+/–	–
Auto news	Journal	www.autochip.vn.ua	–	+	+/–

Table 10. The output value: number of visits

Number of visits	Visiting	Page views	Visiting	Page views
1	1 104 194	1 747 476	35.77%	24.89%
2	248 146	474 620	8.04%	6.76%
3	126 535	267 233	4.10%	3.81%
4	84 933	187 771	2.75%	2.67%
5	64 765	149 478	2.10%	2.13%
6	52 896	124 112	1.71%	1.77%
7	45 104	107 373	1.46%	1.53%
8	39 476	96 071	1.28%	1.37%
9–14	172 195	431 313	5.58%	6.14%
15–25	191 648	504 396	6.21%	7.19%
26–50	239 860	664 731	7.77%	9.47%
51–100	233 159	684 453	7.55%	9.75%
101–200	201 626	630 126	6.53%	8.98%
201+	282 272	950 916	9.14%	13.55%
	3 086 818	7 020 069	–	–

Table 11. Behaviour new visitors and returners

Visitor type	Visiting	Contribution to the total value:
New visitor	1 104 194	35.77%
Returning visitor	1 982 615	64.23%

Table 12. Technology

Name	Value
Visiting	3 086 819 %
Page views	2.27
Average length of visit	00:03:02
% new visits	35.73%
% disclaimer	63.39%

oped and implemented to achieve greater owner-level impact (increased profitability, increased user interest) and user (clarity, simplification of the interface, unification of processes of processing of information resources and expansion of the choice of functionality) e-commerce content systems (Table 12).

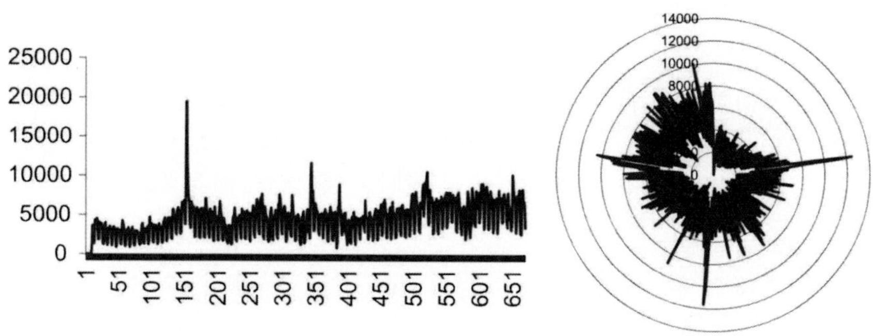

Fig. 3. Daily distribution of the number of visitors

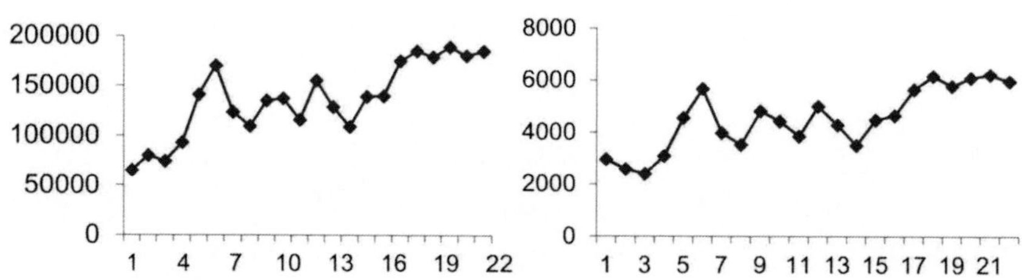

Fig. 4. Monthly distribution of visitors and distribution of average number of monthly visitors

6 Conclusions

The article addresses the problem of developing methods and software for managing information products in the form of a theoretically grounded concept by automating the processing of information resources in ECCS to increase the volume of content sales to a regular user, actively attract potential users and expand the boundaries of the target audience. The scientific novelty of the work is that in the process of theoretical and experimental research such new scientific results are obtained. Content creation, management and maintenance methods were first developed, based on the principles of universality and generality through linguistic content analysis, which enabled the implementation of standardized software tools for processing information resources in e-commerce systems. The structure of ECCS has improved, which is based on the modules of processing information resources and through the distribution of processes of formation, management and maintenance of content, which enables the realization of stages of its life cycle. The model of commercial content lifecycle has further developed, which, thanks to the operational organization of the process of processing information resources in ECCS, increase the efficiency and profitability of the functioning of similar systems. The prospect of further research is to apply the proposed approach to develop recommendations for the design of the ECCS structure by detailing the stages of processing information resources and implementation of subsystems of formation, management and support of commercial content. It allow to effectively simply supporting the content lifecycle at the system developer level (reducing resources and development time, improving system performance).

References

1. EMC, IBM, Microsoft: Content Management Interoperability Services. Appendices. Version 0.5. Hopkinton (2010)
2. Babichev, S., Korobchynskyi, M., Mieshkov, S., Korchomnyi, O.: An effectiveness evaluation of information technology of gene expression profiles processing for gene networks reconstruction. Int. J. Intell. Syst. Appl. **10**(7), 1–10 (2018). https://doi.org/10.5815/ijisa.2018.07.01
3. Babichev, S., Lytvynenko, V., Korobchynskyi, M., Taif, M.A.: Objective clustering inductive technology of gene expression sequences features. Commun. Comput. Inf. Sci. **716**, 359–372 (2017). https://doi.org/10.1007/978-3-319-58274-0_29
4. Bublyk, M., Rybytska, O.: The model of fuzzy expert system for establishing the pollution impact on the mortality rate in Ukraine. In: Computer Sciences and Information Technologies (CSIT), vol. 1, pp. 253–256 (2017). https://doi.org/10.1109/STC-CSIT.2017.8098781
5. Cherednichenko, O., Vovk, M., Kanishcheva, O., Godlevskyi, M.: Studying items similarity for dependable buying on electronic marketplaces. In: CEUR Workshop Proceedings, vol. 2136, pp. 78–89 (2018)
6. Cherednichenko, O., Vovk, M., Kanishcheva, O., Godlevskyi, M.: Towards improving the search quality on the trading platforms. In: 11th SIGSAND/PLAIS. LNBIP, vol. 333, pp. 21–30. Springer, Heidelberg (2018). https://doi.org/10.1007/978-3-030-00060-8_2

7. Chyrun, L., Burov, Y., Rusyn, B., Pohreliuk, L., Oleshek, O., Gozhyj, A., Bobyk, I.: Web resource changes monitoring system development. In: CEUR Workshop Proceedings, vol. 2386, pp. 255–273 (2019)
8. CM: Lifecycle poster (2020). http://www.cmprosold.org/resources/poster/
9. Content, T.: Management possibilities poster (2020). http://metatorial.com/pagea.asp?id=poster
10. Demchuk, A., Lytvyn, V., Vysotska, V., Dilai, M.: Methods and means of web content personalization for commercial information products distribution. Adv. Intell. Syst. Comput. **1020**, 332–347 (2020). https://doi.org/10.1007/978-3-030-26474-1_24
11. Gozhyj, A., Kalinina, I., Gozhyj, V., Vysotska, V.: Web service interaction modeling with colored petri nets. In: International Conference on Intelligent Data Acquisition and Advanced Computing Systems: Technology and Applications, IDAACS, vol. 1, pp. 319–323 (2019). https://doi.org/10.1109/IDAACS.2019.8924400
12. Hackos, J.: Content Management for Dynamic Web Delivery. Wiley, Hoboken (2002)
13. Halvorson, K.: Content Strategy for the Web. New Riders Press, Reading (2009)
14. Kis, Y., Chyrun, L., Tsymbaliak, T., Chyrun, L.: Development of system for managers relationship management with customers. In: Lecture Notes in Computational Intelligence and Decision Making, pp. 405–421. No. 1020 (2020). https://doi.org/10.1007/978-3-030-26474-1_29
15. Kuzmin, O., Bublyk, M.: Economic evaluation and government regulation of technogenic (man-made) damage in the national economy. In: Computer Sciences and Information Technologies (CSIT), pp. 37–39 (2016). https://doi.org/10.1109/STC-CSIT.2016.7589863
16. McGovern, G., Norton, R.: Content Critical. FT Press, Upper Saddle River (2001)
17. McKeever, S.: Understanding web content management systems: evolution, lifecycle and market. In: Industrial Management & Data Systems (MCB UP), vol. 103 (2003)
18. Nakano, R.: Web content management: a collaborative approach. Boston, USA (2002)
19. Papka, R.: On-line News Event Detection, Clustering, and Tracking. Amherst (1999)
20. Rockley, A.: Managing Enterprise Content: A Unified Content Strategy. New Riders Press, Reading (2002)
21. Stone, W.: Plagiarism, duplicate publication and duplicate submission: they are all wrong! In: IEEE Antennas and Propagation, vol. 45 (2003)
22. Sullivan, D.: Invisible web gets deeper (2020). http://searchenginewatch.com/sereport/inproceedings.php/2162871
23. Vysotska, V., Fernandes, V., Emmerich, M.: Web content support method in electronic business systems. In: CEUR Workshop Proceedings, vol. 2136, pp. 20–41 (2018)
24. Woods, R.: Defining a model for content. http://www.contentmanager.net/magazine/inproceedings_785_defining_a_model_for_content_governance.html

Car Parking Data Processing Technique for Smart Parking System as Part of Smart City

Svitlana Popereshnyak(✉) and Iryna Yurchuk

Taras Shevchenko National University of Kyiv, Kyiv, Ukraine
spopereshnyak@gmail.com, i.a.yurchuk@gmail.com

Abstract. The most useful property of any smart system is to have such characteristics as to simulate the process of thinking in order to simplify activities. This Smart parking includes data on location of the car with optimal topology of sensors placement, violation created, and some additional services. The results of it are list of tariffs, list of clients registered in the system, list of registered equipment, list of zones in which the system operates, and list of violations. It includes the following services: active service (for replacing the paper ticketing system and provide instructions to drivers, giving car drivers the ability to use their RFID tag to manage parking), car search service, intelligent parking, regulation, parking security services, availability status, parking lot service maintenance, parking control for automatic door opening, and checking remote access via the Internet and/or GSM network for real-time monitoring. But it is not intended for a specific city, and is a universal solution that can be adapted to any city.

Keywords: Smart parking · Internet speech · Wireless sensor network · RFID · Switch · Sensoracoustic emission signal

1 Introduction

Smart Parking system is a typical component of Smart City. Smart parking is a good illustration of how to make the Internet of Things (IoT) a part of our daily lives. The IoT can be used, for example, to give a driver access to a Smart Parking system through a mobile application to find and reserve a parking space in any parking area. This makes it possible to pay or prepay for a credit card. The Smart Parking system allows you to find free parking, find out if a parking space is still available, and make the decision to reserve and prepay for such parking space. One of the Smart Parking system implementations is parking monitoring, where each parking lot is equipped with a sensor (camera or presence sensor) to detect the presence/absence of vehicles to create an availability map that can be used to manage parking, reservation, and other services. An important aspect of Smart Parking system deployment is creating the optimum topology

S. Babichev et al. (Eds.): ISDMCI 2020, AISC 1246, pp. 183–194, 2021.
https://doi.org/10.1007/978-3-030-54215-3_12

of wireless sensor networks in the parking lots. A software that calculates the optimum wireless sensor network topology will save not only the cost of network equipment but also relieve network engineers from this task, which is an absolute benefit. Such software will also allow you to modify the configuration of wireless devices, as well as to develop and upload any parking scheme described in the correct format as a JSON or XML file. This will allow the application to be used for any network devices and different parking configurations.

2 Relevance of the Topic

In modern car parks, there is no single system for monitoring parking spaces. The main difficulty in creating a single global Smart Parking system is the lack of standardization in software and network tools when building IoT systems. Such a system allows you to monitor in real-time and reserve the right place for the specified time; therefore, it will significantly reduce the time and fuel costs when searching for a parking place. As the deployment of the Smart Parking system requires qualified network engineers and expensive equipment it may seem unprofitable for private owners of parking lots or for state-owned enterprises. On the other hand, an algorithm for solving the problem of optimal sensor arrangement can help automate this process and significantly reduce the cost of Smart Parking system deployment, allowing less time to be spent on network design for engineers.

3 Smart Parking Systems Analysis

Two parking management models have been considered using: parking access (gate monitoring) control and parking space monitoring. Based on entry and exit records, intelligent gate-based parking management can provide important services for drivers, such as the ability to check the availability of parking space and book online. In smart parking management based on monitoring, each parking space is equipped with a sensor (camera sensor or presence sensor) to detect the presence or absence of vehicles to create an accessibility map. The authors of [4,9,11,13,16] propose smart parking managers using RFID technology and embedded systems to help the user check parking availability online with a image detection. [1,2,15] describes a wireless sensor network with a wireless sensor. The authors of [7,12] propose a system with several parking systems to extend scaling services to more than one parking lot. In [14], the authors propose KATHODIGOS. Solution [5] proposes a parking management system called "iParking". In [3], the authors describe a street parking system based on wireless sensor networks. The system can monitor the status of each parking space by deploying a magnetic sensor unit in space. In [6], the optimal placement of static sensors on a network is used to help the agent navigate the area. To study different methods of the correct grid for sensor networks, [8] outputs in each case (square, triangular, and hexagonal) the minimum number of sensors required to ensure full coverage. In divides the investigated landscape into cells and then analyzes how they should be distributed between the cells to avoid network bottlenecks and data loss. Two algorithms are presented in [10], which select locations for a network of sensors with a minimum number of nodes using a mesh model for the terrain and probabilistic model coverage of the sensors.

4 Smart Parking System Prototype Service

A. *Services to be provided by the smart parking system include*:
- Active service: In the current implementation, active RFID technology was used to replace the paper ticketing system and provide instructions to drivers, giving car drivers the ability to use their RFID tag to manage parking.
- Car search service.
- Intelligent parking regulation.
- Parking security services.
- Availability status.
- Parking lot service maintenance.
- Parking control for automatic door opening.
- Checking remote access via the Internet and/or GSM network for real-time monitoring.

B. *The main problems in the implementation of this system include:*
- Sensor selection: Choosing the right detection technology is a complex issue that depends on smart parking efficiency. Size, reliability, adaptation to environmental changes, reliability and cost are all parameters that typically play a key role in selecting the type of sensors for an intelligent parking system.
- Problems with the I2C protocol: One of the major problems is to change the I2C for our application because it is not designed for long-distance travel. Some changes to some I2C parameters are needed, such as impedance and clock speed.
- Issues of energy supply. Another problem was connecting the power supply to all sensors with efficient power consumption.
- Potential disadvantages. When deployed outdoors, ultrasonic sensors are known to be sensitive to changes in temperature and extreme air turbulence. Potential deficiencies associated with internal wireless communications, such as obstacles, reflections, diffraction, and wall penetration, are issues that must be addressed when deploying Smart Parking indoors.
- Effective wireless coverage: For internal and external deployment, effective wireless coverage is an important parameter that determines the lifespan of an intelligent parking system. This is a major problem solved in this system.

5 Smart Parking Network Topology

The smart parking system discussed in this document was built on a sensor method and developed on the basis of a multilayered structure that provides modularity and scalability, as well as providing different services to different users of the parking system. The structure shown in Fig. 1. includes four levels: sensor, network, intermediate and application layer.

A. *Sensor layer*
   This layer defines the platform on which sensor devices are built into the parking lot to detect the presence/absence of a car, and RFID devices located at the car gate and strategic parking points are used to identify vehicles based on the unique mapping between RFID tags and vehicles.
B. *Network layer*
   At this layer, various communication methods were proposed to maintain communication from master and anchor sensors to the gateway and from the gateway to parking users (drivers, remote users and parking owners).
C. *Intermediate layer*
   This is a layer in which situation recognition is accomplished through intelligent algorithms and efficient visualization technologies for service delivery and user convenience.

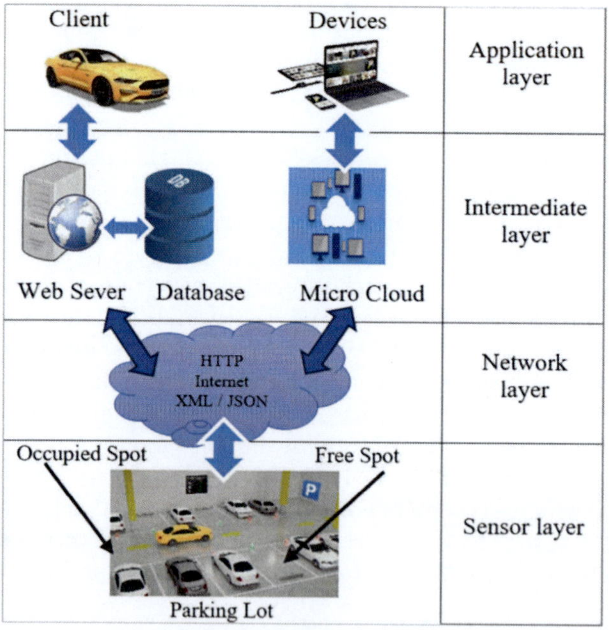

Fig. 1. Network structure

D. *Application layer*
An application layer is a level in which different services are defined and provided to different users. Client devices are connected via the TCP/IP protocol to the parking database. The latest is updated in real time with the status of parking lots.

6 Optimal Sensors Placement Model

Let assume that parking forms a square with $L \times L$ cells. Any cell is either covered or uncovered. The case of partly covered (uncovered) cell is not considered. Every sensor covers all cells which are contained at its sensitivity range S_R and connects to other sensors which are at its connecting range C_R. Optimal placement of sensors and shells is the main problem of their network at parking. We denote by $N = L \times L$ a number of accessible cells. Let C_i be the i-th cell of parking, then its coordinates can be calculated using the following formulas:

$$x(C_i) = \begin{cases} i \mod L, & if\ i \mod L \neq 0 \\ L, & if\ i \mod L = 0 \end{cases}$$

$$y(C_i) = \begin{cases} (i \,\mathrm{div} L) + 1, & if\ i \mod L \neq 0 \\ i \,\mathrm{div} L, & if\ i \mod L = 0 \end{cases}$$

where $(m \,\mathrm{div} n) = q$, $(m \mod n) = r$ if $m = nq + r$. A distance between two cells can be calculated by the following formula:

$$d(C_i, C_j) = \sqrt{(x(C_i) - x(C_j))^2 + (y(C_i) - y(C_j))^2)}.$$

For every C_i, the values x_i, s_i, y_i and z_{ij} can be calculated by using the following formulas:

$$x_i = \begin{cases} 1, & if\ C_i \text{ contains a sensor,} \\ 0, & if \text{ otherwise.} \end{cases} \tag{1}$$

$$s_i = \begin{cases} 1, & if\ C_i \text{ contains a shell,} \\ 0, & if \text{ otherwise.} \end{cases} \tag{2}$$

$$y_i = \begin{cases} 1, & if\ C_i \text{ is covered,} \\ 0, & if \text{ otherwise.} \end{cases} \tag{3}$$

$$z_{ij} = \begin{cases} 1, & if\ C_i, C_j \text{ contain sensors (shells) and } d(C_i, C_j) \leq C_R, \\ 0, & if \text{ otherwise.} \end{cases} \tag{4}$$

So the problem is formulated as follows:

$$\max\{\sum_{i=1}^{N} y_i - \sum_{i=1}^{N}\sum_{j=1}^{N} z_{ij} d(C_i, C_j)\} \tag{5}$$

subject to

$$\sum_{i=1}^{N} x_i = N_S, \tag{6}$$

$$\sum_{i=1}^{N} s_i = N_{Sh}, \tag{7}$$

$$x_i + s_i \leq 1, \ \forall\, i, \tag{8}$$

$$x_i \leq s_j, \ \forall\, i\,, j \text{ such that } d(C_i, C_j) \leq S_R, \tag{9}$$

$$y_i \leq \sum_{j,\ d(C_i,C_j) \leq S_R} x_j, \tag{10}$$

$$x_i + s_i \leq \sum_{j,\ d(C_i,C_j) \leq S_R} x_j, \tag{11}$$

$$z_{ij} \leq x_j, \ \forall i,\ j \text{ such that } j > i \text{ and } d(C_i, C_j) \leq S_R, \tag{12}$$

$$z_{ij} \leq x_j + s_j, \ \forall i,\ j \text{ such that } j > i \text{ and } d(C_i, C_j) \leq S_R, \tag{13}$$

$$z_{ij} \geq x_j + s_j + x_i - 1, \ \forall i,\ j \text{ such that } j > i \text{ and } d(C_i, C_j) \leq S_R, \tag{14}$$

$$x_i \leq \sum_j z_{ij}, \tag{15}$$

$$s_i \leq \sum_j z_{ij}, \tag{16}$$

where N_S be a number of sensors, N_{Sh} – a number of shells, S_R – sensitivity range and C_R – connecting range. Inequality (8) means that there is either a sensor or a shell at the i-th cell. Inequalities (9)–(10) define a parking covering. For (11), if there is either a sensor or a shell at the i-th cell, there is a sensor at an acceptable distance. Inequalities (12)–(16) form function AND: $z_{ij} = 1$ if there is either a sensor or a shell at the j-th cell.

We have to remark that the less distances between sensors and shells are, the more their operating time is. So, for best estimation on operating time favor, the taxicab distance $d_1(C_i, C_j) = |x(C_i) - x(C_j)| + |y(C_i) - y(C_j)|$ have to be used as a distance between cells. If there is a need for cost savings, we have to use Chebyshev distance $d_2(C_i, C_j) = max\{|x(C_i) - x(C_j)|, |y(C_i) - y(C_j)|\}$ as a distance between cells.

7 Requirements for Functions (Tasks) Performed by the System

The developed tool should allow the user to perform the following functions:

- Creating, viewing and editing parking tariff information;
- Creating, viewing and editing information about the areas where the system is used;

- Creating, viewing and editing information about equipment (parking meters, ticket vending machines, ticket dispensers, cameras, tow trucks, etc.) at the sites for placing vehicles;
- Creating, viewing and editing information about additional services that the client can use;
- Display data on the map;
- Create, view, and edit information about system users
- Setting access rights for system users;
- Reviewing data analytics;
- Creating custom criteria analytics.

Input data: data on parking lots; data on areas where the system is applied; data on tariffs for services; data on violations and penalties; list of additional services and their price; list of equipment used in the process of operation; list of employees of control bodies who are users of the system.

Results: list of tariffs; list of clients registered in the system; list of registered equipment; list of zones in which the system operates; list of violations.

8 System Design

A. *Support types requirments*
Smart Parking Manager (SPM) – it is a professional-oriented complex that includes technical, information and software in Fig. 2.

Fig. 2. System components

1. Technical support
The main requirement for technical support is the availability of a device with an Internet connection and a modern browser installed. As the product being created is a web application, it is important to note the server side features. This environment will potentially perform operations on a large amount of information, which will increase over time. Therefore, you need a powerful server with the ability to expand and increase the amount of RAM and storage media on your hard drives.
2. Informational support
SPM is based on the use of a database that contains the information required to perform the duties of the workplace. SPM will store data on tariffs, penalties, penalties, parking areas, employees who are users of the system. The database must be designed to meet the requirements for system response to requests.

B. *Functional design*
In designing the structure of the created tool, the following list of actors was highlighted:
- Parking Service Manager;
- State supervisory authorities representative;
- Law enforcement officer;
- Parking attendant.

All are listed actors should have basic computer skills. The CRUD technique was used to structure the use case (BB) model during design. This abbreviation stands for "Create", "Read", "Update", "Delete". CRUD is the only use that groups operations over homogeneous data. For each operation, it has its own slave flow. When we build an API, we want our models to provide four basic types of functionality. The model must be able to create, read, update and delete resources. Computer scientists often refer to these functions as the acronym CRUD. The CRUD paradigm is common in web application creation as it provides an unforgettable framework for reminding developers how to create full-fledged utility models [17].

C. *Domain System Analysis*
The user information system is based on the information model of the process of interaction of equipment, employees of parking services, vehicles and drivers. Information support should solve the following tasks for its professional activity - to give convenient access to data from various components of the general system. The main function of Smart Parking Software is to provide for the administration of vehicle parking processes and their placement on fine sites (see Fig. 3). Thus, in the system analysis of software, there are two areas of business process research to be distinguished: it is the analysis of monitoring equipment data and process setup.

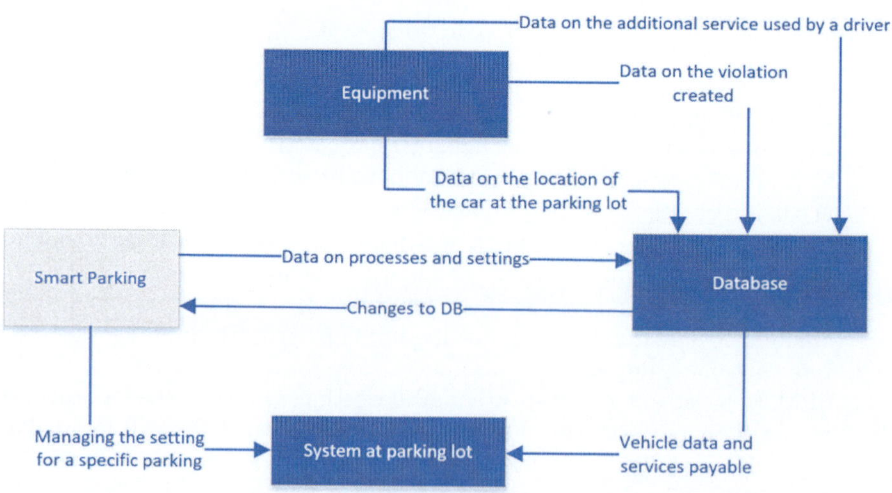

Fig. 3. Smart Parking System place in the parking process as a part of Smart City

There are following ways for data to get into database:

1. From the equipment installed on parking lots and in the city. Smart City technology has sensors installed around the city to collect data.
2. Introduced by mobile users at the parking lot or registration of the offense. The task of the created system - processing, analysis and management of the received data.

D. *Justification of the choice of technology implementation of the software*
The system created will essentially consist of two parts: the first of which is a directly developed web application with a user interface, and the second - a DBMS, which will be the basis of this program - connection and interaction with the database. The completeness of software and user interfaces largely depends on the technology used to develop them. Modern development technologies have reached a considerable level of development in providing interactive, convenient and effective tools both for the development of software systems and for using the results of their work. Therefore, authors chose:

- PostgreSQL DBMS is a freely distributed object-relational database management system (ORDBMS), the most developed of open-source databases in the world and is a real alternative to commercial databases.
- Node.js was selected to develop the server side of the toolbox. An open source platform for executing high-performance web applications written in JavaScript.

Angular framework - Google's framework for creating client applications - was chosen to write the client part. First of all, it aims at developing SPA solutions (Single Page Application), that is, one-page applications.

9 Description of Working with System Modules

After successful authorization, the user goes to the menu and map page. The menu contains all the system functionality that is available for use. By default, the first is a map with designated areas, parking areas, penalty areas and short description equipment. The Projects tab contains information about the areas registered in the system. As the system receives data from a large number of equipment, the user has the opportunity to view and edit all available equipment, this is necessary to go to the page "Assets" (see Fig. 4).

Other tabs include a specific area map (M.A.P.S.) and analytics (Dashboard and Analitics) (see Fig. 5):

This section discusses how to use the created system. This made it possible to understand the capabilities of this system. The system created has a user-friendly and user-friendly interface that allows you to use the system quickly and easily.

10 SWOT-Analysis of the Project

Strengths: Ability to customize network device settings, import and export parking scheme with network topology, portability.

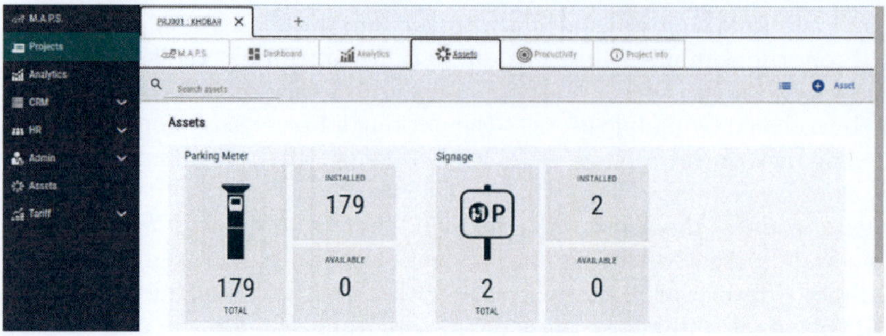

Fig. 4. Equipment in the area overview

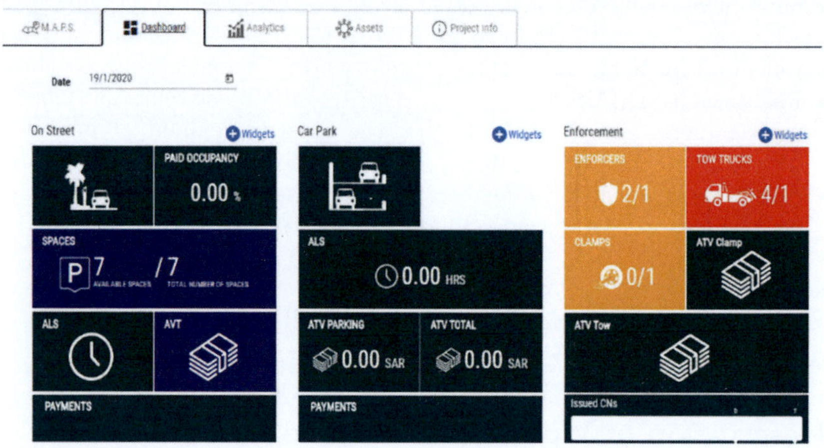

Fig. 5. View area statistics

Weaknesses: High complexity of project implementation due to the need for constant refinement of the latest network technologies.
Opportunities: The rapid growth in demand for Smart Parking deployment tools, the emergence of new technologies, integration with existing popular applications, the rapid growth of market growth.
Threats: Creation of new technical means, competition, changing user needs, market entry of alternative products, slowing down of market growth.

11 Conclusions

Modern rates of development of Smart City technologies indicate the need for active introduction of new methods of work in everyday life. Parking data processing processes are no exception.

To date, systems are being developed that simplify the parking process for users (vehicle drivers), but there is a need to automate the administrative part of this process. This improves the work of both operators and city authorities, who will be able to control and adjust the process of placing vehicles in designated areas. It will also be a major contributor to the fight against parking fraud or fines. The system created solves these problems.

It was built on a sensor method based on a multilayered structure which consists of four levels: sensor to detect the presence/absence of a car, and RFID devices located at the car gate and strategic parking points are used to identify vehicles based on the unique mapping between RFID tags and vehicles; network to maintain communication from master and anchor sensors to the gateway and from the gateway to parking users; intermediate to recognize by intelligent algorithms and efficient visualization technologies for service delivery and user convenience; application layer to make a difference for services are defined and provided to different users.

Smart City technologies should be noted that the system is not intended for a specific city, but is a universal solution that can be adapted to any city. It is already being deployed in many cities around the world, and the system will be able to find a niche and improve performance for cities that implement them.

References

1. Benson, J., O'Donovan, T., O'Sullivan, P., Roedig, U., Sreenan, C., Barton, J., Murphy, A., O'Flynn, B.: Car-park management using wireless sensor networks. In: Proceedings of the 2006 31st IEEE Conference on Local Computer Networks, Tampa, FL, USA, pp. 588–595 (2006). https://doi.org/10.1109/LCN.2006.322020
2. Bi, Y., Sun, L., Zhu, H., Yan, T., Luo, Z.: A parking management system based on wireless sensor network. Acta Autom. Sin **32**, 968–977 (2006)
3. Biagioni, E., Sasaki, G.: Wireless sensor placement for reliable and efficient data collection. In: Proceedings of the 36th Annual Hawaii International Conference on System Science, Big Island, HI, USA (2003). https://doi.org/10.1109/HICSS.2003.1174290
4. Bong, D., Ting, K., Lai, K.: Integrated approach in the design of car park occupancy information system (COINS). IAENG Int. J. Comput. Sci. **35**, 7–14 (2008)
5. Chinrungrueng, J., Dumnin, S., Pongthornseri, R.: Iparking: a parking management framework. In: Proceedings of the IEEE 11th International Conference on ITS Telecommunications (ITST), St. Petersburg, Russia, pp. 63–68 (2011). https://doi.org/10.1109/ITST.2011.6060133
6. Dhillon, S., Chakrabarty, K.: Sensor placement for effective coverage and surveillance in distributed sensor networks. In: Proceedings of the 2003 IEEE Wireless Communications and Networking Conference, New Orleans, LA, USA, vol. 3, pp. 1609–1614 (2003). https://doi.org/10.1109/WCNC.2003.1200627
7. Geng, Y., Cassandras, C.: A new 'smart parking' system infrastructure and implementation. Procedia-Soc. Behav. Sci. **54**, 1278–1287 (2012). https://doi.org/10.1016/j.sbspro.2012.09.842
8. Gu, J., Zhang, Z., Yu, F., Liu, Q.: Design and implementation of a street parking system using wireless sensor networks. In: Proceedings of the 10th IEEE International Conference on Industrial Informatics (INDIN), Beijing, China, pp. 1212–1217 (2012). https://doi.org/10.1109/INDIN.2012.6301241

9. Jian, M.S., Yang, K.S., Lee, C.L.: Modular rfid parking management system based on existed gate system integration. WSEAS Trans. Syst. **7**, 706–716 (2008)
10. Jourdan, D., Roy, N.: Optimal sensor placement for agent localization. ACM Trans. Sens. Netw. (TOSN) **4**, 13:1–13:40 (2008). https://doi.org/10.1109/PLANS.2006.1650596
11. Li, X., Ranga, U.: Design and implementation of a digital parking IoT management system. Technol. Interface J. **10**, 1–11 (2009)
12. Lu, R., Lin, X., Zhu, H., Shen, X.: Spark: a new vanet-based smart parking scheme for large parking lots. In: Proceedings of IEEE INFOCOM 2009, pp. 1413–1421 (2009). https://doi.org/10.1109/INFCOM.2009.5062057
13. Pala, Z., Inanç, N.: Smart parking applications using RFID technology. In: Proceedings of the 1st RFID Eurasia Conference, Istanbul, Turkey, pp. 121–123 (2007). https://doi.org/10.1109/RFIDEURASIA.2007.4368108
14. Samaras, A., Evangeliou, N., Arvanitopoulos, A., Gialelis, J., Koubias, S., Tzes, A.: 89 novel smart parking system based on wireless sensor networks. In: Proceedings of the 1st International Virtual Conference on Intelligent Transportation Systems, Slovakia, pp. 140–145 (2013)
15. Tang, V., Zheng, Y., Cao, J.: An intelligent car park management system based on wireless sensor networks. In: Proceedings of the 2006 1st International Symposium on Pervasive Computing and Applications, Urumqi, China, pp. 65–70 (2006). https://doi.org/10.1109/SPCA.2006.297498
16. Yass, A., Yasin, N., Zaidan, B., Zeiden, A.: New design for intelligent parking system using the principles of management information system and image detection system. In: Proceedings of the 2009 International Conference on Computer Engineering and Applications, vol. 2, pp. 360–364 (2011)
17. Zhang, H., Hou, J.: On the upper bound of α-lifetime for large sensor networks. ACM Trans. Sens. Netw. (TOSN) **1**, 272–300 (2005). https://doi.org/10.1145/1105688.1105693

A Lower Bound for Optimization of Arbitrary Function on Permutations

Sergiy Yakovlev[1(✉)], Oksana Pichugina[1], and Liudmyla Koliechkina[2]

[1] National Aerospace University "Kharkiv Aviation Institute", Kharkiv, Ukraine
svsyak7@gmail.com, oksanapichugina1@gmail.com
[2] University of Lodz, Uniwersytecka Str. 3, 90-137 Lodz, Poland
liudmyla.koliechkina@wmii.uni.lodz.pl

Abstract. A generic constrained permutation-based optimization problem is considered. It is reduced to a Euclidean discrete permutation-based optimization problem, which a discrete optimization problem over vertices of the generalized permutohedron. To this Euclidean combinatorial problem, an equivalent permutation-based optimization problem with convex objective function and convex constraints is formed based on applying the convex extensions theory and a vertex locality of a feasible set. A hybrid approach to solving a polyhedral relaxation of this optimization problem is presented. It jointly uses the penalty method and an offered modification of the conditional gradient method. As a result, a lower bound of the global minimum is found in an arbitrary permutation-based problem.

Keywords: Permutation set · Permutohedron · Convex extension · Penalty method · Conditional gradient method

1 Introduction and Literature Review

Recently, in connection with the expansion of tools of mathematical modeling and computer simulation, as well as with developing relevant high-performance software, interest in solving combinatorial optimization problems, being intractable until then has sharply increased [9,18,22,29]. This class of optimization problems includes so-called Euclidean combinatorial optimization problems (*ECOPs*) [31,33] getting this name due to the fact that their feasible combinatorial domain is explored through its image in Euclidean space. As a result of mapping into Euclidean space, the corresponding optimization problems acquire specific properties making it possible to offer new effective methods for solving these problems [17,36,41,42].

In this paper, we consider a class of ECOPs on so-called vertex-located sets (*VLSs*) [33,39,40]. VLSs are combinatorial sets coinciding with a vertex set of their convex hull after mapping in $\mathbb{R}^n$. ECOPs on VLSs cover a wide class of optimization problems, e.g., permutation-based ECOPs [12,13,20,43] and Boolean ones [7,9,18,22,23,29]. This is due to the fact that a set of permutations, both

S. Babichev et al. (Eds.): ISDMCI 2020, AISC 1246, pp. 195–212, 2021.
https://doi.org/10.1007/978-3-030-54215-3_13

with repetitions and without repetitions, mapped into Euclidean space becomes vertex located. The same is true for the Boolean set [18,29,30,33].

Also, note that there exists a strong connection between general ECOPs and the ones on VLSs. Indeed, on the one hand, an arbitrary discrete set $E \subset \mathbb{R}^n$ can be represented as a union of its vertex-located subsets [33,36]. On the other hand, a lifting of E into $\mathbb{R}^N$, where $N > n$, allows us to move on to considering a VLS in $\mathbb{R}^{n'}$ instead of E [33,36].

It turns out. ECOPs on VLSs possess numerous useful properties making it possible to take a fresh look at the issue of a search of a new optimization approach to their solution. For instance, it concerns an opportunity of applying the convex extensions theory [24,25,32,38,40] in optimization on VLSs.

The aim of this paper is to develop an approach of obtaining lower bounds of the objective function in permutation-based optimization problems based on: a) embedding the permutation set into Euclidean space; b) a reformulation these problems as ECOPs, whose polyhedral relaxation is a convex optimization problem over the generalized permutohedron; c) solving the derived problems despite exponential number of constraints in them.

2 Materials and Methods

2.1 Problem Statement

Let $A = \{a_1, ..., a_n\}$ be a set of arbitrary objects, and Π be a set of permutations of A-elements, i.e., an element $\pi \in \Pi$ has a representation

$$\pi = \langle a_{i_1}, ..., a_{i_n} \rangle, \tag{1}$$

where $\{i_1, ..., i_n\} = \{1, ..., n\} = J_n$. Also, let $\xi : \Pi \to \mathbb{R}^1$.

Let us formulate an optimization problem on Π as follows:

$$\min_{\pi \in \Pi' \subseteq \Pi} \xi(\pi), \tag{2}$$

where Π is the whole search space, Π' is the feasible search space of the problem. In accordance with [13,20], to the problem (2), we will refer to as a generic *permutation-based optimization problem* (*PBP*).

Let $\langle x^*, z^* \rangle = \langle x^*, f(x^*) \rangle$ be an optimal solution of EPBP. Then $\langle \pi^*, z^* \rangle = \langle \pi^*, \xi(\pi^*) \rangle$ is an optimal solution of PBP. Respectively, a lower bound z^l of $\xi(\pi)$ on Π' can found as a lower bound of $f(x)$ on E'.

Establish a bijective map ϕ from Π to a finite set $E \subset \mathbb{R}^n$ as following: to any π in the form of (1), a vector

$$x = \phi(\pi) = (i_1, ..., i_n) \in E \tag{3}$$

is associated.

Then $E = \phi(\Pi)$, $\Pi = \phi^{-1}(E)$. Moreover, an image E' of Π' in $\mathbb{R}^n$ inherits a similar property, namely, $E' = \phi(\Pi')$, $\Pi' = \phi^{-1}(E')$. Respectively, PBP (1) is equivalent to a discrete optimization problem

$$\min_{x \in E' \subseteq E} f(x), \tag{4}$$

where $f : E \to \mathbb{R}^1$ such that $f(x) = \xi(\pi)$, $\forall x = \phi(\pi)$, $\forall \pi \in \Pi$, E is the whole search space, E' is the feasible search space of (4).

By (3), E consists of permutations of the elements of J_n viewed as points in $\mathbb{R}^n$. That is why to the problem (4) we will refer to as a generic Euclidean statement of PBP or an Euclidean PBP (*EPBP*).

Map ϕ, given by (3), is not the only way to embed Π into Euclidean space. If in the scope of PBP elements of A are fully characterized by a finite set of numerical parameters, then EPBP allows a generalization to an optimization problem over a set of Euclidean permutation configurations [33] embedded into $\mathbb{R}^{m \times n}$, where m is the number of these numerical parameters. For instance, if $m = 1$, elements of A are defined by a single parameter, say $g \in \mathbb{R}^1$. This means that for any $i \in J_n$ there exists $a_i \to g_i \in \mathbb{R}^1$, where $g_1, ..., g_n$ are not necessarily different. In this case, PBP is equivalent to EPBP, where

$$x = \phi(\pi) = (g_{i_1}, ..., g_{i_n}) \in E. \tag{5}$$

Here, E is a set of permutations (with repetitions) of elements of a multiset

$$G = \{g_1, ..., g_n\} \subset \mathbb{R}^1 \tag{6}$$

including $k \leq n$ different elements, i.e., G is representable in the form of $G = \{e_1^{n_1}, ..., e_k^{n_k}\}$, where $S(G) = \{e_1, ..., e_k\}$ and $[G] = (n_1, ..., n_k)$ are a base and a multiplicity vector of G. To reflect this, we will use a denotation $E = E_{nk}(G)$ omitting: a) k, if $k = n$; b) G, if $G = J_n$. For example, (3) can be written as $x \in E_n$, while (5) becomes

$$x \in E_{nk}(G). \tag{7}$$

Without loss of generality, assume that a multiset (6) satisfies a condition:

$$g_i \leq g_{i+1},\ i \in J_{n-1}. \tag{8}$$

It is easy to see that (4), (5) is EPBP, which generalizes (3), (4). Further, it will be used as a generic EPBP.

Assume that the constraint $E' \subseteq E_{nk}(G)$ is given analytically, thus, the following discrete optimization problem is considered as EPBP (4):

$$\min f(x), \tag{9}$$

subject to

$$f_{i1}(x) \leq 0, i \in I^1; \tag{10}$$

$$f_{i2}(x) = 0, i \in I^2; \tag{11}$$

$$x \in E, \tag{12}$$

where

$$E = E_{nk}(G), \tag{13}$$

$I^1, I^2 \subset \mathbb{N}$,

$$f : P \to \mathbb{R}^1,\ f_{i1} : P \to \mathbb{R}^1, i \in I^1; f_{i2} : P \to \mathbb{R}^1, i \in I^2; \tag{14}$$

$$P = conv\, E. \tag{15}$$

When the EPBP (9)–(13) (further *Problem 1*) is solved to optimality, the Branch and Bound algorithm design paradigm is one of the most common techniques to accomplish this [7,9,12,18,22,23,29,43]. It assumes three complement stages:

- branching, where Problem 1 is replaced by solving subproblems (9)–(11)

$$x \in E^j \tag{16}$$

  (further *Problem 1.j*) for $j \in J \subset \mathbb{Z}$. Here, $E_{nk}(G) = \underset{j\in J}{\cup} E^j$, $E^j \subset E_{nk}(G)$, $E^j \neq E^{j'}$ for any $j \neq j'$, $j, j' \in J$;
- bounding, where upper and lower estimated bounds on the optimal solution z^* are found. In our case, these are a collection $\{z^{lj}\}_{j\in J}$, where z^{lj} is a low bound of $f(x)$ on $E^{j'} = \{x \in E^j$ subject to (10), (11)$\}$ $(j \in J)$;
- pruning, where all branches unable to produce a better solution than the best one found so far are discarded.

Finding the lower bound z^l of the optimum z^* of PBP enabled an equivalent formulation as EPBP in the form of Problem 1 is a goal of the paper.

Due to $z^l = \min_{j\in J} z^{lj}$, elaboration of the bounds z^{lj} for $j \in J$ results in improving the total lower bound z^l of $f(x)$ on E'. On the other hand, the elaboration is a way to reduce a feasible search space by pruning more alleys' branches.

That is why ways to find lower bounds of the objective function on $E_{nk}(G)$ and its proper subsets is an issue that requires deep consideration.

Note that, for any $j \in J$, Problem 1.j is again a EPBP, which can be represented in the form of Problem 1 complemented by additional constraints:

$$f_{i1}(x) \leq 0, i \in I^{1j};\ f_{i2}(x) = 0, i \in I^{2j},$$

which are singled out E^j from E, where $I^1 \cap I^{1j} = \emptyset$, $I^2 \cap I^{2j} = \emptyset$. Therefore, our contribution will be described for the generic EPBP in the form of Problem 1. Respectively, the lower bound z^l will be found. A standard way to find a lower bound of the objective function in discrete optimization problems is switching to their continuous relaxations [22,23,43].

2.2 Relaxation Techniques for EPBP

Problem 1 allows various such continuous relaxations. For instance, a convex hull of $E_k(G)$ is a permutohedron $P = P_n(G) = conv\, E_n(G)$ [43], while convex hull of $E_{nk}(G)$, $k < n$ is a generalized permutohedron $P = P_{nk}(G) = conv\, E_{nk}(G)$ [27, 33]. In the scope of this paper, we will refer to both of them as permutohedrons. An H-representation of the permutohedron is [31,33,43]:

$$\sum_{i=1}^{n} x_i = \sum_{i=1}^{n} g_i; \tag{17}$$

$$\sum_{i\in\omega} x_i \geq \sum_{i=1}^{|\omega|} g_i, \ \forall \omega \subset J_n, \tag{18}$$

which includes equality (17) and $2^n - 2$ inequalities (18). In turn, these inequalities contain $n-1$ collections corresponding to a specific value of $|\omega| \in J_{n-1}$ of the cardinality $C_n^{|\omega|}$, which are enumerated as a 1-st, 2-nd,..., $n-1$-th collection.

Theorem 1. *[31] The system of constraints* (17) *is irredundant if and only if $n_1 + n_k > 2$, where n_1, n_k are multiplicities of g_1 and g_n, respectively. If $n_1 > 1$, the inequality collections with numbers $|\omega| \in J_{n_1}\backslash\{1\}$ are redundant. If $n_k > 1$, the inequality collections with numbers $|\omega| \in J_{n-2}\backslash J_{n-n_k-1}$ are redundant. Extracting these redundant constraints results in forming an irredundant H-representation of $P_{nk}(G)$.*

Also, $E_{nk}(G)$ is inscribed into a family of hyperspheres centered at a ray $(t,...,t)$, $t \in \mathbb{R}^1\backslash\{0\}$. In these family, a hypersphere $S_r(\overline{\tau})$ centered at $\overline{\tau} = (\tau,...,\tau) \in \mathbb{R}^n$ with a radius r,

$$\tau = \frac{1}{n}\sum_{i=1}^{n} g_i, \ r = \left(\sum_{i=1}^{n}(g_i - \tau)^2\right)^{1/2} \tag{19}$$

is a circumsphere on minimal radius [24].

This results in existence of the following polyhedral-spherical representation of this combinatorial set [24,36,42]:

$$E_{nk}(G) = P_{nk}(G) \cap S_r(\overline{\tau}). \tag{20}$$

Formula (20) allows rewriting (12), (13) as follows:

$$x \in P_{nk}(G), \tag{21}$$

$$x \in S_r(\overline{\tau}). \tag{22}$$

Relaxing any of these two conditions, one comes to a continuous relaxation of Problem 1. One of them is a traditional polyhedral relaxation (9)–(11), (21), which is a nonlinear optimization problem on the permutohedron $P_{nk}(G)$ (further Problem 1.PR). Another one is a spherical relaxation of the form of (9)–(11), (22), which is derived from Problem 1 by relaxing (21). If both these problems were solved, the bound z^l can be chosen as $z^l = \max\{z^P, z^S\}$, where z^P, z^S are global minima of $f(x)$ on the constrained space of the polyhedral and spherical relaxations, respectively. So, it is desirable to solve both of the relaxations to optimality. However, generally, the spherical relaxation is a complicated multi-extremal problem [16]. There exist some approaches to solve it [10,11,14] for specific classes of Problem 1, such as a quadratic problem without additional constraints (11), (12) [10]. In this paper, we restrict consideration to a polyhedral relaxation of Problem 1 and its equivalent reformulation.

Problem 1.PR is solvable to optimality if functions (14) are convex on polytope $P_{nk}(G)$, where

$$P = P_{nk}(G). \tag{23}$$

This is due to, in this case, Problem 1.PR is a convex optimization problem. However, generally, this problem is hard [6]. Respectively, treating it with non-linear programming tools one is able to end at a local optimizer, which yields neither lower nor upper bound on z^* [1,2,21].

We claim that there exists a discrete optimization problem (further a *convexified EPBP*, *CEPBP*, *Problem 2*) equivalent to Problem 1, whose polyhedral relaxation (further *Problem 2.PR*) is a convex optimization problem. It will be shown a way to solve Problem 2.PR to optimality despite the exponential on n number of constraints in it. A value z^{**} will be chosen as the lower bound z^l, where $\langle y^{**}, z^{**}\rangle = \langle y^{**}, F(y^{**})\rangle$ is an optimal solution of Problem 2.PR, $F(x)$ is its objective function,

Problem 2 will be found in the form of:

$$\min F(x), \tag{24}$$

subject to constraints (12), (13):

$$F_{i1}(x) \leq 0,\ i \in J^1; \tag{25}$$

$$F_{i1}(x) \leq 0,\ i \in J^2, \tag{26}$$

where

$$\begin{aligned} &F: P \to \mathbb{R}^1,\ F_{i1}: P \to \mathbb{R}^1, i \in J^1; \\ &F_{i2}: P \to \mathbb{R}^1,\ i \in J^2, \end{aligned} \tag{27}$$

are convex and differentiable on polytope (23).

2.3 Properties of Euclidean Combinatorial Problems on a Vertex-Located Set

Let $E \subset \mathbb{R}^n$ be a finite set of points in Euclidean space, such that

$$E = vert\ P. \tag{28}$$

In accordance with [40], to such a set we will refer to as a vertex located set (*VLS*). Consider a problem (9)–(12), (28), which is a generic constrained Euclidean optimization problem on a VLS E (further *VLP*).

First, it will be justified that any VLP allows an equivalent reformulation (12), (24)–(26) (further referred to as *a convexified VLP, CVLP*), where functions (27) are convex and differentiable on polytope (15), which is a convex hull of a VLS E. Respectively, a polyhedral relaxation of CVLP is a convex optimization problem Then these results related to VLSs will be adapted for $E_{nk}(G)$ and CPBP.

Theorem 2. *[38] For any VLS E and a function* $\varphi : E \to \mathbb{R}^1$, *there exists a differentiable convex function* $\Phi : P \to \mathbb{R}^1$ *such that*

$$\Phi(x) = \varphi(x) \text{ for any } x \in E. \tag{29}$$

For functions satisfying condition (29), the following notation will be used:

$$\Phi(x) \underset{E}{=} \varphi(x). \tag{30}$$

A function Φ defined on polytope (15) is an *extension* of φ from an extension domain E onto P, while φ is a *reduction* of $\Phi(x)$ from P to E [8]. Respectively, if Φ is convex (strongly convex, differentiable) on P, it is a *convex* (strongly convex, differentiable) *extension* of $\varphi(x)$ ($\varphi.CE$) from E onto P [25,32,37,38].

Theorem 2 establishes the existence of a convex extension of any function from an arbitrary VLS E onto its convex hull. Regarding VLP, Theorem 2 states that an unconstrained VLP (9), (12), (28) enables the reformulation as a convexified VLP of the form of (12), (24), (28), where $F(x)$ is a convex extension of $f(x)$ from E onto P.

Now, let us figure out whether this observation is true for VLPs with $I^1 \cup I^2 \neq \emptyset$, i.e., for constrained VLPs. For that, a standard concept of an extension of function (see, for instance, [40]) will be generalized for a case where an extension domain is a proper subset of the domain of a function.

Definition 1. *A function* $\Phi : M \to \mathbb{R}^1$ *is called an extension of* $\varphi : N \to \mathbb{R}^1$ *from an extension domain* $K \subseteq N$ *onto* M, *if* $M \supset N$,

$$\Phi(x) \underset{K}{=} \varphi(x).$$

Theorem 3. *A VLP is equivalent to the following CVLP: find a solution of* (24), *subject to constraints* (12), (28),

$$F_{i1}(x) \leq 0, i \in I^1; \tag{31}$$

$$F_{i2}(x) \leq 0, F_{i3}(x) \leq 0, i \in I^2; \tag{32}$$

where $F(x)$ *is a convex extension of* $f(x)$; $F_{i1}(x)$ *is a convex extension of* $f_{i1}(x)$, $i \in I^1$; $F_{i2}(x)$ *is a convex extension of* $f_{i2}(x)$, $F_{i3}(x)$ *is a convex extension of* $-f_{i2}(x)$, $i \in I^2$, *from* E *onto* P.

Proof. Note that a VLP is equivalent to the following problem:

$$\min \varphi(x), \tag{33}$$

subject to constraints (12),

$$\varphi_{i1}(x) \leq 0, i \in I^1; \tag{34}$$

$$\varphi_{i2}(x) = 0, i \in I^2, \tag{35}$$

where

$$\varphi : E \to \mathbb{R}^1, \ \varphi_{i1} : E \to \mathbb{R}^1, i \in I^1; \ \varphi_{i2} : E \to \mathbb{R}^1, i \in I^2, \tag{36}$$

are reductions of functions (14) from P onto E, respectively.

On the other hand, φ, φ_{i1} are reductions of F, F_{i1} from P onto E ($i \in I^1$), while φ_{i2} are reductions of F_{i2} and $-F_{i3}$ from P onto E ($i \in I^2$). This implies that CVLP is equivalent to a problem (12), (33), (34), (36),

$$\varphi_{i2}(x) \leq 0, -\varphi_{i2}(x) \leq 0, \ i \in I^2,$$

which is the problem (12), (33)–(36).

Corollary 1. *If E is a VLS, then a VLP allows an equivalent reformulation in the form of CVLP.*

Proof. By Theorem 3, this CVLP has the following parameters:

$$J^1 = I^1, \ J^2 = I^2 \cup I^3, \ \text{where } I^3 = \{i+m\}_{i \in I^2},$$
$$m = \max_{i \in I^2} i, F_{i+m,2} = F_{i3}, \ i \in I^2,$$

rest are found from Theorem 3.

In accordance with [33,36], a VLS E enables the following description called its *polyherdral-surfaced representation*:

$$E = P \cap S, \tag{37}$$

where S is a full strongly convex surface given by a strictly convex function $h(x)$, i.e.,

$$S = \{x \in \mathbb{R}^n : \ h(x) = 0\},$$

where *conv* S is a strongly convex body with a boundary S [26]. Moreover, existence of such a representation is a necessary and sufficient condition for vertex locality of E [33].

Coming back to consideration of PBPs, by (20), a set $E_{nk}(G)$ enables a polyherdral-surfaced representation, where P is the permutohedron $P_{nk}(G)$ and a surface S is a hypersphere $S_r(\overline{\tau})$. This implies that $E_{nk}(G)$ is vertex located. Respectively, Theorem 3 holds for the whole class PBPs and states the existence of an equivalent CPBP in the form of (12), (13), (24)–(27) for any PBP (9)–(14).

A polyhedral relaxation of CPBP is a convex optimization program, which can be solved theoretically to optimality by relevant methods [3–5,28,35]. The global minimum can serve a lower bound of z^* that we are looking for.

2.4 A Polynomial CPBP Construction

Theorem 3 establishes the existence of a convexified PBP equivalent to any PBP. Let assume that functions (14) are polynomials. For constructing such CPBPs, approaches to forming convex extensions of polynomials from $E_{nk}(G)$ are required. Some of them are already present in the literature. For instance, in [38, 40], constructive iterative algorithms are offered for forming the convex extensions of arbitrary polynomials in the form of nonsmooth functions. For quadratic and cubic polynomials, a close form of polynomial convex extensions of the same degree is presented in [25, 42]. Respectively, their application to constructing CPBPs results in forming equivalent reformulation of PBPs. Here, we offer a way of building smooth convex extensions of polynomials from a permutation set lying in a positive orthant onto $\mathbb{R}^n_{>0}$.

Theorem 4. *[39] For any monomial*

$$f^+_{\mathbf{s}}(x) = \prod_{i=1}^{n} x_i^{s_i} \tag{38}$$

and any multiset (6) *such that*

$$G \subset \mathbb{R}^1_{>0}, \tag{39}$$

there exists a convex infinitely differentiable extension from $E_{nk}(G)$ onto $\mathbb{R}^n_{>0}$ in the form of:

$$F^+_{\mathbf{s}}(x) \;=\; a\eta^{\mu} \prod_{i=1}^{n} x_i^{s_i-\mu},\ a > 0 \tag{40}$$

where

$$\mathbf{s} = (s_1 ... s_n) \in \mathbb{Z}^n_+,\ \eta = \prod_{i=1}^{n} g_i;\ \mu = \max_{i \in J_n} s_i. \tag{41}$$

Theorem 4 establishes the existence of convex extensions of monomials with positive coefficients from $E_{nk}(G) \subset \mathbb{R}^n_{>0}$ in a class of Laurent polynomials [19]. In a combination with a peculiarity of symmetric functions to be constant on $E_{nk}(G)$ [37, 40], this theorem allows establishing the existence of such Laurent polynomial convex extensions for monomials with negative coefficients. For that, let us assume that a monomial needs to be extended i a convex way:

$$f^-_{\mathbf{s}}(x) = -f^+_{\mathbf{s}}(x), \tag{42}$$

where $f^+_{\mathbf{s}}(x)$ is given by (38).

First, build a monomial symmetric polynomial formed from (38) and define its value on $E_{nk}(G)$:

$$\mathcal{F}_{\mathbf{s}}(x) = a \sum_{y \in E_{n\kappa}(\mathcal{S})} \prod_{i=1}^{n} x_i^{y_i} = \sum_{y \in E_{n\kappa}(\mathcal{S})} f^+_{\mathbf{y}}(x) \underset{E_{nk}(G)}{=} \mathcal{F}_{\mathbf{s}}(g), \tag{43}$$

where $\mathcal{S} = \{s_1, ..., s_n\}$, κ is the number of different elements in $\mathcal{S}$, $g = (g_1, ..., g_n)$, $y = (y_1, ..., y_n)$. Then separate $f_{\mathbf{s}}^-(x)$ from the identity (43) and form convex extensions of polynomials with positive coefficients present in this expression by (40). We get

$$\begin{aligned} f_{\mathbf{s}}^-(x) &= \sum_{y \in E_{n\kappa}(\mathcal{S}) \setminus \{\mathbf{s}\}} f_{\mathbf{y}}^+(x) - \mathcal{F}_{\mathbf{s}}(g) \underset{E_{nk}(G)}{=} \\ \underset{E_{nk}(G)}{=} &\sum_{y \in E_{n\kappa}(\mathcal{S}) \setminus \{\mathbf{s}\}} F_{\mathbf{y}}^+(x) - \mathcal{F}_{\mathbf{s}}(g), \end{aligned} \tag{44}$$

where $F_{\mathbf{y}}^+(x)$ is found by (40) for any $y \in E_{n\kappa}(\mathcal{S})$.

Clearly, on the right-hand side of the expression, there is a convex function, which is, in addition, is a convex extension of $f_{\mathbf{s}}^-(x)$. Thus, the following theorem was proven.

Theorem 5. *For any monomial f defined in $\mathbb{R}^n$ and any multiset G satisfying* (6) *and* (39), *there exists a convex infinitely differentiable extension from $E_{nk}(G)$ onto $\mathbb{R}^n_{>0}$ in the form of: a) if f is of the form* (38), *the extension is given by formula* (40); *b) if f has a form* (42), *then the extension is given by formula*

$$F_{\mathbf{s}}^-(x) = \sum_{y \in E_{n\kappa}(\mathcal{S}) \setminus \{\mathbf{s}\}} F_{\mathbf{y}}^+(x) - \mathcal{F}_{\mathbf{s}}(g),$$

where $F_{\mathbf{y}}^+(x)$ is found by (40) *for any $y \in E_{n\kappa}(\mathcal{S})$.*

Theorem 6. *For an arbitrary polynomial, there exists a convex infinitely differentiable extension from $E_{nk}(G)$ onto $\mathbb{R}^n_{>0}$ induced by a multiset* (6), (39).

Proof. Let $f(x)$ be a polynomial on n variables and G is a multiset (6), (39). $f(x)$ is represented by a finite number of monomials (38) and (42). Applying formulas (40) or (44) depending on signs of the monomials, we obtain a convex extension of $f(x)$ from $E_{nk}(G)$ onto $\mathbb{R}^n_{>0}$ in a form of a Laurent polynomial, which is infinitely differentiable in this orthant.

If a polynomial needed to be extended in a convex way is quadratic, it can be done explicitly as follows.

Theorem 7. *[39] Let a quadratic function be given on a VLS E inscribed in a hypersphere $S_r(\overline{\tau})$ such that*

$$f(x) = (Bx, x) + (b, x), \tag{45}$$

where $B = [b_{ij}]_{n \times n} \in \mathbb{R}^{n \times n}$ is a symmetric matrix, $b \in \mathbb{R}^n$. Then there exists a quadratic convex extension:

$$F(x) = (\widetilde{B}x, x) + (\widetilde{b}, x) + \widetilde{d} \tag{46}$$

of f from E onto $\mathbb{R}^n$, where

$$\widetilde{B} = B + \beta I,\ \widetilde{b} = b - 2\beta\overline{\tau},\ \widetilde{d} = \beta\left(\overline{\tau}^2 - r^2\right), \\ \beta = \sum_{i:\ b_{ii}<0} |c_{ii}| + \sum_{i<j} |b_{ij}|, \tag{47}$$

$\widetilde{B} = \left[\widetilde{b}_{ij}\right]_{n\times n} \in \mathbb{R}^{n\times n}$ is a symmetric positively definite matrix, $\widetilde{b} \in \mathbb{R}^n$, $\widetilde{d} \in \mathbb{R}^1$, I is a unit matrix of the dimension n.

Corollary 2. *Function* (45) *allows constructing a quadratic convex extension from $E_{nk}(G)$ onto $\mathbb{R}^n$ in the form of* (46), (47), *where τ and r are given by* (19).

Corollary 3. *Any quadratic EPBP enables forming an equivalent quadratic CPBP.*

2.5 A Hybrid Method of Convex Optimization on $P_{nk}(G)$

In this section, a hybrid method to solve Problem2.PR is proposed, which enables us to avoid dealing with an exponential number of constraints in polyhedral relaxations of CPBPs. The method is based on combining two well-known optimization techniques, such as the penalty method of nonlinear optimization and conditional gradient methods of convex optimization, with specific properties of linear function on $E_{nk}(G)$ and $P_{nk}(G)$.

Let us assume that, for Problem 1, an equivalent Problem 2 was built. The following hybrid approach to solve its polyhedral relaxation, i.e., Problem 2.PR, is proposed combining the penalty method [2] with the conditional gradient method [2].

Define a family of functions $\psi(x, \alpha_l) : P_{nk}(G) \to \mathbb{R}^1$ depending on a parameter $\alpha_l \in \mathbb{R}^1_{>0}$ and satisfying general requirements to penalty functions [2], $l \in \mathbb{N}$. In accordance with the general scheme of the penalty method [2], let us define a sequence of penalty coefficients $\{\alpha_l\}_{l=1}^{\infty}$ and scalars $\{\varepsilon_l\}_{l=1}^{\infty}$ are defined, such that $\lim\limits_{l\to\infty} \frac{1}{\alpha_l} = 0,\ \lim\limits_{l\to\infty} \varepsilon_l = 0$.

For each l, choose a point x^l that satisfies condition:

$$F\left(x^l\right) + \psi\left(x^l, \alpha_l\right) \le \min_{x\in P_{nk}(G)} \left[F(x) + \psi(x, \alpha_l)\right] + \varepsilon_l, \tag{48}$$

where

$$\psi(x, \alpha_l) = \alpha_l\Big(\sum_{i\in J^1} \left[\max\{0, F_{i1}(x)\}\right]^2 + \sum_{i\in J^2} \left[\max\{0, F_{i2}(x)\}\right]^2\Big). \tag{49}$$

To implement the penalty function method with penalty functions (49), a sequence of auxiliary optimization problems:

$$\Psi(x, \alpha_l) \to \min,\ x \in P_{nk}(G), \tag{50}$$

where

$$\Psi(x, \alpha_l) = F(x) + \psi(x, \alpha_l), \tag{51}$$

need to be solved on $P_{nk}(G)$ for $l = 1, 2, \ldots$. Let

$$x^l = \arg \min_{x \in P_{nk}(G)} \Psi(x, \alpha_k),$$

then the limit sequence $\{x^l\}_{l=1}^{\infty}$ of solutions constructed with respect to condition (48) converges to a solution y^{**} of Problem 2.PR.

Let us focus on an issue of solving the problem (49)–(51) (further referred to as *Problem 3.l*) for a specific value l. It is a problem of optimization convex differentiable function within a polytope given by H-representation (17), (18). An issue is that the number of constraints is, generally, not polynomial in n and reaches the value of $2^n - 1$ for the permutohedron $P_n(G)$ (see Sect. 2.2). Therefore, for high-dimensional problems, applying standard convex optimization methods to solving Problem 3.l becomes problematic.

Remark 1. At the same time, a linear optimization problem on $P_{nk}(G)$ is polynomially solvable. Indeed, the minimum of a linear function $f(x) = \sum_{i=1}^{n} c_i x_i$ on set $P_{nk}(G)$ is attained at point $y^* = (y_1^*, \ldots, y_n^*)$, where $y_{\pi_i}^* = g_i, i \in J_n$, and $\pi = (\pi_1, \ldots, \pi_n)$ is a permutation of the elements of J_n such that $c_{\pi_i} \geq c_{\pi_{i+1}}$, $i \in J_{n-1}$ [40,43]. This peculiarity allows us to offer and apply the following modification of the conditional gradient method [2] for solving Problem 3.l. It is known that, at each step of this iterative method, an auxiliary linear optimization problem is solved on $P_{nk}(G)$ with the gradient of $\psi(x, \alpha_l)$ at point x^l as a vector of the linear objective function. On the permutohedron $P_{nk}(G)$, its solution is reduced to the ordering of the vector's coefficients.

For the iterative process, the following interior point of $\Pi_{nk}(G)$:

$$x^0 = (x_1^0, \ldots, x_n^0), \; x_i^0 = \frac{1}{n} \sum_{j=1}^{n} a_j, i \in J_n$$

is recommended as a starting point.

Note that due to convexity and differentiability of $\psi(x, \alpha_l)$, the above mentioned properties of $P_{nk}(G)$, and a specifics of solving the auxiliary linear programs, one can state that the iterative process converges to the global solution $\langle y^{**}, z^{**} \rangle$ of the polyhedral relaxation Problem 2.PR, which yields the lower bound $z^l = z^{**}$ of z^*.

3 Experiment, Results and Discussion

3.1 Example

Let us illustrate the application of the above results to the formation of a polyhedral relaxation of a EPBP and obtaining the bound z^l.

Consider the following optimization problem (9), where

$$f(x) = 2x_1x_3 - 4x_2x_7 + 3x_1^4x_4^2x_5x_6^3, \tag{52}$$

subject to constraints:

$$\begin{aligned} f_{11}(x) &= x_1^3x_2x_4^2x_5^3 + x_2^2x_5x_7^3 + 5x_3 - x_5 + 3x_7 - 27000 \leq 0, \\ f_{12}(x) &= 2x_2^3 + x_4^3 - 3x_5^3 + 3x_2^2 - -x_3^2 + 2x_6^2 - 7 \leq 0, \end{aligned} \tag{53}$$

where E is a set of permutation with repetitions from elements of $G = \{1, 1, 2, 3, 3, 4, 7\}$ mapped into Euclidean space.

It is required to find a lower bound z^l of $f(x)$ on E.

G is a multiset (6), (8) with a base $S(G) = \{1, 2, 3, 4, 7\}$, multiplicities $(n_1, ..., n_5) = (2, 1, 2, 1, 1)$, and parameters $n = |G| = 7$, $k = |S(G)| = 5$, hence $E = E_{75}(G)$. (52), (53) is a EPBP given in the form of (9)–(13), where $I^1 = \{1, 2\}$, $I^2 = \{\emptyset\}$.

First, let us write explicitly conditions (21), (22).

In accordance with (17), (18), a convex hull of $E_{75}(G)$ is the permutohedron $P_{75}(G)$ defined by relations:

$$\sum\nolimits_{i=1}^{7} x_i = \alpha_7, \tag{54}$$

$$\sum\nolimits_{i \in \omega \subset J_7} x_i \geq \alpha_l, \ l = |\omega| \in J_6, \tag{55}$$

where $\alpha_1 = 1$; $\alpha_2 = 2$; $\alpha_3 = 4$; $\alpha_4 = 7$; $\alpha_5 = 10$; $\alpha_6 = 14$; $\alpha_7 = 21$. By Theorem 1, the collection (55) contains C_7^2 redundant constraints corresponding to α_2. As a result, $P_{75}(G)$ has an irredundant H-representation (54),

$$\sum\nolimits_{i \in \omega \subset J_7} x_i \geq \alpha_l, \ l = |\omega| \in J_6 \backslash \{2\}, \tag{56}$$

of the cardinality $2^n - C_n^2 - 1 = 106$.

By (19), a circumsphere $S_r(\overline{\tau})$ for E has parameters

$$\overline{\tau} = (3, 3, 3, 3, 3, 3, 3), \ r = \sqrt{26}.$$

Its equation is $\sum_{i=1}^{7}(x_i - 3)^2 = 26$ or

$$\sum_{i=1}^{7} x_i^2 - 6\sum_{i=1}^{7} x_i + 37 = 0. \tag{57}$$

As a result, EPBP (9), (12), (52), (53) is equivalent to (9), (52)–(54), (56), (57). Respectively, its polyhedral relaxation is (9), (52)–(54), (56), which is not CPBP due to nonconvexity of $f(x)$, $f_{11}(x)$, $f_{12}(x)$. This EPBP is polynomial, thus it satisfies conditions of Theorem 3 and the transition from this EPBP to an equivalent CPBP is possible and can be made using results of Theorems 4–7 and corollaries from them.

Let us construct convex extensions of nonconvex functions among (52), (53): for quadratic parts – by Theorem 7, for the rest of terms - by Theorems 4, 5, and convex terms, such as linear ones, leave unchanged.

For $f(x)$, the quadratic part is $2x_1x_3 - 4x_2x_7$, for which, by (47), the value β is $\beta = |c_{13}| + |c_{27}| = 6$; parameters in (41) are:

$$\mathbf{s} = (4,0,0,2,1,3,0),\ \mu = 4;\ \eta = \prod_{i=1}^{7} g_i = 504, \eta^{\mu} = 193572384768.$$

Respectively, in accordance with Theorems 4, 6, 7, we have

$$\begin{aligned} F(x) &= \frac{3\eta^4}{x_2^4x_3^4x_4^2x_5^3x_6x_7^4} + 2x_1x_3 - 4x_2x_7 + 6\left(\sum_{i=1}^{7} x_i^2 - 6\sum_{i=1}^{7} x_i + 37\right) = \\ &= \frac{193572384768}{x_2^4x_3^4x_4^2x_5^3x_6x_7^4} + 6\sum_{i=1}^{7} x_i^2 + 2x_1x_3 - 4x_2x_7 - 36\sum_{i=1}^{7} x_i + 222. \end{aligned} \tag{58}$$

For f_{11}, the convexification is required only for first two nonlinear terms, for which $\mu = 3, \eta = 504, \eta^{\mu} = 128024064$. Therefore,

$$F_{11}(x) = \frac{128024064}{x_2^2x_3^3x_4x_6^3x_7^3} + \frac{128024064}{x_1^3x_2x_3^3x_4^3x_5^2x_6^3} + 5x_3 - x_5 + 3x_7 - 27000 \le 0. \tag{59}$$

In f_{12}, the only terms $-3x_5^3$ and $-x_3^2$ require a transformation. By Theorem 5, we have

$$\begin{aligned} F_{12}(x) &= 2x_2^3 + x_4^3 - 3x_5^3 + 3x_2^2 - x_3^2 + 2x_6^2 \\ &- 7 + 3\left(\sum_{i=1}^{7} x_i^3 - \sum_{i=1}^{7} g_i^3\right) + \left(\sum_{i=1}^{7} x_i^2 - \sum_{i=1}^{7} g_i^2\right) \\ &= 3x_1^3 + 5x_2^3 + 3x_3^3 + 4x_4^3 + 3x_6^3 + 3x_7^3 + \\ &+ x_1^2 + 4x_2^2 + x_4^2 + x_5^2 + 3x_6^2 + x_7^2 - 1509 \le 0. \end{aligned} \tag{60}$$

The problem (24), (54), (56)–(60) is the CPBP we looked for. Without (57), it is the corresponding polyhedral relaxation, which is a convex optimization problem. Applying the hybrid method presented in Sect. 2.5 to solve this convex program, we obtain

$$\langle y^{**}, z^{**}\rangle = \langle (1, 4.418, 3.361, 2.642, 3.084, 2.077, 4.418),\ -147.943\rangle .$$

Thus, $z^l = z^{**} = -147.943$ is a lower bound of the global minimum z^* of $f(x)$ on E.

Remark 2. Let us try find an upper bound z^u of z^*. For that, the point y^{**} is projected on set E and yields

$$Pr_E y^{**} = \{y^1, y^2\} = \{(1,7,3,2,3,1,4), (1,4,3,2,3,1,7)\}.$$

These two points are not feasible for the CPBP, because constraint (60) violates at them. Hence, an upper bound z^u is not found on the stage. In order to find it, we recommend to continue the iterative process of the hybrid method from point y^{**} with the objective function $\Upsilon(x, \beta_{l'}) = \Psi(x, \alpha_L) + \beta_{l'}((x - \overline{\tau})^2 - -r^2)$, where L is a number of iterations to achieve y^{**}, $\{\beta_{l'}\}_{l' \in \mathbb{Z}}$ is an increasing sequence of penalty coefficients. If the EPBP is feasible, then the problems' sequence converges and $\Upsilon(x, \beta_{l'}) \underset{l' \to \infty}{\to} x^{**} \in E'$ and yields $z^u = f(x^{**})$.

3.2 Ways to Improve the Lower Bound z^l

Let $K \subseteq \mathbb{R}^n$ be a convex closed set, $E \subset K$, $\phi : K \to \mathbb{R}^1$, $y^{**} = \arg\min_{y \in K} \phi(y)$, for any real-valued function f, $f_+ = \max\{0, f\}$.

Theorem 8. *[37] If $\phi(x)$ is convex and differentiable on K, then $\forall\, x \in K$*

$$\min_{y \in E} \phi(y) \geq \phi(x) - (\nabla \phi(x), x) + \min_{y \in E} \sum_{i=1}^{n} \frac{\partial \phi(x)}{\partial x_i} y_i.$$

Corollary 4. *If $\phi(x)$ is convex and differentiable on K, then*

$$\min_{y \in E} \phi(y) \geq \phi(y^{**}) + \left(\min_{y \in E} \sum_{i=1}^{n} \frac{\partial \phi(x)}{\partial x_i} y_i - (\nabla \phi(x), x) \right)_+ . \tag{61}$$

Theorem 9. *[37] If $\phi(x)$ is strongly convex with a parameter $\rho > 0$ on K, then*

$$\min_{x \in E} \phi(x) \geq \phi(y^{**}) + \rho \min_{x \in E} \|x - y^{**}\|^2. \tag{62}$$

Theorem 10. *[37] If $\phi(x)$ is strongly convex with a parameter $\rho > 0$ and differentiable on K, then $\forall x \in K$*

$$\min_{y \in E} \phi(y) \geq \phi(x) - \frac{1}{4\rho} \|\nabla \phi(x)\|^2 + \rho \min_{y \in E} \left\| y - x + \frac{1}{2\rho} \nabla \phi(x) \right\|^2 .$$

Corollary 5. *If $\phi(x)$ is strongly convex with a parameter $\rho > 0$ and differentiable on K, then $\forall x \in K$*

$$\min_{y \in E} \phi(y) \geq \phi(y^{**}) + (-\frac{1}{4\rho} \|\nabla \phi(y^{**})\|^2 + \rho \min_{y \in E} \left\| y - y^{**} + \frac{1}{2\rho} \nabla \phi(y^{**}) \right\|^2)_+ . \tag{63}$$

Here, a convex extension $F(x)$ of $f(x)$ from E onto $K \supseteq P$ can be used. Respectively, taking $K = P$, all above results can be applied, in particular, $\phi(y^{**}) = F(y^{**}) = z^l$. For set $E = E_{nk}(G)$ under consideration, finding the bound (61) requires solving linear program on E, which can be done easily according to Remark 1. Applying (62), (63) assumes projecting onto E, which is solvable in polynomial time due to reducibility of this quadratic problem to a linear program on E. This is because E is a polyhedral-spherical set [36, 41, 42].

3.3 Discussion

In this study, a general approach of applying convex optimization methods for solving combinatorial optimization problems through their embedding into Euclidean space, reducing to optimization over vertex located sets, convexification of the problems, and solution of their polyhedral relaxation is proposed. These results can be used independently or in Branch and Bound schemes.

A reduction of combinatorial optimization problems to solving a sequence of their convex continuous relaxations is far from the only application of convex optimization in solving combinatorial optimization problems [15, 30, 34]. In combination, all these approaches can be the basis for new highly effective methods.

4 Conclusions

The results on solving the general optimization problem on the permutohedron by applying convex optimization techniques are of independent interest in constrained optimization over combinatorial polytopes with arbitrary functional constraints. On the other hand, the attacked problem can be considered as a relaxation in various solution schemes for permutation-based optimization. Moreover, using these results together with the continuous functional representation theory [24,25], one can single out new classes of combinatorial optimization problems, whose solution by similar hybrid approaches are possible. First of all, it is related to a class of polyhedral-spherical sets that can be represented as an intersection of a combinatorial polytope with a hypersphere. In this class, both permutation-based problems and Boolean problems belong to this class. Using an equation of the circumsphere in a formation of a penalty function, we come to a similar problem statement and, therefore, there exists a possibility of extending results of the paper to solve problems of this class.

References

1. Bazaraa, M.S., Sherali, H.D., Shetty, C.M.: Nonlinear Programming: Theory and Algorithms, 3rd edn. Wiley-Interscience (2006)
2. Bertsekas, D.P.: Nonlinear Programming, 2nd edn. Athena Scientific (1999)
3. Bertsekas, D.P.: Convex Optimization Algorithms, 1st edn. Athena Scientific (2015)
4. Borwein, J., Lewis, A.S.: Convex Analysis and Nonlinear Optimization: Theory and Examples. CMS Books in Mathematics, 2nd edn. Springer (2006). https://doi.org/10.1007/978-0-387-31256-9
5. Boyd, S., Vandenberghe, L.: Convex Optimization, 1st edn. Cambridge University Press, Cambridge (2004)
6. Burer, S., Letchford, A.N.: Non-convex mixed-integer nonlinear programming: a survey. Surv. Oper. Res. Manag. Sci. **17**(2), 97–106 (2012). https://doi.org/10.1016/j.sorms.2012.08.001
7. Butenko, S., Pardalos, P.M., Shylo, V. (eds.): Optimization Methods and Applications: In Honor of Ivan V. Sergienko's 80th Birthday. Springer Optimization and Its Applications. Springer International Publishing (2017). https://doi.org/10.1007/978-3-319-68640-0
8. Christ, M.: The extension problem for certain function spaces involving fractional orders of differentiability. Arkiv för Matematik **22**(1), 63–81 (1984). https://doi.org/10.1007/BF02384371
9. Cook, W.J., Cunningham, W.H., Pulleyblank, W.R., Schrijver, A.: Combinatorial optimization. Wiley-Interscience Series in Discrete Mathematics and Optimization. Wiley, New York (1998)
10. Dahl, J.: Convex Optimization in Signal Processing and Communications, Department of Communication Technology, Aalborg University (2003)
11. Ferreira, O.P., Iusem, A.N., Németh, S.Z.: Concepts and techniques of optimization on the sphere. TOP **22**(3), 1148–1170 (2014). https://doi.org/10.1007/s11750-014-0322-3

12. Gimadi, E., Khachay, M.: Extremal Problems on Sets of Permutations. UMC UPI, Ekaterinburg (2016). (in Russian)
13. Gmys, J.: Heterogeneous cluster computing for many-task exact optimization - Application to permutation problems. Université de Mons (UMONS), University de Lille, Mons (2017)
14. Graf, M., Hielscher, R.: Fast global optimization on the torus, the sphere, and the rotation group. SIAM J. Optim. **25**(1), 540–563 (2015). https://doi.org/10.1137/130950070
15. Horst, R., Tuy, H.: Global Optimization: Deterministic Approaches, 3rd edn. Springer (1996). https://doi.org/10.1007/978-3-662-03199-5
16. de Kierk, E.: The complexity of optimizing over a simplex, hypercube or sphere: a short survey. CEJOR **16**(2), 111–125 (2008). https://doi.org/10.1007/s10100-007-0052-9
17. Koliechkina, L., Pichugina, O.: A horizontal method of localizing values of a linear function in permutation-based optimization. In: Le Thi, H.A., Le, H.M., Pham Dinh, T. (eds.) Optimization of Complex Systems: Theory, Models, Algorithms and Applications. Advances in Intelligent Systems and Computing, pp. 355–364. Springer, Cham (2019). https://doi.org/10.1007/978-3-030-21803-4_36
18. Korte, B., Vygen, J.: Combinatorial Optimization: Theory and Algorithms, 6th edn. Springer (2018). https://doi.org/10.1007/978-3-662-56039-6
19. Lang, S.: Algebra. Graduate Texts in Mathematics, 3rd edn. Springer (2002). https://doi.org/10.1007/978-1-4613-0041-0
20. Mehdi, M.: Parallel Hybrid Optimization Methods for permutation based problems. University des Sciences et Technologie de Lille, Lille (2011)
21. Nocedal, J., Wright, S.: Numerical Optimization. Springer Series in Operations Research and Financial Engineering, 2nd edn. Springer (2006). https://doi.org/10.1007/b98874
22. Papadimitriou, C.H., Steiglitz, K.: Combinatorial Optimization: Algorithms and Complexity. Dover Publications (1998)
23. Pardalos, P.M., Du, D., Graham, R.L.: Handbook of combinatorial optimization. Springer Reference. Springer, New York (2005). https://doi.org/10.1007/b102533
24. Pichugina, O., Yakovlev, S.: Euclidean combinatorial configurations: continuous representations and convex extensions. In: Lytvynenko, V., Babichev, S., Wójcik, W., Vynokurova, O., Vyshemyrskaya, S., Radetskaya, S. (eds.) Lecture Notes in Computational Intelligence and Decision Making, Advances in Intelligent Systems and Computing, pp. 65–80. Springer, Cham (2019). https://doi.org/10.1007/978-3-030-26474-1_5
25. Pichugina, O., Yakovlev, S.: Quadratic optimization models and convex extensions on permutation matrix set. In: Shakhovska, N., Medykovskyy, M.O. (eds.) Advances in Intelligent Systems and Computing IV. Advances in Intelligent Systems and Computing, pp. 231–246. Springer (2019). https://doi.org/10.1007/978-3-030-33695-0_17
26. Pogorelov, A.V.: Extrinsic Geometry of Convex Surfaces. American Mathematical Society, 1st edn. (1973)
27. Postnikov, A.: Permutohedra, associahedra, and beyond. IMRN: International Mathematics Research Notices **2009**(6), 1026–1106 (2009). https://doi.org/10.1093/imrn/rnn153
28. Rockafellar, R.T.: Convex Analysis. Princeton University Press, Princeton (1996)
29. Schrijver, A.: Combinatorial Optimization: Polyhedra and Efficiency. Algorithms and Combinatorics. Springer (2003)

30. Stetsyuk, P.I.: Dual bounds in quadratic extremal problems. A series of scientific publications "Non-differentiable optimization and its applications" dedicated to academician N.Z. Shor, Eureka (2018)
31. Stoyan, Y.G., Yemets', O.: Theory and methods of Euclidean combinatorial optimization (in Ukrainian). ISSE (1993)
32. Stoyan, Y.G., Yakovlev, S.V., Emets, O.A., Valuŏskaya, O.A.: Construction of convex continuations for functions defined on a hypersphere. Cybern. Syst. Anal. **34**(2), 27–36 (1998). https://doi.org/10.1007/BF02742066
33. Stoyan, Y.G., Yakovlev, S.V., Pichugina, O.S.: The Euclidean combinatorial configurations, Constanta (2017)
34. Tawarmalani, M., Sahinidis, N.V.: Convexification and Global Optimization in Continuous and Mixed-Integer Nonlinear Programming: Theory, Algorithms, Software, and Applications. Nonconvex Optimization and Its Applications, 1st edn. vol. 65, Springer (2002). https://doi.org/10.1007/978-1-4757-3532-1
35. Tuy, H.: Convex Analysis and Global Optimization, 2nd edn. Springer (2016)
36. Yakovlev, S., Pichugina, O., Yarovaya, O.: On optimization problems on the polyhedral-spherical configurations with their properties. In: 2018 IEEE First International Conference on System Analysis Intelligent Computing (SAIC), pp. 94–100 (2018). https://doi.org/10.1109/SAIC.2018.8516801
37. Yakovlev, S.V.: Bounds on the minimum of convex functions on euclidean combinatorial sets. Cybernetics **25**(3), 385–391 (1989). https://doi.org/10.1007/BF01069996
38. Yakovlev, S.V.: The theory of convex continuations of functions on vertices of convex polyhedra. Comput. Math. Math. Phys. **34**(7), 1112–1119 (1994)
39. Yakovlev, S.V., Pichugina, O.S.: Properties of combinatorial optimization problems over polyhedral-spherical sets. Cybern. Syst. Anal. **54**(1), 99–109 (2018). https://doi.org/10.1007/s10559-018-0011-6
40. Yakovlev, S.: Convex extensions in combinatorial optimization and their applications. In: Optimization Methods and Applications. Springer Optimization and Its Applications, pp. 567–584. Springer, Cham (2017). https://doi.org/10.1007/978-3-319-68640-0_27
41. Yakovlev, S., Kartashov, O., Pichugina, O.: Optimization on combinatorial configurations using genetic algorithms. In: Proceedings of the Second International Workshop on Computer Modeling and Intelligent Systems (CMIS-2019), CEUR, vol-2353, pp. 28–40 (2019). urn:nbn:de:0074-2353-0
42. Yakovlev, S., Pichugina, O.: On constrained optimization of polynomials on permutation set. In: Proceedings of the Second International Workshop on Computer Modeling and Intelligent Systems (CMIS-2019). CEUR, vol. 2353, pp. 570–580 (2019). urn:nbn:de:0074-2353-0
43. Yemelichev, V.A., Kovalev, M.M., Kravtsov, M.K.: Polytopes. Graphs and Optimisation. Cambridge University Press, Cambridge (1984). translated from the Russian by G. H. Lawden

Method of Structural Functional-Value Modeling of a Complex Hierarchic System

Maksym Korobchynskyi(✉), Mykhailo Slonov, Myhailo Rudenko, and Oleksandr Maryliv

Military-Diplomatic Academy named after Eugene Bereznyak, Kyiv, Ukraine
{maks-kor,ruminik33}@ukr.net,
slonovmu@gmail.com, delavar65000@gmail.com

Abstract. The method of structural functional-value modeling of complex system is offered. It is based on a polynomial approximation of the dependence of the value of subsystems on the level of their functional suitability. The initial data for calculations on the functional-value model of a complex system with serial or parallel interconnection of subsystems are formulated and determined. The choice of an indicator of the structural functional perfection of a complex system and the way of defining the function of the value function of its subsystems are grounded. The order and specific features of structural functional-value calculations of a complex system for serial and parallel subsystems interconnection are developed. The effectiveness of the developed approaches and algorithms is tested in the functional-value modeling of specific complex systems: an educational institution (serial subsystems) and an object monitoring system (parallel subsystems). The minimization of the value of a complex system, provided that it fulfilled its functional purpose at a given level, was carried out by the Lagrange multiplier method. The issues that are solved in the value rationalization of the proposed method include following: ensuring a given level of functional perfection of the system at its minimum value; structural improvement of a complex system due to the choice of a subsystem for which the improvement of the functional perfection of the whole system can be carried out with minimal value. The researches made it possible to formulate the rules of structural rationalization of a complex system. This method is adapted to application at different levels of a priori uncertainty of the source data and can be useful at all stages of the existence of a complex system: development, operation, utilization. In addition, it can be used to investigate poorly formalized and non-formalized complex systems, that is, it enables a qualitative analysis of a complex system to be translated into a quantitative one.

Keywords: Structural functional-value models · Complex hierarchical systems · Functional-value calculations · Approximate functions

S. Babichev et al. (Eds.): ISDMCI 2020, AISC 1246, pp. 213–231, 2021.
https://doi.org/10.1007/978-3-030-54215-3_14

1 Introduction

People are surrounded by technical, organizational, social, environmental systems and etc. Structurally, one of the types of systems is complex systems. A complex system is a managed system in which complexes of machines and groups of people are combined [9]. By another definition, it is a managed system, which is a set of interconnected managed subsystems that are combined for a common purpose [8]. Complex systems are characterized by a large number of components, internal and external connections, and their heterogeneity.

By definition, a structural feature of a complex system is the presence of several subsystems. A set of subsystems defines the content of the task of researching a complex system in a particular case. Each subsystem is characterized by parameters and properties.

The input of a complex system receives a number of input signals, the products whose interaction in the subsystems will provide the output signal, the product. The approximation of the output signals or products of a complex system to a predetermined level according to functional purpose indicates the effectiveness of its functioning. The level of efficiency of functioning of a complex system is controlled by the parameters and properties of subsystems and their structural organization.

A variant of systematic research is functional-value analysis. In the functional-value analysis, there are two generalizing factors. It is the degree of implementation of a complex system of functional purpose and the value of its operation.

One of the directions of functional-value analysis of a complex system is using of approximating polynomial dependences of the value of its subsystems on the level of their functional perfection. The features of this approach for serial or parallel coupling of subsystems in a complex system are discussed in this article.

Functional-value analysis is now a common tool for working with complex systems at all stages of their existence: development, operation, utilization [2,4, 5,7,10,11,13–18]. It aims at finding a rational structure for a complex system. With this structure, the implementation of a given level of functionality of a complex system will be achieved at the expense of minimal value.

Several approaches to functional and value analysis have been developed. The semantic approach is based on the statistical generalization of data on the practice of existence of similar complex systems [3,10,13]. The possibilities of such an approach are limited by the general recommendations in accordance with the implementation of parametric performance control of a complex system.

In [5,15] is considered interval probabilistic efficiency estimation methods. Their branching is determined by the physical heterogeneity of the subsystems of a complex system. The dependence of specific properties and parameters of their functioning on a large number including stochastic influencing factors is taken into account [5]. A promising area of improvement for such methods is the ability to conduct research under uncertainty. The method of the Dempster-Schaefer is actively involved [17]. Unfortunately, at present, in such methods the algorithmic connection between the value of a complex system and its functional perfection has not been worked out.

The limitations, and the inability of analytical formalization of the functioning of complex systems, have led to the proliferation of expert methods of their evaluation [14,16]. Such methods are excellent at solving problems that, in principle, cannot have an algorithmic description. In addition, they allow a parametric description of the operation of a complex system. But the algorithmic correspondence of the level of functional perfection and value of a complex system is not obvious.

The same applies to methods using neural algorithms. They are perfectly adapted to the operation of powerful computational networks [2,7]. However, in this case, the connection between the "parameters-value" of a complex system is not considered analytically.

In [11,18] a method of structural functional-value modeling of a complex hierarchical system with a polynomial approximation of the dependence of its value on the level of functional suitability is proposed. It is based on the relationship between the value and the functionality of a complex system.

But many complex systems are characterized by the parallel interconnection of individual subsystems. Examples of these are the existence of competing businesses, the collection of information about an entity using multiple sources of information etc. A common feature of such systems is the orientation of the functioning of parallel subsystems to achieve the common goal. Questions about the peculiarities of using this method for the study of complex systems with parallel subsystems remain unresolved.

The aim of the research is an algorithmic justification of rational directions of improvement of a complex system with a serial or parallel connection of subsystems with polynomial approximation of their value depending on the level of functional suitability.

2 Materials and Methods

The main assumption when using functional value models is the additivity of value C_i of its individual subsystems relative to the total value $C_\sum$ the whole system [1]:

$$C = \sum_{i=1}^{n} C_i \tag{1}$$

where n is a total number of subsystems.

The decisive rule for the structure of the system in terms of reducing its value is considered a requirement:

$$C = \sum_{i=1}^{n} C_i = \min . \tag{2}$$

Equation (2) shows that the most effective in terms of value is simple ($n \rightarrow 0$) and cheap ($C_i \rightarrow 0$) system.

But each system performs certain functional tasks. Being cheap enough, such a system may not satisfy its functional purpose at all. Therefore, (2) complements the condition of functional perfection of the system. Given the relationship between the value of the system and its functional perfection, the rational existence of a complex system will meet the following conditions [9]:

$$\begin{cases} P_{\sum} = P[C_i] \geq P_s, \\ C_{\sum} = \min \end{cases} \tag{3}$$

where $P_{\sum}$ is a generic indicator of the functional perfection of a complex system; $P[C_i]$ is the functionality that shows the dependence of the system on the task from the invested in it; P_s is a set point of functional excellence.

The Eq. (3) can be considered as a rational functional-value mathematical model of the complex system under study. For its use it is necessary to determine a sufficient level P_s functional excellence of the system. It is desirable that it be regulatory or set point. It is important to find out the value of each subsystem of a complex system depending on its parameters and properties. This will allow you to obtain numerical estimates of the present value of the subsystems.

Note that $C_{\sum}$ depends on the structure of a complex system. It can be characterized by the correctness of its structure, the number of subsystems n, shortcuts l between them, as well as the peculiarities of operation that is maintenance of subsystem life (including their utilization).

If $\Psi(n, l)$ is a function of the value of interconnections between parts of the system and the operating values, so (1):

$$C = \sum_{j=1}^{n} C_j + \Psi(n, l) = \sum_{j=1}^{n} C_j + \Psi_1(l) + \Psi_2(n + l) \tag{4}$$

where $\Psi_1(l)$ and $\Psi_2(l+n)$ are functions of connection values and operating values.

However, if you consider the interconnections as separate subsystems, you can write:

$$\Psi_1(l) = \sum_{j=1}^{l} C_j \tag{5}$$

It is advisable to formalize the function $\Psi(l + n)$. For this purpose, in a simpler case, the level of value per unit of time of operation (depending on the task under consideration - day, year, cycle or period of operation) will be considered proportional to the value of the subsystem. Then:

$$\Psi_2(l + n) = a \sum_{j=1}^{l+n} C_j \tag{6}$$

where a is a estimates the average proportionality factor.

Equation (4), taking into account (5, 6), will be:

$$C_{\sum} = \sum_{j=1}^{l} C_j + \sum_{j=1}^{n} C_j + a \sum_{j=1}^{l+n} C_j = \sum_{j=1}^{l+n} C_j + a \sum_{j=1}^{l+n} C_j = (1 + a) \sum_{j=1}^{l+n} C_j. \tag{7}$$

Thus, (3, 7) is a functional-value model of rational structure of a complex system. Its generalized form is recorded as follows:

$$\begin{cases} P_{\sum} = \prod_{j=1}^{l+n} P_j \geq P_s, \\ C_{\sum} = (1+a) \sum_{j=1}^{l+n} C_j. \end{cases} \tag{8}$$

The solution of (8) is possible only if the functions $P_{\sum}$ and C_j are analytically given and completely adequate to the processes that accompany the existence of the system. The following are approaches to defining them.

Functional excellence of a complex system is determined by the correspondence of the results of its work to the intended purpose. For the widespread assessment of the level of functional excellence, the system develops qualitative and quantitative indicators, which are approved in the regulatory and monographic literature.

With the increase of any system, the complexity and quantity of such indicators greatly differs, making it difficult to describe the exact link between the specific parameters of the system and its functional properties. Therefore, probabilistic indicators are more often used in the analysis of complex systems. They, firstly, allow to take into account the influence on the functional perfection of the parameters of all subsystems, secondly, reflect the stochastic behavior of the large system and, thirdly, allow not to search for exact functional dependencies, but to identify trends of such dependencies [1].

For each subsystem, such an indicator will be the probability P_j of her task at the current values of parameters and properties. The generalized criterion of functional perfection of the whole complex system will be determined by the combination of P_j.

The desired level of functional excellence of a complex system will be $P_{\sum} = 1$. But a probabilistic approach to describing a generalized criterion only allows us to focus on the desired value. The desired minimum level of generalized criterion is considered as such a desirable value. It is set by the normative value P_s. Therefore a rational condition will be:

$$P_{\sum} = \prod_{j=1}^{l+n} P_j \geq P_s. \tag{9}$$

Equation (9) not only sets the required level of efficiency for the whole complex system. It can be used to determine the satisfactory level of performance of an individual subsystem $P_{j,s}$, back if the performance indicators of other subsystems are known:

$$P_{j,s} \geq \frac{P_s}{P_1 P_2 ... P_{j1} P_{j+1} ... P_m}. \tag{10}$$

The possibility of performing (10) answers the question: is it possible to implement a part of the system with a satisfactory value P_j. Detecting all P_j values allows you to proceed to a parametric description of the system and its parts.

From generalizing its characteristics we pass to specific parameters and properties of subsystems. Such ratios can be determined empirically or analytically.

As an example, let's give a ratio for the probability $P_{j,0}$ of detecting an object by its static image. It has the following form [12]:

$$P_{j,0} = \exp[-(\frac{B\acute{m}}{2LR\sqrt{\triangle D}})^2] \tag{11}$$

where B is the coefficient of the shape of the distinguishing feature with the linear size L and the tonal contrast $\triangle D$ of the object.

The object is observed by a species with a resolution R at the denominator of scale $\acute{m}$. The probability $P_{j,0}$ can be considered as a partial criterion for the functional perfection of the species subsystem. Other partial criteria for this task will be the likelihood of the surveillance tool getting into a given area (the subsystem of the media carrier), the likelihood of correctly processing the results of the image decryption (the processing subsystem) etc.

From (11) we can see that to increase P_0 it is necessary to increase the resolution of the image (to improve the quality of the observation tool, to withstand its required modes), the tonal contrast (to select modes and parameters of the receiver according to the contrasting characteristics of the object, background, environment), and the denominator $\acute{m}$ must be reduced (to choose a focal length observation or to reduce the observation distance).

The value of an individual part (subsystem) depends on many factors - the number and perfection of components, the technology of their combination, maintenance and more. It is almost impossible to combine all of these factors into one functional (analytical) dependency.

First, the value of the subsystem consists of the "value" of its parameters - the money spent on reaching the parameters of the required value. According to (11), each combination of design parameters $(R, \acute{m})$ will have its probability value $P_{j,0}$. Thus, it can be argued:

$$C_j = C_j(P_j) \tag{12}$$

where C_j is the value of the j-th subsystem with the level of perfection P_j.

Second, the correct structure of the subsystem implies an increase in functional excellence while increasing the funds invested in it. Therefore, the value equation must meet the following constraints and conditions:

$$\begin{cases} C(P) \geq 0, \\ P_1 > P_2 \rightarrow C(P_1) \geq C(P_2), \\ \lim_{P \to 0} C(P) = 0, \\ \lim_{P \to 1} C(P) = \infty. \end{cases} \tag{13}$$

The analytical constraints and conditions given in (12), (13) are a consequence of the expediency and possibility of physical realization of a complex system. We separately emphasize the additive property of the general value function $C_{\sum}$ with respect to its components C_j:

$$C_{\sum} = \sum_{j=1}^{m} C_j. \tag{14}$$

The requirements (13, 14) satisfy the following approximate expressions:

$$C_j(P) = A_j P \exp(\frac{B_j}{1-P}), \tag{15}$$

$$C_j(P) = \frac{A_j P^{B_j}}{\ln(\frac{1}{P})}, \tag{16}$$

$$C_j(P) = \frac{A_j P}{1-P} \tag{17}$$

where A_j and B_j are constants that can be empirically matched.

The presence of two constants in relations (15) and (16) makes it possible to more accurately and accurately describe true physical dependencies. The factor A_j has a dimension of value. It can be interpreted as the value that the customer can still pay for the functional assignment of the subsystem. The factor B_j characterizes the intensity of the approximation of the system value to the maximum value at $P_j \to 1$.

The rationalization rule will thus be formulated as follows: a complex system must provide a given level of functional perfection P_s with its minimum total value $C_{\sum}$. Let's consider the procedure for using the system of (8) to investigate the standard structural structure of a complex system for serial and parallel subsystem connections.

When using (15), the functional-value model of a rational structure of a complex system (8) will take the following form (excluding operating values):

$$\begin{cases} C_{\sum}(P_j) = \sum_{j=1}^{m} C_j(P_j) = \sum_{j=1}^{m} A_j(P_j) \exp \frac{B_j}{1-P_j} = \min_{0 \le P_j \le 1}, \\ P_{\sum} = \prod_{j=1}^{m} P_j \ge P_s. \end{cases} \tag{18}$$

In specific examples, we consider the features of carrying out functional-value calculations of a complex system with serial or parallel subsystems.

A simpler structure of a complex system with the sequential coupling of its individual subsystems is shown in Fig. 1. The baseline data for functional value calculations is available information on the probabilistic and value properties of P_j, C_j subsystems of a complex system. The result of their implementation is the distribution of the probability of rational functioning of the subsystems which will provide a given level of perfection of a complex system at its minimum value: $P_{\sum} > P_s$, $C_{\sum} = \min$.

If the value dependence for all three parts can be described by (17), (18) and the total value is given by expression (14), then the system of (8) will have the

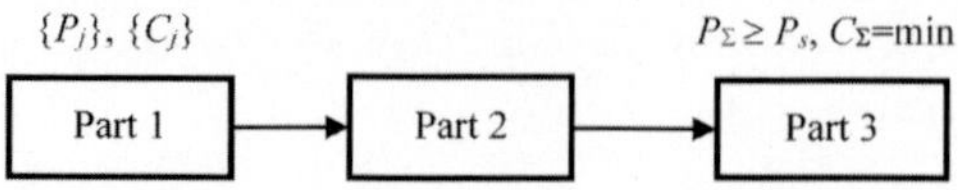

Fig. 1. A complex system of three consecutive elements

following form:

$$\begin{cases} C_{\sum}(P_j) = \sum_{j=1}^{3}(\frac{A_j P_j}{1-P_j}) = \min_{0 \le P_j \le 1}, \\ P_{\sum} = \prod_{j=1}^{3} P_j \ge P_s. \end{cases} \tag{19}$$

Solving (19) means finding the extremum (minimum) of its first equation when the conditions for $P_{\sum}$ and P_j are satisfied. The classical method for finding conditional extremum is the Lagrange multiplier method. For this, the Lagrange function is chosen in the form [6]:

$$f(P_j, \lambda) = \sum_{j=1}^{m}(\frac{A_j P_j}{1 - P_j}) + \lambda(\prod_{j=1}^{3} P_j - P_s) \tag{20}$$

where λ is the Lagrange multiplier.

After finding partial derivatives with respect to the variables P_j, equating (20) them to zero and eliminating the intermediate variables, we obtain such a system of simple algebraic equations:

$$\begin{cases} \frac{A_1 P_1}{1-P_1} = \frac{A_2 P_2}{1-P_2}, \\ \frac{A_1 P_1}{1-P_1} = \frac{A_3 P_3}{1-P_3}, \\ P_{\sum} = P_1 P_2 P_3 \ge P_s. \end{cases} \tag{21}$$

Equation (21) includes three algebraic equations and three variables: P_1, P_2, P_3. Such a system has a unique solution. For small m (in this case $m = 3$) the solution is a direct substitution, with increasing $m (m \ge 4)$ is solved by numerical methods with the help of computer technology. The solution of (21) gives the distribution of the desired probabilities of the levels of excellence of the individual subsystems in terms of the minimum value of the system.

The analysis of functions performed in parallel is appropriate when the operation of parallel components is combined with a common purpose. But this orientation can be achieved in different ways. For example, in a school, as a complex system, subsystems have an educational process, an administrative process, a logistical process etc.

One of the simpler parallel structures is the union of three independent identical parts: Part 1, Part 2, Part 3 (Fig. 2). The system of value rationalization equations for the system under consideration will be:

$$\begin{cases} C_{\sum}(P_j) = \sum_{j=1}^{3} C_j = 3C = \min_{0 \le P_j \le 1}, \\ P_{\sum} = 1 - (1 - P_1)(1 - P_2)(1 - P_3) = 1 - (1 - P)^3 \ge P_s \end{cases} \tag{22}$$

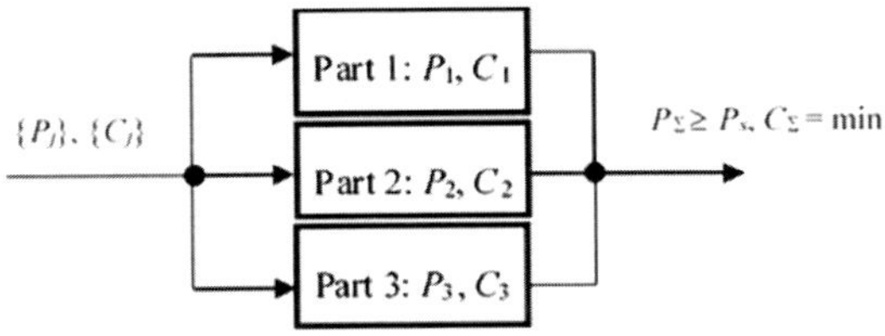

Fig. 2. A complex system of three parts having parallel connection

where C, P is the value and probability of normal functioning of one of the three equal parts of the system during their parallel operation.

If the coupling between P and C can be described by expression (17) and the operation of the parts is independent, then (22) takes the form:

$$\begin{cases} C_{\sum}(P_j) = \frac{3AP}{1-P} = \min_{0 \leq P \leq 1}, \\ P_{\sum} = 1 - (1-P)^3 \geq P_s. \end{cases} \tag{23}$$

Analysis of (23) shows that parallel connections of identical subsystems can only make the system more expensive. Another possible option for parallel pairing is to pair the same purposeful (performing the same function) but with different levels of functional perfection of the parts. In this case, the system of value rationalization will be:

$$\begin{cases} C_{\sum}(P_j) = \sum_{j=1}^{3}(\frac{A_j P_j}{1-P_j}) = \min_{0 \leq P_j \leq 1}, \\ P_{\sum} = 1 - \prod_{j=1}^{3}(1-P_j) \geq P_s. \end{cases} \tag{24}$$

Solution is using the Lagrange multiplier method leads to the following system of algebraic equations:

$$\begin{cases} P_{\sum} = 1 - (1-P_1)(1-P_2)(1-P_3) \geq P_s, \\ \frac{A_1 P_1}{1-P_1} = \frac{A_2 P_2}{1-P_2}, \\ \frac{A_1 P_1}{1-P_1} = \frac{A_3 P_3}{1-P_3}. \end{cases} \tag{25}$$

Equation (25) allows analysis in two directions. First, is the relationship between the values of the approximation coefficients A_j and the values of P_j. Secondly, it is possible to determine the requirements for the required level $P_{j,s}$ behind the functional perfection of the j-th part of the parallel connection at the known levels of this indicator in the other parts.

3 Experiment

The basic principles of functioning of the higher education system are defined by law. At the qualitative level, the regulatory documents set out the requirements for the organization and implementation of the educational process. The developed method allows a quantitative approach to assess the impact of individual units on the process and learning outcomes.

The essence of the approach to compiling a functional-value algorithm in this case is as follows. The educational process according to the regulatory requirements is carried out in the following organizational forms: training sessions (TS), independent work (IW), practical work (PW), control sessions (CS). The learning process as a complex system contains four consecutive subsystems. Each of them is characterized by its functional perfection P_i and conditional value C_i.

The general structural diagram of such a system is presented in Fig. 3.

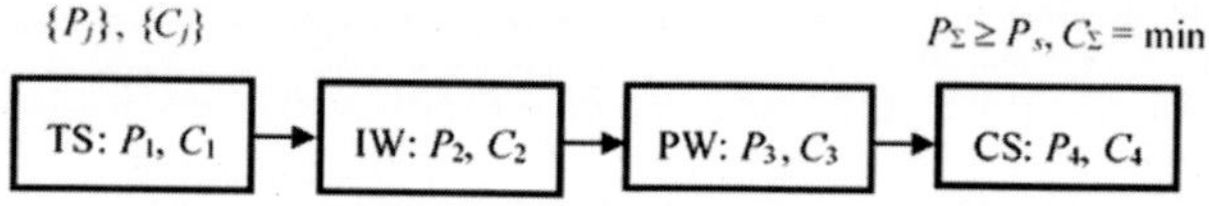

Fig. 3. Structural diagram of interaction of forms of educational process

It will be considered that the functional excellence of the subsystem reflects the likelihood of completing the training task. Its contingent value is determined by measures to ensure the completeness, content and recommended sequence of the educational process.

We will look for the distribution of levels of functional excellence between individual subsystems, provided that the goal of the learning process is achieved at its minimum conditional value. This distribution, when applying the approximation dependence (17), is the solution of a system of equations of the type (25). Only such a system will be of the fourth order, as four subsystems are connected in series. In this case, $P_{\sum}$ is the probability of the graduates matching the given educational and qualification characteristics; $C_{\sum}$ is the conditional value for students to obtain the required level of knowledge in their chosen profession; A_i are empirical constants. P_i is the probability of performing the i-th subsystem of its functional purpose.

The solution of the system of type (25) involves the solution of two problems. First, it is necessary to find out the values of the empirical constants A_i and to determine the value of a given probability of completing the task $P_{\sum}$. The second problem is the direct solution of the system of equations with the determination of the extremum (in this case the minimum) of the function $C_{\sum}(P_i)$.

We solve the first problem. We will consider the contingent value of each subsystem of the learning process as determined by public efforts to implement it. To determine the estimation value of probability P_1 it is necessary to provide preliminary development of the following objective influences on the quality of employment of factors.

The value of P_2 characterizes the functional conformity of the learner's independent work. The objective influential factors of the quality of his independent work are: the program of the discipline; methodical materials; tasks and instructions of the teacher (executors - teachers of the profile department); the level of its material; technical support from the management of the educational institution.

Probability P_3 is a characteristic of functional perfection of practical work. It is provided by carrying out practices and internships (leading department and representatives of the customer, institutions and enterprises that are consumers of graduates of educational institution), games and practical classes (leading department and profile departments).

Control measures are a necessary element of feedback in the learning process. The following types of control are used: entrance (department), current (department), self-control (department), modular (department), as well as control during state certification (leading and profile departments, customer representatives). It is the work in the specified directions that provides the required level of functional perfection of the control sessions P_4.

In the absence of statistics on the value of individual units, a comparative approach can be proposed. Based on the performers of the tasks of each link, one can argue about the significantly different conditional value of the functional perfection of each link. The notional value of the class is the highest. According to the value of the constant A_1, we give the largest value, for example, in 50 units. In the estimating approximation equation of control measures, it is possible to assign $A_2 = 40$ units, practical orientation - $A_3 = 20$ units. In the approximate dependence for the value of student's independent work $A_4 = 10$ units, as such, which is formed almost only by the profile department.

There is an opportunity for an analytical approach to assigning numerical values to the constants A_i. The adjacency matrix [16] will consist horizontally and vertically from the results of A_{ik} expert comparison of A_{ik} with A_{ki}:

$$A_{ik} = \begin{cases} 10, & A_{ik} = A_{ki}, \\ 15, & A_{ik} > A_{ki}, \\ 5, & A_{ik} < A_{ki}, \end{cases} \rightarrow A_i = \sum_{k=1}^{k=4} A_{ik}. \tag{26}$$

Let us dwell on the results of the evaluation comparative approach. We proceed to determine the magnitude of a given probability of fulfilling the task $P_{\sum}$. For the functional perfection probabilities $P_1, \ldots, P_4$ and, respectively, P_s, the desired value is to compare them 1:

$$P_1 \approx P_2 \approx P_3 \approx P_4 \approx P_s \approx 1. \tag{27}$$

But a sufficient level of functional excellence can be considered $P_s = 0.85$. The success of the learning process depends on a large number of probabilistic independent objective and subjective factors (for example, the condition of the teacher at the class, the condition of the student, the audience, samples of logistics, time of the class etc.), which indicates a normal law of density distribution the likelihood of a student successfully learning the material. Therefore, the probability of 0.85 will correspond to almost a reliable level of assimilation. Accept: $P_s = 0.85$.

The solution of (26), (27) and the determination of specific numerical data of the distribution $P_1, \ldots, P_4$ can be made using the Lagrange multiplier method. We introduce the Lagrange function in the form [11]:

$$f(P_i, \lambda) = \sum_{i=1}^{4}(\frac{A_i P_i}{1 - P_i}) + \lambda(\prod_{i=1}^{4} P_i - P_s). \tag{28}$$

Its partial derivatives for the variables P_i and partial extrema are determined by the following equation:

$$\frac{\partial f}{\partial P_i} = \frac{A_i}{(1 - P_i)^2} + \frac{\lambda}{P_i} \prod_{i=1}^{4} P_i = 0. \tag{29}$$

It can be written:

$$\frac{A_i P_i}{(1 - P_i)^2} = -\lambda \prod_{i=1}^{4} P_i = -\lambda P_s. \tag{30}$$

Since for each $i = 1, 2, 3, 4$ the right part of the previous (28–30) is kept constant, we can state the following:

$$\frac{A_1 P_1}{(1 - P_1)^2} = \frac{A_i P_i}{(1 - P_i)^2}, i = 2, 3, 4. \tag{31}$$

Thus, the application of the Lagrange multiplier method leads in the best case to a system of four equations with four variables in the following form:

$$\begin{cases} P_1 P_2 P_3 P_4 \geq 0.85, \\ \frac{50 P_1}{(1-P_1)^2} = \frac{10 P_2}{(1-P_2)^2}, \\ \frac{50 P_1}{(1-P_1)^2} = \frac{30 P_3}{(1-P_3)^2}, \\ \frac{50 P_1}{(1-P_1)^2} = \frac{40 P_4}{(1-P_4)^2}. \end{cases} \tag{32}$$

Systems whose solutions give an equation of order greater than a third have no common solution. They are solved iteratively. The first step is to find the first approximation of $P_1^{(1)}$ of variable P_1, as follows:

$$P_1^{(1)} = \sqrt[4]{P_s} = 0.96. \tag{33}$$

The values of the first approximations of the functional perfection of $P_2^{(1)}$, $P_3^{(1)}$, $P_4^{(1)}$ are calculated by the lower equations of (31), (32) and we check the correspondence of the obtained numerical values to the upper equation of the system. According to the results of the calculations $P_2^{(1)} = 0.982$, $P_3^{(1)} = 0.969$, $P_4^{(1)} = 0.964$. Their validation through functional equation provides a significant over-limit of 0.85. The following approximation of $P_1^{(2)}$ of P_1 (33) is carried out according to the rule:

$$\begin{cases} P_s = \prod_{i=1}^{4}(P_i) = 0,85 > \prod_{i=1}^{4}(P_i^{(1)}), \rightarrow P_1^{(2)} > P_1^{(1)}, \\ P_s = \prod_{i=1}^{4}(P_i) = 0,85 < \prod_{i=1}^{4}(P_i^{(1)}), \rightarrow P_1^{(2)} < P_1^{(1)}. \end{cases} \tag{34}$$

Table 1. Results of iterative calculations of functional conformity

$P_i^{(j)}$	P_1^j	P_2^j	P_3^j	P_4^j	$\prod P_i^j$
$P_i^{(1)}$	0.96	0.982	0.969	0.964	0.881
$P_i^{(2)}$	0.94	0.974	0.952	0.946	0.824
$P_i^{(3)}$	0.95	0.977	0.961	0.955	0.851
$P_{N,i}$	0.05	0.023	0.039	0.045	-
k_i	1.00	2.17	1.28	1.11	-

The results of the three iterations are given in Table 1.

The accuracy of the calculations proved 0.001 and the third iteration will be the last.

In Table 1 in the bottom two lines shows the functional imperfection values of the i-th link of the $P_{N,i}$, and which are entered by the following:

$$P_{N,i} = 1 - P_i \tag{35}$$

as well as their normalized with respect to $P_{N,i}$ values of the imperfection coefficient k_i. Normalization was carried out by:

$$k_i = \frac{P_{N,1}}{P_{N,i}}. \tag{36}$$

As a result of calculations (35), the value rational distribution between the levels of functional perfection of the links of the educational process was obtained. According to the Table 1, the highest requirements of functional perfection are advanced to the value of P_2 - the parameter of functional perfection of independent training of students.

Thus, the least socially valuable way of improving the quality of the educational process is to improve self-study. There are two areas for such work. First, it is an organizational enhancement of the role and responsibility of the profile department for the methodological support of independent work, as well as the management of the educational institution - for the organization and accessibility of its conduct. Second, it is an improvement in the logistical support of self-training.

The purpose of any object's observation is to obtain targeted information about it. The process of observation can be organized in three available ways: expert evaluation (EE), the use of open sources of information (OSI), the use of technical means of accidents (TDA).

Functional excellence levels of EE, OSI and TDA - P_1, P_2, P_3 , their value - respectively C_1, C_2, C_3. The objective of the analysis is to find out the requirements for the levels of functional excellence of the subsystems EE, OSI and TDA provided that the $C_\sum$ values are minimized to improve the functioning of the surveillance system: $C_\sum = \min$.

The description of the dependence of the value of C_j of an individual subsystem on the level of its perfection P_j is polynomial, so we involve a system of equations of type (24) in the solution. The purpose in this case is to find rational ways to improve the process of observation of the object, provided that the requirements to a given level of clarification of the state of the object at minimum conditional values are compulsory. It is this formulation of the question that will ensure that such a process is planned and purposeful.

To determine the empirical constants A_j, we include the mechanism of assembly of the adjacency matrix by (26):

$$A_{jk} = \begin{cases} 10, & A_{jk} = A_{kj}, \\ 15, & A_{jk} > A_{kj}, \\ 5, & A_{jk} < A_{kj}, \end{cases} \rightarrow A_j = \sum_{k=1}^{k=3} A_{jk}. \tag{37}$$

The results of the peer review will be as follows:

$$A_{jk} = \begin{cases} 10..15..15 \\ 5...10...5 \\ 5...15..10 \end{cases} ; A_j = \begin{cases} 35 \\ 20 \\ 30 \end{cases}. \tag{38}$$

We have: $A_1 = 35, A_2 = 20, A_3 = 30$. We enter the value of P_s in the same way as in the previous example and it will assume that $P_s = 0.85$.

We look for the solution of (24) using the Lagrange multiplier method. The search for extremes requires the solution of a system of (25). It will look like:

$$\begin{cases} 1 - (1 - P_1)(1 - P_2)(1 - P_3) = 0.85, \\ \frac{35P_1}{1-P_1} = \frac{20P_2}{1-P_2}, \\ \frac{35P_1}{1-P_1} = \frac{30P_3}{(1-P_3)^2}. \end{cases} \tag{39}$$

The first approximation of $P_1^{(1)}$ to the true value of P_1, which implies the uniformity of P_j values for all components: $P_j = P$, will be equal to:

$$1 - (1 - P_1^{(1)})^3 = 0.85 \rightarrow P_1^{(1)} = 0.469. \tag{40}$$

The first approximations to $P_2^{(1)}$ and $P_3^{(1)}$ are sought by the second and third equations of (39). As a result of the calculations we get: $P_2^{(1)} = 0.607, P_3^{(1)} = 0.507$. We check the calculated values for the correspondence to the first equation of (39). In the following approximations, we use the (34). For the case under consideration, they will be:

$$\begin{cases} 0.85 > 1 - \prod_{i=1}^{3}(1 - P_i^{(1)}), \rightarrow P_1^{(2)} > P_1^{(1)}, \\ 0.85 < 1 - \prod_{i=1}^{3}(1 - P_i^{(1)}), \rightarrow P_1^{(2)} < P_1^{(1)}. \end{cases} \tag{41}$$

The results of the iterative calculations (40), (41) are shown in Table 2.

Table 2. Results of iterative calculations

$P_i^{(j)}$	P_1^j	P_2^j	P_3^j	P_Σ
$P_i^{(1)}$	0.469	0.607	0.507	0.897
$P_i^{(2)}$	0.400	0.533	0.437	0.843
$P_i^{(3)}$	0.410	0.549	0.448	0.853
$P_{N,i}$	0.590	0.451	0.552	-
k_i	1.00	1.308	1.069	-

The accuracy of the calculations is reduced to 0.003 and the third iteration will be the last. In this case, the overall functional excellence of the system will be 0.003 greater than the set value of 0.85.

Subsystems 1, 2, 3 by personal perfection are more likely to be unable to work effectively, but the result of their work together is quite satisfactory. The rational distribution of the functional excellence of the subsystems is proportional to the value of the value factor A_j. That is the increase in the value of an individual component causes the first of all to increase the level of functional excellence of the cheaper components. In the case under consideration, this is to obtain information from open sources of information and through technical means.

4 Results and Discussion

The result of the study is an algorithmic justification of rational directions of improvement of a complex system with a serial or parallel connection of subsystems with polynomial approximation of the dependence of the subsystem value on the level of their functional suitability. This approach in the functional-value modeling of a complex system allowed solving the following questions:

- forecasting the value behavior of a complex system with serial or parallel interconnection of subsystems, provided that the functional tasks with a certain level of perfection are fulfilled and the priority ways of their improvement are determined;
- facilitating the solution of the forward and reverse problems of rationalization of the structure of complex systems with serial or parallel interconnection of subsystems.

Features of application of the developed method for modeling of complex systems with serial or parallel interconnection of subsystems are considered separately. Efficiency of application of the method is supported by functional-value calculations of specific complex systems with minimal a priori information on the functioning of their subsystems.

The value behavior of systems with sequential subsystem coupling was evaluated by (19). For systems with parallel subsystems, this behavior is determined by their structural purpose and evaluated by systems of equations of (19) or (23).

In both cases, the solution was the distribution between the probabilities of the perfect functioning of the individual subsystems at the minimum value of the entire complex system and the specified functional perfection. The introduction of the deficiency coefficient k_i of the subsystem by (36) simplifies the detection of the subsystem, which in terms of value reduction should be improved first and foremost.

The direct and inverse problems of rationalization of the structure of complex systems with serial or parallel coupling of subsystems are proposed to be solved by applying Eqs. (9, 10). This will determine the requirements for the required level of functional excellence of the j-th subsystem with the known levels of this indicator in other subsystems of a complex system.

The research also considered the possibility of parametric control of the value of a complex system. This possibility arises in the presence of empirical or analytical equations of (11), which determine the probability of execution of the subsystem of its functional purpose.

In addition, the article considers approaches to substantiate the choice of coefficients of approximating equations between value and subsystem perfection. The statistical approach involves taking into account the available statistics on the existence of similar systems and subsystems. In the absence of a priori statistics on specific examples, consider the estimation approach and the algorithm for determining priority in pairwise comparisons (37), (38). Such approaches increase the practical significance of the results of functional-value modeling, as well as contribute to the possibility of analytical formalization of non-formalized complex systems.

The conducted researches allow formulating the following rules of rational structural improvement of a complex system.

Rule 1. The rational structure of a complex system (the number of units i and the k connections between them) must provide a given level of functional perfection P_s.

This rule states that the number of links i and the k connections between them should not be excessive:

$$P = P_s \rightarrow i + k = \min . \tag{42}$$

The value of such a system will be minimal. The rule can be called the rule of rational structural structure of a complex system (42).

Rule 2. Complication of a complex system (increasing its value by $\triangle C$) should increase the functional suitability of $\triangle P$.

Based on the entered tags, you can write it as a condition:

$$C_{\sum+1} - \triangle C_{\sum} = C = \min \rightarrow P_{\sum+1} - P_{\sum} = P = \max . \tag{43}$$

The rule aims at rationalizing the improvement of a complex system (43).

Rule 3. A system is constructed correctly if a decrease $(n \rightarrow (n-1))$ or an increase $(n \rightarrow (n+1))$ by any component (link) results in a decrease $(-\triangle C)$ or increase, respectively $(+\triangle C)$ the value of the system:

$$\begin{cases} n \to (n+1) : \triangle C \geq 0, \\ n \to (n-1) : \triangle C \leq 0. \end{cases} \tag{44}$$

That is, in a complex system (44) there are no extra links that do not perform the functionally necessary actions for this system. The results of studies of the method are following:

- the developed method of functional-value analysis of a complex system with the presence of statistics allows the most accurate parametric control of the functional perfection of a complex system at a minimum value. In the absence of such data, it helps to more consciously approach the parametric and structural improvement of a complex system;
- application of this method is appropriate for the study of a complex system at all stages of its existence: development, operation, utilization;
- this method can be applied to the study of weakly formalized and non-formalized complex systems, that is, allows qualitative analysis of complex systems to be translated into quantitative ones.

Unfortunately, the results of the researches presented in the article do not fully cover the peculiarities of the application of this method of functional-value study. Mixed subsystems are not considered, the example of functionally heterogeneous subsystems is not considered for parallel subsystems, and there are limitations to the types of mathematical models the authors worked with. Branching such directions is a priority in further work on improving the capabilities of the developed method.

In addition, the algorithms developed in the examples are not aimed at obtaining specific, physically and mathematically justified values of value and levels of functional excellence. Such structures and calculations only help consciously approach the structure of a complex system.

To a greater extent, these restrictions can be removed when individualizing the application of the developed method for complex systems of a specific purpose, for example, social, administrative, technical, complex systems. In addition, to improve the accuracy of predictive findings from the results of this method, data from previous studies of similar complex systems should be included.

5 Conclusions

Is developed a method of structural functional-value modeling of a complex system. This method is based on a polynomial approximation of the dependence of the value of subsystems of a complex system on the level of their functional perfection. It allows parametric control of the functionality of a complex system with the condition of its minimum value. This method is adapted to application at different levels of a priori uncertainty of the initial data and can be applied at all stages of existence of a complex system: development, operation, utilization. In addition, it is useful for the study of poorly formalized and non-formalized complex systems, that is, allows qualitative analysis of a complex system to be translated into quantitative ones.

The initial data for the calculations by the developed method consist of the choice of a generalized indicator of the structural functional perfection of the system and its components, as well as its value. Choosing as a generalized indicator of the likelihood of fulfilling a functional purpose takes into account the stochastic existence of a complex system. It's allowed to go to parametric control of the level of functional perfection of the system and to use the regulatory requirements for such indicators. The order of structural functional-value calculations of complex system with serial or parallel connection of separate parts for rationalization of its structure is developed. This allowed us to propose a typical solution of a system of equations of functional-value rationalization of a complex system based on the Lagrange multiplier method and to involve an iterative approach for obtaining solutions of systems of equations above the third order.

The efficiency of the developed method algorithms during the research as complex systems of the educational process (sequential interconnection of subsystems) and object monitoring (parallel interconnection of subsystems) were performed. The result of the calculations is the rational value allocation between the levels of functional excellence of the subsystems of each of the systems under study. In determining approximate dependencies, the estimated and priority approaches to substantiate the choice of coefficients of approximating equations between the value and the perfection of the subsystem. These choices allow compensating for the lack of a priori system data are investigated. The introduction of the imperfection factor has simplified the identification of the subsystem, which in terms of value reduction should be improved first and foremost.

References

1. Apollo configuration management manual. D.C. 20 046. US Government Printing Office NASA, Washington (1970)
2. Babichev, S., Korobchynskyi, M., Lahodynskyi, O., Korchomnyi, O., Basanets, V., Borynskyi, V.: Development of a technique for the reconstruction and validation of gene network models based on gene expression profiles. East.-Eur. J. Enterp. Technol. **1**(4(91)), 19–32 (2018). https://doi.org/10.15587/1729-4061.2018.123634
3. Babichev, S., Škvor, J., Fišer, J., Lytvynenko, V.: Technology of gene expression profiles filtering based on wavelet analysis. Int. J. Intell. Syst. Appl. **10**(4), 1–7 (2018). https://doi.org/10.5815/ijisa.2018.04.01
4. Babichev, S., Lytvynenko, V., Škvor, J., Fišer, J.: Model of the objective clustering inductive technology of gene expression profiles based on SOTA and DBSCAN clustering algorithms. Adv. Intell. Syst. Comput. **689**, 21–39 (2018). https://doi.org/10.1007/978-3-319-70581-1_2
5. Beavers, D., Stamey, J.: Bayesian sample size determination for value-effectiveness studies with censored data. PLoS ONE **13**(1), e0190422 (2018). https://doi.org/10.1371/journal.pone.0190422
6. Bertsekas, D.: Constrained Optimization and Lagrange Multiplier Methods. Athena Scientific, Athena (1982)

7. Elrehim, M., Eid, M., Sayed, M.: Structural optimization of concrete arch bridges using genetic algorithms. Ain Shams Eng. J. **10**(3), 507–516 (2019)
8. Glushkov, V.: Introduction to cybernetics. USSR Academy of Sciences, Kyiv (1964)
9. Good, H., Machol, R.: System Engineering. An Introduction to the Design of Large-Scale Systems. McGraw-Hill Book Company, New York (1957)
10. Jaehyun, L., Jinko, K., Woong, K.: Day-ahead electric load forecasting for the residential building with a small-size dataset based on a self-organizing map and a stacking ensemble learning method. Appl. Sci. **9**(6), 12–31 (2019). https://doi.org/10.3390/app9061231
11. Korobchynskyi, M., Slonov, M., Tsybulskyi, S., Dereko, V., Maryliv, O.: Construction of a method for the structural functional-cost modeling of a complex hierarchic system. East.-Eur. J. Enterp. Technol. **14**(103), 11–22 (2020). https://doi.org/10.15587/1729-4061.2020.195379
12. Leachtenauer, J.: Resolution requirements and the Johnson criteria revisited. In: Proceedings of SPIE, vol. 5076, Washington (2003)
13. Muzeyyen, A., Guldmann, J.: Joint values in electricity and natural gas distribution infrastructures: the role of urban factors. Urban Sci. **2**(2), 35 (2018). https://doi.org/10.3390/urbansci2020035
14. Pelin, A., Munteanu, V., Pantea, M., Gligor, D.: Methodological approaches in realizing and applying value-benefit analysis for the investment projects. Ann. Univ. Oradea. Econo. Sci. **2**(1), 156–162 (2009)
15. Ruyi, L., Zhen, W.: Well-posedness of a class of two-point boundary value problems associated with ordinary differential equations. Adv. Differ. Equ. **1**, 1–12 (2018). https://doi.org/10.1186/s13662-018-1510-5
16. Saaty, T.: The us-opecenergy conflict: the payoff matrix by the analytic hierarchy process. Int. J. Game Theory **8**(4), 225–234 (1979)
17. Sentz, K., Ferson, S.: Combination of Evidence in Dempster-Shafer Theory, Sandia National Laboratories, New York, January 2002. https://doi.org/10.2172/800792
18. Slonov, M.: Functional-cost approach to structural improvement of complex system: educational process. Notes V.I. Vernadsky Taurida Nat. Univ. **30(69)**(4), 124–128 (2019). https://doi.org/10.32838/2663-5941/2019.4-1/22

Generalized Approach for Estimating and Forecasting of Dynamical VaR and CVaR Based on Metalog Distribution

Vira Zrazhevska[1(✉)] and Grigoriy Zrazhevsky[2]

[1] National Technical University of Ukraine "Igor Sikorsky Kyiv Polytechnic Institute", Kyiv, Ukraine
vera.zrazhevska@gmail.com
[2] National Taras Shevchenko University of Kyiv, Kyiv, Ukraine
zgrig@univ.kiev.ua

Abstract. The work is devoted to the development of methods for dynamic risk measures VaR and CVaR estimating. As a basic model, a heteroscedastic time series model is considered. The methods proposed in the article are designed for obtaining the forecast estimates of risk measures for volatile time series taking into account the long-range dependence presence. The method of smoothing of the autocorrelation function based on an optimization procedure is used for variance modeling. A metalog distribution is proposed to use for risk measures model residuals estimating. This distribution allows to describe the behavior of the tail part of the distribution with different characteristics. The paper proposes two methods of metalog distribution estimating. The first method is based on an empirical distribution function and the second one on its approximation by sample quantiles. For VaR and CVaR modeling and forecasting, explicit analytical formulas were obtained with different number of members of the metalog distribution. The procedure for obtaining the forecast values of dynamic risk measures VaR and CVaR is formulated as an algorithm. The proposed approach is applied to the time series of the Russian Trading System index for the period 14/10/2005–10/02/2020. For comparison, the forecast of dynamic risk measures is built using well known methods of risk estimation based on the GEV distribution, GPD and historical modeling. Quantitative and qualitative analyzes of the obtained estimates confirmed the high quality of the obtained estimates.

Keywords: VaR and CVaR forecasting · FIGARCH model · Metalog distribution · Long-range dependence · Method of smoothing of the autocorrelation function

1 Introduction and Literature Review

The problem of financial risk forecasting is one of the most actual problems that attract the attention of financiers and mathematicians around the world.

S. Babichev et al. (Eds.): ISDMCI 2020, AISC 1246, pp. 232–245, 2021.
https://doi.org/10.1007/978-3-030-54215-3_15

Numerous economic crises from year to year damage the stable functioning of financial institutions and pose new challenges to mathematicians, demanding further improvement of the existing apparatus and the creation of new approaches, methods and models for risk measures. A clear example of recent times is the COVID 19 pandemic. Among the measures that help to estimate possible losses and, therefore, take timely measures to prevent excessive financial losses, are the VaR (Value-at-Risk) and CVaR (Conditional Value-at-Risk) measures.

The general concept of the measure VaR was first formulated in the works of Harry Markovic in the middle of the XX century. In 1994, J.P. Morgan developed a methodology for VaR estimation and included this measure in the list of mandatory indicators of J.P. Morgan Bank in order to increase the effectiveness of risk management. Since 1999, under the Basel II agreement, VaR has been one of the standard indicators used by most financial institutions.

The shortcomings of the VaR risk measure, in particular the lack of the coherence property, led to the emergence of the CVaR measure in 1999 [1]. The undoubted advantage of CVaR is that this measure allows to quantify the losses in case the VaR limit value is exceeded.

At present, there is already many methods and algorithms for VaR and CVaR estimating and modeling. Thus, in the work [14] the main methods of VaR and CVaR estimation for static risks are analyzed. Most existing methods are based on modeling of a random variable distribution function. In one of the most popular methods, the historical simulation method, the estimation is based on empirical distribution function constructed from historical data. If there is a reasonable assumption about the type of random variable distribution and the ability to estimate its parameters ready-made analytical formulas or the Monte Carlo method can be used. This approach to risk estimation has wide practical application. For example article [9] analyzes two methods for VaR estimation: historical simulation, and Monte Carlo simulation. The authors apply these approaches on six different equities on the Ghana Stock Exchange Market.

If it is not possible to make a reasonable assumption about the type of distribution of all values of a random variable, Extreme Value Theory (EVT) can be used to assess static risks. EVT allows analyzing extreme historical data and evaluate only the tail part of the distribution. The use of Generalized Extreme Value Distribution (GEV) and the use of Generalized Pareto Distribution (GPD) provide analytical formulas for VaR and CVaR. In the article [11], GEV approach is used for VaR estimation as a measurement of gold investment risk in Indonesia.

One of the modern methods of static risk measures estimation is based on the metalog distribution. The main positions of this theory were set out in the work [4]. An advantage of this type of distribution for risk measures estimation is the ability to model the different behavior of the tail of the distribution, depending on the parameters. Being quite new, this method has not yet been widely used in practical assessments of risk measures. Among the latest work on the use of metalog distribution is the prediction of the number of people in the United States who will die as a result of the COVID-19 pandemic in 2020 [2]. In today's fast-paced world, even more relevant than the task of static risks assessing is the dynamic risks modeling and forecasting. The time series analysis is one of the main mathematical tools for this. An analysis of existing models for dynamic VaR and CVaR estimating and forecasting is performed in the article [13]. One of the basic approaches to dynamic risk measures modeling is based on stochastic time series models [6]. The problem of VaR and CVaR modeling in this case is to build a model for variance and to estimate risk measures model residuals. An example of such an approach for CVaR estimation for the Tehran Stock Exchange can be found in the work [12]. The authors used GARCH (1, 1) model for variance modeling and POT theory for CVaR estimation for residuals. The authors of the article [3] proposed a two-stage hybrid method that combines a GARCH filter with the EVT (GARCH-EVT). The method was used to estimate VaR for a long and a short positions applying 10 world stock indices. The article [8] combines the GJR-GARCH model that takes into account the asymmetric shocks of volatility and EVT. The empirical analysis is based on daily average currency exchange rates from the Central Bank of Kenya.

Unpredictable reality generates data that often does not obey standard laws. Therefore, despite the large number of developed methods for assessing and forecasting dynamic and static risk measures, the problem of creating new approaches that best take into account the peculiarities of historical data and can be easily adapted to obtain qualitative forecasts remains relevant. In particular, this applies to data described by time series with the property of long-range dependence. The method of smoothing of the autocorrelation function is among the methods for VaR and CVaR forecasting for volatile financial series with long-range dependence [10]. On the other hand, the unpredictability of the behavior of extreme values requires the use of new types of distributions, such as, in particular, the metallog distribution [4].

The aim of this research is to develop new methods for dynamic risk measures VaR and CVaR modeling and forecasting. These new methods must more fully describe the input data, display their features, thereby improving the quality of forecast estimates. During the period of crisis in the economy and the financial system, a qualitative forecasting of risk measures can maintain the stability of the financial and, in particular, banking sector.

2 Materials and Methods

We consider the continuously distributed time series $\{X_t, t \in \mathrm{T}\}$ defined on the probability space $(\Omega, \Phi_t, \mathrm{P}_t)$ with a set of observations $X_1, ..., X_N$ (Φ_t is the information set containing all available at the time t information about the time series). For a fixed confidence level α dynamic risk measures VaR_α^t is a conditional quantile of the probability distribution function for X_t: $\mathrm{P}_t\left[X_t < -VaR_\alpha^t | \Phi_t\right] = \alpha$. Risk measure $CVaR_\alpha^t$ is defined accordingly to [1] as:

$$CVaR_\alpha^t = E_t[X_{\tilde{t}}\,|X_{\tilde{t}} < -VaR_\alpha^t(t)] = -\frac{1}{\alpha}\int_0^\alpha VaR_\gamma^t\, d\gamma, \quad \alpha < 0.5,$$

where $E_t[\cdot]$ is a conditional expectation defined on Φ_t . The squared time series is considered for variance modulation. Series $\{X_t^2,\, t\, \in\, \mathrm{T}\}$ is assumed to be stationary and long-range dependent: $\exists\, \kappa > 0,\, 0 < \omega < 1 : \lim\limits_{k \to \infty} \rho_k/(\kappa\, k^{-\omega}) = 1$, where $\rho_k = corr(X_t^2, X_{t+k}^2)$ is an autocorrelation function, $k \in N \bigcup \{0\}$.

A model FIGARCH (p, d, q) is used to account the property of a long-range dependence in the modeling of a series [13]:

$$X_t = \mu + \varepsilon_t = \mu + \sigma_t \nu_t \tag{1}$$

$$(1 - b(L)){\sigma_t}^2 = b_0 + (1 - b(L) - \varphi(L)(1 - L)^d){\varepsilon_t}^2 \tag{2}$$

where μ denotes a mean, σ_t^2 is a conditional variance, defined on Φ_t, $\nu_t \sim iid(0,1)$, $LX_t = X_{t-1}$, $a(L) = \sum_{i=1}^{q} a_i L^i$, $b(L) = \sum_{j=1}^{p} b_j L^j$, $\varphi(L) \equiv (1 - a(L) - b(L))(1 - L)^{-1}$, $(1 - L)^d = \sum_{j=1}^{\infty} \frac{\Gamma(j-d)}{\Gamma(-d)\Gamma(j+1)} (-L)^j$ is a fractional difference operator, $\Gamma(\cdot)$ is a gamma function, $d \in (0,\ 1/2)$; $b_0 > 0$; $a_i \geq 0$,$i = 1, 2, \ldots, q$; $b_j \geq 0$, $j = 1, 2, \ldots, p$; $\sum_{i=1}^{q} a_i + \sum_{j=1}^{p} b_j < 1$.

After removing a trend μ from a time series, the model for risk measures based on the FIGARCH(p, d, q) model can be written as [6]:

$$VaR_\alpha\,(X_t) = VaR_\alpha(\nu)\sigma_t, CVaR_\alpha\,(X_t) = CVaR_\alpha(\nu)\sigma_t, \tag{3}$$

where σ_t and ν are defined in (1), (2). $VaR_\alpha(\nu)$ and $CVaR_\alpha(\nu)$ are risk measures model residuals, on the condition $F(\nu_t) \equiv F(\nu)$. Then the forecasting values for S steps ahead for dynamic risk measures can be found as:

$$VaR_\alpha\,(X_{t+S}) = VaR_\alpha(\nu)\sigma_{t+S}, CVaR_\alpha\,(X_{t+S}) = CVaR_\alpha(\nu)\sigma_{t+S}. \tag{4}$$

The following methods are used to estimate $VaR_\alpha(\nu)$ and $CVaR_\alpha(\nu)$.

Historical Simulation Method. According to this method, estimates for risk measures for a random variable X with sample $(X_1, X_2, ..., X_N)$ are found as follow:

$$VaR_\alpha\,(X) = -X_{([N\alpha])}, CVaR_\alpha\,(X) = -\left(\sum_{i=1}^{[N\,\alpha]} X_{(i)}\right) / \left([N\,\alpha]\right), \tag{5}$$

where $X_{(1)} \leq X_{(2)} \leq ... \leq X_{(N)}$ - ordered sample.

Using Generalized Extreme Value (GEV) Distribution. Suppose, that random variable X follows GEV distribution with distribution function $F(x) = \begin{cases} \exp\left(-\left(1 + \xi\,\frac{x-\mu}{\sigma}\right)^{-\frac{1}{\xi}}\right), & \xi \neq 0 \\ \exp\left(-\exp\left(-\frac{x-\mu}{\sigma}\right)\right), & \xi = 0 \end{cases}$. Then the risk measures [5]:

$$
\begin{aligned}
VaR_\alpha(X) &= \begin{cases} -\mu - \frac{\sigma}{\xi}\left((-\ln\alpha)^{-\xi} - 1\right), \ \xi \neq 0 \\ -\mu + \sigma\ln(-\ln\alpha), \ \xi = 0 \end{cases} \\
CVaR_\alpha(X) &= \begin{cases} -\mu - \frac{\sigma}{\alpha\xi}\left(\Gamma(1-\xi, -\ln\alpha) - \alpha\right), \ \xi \neq 0 \\ -\mu - \frac{\sigma}{\alpha}\left(li(\alpha) - \alpha\ln(-\ln\alpha)\right), \ \xi = 0 \end{cases},
\end{aligned} \tag{6}
$$

where $\Gamma(\cdot,\cdot)$ is upper incomplete gamma function, $li(\alpha) = \int \frac{dx}{\ln x}$ is logarithmic integral function.

Using Generalized Pareto Distribution (GPD). Suppose, that a random variable X follows GPD with distribution function:

$$
F(x) = \begin{cases} 1 - \left(1 + \xi\frac{x-\mu}{s}\right)^{-\frac{1}{\xi}}, \ \xi \neq 0 \\ 1 - \exp\left(-\frac{x-\mu}{s}\right), \ \xi = 0 \end{cases}
$$

Then [7] risk measures for losses: $L = -X$, $p = 1 - \alpha$:

$$
\begin{aligned}
VaR_p(L) &= \begin{cases} \mu + s\frac{(1-p)^{-\xi}-1}{\xi}, \ \xi \neq 0 \\ \mu - s\ln(1-p), \ \xi = 0 \end{cases} \\
CVaR_p(L) &= \begin{cases} \mu + s\left(\frac{(1-p)^{-\xi}}{1-\xi} + \frac{(1-p)^{-\xi}-1}{\xi}\right), \ \xi \neq 0 \\ \mu + s\left(1 - \ln(1-p)\right), \ \xi = 0 \end{cases}.
\end{aligned} \tag{7}
$$

Using Metalog Distribution. Suppose that a random variable X with a distribution function $F_X(x)$ follows the metalog distribution [4]. The metalog distribution is determined by the quantile function $M_n(\alpha, \boldsymbol{x}, \boldsymbol{a})$:

$$
M_n(\alpha, \boldsymbol{x}, \boldsymbol{a}) = \begin{cases}
a_1 + a_2\ln\frac{\alpha}{1+\alpha}, & n = 2 \\
a_1 + a_2\ln\frac{\alpha}{1+\alpha} + a_3(\alpha - 0.5)\ln\frac{\alpha}{1+\alpha}, & n = 3 \\
a_1 + a_2\ln\dfrac{\alpha}{1+\alpha} + a_3(\alpha - 0.5)\ln\dfrac{\alpha}{1+\alpha} + & \\
a_4(\alpha - 0.5), & n = 4 \\
M_{n-1} + a_n(\alpha - 0.5)^{\frac{n-1}{2}}, & \text{for odd } n \geqslant 5 \\
M_{n-1} + a_n(\alpha - 0.5)^{\frac{n}{2}-1}\ln\frac{\alpha}{1+\alpha}, & \text{for even } n \geqslant 6
\end{cases}
$$

where $\alpha \in (0,1)$ is probability, $\boldsymbol{a} = (a_1, a_2, ..., a_n)^T$ is a parameter vector defined by:

$$
\boldsymbol{a} = [{Y_n}^T Y_n]^{-1} {Y_n}^T \boldsymbol{x}. \tag{8}
$$

Column vectors $\boldsymbol{x} = (x_1, x_2, ..., x_m)^T$ and $\boldsymbol{\alpha} = (\alpha_1, \alpha_2, ..., \alpha_m)^T$ are coordinates of distribution function $F_X(x)$, $m \geq n$, $\alpha_i \in (0,1)$, $i = 1, ..., m$, at least n of α_i

are distinct. The matrix Y_n is defined as follow:

$$Y_n = \begin{cases} \begin{bmatrix} 1 & \ln\frac{\alpha_1}{1+\alpha_1} \\ \cdots & \cdots \\ 1 & \ln\frac{\alpha_N}{1+\alpha_N} \end{bmatrix}, & n=2 \\ \begin{bmatrix} 1 & \ln\frac{\alpha_1}{1+\alpha_1} & (\alpha_1-0.5)\ln\frac{\alpha_1}{1+\alpha_1} \\ \cdots & \cdots & \cdots \\ 1 & \ln\frac{\alpha_N}{1+\alpha_N} & (\alpha_N-0.5)\ln\frac{\alpha_N}{1+\alpha_N} \end{bmatrix}, & n=3 \\ \begin{bmatrix} 1 & \ln\frac{\alpha_1}{1+\alpha_1} & (\alpha_1-0.5)\ln\frac{\alpha_1}{1+\alpha_1} & (\alpha_1-0.5) \\ \cdots & \cdots & \cdots & \cdots \\ 1 & \ln\frac{\alpha_N}{1+\alpha_N} & (\alpha_N-0.5)\ln\frac{\alpha_N}{1+\alpha_N} & (\alpha_N-0.5) \end{bmatrix}, & n=4 \\ \begin{bmatrix} & (\alpha_1-0.5)^{\frac{n-1}{2}} \\ Y_{n-1}| & \cdots \\ & (\alpha_N-0.5)^{\frac{n-1}{2}} \end{bmatrix}, & \text{for odd } n \geqslant 5 \\ \begin{bmatrix} & (\alpha_1-0.5)^{\frac{n}{2}-1}\ln\frac{\alpha_1}{1+\alpha_1} \\ Y_{n-1}| & \cdots \\ & (\alpha_N-0.5)^{\frac{n}{2}-1}\ln\frac{\alpha_N}{1+\alpha_N} \end{bmatrix}, & \text{for even } n \geqslant 6 \end{cases} \tag{9}$$

Based on the definition of risk measures, we have:

$$VaR_{\tilde{\alpha},n}(X) = -M_n(\tilde{\alpha}, \boldsymbol{x}, \boldsymbol{\alpha}),. \tag{10}$$

$$CVaR_{\tilde{\alpha},n}(X) = -\frac{1}{\tilde{\alpha}}\int_0^{\tilde{\alpha}} M_n(y, \boldsymbol{x}, \boldsymbol{\alpha})\,dy.$$

In this paper the explicit analytical formulas for finding $CVaR_{\tilde{\alpha},n}(X)$ are obtained for the different values of n :

$$CVaR_{\tilde{\alpha},n}(X) = \begin{cases} a_1 + a_2\left(\frac{\ln(1-\tilde{\alpha})}{\tilde{\alpha}} + \ln\frac{\tilde{\alpha}}{1-\tilde{\alpha}}\right), & n=2 \\ CVaR_{\tilde{\alpha},2}(X) + \frac{a_3}{2}\left(1 + (\tilde{\alpha}-1)\ln\frac{\tilde{\alpha}}{1-\tilde{\alpha}}\right), & n=3 \\ CVaR_{\tilde{\alpha},3}(X) + \frac{a_4(\tilde{\alpha}-1)}{2}, & n=4 \\ CVaR_{\tilde{\alpha},n-1}(X) + \dfrac{2a_n}{\tilde{\alpha}(1+n)} & \\ \times\left((-1)^{\frac{n-1}{2}}(0.5)^{\frac{n+1}{2}} + (\tilde{\alpha}-0.5)^{\frac{n+1}{2}}\right), & \text{odd}\, n \geqslant 5 \\ CVaR_{\tilde{\alpha},n-1}(X) + \frac{2^{\frac{2-n}{2}}a_n}{n\tilde{\alpha}}2(2\tilde{\alpha}-1)^{\frac{n}{2}} & \\ \times\, arctg(2\tilde{\alpha}-1) + nG + \ln(1-\tilde{\alpha}) & \\ +(-1)^{\frac{n+2}{2}}\ln\tilde{\alpha}, & \text{even}\, n \geqslant 6 \end{cases} \tag{11}$$

where $G = {}_3F_2\left[1,1,1-{}^{n}\!/_{2};\ 2,2;\ 2\right] + (\tilde{\alpha}-1){}_3F_2\left[1,1,1-{}^{n}\!/_{2};\ 2,2;\ 2-2\tilde{\alpha}\right] + +(-1)^{{}^{n}\!/_{2}}2\tilde{\alpha}\ {}_3F_2\left[1,1,1-{}^{n}\!/_{2};\ 2,2;\ 2\tilde{\alpha}\right]$, ${}_3F_2\left[c_1,c_2,c_3;\ d_1,d_2;\ z\right]$ is generalized hypergeometric function. Coefficients $\boldsymbol{a} = (a_1, a_2, ..., a_n)^T$ are defined from (8).

As we see from (11), the explicit formula for finding $CVaR_{\tilde{\alpha},n}(X)$ for even $n \geq 6$ is quite complex. Therefore, for convenience of practical application, approximate formulas were obtained on the basis of the expansion of $CVaR_{\tilde{\alpha},n}(X)$ in powers of $\tilde{\alpha}$. In particular:

$$\begin{aligned} &CVaR_{\tilde{\alpha},6}(X) = CVaR_{\tilde{\alpha},5}(X) \\ &+ a_6\left({}^1\!/_{12\tilde{\alpha}}\ \ln(1-\tilde{\alpha}) + {}^1\!/_6(\tilde{\alpha}-1) + \left({}^1\!/_4 - {}^{\tilde{\alpha}}\!/_2 + {}^{\tilde{\alpha}^2}\!/_3\right)\ln\tfrac{\tilde{\alpha}}{1-\tilde{\alpha}}\right) + o\left(\tilde{\alpha}^2\right) \\ &CVaR_{\tilde{\alpha},8}(X) = CVaR_{\tilde{\alpha},7}(X) \\ &+ a_8\left({}^1\!/_8 - {}^1\!/_8\tilde{\alpha} + {}^1\!/_{12}\tilde{\alpha}^2 + \left(-{}^1\!/_8 + {}^{3\tilde{\alpha}}\!/_8 - {}^{\tilde{\alpha}^2}\!/_2 + {}^{\tilde{\alpha}^3}\!/_4\right)\ln\tfrac{\tilde{\alpha}}{1-\tilde{\alpha}}\right) + o\left(\tilde{\alpha}^3\right) \end{aligned} \tag{12}$$

The proposed theoretical approach allowed to formulate the algorithm for VaR and CVaR forecasting.

An Algorithm for Dynamic VaR and CVaR Forecast Based on Metalog Distribution.

1. **Statistical data analysis.** Test the time series on volatility and ARCH effect. Test the time series of squares on the long-range dependence: the Hurst parameter estimating with standard methods and obtaining the estimations $\hat{H}_i$, $i = 1, ..., I$.
2. **Variance modeling and prediction** of ${\sigma_t}^2$ with the method of smoothing of the autocorrelation function (MSAF) [10]. According to the MSAF, a regression model is used to find estimates of the autocorrelation function (ACF), which follows from the definition of a long-range dependence:

$$\rho_k = \beta_1 H(2H-1)\, k^{2H-2} + \beta_2 + \upsilon_k, \tag{13}$$

where υ_k $-$ $iid.$, $k_0 \le k \le N$.
According to (13), ACF estimates are: $\tilde{\rho}_k = \tilde{\beta}_1 \hat{H}_m (2\hat{H}_m - 1)\, k^{2\hat{H}_m - 2} + \tilde{\beta}_2$, where $\widehat{H}_m = \frac{1}{I}\sum_{i=1}^{I} \widehat{H}_i$, coefficients estimates $\tilde{\beta}_1$, $\tilde{\beta}_2$ are found with the method of least squares. The estimates $\tilde{\rho}_k$ and the value of the Hurst parameter are corrected with the help of the optimization procedure:

$$\lambda q + (1-\lambda)\frac{1}{k_2 - k_1}\sum_{k=k_1+1}^{k_2} \left(\rho_k - \tilde{\rho}_k\right)^2 \to \min$$

$$\left(\rho_k - \tilde{\rho}_k\right)^2 \le \varepsilon^2 + q, k = k_0, ..., k_1, q \ge 0,$$

where the parameters of the optimization procedure k_1 and $0 \le \lambda \le 1$ are chosen to describe the behavior of the ACF for large values of the argument in the best way. The obtained estimates $\hat{H}_{opt}$, $\hat{\rho}_k = \hat{\beta}_1 \hat{H}_{opt}(2\hat{H}_{opt} - 1)\, k^{2\hat{H}_{opt}-2} + \hat{\beta}_2$, $k_0 \le k \le N$ and the forecasting value $\hat{\rho}_{N+1} = \hat{\beta}_1 \hat{H}_{opt}(2\hat{H}_{opt} - 1)\,(N+1)^{2\hat{H}_{opt}-2} + \hat{\beta}_2$ are used for building the variance model.

To predict the variance, the FIGARCH(p,d,q) model (2) is rewritten as an AR(∞) model for squared process ${\varepsilon_t}^2$:

$${\varepsilon_t}^2 = \gamma_0 + \sum_{i=1}^{\infty} \gamma_i {\varepsilon_{t-i}}^2 + \zeta_t, \tag{14}$$

where $\zeta_t = {\varepsilon_t}^2 - {\sigma_t}^2$ are uncorrelated with mean 0. To find estimates of the autoregressive coefficients $\hat{\gamma}_i$, a reduced system of normal equations is used:

$\sum_{j=1}^{k} \hat{\rho}_{|i-j|}\gamma_j = \hat{\rho}_i$, $i = 1, ..., k$. The value k ($k \leq N$) is determined by the condition of practical convergence of solutions. The coefficients of the system are taken as estimates $\hat{\rho}_i$, for $i = 1, ..., k_1$, evaluated using observed values of the time series. $\hat{\rho}_i$ for $i = k_1 + 1, ..., k-1$ are found by the optimization procedure. $\hat{\rho}_k$ is evaluated by model extrapolation. The order of the reduced autoregressive model (14) $m \leq k$ is determined using information criteria. The built model is used to find variance estimates: $\hat{\sigma}_t^2 = \hat{\gamma}_0 + \sum_{i=1}^{m} \hat{\gamma}_i \hat{\sigma}_{t-i}^2$, $t = 1, ..., N$ and to predict variance: $\hat{\sigma}_{t+s}^2 = \hat{\gamma}_0 + \sum_{i=s}^{m+s} \hat{\gamma}_i \hat{\sigma}_{t-i+1}^2$, $t = N, ...,$ $s = 1, ..., S$. The estimated variance $\hat{\sigma}_t^2$ is used to find the residuals of model (1): $\hat{\nu}_t = X_t/\hat{\sigma}_t$. The residuals $\hat{\nu} = \{\hat{\nu}_t, t = 1, ..., N\}$ are checked to be iid.

3. **Risk measures model residuals $VaR_\alpha(\nu)$ and $CVaR_\alpha(\nu)$ estimating.** Methods based on the metalog distribution (analytical formulas (10)–(12)) are proposed to estimate the measures $VaR_\alpha(\nu)$ and $CVaR_\alpha(\nu)$.
The first method, the Classical Metalog Method (Metalog full), fulfill the parameter estimation of the metalog distribution using the empirical distribution function $\hat{\boldsymbol{\nu}}^* = (\hat{\nu}_{(2)}, ..., \hat{\nu}_{(N)})^T$, $\hat{\boldsymbol{\alpha}} = (\frac{1}{N}, \frac{2}{N}, ..., \frac{N-1}{N})^T$.
Evaluated by (8) and (9) coefficients $\hat{\boldsymbol{a}} = (\hat{a}_1, \hat{a}_2, ..., \hat{a}_n)^T$ are used to estimate $M_n(\tilde{\alpha}, \hat{\boldsymbol{\nu}}^*, \hat{\boldsymbol{a}})$ and find $VaR_{\tilde{\alpha},n}(\hat{\boldsymbol{\nu}})$ by (10), $CVaR_{\tilde{\alpha},n}(\hat{\boldsymbol{\nu}})$ by (11) or (12) for large even n.
The second method, the Quantile Metalog Method (Metalog quant), fulfill the parameter estimation of the metalog distribution using the empirical quantiles. In this case $\hat{\boldsymbol{\nu}}^* = (\tau_{\alpha_1}, ..., \tau_{\alpha_p})^T$, $p < N$. The empirical quantile τ_{α_i} can be found under the following formula:
$$\tau_{\alpha_i} = \begin{cases} \hat{\nu}_{([N\alpha_i]+1)}, & N\alpha_i \notin Z \\ \frac{\bar{\alpha}-\alpha_i}{\bar{\alpha}-\underline{\alpha}}\hat{\nu}_{(\underline{\alpha})} + \frac{\alpha_i-\underline{\alpha}}{\bar{\alpha}-\underline{\alpha}}\hat{\nu}_{(\bar{\alpha})}, & N\alpha_i \in Z \end{cases},$$
where $\underline{\alpha} = \frac{[N\alpha_i]}{N}$, $\bar{\alpha} = \frac{[N\alpha_i+1]}{N}$, $\alpha_i \in (0, 1)$.
Similar to the previous approach, the coefficients are estimated by (8), (9). Risk measures are obtained by (10), (11) or (12). Since the values of the quantiles τ_{α_i} for the tail part of the distribution are the most important to evaluate $VaR_{\tilde{\alpha},n}(\hat{\boldsymbol{\nu}})$ and $CVaR_{\tilde{\alpha},n}(\hat{\boldsymbol{\nu}})$, it is advisable to choose the value of α_i over a non-uniform grid, for example, $\alpha_i = ih$, $h = 0.005$ for $i = \overline{1, 10}$, and $\alpha_i = ih$, $h = 0.01$ for $i = \overline{11, 100}$. Thus, the second proposed method more fully uses the available data to model the tail of the distribution function, which can improve the quality of risk measures estimates.
4. **Estimation and forecasting of dynamic risk measures**. Obtained estimates of variance and risk measures model residuals are used to find estimates of dynamic risk measures $VaR_\alpha\left(\hat{X}_t\right)$, $CVaR_\alpha\left(\hat{X}_t\right)$ by (3). Extrapolation of the model allows to obtain predictive values $VaR_\alpha\left(\hat{X}_{t+S}\right)$, $CVaR_\alpha\left(\hat{X}_{t+S}\right)$ by (4).
5. **Analysis of estimates**. To test the quality of the methods proposed in the work, the predicted values are compared with the estimates based on the observed data. The values of mean error (ME), mean absolute error (MAE),

and root mean square error (MSE) are used for quantitative analysis of the estimates. Qualitative analysis is performed using backtesting. For the VaR estimates, the Kupiec test (LRuc statistics), the Kristoffersen independence test (LRind statistics), the Kupiec and Kristoffersen combined test (LRss statistics) are used. For CVaR estimates, the V - test with statistics V_1, V_2, V, and statistic $bPoE(\alpha)$: $bPoE(\alpha) = 1 - CVaR_{\alpha}^{-1}(X)$ are used [7].

3 Experiment, Results and Discussion

To validate the proposed approach to VaR and CVaR forecasting, time series of returns on a daily basis of the official indicator of the Russian Trading System index (RTS index) is considered. The length of the total sample is 3386 values for the period from 10/14/2005 to 10/02/2020. The confidence level for the risk measures is $\alpha = 0.1$. Method of forecasting is straight, multi-step (S = 5 steps forward), window with accumulation. The length of the initial window is half of the sample. A total number of windows is 337. At each step, the predicted values of the risk measures are compared with the real ones. To obtain the final characteristics of the forecast quality, all S-step forecasts are combined into one.

Five methods are used to estimate the Hurst parameter: the R/S method, the aggregated variance method, the method of absolute values of the aggregated series, the periodogram method, the method of residuals of regression. The analysis of the obtained values confirmes the presence of the long-range dependence for all windows. The average value is $\widehat{H}_m = 0,8639$. The optimal autoregressive order is determined by AIC and HQC information criteria: $m = 26$. Risk measures estimates for model residuels (1) $\hat{\boldsymbol{\nu}}$ are obtained: using formulas (5) (historical method), formulas (6) (GEV method), formulas (7) (GPD method), using the Classical Metalog Method (Metalog full) and Quantile Metalog Method (Metalog quant). Metalog distribution uses $n = 5$. The results of estimation of probability densities for ν and its tail part using the above methods are shown in Fig. 1 and 2 respectively. Note that further for convenience the sample values are multiplied by (-1), $p = 1 - \alpha$.

Table 1. Risk measures model residuals estimates for different windows for RTS index time series

Measure	Method				
	Historical	GEV	GPD	Metalog full	Metalog quant
$VaR_{0.9,5}(\hat{\nu})$ min	0,050522	0,091294	0,047275	0,050834	0,046954
$VaR_{0.9,5}(\hat{\nu})$ max	0,603436	2,676512	0,555759	0,635795	0,546643
$VaR_{0.9,5}(\hat{\nu})$ mn	0,495985	0,7103547	0,4639331	0,473871	0,456269
$CVaR_{0.9,5}(\hat{\nu})$ min	0,093860	0,143814	0,093859	0,107196	0,085901
$CVaR_{0.9,5}(\hat{\nu})$ max	1,434677	11,32127	1,180898	1,984921	1,075892
$CVaR_{0.9,5}(\hat{\nu})$ mn	0,965006	1,179965	0,938788	1,115082	0,880454

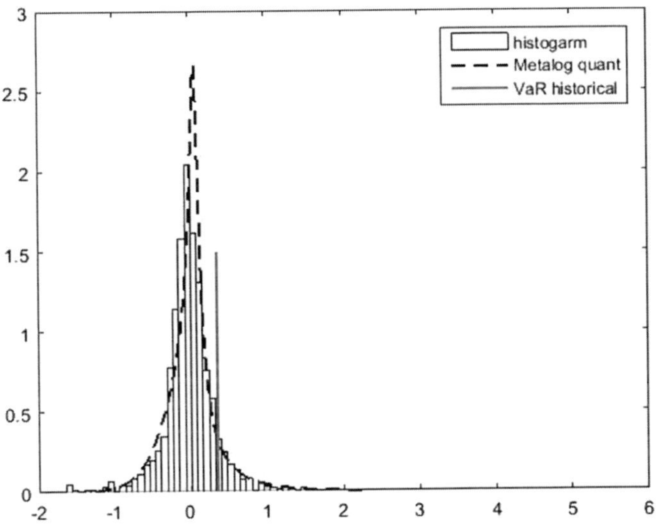

Fig. 1. Empirical probability density function (histogram) and estimated probability density function for ν using Metalog quant

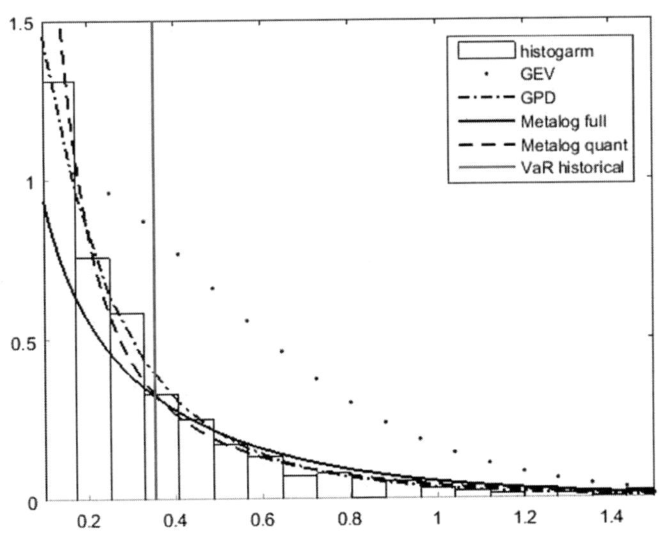

Fig. 2. Tail part for empirical probability density function (histogram) and estimated tail part for probability density function for ν using GEV, GPD, Metalog full, Metalog quant and historical $VaR_{0.9,\ 5}\,(\hat{\boldsymbol{\nu}})$

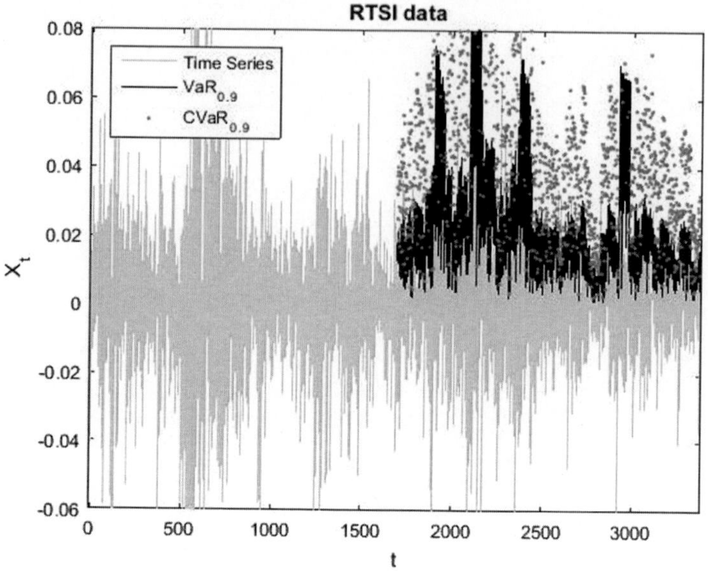

Fig. 3. Historical RTS index time series and dynamic risk estimates using Quantile Metalog Method

Figures 1 and 2 show that the estimates obtained by the Quantile Metalog Method allowed to describe the behavior of the tail part of the function better than the other methods. This leads to the improvement of estimates $VaR_{0.9,\ 5}\left(\hat{\boldsymbol{\nu}}\right)$ and $CVaR_{0.9,\ 5}\left(\hat{\boldsymbol{\nu}}\right)$. Table 1 demonstrates minimum (min), maximum (max) and average (mn) values of the risk measures for 337 windows.

The dynamic risk model is used to find forecast estimates $VaR_p\left(\hat{X}_{t+S}\right)$ and $CVaR_p\left(\hat{X}_{t+S}\right)$. As an example, Fig. 3 shows the historical data of the RTS time series and the estimated values of the dynamic risk measures obtained for 337 windows using Quantile Metalog Method.

A visual comparison of the estimated values with the historical values shows a sufficiently high quality of the obtained estimates: the VaR and CVaR forecast values qualitatively describe the dynamics of the time series and show only a slight delay in the determination of extreme outliers. Prediction errors of risk measures for different methods are given in Table 2.

Table 2 shows that the estimates obtained using the GEV function have the biggest errors, the estimates obtained using the Quantile Metalog Method show the best results. This proves the advantage of the proposed method.

The results of the qualitative analysis are shown in Table 3. For the Kupiec and the Kristoffersen tests the significance level is 0.05. Table 3 shows p-values of statistics. If the forecast estimates for $CVaR_{0.9}\left(\hat{X}_{t+S}\right)$ are good the statistics V, V_1, V_2 are close to zero.

Table 2. The prediction errors of risk measures $VaR_{0.9}\left(\hat{X}_{t+S}\right)$ and $CVaR_{0.9}\left(\hat{X}_{t+S}\right)$ for RTS index time series

Method	Estimate					
	$VaR_{0.9}\left(\hat{X}_{t+S}\right)$			$CVaR_{0.9}\left(\hat{X}_{t+S}\right)$		
	ME ($\times 10^{-3}$)	MAE ($\times 10^{-3}$)	MSE ($\times 10^{-4}$)	ME ($\times 10^{-3}$)	MAE ($\times 10^{-3}$)	MSE ($\times 10^{-3}$)
Historical	3	11.7	3	15	24	1.3
GEV	12	19.4	26	27	38	37.8
GPD	3.2	11.1	2.7	12	23	1.2
Metalog full	3.4	11.3	2.9	21	29	2.0
Metalog quant	2.5	10.9	2.6	11	20	1.0

Table 3. The results of the qualitative analysis of predictive dynamics measures $VaR_{0.9}\left(\hat{X}_{t+S}\right)$ and $CVaR_{0.9}\left(\hat{X}_{t+S}\right)$ for RTS index time series

Method	Estimate					
	LRuc	LRind	LRcc	V	V_1	V_2
Historical	0,0418	0,4743	0,0976	0,024673	−0,00612	−0,04323
GEV	0,0000	0,7330	0,0000	0,031603	−0,00650	−0,06196
GPD	0,7137	0,4123	0,6680	0,022887	−0,00407	−0,04171
Metalog full	0,1738	0,1462	0,1380	0,031372	−0,01151	−0,05184
Metalog quant	0,8388	0,3555	0,6392	0,021145	−0,00393	−0,03836

Table 3 shows the correctness of the estimates obtained by all methods, except the estimates obtained with the GEV function, which corresponds to the results of visual and quantitative analyzes. The results of Table 3 once again confirmes the superiority of the estimates found using the proposed Quantile Metalog Method.

4 Conclusions

The paper considers the problem of dynamic risks measures VaR and CVaR forecast based on the heteroskedastic time series model. Squared time series is considered to be long-range dependent. This specificity of the input data is typical for time series, which describe returns of the different stock exchanges indices, especially for developing countries exchanges. Application of the heteroskedastic time series models reduces the problem of dynamic risks modeling to the problem of variance modeling and risk measures model residuals estimating. The method of smoothing of the autocorrelation function is used to model the variance of the time series. For risk measures model residuals estimating two new parametric methods based on the metalog distribution are proposed: Full Metalog Method and Quantile Metalog Method. The approach based on metalog distribution allows to estimate the tail part of the distribution more precisely which is especially important for qualitative risk measures estimation.

The methods give explicit analytical formulas for CVaR estimating with a different number of members in the metalog distribution n. The choice of parameter n allows to choose formulas for modeling risk measures that are most appropriate for the input data.

To test the proposed methods, the risk measures forecast for the time series of daily return of the Russian Stock Exchange Trading System is built. For comparison, similar model was built and forecast estimates were obtained using widely known methods based on historical modeling, methods using GEV, GPD functions. Conducted quantitative and qualitative analyzes of the obtained estimates confirmed the effectiveness of the new methods proposed in the work on the basis of the metalog distribution.

The proposed approach for dynamic VaR and CVaR modeling and forecasting using metalog distribution can be extended to a multidimensional case and can be applied in portfolio optimization theory.

References

1. Acerbi, C., Tasche, D.: Expected shortfall: a natural coherent alternative to value at risk. Econ. Notes **31**(2), 379–388 (2002). https://doi.org/10.1111/1468-0300.00091
2. Chrisman, L.: Estimating US Deaths from COVID-19 Coronavirus in 2020 (2020). https://lumina.com/estimating-us-deaths-from-covid-19-coronavirus-in-2020
3. Echaust, K., Just, M.: Value at risk estimation using the GARCH-EVT approach with optimal tail selection. Mathematics **8**(1), 114 (2020). https://doi.org/10.3390/math8010114
4. Keelin, T.W.: The metalog distributions. Decis. Anal. **13**(4), 223–293 (2016). https://doi.org/10.1287/deca.2016.0338
5. Khokhlov, V.: Conditional value-at-risk for uncommon distributions. SSRN Electron. J. (2018). https://doi.org/10.2139/ssrn.3200629, https://ssrn.com/abstract=3200629
6. McNeil, A.J., Frey, R., Embrechts, P.: Quantitative Risk Management: Concepts Techniques and Tools. Princeton University Press, Princeton (2005)
7. Norton, M., Khokhlov, V., Uryasev, S.: Calculating CVaR and bPOE for common probability distributions with application to portfolio optimization and density estimation. Ann. Oper. Res. pp. 1–35 (2019). https://doi.org/10.1007/s10479-019-03373-1
8. Omari, C., Mwita, P., Waititu, A.: Using conditional extreme value theory to estimate value-at-risk for daily currency exchange rates. J. Math. Finan. **7**(4), 846–870 (2017). https://doi.org/10.4236/jmf.2017.74045
9. Osei, J., Sarpong, P., Amoako, S.: Comparing historical simulation and Monte Carlo simulation in calculating var. Dama Int. J. Res. **3**(6), 22–35 (2018)
10. Pankratova, N.D., Zrazhevskaja, N.G.: Method of dynamic VaR and CVaR risk measures forecasting for long range dependent time series on the base of the heteroscedastic model. Intell. Control Autom. J. **8**(2), 126–138 (2017). https://doi.org/10.4236/ica.2017.82010
11. Pratiwi, N., Iswahyudi, C., Safitri, R.: Generalized extreme value distribution for value at risk analysis on gold price. J. Phys: Conf. Ser. **1217**, 012090 (2019). https://doi.org/10.1088/1742-6596/1217/1/012090

12. Tabasi, H., Yousefi, V., Tamosaitiene, J., Ghasemi, F.: Estimating conditional value at risk in the Tehran stock exchange based on the extreme value theory using Garch models. Adm. Sci. **9**(2), 40 (2019)
13. Zrazhevska, N.: Construction and application of the classification scheme of dynamic risk measures estimating. Eureka: Phys. Eng. **5**, 67–80 (2016). https://doi.org/10.21303/2461-4262.2016.00162
14. Zrazhevskaja, N.G., Zrazhevskij, A.G.: Classification of methods for risk measures VaR and CVaR calculation and estimation. Syst. Res. Inf. Technol. **3**, 118–125 (2016). https://doi.org/10.20535/SRIT.2308-8893.2016.3.11

Study of the Plot Suitability for Underground Construction: Cognitive Modelling

Nataliya Pankratova[1(✉)], Galina Gorelova[2], and Vladimir Pankratov[1]

[1] Institute for Applied System Analysis, Igor Sikorsky Kyiv Polytechnic Institute, Kyiv, Ukraine
natalidmp@gmail.com, pankratov.volodya@gmail.com

[2] Engineering and Technology Academy of the Southern Federal University, "ITA SFEDU", Taganrog, Russia
gorelova-37@mail.ru

Abstract. It is proposed to use the imitative cognitive modelling to build scenarios for the urban underground development and the evaluating of the megapolises underground space. To analyze the favorableness of urban areas for underground construction the modified method of morphological analysis was selected, which proved to be quite effective in modeling problems whose objects can have a large number of alternative configurations by combining various parameter values. The obtained results are used as initial ones in cognitive modelling for constructing scenarios of evaluating and developing underground urban studies. The corresponding cognitive maps are developed, their structural analysis and stability analysis are carried out, which made it possible to judge the compliance of the cognitive model of reality. The modelling of scenarios for possible processes of the events development in the analyzed complex system is carried out under the influence of various internal and external disturbances and control impulse effects. The results of cognitive modelling scenarios are used to assess the suitability of the selected plot for underground construction.

Keywords: Cognitive modeling · Foresight · Urban underground · Toolkit · Stability · Impulse modeling · Scenarios

1 Introduction

The significant changes that have occurred in recent decades in the life of large cities require a scientific understanding of the new realities and the most likely prospects for their further development. Underground urbanism, which is an integral component of modern megapolises, has already gone beyond the boundaries of individual local objects and is becoming a systemic factor in the existence of large cities. Foresight of future changes in the development of megapolises should be based on a reliable sci-entific and methodological foundation, which should ensure the development of plot and underground development as a whole.

S. Babichev et al. (Eds.): ISDMCI 2020, AISC 1246, pp. 246–264, 2021.
https://doi.org/10.1007/978-3-030-54215-3_16

High tech development, including underground development of megapolises, where innovations are the base for competitiveness, as well as global ecological, economic, demographic challenges have produced a novel world model – sustainable planetary development. The modern concept of sustainable development regarding the city planning should take into account the future needs, which implies the capability to satisfy current needs of society without causing harm to future generations [32]. The important aspect of sustainable development is the potential for timely reaction to possible environmental changes, and minimization of technogenic and ecological impacts. This concept changes the common strategy for engineering projects and replaces the traditional vision of local problems with the position of systemic consideration of large natural-technical and social problems. In the concept of sustainable development of megapolises, underground urbanism is of particular importance, since an effectively planned underground infrastructure improves the quality of life and environmental safety to a much greater extent than a similar functional system on the surface.

The purpose of this paper is to demonstrate the possibilities of sharing the foresight methodology for constructing of scenarios alternatives and the methodologies of cognitive and imitative modelling for creating scenarios of the underground space of megapolises development and the evaluating of underground urbanism. The use of the method of morphological analysis to build alternative heterogeneous design configurations and cognitive and imitative modeling to create their scenarios can become the scientific and methodological basis for constructing a strategic master plan of the "underground city" and contribute to the developing urban underground space of megapolises. The development of megapolises currently actualizes the problems of underground urbanism. One of the main among them is the problem of choosing a construction plot and assessing its suitability for many indicators.

2 Review of Literature

Urban underground development, which is an integral part of modern megapolises, has transcended the scale of separate local objects and became a systemic factor of urban development. Historically established approaches to developing underground space include resource-based, directive-based, city construction-based, and complex approaches. The complex approach is shown in [15] to allow the devising of the most optimal and rational system of utilizing underground space. Prediction of future changes, the capability to swap functions of underground objects, the corresponding urban policy, planning and management of megapolises development should be based on a sound scientific-methodological base, devised to provide systemic surface and underground urban development as a whole [9,22,23].

Various directions of implementing system approach for planning urban surface construction in megacities are known [14,21,30]. Analyzing the trends of the future development of the underground space of megacities are considered the individual projects of underground urban studies that characterize the directions of urban underground construction in the near and medium perspective [10]. So, In Chicago, the second largest economic center in the United States, it is planned to build an underground city with a vertical layout, which will have 100 underground floors. As for the underground development, the studies went no further than general task setting and analysis of research methods [26]. Risks in underground space development which is characterized by space-temporal variability are considered in [6,28].

Scientific and educational personnel from geo-engineering are concentrated in universities and research centers of various profiles (mining, construction, transport), which leads to somewhat different methodological approaches to design problems [14,20,25,31]. A system approach to the planning of underground urban studies, based on the methodology of foresight, as a tool for the concept of sustainable development of megacities was proposed in [23].

The capital of Finland, Helsinki, is perhaps the only city in the world whose underground part was formed not as the sum of independent local objects separately built at different times, but as an underground city built according to a single plan, the development of which began as early as 1972 [33]. Today it is more than 400 underground objects of various sizes and complexity connected with each other with a total area of about 9 million m^2. Helsinki underground is more than 200 km of tunnels of significant sections (up to 21 m^2), in which there are sewage pipes, water pipes, heat supply and a unique system of automatic garbage chutes, which displaced garbage trucks and garbage bins from the city streets.

3 Application of the Foresight and Cognitive Modelling Methodologies for Assessing Suitability of Urban Territories for Underground Construction

We will consider the urban development as a global supersystem in the form of an ordered set of structurally interrelated and functionally interdependent global systems. It is proposed to use the methodologies of foresight and cognitive modelling together: at the first stage, apply the methodology of foresight and use the results in the second stage as the initial data for cognitive modeling, which, based on knowledge and experience, build causal relationships, understand and analyze the behavior of a complex system in a strategic perspective. One of the foresight methodology methods, which is currently efficiently used for generating scenario alternatives, is the modified morphological analysis method (MMAM) [29] which was well proven for the tasks of modelling objects with a multitude of alternative configurations by combining different parameter values of the object. He allowed relying on the identified groups of geological and technogenic factors, to consider a number of solutions and risk groups to assess the feasibility of developing underground urbanism in the areas under consideration.

Study on the basis of morphological analysis method of two plots in Kyiv for construction of underground parking lots is considered. The two underground parking lot plots with different characteristics are investigated. The first construction plot is found at the Shevchenkivsky district at the Peremohy avenue, and the second is also found at the Shevchenkivsky district between the Bulvarno-Kudriavska and Honchara streets. The morphological tables were estimated using the available data of geological and technogenic nature, and the dependencies between parameters were obtained by expert estimation [23].

To build a morphological table for construction plots, some sets of characteristics that affect the result in the same way, in this study, were aggregated into the following parameters: suitability of the plot, the level of dynamic load, the indicator of static load from surface development, the indicator of the static load of the surrounding soil mass, the impact of existing underground facilities, genetic type and lithological composition of soils, estimated soil stable, influence of aquifers and water floods, type of terrain and morphometry, engineering and geological processes, geotechnologies for the construction of underground structures. Studies using the MMAM showed that the second plot refers to the alternative most suitable for underground construction, as evidenced by the absolute probability indicators of the alternative suitability of the plot having values of 0,688 and 0,993 respectively [23]. Therefore, for cognitive modeling, the results obtained for the second plot are used. Cognitive modeling of complex systems (socio-economic, environmental, political, cyberphysical, etc.) and the foresight methodology of possible processes in them, which are used in this study, are developed in [1, 11–13, 24]. Foundation of these works there are articles [2–5, 7, 8, 16–19, 27].

Cognitive modeling of complex systems is a modelling of the structure and behavior of a complex system based on quantitative and qualitative data about it and carried out in stages [11–13, 24]. The main stages: 1) development of a cognitive model in one form or another of mathematical form (from a cognitive map to a vector parametric functional graph); 2) the study of the properties of the model, on the basis of which it is concluded that the properties of the model do not contradict the properties of the studied complex system; 3) scenario analysis carried out by impulse modeling on a cognitive model, which allows to determine the possible development of situations in the system under the influence of external, internal disturbances and control decisions on it. The purpose of such cognitive research is to understand, explain, formally represent a complex system (based on the structuring of theoretical, practical knowledge, statistics and expert data about it) and generate new knowledge about the system, which is necessary for making informed management decisions.

The developed methodology [11–13, 24] and the software [1] of cognitive modeling are a tool for studying the properties of a complex system (stability, connectivity, complexity, etc.) and its behavior. Together with the foresight methodology [12], applied at the first and third stages of cognitive modeling, the results of the study serve to develop recommendations for improving a complex system, to propose desired strategies for developing the system and preventing those that are not desired.

4 Definition of the Possible Development of Processes in a Complex System Based on Cognitive Modelling

In the study of the suitability of the plot for underground construction at the first stage we used cognitive models such as a cognitive map in the form of a sign oriented graph (1) and a functional graph in the form of a weighted sign digraph (3), (4) [16,27]:

$$G = \langle V, E \rangle \tag{1}$$

where G is a cognitive map in which V are concepts, a finite set of vertexes of the cognitive map $V_i \in V, i = 1, 2, \dots k; E = \{e_{ij}\}$ is the set of arcs e_{ij} of the graph, i, j = 1, 2, ... m, reflect the relationship between the vertexes V_i and V_j; the influence of V_i on V_j in the situation under study can be positive (+1) when an increase (decrease) in one factor leads to an increase (decrease) in another, negative (−1) when an increase (decrease) in one factor leads to a decrease (increase) in another, or absent (0). The cognitive map G corresponds to the square matrix of relations A_G in the following form:

$$A_G = \{a_{ij}\} = \begin{cases} 1, \; if \; V_i \; is \; connected \; with \; V_j \\ 0, \; otherwise \end{cases} \tag{2}$$

The ratio a_{ij} can take the value "+1" or "−1". The relation between variables (interaction of factors) is a quantitative or qualitative description of the effect of changes in one variable on others at the corresponding vertexes.

Vector Functional Graph in this case has the form:

$$\Psi = \langle G, X, F(X, E), \Theta \rangle \tag{3}$$

where G is a cognitive map; X is the set of vertex parameters, Θ is the space of vertex parameters; $F(X, E)$ is the arc transformation functional. If

$$F(X, E) = F(x_i, x_j, e_{ij}) = \begin{cases} +\omega_{ij}, \; if \; rising \, / falling \; X_i \\ entails \; rising/falling \; X_j \\ -\omega_{ij}, \; if \; rising/falling \; X_i \\ entails \; falling/rising \; X_j \end{cases} \tag{4}$$

then there is a weighted sign digraph, in which ω_{ij} is the weight coefficient.

At the second stage of cognitive modeling, to study the properties of the cognitive model we used methods of structural stability and perturbation stable analysis [7,27], methods for analyzing model connectivity (simplicial analysis [1]), and graph theory methods [27]. The results of the analysis were compared with the available information on underground construction.

At the third stage of cognitive modeling, to determine the possible development of processes in a complex system and develop development scenarios, we used the impulse process model (modeling the propagation of disturbances in cognitive models) [16]:

$$x_{v_i}(n+1) = x_{v_j}(n) + \sum_{v_j : e = e_{ij} \in E}^{k-1} f(x_i, x_j, e_{ij} P_j(n) + Q_{vi}(n)) \tag{5}$$

where $x(n)$, $x(n+1)$ are the values of the indicator at the vertex Vi at the modelling steps at time $t = n$ and the next $t = n + 1$; $P_j(n)$ is the pulse existed at the vertex V_j at the moment $t = n$; $Q_{Vi}(n) = \{q_1, q_2, ..., q_k\}$ is the vector of external pulses (disturbing or controlling actions) introduced to the vertexes V_i at time moment n.

Table 1. Vertexes V_i of cognitive map G

Code	Name of the vertex	Vertex assignment
V_0	Suitability of the plot	Indicative, target
V_1	The level of dynamic load	Basic, disturbing
V_2	The indicator of static load from surface development	Basic, disturbing
V_3	The indicator of the static load of the surrounding soil mass	Basic, disturbing
V_4	The impact of existing underground facilities	Disturbing
V_5	Genetic type and lithological composition of soils	Basic
V_6	Estimated soil stable	Basic
V_7	Influence of aquifers and water floods	Disturbing
V_8	Type of terrain and morphometry	Basic
V_9	Engineering and geological processes	Disturbing
V_{10}	Geotechnologies for the construction of underground structures	Basic, governing

5 Cognitive Modelling of the Plot Suitability Scenarios for Underground Construction

The First Stage: Cognitive Model Development. In developing of the cognitive model the principle of sequential refinement of the model was used: initially a cognitive map was built (action 1), then not on its basis a model of a weighted sign digraph was constructed and studied (action 2).

Table 2. Fragment of a relation matrix and cognitive map G

Code	Vertex "cause" V_i	Vertex "effect" V_j	Sign of relationship
e_{14}	V_1 The level of dynamic load	V_4 The impact of existing underground facilities	+1
e_{19}		V_9 Engineering and geological processes	−1
e_{110}		V_{10} Geotechnologies for the construction of underground structures	+1
...			
e_{101}	V_{10} Geotechnologies for the construction of underground structures	V_1 The level of dynamic load	−1
e_{104}		V_4 The impact of existing underground facilities	+1
e_{106}		V_6 Estimated soil stable	+1
e_{100}		V_0 Suitability of the plot	+1

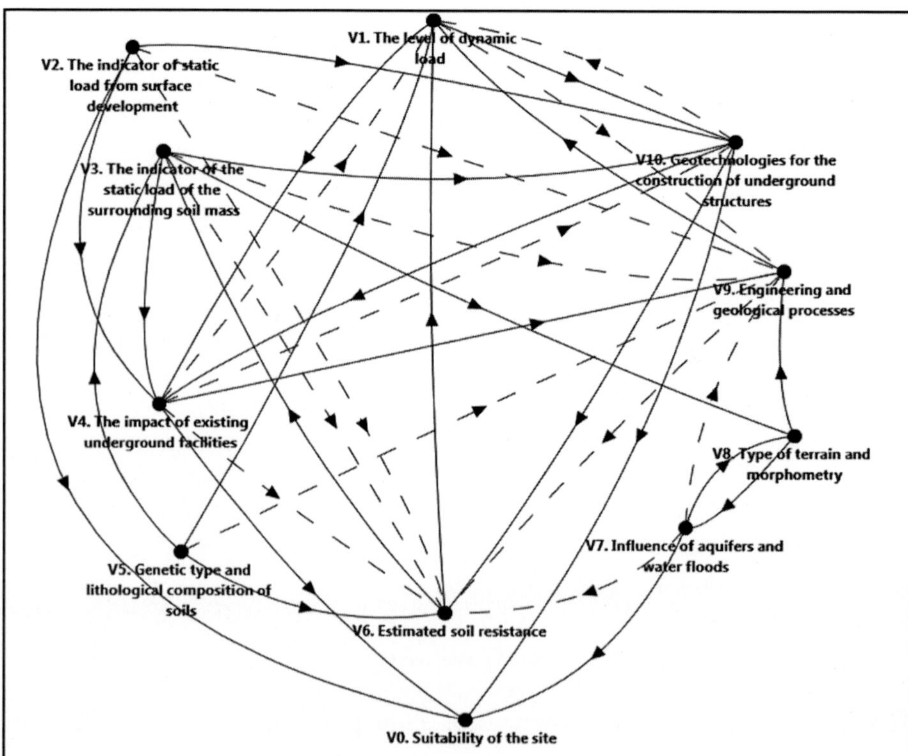

Fig. 1. Cognitive map G

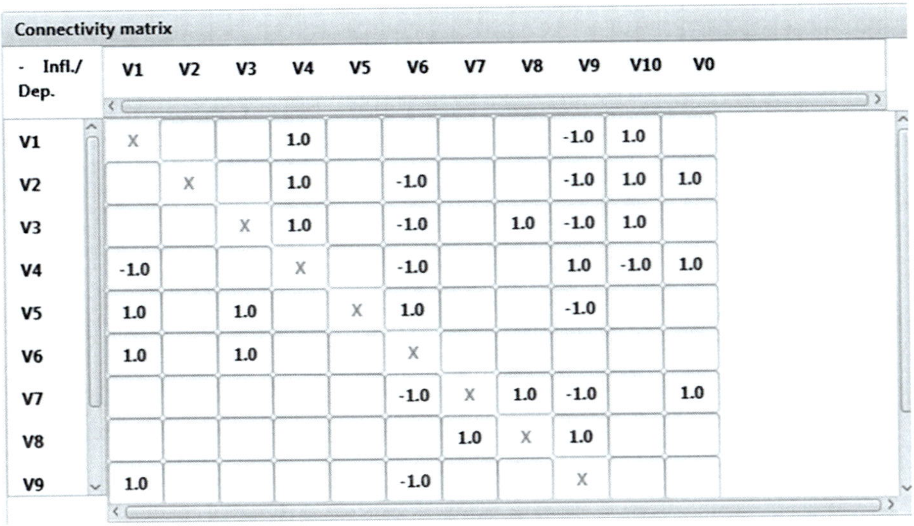

Connectivity matrix

- Infl./ Dep.	V1	V2	V3	V4	V5	V6	V7	V8	V9	V10	V0
V1	X			1.0					-1.0	1.0	
V2		X		1.0		-1.0			-1.0	1.0	1.0
V3			X	1.0		-1.0		1.0	-1.0	1.0	
V4	-1.0			X		-1.0			1.0	-1.0	1.0
V5	1.0		1.0		X	1.0			-1.0		
V6	1.0		1.0			X					
V7						-1.0	X	1.0	-1.0		1.0
V8							1.0	X	1.0		
V9	1.0					-1.0			X		

Fig. 2. Cognitive map G of relationship matrix R_G

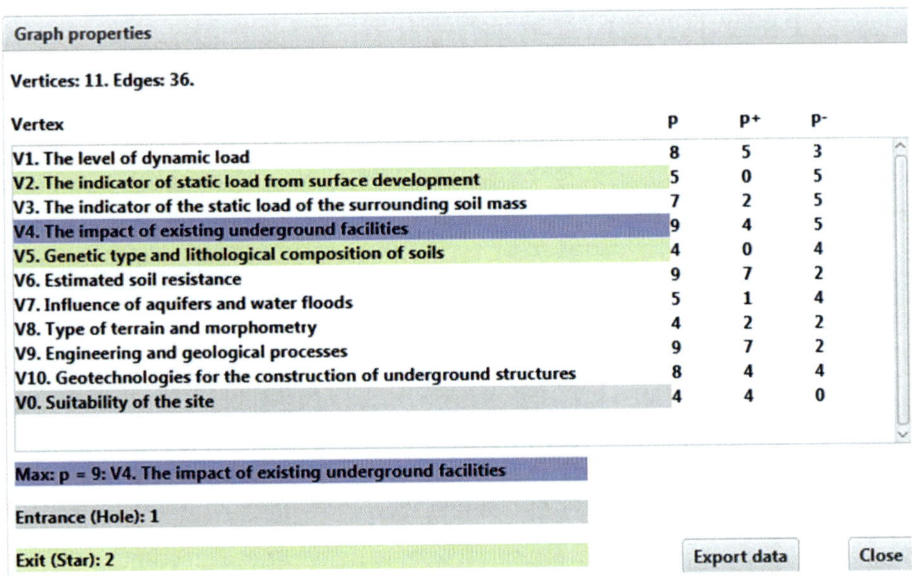

Graph properties

Vertices: 11. Edges: 36.

Vertex	p	p+	p-
V1. The level of dynamic load	8	5	3
V2. The indicator of static load from surface development	5	0	5
V3. The indicator of the static load of the surrounding soil mass	7	2	5
V4. The impact of existing underground facilities	9	4	5
V5. Genetic type and lithological composition of soils	4	0	4
V6. Estimated soil resistance	9	7	2
V7. Influence of aquifers and water floods	5	1	4
V8. Type of terrain and morphometry	4	2	2
V9. Engineering and geological processes	9	7	2
V10. Geotechnologies for the construction of underground structures	8	4	4
V0. Suitability of the site	4	4	0

Max: p = 9: V4. The impact of existing underground facilities

Entrance (Hole): 1

Exit (Star): 2

Export data Close

Fig. 3. Characteristics of the graph G

<u>*Action* 1 :</u> Creation of cognitive map G. Let us call the cognitive map G as "Underground Construction". When developing a cognitive model, it is convenient to present the analyzed and systematized information about the vertexes and their relationships [12] in the form of Table 1 and Table 2.

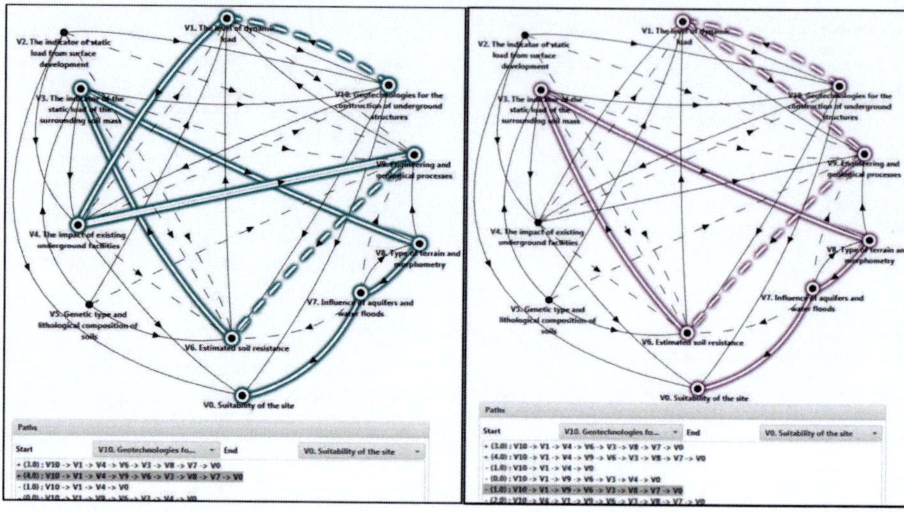

Fig. 4. Ways of the graph G from the vertex V_{10} to the vertex V_0

Cognitive map G constructed using CMLS [11] is shown in Fig. 1. Solid lines correspond to positive arcs "+1", dotted to negative "−1". The relationship matrix RG is shown in Fig. 2.

The Second Stage: Analysis of the Properties of the Cognitive Map ***G*****.** An analysis of the properties of the cognitive map G included: determining the characteristics of the graph, analyzing paths and cycles, analyzing stability to perturbations and structural stability.

1a) Characterization of the graph. In Fig. 3 graph characteristics are given: the number of vertexes and arcs is calculated, the number of incoming (p^+) and outgoing (p^-) arcs is determined.

1b) Definition and analysis of the paths for model G. The study of different paths from top to top on a cognitive map allows us to comprehend how reasonably the cause-effect chains of the cognitive map "Underground Construction" are defined, in which the main characteristics of the plot intended for underground construction are specified. An analysis of the paths makes it possible to interpret their meaning as cause-effect chains, and use the information further to support decisions being made. In addition, the way shows the variety of opportunities to achieve goals, their number may be large and far from obvious to the decision maker. It also becomes clear which vertexes are included in each path, are "touched" on this path, which gives an occasion to evaluate its desirability or non-desirability.

Figure 4 shows examples of paths between the vertexes V_{10} (Geotechnologies for the construction of underground structures) and V_0 (Suitability of the plot); on the left graph of Figs. 4 and 5 there is one of the positive (no negative arcs or an even number of negative arcs) and on the right graph of Fig. 4 and 5 there is one of the negative (odd number of negative arcs) paths.

1c) Analysis of cycles and structural stability of the cognitive map G are presented in Fig. 5.

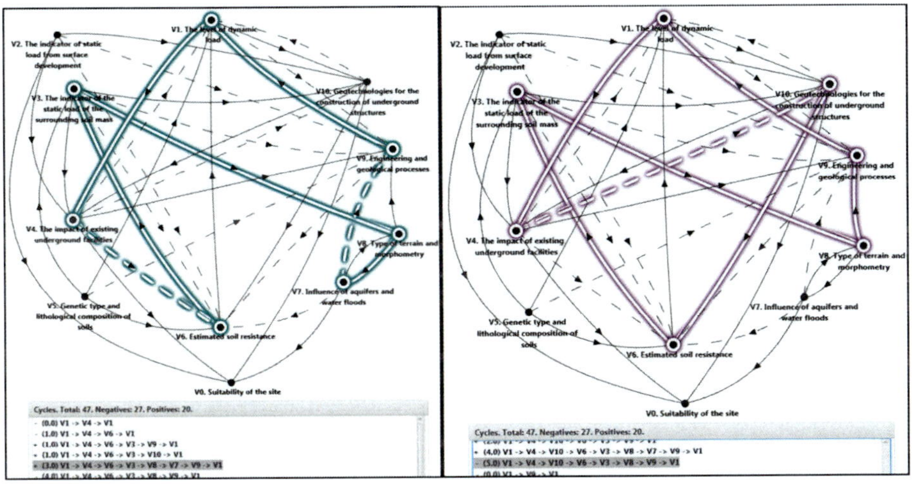

Fig. 5. Cognitive map G cycles

The analysis of the cognitive map cycles showed that this system contains 47 cycles, among them 27 cycles of negative (stabilizing) feedback. This indicates that the analyzed system is structurally stable [3, 4, 7, 12]. Evaluation of this fact as "good" or "bad" is not unambiguous, it can depend both on the goals of management and on the "preferences" of the decision maker, which can upgrade the initial cognitive map.

1d) Analysis of stability to disturbances. This determination is necessary for the analysis of the stability of the system to perturbations and the initial value [7, 27]. The stability criterion $|M| < 1$ is used, where $|M|$ is the maximum modulus eigenvalue of the matrix R_G. Since in this case $|M| = 1.9309 > 1$, the system G is not stable either to perturbation or to the initial value. This indicates that the slightest deviations at the vertexes remove the system from a stable state. Making a decision about how "good"/"bad" is in this case requires further analysis and refinement of the model.

The Third stage. Pulse Modeling, Scenario Analysis. Using the CMLS software system, it is possible to introduce perturbations $Q = \{q_i\}, i = 1, 2, ...k$ of different values (normalized) to any of the vertexes, as well as their combination. Before starting a pulse modelling, it is necessary to develop a plan for a computational experiment. When introducing disturbances to the vertexes, an answer is sought to the question: "What will happen if ...?". In the process of pulse modeling, it is possible to introduce disturbing influences at any modeling step. This makes it possible to change (correct) scenarios in model dynamics, to determine the effects that bring processes closer to the desired ones.

Scenario No. 1. Let the Geotechnologies for the construction of underground structures be improved, the impulse $q_{10} = +1$ is introduced to the vertex V_{10}, the disturbance vector $Q_1 = \{q_1 = 0; q_2 = 0; ...q_{10} = +1\}$. Table 3 presents the results of calculations of impulse processes at the vertexes according to formula (5) at 10 modeling steps, the number of which in this case can be considered sufficient, because trends in situations in the system have already manifested. Figure 6 shows graphs of pulsed processes at the vertexes constructed according to the data in Table 3. The graphs are represented by two figures (a and b) with fewer vertexes in each in order to facilitate their visual analysis.

Table 3. Pulse processes according to scenario No. 1

Step (horizontal) Vertex(vertical)	0.0	1.0	2.0	3.0	4.0	5.0	6.0	7.0	8.0	9.0	10.0
V_1. The level of dynamic load	0.0	0.0	−1.0	−1.0	3.0	−4.0	−10.0	32.0	4.0	−138.0	159.0
V_2. The indicator of static load from surface development	0.0	0.0	0.0	0.0	0.0	0.0	0.0	0.0	0.0	0.0	0.0
V_3. The indicator of the static load of the surrounding soil mass	0.0	0.0	0.0	1.0	0.0	−4.0	2.0	7.0	−21.0	9.0	96.0
V_4. The impact of existing underground facilities	0.0	0.0	1.0	0.0	−1.0	4.0	−3.0	−19.0	35.0	42.0	−180.0
V_5. Genetic type and lithogical composition of soil	0.0	0.0	0.0	0.0	0.0	0.0	0.0	0.0	0.0	0.0	0.0
V_6. Estimated soil resistance	0.0	0.0	1.0	0.0	−4.0	2.0	7.0	−21.0	9.0	96.0	−160.0
V_7. Influence of aquifers and water floods	0.0	0.0	0.0	0.0	0.0	1.0	0.0	−3.0	2.0	4.0	−19.0
V_8. Type of terrain and morphometry	0.0	0.0	0.0	0.0	1.0	0.0	−3.0	2.0	4.0	−19.0	13.0
V_9. Engineering and geological processes	0.0	0.0	0.0	2.0	0.0	−3.0	11.0	2.0	−53.0	54.0	148.0
V_{10}. Geotechnologies for the construction of undeground structures	0.0	1.0	1.0	−1.0	1.0	5.0	−11.0	−4.0	59.0	−51.0	−170.0

Modeling of scenario No. 1 showed that in the case of a system model presented in the form of a cognitive map G, the processes in it will develop in an increasing oscillatory mode with an expected change at the top of V_{10}.

Scenario No. 2. Let four vertexes be excited: there is a negative effect on the system of static load from surface development ($q_2 = -1$) and existing underground facilities ($q_4 = -1$), but geotechnical processes and geotechnologies for the construction of underground structures are assumed to be positive ($q_9 = +1$ *and* $q_{10} = +1$); perturbation vector $Q_2 = \{0; ...q_2 = -1; ...q_4 = -1; ...q_9 = +1; q_{10} = +1\}$. The calculation results are presented in Fig. 7. Analyzing the results of scenario modeling No. 2, we see that the same nature of the growing oscillatory processes is observed as in scenario No. 1. Model G turns out to be impulsively unstable, which also follows from previous results of the cognitive map G study for stability to perturbations. But the model G is structurally stable, and its instability to perturbations can be caused by the fact that all relations a_{ij} are the same, equal to unity. This may not be true. In continuation of the research, an expert survey was conducted, with the help of which the weight coefficients of the arcs were established.

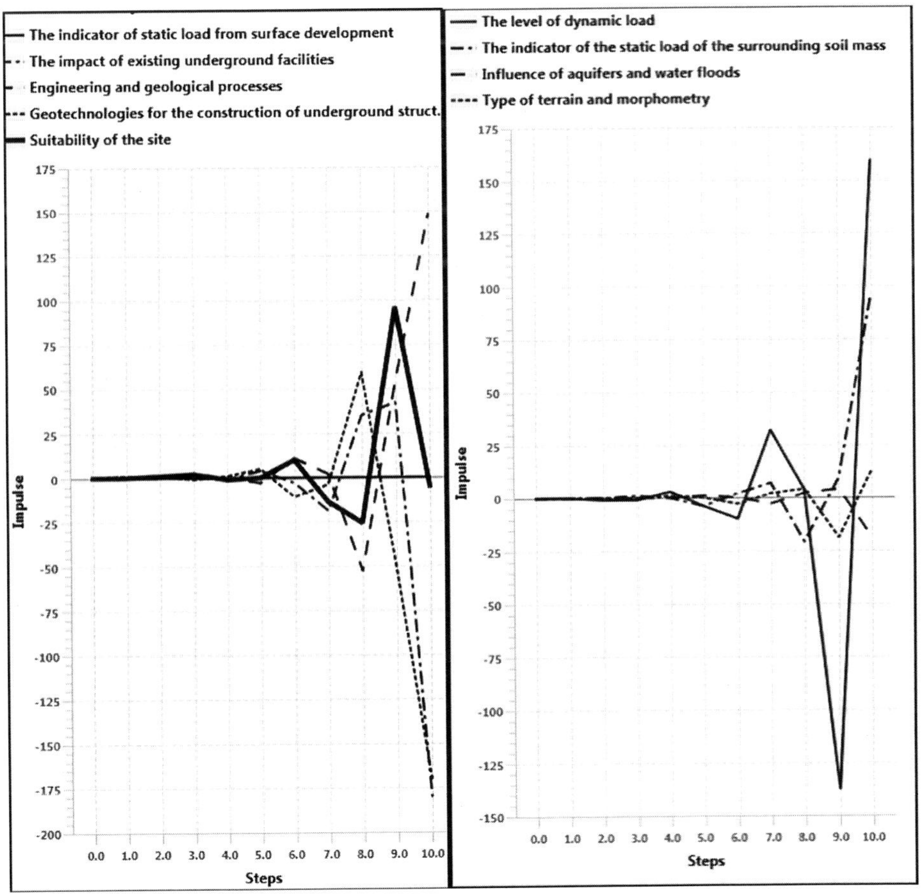

Fig. 6. Graphs of impulse processes according to scenario No. 1

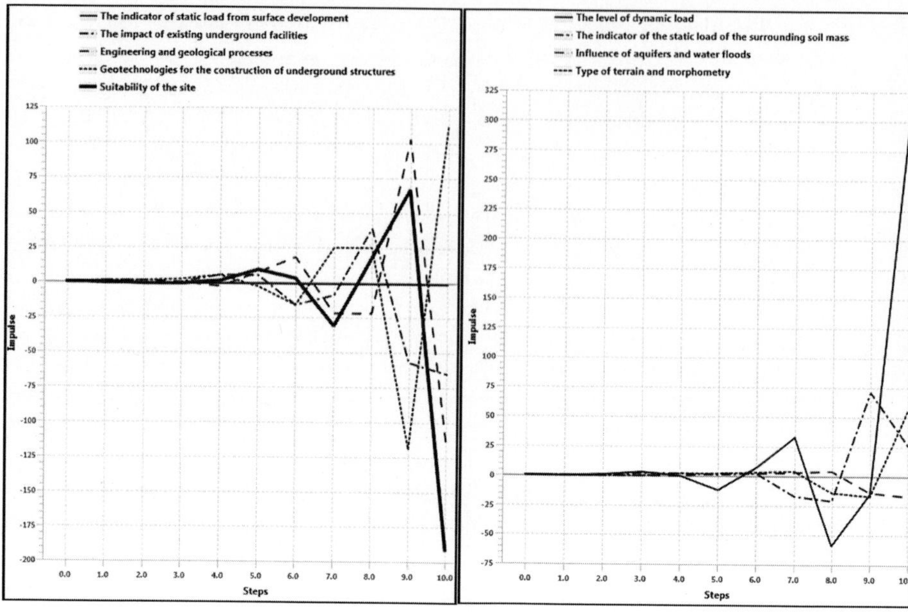

Fig. 7. Graphs of impulse processes according to scenario No. 2

<u>*Action* 2 :</u> Development of a cognitive model of the type of a weighted directed graph Ψ (formulas (3), (4)). Table 4 shows the model data with weights. The corresponding cognitive model Ψ corresponding to Table 4 is shown in Fig. 8. In the course of the study, an analysis was made of the stability of the model to perturbations and according to the initial value. In this case, $|M| = 0.92 < 1$; therefore, the model Ψ is stable to perturbations also with respect to the initial values. Since the structural properties of the models Ψ and G are the same, we pass to pulse modeling. During the study, seventeenth scenarios were analyzed: seven it is when introducing single actions in turn into all seven disturbing vertexes (see Table 1), five it is when introducing single actions into combinations of two vertexes, when introducing single actions into combinations of three vertexes and three scenarios when introducing single actions into combinations of four vertexes. As an illustration of the results, we present the data of pulsed modeling of three scenarios (call them scenarios No. 3, No. 4, No. 5), which allow us to further suggest the main conclusions.

Scenario No. 3. Suppose one of the possible "worst" options, when the level of dynamic load in the system begins to "negatively" affect the entire system ($q_1 = -1$), the perturbation vector is $Q_4 = \{q_1 = -1; q_2 = 0; ...q_{10} = 0\}$. The modelling results are presented in Fig. 9.

As can be seen from the modelling results of scenario No. 3, the negative effect of the dynamic load generates in the system a trend to deterioration in all vertexes, which is especially noticeable in the negative trend at vertex V_0 "Plot

Table 4. Weighting factors of the model Ψ

Code	Name of the vertex	V_0	V_1	V_2	V_3	V_4	V_5	V_6	V_7	V_8	V_9	V_{10}
V_0	Suitability of the plot											
V_1	The level of dynamic load					0.6					−0.7	0.4
V_2	The indicator of static load from surface development	+1				0.2		−0.3			−0.6	0.7
V_3	The indicator of the static load of the surrounding soil mass					0.1		−0.5		0.4	−0.6	0.7
V_4	The impact of existing underground facilities	+1	−0.2					−0.5			0.6	−0.7
V_5	Genetic type and lithogical composition of soil		0.4		0.5			0.3			−0.2	
V_6	Estimated soil stable		0.2		0.7							
V_7	Influence of aquifers and water floods	+1						−0.4		0.6	−0.6	
V_8	Type of terrain and morphometry								0.8		0.2	
V_9	Engineering and geological processes		0.7					−0.3				
V_{10}	Geotechnologies for the construction of undeground structures	+1	−0.7			0.9		0.5				

suitability". The results of the modelling for scenarios No. 4 and No. 5, repeating the conditions of scenarios No. 1 and No. 2, are presented in Fig. 10, 11.

Scenario No. 5: Let there be a negative effect on the system of static load from surface development ($q_2 = -1$) and existing underground facilities ($q_4 = -1$), but geotechnical processes and geotechnologies for the construction of underground structures are assumed to be positive ($q_9 = +1$ and $q_{10} = +1$); perturbation vector $Q_5 = \{0; ...q_2 = -1; ..q_4 = -1; ...q_9 = +1; q_{10} = +1\}$. The modelling results are presented in Fig. 11. Analysis of the 4 and 5 scenarios results showed

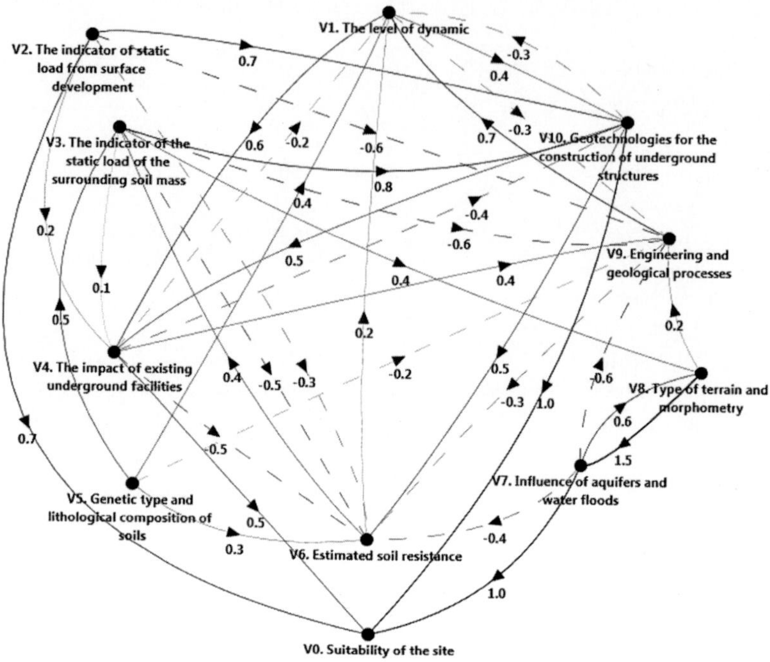

Fig. 8. Cognitive model Ψ

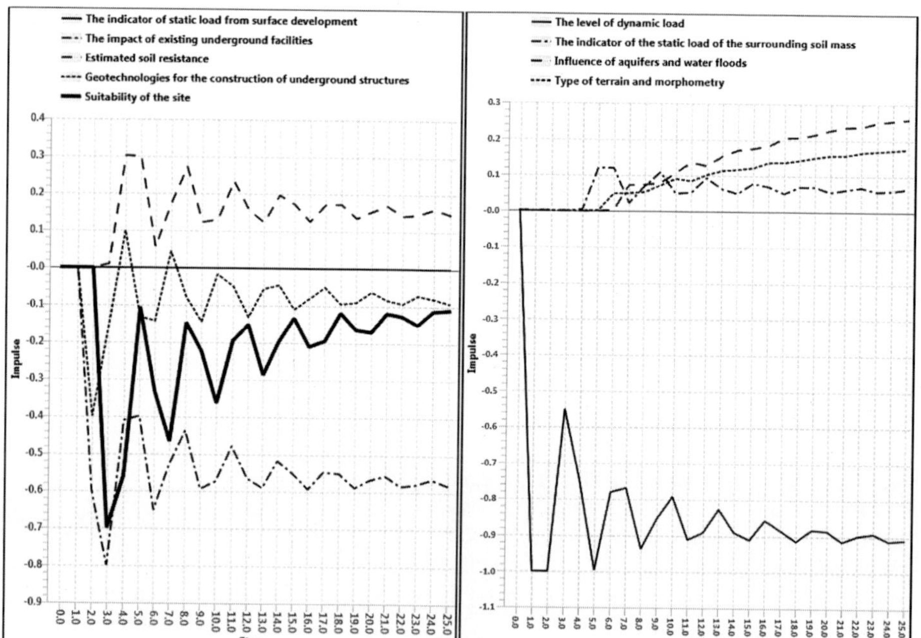

Fig. 9. Graphs of impulse processes according to scenario No. 3

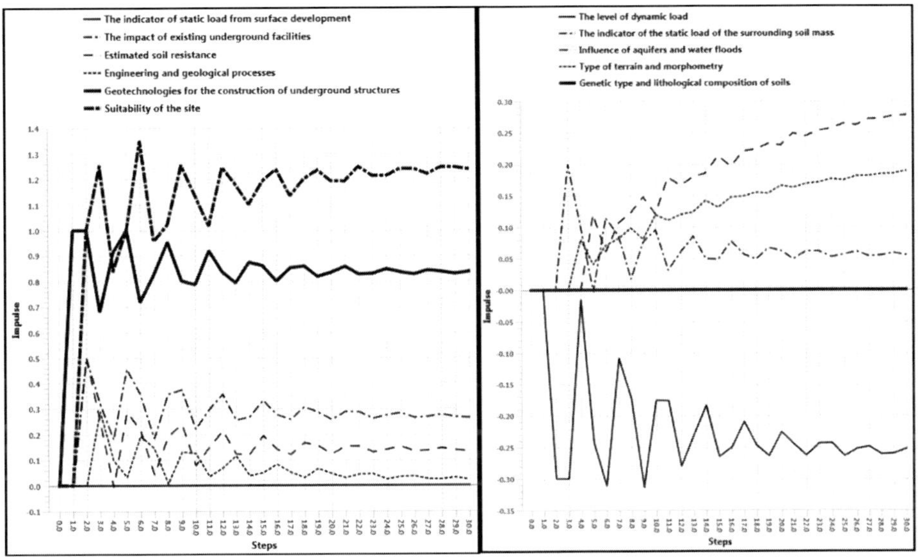

Fig. 10. Graphs of impulse processes according to scenario No. 4

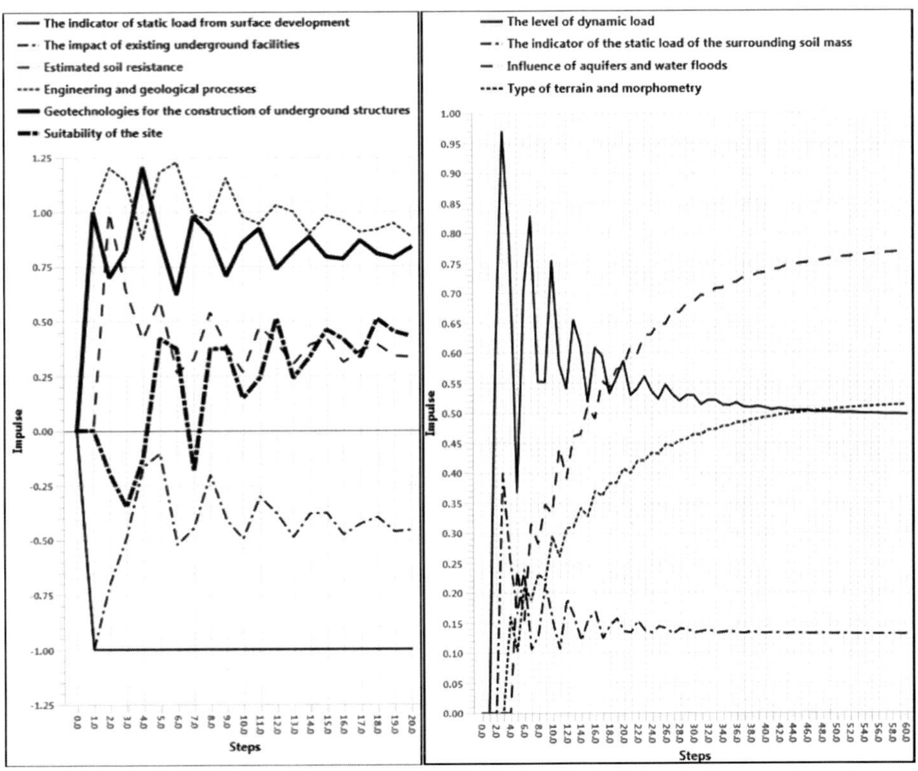

Fig. 11. Graphs of impulse processes according to scenario No. 5

the ability to withstand the negative impacts of the vertexes V_2 (The level of dynamic load) and V_4 (The impact of existing underground facilities) by positive control actions on the vertexes V9 (Engineering and geological processes) and V_{10} (Geotechnologies for the construction of underground structures). Scenario analysis on model Ψ showed that when disturbances are introduced into any of its vertexes, impulse processes are generated that after a certain time stabilize at a certain level; the system is momentarily stable.

6 Conclusions

Study of the plot suitability for underground construction is based on cognitive simulation modeling, which allows to build scenarios in the dynamics of a possible development situation process. Alternatives of the scenarios are constructed using the modified method of morphological analysis which is one of the methods qualitative analysis of the foresight methodology. The obtained results are used as initial ones in cognitive modelling for constructing scenarios of evaluating and developing underground urban studies. The corresponding cognitive maps are developed, their structural analysis and stability analysis are carried out, which made it possible to judge the compliance of the cognitive model of reality. The modelling of scenarios for possible processes of the events development in the analyzed complex system is carried out under the influence of various internal and external disturbances and control impulse effects. The results of the conducted cognitive modeling make it possible to judge that the cognitive models G and Ψ, which systematize and structure various information about the underground construction system, correspond to the real system and can be used to anticipate the possible processes of situations in the system under the influence of various disturbing and controlling factors. The developed methodology and tools made it possible to combine the assessment of the impacts and relationships of geological factors, technogenic and structural-functional types for the analysis of the plot suitability of underground construction.

But when solving construction problems on a specific plot, it is necessary to introduce quantitative characteristics of vertexes into the research scheme; it may also be necessary to increase the number of vertexes and adjust the structure of the cognitive model.

The synthesis of methodologies for cognitive modeling and foresight can become the scientific and methodological basis for constructing a strategic master plan for the "underground city" and contribute to the development of a mechanism for assessing the socio-economic efficiency of developing urban underground space in megapolises.

References

1. Program for cognitive modeling and analysis of socio-economic systems at the regional level, certificate of state registration of computer programs N 2018661506 (2018)

2. Abramova, N., Avdeeva, Z.: Cognitive analysis and management of the development of situations: problems of methodology, theory and practice. Prob. Control **3**, 85–87 (2008)
3. Atkin, R.: Combinatorial Connectivies in Social Systems. An Application of Simplicial Complex Structures to the Study of Large Organisations. Interdisciplinary Systems Research (1997)
4. Atkin, R., Casti, J.: Polyhedral Dynamics and the Geometry of Systems ons. International Institute for Applied Systems Analysis, Laxenburg (1997)
5. Avdeeva, Z., Kovriga, S.: Scientific foresight and cognitive modeling of socio-economic systems. In: 18-th IFAC Conference on Technology, Culture and International Stability, vol. 51, pp. 145–149 (2018). https://doi.org/10.1016/j.ifacol.2018.11.264
6. Bondarik, G.: Extrapolation, Interpolation, and Smoothing of Stationary Time Series. Nedra, Moscow (1981)
7. Casti, J.: Connectivity, Complexity, and Catastrophe in Large-Scale Systems. Wiley, Chichester/New York/Brisbane/Toronto (1979)
8. Eden, C.: Cognitive mapping. Eur. J. Oper. Res. (36), 1–13 (1998)
9. Gayko, G.: Developing underground space in the concept of sustainable development for large cities. Geotechnologies **1**, 60–64 (2018)
10. Gilbert, P., et al.: Underground Engineering for Sustainable Urban Development (2013). https://doi.org/10.17226/14670
11. Ginis, L., Gorelova, G., Kolodenkova, A.: Cognitive and modelling modeling of development of regional economy system. Int. J. Econ. Finan. **5**, 97–103 (2016). https://doi.org/10.2991/itsmssm-17.2017.12
12. Gorelova, G., Pankratova, N. (eds.): Innovative Development of Socio-Economic Systems Based on Foresight and Cognitive Modelling Methodologies. Naukova Dumka, Kiev (2015)
13. Gorelova, G., Pankratova, N.: Scientific foresight and cognitive modeling of socio-economic systems. In: 18-th IFAC Conference on Technology, Culture and International Stability, vol. 51, pp. 145–149 (2018). https://doi.org/10.1016/j.ifacol.2018.11.264
14. Kartosiya, B.: Developing underground space of large cities. new tendencies. Min. Inf.-Anal. Bull. (1), 615–629 (2015). https://doi.org/10.29202/nvngu/2019-3/14
15. Konykhov, S.: Systematization of approaches for developing underground city space. Vestnik MGSU (4), 56–61 (2010)
16. Kulba, V., Kononov, D., Kovalevsky, S., Kosyachenko, S., Nizhegorodtsev, R., Chernov, I.: Scenario Analysis of the Dynamics of Behavior of Socio-Economic Systems. IPU, Moscow (2002)
17. Langley, P., Laird, J., Rogers, S.: Cognitive architectures: research issues and challenges. Cogn. Syst. Res. **10**(2), 141–160 (2006). https://doi.org/10.1016/j.cogsys.2006.07.004
18. Maksimov, V.: Cognitive technology - from ignorance to understanding. In: 1st Work Cognitive Analysis and Management of the Development of Situations, (CASC 2001), pp. 4–18 (2001)
19. Maximov, V., Kovriga, S.: Application of structure-objective analysis of socio-economic situations development. Probl. Control **3**, 39–44 (2005)
20. Owen, C.: A critical role for design technology. Des. Manag. J. **3**(2), 10–18 (1993)
21. Owen, C., Bezerra, C.: Evolutionary structured planning. a computer-supported methodology for the conceptual planning process. In: Artificial Intelligence in Design 2000, pp. 287–307. Kluwer Academic Publishers, Dordrecht (2000)

22. Pankratova, N., Gayko, G., Kravets, V., Savchenko, I.: Problems of megapolises underground space system planning. Autom. Inf. Sci. **48**(4), 32–38 (2016). https://doi.org/10.1615/JAutomatInfScien.v48.i4.40
23. Pankratova, N., Savchenko, I., Haiko, H., Kravets, V.: System approach to planning urban underground development. Int. J. Inf. Content Process. **6**(1), 3–17 (2019)
24. Pankratova, N., Gorelova, G., Pankratov, V.: Strategy for the study of interregional economic and social exchange based on foresight and cognitive modeling methodologies. In: 8th International Conference on Mathematics. Information Technologies in Education, Shatsk, Ukraine, 2–4 June, pp. 136–141 (2019)
25. Powell, J.: An application of a network-based futures method to strategic business planning. J. Oper. Res. Soc. **48**, 857–872 (1997)
26. Resin, V., Popkov, Y.: Large Cities Development in Conditions of Transition Economy (System Approach). Knizhnyi dom "Librokom", Moscow (2013)
27. Roberts, F.: Graph Theory and its Applications to Problems of Society, Society for Industrial and Applied Mathematics. In: Society for Industrial and Applied Mathematics, Philadelphia (1978). https://doi.org/10.7551/mitpress/2946.001.0001
28. Saługa, P.: Ocena ekonomiczna projektów i analiza ryzyka w górnictwie [Economic Evaluation and Risk Analysis of Mineral Projects]. nr.152, Wyd. IGSMiE PAN, Kraków (2009)
29. Savchenko, I.: Estimating the solution sensitivity in application of the modified morphological analysis method. Cybern. Syst. Anal. **52**(5), 782–790 (2016). https://doi.org/10.1007/s10559-016-9879-1
30. Sterling, R., Admiraal, N., Bobylev, H., Parker, J., Godard, I., Vähäaho, X.S., Hanamura, T.: Sustainability issues for underground spaces in urban areas. ICE. - Urban Des. Plan. **165**, 241–254 (2012). https://doi.org/10.1680/udap.10.00020
31. Tajdus, A., Cala, M., Tajdus, K.: Geomechanika w budownictwie podziemnym. Projektowanie i budowa tuneli, AGH (2012)
32. Vernadsky, V.: Biosphere and Noosphere. Iris Press, Moscow (2012)
33. Vähäaho, I.: Underground space planning in Helsinki. J. Rock Mechan. Geotech. Eng. **6**, 387–398 (2014). https://doi.org/10.1016/j.jrmge.2014.05.005

Risk Assessment of Innovative Projects Based on Fuzzy Modeling

Yuri Samokhvalov(✉)

Taras Shevchenko National University of Kyiv, 64, Volodymyrska Street, Kyiv 01033, Ukraine
yu1953@ukr.net

Abstract. This article is devoted to the consideration of innovative risk assessment approach. The approach includes two stages. At the first stage, as a result of fuzzy network modeling, based on fuzzy linguistic assessments of duration and cost of the job, the size of initial investments and the projected duration of the initial period of the innovative process, using β-distribution, are determined. On the second, random scenarios are simulated and the possible risk is calculated. This procedure combines fuzzy estimates and the Monte Carlo simulation. Fuzzy estimates of the model parameters of possible results allow us to take into account the uncertainty of the initial information in the process of simulating random values. In this case, for modeling random variables, the Gauss membership function is used, which determines the distribution of expectations of input variables.

Keywords: Innovative project · Risk · Net present value · Fuzzy sets · Linguistic assessment · Membership function · The Monte Carlo method

1 Introduction and Literature Review

Recent decades have brightly demonstrated the scientific and practical importance of innovative development as the most important factor in the economic growth of national economies and all business entities [22]. A characteristic feature of investment engineering is that the results are predictive in nature. Moreover, the more distant the forecast horizon, the less its accuracy.

Uncertainty of the predicted results leads to the risk that the goals, which are set in the project, may not be achieved fully or partially. Incorrect decisions regarding long-term investments can have especially serious consequences. Therefore, risk assessment is one of the main components of investment analysis, when making decisions, which are regarding the implementation of innovative projects. Moreover, the main task of the project risk analysis is to provide reliable information based on which decisions can be made on the feasibility of implementing or financing projects. Risk assessment of innovative (venture) projects is characterized by the fact that such projects are aimed at developing

S. Babichev et al. (Eds.): ISDMCI 2020, AISC 1246, pp. 265–281, 2021.
https://doi.org/10.1007/978-3-030-54215-3_17

and implementing a new idea, technology or product. Consequently, no statistical information is available about the object of study, and it is rarely possible to draw an analogy with such projects. That's why, when innovative projects are evaluating, a combination of expert and quantitative methods is necessary. This is due to the fact that persons involved in the innovative process, are forced to be guided not by data and calculations confirmed by previous practice, but by their subjective feelings and assessments, including regarding risks [6,9].

The Monte Carlo method is widely used for risk assessment in modern innovative practice [5,14,20,31,36]. This method allows taking into account the influence of uncertainty on the effectiveness of an innovative project and bringing its results closer to reality as much as possible. This makes Monte Carlo method one of the best ways to evaluate investment projects. However, despite the theoretical attractiveness and advantages of the Monte Carlo method, the application of this method is associated with certain difficulties in practice. These are because of the significant uncertainty conditions of the innovative process and its result, the accuracy of the simulation largely depends on the reliability of the initial data and the parameters of the forecast model. This can lead to an incorrect risk assessment, therefore, to the making a wrong investment decision. So, such an apparatus is needed that would adequately describe the initial data of the project and, as a result, increase their reliability in such conditions. Such an instrument can be the apparatus of the fuzzy set theory.

Fuzzy modeling is one of the most active, promising areas of applied research in the field of analysis and risk assessment of innovative, and investment projects recently [1,2,7,10,24,32]. In particular, studies have been devoted to the use of fuzzy sets in the Monte Carlo method [16,21,26,34,35]. So, Fuzzy Monte Carlo Simulation (FMCS) is considered for modeling scenarios of a sustainable society (3S Simulator) in [34]. In addition, FMCS, which combines the classical Monte Carlo theory of probability with the theory of probability of fuzzy arithmetic, was discussed in [16]. Moreover, this allows one to obtain a result not with one probability distribution, but with a family of distributions. FMCS, which uses both the probability density function (pdf) and the probability distribution (for example, fuzzy numbers) for modeling the uncertainty/inaccuracy associated with the input parameters and output data, was proposed in [35]. Fuzzy Monte Carlo method, which combines Fuzzy Logic and a Monte Carlo simulation to assess the sustainability of urban vehicles, was considered in [21]. Finally, FMCS for risk analysis of the construction projects was discussed in [26]. In this method, a fuzzy cumulative distribution function is constructed to represent the uncertainty.

The main point of the considered FMCS is the presentation of parameters and variables only by triangular fuzzy numbers. Such numbers are used when the predicted values of parameters or variables are known. However, intervals of possible values of a random variable are often known in practice. In this case, trapezoidal fuzzy numbers specify such parameters. In addition, these jobs do not provide mechanisms for substantiating the size of investments and their payback periods.

The article considers an approach combining fuzzy network modeling and fuzzy Monte Carlo simulation. This allows justifying the time to prepare the production of new products and the amount of initial investment based on fuzzy estimates and modeling random variables, use triangular and trapezoidal fuzzy numbers given in the linguistic form.

2 Innovative Projects and Risks

An innovative project (IP) is understood as an investment project that has the following distinctive features [8]. These projects contain significant expenses for research, development and experimental job, as well as for manufacturing and testing of new products. An increased level of possible risks is another important feature of IP. Since innovative activity is associated with the creation of new or modification of existing goods or services, ceteris paribus, the implementation of an innovative project has a higher degree of uncertainty than the implementation of a conventional investment project. In addition, the uncertainty is also caused by the influence of both economic factors (market fluctuations, prices, exchange rates, inflation, etc.) that are not dependent on the efforts of investors and non-economic factors (climatic and environmental conditions, political relations, etc.), which are not always amenable to accurate assessment [23]. On the other hand, underestimating the risks of innovative projects leads to serious consequences associated with the loss of time, money and reputation.

Another important difference between IP and investment is a long time between the end of the investment part of the project and the earning of the first income. Further, investments in any project are associated with a certain risk of not receiving the expected profit from the innovative project. The practice of innovative activity shows that investment and innovative risks can be caused by the following factors [13]: too large assumptions regarding the prospects of innovative; the presence of corporate stress associated with the implementation; an insufficient degree of distinguishability of the new development; the risk of unexpected technical problems; unforeseen expenses of an innovative project.

Other risks can be highlighted, for example, the risk associated with a change in a firm's position in the market, or with its market "strength", as well as with a change in its status, which entails a drop in business and social reputation and a loss of administrative resources. It can be bankruptcy, mergers and absorption, force majeure. These risks are much less common and therefore they are pointed as the "other" group.

With this in mind, the innovative risk is the result of the combined action of these factors determining various types of risks: scientific, technical, economic, political, business, social, environmental and others. A more detailed classification of innovative risks can be found in [17].

3 An Innovative Project Performance Indicator

An efficiency mark of innovative projects is one of the main elements of investment analysis. As larger, the innovative project and more significant changes it causes in the business, then more accurate should be the calculation of cash flows and methods for evaluating such projects performance indicator [12].

To assess the assessment of the effectiveness of innovative projects, the net present value (NPV) of the project is most often used [19]. It reflects the investor's profit (value-added of investments) that he expects to receive from the implementation of the project after the cash inflows cover his initial investment costs and the periodic cash outflows associated with the implementation of such a project. This indicator is used to evaluate project performance indicator and to select the best option when comparing alternative projects. The net present value is calculated by discounting (bringing to the current value, i.e., at the time of investment) the expected cash flows (both income and expenses)[19]:

$$NPV = \sum_{i=1}^{n} \frac{CF_i}{(1+p)^i} - IC, \tag{1}$$

where n is a total number of periods (calculation horizon); p is an interest rate or a discount rate; CF_i is the cash inflow in the i-th year; IC are the investments in the project.

The calculation of NPV in practice is based on the annual cash inflow CF_i (Cash Flow). Cash inflow means the total income received in each time period minus the costs associated with its obtainment during the implementation of the investment project. CF_i is a generalized indicator that includes many components: funds from the implementation of core, investment and financial activities. It acts as a certain random number obeying the normal distribution law.

The interest rate p determines the investor's income from the implementation of an innovative project. The choice of interest rate depends on both external factors (economic situation, inflation rate) and internal factors related to the creditworthiness of the company, assessment of its near and distant future, etc. Various methods and approaches have been developed for calculating and justifying the interest rate [3]. An analysis of its possible options allows the investor to take as a rate for discounting some acceptable level of profitability for him.

Initial investments in innovative projects IC (Invest Capital) are related to the preparation of production for the release of new or improved products. The purpose of preparing a new product is the creation and organization of the new competitive products release, the transfer of the production process to a higher technical and social-economic level, ensuring the efficient job of the company.

The timing risks are of particular importance for such projects since failure to meet with the deadlines can lead to additional costs (delayed payments, loss of interest, increased project expenses, etc.). Therefore, to make an investment decision, a reasonable assessment of the initial investment is necessary, which to some extent compensates for such risks. Given these provisions, the proposed approach involves two steps. The first calculates the amount of initial investment, and the second simulates the possible values of NPV. A common and important task that is solved at these stages is to ensure the reliability of the source data, on which the accuracy of the results of the innovative design largely depends. Given the significant uncertainty of the innovation process, the initial data and model parameters are set by fuzzy linguistic values. This will allow us to reduce the psychological uncertainty of a person in the correctness of his estimates, under these conditions, since it is easier for him to give a fuzzy estimate to a specific parameter than to assign a measure to him. Therefore, first, consider the processing mechanisms of fuzzy linguistic quantities, and then the indicated steps.

4 Approximation of Fuzzy Linguistic Quantities by Numerical Intervals

A fuzzy linguistic quantity means a numerical quantity, which is expressed using the modality "approximately". Such quantities are represent by statements "the number is approximately equal T" or "the number is approximately in the range from T_1 to T_2" and are approximated by numerical intervals, the boundaries of which are the transition points of the membership functions (MF) of the corresponding fuzzy sets. To determine the distance between the transition points, we use the algorithm that was considered in [4]. This method is based on statistics data, that were obtained as a result of the experiment. These data represent the views of experts on the distances between transition points for various numbers and are given in Table 1 [4].

Table 1. The distance between the transition points

Number: x	The distance between the transition points: $b(x)$
1, 2, 3, 4, 6, 7, 8, 9	0,46x
10, 20, 30, 40, 60, 70, 80, 90	(0,357–0,00163)x
35, 45, 55, 65, 75, 85, 95	21 (0,213–0,00067x)x
5	2, 8
15	6, 48
25	6, 75
50	24
Other two-digit numbers	$\frac{1}{2}(b([\frac{x}{10}] \cdot 10 + 5) + b(x - [\frac{x}{10}] \cdot 10))$

Let a fuzzy linguistic quantity "the number is near the number T" be given, where T is a natural number. If $T \in [1, 99]$, then $b(T)$ could be found according to the Table 1, in which $[a]$ is an integer part of the number a. Otherwise, let its least significant digit has an order of q. We divide the possible values q into the residue classes modulo 3. As a result, we obtain three classes $M_d, d \in \{0, 1, 2\}$, where $d = q\text{mod}(3)$. In this case the value $b(T)$ also depends on the class M_d, to which the number T belongs.

Let r_q be the numeral that is in the q-th place of the number T. Then:

1. If $T \in M_0$ (for example, 300, 300000 etc.), then $b(T) = b(x) \cdot 10^{q-2}$, where $x = r_q \cdot 10$ and $b(x)$ is taken from Table 1.
2. If $T \in M_1$ (for example, 101, 202000, 15000 etc.), then two options are possible:
   a) if $r_{q+1} = 0$, then $b(T) = b(x) \cdot 10^{q-1}$, where $x = r_q$;
   b) if $r_{q+1} \neq 0$, then $b(T) = b(x) \cdot 10^{q-1}$, where $x = r_{q+1} \cdot 10 + r_q$.
3. If $T \in M_2$ (for example, 2030, 2140 and etc.), then two options are possible:
   a) if $r_{q+1} = 0$, then $x = r_q \cdot 10$; $b(T) = b(x) \cdot 10^{q-2}$;
   b) if $r_{q+1} \neq 0$, then $x = r_{q+1} \cdot 10 + r_q$; $b(T) = b(x) \cdot 10^{q-1}$;

As a result, the value $b(T)$ will be obtained. Then the transition points $\alpha = T - \frac{b(T)}{2}$ and $\beta = T + \frac{b(T)}{2}$ are calculated. These values define the boundaries of the confidence interval of numbers, which are approximately equal to the number T.

Now consider the case when "the number is approximately in the range from T_1 to T_2". In this case, the values $b(T_1)$ and $b(T_2)$ are found first, and then the boundaries of the confidence interval of this fuzzy linguistic quantity are calculated: $\alpha = T_1 - \frac{b(T_1)}{2}$, $\beta = T_2 + \frac{b(T_2)}{2}$.

For better understanding such calculations, let's consider an example like: determine the confidence interval of the fuzzy set "the number is approximately in the range from 90 to 137." In this case, firstly, we find the values $b(90)$ and $b(137)$. The value $b(90) \approx 32$ is determined according to Table 1, and the value $b(137)$ is calculated as follows.

The least significant digit of 137 is in the discharge of units ($q = 1$), therefore, $r_q = r_1 = 7$, $r_{q+1} = r_2 = 3$, is a digit whose order is unit higher than the order of the least significant digit of the number 137. After dividing q by 3, the remainder gives 1, therefore, the number 137 belongs to the equivalence class M_1, so $d = 1$. Since $r_{q+1} \neq 0$, then according to item 2b we have that $x = r_{q+1} \cdot 10 + r_q = r_2 \cdot 10 + r_1 = 37$ and $b(137) = b(37)$, and $b(37)$ is calculated by the formula:

$$b(37) = \frac{1}{2}(b([\frac{37}{10}] \cdot 10 + 5) + b(37 - [\frac{37}{10}] \cdot 10)) = \frac{1}{2}(b(35) + b(7)),$$

where $b(35)$ and $b(7)$ are found in Table 1: $b(35) = 6.63$ and $b(7) = 3.22$. Then $b(137) = \frac{1}{2}(6.63 + 3.22) \approx 5$. As a result, we get $\alpha = 74$, $\beta = 139.5$.

5 Initial Investment Estimation

Determining the initial investment is one of the main tasks of the preliminary innovative research. The discounted payback period depends on the correct assessment of the size of initial investment, which is a key indicator of assessing the investment attractiveness of an innovative project. For correct assessing the size of initial investments, it is necessary to know their structure, or, which is the same, the articles financing the jobs on preparing a new production. By the type and nature of the job, the preparation of production includes the following stages: conducting research and development job; technological preparation of production; organizational and economic preparation of production [18].

Scientific research and development job are designed to develop new and improve existing products and processes, ways and methods of organizing and managing production. The result is a prototype of a new product. Technological preparation of production consists in choosing the most rational and economical technology route that determines the sequence of basic operations (job) to create new products. Organizational and economic preparation of production includes the development of a project the organization of primary and auxiliary production, labor and wages, and their adaptation to the conditions for manufacturing new products. Given the significant amount of jobs involved in the preparatory phase of innovative research and their planning in the face of uncertainty, it is proposed to evaluate the initial investment by the method [28,29], which combines formal procedures and fuzzy expert estimates. In addition, this method also allows predicting the most likely duration of the preparatory phase. It will make possible to estimate the payback period of investments relative to the start of the innovative process that is more informative for the investor than the assessment of only the investment period. For the purpose of a holistic perception of the material, we consider it in the context of job [29].

Suppose that the preparation of production require a certain number of jobs $s_1, s_2, ..., s_m$. There may be jobs of a scientific, technical, technological and organizational nature among them. These jobs will be called *basic*. First, the list of main jobs is supplemented by *intermediate jobs* $s_{m+1}, ..., s_{m+n}$, the implementation of which may be necessary or useful for the implementation of the main jobs. Then, for each job (goal) s_i $(i = \overline{1, m+n})$, the experts that formulate the conditions for its achievement $s_{i1}, s_{i2}, ..., s_{ik}$ and give the duration estimates of the job after fulfilling the given condition are invited. The estimates of the duration and cost of job s_i are given by the statements "the duration (cost) of job s_i is approximately equal to T" or "the duration (cost) of job s_i is approximately in the range from T_1 to T_2". Then these estimates are approximated by intervals of time $[t'_i, t''_i]$ and cost $[w'_i, w''_i]$.

In order to increase the objectivity of expert assessments, they are coordinated. As a result, expert evaluations $[t'_{ij}, t''_{ij}]$ and $[w'_{ij}, w''_{ij}]$ are led to consensus estimates $[t'_i, t''_i]$, $[w'_i, w''_i]$. The coordination of interval expert evaluations is described in detail in [30], therefore, these issues are omitted. As a result, for each job s_i several conditions for its implementation will be obtained, as well as estimates of time and cost, in accordance with the number of experts involved. After this, the job s_i $(i = \overline{1, m+n})$ is divided into sets $M_0, M_1, ..., M_p$, where $M_i \cap M_j = \varnothing$. The set M_0 includes jobs that have only unconditional estimates of the execution time, and the set M_i includes jobs whose conditions are only jobs from the sets $M_0, M_1, ..., M_{i-1}$ $(i = \overline{1, p})$. In addition, for each job s_i $(i = \overline{1, m+n})$, a "shifted" estimate of its duration is relative to the time when the condition is fulfilled and the maximum execution time are calculated. The shifted estimates are calculated as follows:

Case 1. Let s_i be an unconditional job, and let $t_{ik} = [t'_{ik}, t''_{ik}]$ $(k = \overline{1, m_i})$ be estimates of its execution time given by m_i experts. Then these estimates will be shifted estimates of the time taken to complete the job s_i, and $t_i^{max} = \max_k t''_{ik}$ will be their maximum duration.

Case 2. Let $s_{i1}, s_{i2}, ..., s_{ik_i}$ be the jobs that are the condition for the job s_i, to be completed, let $t_{ik} = [t'_{ik}, t''_{ik}]$ $(k = \overline{1, m_i})$ be the time estimates for its duration after this condition is met, and let t_{ij}^{max} be the maximum job $s_{ij}(j = \overline{1, k_i})$ duration. Then the boundaries of the shifted estimates of the execution time of the job s_i are calculated as follows:

$$t'_{ik} = t'_{ik} + t_{iu} \quad and \quad t''_{ik} = t''_{ik} + t_{iu},$$

where t_{iu} is the maximum execution time of fulfillment of the condition s_{i1}, $s_{i2}, ..., s_{ik_i}$.

Moreover, the time for fulfilling this condition depends on the logic of the performance of jobs s_{ij}. If jobs are arranged in a sequence of the form "end-start", then the maximum execution time of condition $s_{i1}, s_{i2}, ..., s_{ik_i}$ is calculated as the sum of t_{ij}^{max} values. If jobs are performed in parallel, then the maximum value of t_{ij}^{max} is taken as the maximum time for fulfilling the condition $s_{i1}, s_{i2}, ..., s_{ik_i}$. In other cases, the time t_{iu} is calculated according to the corresponding job schedule s_{ij}.

Estimation of the Most Probable Duration of the Initial Period. Let $[t', t'']$ be an estimate of the duration of job s. Then the probability $p(t|t', t")$ that the job s will be completed by time $t' \le t \le t"$ is calculated by the formula:

$$p(t|t', t'') = \begin{cases} 0, & t < t' \\ F(t), & t' \le t \le t'' \\ 1, & t > t'' \end{cases} \tag{2}$$

where $F(t) = 12 \int_0^x y \cdot (1 - y)^2 dy$ is a β-distribution function with parameters $\alpha = 2$ and $\beta = 3$, $x = \dfrac{t - a}{b - a}$ is a zoom variable $(0 \le x \le 1)$.

D. Golenko proposed the function $F(t)$ in [11]. It describes the β-distribution in the interval $[a, b]$ and requires only an optimistic estimate a and a pessimistic estimate b.

Let $p_i(t|t'_{ij}, t''_{ij})$ be the probability of the fact that by the time moment $t'_{ij} \le t \le t''_{ij}$ the job s_i will be completed, where $[t'_{ij}, t''_{ij}]$ is the estimate of its duration, which was given by j-th expert. Then the probability $p_i(t)$ of the fact that by the time moment t the job s_i will be completed, we will calculate by the formula [30]:

$$p_i(t) = \sum_{j=1}^{m_i} \lambda_{ij} p_{ij}(t|t'_{ij}, t''_{ij}) p_{ij1}(t) p_{ij2}(t) ... p_{ijk_j}(t), \tag{3}$$

where ij_1, ij_2 ,..., ijk_j are the numbers of intermediate jobs that the j-th expert determined as a condition for the job s_i completion; λ_{ij} is this weight coefficient, whereas $\sum_{j=1}^{m_i} \lambda_{ij} = 1$; t'_{ij}, t''_{ij} are the boundaries of the "shifted" estimate of the time taken to complete the job s_i after the fulfillment of the condition; m_i is a number of experts who have evaluated the job s_i. This formula gives an average probability estimate taking into account the weights of experts. Moreover, for unconditional job, the expression $\lambda_{ij} p_{ij}(t|t'_{ij}, t''_{ij}) p_{ij1}(t) p_{ij2}(t)...p_{ijk_j}(t)$ in (3) changes to $\lambda_{ij} p_{ij}(t|t'_{ij}, t''_{ij})$.

Let $s_k \in M_p$ be the final job of the preparatory phase. Then, using (3), it is possible to obtain a forecast of the most probable time duration t of this stage. As such time, it is customary to consider the median of distribution $p_k(t) = 0.5$ or $t = p_k^{-1}(0.5)$.

The expected estimate of the initial investment. Let $[w'_{ij}, w''_{ij}]$ be an estimate of the cost of job s_i given by the j-th expert. Based on this assessment, the expected cost of job s_i can be calculated by the formula:

$$w_{exp}^{ij} = \frac{w'_{ij} + w_p^{ij} + w''_{ij}}{3}, \tag{4}$$

where w_p^{ij} is the most probable estimate of the cost of job s_i.

According to [15], costs have a triangular distribution in the interval $[w'_{ij}, w''_{ij}]$ with a function

$$F(w) = \begin{cases} \dfrac{2(w - w'_{ij})^2}{(w''_{ij} - w'_{ij})^2}, w'_{ij} \le w \le \dfrac{w'_{ij} + w''_{ij}}{2} \\ 1 - \dfrac{2(w''_{ij} - w)^2}{(w''_{ij} - w'_{ij})^2}, \dfrac{w'_{ij} + w''_{ij}}{2} \le w \le w''_{ij} \end{cases}. \tag{5}$$

Then, the most probable estimate of the cost of job s_i is $w_p^{ij} = F^{-1}(0.5)$. Taking into account the assessments of all experts, the expected cost of job s_i is calculated as a weighted sum of

$$w_{exp}^{i} = \sum_{j=1}^{m_i} \lambda_{ij} \cdot w_{exp}^{ij}, \tag{6}$$

where λ_{ij} is the weight coefficient of the j-th expert ($\sum_{j=1}^{m_i} \lambda_{ij}$=1), m_i is the number of experts who evaluated the job of s_i. And finally, knowing the expected cost of the jobs w_{exp}^i, the amount of initial investment is calculated by the formula

$$IC = \sum_{j=1}^{k} w_{exp}^{ij}, \tag{7}$$

where k is the number of planned jobs.

6 Simulation of Random Scenarios

Risk assessment using the Monte Carlo method is performed using a model of possible outcomes, i.e. a mathematical model for determining the indicator of the effectiveness of an innovative project (1), which is exposed to a series of simulation runs using a computer. The cash flow CF_i is the key in this model. It is a random variable, based on which possible scenarios (project results) are built in the process of simulation. Since the values of this parameter are predictive, so it is estimated by fuzzy statements like "expected cash flow CF_i is approximately equal to C_i" or "expected cash flow CF_i is approximately in the range from CF^i_{min} to CF^i_{max}". Here CF^i_{min} and CF^i_{max} are values of cash inflows, which are calculated respectively in extremely favorable (excluding risk) and most unfavorable (taking into account all possible risks) conditions. Note that the values C_i, CF^i_{min} and CF^i_{max} can be obtained by one of the known methods [25]. Let's consider two cases:

Case 1. Let parameter CF_i be given a fuzzy linguistic assessment "the expected cash inflow is approximately equal to C_i". So, let build the membership function of this fuzzy set $\tilde{C}_i$. As such a function, will use the Gauss function [4]:

$$\mu_{\tilde{C}_i}(x) = e^{-a(x-C_i)^2}, \tag{8}$$

where $a = -\dfrac{4\ln 0.5}{b^2}$, and b is the distance between the transition points of function $\mu_{\tilde{C}_i}(x)$. This distance is calculated by the algorithm described above.

The Gauss function has an unbounded carrier since on the left and on the right it asymptotically tends to zero. However, in practice, the carrier of this function can be considered as limited by points $x = C_i \pm 3\sigma$, at which its value is approximately equal to 0.01. Therefore, in practice, we can assume that the value of the function, that is equal to 0.01, corresponds to the complete non-belonging of the element to the fuzzy set $\tilde{C}_i$. If we go from σ to b then the boundaries of this interval will be $x = C_i \pm \dfrac{k \cdot b}{2}$, where $k \approx 2.5$ is the scaling factor. Let denote these boundaries by C^i_{min} and C^i_{max}.

The membership function $\mu_{\tilde{C}_i}(x)$, built in this way, is a function of the probability density of the normal law and determines the distribution of expectations for cash inflows in the i-th year. We will use this function to model random CF_i values.

Let the initial investment amount IC be given, the interest rate p, for each period i the intervals $[C^i_{min}, C^i_{max}]$ of possible values of the parameter CF_i and membership function $\mu_{\tilde{C}_i}(x)$ are known. Next, let r be a random number. If $r \in [0.01, 0.5]$, then the value of parameter CF_i is the root of equation $\mu_{\tilde{C}_i}(x) = 2r$, which lays in the interval $[C_i, C^i_{max}]$. If $r \in [0.5, 0.99]$, then the root of equation $\mu_{\tilde{C}_i}(x) = 2(1-r)$, which belongs to the interval $[C_i, C^i_{max}]$, is taken. As a result, the set of values $CF_i \in [C^i_{min}, C^i_{max}]$ is obtained.

Note that in this case, to simulate all the values of CF_i, one random number r was used. However, you can use your own random number for each CF_i.

Case 2. Let parameter CF_i be given a fuzzy linguistic assessment "the expected cash inflow is approximately in the range from CF^i_{min} to CF^i_{max}." Let also $[CF^i_1, CF^i_2]$ be the confidence interval of this estimate. Then the boundaries of the parameter carrier CF_i of this fuzzy set are calculated using formulas $C^i_{min} = CF^i_{min} - \frac{k \cdot b_1}{2}$ and $C^i_{max} = CF^i_{max} + \frac{k \cdot b_2}{2}$, where b_1, b_2 are the distances between the transition points of the membership functions of the fuzzy set "number is approximately equal to CF^i_{min}" and "number is approximately equal to CF^i_{max}", respectively, and also $k \approx 2.5$. In this case, $\frac{b_1}{2} = CF^i_{min} - CF^i_1$, and $\frac{b_2}{2} = CF^i_2 - CF^i_{max}$.

Next, let $C_i = \frac{CF^i_{min} + CF^i_{max}}{2}$ and build the MF of the fuzzy set $\tilde{C}_i \triangleq$ "the expected cash inflow is approximately equal to $\tilde{C}_i$". Since the $\frac{b_1}{2} \neq \frac{b_2}{2}$, therefore, this function consists of two Gaussian functions:

$$\mu_{\tilde{C}_i}(x) = \begin{cases} \mu^1_{\tilde{C}_i}, \; x \in [C^i_{min}, C_i] \\ \mu^2_{\tilde{C}_i}, \; x \in [C^i, C^i_{max}] \end{cases}, \tag{9}$$

where $\mu^1_{\tilde{C}_i} = e^{-\frac{(x-C_i)^2}{2\sigma_1^2}}$, $\mu^2_{\tilde{C}_i} = e^{-\frac{(x-C_i)^2}{2\sigma_2^2}}$.

In these functions, parameters σ_1 and σ_2 are found from the equations

$$e^{-\frac{(C^i_{min}-C_i)^2}{2\sigma_1^2}} = 0.01 \;\; and \;\; e^{-\frac{(C^i_{max}-C_i)^2}{2\sigma_2^2}} = 0.01.$$

In this case, in the process of modeling the values $CF_i \in [C^i_{min}, C^i_{max}]$, if $r \in [0.01, 0.5]$, then the value of parameter CF_i is the root of equation $\mu^1_{\tilde{C}_i}(x) = 2r$, and if $r \in (0.5, 0.99]$, then the root of equation $\mu^2_{\tilde{C}_i}(x) = 2(1-r)$. Then, the NPV is calculated. We obtain statistics on random scenarios NPV, repeating this calculations for different numbers r_i many times. After that, the investment risk is estimated. It is the probability of a scenario with a negative NPV as an unfavorable project result for the investor. This probability is defined as the ratio of the number of such scenarios to their total number:

$$P(NPV < 0) = \frac{k}{m},$$

where k is a number of negative values NPV, m is a number of performed experiments.

If the investor accepts the risk, the discounted payback period is calculated. This indicator is calculated by the formula [3]:

$$DPP = min\, n, \; at\; which \sum_{i=1}^{n} CF_i > IC, \tag{10}$$

where $CF_i = C_i$.

Then, taking into account the duration T of the preparatory stage, the return on investment will occur in $T + n$ units of time.

In conclusion, note that if the task is to choose the preferred option when comparing alternative projects, in this case, the criteria are the most probable duration of the preparatory phase, the investment size, the payback period, the expected value of NPV and the probability of receiving a negative result from the implementation of an innovative project (investment risk). Such a choice can be effectively made by the method of [27].

7 Practical Implementation

Consider an example illustrating the proposed approach to the risk assessment of innovative projects. Let the investor have to decide on the feasibility of releasing new products. This approach includes two stages: calculation of the initial investment and modeling of random scenarios.

Calculation of the initial investment. Suppose that it is necessary to carry out two main jobs s_1 (to master the production of products A), s_2 (to release a prototype) and one intermediate job s_3 (to conduct scientific research) at the initial stage of innovative research. Also, let each job be evaluated by two experts. The results of the evaluation of jobs are shown in the Table 2.

Table 2. The expertise results

Works	s_1		s_2		s_3	
Layers	M_2		M_1		M_0	
Experts	1	2	1	2	1	2
Weights	0.6	0.4	0.3	0.7	0.4	0.6
Conditions	(s_2, s_3)	s_2	s_3	–	–	–
Cost estimates	[20, 30]	[10, 30]	[10, 50]	[20, 40]	[10, 20]	[10, 30]
Time estimates	[1, 3]	[2, 4]	[1, 4]	[2, 3]	[1, 3]	[2, 3]
Shifted time estimates	[8, 10]	[9, 11]	[4, 7]	[2, 3]	[1, 3]	[2, 3]
Max time	11		7		3	

According to this table, jobs s_1 and s_2 are conditional, and job s_3 is unconditional. Condition (s_2, s_3) means that job s_1 is performed after jobs s_2 and s_3, and time evaluation, for example, [1, 3] means that after fulfilling this condition, from one to three units of time are needed to complete job s_1. In addition, the confidence intervals of the corresponding linguistic estimates are used as estimates of the cost and duration of the jobs s_1, s_2, and s_3. For the convenience of calculations, the boundaries of these intervals are rounded to the nearest integer. For example, the confidence interval of the fuzzy linguistic assessment "the cost of job s_3 is approximately in the range from 13 to 17" is the interval [9.84, 20.39]. And taking into account the rounding of its borders, we get the estimate [10, 20]. The data presented in the table allow obtaining a forecast of initial investments, as well as the most probable duration of the t_p preparatory stage of the innovative research. The duration of the preparatory phase is determined by the completion time of the job s_1. For this job, according to (2) and (3), the probability distribution function has the form: $p_1(t) = 0.6p_{11}(t|8,10)p_2(t)p_3(t) + 0.4p_{12}(t|9,11)p_2(t)$, where $p_2(t) = 0.3p_{21}(t|4,7)p_3(t) + 0.7p_{22}(t|2,3)$, $p_3(t) = 0.4p_{31}(t|1,3) + 0.6p_{32}(t|2,3)$.

Then, $t_p = p_1^{-1}(0.5)$. Given the form of function $p_1(t)$, it is difficult to obtain an analytical expression for $p_1^{-1}(x)$. So, in this case, we calculate the median in an approximate way. That is, we find such $t \in [0, 11]$ that inequality $|p_1(t)-0.5| \leq \varepsilon$ holds, where ε is the calculation error. Let $\varepsilon = 0.01$, then $t_p \approx 9$, at which the probability will be $p_1(t_p) \approx 0.503$.

The expected size IC of initial investments is calculated by the formula (7): $IC = \sum_{i=1}^{3} w_{exp}^i$, where w_{exp}^i is expected cost of job s_i. Using estimates of the cost of jobs s_1, s_2, s_3 and formulas (4), (6) we get that $IC = 65$. An inverse function of the form is used, when calculating the most probable cost w_p^{ij} in formula (4):

$$F^{-1}(x) = \begin{cases} w'_{ij} + (w''_{ij} - w'_{ij})\sqrt{\dfrac{x}{2}}, \ x \leq \dfrac{1}{2} \\ w''_{ij} + (w''_{ij} - w'_{ij})\sqrt{\dfrac{(1-x)}{2}}, \ x > \dfrac{1}{2} \end{cases}.$$

Simulation of random scenarios. Table 3 shows the initial simulation data for four innovation periods.

Table 3. The initial simulation data

Periods (i)	1	2	3	4
Estimate CF_i: (C_i)	≈22	≈27	≈33	≈18
Interval of possible values CF_i: $[C_{min}^i, C_{max}^i]$	[17, 27]	[21, 33]	[25, 41]	[13, 23]
Interest rate (p)	12			
Investments (IC)	65			

In this table, estimates CF_i are given by fuzzy linguistic expressions like "expected cash flow is approximately equal to 22, 27, 33, 18". In addition, for the convenience of calculations, the boundaries C_{min}^i, C_{max}^i, are rounded to the nearest integer. Based on these data, for each investment period i membership functions (8) of fuzzy sets $\tilde{C}_i \triangleq$ "expected cash flow approximately equal C_i" are constructing:

$$\mu_{\tilde{C}_1}(x) = e^{-\dfrac{4\ln 0.5(x-22)^2}{16}}, \ \mu_{\tilde{C}_1}(x) = e^{-\dfrac{4\ln 0.5(x-27)^2}{26.6}},$$

$$\mu_{\tilde{C}_1}(x) = e^{-\dfrac{4\ln 0.5(x-33)^2}{36}}, \ \mu_{\tilde{C}_1}(x) = e^{-\dfrac{4\ln 0.5(x-18)^2}{13}}.$$

In one run of the forecast model, the values $CF_i(i = \overline{1,4})$ are modeled by one random number r. If $r \in [0.01, 0.5]$, then values CF_i are the roots of equations $\mu_{\tilde{C}_1}(x) = 2r$, which are in the intervals $[C^i_{min}, C_i]$. If $r \in [0.5, 0.99]$, then the roots of equations $\mu_{\tilde{C}_1}(x) = 2(1-r)$ are taken, which belong to the intervals $[C^i, C^i_{max}]$. As a result, the set of values $CF_i \in [C^i_{min}, C^i_{max}]$ is obtained. Then, the indicator $NPV = \sum_{i=1}^{4} \frac{CF_i}{(1+p)^i} - IC$ is calculated. 1000 model runs were made. The result of the statistical analysis of the obtained data is given in Table 4.

Table 4. Summary statistics of innovative risk assessment

Indicator	Value
The most likely duration of the preparatory phase	9
Investment size	65
Minimum value of NPV	0.02
Maximum value of NPV	27.8
Expected value of NPV	11
The number of cases when $NPV < 0$	80
The probability of a negative result from the implementation of the innovative project (investment risk)	0.08

The following scale can be used for verbal expression the risk assessment of a project [33] (Table 5).

Table 5. Verbal risk assessments

Quantitative assessment	Verbal assessment
Less than 0.01	Low
0.01–0.1	Average
Over than 0.1	High

Since the risk is medium (we will consider it as acceptable for the investor), the discounted payback period is calculated, i.e. refund period taking into account the time value of money (discount rate). This period is calculated by the formula (8):

$$DPP = min\, n,\ at\ which \sum_{i=1}^{n} \frac{CF_i}{(1+p)^2} > IC.$$

In this case, we get $DPP = 3$ If the unit of time is considered a year, then taking into account the duration of the preparatory phase, the return on investment will occur in about 12 years. Note that if the task is to choose the preferred option when comparing alternative projects, so in this case the criteria may be the duration of the preparatory phase, the amount of investment, the payback period, the expected value of, and the probability a negative result. Such a choice can be effectively made with the method, given in [27].

8 Conclusions

This article considers the approach to risk assessment of innovative projects. The risk assessment involves two stages. At the first stage, as a result of fuzzy network modeling, based on fuzzy linguistic assessments of duration and cost of the jobs, the size of initial investments and the projected duration of the initial period of the innovative process. To calculate the volume of investments, a triangular distribution function is used, the function of the β-distribution of the time taken to complete the jobs for the duration of the initial period. This function allows us to build the β-distribution of a random variable over the entire domain of its definition, based on a set of approximate point and interval expert estimates. This makes it possible to estimate the payback period of investments relative to the start of the innovation process, which is more informative for an investor than the assessment of only the investment period.

In the second stage, random scenarios are modeled and the possible risk is calculated. Simulation of random scenarios is performed by the Monte Carlo method using the model for determining the NPV indicator of the effectiveness of an innovative project. Fuzzy estimates of the model parameters are used during modeling, which allows taking into account the uncertainty of the initial information in the process of modeling random variables. In this case, to model random variables, the Gauss membership function is used, which determines the distribution of expectations of input variables. Investment risk is defined as the probability of a project outcome unfavorable to the investor.

In general, the use of this approach allows us to determine the volume of initial investments, and the duration of the initial period of the innovation process, the risk of investments, and their payback period. These indicators can be used as criteria when it is necessary to choose the preferred option when comparing alternative innovative projects.

The proposed approach does not pretend to be complete and can be used as the basis at the development of the appropriate methodological apparatus for risk assessment of various innovative projects.

References

1. Abreu, A., Martins, J., Calado, J.: Fuzzy logic model to support risk assessment in innovation ecosystems. In: 13th APCA International Conference on Automatic Control and Soft Computing (CONTROLO), Ponta Delgada, Azores (2018). https://doi.org/10.1109/CONTROLO.2018.8514281

2. Baranova, M., Khaibullina, G., Galiaksarova, G.: The use of fuzzy logic for the analysis of an investment project. Sci. Contrib. Future **6**(4), 21–24 (2016)
3. Belyi, E., Romanova, I.: Methods of financial and commercial settlements, tutorial, UlSU, Ulyanovsk (2015)
4. Borisov, A., Krumberg, O., Fedorov, I.: Decision making based on fuzzy models: examples of use, knowledge, Riga (1994)
5. Crowey, M., Gailay, D., Mark, R.: Basics of Risk Management. Publishing House Yurayt, Moskov (2016)
6. Deptula, A., Knosala, R.: Risk assessment of the innovative projects implementation. Manag. Prod. Eng. Rev. **6**(4), 15–25 (2015). https://doi.org/10.1515/mper-2015-0032
7. Doskocil, R.: An evaluation of total project risk based on fuzzy logic. Bus. Theory Pract. **17**(1), 23–31 (2015). https://doi.org/10.3846/btp.2016.534
8. Dosuzheva, E.: Innovation and investment projects, their features and main forms of implementation. Internet-jornal "Sci. Sci." **7**(2) (2015). http://naukovedenie.ru/PDF/125EVN215.pdf
9. Gelfer, L.: Risk analysis features of venture investment project. Russ. Entrep. **8**(8), 105–110 (2007)
10. Glushenko, S.: The use of a fuzzy logic mechanism for assessing the risk of investment and construction projects. Vestnik RINH **47**(3), 58–67 (2014)
11. Golenko-Ginzburg, D.: Stochastic Network Models for Development Planning and Management. Scientific Book, Voronezh (2010)
12. Goncharenko, L.: Innovation Policy: Textbook. KNORUS, Moscow (2011)
13. Gracheva, M., Lyapina, S.: Risk Management in Innovation. UNITY-DINA, Moscow (2010)
14. Gruzin, N.: Modern approaches to the assessment of project risks of organizations. Internet J. "Sci." **7**(6), 1–7 (2015). https://doi.org/10.15862/68EVN615
15. Hulett, D.: Integrated Cost-Schedule Risk Analysis, New edn. Ashgate Publishing, Surrey (2011)
16. Kim, Y.: Monte Carlo vs. fuzzy monte carlo simulation for uncertainty and global sensitivity analysis. Sustainability **9**(4), 539 (2017). https://doi.org/10.3390/su9040539
17. Konovalova, O.: Modern approaches to the classification of innovative risks. Vestnik Universiteta (9), 59–65 (2015)
18. Korshunov, V.: Economics of Organization (Enterprise). Theory and Practice. Yurayt Publishing, Moscow (2014)
19. Kosov, V., Livshits, V., Shakhnazarov, A.: Guidelines for the evaluation of investment projects, economics, Moscow (2000)
20. Kuzina, S., Kuzin, P.: Risk assessment of an innovative project by simulation method (monte carlo method), bulletin TVSU. Ser. "Econ. Manag." (Penza State Univ. Penza Russ) (2), 257–264 (2014)
21. Lambros, K., Mitropoulos, P., Prevedouros, X., Yu, E., Nathanail, A.: A fuzzy and a monte carlo simulation approach to assess sustainability and rank vehicles in urban environment. Transp. Res. Procedia **24**, 296–303 (2017). https://doi.org/10.1016/j.trpro.2017.05.121
22. Makhovikova, G., Efimova, N.: Innovation Management. Eksmo, Moskow (2010)
23. Matveeva, L.: Management of investment projects in conditions of risks and uncertainty: a training manual. Yurayt Publishing House, Moscow (2018)
24. Narendra, M.: Risk assessment is fuzzy business-fuzzy logic provides the way to assess off-site risk from industrial installations. Risk **206**, 218–231 (2004)

25. Prodchenko, I.: Theoretical Foundations of Financial Management: Educational and Methodological Complex. MIEMP, Moscow (2013)
26. Sadeghi, N., Fayek, A.R., Pedrycz, W.: Fuzzy Monte Carlo simulation and risk assessment in construction. Comput.-Aided Civ. Infrastruct. Eng. **25**(4), 238–252 (2010). https://doi.org/10.1111/j.1467-8667.2009.00632.x
27. Samokhvalov, Y.: Distinctive features of using the method of analysis of hierarchies in estimating problems on the basis of metric criteria. Cybern. Syst. Anal. **40**, 639–642 (2004). https://doi.org/10.1007/s10559-005-0002-2
28. Samokhvalov, Y.: Development of the prediction graph method under incomplete and inaccurate expert estimates. Cybern. Syst. Anal. **54**, 75–82 (2018). https://doi.org/10.1007/s10559-018-0008-1
29. Samokhvalov, Y.: Construction of the job duration distribution in network models for a set of fuzzy expert estimates. In: Lytvynenko, V., Babichev, S., Wojcik, W., Vynokurova, O., Vyshemyrskaya, S., Radetskaya, S. (eds.) Lecture Notes in Computational Intelligence and Decision Making. ISDMCI 2019. Advances in Intelligent Systems and Computing, vol. 1020. Springer, Cham (2020). https://doi.org/10.1007/978-3-030-26474-1_8
30. Samokhvalov, Y.: Matching of expert estimates in preference relation matrices. Upravlyayushchie Sistemy i Mashiny (6), 49–54 (2002)
31. Sazonov, A., Sazonova, M.: Application of the Monte Carlo method for modeling economic risks in projects. Sci. Mod. **43**, 228–232 (2016)
32. Tah, H., Carr, V.: A proposal for construction project risk assessment using fuzzy logic. Constr. Manag. Econ. **18**(4), 491–500 (2000). https://doi.org/10.1080/014461900050024905
33. Tsarev, V.: Assessment of the Economic Efficiency of Investments. SPB, Peter (2004)
34. Ullah, A.: A fuzzy Monte Carlo simulation technique for sustainable society scenario (3s) simulator. In: Matsumoto, M., Masui, K., Fukushige, S., Kondoh, S. (eds.) Sustainability Through Innovation in Product Life Cycle Design. EcoProduction (Environmental Issues in Logistics and Manufacturing). pp. 601–618. Springer, Singapore (2017). https://doi.org/10.3390/su10061901
35. Ullah, S., Shamsuzzaman, M.: Fuzzy Monte Carlo simulation using point-cloud-based probability–possibility transformation. Simul.: Trans. Soc. Model. Simul. Int. **89**(7), 860–875 (2013). https://doi.org/10.1177/0037549713482174
36. Vorontsovsky, A.: Investment Management: Investments and Investment Risks in the Real Sector of the Economy. Yurayt Publishing House, Moskow (2020)

Theoretical and Applied Aspects of Decision-Making Systems

Assessing Forest Fire Dynamics in UAV-Based Tactical Monitoring System

Volodymyr Sherstjuk[1(✉)], Maryna Zharikova[1], Irina Dorovskaja[2], and Vasyl Sheketa[3]

[1] Kherson National Technical University, Kherson, Ukraine
vgsherstyuk@gmail.com, marina.jarikova@gmail.com
[2] European University, Kryvyi Rih Branch, Kryvyi Rih, Ukraine
irina.dora07@gmail.com
[3] Ivano-Frankivsk National Technical University of Oil and Gas, Ivano-Frankivsk, Ukraine
vasylsheketa@gmail.com

Abstract. This work presents a new method to assess the forest-fire parameters and fire front based on uncertain observations in a forest fire-fighting monitoring system, which jointly uses unmanned aerial vehicles and remote sensing. The proposed method is based on analyzing the consecutive image frames captured by optical and infrared cameras. The algorithm of the analysis of images obtained from different points of view simultaneously but separately by both types of cameras is described. Analysis of the optical images is divided into flame and smoke channels. It allows reducing the noise and eliminating cameras' displacement effects in remote sensing images occurring due to vehicles' vibrations. As the result of the analysis of images captured from different observation points, a semi-volumetric representation of the fire front can be evaluated, and various fire areas can be classified by intensity based on approximate flame height estimations. This result allows decision-makers to obtain the necessary information for decision-makers to correctly assess the dynamics of a forest fire. The proposed method has enough performance for assessing forest fires in real-time within the fire response decision support system.

Keywords: Forest fire monitoring · Remote sensing · Uncertain observations · Image processing · Image fusion · Fire front · Fire line intensity · Fire classification

1 Introduction

The problems of forest fires are increasing throughout the world due to the significant growth of their number and scale. Forest fires' effects are often dramatic, so they need a timely and efficient response. However, forest fire response operations

S. Babichev et al. (Eds.): ISDMCI 2020, AISC 1246, pp. 285–301, 2021.
https://doi.org/10.1007/978-3-030-54215-3_18

are usually difficult and expensive. They rely primarily on active fire observations and decision-makers' experience. Since flame and smoke significantly affect visibility while heat does not allow to approach the fire front, decision-makers forced to plan and perform response operations in conditions of uncertainty, time shortage, and heavy responsibility. The most important element of the forest fire response operations is the correct assessment of fire dynamics, which allows decision-makers to predict the development of the forest fire, identify the most vulnerable areas of the terrain, distribute fire fighting forces, and plan their activity properly. Real-time decision support systems (DSSs) based on such modern techniques as remote sensing, image processing, and geographical information systems (GIS) [2,14], can help optimize solutions related to forest fire response operations.

Online monitoring system tools is one of the main parts of such DSS that allow observing forest fire and estimating the dynamics of its parameters and other factors relevant to the fire response operation. Due to inaccuracy and the danger of direct observations, indirect monitoring based on remote sensing techniques must be used. Since unmanned aerial vehicles (UAV) can hover closely to the forest fire perimeter without endangering, they are the most suitable carriers for remote sensors such as optical and infrared cameras. Thus, it is not surprising that UAVs have undergone a great surge in their use for forest fire fighting tasks in recent years [3]. Using UAVs, one can cover large terrains and observe forest fire more detailly. Solving a range of fire detection, monitoring, and ever extinguishing tasks, UAVs can provide decision-makers with more reliable information for the proper planning of response operations. Thus, the topic of our interest is the study of approaches to forest fire assessments using multi-UAV based online monitoring tools.

2 Related Works

Several works are emphasizing the significant success of recent years in the field of UAV-based forest fire fighting [13]. Their emphasis has gradually shifted from the use of single expensive and complex UAV to the cooperative use of small and even large groups of low-cost UAVs [19]. Multi-UAV systems can be a suitable solution for remote sensing-based online forest fire monitoring.

Advances in the field of cooperative use of UAVs and data processing techniques have broadened the horizons of remote sensing of forest fires and acquired high-resolution images more easily available [17]. Remote sensor systems that were previously restricted in use due to high cost or technological limits are now becoming widely available and affordable [5]. Modern UAVs are quite flexible in hosting different kinds of sensors, but the most commonly used ones are the electro-optical and infrared cameras. Recently, their capabilities and quality have been improved, while their cost has been significantly decreased [27].

The geometrical fire parameters such as the height of the flame should be measured by adopting a 3D vision-based instrumentation with visual cameras and navigation sensors [18]. There are emerging 3D mapping technologies such as Light Detection and Ranging (LiDAR) that can be a vest-effective means to obtain 3D information on forest fire attributes with high spatial resolution [6]. In practice, 3D information can be affected by numerous potential sources of inaccuracy and errors. For example, duct and smoke may occlude the fire scene, while sensor noise may affect the measurement of forest fires.

Despite huge progress in the acquisition of high-resolution imagery and 3D data processing, there is a lack of works dealing with processing inaccurate information obtained from the sensors in 3D space. Unfortunately, high-quality images from these cameras are exposed to distortions [11]. Currently, UAVs are most often used for forest fire detection where the effects of such distortions are not significant [26]. Images obtained by UAVs during forest-fire monitoring are needed to be represented as data flows of heavy volumes analyzed in real-time. Nonetheless, some works concerning the forest fire monitoring are devoted to observing the fire parameters at specific points in space [28], whereas monitoring the fire-front dynamics requires simultaneous observation from several points as well as the processing of several data streams by a ground station at once. Therefore, it is very sensitive to inaccuracy and inconsistency of remote sensing data, which leads to a significant level of uncertainty of determining the fire front and the parameters of its spreading and ultimately contradicts the requirements of the credibility of decision making [22]. This becomes especially unpleasant if DSS is based on a forest fire spreading model. To date, such issues are almost not studied [20].

This paper dwells on the forest fire assessment for diagnosing the situation within DSS based on multi-UAV online monitoring using image processing and analyzing techniques. We assume that a decision-maker does not require precise estimations of the parameters of the forest fire but a prediction of its dynamics only. Thus, using image processing and analyzing remote sensing data streams, we can find out the location of various fire sites as well as evaluate their intensity and dynamics. It is also important to classify fire sites concerning their intensity degree using the flame height estimations [1]. These tasks are often solved along with the task of the fire-front recognition [21], which representation can be vague and blurred.

This work aims to develop a method to assess forest fire dynamics and classify it based on fireline intensity and evaluated flame length.

3 Materials and Methods

A proposed method is aimed at a tactical multi-UAV online monitoring system developed according to [22]. Monitoring UAVs have hovering capabilities and are equipped with both a thermal infrared camera and a high-resolution 16-Mp electrooptical camera with pan and tilt units providing scalability of images. They are also equipped with a GPS receiver and inertial measurement unit for self-localization.

Such UAVs can capture images flying towards the fire and hovering around it. During the monitoring, they perform onboard only stabilization, geo-localization, and geo-rectification of images (Fig. 1). Their sensors capture and transmit images to the ground control station. The digital elevation model represents cartographic dataset describing terrain surface, so the captured images are transformed into a stream of geo-mapped frames, where each frame is complemented with the coordinates of the upper left corner of the frame and scale value. This makes it possible to merge images taken from different positions of observation.

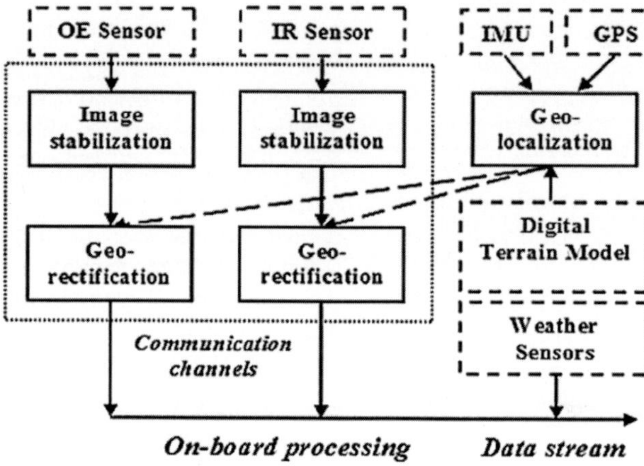

Fig. 1. Onboard image processing

Suppose that that fire spreads in three-dimensional space C above the considered terrain. Space C is discretized by a grid $D = \{d_{xyz}\}$ that approximates the terrain by isometric cubic cells d_{xyz} of size $\delta \times \delta \times \delta$. It constitutes a certain GIS layer, where each spatial object's location is discrete and bounded to a specific cell. Each cell is considered as a spatial object of a minimal volume associated with certain attributes. Thus, each cell $d_{ijk} \in D$ is related to the coordinates (i, j, k) and has attribute values $d_{ijk} = \{a_{1ijk}, \ldots, a_{nijk}\}$.

3.1 Image Analyzing Process

We consider that each UAV should sequentially transmit the images captured by the electro-optical and infrared cameras to the ground center to analyzing them [15]. Thus, the images obtained from different points of view will be received by the ground center in separate data streams and analyzed within optical and infrared channels in parallel as depicted in Fig. 2.

Image Analysis in Infrared Channel. Infrared cameras allow capturing image frames in challenging conditions such as smoke-filled or low-light environments. A pixel is black in the image corresponding to each area, where heat is not radiated. The increase in heat radiation is determined by changing the color of pixels from black to white. Thus, burning areas will be represented as white areas in the image, while free of burning areas will be represented as black areas. Nevertheless, a majority of pixels have grey colors.

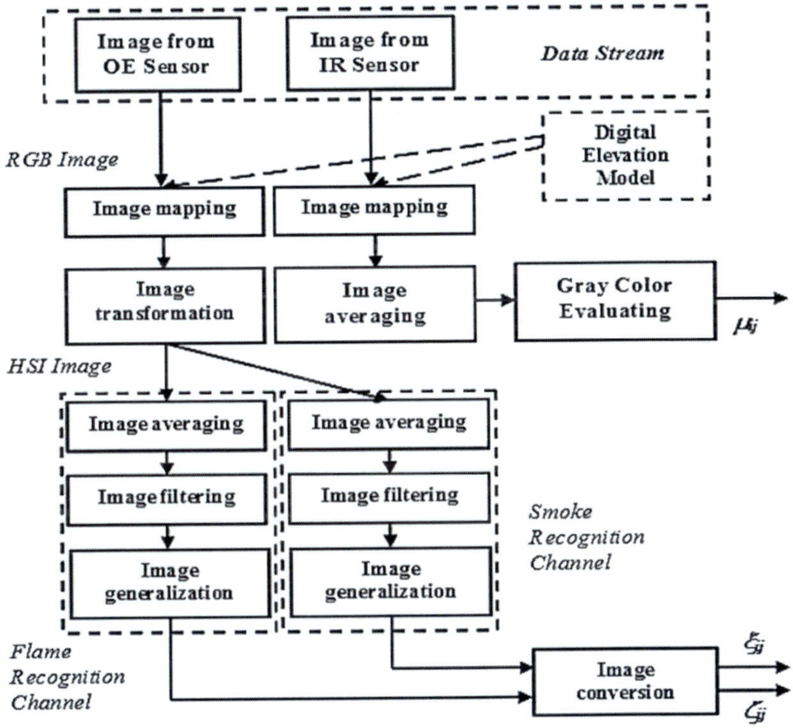

Fig. 2. The image analysis at the ground center

In the infrared channel of each data stream, the image analysis consists of three stages:

1. *Image mapping.* During this process, all pixels of the 2D image frame will be transformed into three-dimensional space C using Digital Elevation Model and corresponding projection function, then they should be mapped matching a certain set of pixels to some cell $d_{ijk} \in D$.
2. *Image averaging.* At this stage, the values of the brightness of all pixels within the cell $d_{ijk} \in D$ will be averaged taking an average brightness B_{ijk}.
3. *Gray color evaluation.* Using an order of grey colors (Fig. 3), pixel colors must be transformed from the ordinal color scheme, i.e. from black to white, to a numerical scale ranged from 0 to 1. As a result of this transformation, each cell $d_{ijk} \in D$ will be associated with a certain "degree of grayness" μ_{ijk} based on its average brightness B_{ijk}.

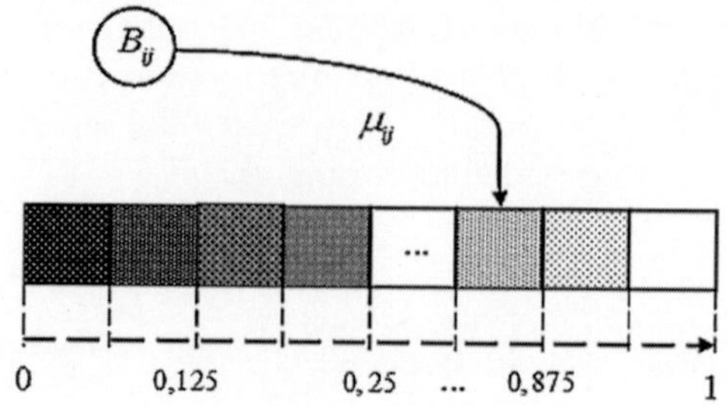

Fig. 3. The partial order of grey colors

Image Analysis in Optical Channels. The image frames captured by optical cameras are analyzed in two separate channels providing flame [9] and smoke [8] recognition. These channels work similarly but have different color settings within the *HSI* model. For example, in the flame recognition channel, for conventional fires (from dark red to light yellow) color ranges can be defined as $\rho_1 = [0° - 60°, 0.4 - 1.0, 0.5 - 1.0]$ (daytime) and $\rho_2 = [0° - 60°, 0.2 - 0.85, 0.5 - 1.0]$ (nighttime and bad weather conditions) while for well-developed fires (from light blue to white) they can be $\rho_3 = [220° - 260°, 0.4 - 1.0, 0.5 - 1.0]$ and $\rho_4 = [220° - 60°, 0.2 - 0.85, 0.5 - 1.0]$ respectively [12]. In the smoke recognition channel, the color ranges can be defined as $\rho_5 = [355° - 5°, 0 - 0.05, 0.5 - 0.85]$ (gray) and $\rho_6 = [230° - 240°, 0 - 0.05, 0.5 - 0.85]$ (silver).

Furthermore, there are two basic center colors defined in the flame recognition channel such as $\pi_1 = [30°, *, *]$ (bright orange) and $\pi_2 = [240°, *, *]$ (blue), and two basic center colors defined in the smoke recognition channel such as $\pi_3 = [235°, *, *]$ (silver) and $\pi_3 = [0°, *, *]$ (white) [7].

The whole process of analysis can be represented as consisting of six following stages:

1. *Image mapping.* As a result of the mapping process, the image frame will be oriented, transformed, and then mapped in a certain way, so that each cell $d_{ijk} \in D$ becomes imposed by a matrix m_{ijk} of pixels having a size $n \times n \times n$ that uniquely corresponds to this cell.
2. *Image transformation.* During this transformation, each *RGB* pixel of the image frame should be converted within the *HSI* color model, in which H, S, and I traditionally mean hue, saturation, and intensity. In this model, the H component stands for a dominant color, while the S component indicates a degree to what this color is diluted with the white color, as well as the I component denotes its brightness. All components of *HSI* pixel should be regularized to accept the following ranges: $0 \le H \le 360°$, $0 \le S \le 1$, and $0 \le I \le 1$ (0 corresponds to black, 1 corresponds to white). Such regularization provides a computational convenience. The image transformation stage allows the I component for all pixels to be separated from the color information to speed up the subsequent analysis.

3. *Image averaging.* At this stage, the values of H, S, and I components within the cell $d_{ijk} \in D$ should be averaged, so that the cell d_{ijk} is associated with the mean values H_{ijk}, S_{ijk}, and I_{ijk} within the matrix m_{ijk} of pixels having a size $n \times n \times n$ that corresponds to this cell.
4. *Image filtering.* Such filtering is performed separately for the flame and smoke recognition channels. During this process, the cells having the *HSI* values not being in the proper ranges for the correspondent filters ($\rho_1 - \rho_6$) should be eliminated and their colors should be replaced by black color to speed up the consequent computations.
5. *Image generalization.* At this stage, the value of the H component for each cell should be substituted by a degree of similarity to the correspondent center color within the *HSI* color model. In the flame recognition channel, the center colors π_1 and π_2 should be used, while in the smoke recognition channel, such center colors should be π_3 and π_4. Similarity degrees defined in both flame (ξ_{ijk}) and smoke (ζ_{ijk}) recognition channels should be normalized within the range [0, 1] for all cells d_{ijk}.
6. *Image conversion.* As a result of this process, two different images consisting of grayscale pixels will be formed separately in the flame and smoke recognition channels.

3.2 Final Image Fusion and Analysis

At the next phase, the flows of values ξ_{ijk}, ζ_{ijk}, and μ_{ijk} associated with a certain cell d_{ijk} should be merged to build a lightweight image of the process, which can be subjected to the final analysis as depicted in Fig. 4.

The final image analysis consists of the following stages:

1. *Preparatory filtration.* As a result of such filtration, all cells that have inappropriate (low) values of intensity or saturation must be filtered out.
2. *Average merging.* At this stage, images observed from different viewpoints $k = \{1, m\}$ received from different UAVs, and processed separately in different channels, should be finally combined. It means that the flows of the values ξ_{ijk}, ζ_{ijk}, and μ_{ijk} calculated on the basis of the analyzed images for each observation point, will be averaged concerning the correspondent coincidental cell by simple arithmetic mean and then merged.
3. *Image cleaning.* At this stage, the foreground and background will be segregated from the image. A set of threshold values $\{\beta_1, \beta_2, \beta_3\}$ should be defined so that the analyzed terrain will be broken down to the background (i.e., all cells having $\xi_{ijk} \leq \beta_1$), the foreground (all cells having $\xi_{ijk} \geq \beta_2$), and suspicious areas that fulfill the condition $\zeta_{ijk} \leq \beta_3$ due to the exposure of smoke or precipitation.
4. *Image fusion.* At this stage, a set of interjacent coefficients $\{\lambda_1, \lambda_2, \lambda_3, \lambda_4\}$ satisfying the conditions $\lambda_1 + \lambda_2 = 1$ and $\lambda_3 + \lambda_4 = 1$ should be defined. Based on these coefficients as well as on the "day/night" feature, the value η_{ijk} should be assessed for each foreground cell d_{ijk}. We suppose that in the daytime, the condition $\eta_{ijk} = \lambda_1 \cdot \mu_{ijk} + \lambda_2 \cdot \xi_{ijk}$ must be satisfied for the

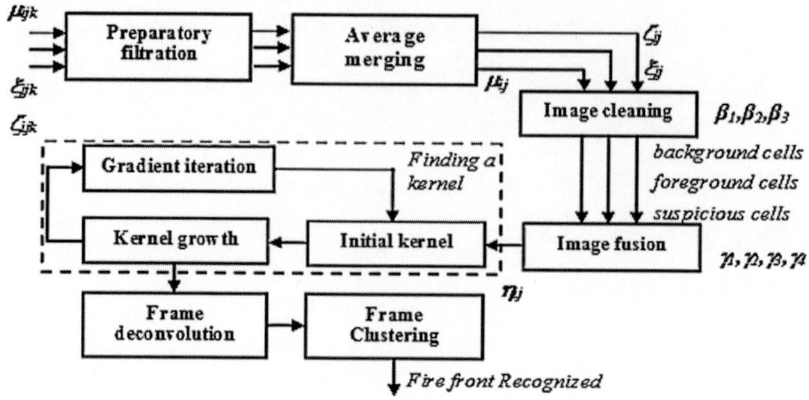

Fig. 4. Final image fusion and analysis

suspicious cells while the condition $\eta_{ijk} = \xi_{ijk}$ must be satisfied for the other foreground cells. At the same time, in the nighttime and bad weather, the conditions $\eta_{ij} = \mu_{ij}$ and $\eta_{ijk} = \lambda_3 \cdot \mu_{ijk} + \lambda_4 \cdot \xi_{ijk}$ must be satisfied with the suspicious cells and the other foreground cells respectively. The choice $\lambda_1 = -\zeta_{ijk}$ and $\lambda_4 = \zeta_{ijk}$ is the most convenient. As a result, we obtain a grayscale frame Φ that takes a certain value η_{ijk} for all cells $d_{ijk} \in D$.

5. *Finding a frame kernel.* The grayscale frame Φ is analyzed to find its vague kernel based on a linear convolution of the kernel K with certain latent Λ and noise N, such that $\Phi = \mathrm{K} \otimes \Lambda + \mathrm{N}$. The initial kernel is determined by a statistically continuous spot of the brightest cells [23]. The gradient method is used to increase it at each step under the assumption that the noise model is Poisson [24].
6. *Frame deconvolution.* Using the frame kernel determined in the previous step, a standard non-blind deconvolution method [25] is performed using the Richardson-Lucy algorithm, which optimizes iteratively the likelihood function of the Poisson noise model and has a high performance [10].
7. *Image projection.* At the final stage, the three-dimensional structure of the observed space is projected into a two-dimensional one. To do this, the values η_{ijk} for all cells $d_{ij0}, \ldots d_{ijk}, \ldots d_{ijn}$ arranged vertically from 0 to n, should be averaged using the simple arithmetic mean, and then the resulted value φ_{ij} will be assigned to the projected cell $d_{ij} \in D'$ lying "at the ground" of the corresponding vertical.

The last conversion is necessary to evaluate the current position of the forest fire front at the terrain, while a three-dimensional representation of the space is necessary to determine the height of the flame.

Thanks to the proposed multi-stage processing algorithm, the proposed image analyzing method can reduce noise and eliminate cameras' displacement effects in remote sensing images, which are the result of vibration and other distortions onboard the UAV.

3.3 Fire Front Evaluation

As a result of performing the image projection, we obtain a distribution of the values φ_{ij} represented by cells' colors over the discretized terrain. Using the results presented by authors in [31], we can break down a set of cells belonging to a projection plane D', into the subset of "free of fire", "burning", and "burnt" cells. However, due to the uncertainty of observations, the boundaries between these subsets are usually blurred.

To overcome the uncertainty, we represent D' as a fuzzy rough soft set of cells, where rough sets allow us to blur the soft set. Thus, we can determine the upper and lower approximations of the rough set based on the value φ_{ij} for each cell, and a boundary region that is a subset of cells with the degrees of membership in the rough set is unknown. Using the fuzzy sets, we can approximate boundary regions with membership degrees of cells equal to their degrees of burning possibility represented by the cells' average colors.

Thus, based on values φ_{ij} calculated as the result of the image analysis, we first determine the rough set B of already burnt cells and the rough set F of free of fire (non-burning) cells and consider all cells lying between the boundary regions of these rough sets (i.e., inside their intersection) as obviously burning cells. Next, we assess degrees of membership of the cells from the boundary regions using degrees of their burning possibility obtained by image processing. As a result, we represent the discretized terrain containing the fire front as the fuzzy rough soft set. Figure 5 depicts a three-dimensional representation of the fire front, and Fig. 6 shows a two-dimensional (flat) projection of the fire front on the terrain map.

3.4 Fire Intensity Assessment

Evaluation of the forest fire front as proposed in the previous section shows a clear picture of the ongoing processes to decision-maker that is enough to understand a direction and a rate of fire spreading. At the same time, due to the influence of spatial and temporal variability of weather, fuels, and topography conditions, there is a segmentation of the fire front and multiplicity of fire spots that essentially complicate decision-making. Thus, it is necessary to classify different sections and segments of the fireline by their type and intensity to obtain a correct assessment of the fire dynamics and plan response operation adequately. Such classification is important because the forest fire can be surface or crown, and they demonstrate different behavior.

Fireline intensity usually represents the rate of heat released from a segment or portion of the fire perimeter (i.e., head, flank, back, or rear) [4]. It is known that flame size is the main visual manifestation of fireline intensity [1]. We can estimate fireline intensity both directly using infrared remote sensing and indirectly by assessing the flame length/height. Since the fireline intensity uniquely determines fire impacts, short-term fire effects, and fire suppression difficulty, we must take into account both direct and indirect estimation of fireline intensity to classify different segments of the fire front.

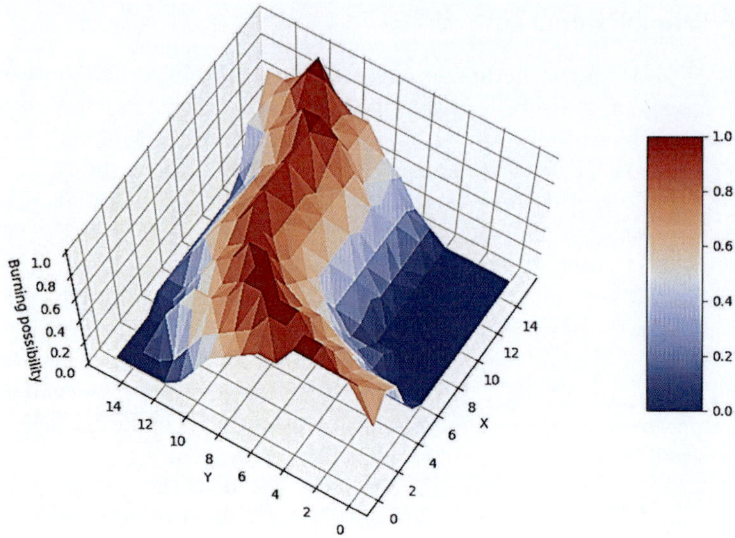

Fig. 5. Three-dimensional representation of the forest fire front

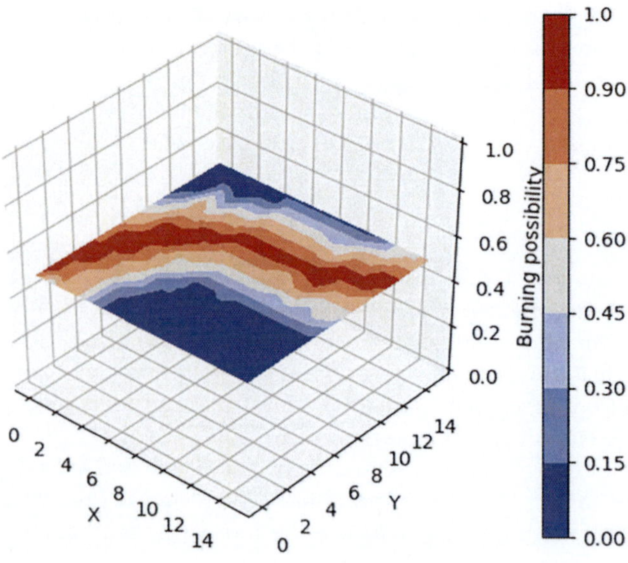

Fig. 6. Two-dimensional projection of the forest fire front assessed by the possibility of burning

Let us build an assessment of the heating rate based on [30] as follows. Let $\preceq$ be a partial order that implies a certain $(m+1)$-leveled ordinal scale $\mathrm{M} = \{\mu_0, \mu_1, \ldots, \mu_m\}$ such that $\mu_0 \preceq \mu_1 \preceq \cdots \preceq \mu_m$. Suppose ϑ is a function that uniquely takes each value of averaged "degree of grayness" μ_{ij} within some cell $d_{ij} \in D'$ onto a certain level of the scale M, such that $\vartheta : \mu_{ij} \rightarrow$ M. Thus, we define a qualitative estimation of the heating rate. Let μ_i, $i = 1..m$ be threshold values that can be used as the first feature to define fire intensity assessment.

The second feature is a maximum flame length. This feature can be estimated based on the volumetric structure of the observed space obtained after the frame deconvolution stage. Since each column of cells $d_{ij0}, \ldots d_{ijk}, \ldots d_{ijn}$ is arranged vertically and all cells are indexed from 0 to the certain value k within this column, we can divide it using the values η_{ijk} onto the cells that contain the flame, cells that contain smoke, and cells that do not contain flame or smoke as shown in Fig. 7. Obviously, the maximal index $s \leq k$ of the "highest" cell d_{ijs} containing the flame determines its length as $s \times \delta$. Accordingly, this value should be mapped together with the value μ_{ij} to the correspondent cell $d_{ij} \in D'$ lying "at the ground" on the projection plane. Using both heating rate and the flame length, we can assess fire intensity within the cell $d_{ij} \in D'$ based on the rules presented in Table 1.

Thus, the fire intensity estimation can be assigned to the various cells that constitute certain segments and sections of the fire front. Such estimations can be highlighted by the cell color on the terrain map allowing decision-makers to assess the forest fire dynamics correctly.

Table 1. Four-levels fire intensity assessment

Fireline intensity level	Maximum flame length, m	Fire Intensity Class
μ_0	< 1.5	Low
μ_1	1.5–2.5	Moderate
μ_2	2.5–3.5	High
μ_3	> 3.5	Very High

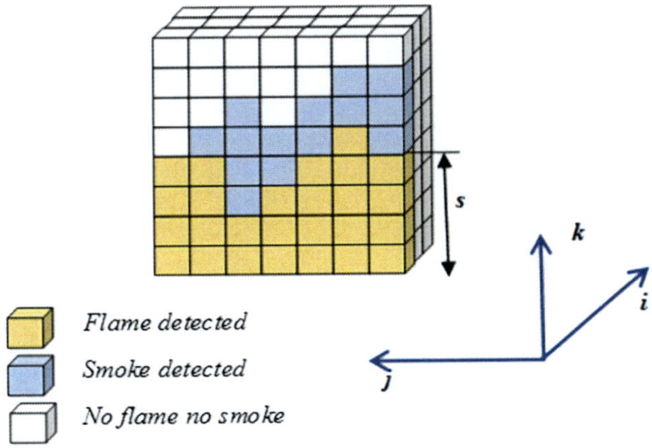

Fig. 7. Estimation of the flame length

4 Results and Discussion

The prototype of the tactical forest-fire fighting system has been implemented for the Tsurupinsk forestry (Kherson region, Ukraine). UAV onboard control system has been developed based on microcontroller STM32F429 (180 MHz Cortex M4, 2 Mb Flash/256 Kb RAM internal, QSPI Flash N25Q512). The ground control center prototype uses two servers HP ProLiant ML350 (Intel Xeon E5-2620, 8 cores up to 3 GHz) and several client computers based on the Pentium i5-7400 3 GHz processor and 16 GB RAM.

The proposed method of image processing, fire front evaluation, and fire intensity classification has been implemented using Visual C++. The system has been tested in laboratory conditions with the multi-UAV team (3 monitoring octocopters). It used the approximate fire-spreading model [29] that allows calculating the rate of fire spread and obtaining the other fire parameters necessary for decision-making. All UAVs' cameras have been precisely calibrated.

During the simulations, we had two descriptions of the forest fire front and two estimations of the fire intensity at each time: the first pair is the result of the performing of the proposed method of an image analyzing, while the second is the result of the prediction based on the approximate fire spreading model. Thus, we can compare periodically the model-based predictions with observation-based assessments. Using Fuzzy Neuron Network (FNN), we can organize the feedback loop through the assessment process and improve the assessment performance. All deviations between model-based predictions and observation-based assessments need to be detected, the number and value of such deviations determine the accuracy of the classification and the quality of the assessment model. FNN included in the feedback loop can significantly reduce those deviations [16].

The image processing and the correspondent calculations need to be performed as fast as possible to neglect the influence of some random factors. Thus, we must minimize the time difference between the observation moment and the prediction moment based on the fire spreading model. To speed up the system, the case-based methods have been also involved in the model justification process as it is shown in Fig. 8.

The case-based system allows us to obtain quickly the required threshold values $\{\beta_1, \beta_2, \beta_3\}$ and coefficients $\{\lambda_1, \lambda_2, \lambda_3, \lambda_4\}$ needed for the image analyzing process for various terrain areas, vegetation types, and weather conditions. The case base store such values concerning the similarity to environmental conditions, so they can be quickly retrieved if the search parameters match the case description.

While FNN creates a local feedback loop and allows mutual adjustment of the parameters of the fire spreading model and the coefficients of the neural network, the case-based system creates a deep feedback loop that includes the entire process of image analysis and fire dynamics assessment. This ensures the necessary robustness and stability of the image analysis, fire front evaluation, and fire intensity estimation process. Thus, the classification results are also reliable and robust.

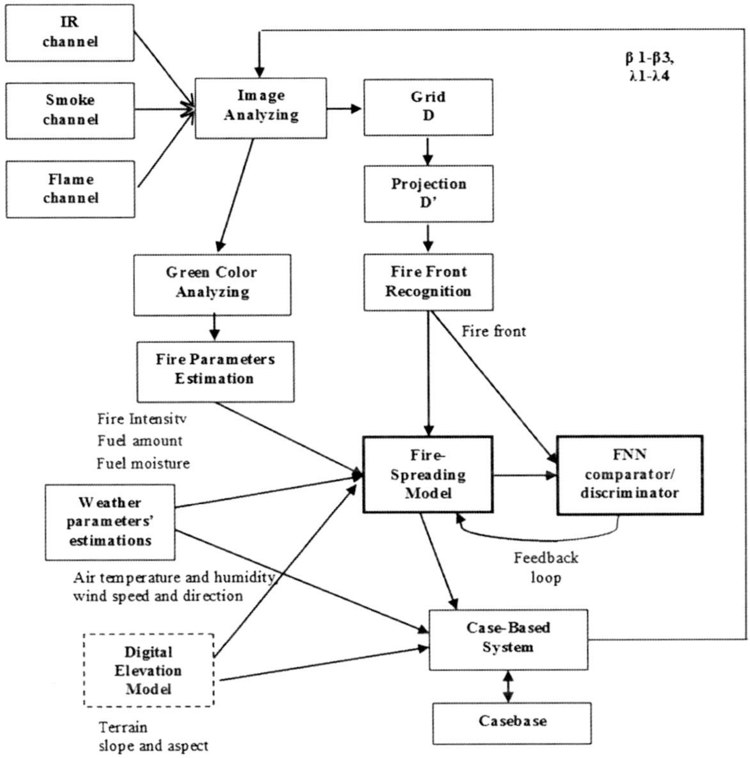

Fig. 8. Adjusting the forest fire spreading model

The performance of the proposed method has been studied during the simulation. The results of the simulation have shown that the proposed method can achieve the accuracy of classification up to 99% for different terrains and weather conditions but it depends on the cell size (Fig. 9).

During the simulation, the cell size has been varied from 5 m to 50 m. The number of levels within the discretized gray color scale in image processing algorithms has been also varied. There is an inverse relationship between classification accuracy and analyzing time.

The proposed method can provide acceptable performance with the cell size 5 m and above and the 16-levels color discretization. Changing the threshold values β_1 and β_2 speeding up computing but reducing accuracy.

The main problem that we address in the paper is the assessment of forest fire dynamics in UAV-based real-time decision support systems for wildfire response operations. Since the conditions, in which forest fires occur and spread are unpredictable, and the laws of their dynamics remain not fully explored, the online monitoring based on remote sensing and image processing is the only way to assess the fire dynamics and make forest fire response decisions more grounded.

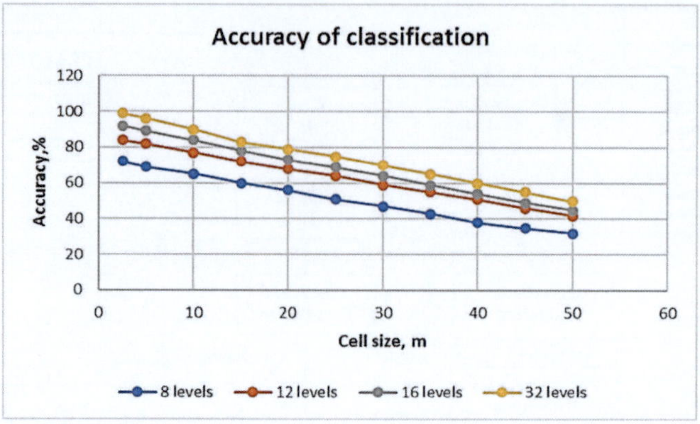

Fig. 9. Influence of the cell's size on the accuracy of classification

In this study, we propose to split the information obtained by two types of cameras from the different points of view and analyze it in the flame and smoke channels. It reduces observation errors due to UAV movement, vibration, drifting, changes in speed, and hovering. Taking into account uncertainty of fire front recognition we use fuzzy rough soft sets that allow clustering recognized fire front with the use of soft and rough sets and representing the boundary regions of the rough sets as the fuzzy sets. The developed system can provide the required credibility and efficiency of fire front recognition and can be used for solving the practical forest fire-fighting problems in near-real-time (Fig. 10).

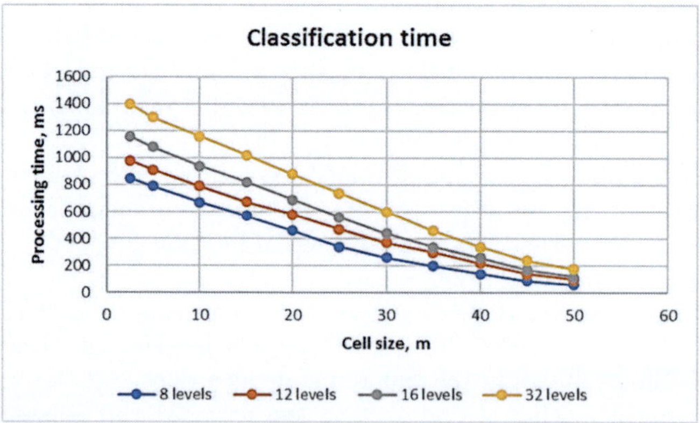

Fig. 10. Influence of the cell's size on the classification time

5 Conclusions

The method of forest fire assessment for diagnosing the situation within DSS based on multi-UAV online monitoring using image processing and analyzing techniques is proposed. This method allows using uncertain observations captured by UAVs' cameras during the forest fire monitoring. The described algorithm performs analyze of the image obtained from the different points of view simultaneously but separately from both types of cameras as well as in the flame and smoke channels. It reduces noise and eliminates cameras' displacement effects occurring due to UAVs' vibrations. Using image processing and analyzing remote sensing data streams, the proposed algorithm can find out the location of various fire sites, evaluate their intensity and dynamics, classify them concerning their intensity degree using the flame height estimations based on the semi-volumetric information. The proposed method improves image quality and accuracy of the fire front evaluation as well as fire intensity assessment. It provides enough performance to fulfill the requirements of efficiency within a real-time forest-fire response decision support system.

References

1. Alexander, M., Cruz, M.: Fireline intensity. In: Manzello, S. (ed.) Encyclopedia of Wildfires and Wildland-Urban Interface (WUI) Fires. Springer (2018). https://doi.org/10.1007/978-3-319-51727-8_52-1
2. Allison, R., Johnston, J., Craig, G., Jennings, S.: Airborne optical and thermal remote sensing for wildfire detection and monitoring. Sensors **18**(8), 1310 (2016). https://doi.org/10.3390/s16081310
3. Ambrosia, V., Zajkowski, T.: Selection of appropriate class UAS/sensors to support fire monitoring: experiences in the united states. In: Handbook of Unmanned Aerial Vehicles. pp. 2723–2754. Springer, Netherlands (2015). https://doi.org/10.1007/978-90-481-9707-1_73
4. Andrews, P.: The Rothermel surface fire spread model and associated developments: a comprehensive explanation. General Technical report RMRS-GTR-371. USDA Forest Service, Rocky Mountain Research Station, Fort Collins (2018)
5. Aydin, B., Selvi, E., Tao, J., Starek, M.: Use of fire-extinguishing balls for a conceptual system of drone-assisted wildfire fighting. Drones **3**(1), 17 (2019). https://doi.org/10.3390/drones3010017
6. Behera, M., Roy, O.: Lidar remote sensing for forestry applications: the Indian context. Curr. Sci. **83**(11), 1320–1328 (2002)
7. Bugaric, M., Jakovcevic, T., Stipanicev, D.: Computer vision based measurement of wildfire smoke dynamics. Adv. Electr. Comput. Eng. **15**(1), 55–62 (2015). https://doi.org/10.4316/AECE.2015.01008
8. Chen, T., Yin, Y., Huang, S., Ye, Y.: The smoke detection for early fire-alarming system base on video processing. In: Proceedings of 2006 International Conference on Intelligent Information Hiding and Multimedia, pp. 427–430. Pasadena (2006). https://doi.org/10.5555/1193214.1193962
9. James, M.: Flame and smoke estimation for fire detection in videos based on optical flow and neural networks. J. Res. Eng. Technol. **3**(8), 324–328 (2014). https://doi.org/10.15623/ijret.2014.0308050

10. Kim, Y.J., Kim, E.G.: Image based fire detection using convolutional neural network. J. Korea Inst. Inf. Commun. Eng. **20**, 1649–1656 (2016). https://doi.org/10.6109/jkiice.2016.20.9.1649
11. Leblon, B., San-Miguel-Ayanz, J., Bourgeau-Chavez, L., Kong, M.: Remote sensing of wildfires. In: Land Surface Remote Sensing. pp. 55–95. Elsevier (2016). https://doi.org/10.1016/B978-1-78548-105-5.50003-7
12. Luo, Y., Zhao, L., Liu, P., Huang, D.: Fire smoke detection algorithm based on motion characteristics and convolutional neural networks. Multimedia Tools Appl. **77**(12), 15075–15092 (2018). https://doi.org/10.1007/s11042-017-5090-2
13. Merino, L., Caballero, F., Martínez de Dios, J., Maza, I., Ollero, A.: An unmanned aircraft system for automatic forest fire monitoring and measurement. J. Intell. Robot. Syst. **6**(1-4), 533–548 (2012). https://doi.org/10.1007/s10846-011-9560-x
14. Merino, L., Martinez de Dios, J., Ollero, A.: Cooperative unmanned aerial systems for fire detection, monitoring, and extinguishing. In: Handbook of Unmanned Aerial Vehicles, pp. 2693–2722. Springer, Netherlands (2015). https://doi.org/10.1007/978-90-481-9707-1_74
15. Mueller, M., Karasev, P., Kolesov, I., Tannenbaum, A.: Optical flow estimation for flame detection in videos. IEEE Trans. Image Process. **22**(7), 2786–2797 (2013). https://doi.org/10.1109/TIP.2013.2258353
16. Qiang, Y., Pei, B., Zhao, J.: Forest fire image intelligent recognition based on the neural network. J. Multimedia **9**(3), 449–455 (2014). https://doi.org/10.4304/jmm.9.3.449-455
17. Qin, H., Cui, J., Li, J., Bi, Y., Lan, M., Shan, M., Liu, W., Wang, K., Lin, F., Zhang, Y., Chen, B.: Design and implementation of an unmanned air vehicle for autonomous firefighting missions. In: Proceedings of the IEEE International Conference on Control and Automation, pp. 62–67. Kathmandu (2016). https://doi.org/10.1109/ICCA.2016.7505253
18. Rossi, L., Molinier, T., Akhloufi, M., Tison, Y., Peri, A.: A 3D-vision system for the measurement of the rate of spread and the height of fire fronts. Meas. Sci. Technol. **21**(10), 1–12 (2010). https://doi.org/10.1088/0957-0233/21/10/105501
19. Shahbazi, M., Theau, J., Menard, P.: Recent applications of unmanned aerial imagery in natural resource management. GISci. Remote Sens. **51**(4), 339–365 (2014). https://doi.org/10.1080/15481603.2014.926650
20. Sharifi, F., Mirzaei, M., Zhang, Y., Gordon, B.: Cooperative multi-vehicle search and coverage problem in an uncertain environment. Unmanned Syst. **3**(1), 35–47 (2015). https://doi.org/10.1109/CDC.2011.6161448
21. Sherstjuk, V., Zharikova, M.: Fire-front recognition in UAV-based forest-fire monitoring system using fuzzy rough soft sets. In: Proceedings of 2019 IEEE 2nd Ukraine Conference on Electrical and Computer Engineering (UKRCON), pp. 1091–1096. Lviv (2019). https://doi.org/10.1109/UKRCON.2019.8879829
22. Sherstjuk, V., Zharikova, M., Sokol, I.: Forest fire-fighting monitoring system based on UAV team and remote sensing. In: Proceedings of 2018 IEEE 38th International Conference on Electronics and Nanotechnology, pp. 663–668, Kyiv (2018). https://doi.org/10.1109/ELNANO.2018.8477527
23. Simoncelli, E.: Statistical modeling of photographic images. In: Handbook of Image and Video Processing, pp. 431–441. Academic Press (2005). https://doi.org/10.1016/B978-012119792-6/50089-9
24. Tang, C., Wu, J., Hou, Y., Wang, P., Li, W.: A spectral and spatial approach of coarse-to-fine blurred image region detection. IEEE Signal Process. Lett. **23**(11), 1652–1656 (2016). https://doi.org/10.1109/LSP.2016.2611608

25. Wang, N.C., Ding, J.J., Chen, L.A., Chang, R.: Efficient image deblurring via block-wise non-blind deconvolution algorithm. In: Proceedings of 10th International Conference on Information, Communications and Signal Processing, pp. 1–5. Singapore (2015). https://doi.org/10.1109/ICICS.2015.7459884
26. Yuan, C., Liu, Z., Zhang, Y.: Aerial images–based forest fire detection for firefighting using optical remote sensing techniques and unmanned aerial vehicles. J. Int. Robot. Syst. **88**, 635–654 (2017). https://doi.org/10.1007/s10846-016-0464-7
27. Yuan, C., Zhang, Y., Liu, Z.: A survey on technologies for automatic forest fire monitoring, detection and fighting using uavs and remote sensing techniques. Can. J. For. Res. **45**(7), 783–792 (2015). https://doi.org/10.1139/cjfr-2014-0347
28. Zhang, T., Li, Q., Zhang, C., Liang, H., Li, P., Wang, T., Li, S., Zhu, Y., Wu, C.: Current trends in the development of intelligent unmanned autonomous systems. Front. Inf. Technol. Electron. Eng. **18**(1), 68–85 (2017). https://doi.org/10.1631/FITEE.1601650
29. Zharikova, M., Sherstjuk, V.: Development of the model of natural emergencies in decision support system. East. Eur. J. Enterp. Technol. **4**(1), 62–69 (2015). https://doi.org/10.15587/1729-4061.2015.37801
30. Zharikova, M., Sherstjuk, V.: Threat assessment method for intelligent disaster decision support system. In: Advances in Intelligent Systems and Computing, vol. 512, pp. 81–99 (2016). https://doi.org/10.1007/978-3-319-45991-2_6
31. Zharikova, M., Sherstjuk, V., Wójcik, W., Syzdykpayeva, A., Muslimov, K.: A model of destructive processes based on interval fuzzy rough soft set. Przeglad Elektrotechniczny **4**, 132–137 (2019). https://doi.org/10.15199/48.2019.04.23847

Ontology and Analytic Hierarchy Process in the Information and Analytical Systems

Oleksandr Nesterenko(✉)

Nationale Academie of Management, Kyiv, Ukraine
aleksandrnesterenkoua@gmail.com

Abstract. In our days, problems relating to decision making support in the information and analytical systems are caused mainly by the need to handle a large volume of diverse information and the occurrence of a great number of alternatives of alternatives and multi-objective criteria when choosing them. Application of ontological descriptions in the decision making support chain ensures the dynamic formation of corresponding sets of alternatives and criteria based on the properties of concepts of the domain areas for which relevant decisions are made. Inclusion of ontological models in the information and analytical system environment allows effectively apply a method of hierarchy analysis as a systematic procedure for hierarchic representation and analysis of elements which establish the core of a problem. In this case, validity of a decision depends entirely on the correctness and relevance of ontological domain model, while the objectivity of a method is ensured by fixing the transitive consistency of expert judgment, which eliminates subjectivity and supports the principle of impartiality and justice.

Keywords: Ontology · Analytic hierarchy process · Decision make support · Information and analytical system

1 Introduction and Literature Review

Decision making support is an integral functional component of advanced information and analytical systems (IAS), applied in the administrative management field particularly at the governmental level. According to the specified requirements, a solution depends on the complexity of presentation and perception of properties and functionality of composite objects and domain area (DA) processes relating to the problems under consideration. At the same time, as long as nowadays the new data sources appear and the streaming data and distributed data storage are used, much more data can be processed in IAS than before. Therefore, such systems should contain in their operating environment not only the tools for processing and analyzing data, used in the decision making process, but also the knowledge management tools to ensure processing of certain

S. Babichev et al. (Eds.): ISDMCI 2020, AISC 1246, pp. 302–314, 2021.
https://doi.org/10.1007/978-3-030-54215-3_19

opinions, propositions and statements that carry the object representations and perceptions of DA.

Usually, three stages are distinguished in the decision making process: information retrieval and task setting stage; a stage when a set of alternatives is built; and a stage of choosing the best alternative. Thus, as is known, in the most general view a decision making task is characterized a next cortege: $<\mathcal{A}, E, S, T>$, where $\mathcal{A}$ are the possible alternatives, E - decision-making environment, S – system of advantages of person that makes decision (PDM). It is needed to make some action of T with the great number of alternatives $\mathcal{A}$: to find the most acceptable alternative, arcwise to put in order the list of possible alternatives, etc.

Mathematically, such tasks are described by a set of alternatives $\mathcal{A}$ according to which the values of certain indicators (criteria) are specified for each of them. Solution of a problem is regarded an alternative with the best (in total) values of criteria, which in the general case are characterized by different importance.

At the same time in modern terms the environment of decision-making E is characterized of the joint processing of interrelated diverse information, its integration and interaction with other systems and subsystems, which are different in purpose, is associated with their representation by certain information models [1,10,13,14]. In practice, each such model reflects a certain set of knowledge, which describes the properties of objects and processes being analyzed.

Problem in solving the practical tasks relating to decision making support, id east determination of the system of advantages S and implementation of certain action T is above the great number alternatives $\mathcal{A}$, is often a numerous sets of alternatives and multi-objective criteria when choosing them. One of the recognized approaches to overcome this problem is the Analytic Hierarchy Process (AHP) [17]. It is commonly known that analysis of a problem, relating to decision making process in AHP, begins with the construction of a hierarchic structure that includes the purpose, criteria, alternatives and other factors which influence the choice.

Each element of the hierarchy can represent various aspects of DA, notably that material as well as non-material factors, measurable quantitative parameters as well as qualitative characteristics, objective data as well as subjective expert estimates can be taken into consideration at that time. It is at this stage that the effectiveness of a method is determined, since availability of an identical information model and body of knowledge attributed to DA makes it possible to clearly define a list of alternatives and restrictions as well as to minimize the lack of AHP application, which is associated with the relations of consistency as a Q-factor of expert estimates [4].

In addition, something must be done to overcome the following drawbacks of the method: (1) it is difficult to propose a universal model of quality evaluation solution; (2) existing model is difficult to reuse. As a result, many experts have to spend a lot of time updating the model.

Among existing approaches to the representation of information models in the analytical system environment, a certain set of ontologies is considered to be the most adequate representation [3,5,6,11,12,15,18,21], and each model or combination of models can be represented and defined by distinct taxonomy.

Any ontology contains informational descriptions based on an object-oriented formalization procedure and on descriptions of interpretational functions, which (based on ontology) control the supply of information resource for its further processing. Such structure reflects the comprehension of a problem by PDM. Equally, analysis of the decision-making situation in AHP is similar to the procedures and methods of argumentation normally used by a person at the intuitive level. These distinctive features of two methods under consideration represent the factor that unites them.

The same time period has also seen marked growth in corporate structural complexity. This lack of awareness of the corporate data landscape impacts the ability to govern data, which in turn impacts overall data quality within organizations. Not surprisingly, many researchers and practitioners offer tools and methods to better manage data and processes, including the use of ontology-based related data.

At the same time, due to the complexity of business processes, the main idea in decision making is to use expert knowledge to help the user in making the best decision in terms of multicriteria. There are many researchers are considering AHP for this purpose, which is well suited for hierarchical data structures, in particular those that have been formed in terms of ontologies.

However, it should be noted that in many ontology- and AHP-based sharing research and projects mostly interconnected integration is not observed. Mainly, it is an ontology based preliminary data preparation and then separately used by AHP to obtain certain estimates. Owing to the known drawbacks of AHP, it is often the case that search for a multicriteria analysis method best suited to the problem is expanded or modified by AHP, or other methods. For example, it is proposed in [20] concerning the development of a software selection decision support system.

The study [14] presents a generic ontology-based architecture using a multi-criteria decision making technique to design a personalized route planning system. Initially, an ontology builds a general user-oriented architecture. A user-centric abstract model has a domain-specific ontology. From the built models stand out criteria that are weighted and evaluated by the AHP.

In [22] combining AHP and the scores of evaluation features obtained by ontology querying a comprehensive evaluation is realized. An ontological request is performed to obtain information on the quality of the parts. It is then proposed to quantify the degree of reuse parts by AHP.

In [2] the authors present a method for using information from WordNet ontology (general purpose ontology), and their related terms and relationships from domain ontology. The web pages ranking module uses AHP algorithm.

The purpose of paper [9] is to propose a way to ontologically represent the AHP method, which not only specifies the concepts and their relationships that are necessary for applying the AHP method, but also realizes the priority assessment and consistency evaluation mechanisms in the AHP method through corresponding reasoning rules. The result is an ontology called AHP Ontology.

As can be seen from the analysis of existing solutions, the issues of using ontologies and AHP in administrative management are very limited. Among the few examples is work [19] that offers conceptual modeling of requirements of government to citizens (G2C) service provision. An ontology model developed represents requirements of service provision in a web service environment. Hereafter, AHP is used to determine the pre-selected services for contributing to quality features.

The main conclusion is that such approaches allow us to find acceptable solutions only if the state of DA is clearly defined and its mathematical description is presented in the form of defined sets of concepts and their properties.

This part of the problem, which is still poorly understood in the field of administrative management, can be solved by combining a process of ontological model-based domain formalization and information structuring with AHP-based multicriteria analysis.

The aim of the research is the proposes a technique based on a joint use of ontology and analysis of hierarchies methods with a view to ensure the correctness and effectiveness of the decisions supported.

2 Materials and Methods

2.1 Core of Ontological Methodology in the IAS Operating Environment

Considering the core of ontological methodology, it should be noted that it is based on the object-oriented approach, in which DA is represented as a set of objects interacting with each other by means of semantic binding of propositions, statements and opinions. An object is understood as a certain entity (real or abstract) characterized by a status, behavior and identity.

A status of an object is determined by a list of all possible properties and values of each of such properties. Behavior of an object (or its functionality) characterizes how an object interacts or is exposed to interactions from the side of other objects, manifesting its identity. Behavior of an object is realized in the form of functions, which are called methods. In this case, the structure of an object is accessible only through its methods, which all together form the interface of an object. Finally, the identity of an object is characterized by its properties that distinguish this object from the other objects.

While forming a correct operating environment IAS, two types of hierarchic object relationships are in particular interest: 1) the links, denoting equitable relations between objects; an object cooperates with others through the connecting links; 2) the aggregation, describing a ratio of the whole and the parts that lead to the corresponding taxonomy (hierarchy of objects).

In the general case, the computer ontology of some DA is formally represented as an ordered triple $O = <X, R, F>$, where X, R, F – are the finite sets of correspondingly: X – concepts (notions, terms) of a domain, R – relations between them, F – interpretation functions X and/or R [3,15,18].

Then the ontology of certain DA shall be considered as a certain not empty set of objects satisfying the following requirements:

1. Objects are organized in the form of a hierarchic structure of a finite set of concepts that describe a given DA.
2. Structure can be represented by a set of bipartite graphs, where: the vertices are concepts, and the arcs are semantic relations between them.
3. Concept and relation are interpreted in accordance with the generally valid functions of interpretation from the electronic sources relating to a given DA.
4. Definition of concepts and relations shall be derived from the axioms and restrictions of their scope.
5. Interpretation functions and axioms are described in the formal theory language.

The process of formation of the IAS operating environment requires setting of a definition of thematic ontology with the most complete set of concepts and conceptual relations, and the interpretation function with the added axioms, definitions and restriction of the subject of a particular system, that is, there is: $R \neq \oslash, F \neq \oslash$.

Thematic ontology is a formal representation of conceptual knowledge about the subject domain and can be represented by a certain system of information resources. Building process of such information system can be represented as a composition of certain propositions, opinions, statements, terms (concepts) and relations between them, while its result can be represented as a foundation for building ontological knowledge base in a given DA, described in a declarative form [1,3]. In this case, description of all components is given in a formal language, which can be interpreted by some procedure (algorithm).

Formal model scheme of thematic ontology is described by a quartet [15]:

$$O = <X, R, F, A(D, R_S)>, \tag{1}$$

where X is the concepts set; $X = \{X_1, X_2, ..., X_i, X_n\}$, $i = \overline{1,n}$, a finite set of concepts (notions) defined in the IAS operating environment;

$R = \{R_1, R_2, ..., R_k, ..., R_m\}$, $R : X_1 \times X_2 \times ... \times X_n$, $k = \overline{1,m}$, $m = CardR$ - is a finite set of semantically significant properties (relations) among the DA concepts. They define a type of interaction between concepts. In the general case, relations are divided into universally significant (of which, as a rule, a partial order relation is singled out) and the specific relations of a given DA;

$F : X \times R$ - is a finite set of interpretation functions defined on concepts and/or relations. A glossary, compiled for a set of X concepts, is a particular case of a definition of a set of F interpretation functions. In the general case, definition of X_i concept includes a subset of $\{x_{i-1}\}$, concepts, through which the X_i, relations (connecting X_i and $\{x_{i-1}\}$, are determined, and a set of attributes (indicators) inherent in $X_i \mid i = \overline{1,n}$;

A – is a finite set of axioms used to record the identically true propositions (definitions and restrictions) - tautologies in the terms of DA subjects;

D – is a set of additional definitions of concepts (notions) in the terms of DA subjects; R_S – is a set of restrictions, which define certain properties of DA concepts and can be interpreted as criteria in defining the areas of conceptual structures (concepts, notions, propositions, statements) of a certain subject – $R_S : R_S \subset R$.

2.2 Properties of Ontology Objects in Multi-criterion Decision Making Tasks

When forming ontologies in the IAS operating environment, we define R_S a set of restrictions, as the one that allows selecting such subset A from a set of X concepts that it can be decomposed into subsets $\mathcal{A}_i = \{a_{i1}, ..., a_{in}\}$, which intersect and which we call a set of characteristics alternatives ($\prod_{i=1}^{n} \mathcal{A}_i \neq 0$, where 0 is an empty set) (1). All the a_{in} elements of each $\mathcal{A}_i$ set shall have the property of a certain advantage, enabling selection of the necessary tautology at the problem solving stages related to the decision making support. That is, a lot of restrictions in such tasks allow building a set of alternative concepts based on the definition of ontology taxonomic structure (Fig. 1).

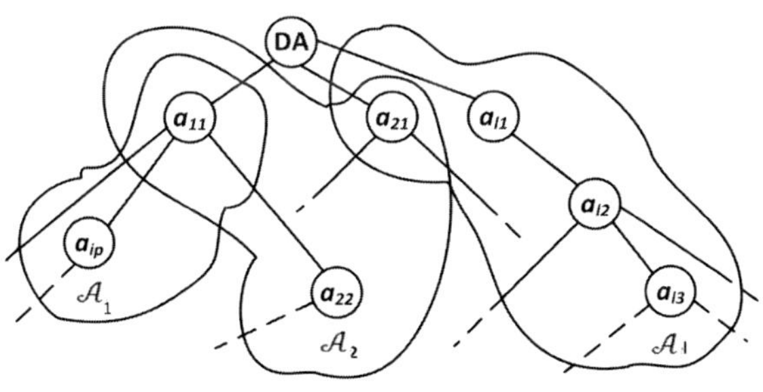

Fig. 1. General scheme allocation characteristics of alternatives on ontological data of the domain area

The properties of ontology objects can be used as criteria, according to which experts can choose alternative from a set of possible alternatives. In the ontological representation, criteria constitute a certain subset in a set of properties R. Taxonomic structure of ontology provides the selection of a certain set of alternatives that define some tasks relating to choice [8]. Each of the ontology elements used in the decision making tasks has a certain set of criteria (properties).

Some set of criteria can be represented in the form of a certain evaluation function $\mathcal{C}$, that takes values on some set of estimates $\mathcal{O}$, or in the form of a rule by which the "best" alternative is chosen. In this case, the "best" alternative corresponds to the maximum or minimum value of the evaluation function, depending on the content of the criterion. If $\mathcal{A} = \{a_i, ..., a_n\}$ is a set of alternatives or solutions, then $\mathcal{C} : \mathcal{A} \rightarrow 0$ [7].

Occurrence of several criteria turns the decision-making task into multi-criteria one. A multi-criteria task has a set of $q > 1$ criteria such that $\mathcal{C}_1, ..., \mathcal{C}_q$, $\mathcal{C}_q : \mathcal{A} \rightarrow \mathcal{O}_r$. Where $\mathcal{O}_r$ is a set of values of the function $\mathcal{C}_q$. Sometimes it may be convenient to consider several criteria in the form of a single vector criterion or vector estimate $\mathcal{C}(a) = (\mathcal{C}_1(a), ..., \mathcal{C}_q(a))$ of alternative a.

Thus, a multi-criteria decision making task is determined by a set of possible solutions $\mathcal{A}$, vector criterion $\mathcal{C}$ and a ratio of preferences in a set $\mathcal{A}$. The aim of solution is to search for "optimal" (to some extent) alternative A or a group of alternatives with consideration of the preference relations based on the vector criterion $\mathcal{C}$.

This is ensured by arranging ontology object-concepts on the basis an integrated processing of properties that characterize them. For this we use weight, ball and linguistic scales. Each such scale defines the values of the criteria characterizing the properties objects of DA thematic ontology.

2.3 Combination of Ontology and Analytic Hierarchy Process

Decision of a problem at decision-making stage is a process of gradual setting of priorities. At the first stage, the most important elements of the problem are identified; the second stage is the best way to verify observations, to test and evaluate the elements; and the next stage can possibly be evolving of a way to apply a decision and to assess its quality.

The process can be carried out over a sequence of hierarchies: in such case, the results obtained in one of them are used as inputs for study of the following one. The entire process shall be checked and re-evaluated until confidence that the process has covered all the important characteristics required for presentation and solution of the problem.

This multistage method is inherent in the AHP theory and involves systematization of a task solving process on the principles of identity and decomposition, discrimination, comparative opinion and synthesis.

When applying AHP, the first stage has provision for structuring of a problem in the form of hierarchy or network based on the principle of identity and decomposition. Typically, a hierarchy is built from the top (i.e. the targets from the management point of view), through intermediate levels (criteria and sub-criteria, on which the subsequent levels depend) to the lowest level (hierarchy leaves), which is usually a list of alternatives. Here, alternatives are understood by some objects in the subject area that are evaluated relative to the achievement of the goals specified at the top of the hierarchy.

Thus, this hierarchical structure is a graphical representation of the problem in form inverted tree, where each element, with the exception of the uppermost one, depends on one or more elements located above. Using information available in the database, built on the ontological model considered above, such a hierarchy can be formed on the basis of the R_S and $\mathcal{A}$.

After the hierarchical reproduction of the problem the issue of setting criteria priorities and evaluating each of the alternatives by criteria is identified, identifying the most important of them.

In AHP it is necessary to find out important elements of a problem (objects) in terms of their influence w_i (weight) on a common characteristic (criterion). For this purpose experts perform pairwise comparisons y_{ij}, which lead to a square table - a matrix form with the property of inverse symmetry, i.e. $y_{ij} = 1/y_{ij}$, where i and j indices refer to a row and to a column, respectively. Then "solutions" of such square inversely symmetric matrices are carried out. The meaning of such calculations is that they determine the way for quantitative estimation of relative importance of the factors in a problem being analyzed (sets of local priorities). Further on, while solving a problem, attention shall be focused on factors with the largest values of importance.

It is important to understand that the values $w_1, w_2, ..., w_n$ are not known in advance, and they are formed based on the subjective opinions made by the experts who evaluate their number using a special scale of relative importance (value); its structure is proposed by the author of the AHP ideology.

Starting with the left element in the matrix, an expert has a question: at what extent is this element more important than an element at the top? If the first element is more important than the second, then an integer from the scale of relative importance is used, and otherwise an inverse value is used. (When an element is compared with itself, the ratio is equal to one). The relations, that are reverse to each other, are recorded in the symmetric positions of matrix. Therefore, experts normally deal with positive inverse symmetric matrices, and only $n(n-1)/2$ opinions are necessary, where n is the total number of compared elements. It is not assumed that opinions are fully coordinated and they are not subject for coordination, except for the requirements of inverse symmetry.

A very useful indicator in the AHP theory is the so-called consistency index (CI), which provides degree of violation made by experts in relation to numerous and transitive (ordinal) consistency. That is, this quantitative assessment of the inconsistency of the results of comparisons arising from subjective views (and errors) of experts. In every matrix and for the whole hierarchy CI can be obtained by certain calculations. The CI value is positive, and the less contradiction exists in comparisons, the less is the CI value.

When CI is divided by a number corresponding to the random consistency of a matrix of the same order, the compatibility relation (CR) will be obtained. CR value shall be not more than 10% in order to be acceptable. In some cases, this value can be tolerated up to maximum 20%. If the CR goes beyond these limits, the expert participants should investigate the task again and check their opinions.

To prevent inconsistencies of matrices due to the influence of various factors and properties of evaluated alternatives on expert opinions, the extreme subjectivity shall be avoided. Within such frames the DA ontology shall be built; and it shall clearly define all the DA factors and make the basis for expert estimates.

It is the ontological model that enables a group of experts interacting among themselves in order to solving a task, modifying their opinions and, as a result, forming their own opinion in a rational way.

Thus, the acceptable consistency relations are provided by iteratively repeating the AHP stages and using the knowledge about DA, formed on the basis of ontological model, and giving more attention to certain stages comparing to others in some situations.

3 Experiment, Results and Discussion

Let's consider an example of a decision related to choice (ranking) standard (industrial) software of the class ERP.

The algorithm for performing the actions in the conditions of using the proposed integrated method is shown in Fig. 2.

First of all, there is a list well-known and widely used standard industrial software solutions of the class ERP (hereinafter - ERP-systems). Given the large number of such solutions on the market, it is advisable to select only a few. This choice is made by the manager on the analysis of ontological data domain based. Their groups (short list) formation is carried out on the basis of the ontology of the domain "Software quality", built in accordance with the standards ISO/IEC 12119 and ISO/IEC 9126 [16].

Next, you need to define a set of characteristics of ERP-systems for use as criteria for ranking ERP-systems from the list of alternatives. Characteristics are desirable to be grouped together in such a way that the presence of each group characteristics is a prerequisite for choosing an ERP-system. At the same time, an inadequate level or even lack of any characteristic in the middle of the group may not be critical in the final selection of the best ERP-system. (2). The set characteristics and their groups formation is carried out on the basis of the ontology. This is done by distinguishing from the ontology sets characteristics $\mathcal{A}$, which are formed taking into account the constraints R_S, which are the peculiarities of ERP-system business solutions.

As a result, in this example 6 groups of characteristics were selected: positioning ERP- system on the world market, positioning ERP-As a result, in this example 6 groups of characteristics were selected: positioning ERP- system on the world market, positioning ERP-system in the domestic market, technological characteristics, supplier's infrastructure in the local market, functional characteristics, and others. (3). Attributable descriptions (properties) of criteria can be presented in the ontological database in form of frames, in slots containing the corresponding numerical or linguistic data. Further, according to the AHP algorithm, to accomplish this task, you must:

(i) build a dominant hierarchy of criteria - those properties of alternatives that have a significant impact on the task;
(ii) make expert comparisons of alternatives with an assessment of the superiority of one alternative over the other on each criterion on a Saati scale;
(iii) summarize the estimates obtained using a scalar (linear) convolution, taking into account the importance (weight) of the criteria and, possibly, the competence (weight) of the experts.

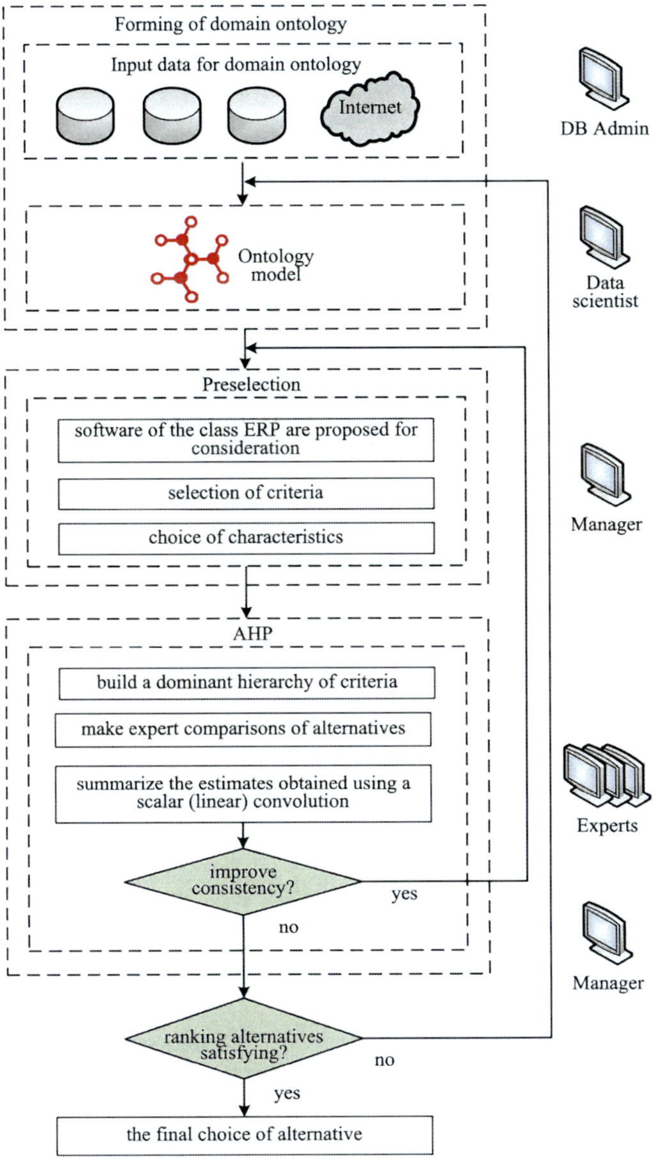

Fig. 2. A structural block chart of the algorithm to performing the actions in the conditions of using the proposed integrated method

In the end, this will allow you to get a total rating (rating) for each alternative and thus rank them.

Next, a group of experts who have knowledge and experience in implementation, support (maintenance) and operation of ERP-systems, select the best ERP-system in steps AHP algorithm (Fig. 3).

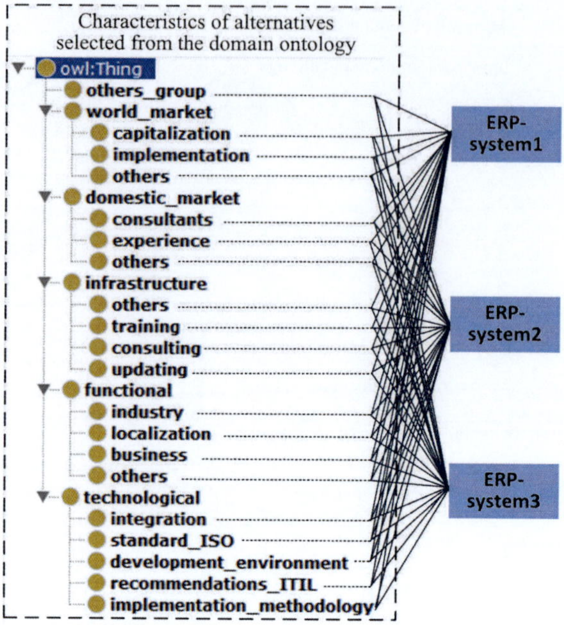

Fig. 3. Hierarchical reproduction of the task selecting an ERP-system using the domain area ontology

The presence of an attribute component in an ontological basis for each concept provides an adequate evaluation by experts in pair comparisons, making it impossible to make unreasonable marks. Thus, the necessary informational support for the experts to solve the problem of comparing alternatives may consist in the application of an ontological model of DA based on interpretive selection functions constructed by means of hyperrelationships over the concepts of the ontological taxonomic structure and their properties. It is an ontological model that provides a group of experts to interact with each other when solving a task, to modify their judgments and, as a result, to form their own opinion in a rational way.

In order to improve consistency, the search for additional information and viewing of the data, used for building the scale, can be recommended. A set of additional definitions of concept (notions) in the terms of DA subjects can be viewed as a source.

An important component of the proposed algorithm is to help experts immediately support the transitivity of their judgments, which will ensure consistency in the estimates of the various alternatives at the overall preference level.

On the other hand, according to the results of various decision-making tasks, certain corrections to the ontological model shall be made in the given DA aiming to bring it (in terms of adequacy) as close to the domain area as possible, as

AHP allows acquire new knowledge relating to DA, verifying the validity and consistency of the expert opinions, which also are the basis for ontological model.

4 Conclusions

Thus, it can be concluded that the use of ontological descriptions in decision making support process ensures the dynamic formation of corresponding sets of criteria for information and analytical systems based on the properties of concepts of those domain areas for which the relevant decisions are made. Ontological modeling ensures the decomposition of opinion-problem into simpler constituent parts – tautologies, as well as further processing of opinion sequence by experts on the basis of a certain preference property. As a result, the numerical expression for the relative degree (intensity) of interaction of elements in the hierarchy can be found.

Integrating the use of ontologies and AHP to support expert decision-making allows the same technology to be applied at different stages of the process: to determine the priority (weight) of criteria, the competence (weight) of experts, and to evaluate alternatives in-house. The proposed algorithm significantly simplifies expert activities by unifying the process of evaluating alternatives, provides information support to the transitional consistency of expert opinions, and promotes cardinal consistency.

Proceeding from the foregoing, it can be noted that the inclusion of ontological models in the IAS environment makes it possible to apply the hierarchical analysis method quite effectively as a systematic procedure for hierarchical representation and analysis of the elements, which determine the essence of a problem. At the same time, the supportability of a decision depends entirely on the correctness and adequacy of domain area ontological model, and the objectivity of method is ensured by fixing the transitive coherence of expert opinions that excludes the subjectivism and supports the principle of their impartiality and correctness.

References

1. Diets, J.L.: Enterprise Ontology Theory and Methodology. Springer, Heidelberg (2006)
2. El-Dsouky, A., Ali, H., Rashed, R.: Ranking documents based on the semantic relations using analytical hierarchy process: query expansion and ranking processs. Inf. Retrieval Manag.: Concepts Methodol. Tools Appl. (4), 1841–1859 (2018). https://doi.org/10.4018/978-1-5225-5191-1.ch084
3. Guarino, N.: The ontological level: revisiting 30 years of knowledge representation. In: Conceptual Modelling: Foundations and Applications. Essays in Honor of John Mylopoulo, pp. 52–67 (2009)
4. Ishizaka, A., Labib, A.: Analytic hierarchy process and expert choice: benefits and limitations. OR Insight **22**(4), 201–220 (2009). https://doi.org/10.1057/ori.2009.10

5. Järvenpää, E., Siltala, N., Hylli, O., Lanz, M.: The development of an ontology for describing the capabilities of manufacturing resources. J. Intell. Manuf. **30**(2), 959–978 (2019). https://doi.org/10.1007/s10845-018-1427-6
6. Kryivyiy, S.: Formalized ontological models in scientific researchs. USiM **3**, 4–15 (2016)
7. Larichev, O.: Theory and decision-making methods, as well as a chronicle of events in magical countries, Logos, Moskva (2006)
8. Larichev, O., Olson, D.L.: Multiple Criteria Analysis in Strategic Siting Problems. Kluwer Academic Publishers, Boston (2001)
9. Liao, Y., Loures, E.R., Canciglieri, O., Panetto, H.: A novel approach for ontological representation of analytic hierarchy process. Adv. Mater. Res. **988**, 675–682 (2014). https://doi.org/10.4028/www.scientific.net/AMR.988.675
10. Lytvyn, V.: Knowledge bases of intellectual decision support systems. Vydavnycztvo Lvivskoyi politekhniky, Lviv (2011)
11. Munir, K., Anjum, M.: The use of ontologies for effective knowledge modelling and information retrieval. Appl. Comput. Inform. **14**(2), 116–126 (2018). https://doi.org/10.1016/j.aci.2017.07.003
12. Nesterenko, O., Trofymchuk, O.: Patterns in forming the ontology-based environment of information-analytical activity in administrative management. East.-Eur. J. Enterp. Technol. **101**, 33–42 (2019). https://doi.org/10.15587/1729-4061.2019.180107
13. Nesterenko, O.: Informational approach to management providing in automated information-analytical systems of government bodies. Reiestratsiia zberihannia i obrobka danykh **10**(3), 46–55 (2008)
14. Niaraki, A., Kim, K.: Ontology based personalized route planning system using a multi-criteria decision making approach. Expert Syst. Appl. **36**(2), 2250–2259 (2009). https://doi.org/10.1016/j.eswa.2007.12.053
15. Palagin, A., Kryivyiy, S., Petrenko, N.: Ontological methods and means of processing subject knowledge. Publishing house of VNU them V, Dahl, Lugansk (2012)
16. Polizhuk, V., Netesin, I., Karpovich, A.: How to choose ERP by Saati method. Informatsionnyie tehnologii dlya menedzhmenta (3–4), 10–14 (2014)
17. Saati, T.: Decisions making. Hierarchy analysis method. Radio i svyaz, Moskva (1989)
18. Stryzhak, O.: Tools for ontological integration and support of distributed spatial and semantic informational resources. Ekolohichna bezpeka ta pryrodokorystuvannia (12), 166–178 (2013)
19. Sun, L., Mushi, C.J., Alsoud, A.: Conceptual modelling for requirements of government to citizen service provision. In: Information Systems Development - Business Systems and Services: Modeling and Development, pp. 747–758 (2011). https://doi.org/10.1007/978-1-4419-9790-6_60
20. Wasielewska, K., Ganzha, M., Paprzycki, M., et al.: Multicriteria analysis of ontologically represented information. In: AIP Conference Proceedings, vol, 1629, pp. 281–295 (2014). https://doi.org/10.1063/1.4902284
21. Weisfeld, M.: The Object-Oriented Thoughn Process. Addison-Wesley Professional, Boston (2013)
22. Xu, F.X., Liu, X.H., Chen, W., Zhou, C., Cao, B.W.: An ontology and AHP based quality evaluation approach for reuse parts of end-of-life construction machinery. Math. Probl. Eng. **1**, 1–12 (2018). https://doi.org/10.1155/2018/3481030

Dynamic Bayesian Networks Application for Evaluating the Investment Projects Effectiveness

Volodymyr Lytvynenko[1(✉)], Oleksandr Naumov[2], Mariia Voronenko[1], Jan Krejci[3], Larisa Naumova[2], Dmytro Nikytenko[4], and Nataliia Savina[4]

[1] Department of Informatics and Computing Technology, Kherson National Technical University, Kherson, Ukraine
immun56@gmail.com, mary_voronenko@i.ua

[2] University of State Fiscal Service of Ukraine, Irpin, Ukraine
abnaumov75@gmail.com, lnnaumova46@gmail.com

[3] Jan Evangelista Purkyne University in Usti nad Labem, Ústí nad Labem, Czech Republic
jan.krejci@ujep.cz

[4] National University of Water and Environmental Engineering, Rivne, Ukraine
{d.v.nikytenko,n.b.savina}@nuwm.edu.ua

Abstract. In this paper, we propose a methodology for using dynamic Bayesian networks (DBN) in the tasks of assessing the success of an investment project. The methods of constructing DBN, their parametric learning, validation and scenario analysis of "What-if" are considered. A dynamic Bayesian model has been developed for scenario analysis and forecasting the success of an investment project. The model takes into account the time component and is designed in collaboration with expert economists in the selection and quantification of input and output variables. Now, using the dynamic Bayesian model, it is possible with a certain degree of probability to assess the degree of success of the capital investment, without incurring monetary and temporary losses. This will greatly facilitate the investment forecast for identifying profitable investment sources. This is the advantage of the proposed approach.

Keywords: Investment project · Bayesian networks · Structural learning · Parametric learning · Sensitivity analysis · Validation

1 Introduction

The task of adequately assessing the attractiveness of a project related to capital investment is to determine how much future revenues justify today's costs. Thus, the problem of making a decision on investments is to evaluate the plan for the expected development of events from the point of view of how much the content of the plan and the likely consequences of its implementation correspond to the expected result [17,30].

S. Babichev et al. (Eds.): ISDMCI 2020, AISC 1246, pp. 315–330, 2021.
https://doi.org/10.1007/978-3-030-54215-3_20

The formation of a system for assessing the quality of an investment project, which makes it possible to predict its ability to become more successful over time, contributes to an increase in investment, which will positively affect the development of the economy as a whole, is dynamic. Socio-economic processes that control investment, all investment activities in general, like many phenomena occurring in society, are dynamic. Performance indicators of an investment project, such as internal profitability, project implementation period, payback period, are variables that are observed at successive points in time. The target indicator characterizing the degree of success of the investment project will also change at each time interval.

This analyzed social phenomenon is characterized not only by current values but also by the results of previous periods. To present and analyze economic phenomena that change over time, we need to use a model that takes into account the relationship between the values of all parts of the model at different points in time. In this paper, to solve this problem, such probabilistic graphic models as DBNs are used.

A comparative analysis of the research results determines the novelty of the results: having a set of statistical data available, we still use the probabilistic approach of Dynamic Bayesian networks, the use of which allows us to trace the results several time steps forward; according to the results of the experiments, tendencies were successfully identified in which the indicators of innovative and financial development of the country will improve; the conditions under which the investment project will bring profit to all its participants are determined.

In the "Problem Statement" section, we present the necessity of a support tool for investment solutions. In the section "Related Work" we show what research in the field of investments was carried out earlier. The section "Proposed Approach" presents an input set of indicators and their purpose, as well as a brief overview of the methods that are used in the work. In the "Experimental Results" section, the original static Bayesian model is presented, we also demonstrate the scenario analysis of "What-if" of the designed Bayesian network, considering the time components, such as the period of the project and its payback period. The chapter "Discussion" presents an analysis of the results of the study and shows the conditions under which the investment project will be successful. The chapter "Conclusion" summarizes the results and formulates promising areas of our research.

2 Related Works

The most important property of capital is the ability to make a profit for its owner, long-term investment of capital or investment is one form of using this opportunity. The authors of [7,19,21] show various types of investment instruments and also tips for choosing companies, in understanding their accounts, managing and evaluating the effectiveness and risks of investing money on the example of how some companies met with the financial crisis and successfully overcome it.

The authors of [24] have shown that the rate of accumulation of physical capital is an important determinant of economic growth through a large sample of countries. The decline in investment is a concern given the close relationship between investment rates and economic growth, as noted in recent studies, such as [3,4,12].

Another approach suggested by [29,32] emphasizes the importance of integrating financial instruments in the economy and high-interest rates to stimulate growth. In this view, investment is positively dependent on the real interest rate as opposed to the neoclassical theory. The reason for this is that raising interest rates increases the value of the financial resources raised by the financial intermediaries and thus forms the funds for investment. In [13,23,28] investment incentives and discounting of depreciation deductions are studied, the value of expectations when investing money and its consequences are analyzed.

Bayesian networks are a widely used modeling tool that allows you to design economic models of control over re-source management, as shown in [8,9,11]. In [10,22] proposed the use of object-oriented BN for the design of agroeconomic models. As a result, the authors achieved some positive results in controlling the management of groundwater in Spain.

Dynamic Bayesian networks have found application in the diagnosis of diseases [33] and forecasting weather conditions [27]. It is interesting the using of Dynamic Bayesian networks in recognizing handwritten Arabic words in [18]. The authors achieved the goal that they set for themselves, but the question remains whether this dynamic model can be applied to other languages, and not just Arabic. In [31], the authors used Dynamic Bayesian networks to predict passenger flow. The problem was solved, however, in their study it remained unclear whether other areas of application for the developed model could be found.

The tasks of finding conditions for the success of an investment project in the current economic situation for a particular country, taking into account the time component, have not yet been resolved. Also, until now, insufficient attention has been paid to detecting growth trends in the region's financial indicators in the presence of incomplete data in a changing situation over time.

The aim of the work is development of a model that allows predicting the success of the investment project in time, which will contribute to a gradual increase in investment based on statistical indicators using a dynamic Bayesian network model.

3 Problem Statement

Ukraine has a certain investment attractiveness. Factors such as natural resource potential, allowing to develop various types of industries, favorable geographical position; the presence of a developed processing sector, developed transport infrastructure; significant domestic consumer market; the presence of scientific organizations for the innovative development of production, give our country an investment advantage. However, there are a number of factors to restrain investment activity in Ukraine, such as the risk of changes in the political and socio-economic situation in the country, the lack of a securities market, the lack of targeted state support for innovative investments, etc. These factors largely determine the investment climate and investment activity in the economy as a whole, as well as the opportunities and risks of investing in specific projects in Ukraine. Figure 1 shows the concepts that investors guide in evaluating investments.

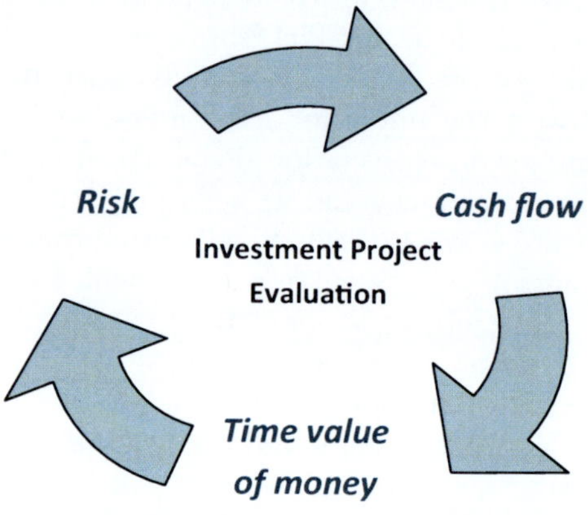

Fig. 1. Basic concepts used in the assessment of investments

A mathematical model that, when analyzing financial, investment and economic enterprises would help in advance to determine whether an investment project will be successful, requires input data. Bayesian network methods, with a certain degree of probability, make it possible to achieve the goal. Moreover, Bayesian networks are a powerful and flexible tool for analyzing economic data and research results. The methods for developing an investment risk forecasting tool can be divided into 2 approaches: regression models and machine learning algorithms. BN risk modeling has advantages over regression based approaches, and in this article we will focus on three of them:

- creating network structures that make it easy to establish relationships between variables,
- the ability to apply Bayes' theorem to assess risk at an individual level,
- easy transformation of Bayesian models into decision-making models.

Having input indicators, such as the result of the financial activity of the enterprise (UAH); cash balance at the end of the project (UAH), project implementation period (quarter), internal profitability of the enterprise (%), etc. it is necessary to build a dynamic Bayesian network that allows determining the influence of the input data on the integral indicator of the effectiveness of the investment project.

4 Materials and Methods

Among the criteria for making a decision on investing, one can single out: the indicator of net present value, the index of profitability, the indicator of internal rate of return and the payback period (Table 1). If the net present value of the NPV investment project is positive, then the project is effective at this projected discount rate, that is, we can consider the issue of its adoption and financing. The greater the NPV, the more effective the project.

Table 1. Matrix of indicators of business activity

Number	Indicators	Description
1	X1	Volume of investments of the enterprise, UAH
2	X11	Own funds of the enterprise, UAH
3	X12	Cash investor, UAH
4	X2	The result of the financial activities of the enterprise, UAH
5	X21	Net profit of the enterprise, UAH
6	X22	Depreciation costs of the enterprise, UAH
7	X23	Costs of the enterprise for paying for a loan, UAH
8	X24	Balance of funds of the enterprise at the end of the project, UAH
9	X3	T - Project implementation period, quarter
10	X4	r - Discount rate,%
11	X31	NPV - Net present value ($NPV > 0$)
12	X32	PI - Profitability Index ($PI>1$)
13	X33	PBP - Payback term ($PBP < T$), quarter
14	X41	IRR - Internal rate of return ($IRR > r$),%

The yield index PI is closely related to NPV, namely: if NPV is positive, then $PI > 1$ and vice versa. So if $PI > 1$, then the project is effective, and if $PI > 1$ is not effective. Unlike NPV, the profitability index is a relative indicator. The value of the IRR internal rate of return should exceed the rate of return on capital acceptable to the investor, and should be higher than the bank rate for long-term loans (per quarter), otherwise it would be more profitable to invest in a bank than to spend them on financing a project. In the case when the IRR is equal to or exceeds the rate of return on capital necessary for the investor, then investments in this project are effective and the feasibility of financing it can be considered.

The acceptable payback period for a PBP project depends on the length of the period during which the project is completed. That is, if the payback period of a certain project exceeds the level set by the investor, then the project is considered acceptable for financing. If a comparison of alternative investment projects for NPV and IRR leads to opposite results, NPV should be preferred. An important role is also played by the discount rate, which contains uncertainties and risks associated with, for example, deadlines for payments, jumps in inflation, designers' mistakes, unforeseen expenses or penalties for breach of contractual obligations. All the indicators discussed above are closely interconnected, the nature of which is shown in Fig. 3. None of the criteria discussed above alone is sufficient to make a decision on investing money. The decision on investment should be made taking into account the values of all the listed criteria and the interests of all participants in the investment process.

In our study, we use a set of statistical data, consisting of 14 indicators for 11 enterprises for the period 2017–2019 (Table 2). Table 2 contains statistics for key nodes only. In our study, each quarter is one time step. These indicators most fully reflect the financial, economic and business activity of the enterprise. The resulting indicator Y is an integral indicator of the success of an investment project.

Table 2. Statistical data for the period 2017–2019

Period of time	X1	X11	X2	X21	X3	X31	X4	X41	Y
1 quarter 2017	530860	48860	0	1654435	15	12479	12	16	0
2 quarter 2017	180000	20000	−179750	216191	15	11937	8	20	1
3 quarter 2017	2117392	117392	117392	2555565	15	89534	10	18	1
4 quarter 2017	4702502	646702	4702502	4049603	15	297771	12	24	0
1 quarter 2018	1800800	300800	0	2119018	15	64701	12	17	0
2 quarter 2018	411000	0	0	958863	20	672840	24	24	0
3 quarter 2018	456000	176000	0	392712	15	7022	12	15	0
4 quarter 2018	1000000	0	349660	2034038	15	128964	12	20	0
1 quarter 2019	260000	0	0	410054	15	38300	12	22	1
2 quarter 2019	1800000	0	−595200	4722300	40	486390	6	59	1
3 quarter 2019	3838000	93000	−3210000	2840170	12	164410	12	14	1
4 quarter 2019	19600	0	−3920	34786	15	11185	4	43	0

4.1 Applicable Methods

Formally, A pair $<G, B>$ called a Bayesian network (BN), when the first part of G is a acyclic directed graph corresponding to random variables [2,5]. When each variable is autonomous of its parents in G, so a graph is written as a composition of autonomous conditions. The second part of the pair, B, is the composition of parameters defining the network. It composed of parameters $Q_{x^i|pa(X^i)} = P(x^i|pa(X^i))$ for each possible x^i value from X^i and $pa(X^i)$ from $Pa(X^i)$, where $Pa(X^i)$ means the variable X^i parents set in G . Each variable X^i in graph G is suggested as a vertex. If we consider more than only one graph, then we use the notation to identify the parents $Pa^G(X^i)$ in graph G [5]. The BN's cumulative probability B is determined by the equation:

$$P_B(X^1, ..., X^N) = \prod\nolimits_{i=1}^{N} P_B(X^i|Pa(X^i)).$$

The BN represents a model for getting probabilistic dependencies, as well as the absence of these dependencies. At the same time, the $A \rightarrow B$ interdependence can be causal, when event A causes B to occur. So that is, at the time when there is a mechanism whereby the value accepted by A affects the value adopted by B. When all BN's connections are causal, so BN is called causal.

Markov condition is fulfilled: each network variable is independent of all other variables except the parent precursors of this variable [6]. The construction of the BN structure and the conditional probability tables can be based on:

- obtain information from subject matter experts,
- obtaining statistical information.

The structure of the network is most often determined by domain experts, although there are data-based structural learning methods for BN. Probability tables, on the contrary, are often generated on the basis of data by statistical methods. The goal of parametric learning is to find the most likely variables θ that explain the data. Let $D = \{D_1, D_2, \ldots, D_N\}$ be training data, where $D_1 = \{x_1[l], x_2[l], \ldots, x_n[l]\}$ consists of instances of Bayesian network nodes. The learning parameter is quantified by the logarithmic likelihood function, denoted as $L_D(\theta)$ [16]. Sensitivity analysis of a Bayesian network basically amounts to establishing, for each of each of the network's parameters, a function expressing an output probability in terms og the parameter under study [11,22].

For output probabilities, we shall consider posterior marginal probabilities of the form $y = p(a|e)$, where a is a value of a variable A and e denotes the evidence available. Each of the network's parameters is of the form $x = p(b_i|\pi)$, where b_i is a value of a variable B and π- is an arbitrary combination of values of the set of parents $\prod = pa(B)$ of B. We will mark $p(a|e)(x)$ to denote the function expressing the posterior marginal $p(a|e)$ in terms of the parameter x. Hereinafter, we will assume that in a sensitivity analysis, upon varying a parameter $x = p(b_i|\pi)$ each of the other probabilities $p(b_j|\pi)$ is co-varied accordingly. The function y(x), yielded by a sensitivity analysis is a quotient of two linear functions in x [16,27].

4.2 Dynamic Bayesian Networks

The concept of DBN, first introduced by Dean and Kanazawa [15] in 1988, is an extension of the Bayesian network (BN) [14,20] to simulate dynamic systems that change over time. A DBN contains the same basic DAG structure, but adds time arcs to capture dependencies between nodes that have some time delay. In fact, for DBN design we only need to know the static structure and set of time arcs. Then we can repeat the structure as many times as necessary to be able to simulate any period of time. In our study, the GeNIe 2.4 Academic software environment was used. In this software environment, DBNs are represented as shown in Fig. 2.

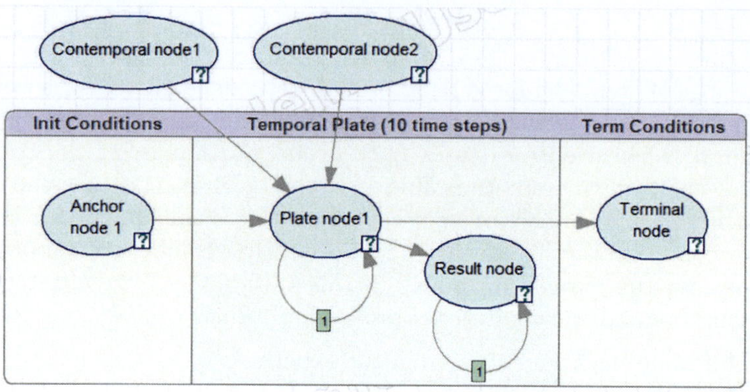

Fig. 2. DBN representation example in GeNIe 2.4

The presented network, as an example of a DBN, has two normal nodes, two temporary nodes and a temporary plate consisting of two-time steps. Straight bold arrows mean normal (static) arcs, while bright arrows with marks on them show temporary arcs. The time arc with the tag "1" shows the dependence of the child node in time t on the status of the parent node in time [15].

Thus, the trend of probability fluctuations for each node in all time steps can be analyzed in detail. In the DBN area, the actual observation of an event is called evidence. By providing such evidence and updating the network, the entire network and node confidence will be changed [15].

5 Experiment, Results and Discussion

Figure 3 shows the initial static BN model aimed at predicting the success of an investment project. The final accuracy of this network is 61%.

On the basis of the static BN we will build a dynamic BN, which makes it possible to show the effectiveness of investment, taking into account time parameters. Designed DBN has qualitative and quantitative components. A quality component is a network structure, its conceptual model, extracted from the knowledge of experts. The quantitative component is the conditional probability distribution, which determines the exact relationship between states and variables. We take the static Bayesian network of 14 nodes (see Fig.4), after

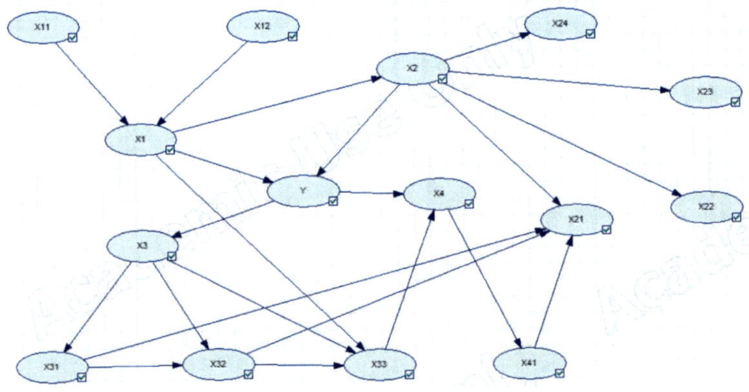

Fig. 3. Initial static BN model aimed at predicting the success of an investment project

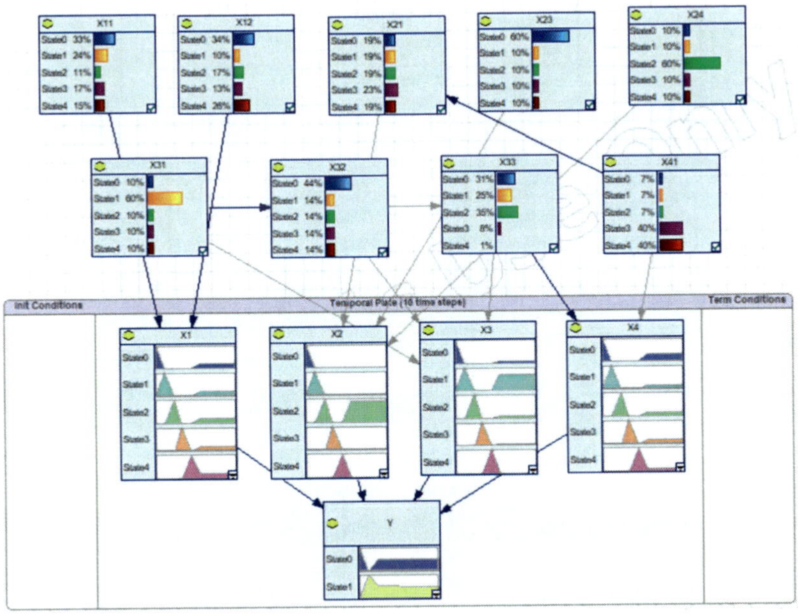

Fig. 4. Dynamic BN for determine the effectiveness of investment

re-verification, the accuracy of the initial model reached 61%. Next, we determine the normalized and terminal nodes, and then rebuild the network, taking into account the presence of the time domain. We assume that the nodes X11, X12, X21–X24, X41 are normalized static, whose values do not change with time when adding time arcs. On nodes, X1–X4 we set temporary arcs and now previous probabilities are taken into account on key variables X1–X4. After learning the parameters, you need to set time periods for all key variables. The resulting dynamic network will look like that shown in Fig. 4.

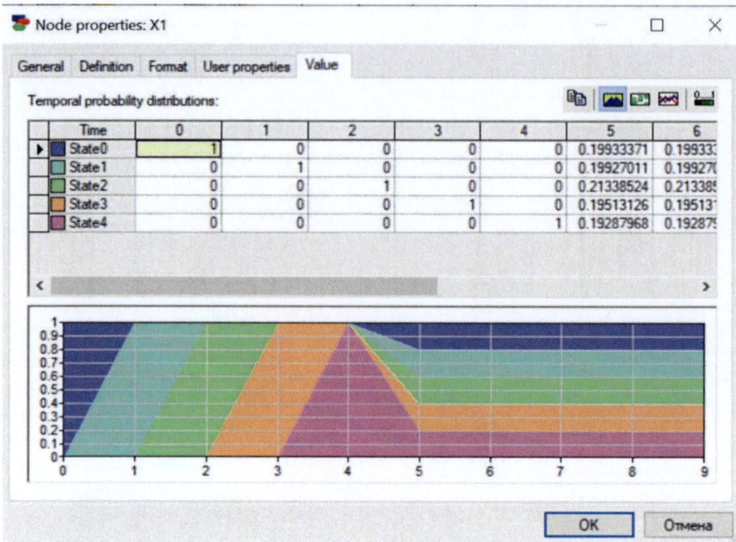

Fig. 5. The conditions of maximization the volume of investments

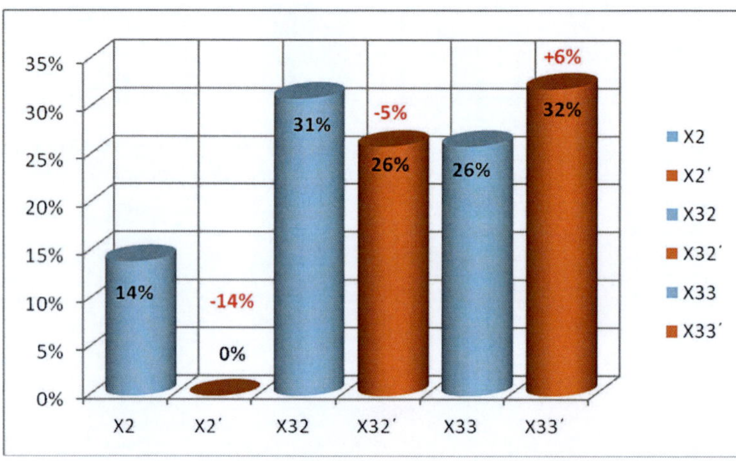

Fig. 6. The consequences of reducing financial activity to a minimum

Considering only the economic aspect of investing to a greater extent, it turns out to be wrong, since the external economic situation is characterized by general instability, the imperfection of the legislative system and underdeveloped market relations (primarily the capital market). This is the reason why it is necessary to take into account the temporary component of investment activity. Under the conditions of the so-called "perfect competition", the criterion for the effectiveness of an investment project is the level of profit received on invested

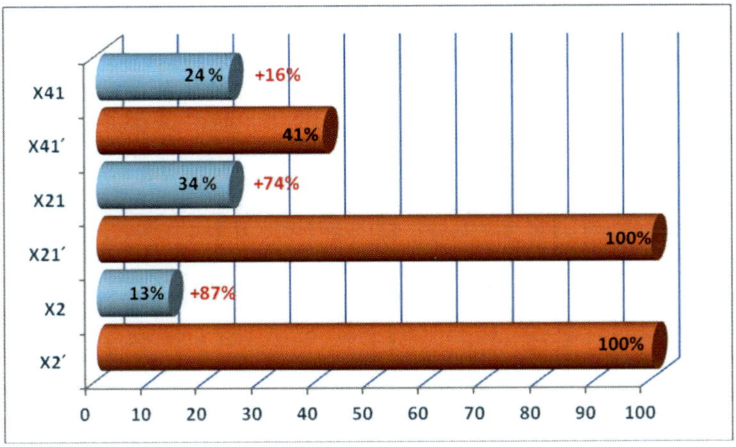

Fig. 7. Initial static BN model aimed at predicting the success of an investment project

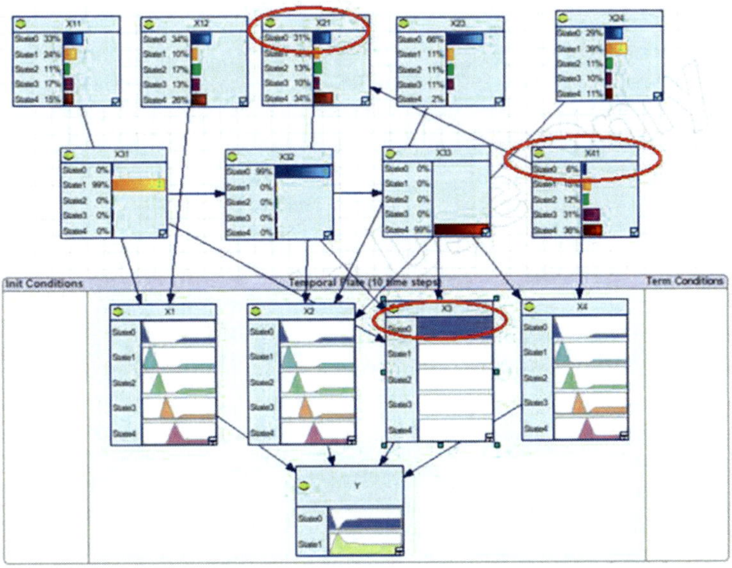

Fig. 8. Conditions under which the internal rate of return of IRR will decrease in the next time step

capital. For a "perfect investor," that is, a person who is interested in striving for the highest possible rate of return, all investment projects that provide the same levels of return are attractive. The process of structural and parametric design of a network, based only on expert estimates, as for example in [1], takes a long time and is time-consuming even for small networks. Also, any static BN, such as in [26], which models spatial dependencies, does not make a time forecast. Unlike a system that uses input from a numerical model as an input to a static network to predict the values of key variables, the presented DBN uses the observed values of key variables to predict future values of the same variables [25]. During the experiments, it was possible to identify conditions that contribute to improving the efficiency of the investment project at each time step.

1) To maximize the volume of investments (i.e., X1 increases by 86% compared to the initial state), at the previous time step, you need to take care of increasing the amount of your own investments by 85% (that is, create conditions for the growth of X11 from 15% to 100%), as well as increasing the number of investments of investors by 74% (that is, it is necessary to ensure the growth of X12 from 26% to 100%), as shown in Fig. 5.
2) The results of the experiment show that as soon as the financial activity are reduced to a minimum (i.e., X2 is reduced by 14% compared to the initial state), then at the next time step this will necessarily lead to a decrease in the PI index of profit by 5% (X32 will decrease from 31% to 26%) and increase the payback period by 6% (X33 will increase from 26% to 32%), as shown in Fig. 6.
3) For the financial results at the next time step to be maximum (i.e., that is, X2 will increase by 87%), it is necessary at the current time to increase the level of net profit by 76% (X21 will increase from 24% to 100%) and increase the internal rate of return of IRR by 16% (X41 will increase from 25% to 41%), as shown in Fig. 7.
4) A decrease in net profit of only 4% now (that is, provided that X21 is reduced from 27% to 31%) will lead to a decrease in the internal rate of return of IRR at the next time step by 23% (X41 will decrease from 29% to 6%), as well as a decrease discount rates to a minimum (X3 minimum), as shown in Fig. 8.
5) In order for the investment project to be implemented in the shortest possible time (that is, the payback period X4 decreased from 19% to the minimum), it is necessary to reduce the PI index by 15% at the current time step (X32 will decrease from 31% to 16%) and achieve growth the payback period of the investment project by 15% (X33 will increase from 26% to 41%), as shown in Fig. 9.
6) With the maximum implementation period of the investment project, net profit will increase by 19% (X21 will increase from 26% to 45%), and the profitability index will increase by 13% (X32 will increase from 16% to 29%), as shown in Fig. 10.
7) If at the moment the investment project has maximum investments (investment volume X1 = 100%) and the longest implementation period (X4 = 100%), then at the next time step this will lead to an increase in net profit by 21% (that is, X21 will increase from 24 to 45%) and an increase in the profitability index by 16% (X32 will increase from 16% to 32%), as shown in Fig. 11.

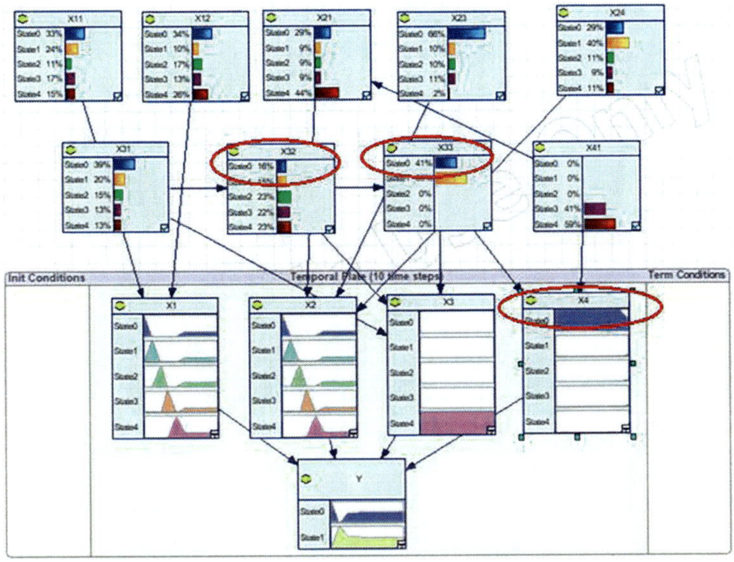

Fig. 9. Conditions under which the investment project will be implemented as soon as possible

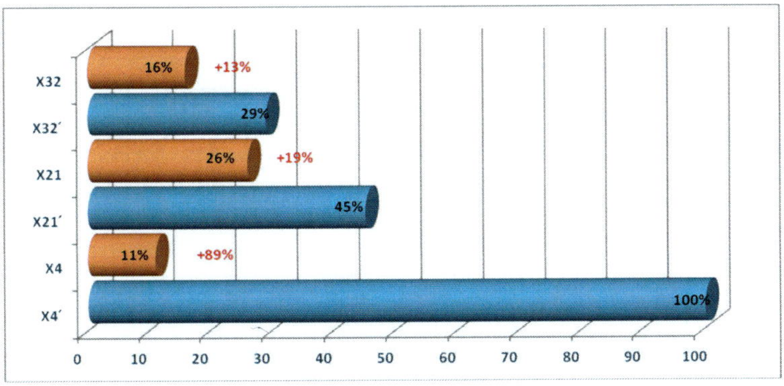

Fig. 10. Net profit increase terms

Previously, to achieve the goal of ascertaining the success of an investment project, an audit of the enterprise and analysis of the economic situation were necessarily carried out. This entailed additional time and material costs. Now, using the dynamic Bayesian model, it is possible with a certain degree of probability to assess the degree of success of the capital investment, without incurring monetary and temporary losses. This will greatly facilitate the investment forecast for identifying profitable investment sources. This is the advantage of the proposed approach.

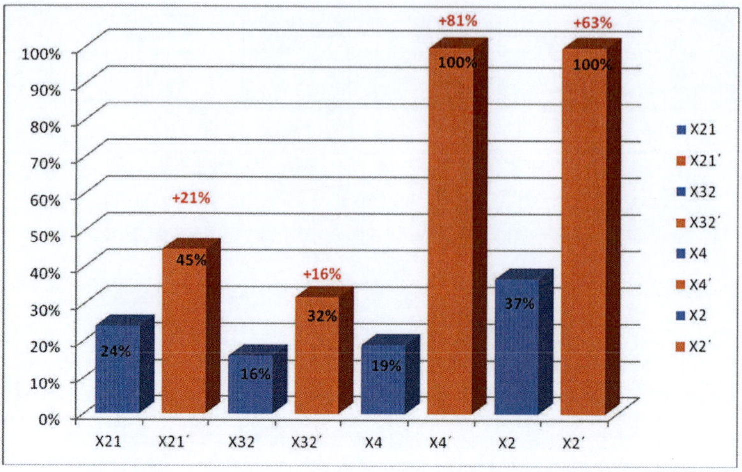

Fig. 11. Reasons for the increase in the index of profitability

6 Conclusions

In the work, a dynamic BN is designed for evaluating the success of an investment project at each time step. It is well known that there are periods of decline in investment activity and there are periods of recovery when investors need to be prepared to increase economic and financial risks, the emergence of new investment opportunities, etc. In such cases, it is necessary to ensure a quick response, take care of additional reserves.

In order for the investment project to be implemented in the shortest possible time, it is necessary to reduce the PI index by 15% at the current time step and achieve growth the payback period of the investment project by 15%. For the financial results at the next time step to be maximum, it is necessary at the current time to increase the level of net profit by 76% and increase the internal rate of return of IRR by 16%.

The scope of the developed DBN is very extensive, for example, audit, economics and finance. The model can be used to assess investment risks, in order to increase the investment potential of the country, both internal and external. The end users of the presented DBN are: investors whose goal is to increase the level of investment, while striving to reduce the risk of capital investment. In the field of investment, monitoring is an additional opportunity to provide investors with timely information.

The future area of the authors' research is implementation of this model on real data.

References

1. Abramson, B., Brown, J., Edwards, W., Murphy, A., Winkler, R.: Hailfinder: a Bayesian system for forecasting severe weather. Int. J. Forecast. **12**(1), 57–71 (1996). https://doi.org/10.1016/0169-2070(95)00664-8
2. Acid, S., Campos, L., Castellano, J.: Learning Bayesian network classifiers: searching in a space of partially directed acyclic graphs. Mach. Learn. **59**(3), 213–235 (2005). https://doi.org/10.1007/s10994-005-0473-4
3. Barro, R.: Economic growth in a cross section of countries. Q. J. Econ. **106**(2), 407 (1991). https://doi.org/10.2307/2937943
4. Ben-David, D.: Convergence clubs and subsistence economies. J. Dev. Econ. **55**(1), 155–171 (1998). https://doi.org/10.1016/s0304-3878(97)00060-6
5. Bidyuk, P., Gozhyj, A., Kalinina, I.: Probabilistic inference based on LS-method modifications in decision making problems. In: Lecture Notes in Computational Intelligence and Decision Making. ISDMCI 2019, vol. 1020, pp. 422–433 (2019). https://doi.org/10.1007/978-3-030-26474-1_30
6. Bidyuk, P., Matsuki, Y., Gozhyj, A., Beglytsia, V., Kalinina, I.: Features of application of Monte Carlo method with Markov chain algorithms in Bayesian data analysis. In: Advances in Intelligent Systems and Computing IV. CSIT 2019, vol. 1080, pp. 361–376. Springer, Cham (2020). https://doi.org/10.1007/978-3-030-33695-0_25
7. Block, S.: Are real options actually used in the real world? Eng. Econ. **3**(52), 255–267 (2007). https://doi.org/10.1080/00137910701503910
8. Brouwer, R., De-Blois, C.: Integrated modelling of risk and uncertainty underlying the cost and effectiveness of water quality measures. Environ. Model Softw. **23**(7), 922–937 (2008). https://doi.org/10.1016/j.envsoft.2007.10.006
9. Cai, X., McKinney, D., Lasdon, L.: Integrated hydrologic-agronomic-economic model for river basin management. J. Water Resour. Plan. Manag. **129**(1), 4–17 (2003). https://doi.org/10.1061/(asce)0733-9496(2003)129:1(4)
10. Carmona, G., Varela-Ortega, C., Bromley, J.: The use of participatory object-oriented Bayesian networks and agro-economic models for groundwater management in spain. Water Resour. Manag. **25**(5), 1509–1524 (2011). https://doi.org/10.1007/s11269-010-9757-y
11. Castelletti, A., Soncini-Sessa, R.: Bayesian networks and participatory modelling in water resource management. Environ. Model Softw. **22**(8), 1075–1088 (2007). https://doi.org/10.1016/j.envsoft.2006.06.003
12. Chari, V., Kehoe, P., McGrattan, E.: Sticky price models of the business cycle: can the contract multiplier solve the persis-tence problem. Econometrica **68**, 1151–1179 (2000). https://doi.org/10.1111/1468-0262.00154
13. Cox, J., Ross, S., Rubinstein, M.: Option pricing: a simplified approach. J. Financ. Econ. **7**(3), 229–263 (1979). https://doi.org/10.1016/0304-405x(79)90015-1
14. De Campos, L., Castellano, J.: Bayesian network learning algorithms using structural restrictions. Int. J. Approx. Reason. **45**(2), 233–254 (2007). https://doi.org/10.1016/j.ijar.2006.06.009
15. Dean, T., Kanazawa, K.: Probabilistic temporal reasoning. In: Proceedings of the National Conference on Artificial Intelligence (AAAI - 1988), pp. 525–529. AAAI Press/The MIT Press (1988)
16. Dempster, A., Laird, N., Rubin, D.: Maximum likelihood from incomplete data via the EM algorithm. J. Roy. Stat. Soc. **39**, 1–38 (1977)

17. Dixit, A.: Investment and hysteresis. J. Econ. Perspect. **6**(1), 107–132 (1992). https://doi.org/10.11257/jep6.1.107
18. Ghanmi, N., Awal, A., Kooli, N.: Dynamic Bayesian networks for handwritten Arabic word recognition. In: 1st International Workshop on Arabic Script Analysis and Recognition (ASAR) (2017). https://doi.org/10.1016/j.ijar.2006.06.00910.1109/asar.2017.8067769
19. Graham, J., Harvey, C.: The theory and practice of corporate finance: evidence from the field. SSRN Electron. J. (2000). https://doi.org/10.1016/j.ijar.2006.06.00910.2139/ssrn.220251
20. Heckerman, D., Geiger, D., Chickering, D.: Learning Bayesian networks: the combination of knowledge and statistical data. Mach. Learn. **20**(3), 197–243 (1995). https://doi.org/10.1023/a:1022623210503
21. Helfert Erich, A.: Financial Analysis: Tools and Techniques: A Guide for Managers. McGraw Hill, New York (2001). https://doi.org/10.1036/0071395415
22. Jakeman, A., Letcher, R.: Integrated assessment and modelling: features, principles and examples for catchment management. Environ. Model Softw. **18**(6), 491–501 (2003). https://doi.org/10.1016/s1364-8152(03)00024-0
23. Lawrence, H.: Investment incentives and the discounting of depreciation allowances. In: Effects of Taxation on Capital Accumulation, pp. 296–304 (1987)
24. Levine, R., Renelt, D.: A sensitivity analysis of cross-country growth regressions. Am. Econ. Rev. **4**(82), 942–963 (1992)
25. Lytvynenko, V., Savina, N., Krejci, J., Fefelov, A., Lurie, I., Voronenko, M., Lopushynskyi, I., Vorona, P.: Dynamic Bayesian networks in the problem of localizing the narcotic substances distribution. In: AISC, vol. 1080, pp. 421–438. Springer(2019)
26. Lytvynenko, V., Savina, N., Voronenko, M., Doroschuk, N., Smailova, S., Boskin, O., Kravchenko, T.: Development, validation and testing of the Bayesian network of educational institutions financing. In: The crossing point of Intelligent Data Acquisition and Advanced Computing Systems and East and West Scientists (IDAACS-2019) (2019). https://doi.org/10.1109/IDAACS.2019.8924307
27. Mbuvha, R., Jonsson, M., Ehn, N., Herman, P.: Bayesian neural networks for one-hour ahead wind power forecasting. In: IEEE 6th International Conference on Renewable Energy Research and Applications (ICRERA) (2017). https://doi.org/10.1109/icrera.2017.8191129
28. McDonald, R., Siegel, D.: The value of waiting to invest. Quart. J. Econ. **101**, 707–728 (1986)
29. McKinnon, R.: Money and Capital in Economic Development. Brookings Institution, Washington (1973)
30. Purvis, A., Boggess, W., Moss, C., Holt, J.: Technology adoption decisions under irreversibility and uncertainty: an ex ante approach. Am. J. Agric. Econ. **77**(3), 541–551 (1995). https://doi.org/10.2307/1243223
31. Roos, J., Bonnevay, S., Gavin, G.: Dynamic Bayesian networks with Gaussian mixture models for short-term passenger flow forecasting. In: 12th International Conference on Intelligent Systems and Knowledge Engineering (ISKE) (2017). https://doi.org/10.1109/iske.2017.8258756
32. Schmitt, H., Shaw, E.: Financial deepening in economic development. Am. J. Agric. Econ. **56**(3), 670 (1974). https://doi.org/10.2307/1238641
33. Xie, H., Shi, J., Lu, W., Cui, W.: Dynamic Bayesian networks in electronic equipment health diagnosis. In: Prognostics and System Health Management Conference (PHM-Chengdu) (2016). https://doi.org/10.1109/phm.2016.7819945

Development of a Fuzzy Inference Model for the Management of a Marine Engine

Sergii Babichev[1,2(✉)], Liliya Strielkovskaya[3], Oleksandr Zaitsev[4], and Orest Khamula[2]

[1] Jan Evangelista Purkyně University in Ústí nad Labem, Ústí nad Labem, Czech Republic
sergii.babichev@ujep.cz
[2] Ukrainian Academy of Printing, Lviv, Ukraine
khamula@gmail.com
[3] Kherson State Maritime Academy, Kherson, Ukraine
stliliya8810@gmail.com
[4] Military-Diplomatic Academy named after Eugene Bereznyak, Kyiv, Ukraine
a.zaysev@gmail.com

Abstract. The paper presents the results of the research concerning development of a fuzzy inference system to determine the current state of marine engine based on criteria established by the manufacturer. The practical implementation of the proposed model has been carried out for marine engine Wärtsilä 6L46C, which is one of the most used types of current marine engines. Implementation of the fuzzy inference system was performed based on R software using Mamdani inference algorithm. The gaussian and triangular membership functions were used for input and output variables respectively during the simulation process. The results of the simulation have shown that value of the output parameter (technical state of engine) was changed within the range from 0.15 to 0.893 during variation of the input parameters from minimal to maximal values. This fact indicates the adequate of the proposed fuzzy inference system operation.

Keywords: Fuzzy logic · Decision-making system · Membership function · Fuzzy rules · Marine engine · Technical diagnostics

1 Introduction

Development of an automated decision-making systems in order to determine the most effective modes of marine engine work, which allows us to maintain the reliability of the technical objects for a specified time while maintaining technical and economic indicators, is one of the currents aspects of modern complex technical objects. A solution of this problem is possible only in the case if the automatic control system is able to track not only the set of parameters, but also take into account the state of the technical object in the real time. Since the technical condition of complex objects is determined by a large number of factors,

S. Babichev et al. (Eds.): ISDMCI 2020, AISC 1246, pp. 331–340, 2021.
https://doi.org/10.1007/978-3-030-54215-3_21

assessing the current state of the investigated objects is a complex multifactorial task. Nowadays, in most cases, the solution of the problem of determining the technical condition of a marine engine is reduced to establishing the share values of operational factors (load, time modes, et al.) considering the personal experience of staff members. In other words, the control task is divided into two sub-tasks: determination of the systems parameters values by the operator based on his experience and setup the automated control system considering the determined before parameters. In this case, the operation of marine engine has a several of specific features consisting in the fact that a decision concerning the choice of an optimal operational mode is made by a very limited number of persons. As a result, ensuring the trouble-free operation of such complex objects as marine engines depends on the knowledge and experience of several persons. This fact increases the risk of the making wrong decision. According to hereinbefore, an important task of the technical operation of a marine engine is the development of techniques to generalize and formalize the experience of a large number of specialists in the fields of the marine engines operation, presenting this experience as a system which allows us to make decisions in accordance with the current situation considering the current state of the technical object.

1.1 Reviews of Literature

There are a lot of technical diagnostic methods to evaluate the current state of the complex objects nowadays [7,9,10,21,25]. So, the paper [9] presents a technique of data fusion obtained from the various sensors. The technique is based on adaptive models, which are the combinations of fuzzy neural systems and network models implementing Dempster-Shafer computational methodology of evidence combining. In the paper [10], the authors present the results of the research concerning a non-destructive diagnostic technique used for preliminary determination the location and size of delamination in composite coatings of wind turbine blades. The proposed by the authors technique of damage detection is based on the analysis of the ten first mode shapes of bending vibrations, which correspond to displacements of rotor blades perpendicular to the rotor plane. The results of the research concerning diagnostics of a compressor of marine large two-stroke diesel engines are presented in [21]. The authors have shown that for this type of engine, the compressor mass flow rate and isentropic efficiency empirical models are preferred for both the working cycle dynamic simulation research and the design and testing of control and diagnostics algorithms due to their compact and simple structures, and satisfactory prediction accuracy. In [7] the author has developed the measurement and diagnostic system to assess the technical condition of machinery and equipment in operation on board the ship. The author has shown that modernization of measuring equipment, significantly expanded the research capabilities, which resulted in increasing the scope, improving quality, and speeding up the research and development work in the field of safety of operation and diagnostics of engine rooms and ship operation. The detailed description of the experiment setups and principle of the three optical diagnostic techniques for marine engine are provided in the paper

[25]. The authors have shown that the liquid and vapor penetrations decrease with increasing the ambient density and increasing the ambient temperature can reduce the liquid penetration, but has no obvious effect on the vapor penetration.

However, the detail analysis of hereinbefore described papers has shown that the most of the existing methods of technical diagnostic have used a small number of input parameters, which in many cases are not enough to determine the technical condition of such a complex object as a marine engine is. In this case, to evaluate the technical state of complex objects it is reasonable to apply techniques which are based on complex use of various parameters determined the conditions of this object functioning. These techniques include the following ones: statistical techniques including Bayesian networks which are based on evaluation of a priori probabilities interconnections of both input and output parameters of the system between each other [6,16,20]; techniques based on fuzzy logic, which are applied nowadays in various fields of scientific research [4,5,12,23]; current techniques to diagnostics diesel internal combustion engines which are based on modern methods of complex data processing [11,15,18,24].

The problems concerning monitoring the output parameters of marine engine is currently solving by various manufacturers of marine internal combustion engines. So, the largest manufacturer of marine diesel engines MAN B&W Diesel has developed a computer diagnostic system CoCoS EDS [8]. The software is designed in such a way that it allows monitoring the main engine through the standard interface which is integrated with the central control system. FAKS2i and MoniTrend (Data Analysis System) softwares are produced by Wartsila company [1,2]. The algorithm of FAKS2i program in some aspects is similar to CoCoS EDS software. MoniTrend package analyzes the data received from the centralized control and monitoring system and detects the current trend in real-time.

However, it should be noted that the systems presented hereinbefore are able to service only some of the new diesel engines for which they were created. At the same time, there are a large number of ships which have not such systems and which are now in operation and will be used for several tens of years. Considering the increasing requirements for the safety of navigation and increased competition in the maritime industry becomes urgent to reduce the likelihood of making an incorrect or economically unjustified decision during the operation of a marine engine. This problem can be solved by introducing at these ships universal decision making systems, which are able to develop recommendations related to the implementation of operational measures or the choice of operating modes of the main and auxiliary engines on the basis of the information received of the system. It can be achieved by using current techniques, models and information technologies which are used successfully in various fields of scientific research nowadays [3,14,19,22]. Taking into account existing uncertainty, incompleteness, ambiguity, and also the fuzziness of the obtained information concerning the state of the examined marine engine, the use of the fuzzy logic technique is one of the promising directions for solving the problem of determining the marine engine technical state.

The aim of the paper is development of a fuzzy inference model for determining the current state of a marine internal combustion engine based on criteria established by the manufacturer.

2 Materials and Methods

In the case of the decision making system of marine engines the mathematical model can be represented as follows:

$$y = f(x_1, x_2, ..., x_n), \tag{1}$$

where y is the output parameter characterized a technical state of the investigated system; $\overline{x_1, x_n}$ are the input parameters. These parameters are controlled during the marine engine operation. The main task in this case consists in finding adequate compliance between set of input parameters and output variable. To solve this task, we propose to use a fuzzy inference system based on Mamdani inference algorithm. This choice is determined by the following reasons: in the case of marine engine testing the values of the input parameters functions do not require any scaling, all membership functions can be isotropic, resulting membership function can be represented as a simple set and the implementation of the defuzzification process does not require the use of a special functional. In this case the use of Mamdani fuzzy inference algorithm is reasonable. Structure block chart of a fuzzy logic model is presented in Fig. 1.

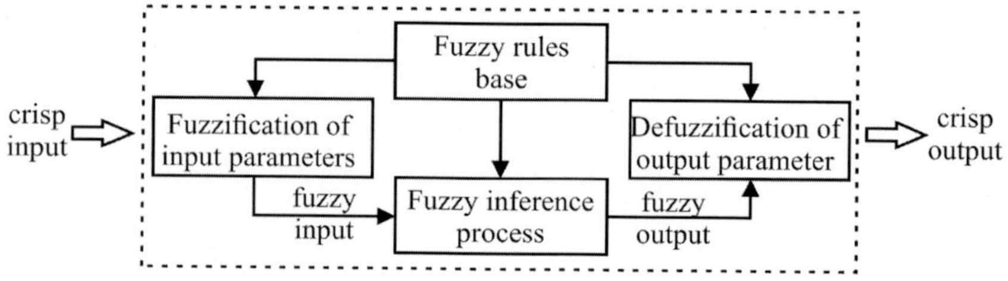

Fig. 1. A structural block chart of a fuzzy logic model

Implementation of this process assumes the following stages:

Stage 1. Setup of the system.

1. Transformation of input $\overline{x_1, x_n}$ and output y variables into linguistic estimates with corresponding membership functions;
2. form a basic term-set with the corresponding membership functions for each of the terms: $A = a_1, a_2, ..., a_i$;
3. formation of fuzzy rules agreed between input and output variables:

$$\bigcup_{k=1}^{m}[\bigcap_{i=1}^{n}(x_i = a_i^k),\ for\ \omega_k] \rightarrow D = d_k, \tag{2}$$

where $k = \overline{1,m}$ is the number of logical statements; $i = \overline{1,n}$ is the number of used terms; $D = d_1, d_2, ..., d_m$ is a set of fuzzy output variables.

Stage 2. *Fuzzification* or establishing a correspondence between the crisp values of the input variables and appropriate values of the membership functions. In other words, it is necessary to determine the $\mu^{a_i^k}(x_i)$ - value of membership function of variable x_i for term a_i^k.

Stage 3. Fuzzy inference process. Implementation of this stage assumes the following steps:

1. *aggregation* or determining the degree of truth of the conditions for each of the rules. The clipping levels for the prerequisites of each of the rules are found using the *min* operation at this step:

$$\alpha_k = \bigwedge_{i=1}^{n} [\mu^{a_i^k}(x_i)]; \tag{3}$$

2. *activation* or finding the degree of truth for each of the fuzzy rules of knowledge base, i.b. definition of truncated membership functions of fuzzy sets:

$$\mu'_k = (\alpha_k \wedge \mu_k(D)); \tag{4}$$

where $\mu(D)$ and $\mu'(D)$ are the truncated membership functions for input variables and fuzzy rules respectively;

3. *accumulation* or finding the membership function of the resulting fuzzy subset for the output variable:

$$\mu_\Sigma(D) = \bigvee_{k=1}^{m} [\mu'_k(D)] \tag{5}$$

Stage 4. *Defuzzification* or determining crisp value of the output variable. We used the mass center technique (mass center of the resulting membership function) to calculate the crisp value of the output variable:

$$y = \frac{\int_{min}^{max} x \cdot \mu(x)dx}{\int_{min}^{max} \mu(x)dx} \tag{6}$$

Thus, the implementation of fuzzy inference system assumes formation of set of both the input and output systems parameters at the first step. Then, formalization of these parameters into linguistic estimates with corresponding membership functions. The next step involves formation of a set of fuzzy rules agreed between input and output parameters. Finally, it is necessary to evaluate the model using the test data.

3 Experiment, Results and Discussion

The practical implementation of the hereinbefore presented model was performed for a Wärtsilä 6L46C marine engine [2] using *sets* package [17] of *R* software [13]. The following seven input parameters were selected based on analysis of boundary values considering actuation of signalization:

- x_1 - air temperature after air cooler;
- x_2 - pressure to the engine (lubricating oil system);
- x_3 - pressure of circulation pump;
- x_4 - temperature before engine, nominal (lubricating oil system);
- x_5 - pressure to the engine (high temperature circuit);
- x_6 - temperature after cylinders (high temperature circuit);
- x_7 - pressure to the engine (low temperature circuit).

The terms of linguistic variables were setupped as follows: "normal", "admissible", "non-admissible". Output parameter y - "technical state" had the following terms: "good", "acceptable", "permissible" and "non-acceptable". The ranges of these parameters changes were determined based on the design values of the auxiliary systems presented in the marine engine manual. The gaussian and triangular membership functions were used for input and output variables respectively. Table 1 and Table 2 present the parameters of the appropriate membership functions which were determined based on analysis the corresponding variables boundary values. The concentration coefficient σ in the cases of gaussian functions use was determined empirically during the simulation process.

Table 1. Parameters of gaussian membership functions for input variables

Parameter	Coefficient	Terms		
		Normal	Admissible	Non-admissible
x1	c	325.15	345.65	360.65
	σ	12	7	10
x2	c	0.35	0.25	0.1
	σ	0.03	0.03	0.06
x3	c	0.079	0.064	0.025
	σ	0.006	0.084	0.015
x4	c	343.65	352.15	363.15
	σ	6	4	10
x5	c	0.4	0.26	0.1
	σ	0.08	0.06	0.1
x6	c	366.65	380.65	393.15
	σ	4	6	10
x7	c	0.06	0.06	0.1
	σ	0.6	0.6	1

The charts of the membership functions for both the input and output variables are presented in Fig. 2. Causal relationships between selected input parameters and the technical state of the examined engine are formalized as a set of fuzzy rules. To create a set of fuzzy rules, an expert assessment of various

Table 2. Parameters of triangular membership functions for output variable

Parameter	Coefficients	Terms			
		Good	Acceptable	Permissible	Non-acceptable
y	a	0.8	0.5	0.3	0
	b	0.9	0.65	0.4	0.15
	c	1	0.8	0.5	0.3

combinations of both the input and output parameters was performed with a concurrent assessment of appropriate membership functions values. The weights of all the rules were assigned as 1. Since the totality of fuzzy rules describes the marine engine management strategy, special attention was attended to their consistency. The simulation results have shown that the least approximation error is achieved in the case of the use of 68 of fuzzy rules.

The results of the obtained fuzzy inference model operation for various combination of the input parameters are presented in Table 3. As it can be seen, the value of the output parameter (technical state of the marine engine) decreases from 0.893 to 0.15 during the variation of the input parameters values within the

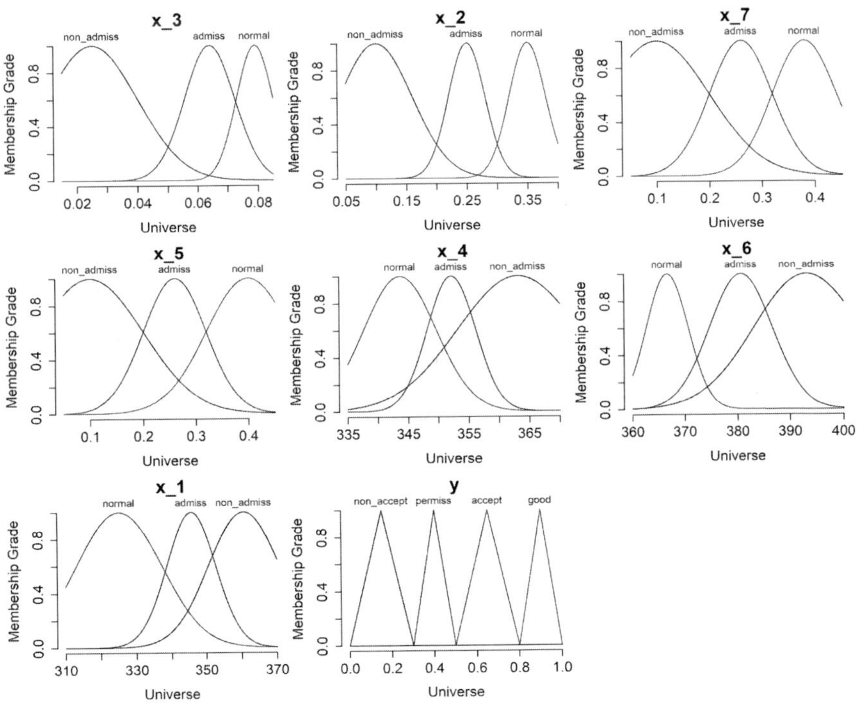

Fig. 2. The charts of the membership functions for input and output parameters

Table 3. The results of the fuzzy inference system operation for various combinations of the input parameters

Case	x_1	x_2	x_3	x_4	x_5	x_6	x_7	y
1	325	0.35	0.078	345	0.415	368	0.398	0.893
2	325	0.35	0.06	345	0.415	368	0.398	0.887
3	313	0.25	0.058	350	0.38	356	0.349	0.768
4	320	0.25	0.079	352	0.26	360	0.36	0.688
5	346	0.24	0.063	353	0.3	379	0.22	0.491
6	342	0.35	0.05	351.5	0.2	365	0.2	0.407
7	318	0.3	0.078	350	0.1	367	0.38	0.158
8	344	0.25	0.025	352	0.26	380	0.28	0.15

appropriate ranges. So, in the first case, when the values of all input parameters correspond to normal states, the value of the output variable is 0.893, what, by the experts' mind, corresponds to the good state of the marine engine. In the second case, the value of the parameter x_3 was not normal, but it was acceptable by the experts' mind. The value of the output parameter in this case was 0.887, that is less in comparison with the first case. However, the value of this parameter corresponds to *good* state of the marine engine too. In the cases from 3 to 6 the input parameters values were changed within the *admissible* range. The output parameter values were decreased within the range from *acceptible* to *permissible*, that is aggreed with the exprets' mind. In the last two cases (*7th* and *8th*) one of the input parameters (x_5 in the *7th* case and x_3 in the *8th* one) was not acceptable. Other input parameters were normal or admissible. The value of the output parameter in these cases was not acceptable too (0.158 and 0.15 respectively). To the experts' mind, the technical state of the marine engine in these cases is critical and an alarm is triggered in the engine room. The analysis of the obtained results indicates about adequately of the proposed fuxxy inference model operation.

4 Conclusions

In this paper, we have presented the results of the research concerning development of fuzzy inference model on order to determine the technical state of marine engines. A Wärtsilä 6L46C marine engine was used as the experimental one during the simulation process. The seven input parameters have been selected based on analysis of boundary values considering actuation of signalization during the engine expluatation. The terms of linguistic variables have been setupped as follows: "normal", "admissible" , "non-admissible". Output parameter y - "technical state" had the following terms: "good", "acceptable", "permissible" and "non-acceptable". The ranges of these parameters changes have been determined based on the design values of the auxiliary systems presented in the marine engine manual. The gaussian and triangular membership functions have been used for input and output variables respectively. The set of fuzzy rules containes 68 of rules.

The results of the simulation have shown high level of adequately of the obtained model operation. In the case of normal value of the all input parameters, the output variable corresponded to good technical state of the marine engine. When the values of the input parameters were varied within the admissible range, the values of the output parameter were changed within the range from acceptible to permissible, that is aggreed with the exprets' mind. In the cases, when the value of one of the input parameters was not admissible, the value of the output parameter was not acceptable too. To the experts' mind, the technical state of the marine engine in these cases is critical and an alarm is triggered in the engine room.

References

1. Wartsila. https://www.wartsila.com/
2. Project guide for marine applications. Wärtsilä finland oy, marine p.o. box 252, fin-65101 vaasa, finland (2001). https://www.wartsila.com/
3. Babichev, S.: An evaluation of the information technology of gene expression profiles processing stability for different levels of noise components. Data **3**(4), 48 (2018). https://doi.org/10.3390/data3040048
4. Babichev, S., Lytvynenko, V., Gozhyj, A., Korobchynskyi, M., Voronenko, M.: A fuzzy model for gene expression profiles reducing based on the complex use of statistical criteria and Shannon entropy. In: Advances in Intelligent Systems and Computing, vol. 754, pp. 545–554 (2019). https://doi.org/10.1007/978-3-319-91008-6_55
5. Babichev, S., Barilla, J., Fišer, J., Škvor, J.: A hybrid model of gene expression profiles reducing based on the complex use of fuzzy inference system and clustering quality criteria. In: 2019 Conference of the International Fuzzy Systems Association and the European Society for Fuzzy Logic and Technology (EUSFLAT 2019). Atlantis Press, August 2019. https://doi.org/10.2991/eusflat-19.2019.20
6. Bidyuk, P., Gozhvi, A., Kalinina, I.: Modeling military conflicts using Bayesian networks. In: Proceedings of the 2018 IEEE 1st International Conference on System Analysis and Intelligent Computing, SAIC 2018, p. 8516861. Institute of Electrical and Electronics Engineers Inc. (2018). https://doi.org/10.1109/SAIC.2018.8516861
7. Charchalis, A.: Measurement and diagnostic system for marine engines. J. Konbin **49**(3), 33–48 (2019). https://doi.org/10.2478/jok-2019-0049
8. Diesel, M.: User's guide CoCos EDS (engine diagnostics system) version 1.70 (2008). https://seatracker.ru/viewtopic.php?t=380
9. Dolgiy, A., Kovalev, S., Sukhanov, A., Styskala, V.: Hybrid fuzzy neural model based dempster-shafer system for processing of diagnostic information. In: Lecture Notes in Electrical Engineering, vol. 554, pp. 24–33 (2020). https://doi.org/10.1007/978-3-030-14907-9_3
10. Dolinski, L., Krawczuk, M., Zak, A.: Damage detection in the wind turbine blade using root mean square and experimental modal parameters. In: Lecture Notes in Mechanical Engineering, pp. 728–742 (2020). https://doi.org/10.1007/978-981-13-8331-1_57

11. Galiullin, L., Galiullin, I.: Fault diagnostic method for IC engines. Int. J. Intell. Syst. Appl. **11**(8), 2273–2279 (2019)
12. Hu, Z., Bodyanskiy, Y., Tyshchenko, O., Samitova, V.: Possibilistic fuzzy clustering for categorical data arrays based on frequency prototypes and dissimilarity measures. Int. J. Intell. Syst. Appl. **9**(5), 55–61 (2017). https://doi.org/10.5815/ijisa.2017.05.07
13. Ihaka, R., Gentleman, R.: R: a language for data analysis and graphics. J. Comput. Graph. Stat. **5**(3), 299–314 (1996). http://www.tandfonline.com/, https://doi.org/10.1080/10618600.1996.10474713
14. Kanishcheva, O., Vysotska, V., Chyrun, L., Gozhyj, A.: Method of integration and content management of the information resources network. In: Advances in Intelligent Systems and Computing, vol. 689, pp. 204–216 (2019). https://doi.org/10.1007/978-3-319-70581-1_14
15. Liu, J., Ma, B., Jiang, Z.: A study of probabilistic diagnosis method for three kinds of internal combustion engine faults based on the graphical model. Int. J. Intell. Syst. Appl. art. no. 8156450 (2019). https://doi.org/10.1155/2019/8156450
16. Luo, X., Kareem, A.: Bayesian deep learning with hierarchical prior: predictions from limited and noisy data. Struct. Saf. **84**, art. no. 101918 (2020). https://doi.org/10.1016/j.strusafe.2019.101918
17. Meyer, D., Hornik, K., Buchta, C.: Sets, generalized sets, customizable sets and intervals (2017). https://cran.r-project.org/package=sets
18. Monieta, J.: Preliminary investigations of marine diesel engines turbochargers diagnostic. Appl. Cond. Monit. **10**, 81–90 (2018). https://doi.org/10.1007/978-3-319-62042-8_7
19. Naum, O., Chyrun, L., Vysotska, V., Kanishcheva, O.: Intellectual system design for content formation. In: Proceedings of the 12th International Scientific and Technical Conference on Computer Sciences and Information Technologies, CSIT 2017, vol. 1, pp. 131–138. Institute of Electrical and Electronics Engineers Inc. (2017). https://doi.org/10.1109/STC-CSIT.2017.8098753
20. Shafay, A., Sultan, K.: Bayesian inference based on multiply type-II censored samples of sequential order statistics from pareto distribution. J. Test. Eval. **48**(6), art. no. JTE20170699 (2020). https://doi.org/10.1520/JTE20170699
21. Shen, H., Zhang, C., Zhang, J., Yang, B., Jia, B.: Applicable and comparative research of compressor mass flow rate and isentropic efficiency empirical models to marine large-scale compressor. Energies **13**(1), art. no. 47 (2019). https://doi.org/10.3390/en13010047
22. Tkachenko, R., Doroshenko, A., Izonin, I., Tsymbal, Y., Havrysh, B.: Imbalance data classification via neural-like structures of geometric transformations model: local and global approaches. In: Advances in Intelligent Systems and Computing, vol. 754, pp. 112–122 (2019). https://doi.org/10.1007/978-3-319-91008-6_12
23. Zadeh, L.A., Abbasov, A.M., Shahbazova, S.N.: Fuzzy-based techniques in human-like processing of social network data. Int. J. Uncertainty Fuzziness Knowl. Based Syst. **23**, 14–17 (2015)
24. Zadrg, R., Tomasz, K.: Ranking of toxic compound concentrations as diagnostic parameetrs of marine internal combustion engine. Pol. Marit. Res. **25**(1), 234–242 (2018). https://doi.org/10.2478/pomr-2018-0047
25. Zhang, W., Li, X., Huang, L., Feng, M.: Experimental study on spray and evaporation characteristics of diesel-fueled marine engine conditions based on optical diagnostic technology. Fuel **246**, 454–465 (2019). https://doi.org/10.1016/j.fuel.2019.02.065

Integral Criterion of Environmental Safety as an Indicator of the Effectiveness of a Motor Transport Enterprise Management

Serhii Kolomiiets[1](✉) and Anna Kolomiiets[2]

[1] National Transport University, Kyiv, Ukraine
s.kolomiiets@ntu.edu.ua
[2] Taras Shevchenko National University of Kyiv, Kyiv, Ukraine
kolomietsa@fit.knu.ua

Abstract. In the course of the study, a set of criteria for assessing the environmental safety of motor transport enterprises was developed, which consists of ten separate criteria for evaluating stationary and mobile sources of harmful effects of the enterprise, determined by the results of the analysis of significant environmental aspects of individual technological processes of maintenance and repair of vehicles and grouped into factors of resource costs, emissions, and waste, and integral criteria, which forms a single function of the goal, which determines the level of environmental safety of motor transport enterprises. To assess the level of environmental safety of motor transport enterprises, as well as vehicles serviced by them, the most informative criteria are identified, which are able to compare a large number of possible schemes for organizing the maintenance and repair process and options for the fleet of vehicles for various purposes in terms of the level of environmental pollution, waste generation and the cost of operating materials. This should take into account the universality of criteria for comparing different schemes of the same type and different types of motor transport enterprises, taking into account the operating conditions of the fleet of vehicles, as well as for assessing their compliance with modern environmental safety requirements. The study was formed by the method of formation of an integral criterion of ecological safety of motor transport enterprise, which is based on the method of expert estimations of the criteria of harmful emissions, waste, and costs of operational materials. The regularities of changes in individual, group and integral criteria of ecological safety of motor transport enterprises are determined according to the results of the study of the influence of the fleet structure and technological regimes on energy consumption, emissions and waste in the technological processes of recovery of the vehicle, depending on the proposed options to improve the environmental safety of the enterprise.

Keywords: Motor transport enterprises · Environmental safety · Fuel and resource consumption · Emissions · Waste

S. Babichev et al. (Eds.): ISDMCI 2020, AISC 1246, pp. 341–356, 2021.
https://doi.org/10.1007/978-3-030-54215-3_22

1 Introduction

The motor transport complex is one of the main sources of environmental pollution. Levels of air pollution with nitrogen and carbon oxides, hydrocarbons and other harmful substances in most major cities, according to the meteorological agency in recent years, exceed the maximum permissible concentrations several-fold. Motor transport accounts for about a third of the total emissions of harmful substances into the atmosphere, and in megacities this value reaches 85–90%, which creates significant problems for the health of the population and the state of the environment as a whole [13].

The current level of environmental safety of the motor transport complex as a whole and its main elements is determined by the level of energy and resource consumption, environmental pollution by emissions and waste during the operation of vehicles and their recovery. At the level of individual motor transport enterprises, the quality and timeliness of maintenance and repair work has a significant impact on the level of environmental safety of vehicles, since technical failures of individual vehicle systems can lead to a significant increase in fuel consumption and emissions of harmful substances. In addition to providing a given level of technical condition of vehicles, the current tasks of the functioning of motor transport enterprises include monitoring of energy and resource consumption, emissions and waste in technological processes of restoring the operability of vehicles to determine the current level of environmental safety of the enterprise and constant implementation of measures aimed at reducing fuel and resource consumption, environmental pollution by emissions and waste.

A great deal of scientific research is devoted to the issue of environmental pollution by road. In particular, the study of G. Pontikakis [17], V. Khrutba [12], P. Levkovets [14] and others is devoted to improving the environmental performance of individual vehicles. In the works of V. Matechyk [15], V. Bilichenko [5], I. Gritsuk [11] the methods of estimation and prediction of the harmful influence of the operation of vehicles and the organization of their maintenance process on the level of environmental pollution are proposed. Issues of increase of efficiency of work of the motor transport enterprise are revealed in the works of I. Fedotova [9], Yu. Petrenko [16], V. Matechyk [10] and others. At the same time, the question of the impact of technological processes of restoring the efficiency of vehicles on the level of environmental safety and management efficiency of motor transport enterprises has not been investigated yet, which determines the need for this study and defines its purpose and objectives.

The aim of the research is study is the regularities of changes in individual, group and integral criteria of ecological safety of motor transport enterprises as an indicator of the effectiveness of management of a motor transport enterprise according to the results of the study of the influence of the fleet structure and technological regimes on energy consumption, emissions and waste in the technological processes of recovery of the vehicle, depending on the proposed options to improve the environmental safety of the enterprise.

2 Generalized Assessment of the Level of Environmental Safety of a Motor Transport Enterprise

Decision-making on the choice of features of the activity of a motor transport enterprise (modes of movement of vehicles in separate technological cycles, type of vehicles, characteristics of technological operations, frequency of main technological impacts during the life cycle) is based on a comparative assessment of indicators of various properties of the monitoring system of the technological process of motor transport enterprises.

To assess the impact of motor transport enterprises on the environment, as a rule, not one, but a whole set of separate criteria $\{K_i\}$ is considered, which form vector criteria $K = \{K_1, K_2, \cdots, K_i\}$. Vector criteria K is reduced to various group and integral criteria. To assess the level of environmental safety of motor transport enterprises, as well as vehicles serviced by them, the most informative criteria are needed, which are able to compare a large number of possible schemes for organizing the maintenance and repair process and options for the fleet of vehicles for various purposes in terms of the level of environmental pollution, waste generation and the cost of operating materials. At the same time, these criteria should be universal for comparing different schemes of the same type and different types of motor transport enterprises, taking into account the operating conditions of the fleet of vehicles, as well as for assessing their compliance with modern environmental safety requirements [13].

For an integrated estimation of influence of separate technological processes of motor transport enterprises and structure of the fleet of vehicles on the environmental safety level of these enterprises, and also the choice of effective ways of increasing a preliminary analysis of environmental aspects of selected technological processes of technical maintenance and repair of vehicles. This approach is acceptable for choosing a particular option of the organization of the production process of an enterprise or the composition of the fleet of vehicles in certain conditions, each of which is characterized by a certain set of values of indicators.

The approach is based on the tasks of determining the generalized management function of motor transport enterprises and the impact on the environment. The tasks of defining this function, investigating and assessing its extremes are divided into separate parts, for which partial solutions are determined, and then the results obtained are synthesized. It is difficult to use analytical methods to identify the value structure of individual indicators. These values can be considered together with the impact on the environment in an informal procedure that does not have the logical and theoretical basis inherent in the method of decision analysis.

At the first approximation, the level of ecological safety of motor transport enterprises in comparison with the base object can be evaluated on the totality of indicators $\overline{\chi} = \|\chi_1, \chi_2, \cdots, \chi_n\|$, $n = 5 \ldots 50$ and as the result of a pairwise comparison of the values of the meters of single indicators of basic (B) and the object that is considered, (A) in the form:

$$\begin{cases} \overline{\chi_A} > \overline{\chi_B} \\ \overline{\chi_A} = \overline{\chi_B} \\ \overline{\chi_A} < \overline{\chi_B} \end{cases} \tag{1}$$

where $\overline{\chi_A} \leq \overline{\chi_0}$ and $\overline{\chi_0}$ is the constraint vector.

It is assumed that all values of the considered single indicator meters are bounded at the top and bottom:

$$[\overline{\chi_i^{\partial}}] < \overline{\chi_i} \leq \overline{\chi_0}, \quad i = \overline{1,\ n} \tag{2}$$

Here, $[\overline{\chi_i^{\partial}}]$ is the vector of the values of single indicator meters given in the normative and technical documentation.

The upper limit is related to both technical capabilities and economic feasibility. The lower limit is associated with the fact that the indicators of the object considered must meet the current standards and requirements. However, when simultaneously comparing the values of meters more than 10–12 indicators, the level of environmental safety of motor transport enterprises is usually difficult to characterize in the above expression as exceeding the basic or inferior to the basic. Therefore, to describe the advantages, it is necessary to use a target function $\varphi\left[\vartheta\left(A_i\right)\right]$ that defines the order of granting an advantage and characterizes the desired integral criterion [18]. This function provides a χ real number corresponding to $\vartheta\left(\chi\right)$ each value, and χ preferably χ' only when $\vartheta\left(\chi\right) > \vartheta\left(\chi'\right)$, and χ equivalent χ' only in the case when $\vartheta\left(\chi\right) = \vartheta\left(\chi'\right)$.

If the function is adaptive, then for $i = 1,\ \ldots,\ n$ $(n > 3)$ single indicators, the target function for a single indicator type can be represented as:

$$\vartheta\left(\chi_1,\ \ldots,\ \chi_n\right) = \sum_{i=1}^{n} \gamma_i \cdot \vartheta_i\left(\chi\right) \tag{3}$$

where ϑ_i is the partial function of the interaction, expressed by the values in a dimensionless form; γ_i is the weight coefficients that characterize the value relations between the meters and satisfy the condition:

$$\sum_{i=1}^{n} \gamma_i = 1 \tag{4}$$

Depending on the direction of the priority vector, when using the appropriate scheme for organizing the production process of motor transport enterprises or the composition of the fleet, the target function of j of the enterprise will be considered more preferable if:

$$\varphi_j = \varphi\left[\vartheta\left(A_j\right)\right] = \left[\sum_{i=1}^{n} \vartheta_i\left(\chi_{ij}\right)\right]_{j \in J} \to \min \tag{5}$$

where j is a variant of the scheme of organization of the production process of the enterprise or the composition of the fleet; n is the number of the meter.

The problem of identifying a function φ_j and quantifying its values is divided into separate parts, for which partial solutions are found, which are then synthesized, indexed, and analyzed. The most common method is expert assessments, according to which each expert performs ordering of particular goals or meters in accordance with the established main goal. The first important meter of a single indicator gets the appropriate rank, r_1 and so on. Each meter in the ranked series is assigned a quantitative score a_i (points). Therefore, the assessment of the level of environmental safety will be based on the following methodological principles [13]:

1. Determining the number of experts who will participate in the assessment of coefficients, which is embodied in the formula:

$$N = (\frac{3}{\xi} + 5) \cdot 0,5 \tag{6}$$

   where N is the minimum number of experts; ξ - the margin of error of the examination results $(0 < b < 1)$.
2. Using the method of expert assessments, the main criteria for assessing the level of environmental safety of motor transport enterprises are selected.
3. The weight of the coefficients is calculated using the formula:

$$w_i = \frac{\sum_{j=1}^{n} x_{ij}}{\sum_{i=1}^{m} x_{ij} \sum_{j=1}^{n} x_{ij}} i = \overline{1,m}, j = \overline{1,n} \tag{7}$$

   where x_{ij} is the estimate of the factor and obtained from the j-th expert; n is the number of experts; m is the number of coefficients to be assessed.
4. The results of such an assessment should be processed not only quantitatively, but also qualitatively, since they may not coincide. Therefore, we proposed to determine the degree of consistency of expert opinions using the following formulas:

$$W = \frac{12 \cdot S}{n^2(m^3 - m)}, \quad S = \sum_{i=1}^{m} (\sum_{j=1}^{n} x_{ij} - \frac{1}{2} n(m+1))^2 \tag{8}$$

   where W is the concordance coefficient (sufficient for a value of ≥ 0.5); S is the sum of squares of deviation of all ratings of ranks of each coefficient from the average value; n is the number of experts; m is the number of coefficients to be assessed.
5. Coefficients that experts believe are the most important are represented as a matrix a_{ij}, where the coefficients $a_i, i = \overline{1, m}$ are fixed in the rows, and the values of the columns reflect their specific value.
6. Find the adjusted indicators (comparing the available values with the values for the most successful period of activity of road transport companies or

reference values for the industry), which are in the range from zero to one and is calculated using the formula:

$$b_{ik} = \frac{a_{ik}}{\max a_{ik}} \tag{9}$$

7. The components of the integral indicator E_k are found as a weighted average, and the aggregate indicator is found as the arithmetic mean of the indicators:

$$E_k = \sum_{i=1}^{m} w_i \cdot b_{ik} \tag{10}$$

where w_i is the weight of i-th indicator; m is the number of indicators.

3 Determination of the Environmental Aspects of Individual Technological Processes of the Stage of Vehicle Operability Restoration

The system of vehicle maintenance and repair is a complex multi-component structure that must also be considered from the point of view of environmental impact: energy consumption and environmental pollution by harmful emissions and waste, which in turn increases the environmental risk – the probability of deliberate or accidental, gradual and catastrophic anthropogenic changes [5,6,14,17]. The environmental impact of a certain process stage of the recovery vehicles is the use of operating materials, fluids, oil products, natural gas, atmospheric air, which is accompanied by negative processes, namely [7,12,15]: atmospheric pollution; water pollution; land and soil pollution; noise, electromagnetic and vibration effects; release of unpleasant odors into the atmosphere; release of toxic waste; thermal pollution.

The impact of vehicles on the environment is evident:

1. while vehicles are moving;
2. during maintenance;
3. when the infrastructure that provides its operation is functioning.

Sources of emissions of pollutants in the processes of restoring the operability of vehicles at motor transport enterprises are [1,2,7,12,15]: brake pad regeneration area – boring on lathes and processing on a grinding machine of brake linings (these sources are equipped with dust cleaning plants); lathes, milling machines, arc welding, gas cutting, grinding machine is equipped with a closed-cycle aspiration system; maintenance shop, aggregate department, washing of parts and aggregates in the washing machine, in baths where degreasing agents are used; pneumatic equipment repair department - washing of parts in diesel fuel; body section-arc and gas cutting; carpentry area, woodworking machines; forge; plastics area -injection molding of products made of nylon, polystyrene; separation of arc welding in argon medium; coppersmith department- soldering station for units and parts; electric area – soldering with a tin-lead solder in the repair; fuel equipment repair area where the fuel equipment is washed, repaired, tested and adjusted; an area of engine running-in that has an exhaust pipe through which fuel combustion products are released into the atmospheric air when the engine is running; the batteries are charged in the battery compartment; grinding and turning department, where a cooling agent – emulsor is used for metalworking machines during metal working; paint department – paint application and drying, storage of paint materials in the cabinet; welding area – welding and gas cutting; spraying area, the plasma-spraying using welding wire; motor area -grinding machine;department of impregnation of generator windings- impregnation with varnish and drying oven;vulcanization department, where the following technological operations are carried out: production of rubber parts on vulcanization presses, vulcanization of repair sites for chambers, tires, roughing of repair sites on a grinding machine equipped with a dust cleaning plant; maintenance shop – fuel equipment repair; sharpening machines in the technical maintenance areas in the aggregate department; gas station-diesel fuel tanks, vehicle refueling stations. Thus, the cost of resources, emissions of harmful substances and waste generated are the main elements of the harmful impact of motor transport enterprises on the environment. Since the environmental impacts of individual technological processes of the vehicle recovery stage are various and numerous, it is necessary to determine the most significant environmental aspects of individual technological processes of motor transport enterprises. The materiality of certain aspects is assessed in accordance with the methodology set out in section 6 of ISO 14001 [3,4]. The primary ranking of environmental aspects to be included in the "Register of environmental aspects" is based on materiality criteria. Aspects are assessed on a three-point scale from 1 to 3, from less impact to more. The scheme for assessing the environmental impact of motor transport enterprises is shown in Fig. 1.

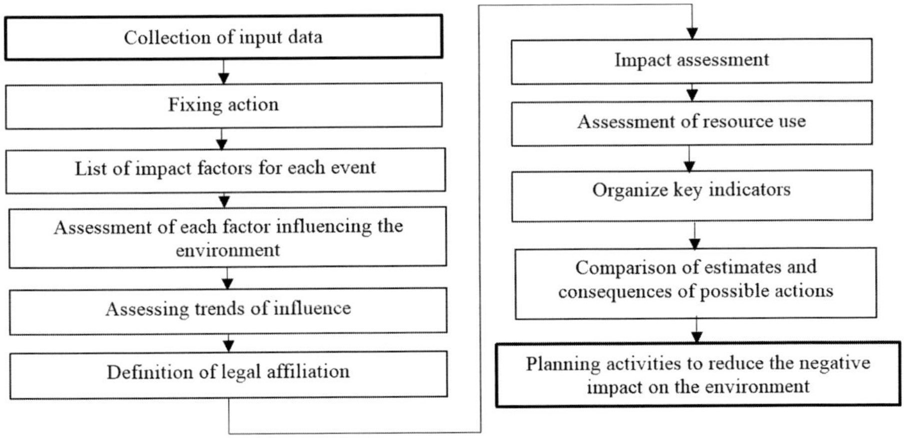

Fig. 1. The scheme for assessing the environmental impact of motor transport enterprises

The assessment of the materiality of aspects is defined as the sum of: the risks of environmental impact (probability, intensity of impact and severity of consequences); the scale of impact; compliance with legal and regulatory documents; public significance and interest of the parties; financial costs for bringing to standards; the ability to apply technology multiplied by the ability to manage the aspect. Thus, those aspects that cause the most problems and that transport companies have the ability to influence will become more important.

All significant aspects of objects, activities, products and services that have a negative impact on the environment or a positive effect are entered in the register of significant environmental aspects of the enterprise. Based on the results of the analysis of the cost aspects of certain types of energy resources, emissions of certain harmful substances and the formation of certain types of waste, 10 most important individual criteria for environmental safety have been identified, which should be grouped into group and integral criteria for environmental safety (Fig. 2).

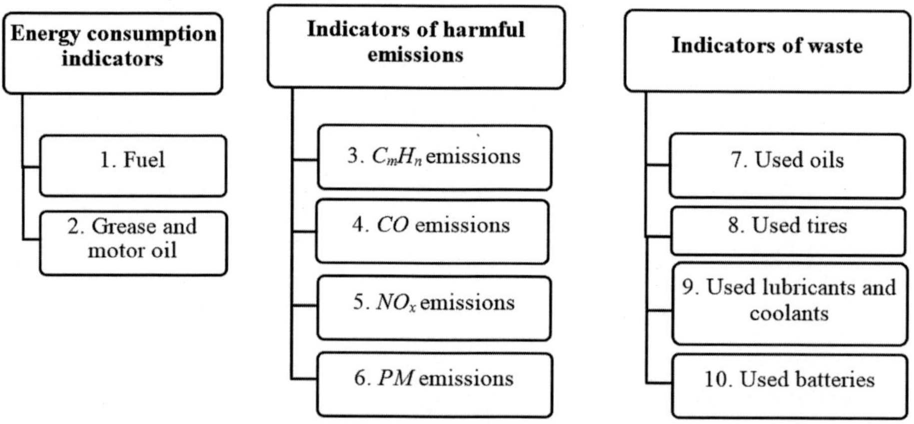

Fig. 2. Criteria for assessing the adverse impact of road transport enterprises

For example, individual criteria for energy costs, emissions of harmful substances, and waste generation are reduced to the corresponding group criteria of K_{BE}, K_{BH}, K_{BB}:

$$K_{BE} = \sum_{i=1}^{3} \beta_i \cdot \frac{E_i}{E_{i\,b}} \tag{11}$$

$$K_{BH} = \sum_{i=1}^{4} \delta_i \cdot \frac{C_i}{C_{i\,b}} \tag{12}$$

$$K_{BB} = \sum_{i=1}^{7} \gamma_i \cdot \frac{B_i}{B_{i\,b}} \tag{13}$$

where $E_i, E_{i\,b}$ are the values of individual i criteria for energy and resource consumption of the studied motor transport enterprise and their desired level; $C_i, C_{i\,b}$ are the value of individual i criteria for emissions of harmful substances of the studied motor transport enterprise and their desired level; B_i, B_{ib} are the value of individual i criteria for waste of the studied motor transport enterprise and their desired level; β_i , δ_i, γ_i are the weight coefficients of individual types of costs, emissions, and waste, respectively ($\sum \beta_i = 1$; $\sum \delta_i = 1$; $\sum \gamma_i = 1$).

Based on the group criteria for energy resources costs K_{BE}, emissions of harmful substances, K_{BH} and waste generation K_{BB}, an integral criterion for environmental safety of motor transport enterprises is formed, which can be presented as follows:

$$K_{EB} = \alpha_1 \cdot K_{BE} + \alpha_2 \cdot K_{BH} + \alpha_3 \cdot K_{BB} \tag{14}$$

where α_1, α_2, α_3 are the weight coefficients of the components of environmental safety indicators of motor transport enterprises ($\sum \alpha_i = 1$).

The values of the weight coefficients of the group and integral criteria for environmental safety of motor transport enterprises were determined by the method of expert assessments, taking into account research and assessments of other authors, technical characteristics of vehicles and the technological process of the enterprise.

By the inverse value of the integral criterion, a one-dimensional target function is formed to assess the level of environmental safety of motor transport enterprises:

$$R_{EB} = \frac{1}{K_{EB}} \Rightarrow 1 \tag{15}$$

On the basis of various options of situation studies and taking into account the excess of sanitary and hygienic standards, the value of the stages of the level of the environmental safety is proposed, which are shown in Table 1, which allow determining the zones of safe operation of the technological process of motor transport enterprises.Taking into account the developed criteria, the model for

Table 1. The degree of environmental safety of the motor transport enterprises

Degrees of ecological safety level	The value of the integral index
Completely safe	(1,00–0,80)
Safe	(0,80–0,63)
Medium safe	(0,63–0,37)
Dangerous	(0,37–0,2)
Particularly dangerous	(0,2–0,00)

assessing the level of environmental safety of motor transport enterprises was expanded. At the stage of calculations for a comprehensive assessment, it is necessary to introduce the definition of the level of environmental safety criteria for motor transport enterprises, depending on the level of energy and resource consumption, the formation of harmful substances and waste.

4 Determination of the Existing Level of Environmental Safety of the Bus Fleet No. 2 "Kyivpastrans"

The assessment of the level of environmental safety of the Bus fleet No. 2 "Kyivpastrans" will be based on the method of expert assessments [13].

Each of the experts assessed the significance of each environmental safety criterion, among those listed in the previous section, in order. The estimates of 18 experts are taken into account. The most significant criterion is assigned a rank of 10. Then put the ranks in order. If the expert concludes that the weight of some criteria is the same, it gives them the same rank. However, the amount of ranks assigned by experts is different. The results of expert assessment of individual criteria for energy costs (E1, E2), emissions (C3, C4, C5, C6) and waste (B7, B8, B9, B10) are shown in Table 2.

Table 2. Results of expert assessment of individual criteria for environmental safety of motor transport enterprises

Expert number	E1	E2	C3	C4	C5	C6	B7	B8	B9	B10
1	9	6	5	7	3	4	2	10	1	8
2	8	6	2	7	1	5	4	10	3	9
3	9	5	6	10	4	1	3	8	2	7
4	9	6	2	7	3	4	5	10	1	8
5	10	6	5	7	4	1	3	9	2	8
6	9	10	4	6	3	5	1	8	2	7
7	8	6	5	9	2	1	4	10	3	7
8	10	5	1	7	4	3	6	9	4	8
9	7	9	5	10	4	1	3	8	2	6
10	8	7	1	6	5	5	2	9	3	10
11	10	6	4	9	4	3	2	7	1	8
12	9	5	6	7	4	2	3	8	1	10
13	8	5	5	10	1	3	1	7	2	9
14	9	4	4	10	5	6	2	8	3	7
15	10	5	6	7	4	2	3	9	1	8
16	10	4	3	8	5	6	1	9	2	7
17	8	4	6	7	5	2	3	9	1	10
18	8	6	5	10	4	3	1	7	2	9
Total points	159	107	75	144	62	57	49	155	36	146
Rank rate	1	5	6	4	7	8	9	2	10	3
Coefficient weight	0,1606	0,1080	0,0757	0,1454	0,0626	0,0575	0,0494	0,1565	0,0363	0,1474

Using the methodology for assessing the level of environmental safety of motor transport enterprises given in Sect. 2 and the statistical information of the Bus fleet No. 2 CE "Kyivpastrans" given above, the values of individual criteria for environmental safety of the Bus fleet No. 2 CE "Kyivpastrans" for 2014–2017 were calculated. (Table 3, Fig. 3).

Table 3. The value of individual and integral criteria of environmental safety for the BF No. 2 CE "Kyivpastrans" for 2014–2017

Coefficient	2014	2015	2016	2017
Indicators of energy consumption				
1. Fuel consumption	0,49	0,49	0,65	0,67
2. Consumption of motor oil and grease	0,51	0,5	0,68	0,68
Indicators of emissions of harmful substances				
3. CmHn emissions	0,39	0,39	0,45	0,46
4. CO emissions	0,4	0,4	0,49	0,51
5. NOx emissions	0,33	0,33	0,42	0,44
6. PM emissions	0,31	0,31	0,39	0,41
Indicators of waste generation				
7. Used oils	0,4	0,4	0,45	0,46
8. Used tires	0,55	0,55	0,57	0,58
9. Used lubricants and coolants	0,48	0,48	0,52	0,53
10. Used batteries	0,42	0,42	0,44	0,45
The integral indicator	0,445	0,444	0,527	0,541

Integral criterion of environmental safety calculated taking into account the weight coefficients determined on the basis of expert assessment in Table 2.

Based on the analysis of the data in Fig. 3 and Table 3, it can be concluded that at present the level of environmental safety of the Bus fleet No. 2 CE "Kyivpastrans" corresponds to the level of average safety (in accordance with Table 1), even taking into account the fact that in 2016 about 20 EURO-4 class buses were purchased.

5 Forecasting Changes in the Level of Environmental Safety of the Bus Fleet No. 2 CE "Kyivpastrans"

Based on the analysis of the current state of environmental impact from the production activities of the Bus fleet No. 2 of CE "Kyivpastrans" and research on resource and energy saving and reducing the negative impact on the environment during the operation of vehicles [8,10,11,16], in order to increase the level of environmental safety, four options for changing the basic characteristics of the enterprise's activities are proposed:

- 1-basic option without any changes (for comparison);

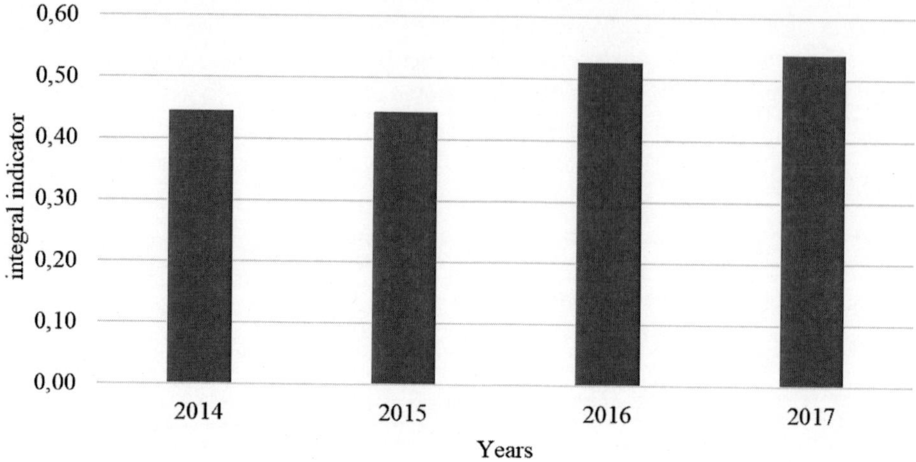

Fig. 3. Changes over the years in the integrated indicator of environmental safety of the BF No. 2

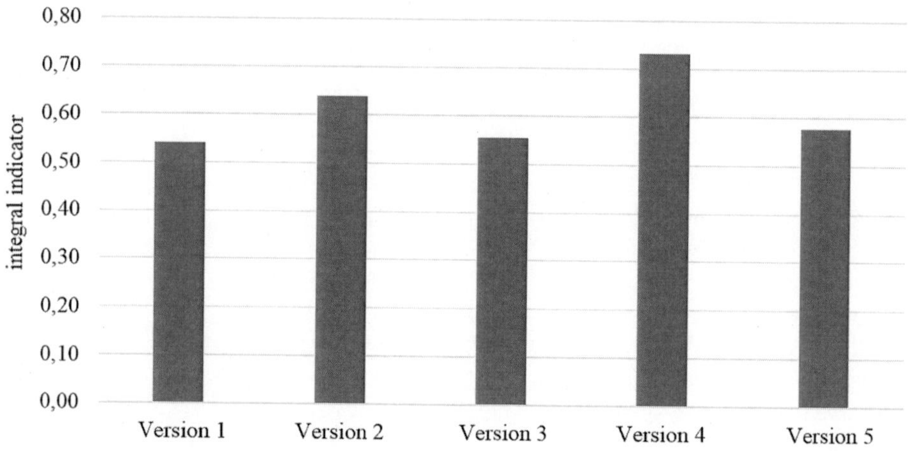

Fig. 4. Comparative assessment of the level of environmental safety in various options of the organization of production activities

- 2-using a vehicle engine thermal preparation system and a catalytic converter;
- 3-increasing the speed of technological traffic on the territory of the Bus fleet No. 2 CE "Kyivpastrans" without changing the flow patterns of technological flows;
- 4-replacing the entire fleet of the BF No2 with EURO-5 ecological class vehicles while maintaining passenger capacity and the number of vehicles;
- 5-improving the patterns of movement of technological flows of vehicles on the territory of the Bus Fleet No. 2 CE "Kyivpastrans" (Fig. 4, Table 4).

Table 4. The significance of individual and integral criteria for environmental safety of the Bus fleet No.2 "Kyivpastrans" in various options of industrial activity organization

Coefficient	Version 1 (basic version)	Version 2	Version 3	Version 4	Version 5
Indicators of energy consumption					
1. Fuel consumption	0,67	0,8	0,68	0,85	0,73
2. Consumption of motor oil, grease	0,68	0,7	0,69	0,73	0,72
Indicators of emissions of harmful substances					
3. CmHn emissions	0,46	0,60	0,48	0,74	0,51
4. CO emissions	0,51	0,64	0,52	0,73	0,54
5. NOx emissions	0,44	0,59	0,46	0,77	0,48
6. PM emissions	0,41	0,53	0,43	0,79	0,49
Indicators of waste generation					
7. Used oils	0,46	0,59	0,48	0,61	0,52
8. Used tires	0,58	0,64	0,59	0,69	0,51
9. Used lubricants and coolants	0,53	0,65	0,56	0,66	0,58
10. Used batteries	0,45	0,52	0,46	0,67	0,55
The integral indicator	0,541	0,640	0,554	0,732	0,578

An option of using the thermal preparation system for the vehicle engine and the catalytic converter is described in theses [10,11]. From the obtained results of assessing the impact of the developed heating system to the fuel consumption and harmful emissions of the vehicle during its movement in the riding cycle shows that it is the warm-up phase of engine and catalytic converter in the initial phase of movement of the vehicle in riding a cycle in which there is the greatest intensity of the fuel consumption and emissions of harmful substances causes a significant impact on these figures. Thus, the use of the heating system in idle mode can reduce fuel consumption by 32.7%, carbon monoxide emissions by 59.6%, hydrocarbon emissions by 56.1%, nitrogen oxide emissions by 93.5%, and reduce the overall engine heating time by 17.8% [10]. In a heating mode, fuel consumption is reduced by 12.4%, carbon monoxide emissions by 56.3%, hydrocarbon emissions by 37.8%, and nitrogen oxide emissions by 22.1%. In other words, the use of a catalytic converter heating system reduces the total emissions of nitrogen oxides during heating in motion, which increases when using additional engine heating only [11].

The option to increase the rate of technological movement on the territory of the Bus fleet No. 2 CE "Kyivpastrans" without changing the traffic pattern is associated with the specifics of the operation of the enterprise. Today, a speed limit of 5 km/h is set for traffic on the territory of the enterprise. After analyzing the area of the enterprise, the layout of the main production sites, the routes of technological traffic, as well as the number of vehicles that are serviced by motor transport enterprises, it turned out that there are all opportunities to increase the speed in technological traffic to 15 km/h. This, in turn, will allow buses to move through the territory of the enterprise not in the first, but in the second gear, which will reduce fuel consumption and pollutants emissions in the technological cycle of vehicle maintenance [9].

The option of replacing the entire fleet of the Bus fleet No. 2 "Kyivpastrans" with EURO-5 ecological class vehicles while maintaining passenger capacity and the number of vehicles is based on the research of many scientists and development strategies of both individual motor transport enterprises and the transport and road complex of Ukraine as a whole [16]. Today, the company operates 40 EURO-2 class buses, 168 EURO-3 class buses and 35 EURO-4 class buses. The option of improving the flow pattern of technological flows of vehicles on the territory of the BF No. 2 is considered in the thesis, which justifies the option of eliminating the centralized parking space for buses of motor transport enterprise, creating a reserved area on this site, and organizing parking places directly near the exits and arrivals in the areas of daily maintenance, technical maintenance, and routine repairs.

Using the methodology for assessing the level of environmental safety of motor transport enterprises given above and proposed options for improving the level of environmental safety of the Bus fleet No. 2 CE "Kyivpastrans", which were given earlier, the values of individual criteria for environmental safety of the Bus fleet No. 2 CE "Kyivpastrans" were calculated (Table 4). Integral criterion of environmental safety calculated taking into account the weight coefficients determined on the basis of expert assessment in Table 2.

Analysis of the level of ecological safety of the Bus fleet No. 2 CE "Kyivpastrans", according to the integral criterion and its prediction during the implementation of all recommended actions evidence a significant ecological and economic effect and increase of the level of ecological safety of the enterprise due to implementation of the proposed measures (Fig. 4).

Based on the analysis of the data in Table 4 and Fig. 4, we can conclude that today the level of environmental safety of the Bus fleet No. 2 of CE "Kyivpastrans" corresponds to the level of average safety (in accordance with Table 1), but due to the implementation of options 2 and 4, the level of environmental safety of the enterprise will increase to a safe level.

6 Conclusions

Thus the complex criteria of an estimation of ecological safety of motor transport enterprises, different from the existing ones with a degree of differentiation of assessment of individual stationary and mobile sources of harmful influence of the enterprise, grouped by the factors of resources, emissions, and waste and integration of all of the harmful factors in a single function target, which uniquely identifies the level of ecological safety of motor transport enterprises was perfected.

The regularities of changes in individual, group and integral criteria of environmental safety of motor transport enterprises are obtained, which differ from the existing ones by taking into account the influence of the structure of the vehicle fleet of parameters and technological processes of the enterprise on the amount of energy and resource consumption, emissions and waste.

A set of criteria for assessing the environmental safety of motor transport enterprises has been developed, which consists of ten separate criteria for assessing stationary and mobile sources of harmful effects of the enterprise, determined by the results of the analysis of significant environmental aspects of individual technological processes of maintenance and repair of vehicles and grouped into factors of resource costs, emissions, and waste, and integral criteria, forms a single function of the goal, which determines the level of environmental safety of motor transport enterprises.

To assess the level of environmental safety of motor transport enterprises, as well as vehicles serviced by them, the most informative criteria are identified, which are able to compare a large number of possible schemes for organizing the maintenance and repair process and options for the fleet of vehicles for various purposes in terms of the level of environmental pollution, waste generation and the cost of operating materials. This should take into account the universality of criteria for comparing different schemes of the same type and different types of motor transport enterprises, taking into account the operating conditions of the fleet of vehicles, as well as for assessing their compliance with modern environmental safety requirements. The study was formed by the method of formation of an integral criterion of ecological safety of motor transport enterprise, which is based on the method of expert estimations of the criteria of harmful emissions, waste, and costs of operational materials.

References

1. On approval of the regulation on the maintenance and repair of road vehicles of road transport, March 1998. https://zakon.rada.gov.ua/laws/show/z0268-98
2. ISO 3929:2003 road vehicles - measurement methods for exhaust gas emissions during inspection or maintenance, September 2003. https://www.iso.org/ru/standard/36351.html
3. ISO 14001:2004 environmental management systems—requirements with guidance for use, November 2004. https://www.iso.org/standard/31807.html
4. ISO 14004:2004 environmental management systems—general guidelines on principles, systems and support techniques, November 2004. https://www.iso.org/standard/31808.html
5. Bilichenko, V., Tsymbal, V.: Research of influence of specialization level of permanent repair posts on efficiency of autoservice enterprises functioning. Bull. Polytech. Inst. Iasi **2**, 45–58 (2014)
6. Borschevsky, O.: Some issues of environmental compliance of vehicles of categories. Sci. Prod. J. **3**, 14–19 (2012)
7. Brannigan, C., Skinner, I., Gibson, G., Kay, D.: Report on the implementation of directive 1999/94/EC relating to the availability of consumer information on fuel economy and CO2 emissions in respect of the marketing of new passenger cars. Technical report, vol. 2 (2011)
8. Delphi: Worldwide 2018–2019 heavy duty and off–highway vehicles. https://www.delphi.com/innovations.emissionstandards
9. Fedotova, I.: Estimation of the level of ecological safety of a motor transport enterprise. Econ. Transp. Complex **29**, 30–40 (2017). https://doi.org/10.4271/2017-26-0364

10. Gritsuk, I., Mateichyk, V., Tsiuman, M.: Reducing harmful emissions of the vehicular engine by rapid after-start heating of the catalytic converter using thermal accumulator. SAE Technical Paper, p. 12 (2018). https://doi.org/10.4271/2018-01-0784
11. Gritsuk, I., Volkov, V., Mateichyk, V.: The evaluation of vehicle fuel consumption and harmful emission using the heating system in a driving cycle. SAE Int. J. Fuels Lubr. **10**, 236–248 (2017). https://doi.org/10.4271/2017-26-0364
12. Khrutba, V., et al.: Peculiarities of implementation of the environmental management system of motor transport enterprises on the urban territories. J. Environ. Manag. Tour. **5**, 1094–1104 (2019). https://doi.org/10.14505//jemt.10.5(37).16
13. Kolomiiets, S.: Methods of integral criterion of ecological safety of motor transport enterprise. Problems of Chemotomology. Theory and Practice of Rational Study of Traditional and Alternative Fuels and Lubricants, Center for Educational Literature (2017)
14. Levkovets, P.: Estimation of the quality of management of efficiency of functioning of motor transport enterprises. Bull. Natl. Transp. Univ. **10**, 204–211 (2005)
15. Mateichyk, V., Gorid'ko, N., Kolomiiets, S.: Developing operating procedures of environmental management system at a motor transport enterprise. Systemy środki tranaportu samochodowego. Efektywność i bezpieczeństwo. Wybrane zagadnienia (11), 57–62 (2017)
16. Petrenko, Y., Shilova, T., Kirichenko, A.: Stages of ecological project on reduction of negative influence of motor transport enterprise. Bull. KhNADU **69**, 91–94 (2015)
17. Pontikakis, G., Stamatelos, A.: Mathematical modelling of catalytic exhaust systems for EURO-3 and EURO-4 emissions standards. Proc. Inst. Mech. Eng. **215**, 1005–1015 (2001). https://doi.org/10.1243/0954407011528572
18. Yalovy, G., Tsesliv, A.: Estimation of the state of integration enterprises. Econ. Bull. Natl. Tech. Univ. Ukraine "Kyiv Polytechnic Institute" (10), 236–240 (2013)

Supporting the Decision-Making About the Possibility of Donation and Transplantation Based on Civil Law Grounds

Tetiana Hovorushchenko[1(✉)], Alla Herts[2], Yelyzaveta Hnatchuk[1], and Oleg Sachenko[3]

[1] Khmelnytskyi National University, Khmelnytskyi, Ukraine
govorushchenko@gmail.com, hnatchuk@gmail.com

[2] Ivan Franko National University of Lviv, Lviv, Ukraine
agerc@gmail.com

[3] Ternopil National Economic University, Ternopil, Ukraine
olsachenko231@gmail.com

Abstract. This study is focused to current task of supporting decision-making concerning the possibility of donation and transplantation based on civil law grounds. Methods of supporting the decision-making about the possibility of donation and transplantation based on civil law grounds have been developed. These methods help to the transplantation clinic staff to correctly form the Register of donors (living donors and posthumous donors) and Lists of persons for transplantation (from the living and posthumous donors or solely posthumous donors), taking into account all current civil law grounds. The developed methods standardize the procedure of including donors in the Register and recipients in the Transplant List, minimizing the impact of subjectivism and the human factor herewith. Thus, the developed methods of supporting the decision-making about the possibility of donation and transplantation minimize the damage that the whole medical transplantation system may suffer and leverage the occurrence for its improvement. The perspective directions of the future authors' work are the design and development of the module (subsystem) of the Unified State Information System for Organ and Tissue Transplantation for supporting the decision-making about the possibility of donation and transplantation.

Keywords: Supporting the decision-making · Production rules · Unified State Information System for Organ and Tissue Transplantation · Donation · Transplantation · Civil law grounds

1 Introduction

Nowadays, transplantation of human organs and (or) tissues is effective and usually the only means of saving lives and restoring health in some cases. It is used

S. Babichev et al. (Eds.): ISDMCI 2020, AISC 1246, pp. 357–376, 2021.
https://doi.org/10.1007/978-3-030-54215-3_23

throughout the civilized world when medical or surgical methods of treatment are not possible, however, the patient has the potential for further healthy activity. In developed countries, organ transplantation has already begun to significantly affect the life expectancy and quality of people life, as successful transplantation can prolong a person's life sometimes by more than 25 years. Today, there are more than 1 million people with transplanted organs in the world that live an active lifestyle and even play sports [13].

According to the Law of Ukraine "On the application of transplantation of anatomical materials to human", transplantation is a special method of treatment that involves the transplant of an organ or other anatomical material (taken from humans or animals) to the recipient. *Transplantation* is the transfer (engraftment) of cells, tissues or organs from a donor to a recipient in order to restore their function(s) in the body [1]. Transplantation as a method of treatment is used solely in the presence of medical indications, as determined by the Concilium of physicians of the relevant health care institution or scientific institution, and if the use of other means and methods for maintaining the life, restoring or maintaining the health does not produce the desired results. *Donation* is the voluntary transfer of blood and other organs to help other people to heal and recover. There are 2 types of donation: after life (with the consent of the person over life or family members after his or her death) and over life (with the consent of the donor).

Every year, almost 2,000 Ukrainians need kidneys transplant, 1,500 - liver transplant, 1,000 - heart transplant, 600 - bone marrow transplant [21]. In developed countries, organ transplantation is the standard of treatment for many diseases of the kidneys, heart, liver, lungs, intestines, etc. Ukraine is lagging behind by 20–25 years in the development of organ transplantation due to lack of funding the development of the transplantation system, the lack of a single register of donors, the ratio of presumption of consent and presumption of disagreement, the prospects of developing a transplant-coordination service, the problems of training and preparing the doctors, the lack of a single electronic system, etc.

The level of organ transplantation determines the overall level of medical development in the country. According to open data from the Ministry of Health of Ukraine, the posthumous organ donation rate in Ukraine is the lowest among European countries and amounts to 0.15 cases per 1 million person [13] (for comparison - Spain has the highest organ donation rate worldwide (39.7 per million persons in a year) [17]).For example, just over one hundred kidney transplants are done in our country during the year, while the total estimated need for such transplantation is 13.5 thousand per year. The average annual number of liver transplants performed in Ukraine does not exceed 10 and has been decreasing in recent years, but about 2.5 thousand patients require such a transplant. From 2001 to 2019, only 5 heart transplants were performed in Ukraine, although the need is around 200 operations annually [13]. According to the public organization "All-Ukrainian Donation Platform "iDonor" [2], more than three thousand patients die in Ukraine every year without a transplant of an organ.

Organ transplantation is a very specific field of medicine, which, in addition to the clear regulation by the regulatory documents, must be provided with a developed and implemented Transplantation System. The Transplantation System, which operates in the country, ensures the clear and coordinated work and interaction of the specialists-transplantologists [13]. According to the Law of Ukraine "On the application of transplantation of anatomical materials to human", transplantation activity should be based solely on the basis of the Unified State Information System for Organ and Tissue Transplantation (USIST). This information system is still in the process of being created. Its test launch will take place in April 2020, and a full launch - from January 1, 2021. The System will contain a number of data, including donor data and patient data. Direct data handling will be semi-automatic. In minds of the authors, **a significant and actual** for USIST may be a successfully developed and implemented module (subsystem) of supporting the decision-making about the possibility of donation and transplantation based on the civil law grounds, which will help to decide about the possibility or impossibility of donation and transplantation in one case or another taking into account civil law grounds.

2 Literature Review

Today, many researchers around the world are paying attention to the problem of developing the information system for organ and tissue transplantation.

Let's consider basic general information technology decisions for medical law in general and for the field of donation and transplantation in particular.

In [8] the application of information technology to in-service education for organ transplantation professionals has been suggested. Authors of [8] concluded, that it is necessary to develop a theoretical construct to create a complete curriculum framework and to establish complete fit-in professional and ethical organ transplantation learning programs based on informatics technology to increase the interdisciplinary exchange of experience.

Authors of [15] established a system for collecting, collating, analyzing, distributing and reacting to serious adverse events and reactions (SAE/R) in unrelated HPC donors. The system has a robust governance structure, formal feedback to the World Marrow Donor Association membership and transparent information flows to other agencies, specialist physicians and transplant programs and the general public.

The Global Database on Donation and Transplantation is one of the main areas of the Global Observatory on Donation and Transplantation (GODT), which is the information platform designed to implement some of the requirements of the World Health Assembly Resolution WHA57.18. The objectives are to collect, analyze, and disseminate information on legislative and organizational frameworks and annual aggregated data on donation and transplantation activities of the World Health Organization (WHO) Member States. Another objective is to establish the transplantation network with the national health authorities. The database provides a working tool that helps to exchange information and be a reference for professionals involved, promoting international cooperation through the website [14].

Authors of [19] proposed the new content and infrastructure for harmonized documentation forms in the domain of organ transplantations enabling worldwide reuse and exchange.

The article [10] presents a medical relational model for the extraction of logical rules from medical law, required to design a medical decision support system that facilitates the process of exchanging data electronically with minimum human intervention. The division of medical law into small concept classes makes it easier to formalize the legal text of medical law into logical rules.

The paper [11] presents a mobile decision support system prototype designed for healthcare providers of an emergency department to identify and follow the correct procedure based on applicable laws to mitigate the chance of incorrect actions being taken, along with the associated risks and preserve both the patients' and their own rights.

Authors of [7] described the process of assessing the sufficiency of information for conducting the surrogate motherhood procedure and developed a conceptual scheme of the intelligent decision support system in the legal regulation of surrogate motherhood.

Let's also consider the basic cases of the implementation and use of information technology to support donation and transplantation through the example of different countries of the world.

So, the authors of [12], using the United Network for Organ Sharing database, showed that implementation of the Kidney Allocation System (KAS) changed how kidneys are allocated and the information on which organ utilization decisions are based. Moreover, after KAS implementation, patient survival for recipients of SCD kidneys was significantly worse.

Paper [20] describes a new prospective-designed version of the Korean Organ Transplantation Registry (KOTRY) - 59 centers are participating in KOTRY, the estimated annual enrollment is more than 2000 cases. The authors of [9] concluded that the processes used to establish KOTRY provide a good model for launching new nationwide transplant cohort studies.

Paper [4] describes how Nagasaki University Hospital established and implemented a Donor Advocacy Team (DAT) program: a risk management program for initiation in the event of serious, persistent, or fatal impairment of an organ, tissue, or cell transplantation from a living donor.

According to the legal framework of Mexico, the National Transplant Center is responsible for coordinating National Subsystem of donation and transplantation, and one of its functions is to integrate and backup information regarding donation and transplantation through the National Transplant Registry System. Distribution and allocation processes are a key element to provide a fair distribution for those patients waiting for that organ; thus the electronic tool is proposed, that aims to support the decision of the donation and/or transplants coordination committee by providing the necessary elements to make this process more efficient [18].

Organ transplantation is vulnerable to serious adverse reactions and events, which require a system for their monitoring and management, as required of EU Member States by Directive 2010/53/EU. A management system was implemented in Poland using modern network technologies through the following steps: the development of a catalogue of events and reactions, the preparation and implementation of the network module, the operational procedures, the evaluation system [3].

The Agence de la Biomedecine (France) maintains a national information system used to Coordinate and administer organ procurement and transplant activities and to perm the evaluation of organ transplantation activities. The paper [16] describes the core uses and functional requirements of Cristal, the donor, recipient and allocation software of the information system and its evolution in the new system.

The Etablissement francais des Greffes (EfG) is a state agency dealing with Public Health issues related to organ, tissue and cell transplantation in France. The evaluation of organ retrieval and transplantation activities, one of its missions, is supported by a national information system (EfG-IS). The EfG-IS is moving towards a new n-tier architecture comprising a terminology server for end-stage diseases, organ failure, dialysis and transplantation (EfG-TS). Paper [9] describes the methodological approach used to build the terminological server on "sound ontological foundations".

3 Problem Statement

For Ukraine, **the actual task** is the development of the Unified State Information System for Organ and Tissue Transplantation. For the successful and full functioning of the USIST, a module (subsystem) for supporting the decision-making about the possibility of donation and transplantation is required, which will provide a conclusion on the possibility or impossibility of donation and transplantation taking into account the civil law grounds. As shown by the review of the literature, neither the known information systems in the field of donation and transplantology (papers [7] - [17]), nor the known systems in the field of medical law (papers [18] - [20]) propose the method or subsystem of supporting the decision-making about the possibility of donation and transplantation based on the civil law grounds.

Given the proven actuality and necessity, **the aim of this research** is the design and development of the module (subsystem) for supporting the decision-making about the possibility of donation and transplantation.

Such a module, like any decision-making support system (DMSS), can be represented as a formal system:

$$R = \langle T, P, M \rangle \tag{1}$$

where R is the set of alternatives that are generated by module; T is the set of the basic elements (knowledge base); P is the set of rules by which alternatives are generated for objects with T; M are methods used in data processing.

The DMSS inputs (set T) are attributes and factors that describe the state of the object and the processes that require decision-making, as well as knowledge of experts of the subject domain (all this information is contained in the section of data of knowledge base). The outputs of the DMSS (set R) are the results of the data analysis, on the basis of which the decisions are generated, as well as the decisions themselves. Then the relationship between the input and output parameters is a mathematical description of DMSS:

$$R = M(T) \tag{2}$$

where M is a method that allows to parameters of T to match an alternative of R.

In order to achieve the aim of the research, **the following tasks** must be solved:

- conducting the analysis of subject domain (attributes, factors, knowledge of experts) to fill the section of data of knowledge base and to form the production rules for the section of rules of the knowledge base;
- developing the production rules (subset PRD of set P) and method (subset Md of set M) of supporting the decision-making about the possibility of donation based on civil law grounds;
- developing the production rules (subset PRT of set P) and method (subset Mt of set M) of supporting the decision-making about the possibility of transplantation based on civil law grounds;
- design of module (subsystem) for supporting the decision-making about the possibility of donation and transplantation of the USIST (as decision-making support system).

4 Civil Law Grounds of the Donation and Transplantology

The first task of the study is to analyze the subject domain for filling the section of data of the knowledge base and for forming the production rules for the section of rules of the knowledge base. So let's conduct review the civil law grounds of the donation and transplantology in Ukraine. The legal facts, that are the basis for the emergence of civil legal relationships for the transplantation, must be satisfied in the aggregate (set) [5].

The emergence of law relationships for the posthumous donation is only possible if the donor dies. According to Art. 52 of the Law of Ukraine "On the application of transplantation of anatomical materials to human", a person is considered to have died since the moment of death of his brain or his biological death.

According to Art. 13 of the Law of Ukraine "On the application of transplantation of anatomical materials to human", transplantation as a method of treatment is used only in the presence of medical indications, if it is impossible to preserve life or restore the health of the patient by other methods of treatment.

Transplantation becomes possible only if the organs or other anatomical materials of the donor and recipient are compatible. The most complex and debatable issue is the consent of the donor (or family members after his or her death) on the removal of organs over life or after death. Civil Code of Ukraine (Part 3 of Article 290), Law of Ukraine "Fundamentals of the legislation of Ukraine on health care" (Part 3 of Article 47), Law of Ukraine "On the application of transplantation of anatomical materials to human" (Part 4 of Article 14 and Part 1 of Article 16) stipulate that any adult legal capable person may give written consent or disagreement to become a donor of anatomical materials. Donation of organs over life is forbidden to pregnant women, foreigners and stateless persons who are illegally staying in Ukraine, sick persons, people being held in prisons, persons who have previously donated an organ or part of it for transplantation. Posthumous donation of organs is forbidden to orphans children, veterans of anti-terrorist operation and/or the operation of the joint forces, as well as to whose personalities of which are not identified. In addition, the organs cannot be used, if law enforcement, the medical examiner or the court oppose it. The contract on transplantation of organs and other anatomical materials and contract on the donation can only be signed if the damage to the donor, which will be caused by the transplant, is less than the one that threatened the recipient. The contract on transplantation of organs and other anatomical materials is legal if there is a written consent of the objectively informed donor and the recipient, as well as parents or other legal representatives of the recipient, if the recipient is a person between the ages of 15 and 18. This condition is not mandatory only in urgent cases where there is a real threat to the life of the recipient. Transplantation of any organs and tissues other than those authorized for transplantation by the Ministry of Health of Ukraine is illegal. Transplantation by health care institutions or by scientific institutions that are not authorized to transplant it is illegal. The removal from a living donor of a different transplant other than one of the paired organs or part of an organ or part of the tissue is illegal.

Transplantation and donation are inextricably interlinked, as transplantation is a method of transplanting a donor's organ or tissue (transplantant) to a recipient. Therefore, a donor is required for transplantation of transplantant to a recipient. So, two persons are always involved in the transplant operation - the donor and the recipient. Therefore, further work will be logically divided into 2 parts - supporting the decision about the possibility of organ donation and supporting the decision about the possibility of transplantation (based on civil law grounds).

5 Supporting the Decision-Making About the Possibility of Donation and Transplantation Based on Civil Law Grounds

Given the above-mentioned civil law grounds of donation and transplantology, we will first develop production rules for supporting the decision-making about the possibility of donation and transplantation based on civil law grounds.

In [6,7], the authors represented methods for supporting the decision-making based on the use of ontologies. The ontology of the subject area for the task of supporting the decision-making about the possibility of donation and transplantation has not been developed, so this approach cannot be used.

Production rules for supporting the decision about the possibility of organ donation based on civil law grounds (subset $PRD = \{prd1, ..., prd23\}$) can be divided into the following sections: $\{prd1, ..., prd5\}$ - rules for determining the possibility of donation after life, $\{prd6, ..., prd9\}$ - rules for determining the donor's brain death, $\{prd10, prd11\}$ - rules for determining the biological death of donor, $\{prd12, ..., prd22\}$ - rules for determining the possibility of donation over life. The production rules $\{prd1, ..., prd22\}$ are based on the verification of the fulfilment/non-fulfilment of each civil law ground separately (or several at the same time - if it's necessary) and, if the appropriate civil law ground is fulfilled, the corresponding counter is incremented - k in the rules for determining the possibility of donation after life, i in the rules for determining the donor's brain death, j in the rules for determining the biological death of the donor, l in the rules for determining the possibility of donation over life. According to the rule $\{prd23\}$, the values of the counters k, i, j, l are analyzed and, depending on their values, a conclusion is made about the possibility or the impossibility of donation by the Person-potential donor and, if donation is possible, the type of donation – over life or after life – is determined.

Production rules for supporting the decision-making about the possibility of organ donation based on civil law grounds (subset $PRD = \{prd1, ..., prd23\}$ of set P):

1. if (there is a consent of an adult legal capable Person-potential donor to remove organs after death) or ((there is no disagreement of an adult legal capable Person-potential donor to become a donor after death) and (there is a consent of family members after the death of an adult legal capable Person-potential donor (spouse, children, parents, siblings) to remove organs after death)) or (there is a consent of legal representatives of a minor Person-potential donor to remove organs after death) or (there is a legal consent of legal representatives of the incapable of the Person-potential donor to remove organs after death), then $k = k + 1$;
2. if the Person-potential donor is not an orphan child, then $k = k + 1$;
3. if the Person-potential donor is not a veteran of anti-terrorist operation and/or the operation of the joint forces, then $k = k + 1$;
4. if Person-potential donor is identified, then $k = k + 1$;
5. if law enforcement, the medical examiner or the court don't oppose the posthumous donation of the Person-potential donor, then $k = k + 1$;
6. if there has been a complete and irreversible cessation of all brain function of the Person-potential donor (with a working heart and artificial lung ventilation), then $k = k + 1$;
7. if the complete and irreversible cessation of all brain function of the Person-potential donor is attested by the Concilium of physicians, then $i = i + 1$;

8. if the physicians, which will be involved in the removal of the anatomical materials and their transplantation were not included in the Concilium of physicians who ascertained the death of the brain of the Person-potential donor, then $i = i + 1$;
9. if the transplant-coordinator of the healthcare facility wasn't included in the Concilium of physicians who ascertained the death of the brain of the Person-potential donor, then $i = i + 1$;
10. if there has been an irreversible cessation of circulation and respiratory function in the Person-potential donor, then $j = j + 1$;
11. if early and/or late cadaveric changes occurred in the Person-potential donor, then $j = j + 1$;
12. if (there is a consent of an adult legal capable Person-potential donor for the removal of his organ (which is allowed for transplantation by the Ministry of Health of Ukraine of one of the paired organs or part of the organ) over life) or (there is a consent of the legal representatives of the minor Person-potential donor for the removal of his organ (which is allowed for transplantation by the Ministry of Health of Ukraine of one of the paired organs or part of the organ) over life) or (there is a consent of legal representatives of the incapable Person-potential donor for removal of its organ (which is allowed for transplantation by the Ministry of Health of Ukraine of one of the paired organs or part of the organ) over life), then $l = l + 1$;
13. if the transplant harm to the Person-potential donor is less than that which threatened the Recipient, then $l = l + 1$;
14. if the Person-potential donor is not a pregnant woman, then $l = l + 1$;
15. if the Person-potential donor is not a foreigner illegally staying on the territory of Ukraine, then $l = l + 1$;
16. if the Person-potential donor is not a stateless person illegally staying in the territory of Ukraine, then $l = l + 1$;
17. if the Person-potential donor does not suffer from severe mental disorders, then $l = l + 1$;
18. if the Person-potential donor does not have any diseases that can be transmitted to the Recipient, then $l = l + 1$;
19. if the Person-potential donor is not held in prison, then $l = l + 1$;
20. if the Person-potential donor has not previously provided the organ or its part for transplantation, then $l = l + 1$;
21. if one of the paired organs or part of the organ or part of the tissue will be removed from the Person-potential donor (for the purpose of further transplantation), then $l = l + 1$;
22. if law enforcement, the medical examiner or the court don't oppose the lifetime donation of the Person-potential donor, then $l = l + 1$;
23. if $k = 5$ and $i = 4$ or $j = 2$, then *Organ donation after life (posthumous) is possible*, else if $l = 11$, then *Organ donation over life is possible*, else *Organ donation is not possible.*

The decision tree about the possibility of organ donation based on civil law grounds is represented in Fig. 1.

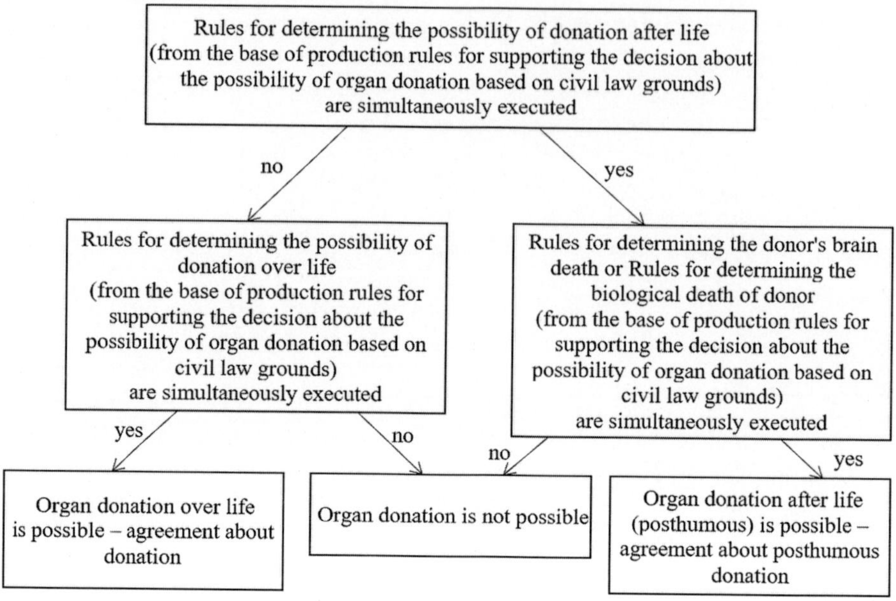

Fig. 1. The decision tree about the possibility of organ donation based on civil law grounds

Method of supporting the decision-making about the possibility of donation based on civil law grounds (subset *Md* of set *M*) consists of the next steps:

1. to set that the counters k, i, j, l are equal 0;
2. by the method of conducting a search in width in the forward direction in the subset {*prd1, ..., prd22*} of the base of production rules for supporting the decision-making about the possibility of organ donation, each rule for the Person-potential donor is verified and the calculation of relevant counters is conducted;
3. according to the rule {*prd23*}, the received values of the counters k, i, j, l are analyzed, and a conclusion about the possibility or the impossibility of donation by the Person-potential donor is made; if the donation is possible, the type of donation - over life or after life - is determined;
4. if the conclusion is made about the possibility of donation after life, then the data of the Person-potential donor shall be entered in the Register of posthumous donors; otherwise, if the conclusion is made about the possibility of donation over life, then the data of the Person-potential donor shall be entered in the Register of living donors; otherwise, if the conclusion is made about donation impossibility, then the data of the Person-potential donor shall not be entered in any of the Donor Registers.

Production rules for supporting the decision about the possibility of transplantation based on civil law grounds {*prt1, ..., prt5*} are based on the verification of the fulfilment/non-fulfilment of each civil law ground separately (or several at the same time - if it's necessary) and, if the appropriate civil law ground is fulfilled, the counter m is incremented. According to the rule $\{prt6\}$, the value of the counter m is analyzed and, depending on its value, a conclusion is made about the possibility or the impossibility of transplantation for the Person-recipient.

Production rules for supporting the decision-making about the possibility of transplantation based on civil law grounds (subset $PRT = \{prt1, ..., prt6\}$ of set P):

1. if the Person-recipient has a disease through which his/her life cannot be saved and/or his/her health cannot be restored by other (non-transplant) treatments, then $m = m + 1$;
2. if the medical indications for transplantation into the Person-recipient were established by Concilium of physicians, then $m = m + 1$;
3. if (there is a written consent of the adult Person-recipient) or ((there is a written consent of the Person-recipient age 15 to 18) and (there is a written consent of the parent or other legal representative of the Person-recipient age 15 to 18)) or (there is a written consent of the parents or other legal representatives of the Person-recipient under the age of 15) or (there is an urgent case with a real threat to the Person-recipient's life), then $m = m + 1$;
4. if the Person-recipient needs an organ or tissue, the transplantation of which is authorized by the Ministry of Health of Ukraine, then $m = m + 1$;
5. if a future transplant will be carried out by a healthcare institution or scientific institution, which is authorized to carry out it, then $m = m + 1$;
6. if $m = 5$, then "Transplantation is possible" else "Transplantation is not possible".

Method of supporting the decision-making about the possibility of transplantation based on civil law grounds (subset Mt of set M) consists of the next steps:

1. to set that the counter m is equal 0;
2. by the method of conducting a search in width in the forward direction in the subset $\{prt1, ..., prt5\}$ of the base of production rules for supporting the decision-making about the possibility of transplantation, each rule for the Person-recipient is verified and the calculation of relevant counters is conducted;
3. according to the rule $\{prt6\}$, the received value of the counter m is analyzed, and a conclusion about the possibility or the impossibility of transplantation for the Person-recipient;
4. if the conclusion is made about the impossibility of transplantation, then the data of the Person-recipient shall not be entered in the List of persons for transplantation, otherwise, if the conclusion is made about the possibility of transplantation, then if the Person-recipient needs a transplant of one of the paired organs or part of the organ or part of the tissue, then the data of the Person-recipient shall be entered in the List of persons for transplantation

from the living donors or the posthumous donors, otherwise, if the Person-recipient needs a transplant of one of the nonpaired organs, then the data of the Person-recipient shall be entered in the List of persons for transplantation from the posthumous donors.

On the basis of two developed methods of supporting the decision-making about the possibility of donation and transplantation based on civil law grounds, the authors will design a module (subsystem) of supporting the decision-making about the possibility of donation and transplantation. Structure of such module is represented in Fig. 2.

Therefore, the data gathering is conducted, the required data (set T) is selected from the data sources, after which the section of data of the knowledge base is filled. Further on, to the data from the set T is applied one of two methods from the set M (the method of supporting the decision-making about the possibility of donation based on civil law grounds or the method of supporting the decision-making about the possibility of transplantation based on civil law grounds), which allows to the parameters of the set T to match the alternative R. In step 1 of the corresponding method, processing (analyzing) the data is performed using the developed production rules (from the set P), which are contained in the section of rules of the knowledge base. In step 2 of the corresponding method, making the decision is performed, in step 3 of the corresponding method, implementing the decision R is performed. In addition, the generated conclusion (decision) R is also made known to the user of the module (subsystem).

The example of the use (implementing the decision) of a future module (subsystem) of supporting the decision-making about the possibility of donation and transplantation in the Unified State Information System for Organ and Tissue Transplantation is represented in Fig. 3.

6 Experiment and Results

Let's look at the functioning of the developed methods of supporting the decision-making about the possibility of donation and transplantation.

For example, let's consider the case 1. A 25-year-old legal capable man (citizen of Ukraine) wants to become a donor for life. He wants to give his kidney to a sister with kidney failure. The man signed consent. The examination of the man and his sister revealed that the woman without a kidney transplant will die soon, and the man has two healthy kidneys, so he can live fully with one kidney. In addition, doctors found that man had no severe mental disorders, had no diseases that could be transmitted to the recipient, had not previously provided any organs or parts of them for transplantation. The lawyers of the clinic also found that he was not sentenced to imprisonment; law enforcement, the medical examiner or the court don't oppose his lifetime donation. The man is not a veteran of anti-terrorist operation and/or the operation of the joint forces.

In accordance with the method of supporting the decision-making about the possibility of donation based on civil law grounds, each rule from the subset

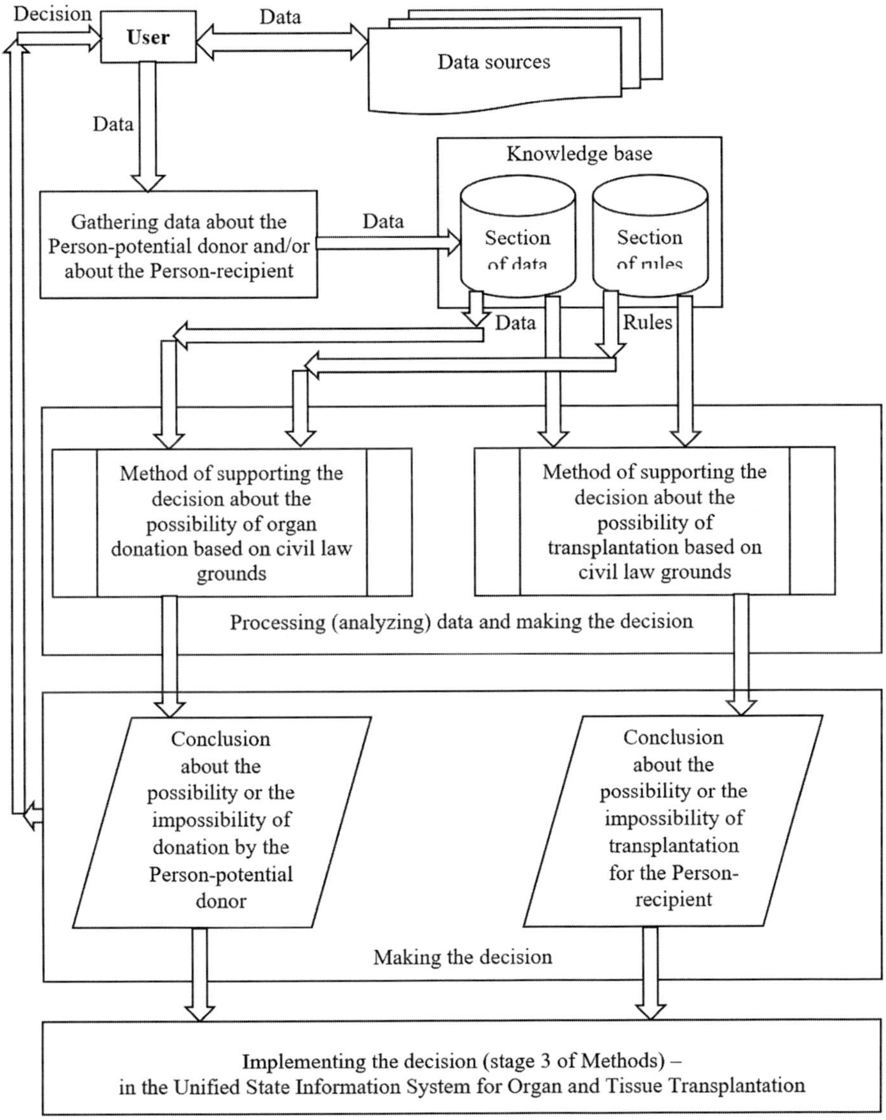

Fig. 2. Structure of module (subsystem) of supporting the decision-making about the possibility of donation and transplantation

{*prd1*, ..., *prd22*} of the production rule's base for supporting the decision-making about the possibility of organ donation was verified. As a result, the following values of counters were received: $k = 4$, $i = 0$, $j = 0$, $l = 11$. According to rule {*prd23*}, an analysis of the received counters was made, and a conclusion was made about the possibility of donation over life by this man, so his data is entered in the Register of living donors.

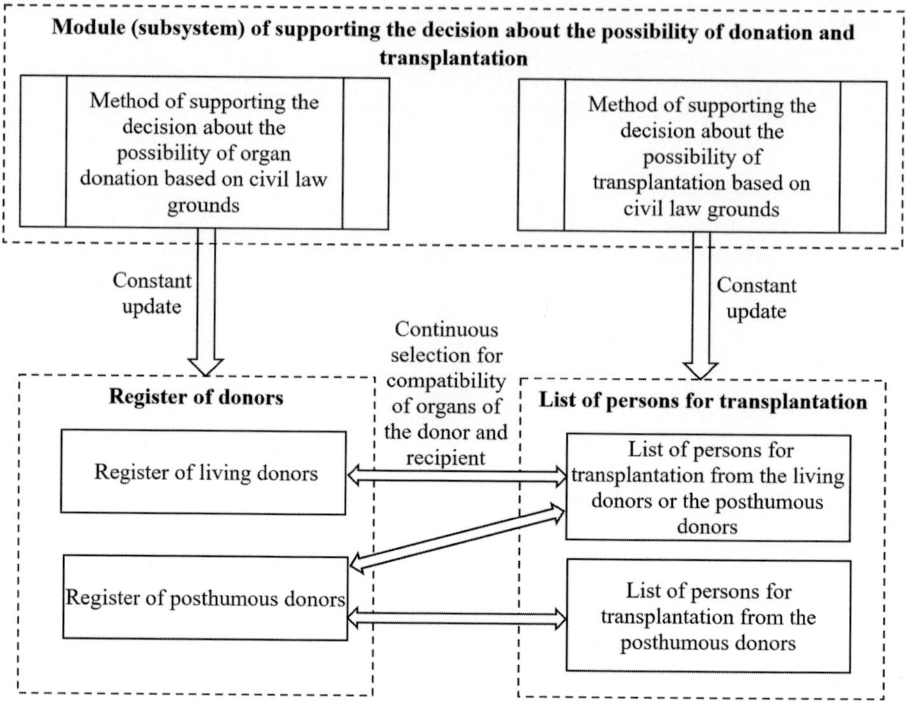

Fig. 3. The example of the use of a future module (subsystem) of supporting the decision-making about the possibility of donation and transplantation in the Unified State Information System for Organ and Tissue Transplantation

This man's 22-year-old sister is ill with kidney failure in an incurable form, i.e. only a kidney transplant can save her life. This fact was attested by the Concilium of physicians. She gave written consent for a kidney transplant. Transplantation of a kidney is authorized by the Ministry of Health of Ukraine. Future kidney transplantation will be performed by a clinic that is eligible for such operations. According to the method of supporting the decision-making about the possibility of transplantation based on civil law grounds, each rule from the subset $\{prt1, \ldots, prt5\}$ of the production rule's base for supporting the decision-making about the possibility of transplantation was verified. As a result, the following counter value was received: $m = 5$. According to rule $\{prt6\}$, a decision was made about the possibility of transplantation for this woman. Since a woman needs a transplant from one of the paired organs, her data is entered in the List of persons for transplantation from the living donors or the posthumous donors.

Since the kidney of the donor is compatible with the recipient, surgery of transplantation of the kidney from the man to his sister was approved.

Let's consider case 2. A 27-year-old man (citizen of Ukraine), whose identity was immediately established under the driver's license, was involved in the accident. As a result of the accident, the patient's brain was severely affected, resulting in the complete and irreversible cessation of all brain functions (with a working heart and artificial lung ventilation). Brain death was recorded by the Concilium of physicians, which did not include transplantologists and transplant-coordinator of the clinic. The clinic staff wanted to enter the organs of this man, which were not injured in the accident, in the Register of posthumous donors. His disagreement to become a donor after his death was not recorded. The wife was ready to write consent for the removal of the organs from her deceased husband. The lawyers of the clinic also conducted a check on the man and found that law enforcement, the medical examiner or the court don't oppose his donation after life. The man is the veteran of the anti-terrorist operation in Ukraine. In accordance with the method of supporting the decision-making about the possibility of donation based on civil law grounds, each rule from the subset $\{prd1, ..., prd22\}$ of the production rule's base for supporting the decision-making about the possibility of organ donation was verified. As a result, the following values of counters were received: $k = 4,\ i = 4,\ j = 1,\ l = 9$. According to rule $\{prd23\}$, an analysis of the received counters was made, and a conclusion was made about the impossibility of organ donation after life by this man, so his data cannot be entered in the Register of posthumous donors.

Let's consider case 3. A 16-year-old teenager has kidney disease. He is undergoing hemodialysis procedures. His health can only be restored by kidney transplantation. This fact was attested by the Concilium of physicians. The parents of the teenager provided written consent for a kidney transplant for their son. However, the teenager categorically objects to the operation, so he does not sign a written consent for the transplant. His health current state is such that there is no real threat to life. Transplantation of a kidney is authorized by the Ministry of Health of Ukraine. Future kidney transplantation will be performed by a clinic that is eligible for such operations. According to the method of supporting the decision-making about the possibility of transplantation based on civil law grounds, each rule from the subset $\{prt1, ..., prt5\}$ of the production rule's base for supporting the decision-making about the possibility of transplantation was verified. As a result, the following counter value was received: $m = 4$. According to rule $\{prt6\}$, a decision was made about the impossibility of transplantation for this boy, so his data cannot be entered in the List of persons for transplantation.

Thus, using the developed production rules and methods of supporting the decision-making about the possibility of donation and transplantation based on civil law grounds in the 3 above described cases showed that in one case the decision was made as to the possibility of both donation and transplantation, and therefore kidney transplant surgery was allowed, and in two other cases, it was decided that donation and transplantation would be impossible. These decisions were made not only on medical grounds but also on civil law grounds.

So, the developed methods of supporting the decision-making about the possibility of donation and transplantation based on civil law grounds provided the opportunity to perform a quick check of all civil law grounds for entering the person in the Register of donors or List of persons for transplantation, respectively. Such a check allows making unmistakable (in terms of civil law grounds) decisions about donation and transplantation. Given the gravity of such decisions, it's necessary to minimize the impact of subjectivism and the human factor during their making that is also provided by the developed methods.

The authors verified the correctness of the decision on the possibility of donation and the possibility of transplantation of 100 performed organ or tissue transplantations in Ukraine in 2010–2019. The results of such verification are represented in Fig. 4.

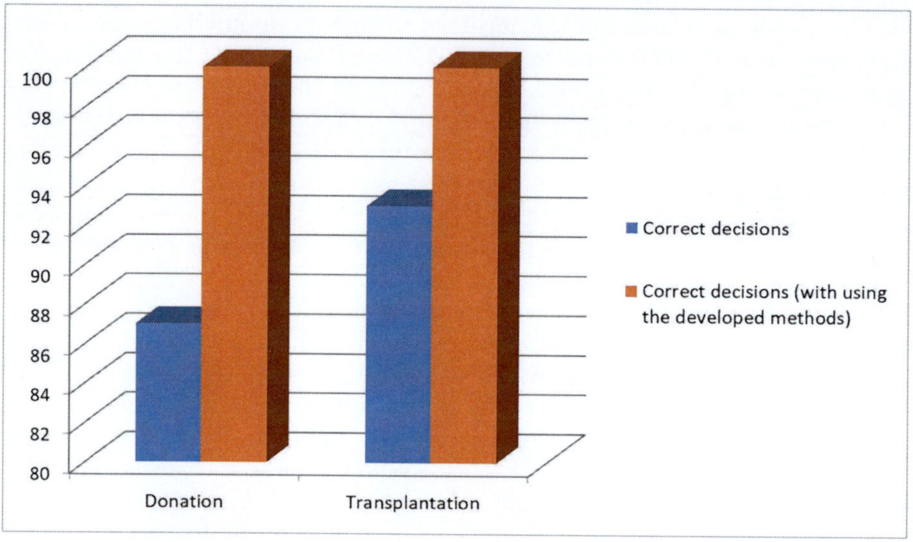

Fig. 4. Correctness of the decision on the possibility of donation and transplantation for 100 performed organ or tissue transplantation in Ukraine and the world in 2019

Obviously, only 87 decisions on donation and only 93 on transplantation were legally correct given the civil law grounds, which are represented in Sect. 4 of this paper. Using the developed rules and methods of supporting the decision-making about the possibility of donation and transplantation based on civil law grounds gives 100-percent correctness of the made decisions in terms of the civil law grounds (all 100 decisions would be correct if the developed rules and methods of supporting the decision-making about the possibility of donation and transplantation would be applied).

In addition, the authors have analyzed the materials of 40 court cases initiated against medical clinics that used donation and transplantation, in 30 of which the court ruled that the donation and/or transplantation was unlawful. An analysis of the data on these cases by the developed methods of supporting the decision-making about the possibility of donation and transplantation based on civil law grounds showed that if the developed methods will be used before the operations, all correct and lawful decisions would be made - Fig. 5.

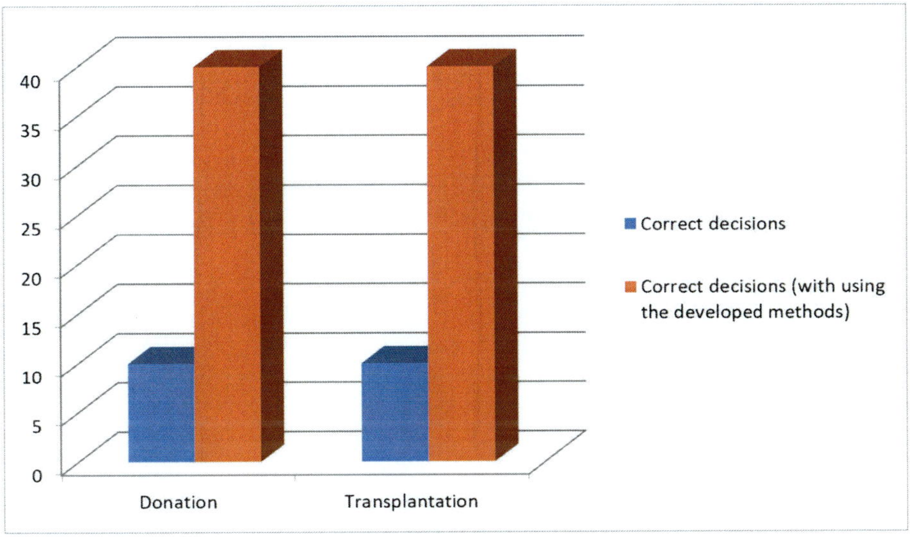

Fig. 5. Correctness of the decision on the possibility of donation and transplantation in 40 court cases initiated against medical clinics that used donation and performed transplantation

The described results proved the workability, effectiveness and relevance of the developed methods of supporting the decision-making about the possibility of donation and transplantation based on civil law grounds.

7 Discussion

Among the reasons holding back the development of human anatomical material transplantation in Ukraine is the unpreparedness of the Unified State Information System for Organ and Tissue Transplantation (USIST), and there is also a lack of a method or subsystem for supporting the decision-making about the possibility of donation and transplantation based on civil law grounds, which is why clinics do not always make correct decisions about the possibility/impossibility of donation and/or transplantation. According to the authors, the functioning of the USIST will not be complete without a successfully developed and implemented module (subsystem) for supporting the decision-making about the possibility of donation and transplantation, which will help to decide on the possibility or impossibility of donation and transplantation in the certain case, taking into account existing and current civil law grounds.

The performed experiments and the obtained results show that the developed production rules and methods of supporting the decision-making about the possibility of donation and transplantation based on civil law grounds help to avoid incorrect (unlawful) decisions regarding the possibility/impossibility of donation and/or transplantation. The use of the developed rules and methods increased the level of correctness of the made decisions on donation by 13% (from 87 to 100 correct decisions) and on transplantation by 7% (from 93 to 100 correct decisions). Therefore, the module (subsystem) for supporting the decision-making about the possibility of donation and transplantation of USIST-system will allow: rapid and automatic verification of all civil law grounds for entry of the person's data in the Register of donors or Lists of persons for transplantation, respectively; making unmistakable (in terms of civil law grounds) decisions on donation and transplantation; minimizing the impact of subjectivism and the human factor in making decisions about donation and transplantation. Therefore, the important and urgent task now is to implement the designed module (subsystem) for supporting the decision-making about the possibility of donation and transplantation.

8 Conclusions

In this paper, the actuality and importance of the process of supporting the decision-making about the possibility of donation and transplantation based on the civil law grounds are proved. This realized process will help to decide about the possibility or impossibility of donation and transplantation, in any case, taking into account all existing civil law grounds.

A review of the literature was conducted. It proved that neither the known information systems in the field of donation and transplantology, nor the known systems in the field of medical law propose the method or subsystem of supporting the decision-making about the possibility of donation and transplantation based on the civil law grounds. The review of the civil law grounds of donation and transplantology in Ukraine was conducted.

Methods of supporting the decision-making about the possibility of donation and transplantation based on civil law grounds have been first time developed. These methods help to the transplantation clinic staff to correctly form the Register of donors (living donors and posthumous donors) and Lists of persons for transplantation (from the living and posthumous donors or solely posthumous donors), taking into account all current civil law grounds. The developed methods standardize the procedure of including donors in the Register and recipients in the Transplant List, minimizing the impact of subjectivism and the human factor herewith. Thus, the developed methods of supporting the decision-making about the possibility of donation and transplantation minimize the damage that the whole medical transplantation system may suffer and leverage the occurrence for its improvement.

Experiments have shown that the developed methods are workable and effective in determining the possibility of donation and transplantation based on all currently existing civil law grounds in the complex.

The perspective direction of the future authors' work is the development of the module (subsystem) of the Unified State Information System for Organ and Tissue Transplantation for supporting the decision-making about the possibility of donation and transplantation - on the basis of the structure of such the module (subsystem), which is developed in this paper.

References

1. World Health Organization: Transplantation. https://www.who.int/topics/transplantation/en
2. Non-Government Organization "All-Ukrainian Donor Platform "iDonor" (2020). https://idonor.org.ua/
3. Czerwinski, J., Kalicinski, P., Danielewicz, R.: Serious adverse events and reactions in organ donation and transplantation: a web-net tool-based nationwide system for reporting and monitoring. Ann. Transplant. **20**, 243–248 (2015). https://doi.org/10.12659/AOT.893009
4. Eguchi, S., Soyama, A., Nagai, K., Miyazaki, Y., Kurihara, S., Hidaka, M., Ono, S., Adachi, T., Natsuda, K., Hara, T.: The donor advocacy team: a risk management program for living organ, tissue, and cell transplant donors. Surg. Today **47**(8), 980–985 (2017). https://doi.org/10.1007/s00595-017-1468-z
5. Herts, A.: The peculiarities of civil-legal regulation of transplantation in Ukraine and Europe. Baltic J. Eur. Stud. **8**(1), 33–48 (2018). https://doi.org/10.1515/bjes-2018-0003
6. Hovorushchenko, T.: Information technology for assurance of veracity of quality information in the software requirements specification. Adv. Intell. Syst. Comput. **689**, 166–185 (2018). https://doi.org/10.1007/978-3-319-70581-1_12
7. Hovorushchenko, T., Herts, A., Hnatchuk, Y.: Concept of intelligent decision support system in the legal regulation of the surrogate motherhood. CEUR-WS **2488**, 57–68 (2019)
8. Hsieh, M., Shih, F., Sheu, S., Wang, S., Shih, F.: Using an informatics education strategy to resolve the dilemma of teaching transplantation in medical institutions: multidisciplinary medical team perspectives. Medicine **97**(43), e12809 (2018). https://doi.org/10.1097/MD.0000000000012809
9. Jacquelinet, C., Burgun, A., Delamarre, D., Strang, N., Djabbour, S., Boutin, B., Le Beux, P.: Developing the ontological foundations of a terminological system for end-stage diseases, organ failure, dialysis and transplantation. Int. J. Med. Inform. **70**(2), 317–328 (2003). https://doi.org/10.1016/s1386-5056(03)00046-7
10. Khan, I., Sher, M., Khan, J., Saqlain, S., Ghani, A., Naqvi, H., Ashraf, M.: Conversion of legal text to a logical rules set from medical law using the medical relational model and the world rule model for a medical decision support system. Inform.-Basel **3**(1), 2 (2016). https://doi.org/10.3390/informatics3010002
11. Khodambashi, S., Gulla, J.A., Abrahamsson, P., Moser, F.: Design and development of a mobile decision support system: guiding clinicians regarding law in the practice of psychiatry in emergency department. In: 2017 IEEE 30th International Symposium on Computer-Based Medical Systems (CBMS), pp. 67–72. IEEE (2017). https://doi.org/10.1109/CBMS.2017.77
12. King, C., Perkins, J., Johnson, C., Blosser, C., Leca, N., Sibulesky, L.: Utilization of standard criteria donor and expanded criteria donor kidneys after kidney allocation system implementation. Ann. Transplant. **23**, 691–703 (2018). https://doi.org/10.12659/AOT.91050
13. Koval, D., Rudenko, K.: Translation of bodies = saving human lives (2019). https://www.civic-synergy.org.ua/wp-content/uploads/2018/04/Transplantatsiya-organiv-poryatunok-lyudskyh-zhyttiv_rol-transplant-koordynatora.pdf
14. Matesanz, R., Mahillo, B., Alvarez, M., Carmona, M.: Global observatory and database on donation and transplantation: world overview on transplantation activities. Transplant. Proc. **41**(6), 2297–2301 (2009). https://doi.org/10.1016/j.transproceed.2009.05.004

15. Shaw, B., Chapman, J., Fechter, M., Foeken, L., Greinix, H., Hwan, W., Philips-Johnson, L., Korhonen, M., Lindberg, B., Navarro, W.: Towards a global system of vigilance and surveillance in unrelated donors of haematopoietic progenitor cells for transplantation. Bone Marrow Transplant. **48**(12), 1506–1509 (2013). https://doi.org/10.1038/bmt.2013.104
16. Strang, W., Tuppin, P., Atinault, A., Jacquelinet, C.: The French organ transplant data system. Connect. Med. Inform. Bio-Inform. **116**, 77–82 (2005)
17. Tackmann, E., Dettmer, S.: Measures influencing post-mortem organ donation rates in Germany, The Netherlands, Spain and the UK: a systematic review. Anaesthesist **68**(6), 377–383 (2019). https://doi.org/10.1007/s00101-019-0600-4
18. Tenorio, S., Daza, L., Estrada, S.: Electronic tool for distribution and allocation of heart on donation and transplantation in Mexico. Transplant. Proc. **48**(2), 564–567 (2016). https://doi.org/10.1016/j.transproceed.2016.03.018
19. Varghese, J., Schulze Suenninghausen, S., Dugas, M.: Standardized quality assurance forms for organ transplantations with multilingual support, open access and UMLS coding. Stud. Health Technol. Inform. **212**, 15–22 (2015)
20. Yang, J., Jeong, J., Lee, J., Kim, Y., Paik, H., Kim, J., Park, H., Kim, M.: Design and methods of the Korean organ transplantation registry. Transplant. Direct **3**(8), e191 (2017). https://doi.org/10.1097/TXD.0000000000000678
21. Yurchenko, N.: Organ transplants are thousands of saved lives (2016). http://www.golos.com.ua/article/278750

Application of Game Theory for Decisions Making on the Development of IT Products

Viktor Morozov(✉), Olga Mezentseva, and Maksym Proskurin

Taras Shevchenko National University of Kyiv, Kyiv, Ukraine
knumvv@gmail.com, olga.mezentseva.fit@gmail.com, proskuryn69@gmail.com

Abstract. The article presents the results of research related to the problems of development of start-up projects in the IT-industry. Such projects are based on innovation and need to be formalised as independent enterprises for their implementation. Under market conditions, such enterprises are forced to develop modern IT products to support small and medium-sized businesses based on the interaction with numerous potential customers, who subsequently become regular ones. At the same time, the type of SaaS company is determined to a large extent by its "average client". This is the topic used in many empirical rules of SaaS metrics. We proposed a hypothesis about the possibility of applying game theory to improve the efficiency of customer interaction with SaaS companies, based on a model that takes into account the mutual influence and interaction of the business owner, developers and customers. The development of innovative IT products should be carried out with the fullest satisfaction of the interests and needs of all key stakeholders. In this article, we have proposed a mathematical description of the model and method of modelling these interactions using game theory. It allows building corresponding formal models. For modelling and testing of the offered models, the heuristic algorithm with optimisation function is offered. For carrying out of chains training, the Python functionality with users activity datasets processing is used. An analysis was made, and conclusions were reached on the effectiveness of the proposed approach.

Keywords: Projects · Innovation · IT product · Information interaction · Start-up · Decision making · Game theory · Modelling · Stakeholders

1 Introduction

Modern experience in development and implementation of IT products is connected with numerous discoveries in innovative start-up projects. The number of such projects is steadily growing all over the world. Usually, such start-ups are designed as small enterprises, which create favourable conditions for the development of innovation. Indeed, the concept of "Start-up" – a new and rapidly

S. Babichev et al. (Eds.): ISDMCI 2020, AISC 1246, pp. 377–394, 2021.
https://doi.org/10.1007/978-3-030-54215-3_24

growing company, seeks to meet the needs of the market by developing a business model based on an innovative idea [1]. The projects pursued by start-ups are very risky, and their survival rate is quite low. A recent study by Small Biz Trends [27] shows that 90% of new start-ups fail, and CB Insights [4] has developed a diagram of the top 20 reasons for start-up failure by analysing more than one hundred real cases (Fig. 1).

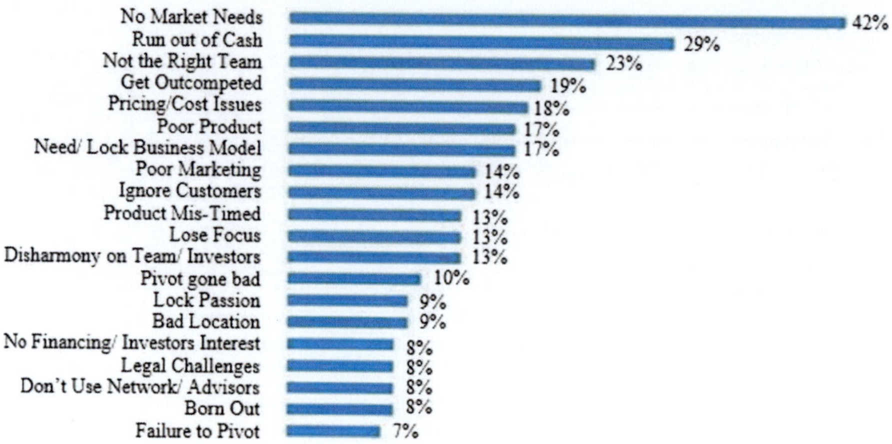

Fig. 1. Reasons for start-up failure.

U. S. bank research shows that 82% of enterprises that failed cited cash flow problems as one of the factors behind their failure [9]. The average small business has enough money streams to cover 3–4 weeks of expenses, and many are struggling with overdue invoices, because some of them exceed 30 days. Among all the small enterprises that started their activities in 2014 [31]:

- 80% held out until year two (2015);
- 70% held out until the third year (2016);
- 62% lived to the fourth year (2017);
- 56% until the fifth year (2018).

With those numbers in mind, a bit more than half of all start-ups actually survive to their fourth year, while the start-up failure rate at four years is about 44%.

Companies with the SaaS (Software as a Service) distribution model [17] and B2B (Business to Business) transactions [28] have problems with long sales cycles, as the decision of the client is collective and depends on many factors, so the quality of customer service cannot be an individual indicator. Besides, the B2B client experience is characterised by numerous interactions with the company and includes a wide range of contact persons.

The practical value of this general principle is the subscription model with recurring fees (revenue). Unlike licensed software customers, SaaS customers easily enter into their relationship with SaaS vendors through multiple repeat purchases. Each new SaaS client brings a new recurring revenue stream and joins a large cohort of clients to create a total recurring revenue stream for SaaS.

Game theory can be used to simulate the relationship between the business owner, developers and customers [22]. Its use is justified by the nature of the relationships of these stakeholders. At the same time, the game theory itself is presented as a mathematical method of search and implementation of optimal solutions of behavioural policy in situations similar to the so-called "game situations". The game is understood as a process of competitive struggle of two or more parties for the realisation of their interests and profit. Each player has a clearly defined goal (the so-called "winning the game" - the target function) and uses a set of stable, variable and unstable data characteristic of them, and data chains that can lead to victory or loss in the game, which depends on the behaviour (data) of other players. Game theory will help us to build effective strategies and tactics in management, applied marketing and PR, allowing us to select the best strategies for IT products, taking into account the views of other players participants, their resource and potential and their possible actions, taking into account the existing risks.

In this aspect, game theory can be used in decision making to build IT product development programs and strategies more effectively. Game theory is a section of operations research, and it is not related to the theory of decision making. The main value of game theory is for studying, modelling and formation of a choice of the behaviour of artificial intelligent agents and in studying of information systems, processes and features of stakeholder engagement.

Under market conditions, the equilibrium development factor is very important during work, especially in IT projects. Participants benefit from maintaining this balance because any change will make their position worse. These works by J. Nash [8,23] have made a serious contribution to the development of game theory, and completely new mathematical tools for economic modelling have been proposed. Nash proves that the approach to competition proposed by A. Smith, when everyone pursues only his interests (acting for himself), is not the best. He argued that strategies are more optimal when everyone tries to do better for themselves, taking into account the interests of others.

The hypothesis is that players behave in a way that allows them to achieve maximum aggregate value (the "economic person" model of behaviour). However, in practice real behaviour of people often does not correspond to it because of irrationality, elements of "double bind" in the control system, modelling of discussion, other irrational motives of players (including altruism). In our case, game theory can be used as a reasonable ideal model by analogy with the same models in physics.

Thus, it is relevant to create an effective game model to analyse stakeholder engagement, in terms of determining the strategic directions of development of a particular IT product. This will allow us to build the content of the start-up development project, model and obtain the following datasets for analysis and appropriate decisions on IT product development of such projects. At the same time, start-ups will be provided with a strong competitive advantage, which will be the key to business success. As a whole, it will reduce a strong competitive advantage and the key to success to shorten sales cycle [5] and give the product a higher efficiency.

2 Analysis of Recent Research and Publications

Application of the project approach to start-up development and effective use of products of these projects at the creation of modern IT were considered in research papers [3,11,20,21,25,26,30]. Lean Start-up methodology, based on an Agile approach, is described in [25]. Methods of Lean Production of IT products and effective development of products of the project activity are offered there, but methods of promotion and development of such products are not given. In [11], the basic technological processes of effective management of medium and big projects are defined, but not all of these processes are necessary for the management of start-up projects. In [20,21,30] models and methods of proactive product management of the complex IT projects are considered. However, the interaction of clients with IT product is not observed, and the definition of directions of product development for the purpose of increase of clients' loyalty is not specified. In [21], new models of innovation management in projects based on convergence, combining different methodologies are proposed. In [26], questions of construction of effective project teams that can essentially influence the promotion of IT product among subscribers are considered. However, it is not specified what methods can be used for the accounting of interactions of the owner of a product, clients and developers. Such an approach can be used for further development of start-up projects. However, assessment of customer value to the products of such projects is not made.

Reviewing processes of product development of IT projects, it is necessary to consider directions of development of the start-up companies with Software-as-a-Service (SaaS) [3] distribution model and Business to Business (B2B) [2] transactions. As it has already been noticed above, models and methods of the game theory are best suited to such business models for the effective accounting of interactions. The following major scientific papers also address these issues [7,10,13,16,19,24]. As it is noted in these papers, such development should be based on new tools allowing to estimate productivity (value) of the interaction of clients with IT product, and as a feedback, such tools allow to define the basic directions of further improvement of innovative IT products. Here we can use well-proven customer journey maps. However, it is not specified how to process quite big data on numerous metrics analytically.

The analysis of the considered information sources [1–31] allows us to conclude, that for management of development of innovative commercial products of IT projects there are no integrated models and the methods that would allow to react to various interactions of various classes of clients with the least expenses. And also, based on the analysis of such interactions, to identify directions of the further improvement and development of such products for the increase of economic efficiency.

This significantly reduces the opportunities for effective management of start-up projects development. The possibilities of proactive management of innovative IT products development based on game theory methods, which would allow taking into account dynamic interaction on product development processes and further management of the start-up project, have not been investigated sufficiently.

3 Setting the Objective

The goal of the paper is to substantiate and develop a model to assess the dynamic interaction of the three main types of stakeholders. The evaluation of such engagement will be carried out using a method based on game theory. The assessment results will be used to develop a strategy to improve the IT product of the project with maximum efficiency This will provide an opportunity to improve the quality of development management of such projects and will increase commercial efficiency and extend the life cycle of the product.

The following tasks have been defined to achieve the objectives:

- to identify the key stakeholders that make up the elementary basis of the development management processes for commercial products of IT projects;
- to build a conceptual model of the game mutual influence of stakeholders based on information interaction;
- to carry out a formal description of the elements and to build a mathematical model of the stakeholder engagement process and to investigate the obtained model;
- conduct modelling and research on the results, analyse them and identify key conclusions.

4 Materials and Methods

4.1 Building a Conceptual Model of Stakeholder Engagement

As noted above, game theory is one of the effective tools to assess the quality of interaction between customers, business owners and development team with IT product and among themselves. In this case, the tools of visualisation of such interaction between the consumer and a product or service can be used. Creation of such models is simultaneously a process of analysis and a method for generating ideas for improving a product or service.

Note the characteristic conditions for modelling the game as a mathematical model that describes the chosen approach:

- availability of two or more participants;
- uncertainty of participants' behaviour, multiple options available to each of them;
- different interests of the participants (partial or absolute mismatch);

– the interconnection of strategies among the participants, because the result obtained by each of them depends on the behaviour of all together.

The conceptual model for stakeholder engagement in the development projects of IT start-ups may look as shown in Fig. 2.

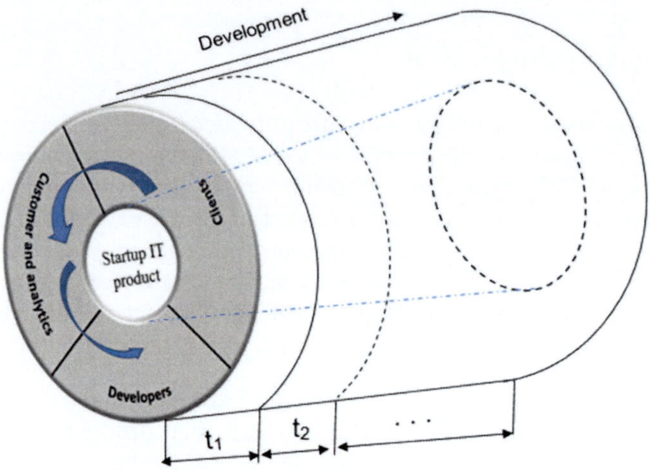

Fig. 2. The conceptual model "Growing cone" of stakeholder engagement in IT start-ups.

As shown in Fig. 2, we have selected three key stakeholders to conduct the research. They interact over time and, despite their different interests, must find points of compromise in order to develop the IT product collaboratively. The value of the product grows with time (circle in the middle) because its content is complemented by new features. Such additional features are defined according to the results of the game modelling of three stakeholders.

Start-ups with a SaaS distribution model often face a high risk of failure due to high initial costs, which entails losses in the first years of the company's activity. "Software as a Service" significantly outperforms the business of selling software licenses but has advantages and disadvantages that adapt to each particular type of service. The essence of this model of distribution in terms of cash flows is that before and during the launch of the business acquisition of new customers requires high costs, which are gradually returned over a long period. The faster a company tries to grow, the higher its losses, and the more it spends on the start, the further the break-even point of this business is postponed. In our opinion, this problem can be solved in several ways, considering: the quality of the developer's service, the possibility of its full automation and adjustment of the model of service delivery, etc. In our article, we will use as an example of a B2B SaaS business with low-touch sales model in the logistics sector. This is due to the following aspects:

- there is no regular personal communication with the representatives of the software development company;
- the main channels to attract new potential customers are the software website, email marketing, direct advertising and a free trial version, which is designed to be very easy and quick to start using;
- B2B sales are made by monthly subscription with a price range of 20–500 USD per month.

Thanks to a retrospective analysis of the positive experience of the companies Basecamp [6] (main products: Basecamp, Ruby on Rails, Ta-Da List and CRM-system Highrise) and Atlassian [18] (main products: JIRA, Trello, Confluence), we came to the conclusion that the adjustment of the SaaS model is based on the needs not only of customers but also of the main participants of this business: developers and owners. In the course of the adjustment, each of them makes their proposals and limitations, trying to find a way out - the most convenient and profitable strategy for all. In a case, if convenience (benefit) will be only on the client-side, such model quickly goes to a loss. Also, the unsatisfied inquiries from developers will not allow to automate and improve the software as it is required by such business, for its scaling etc.

Such conditions and requirements for solving the problem of this business can be modelled as a cooperative game with a non-zero sum in the field of game theory with its further automation for easy decision making in the management of start-up projects in the SaaS field. In other words, how to create an optimal strategy of behaviour in the market in order to reach the break-even point as soon as possible, making it the basis for decision making in management.

4.2 Construction of Local Models of Stakeholder Engagement

The main concept in the theory of cooperative games is the concept of negotiation set because it is on this set is negotiating about the common strategy of the players. Thus, the competitive cooperation of the companies can be coordinated at the negotiation set, and the appropriate coefficients of influence on the characteristics of the company (weighty) can be determined. For SaaS business, we have three players: software developers, customers and business owners (entrepreneurs).

One player is defined as the main player, while the other two are subordinated to the negotiation strategy. Each player has its optimisation function, and negotiations are built up between them. Let us consider the conditions and requirements of these players.

A. The company's ***customers*** like SaaS because it meets their needs and requirements in the right amount. In contrast to licensed software, customers do not spend money due to hardware failures, errors and cyber-attacks. The effectiveness of SaaS for the customer significantly exceeds the performance of the company's own IT department. Also, using SaaS is usually less expensive than using software and goods with other payment models. Therefore, the main focus for clients at any further scaling of interaction is to reduce their costs when they use the service. The cost reduction, in this case, includes not only the reduction of the monthly subscription fee, thus MRR (Monthly Recurring Revenue) for the owner and the payment for software updates, but also the reduction of tax payments, for the use of roads and vehicles in transportation (environmental tax), etc. For the developer, this, in turn, means constantly improving route optimisation algorithm (for example, parcel delivery tasks for a logistics company). The target function for the client can be summarized as shown below:

$$(MRR, T_{ec}, UP_s) \rightarrow min \tag{1}$$

where MRR is the monthly subscription fee; T_{ec} is the tax costs on product updates and modifications; UP_s is the cost of updating the basic software (price increase).

B. *Owners and investors* prefer the SaaS distribution model for the reason that revenue usually comes in regularly. This makes it much easier to predict cash flows in SaaS companies. For this reason, SaaS companies are considered one of the fastest-growing in IT history.

The link between the SaaS CLTV (Customer Lifetime Value) and the SaaS valuation comes from the subscription model, where the company's revenue is defined as the sum of individual customer revenue streams.

In this case, a financially acceptable method of assessing the value of the company is to calculate the accumulated amount of its future cash flows, or net present value:

$$NPV_{SaaS} = CLV_{LTV} \times K_{LTV} \tag{2}$$

where: CLV is the average time spent by a SaaS client with IT product; K_{LTV} is the similar metric for the number of customers with a lifetime value, which is the forecasted number of new customers acquired by the SaaS owner during the lifetime of that IT product version.

As a result, we get the following formula to consider the interests of IT business owners:

$$K_{LTV} = \frac{K_1}{(1+i)} + \frac{K_2}{(1+i)^2} + ... + \frac{K_n}{(1+i)^n} \tag{3}$$

where K_{LTV} is the current value of all new customers that the company will ever acquire at a discount on how long it takes to acquire them; K_n is the number of new clients acquired during the period n;

The above formula fully covers the three critical questions asked by SaaS VC (Venture Capital) investors:

- What is the value to the customer CLV?
- How big is this market K_n?
- How fast can you get there K_{LTV}?

Consider a generalised metric that describes the average SaaS client as follows:

$$CLV = LARR_{LTV} - LACS_{LTV} - LCAC \tag{4}$$

$$CLTV = \frac{(ARR_1 - ACS_1)}{(1+i)} + \frac{(ARR_2 - ACS_2)}{(1+i)^2} + \cdots + \frac{(ARR_n - ACS_n)}{(1+i)^n} - CAC \quad (5)$$

where: $LARR_{LTV}$ is the average lifetime recurring revenue per client; $LACS_{LTV}$ is the average lifetime recurring service cost per customer; $LCAC$ is the average cost of acquisition a client; ARR_n and ACS_n is the average client recurring revenue or service cost for the period n; $CLTV$ is the client's value to the business for the whole period of their interaction; CAC is the cost of acquisition a client;i is he cost of capital, which for VC funded start-ups is 30%.

In the continuous model, the net value of the company's periodic income NPV_{SaaS} will be determined by the following formula:

$$NPV_{SaaS} = TRR_{LTV} = \int TRR\ (t) \times e^{-it} dt \rightarrow max \quad (6)$$

where: TRR_{LTV} is the total regular income; t is the life cycle time; i is the cost of capital.

If a SaaS company grows rapidly, rising acquisition costs may outpace the accumulation of recurring deposits, which makes profitability impossible.

As stated above, regarding the profitability of the SaaS business model in the IT industry, the time to make a profit comes when the profit rate is zero:

$$[LARR - LACS] \times C(T_{profit}) - CAC \times \Delta C_{new}(t_{profit}) = 0 \quad (7)$$

where: $C(T_{profit})$ is the cost of making a profit; $\Delta C_{new}(t_{profit})$ is the increase/decrease in costs of attracting profitable customers.

C. The SaaS model is convenient for ***developers***, because it has the most acceptable way of payment. Therefore, their interests are aimed at increasing the expenses for the owner to perform the ordered work and reduce labour costs unrelated to the costs of automation. Also, this function is completely opposite to the needs of customers. Thus, developers, interacting in the cooperative game, are competitors and opponents of customers for each criterion:

$$(MRR, T_{ec}, UP_s) \rightarrow max \quad (8)$$

In general, the restrictive functions of the cooperative game for SaaS ***customers***, ***owners (investors)*** and ***developers*** will look like this:

$$\begin{cases} NPV_{SaaS} = TRR_{LTV} = \int TRR\ (t) \times e^{-it} dt \rightarrow max, \\ [ARR - ACS] \times C(T_{profit}) - CAC \times \Delta C_{new}(t_{profit}) \rightarrow min, \\ ARR_{LTV} = (MRR, T_{ec}, UP_s) \rightarrow max \end{cases} \quad (9)$$

4.3 Building a Model of Game Situations

To describe such a cooperative game we use a simplified model of the company operating on the market: Let the state of the company be characterized at every moment t by two values: $N_{(t)}$ is the volume of subscriptions provided and $U_{(t)}$ is the capital of the company with knowledge of U_0 at the beginning of surveillance of the company. The functioning of a company is modelled by the following processes that take place in such a sequence:

1. The company receives a new flow of risks from new subscriptions: intensity λ. The company develops $N_{(t)}$ products, making a profit as a random variable with a mathematical expectation of a during the time t. The capital of the company increases by this magnitude $\lambda \cdot a \cdot t$.
2. Part of subscriptions sold stops coming with the time t, as the customer's churn. Moreover, if this process has an intensity of μ, and the volume of all subscriptions $N_{(t)}$, the loss of profit equals $\mu \cdot a \cdot t$.
3. With an intensity of $\lambda_1 N_{(t)}$ there is additional income from software updates. These revenues are a random variable with a mathematical expectation of b. The capital of the company from this increases by this magnitude $(\lambda_1 \cdot b \cdot \lambda \cdot t)/\mu$.
4. With the intensity $\mu_1 N_{(t)}$ there is an unexpected loss of profit as a random variable with mathematical expectation c. That is, the capital of the company from this decreases by the magnitude $(\lambda_1 \cdot c \cdot \lambda \cdot t)/\mu$.

For these assumptions, it is shown [9,25] that the mathematical expectation of the company's capital $U_{(t)}$ has a form:

$$M[U_{(t)}] = \frac{U_0 + \lambda t(a\mu + b\lambda_1 - c\mu_1)}{\mu} \tag{10}$$

In this case, the growth rate of the company capital K equals:

$$K = \frac{dM[U_{(t)}]}{dt} = \frac{\lambda(a\mu + b\lambda_1 - c\mu_1)}{\mu} = \lambda(a + \frac{b\lambda_1 - c\mu_1}{\mu}) = \lambda(a + \delta) \tag{11}$$

where

$$\delta = \frac{b\lambda_1 - c\mu_1}{\mu} \tag{12}$$

Let us assume that there are three players in the market is being investigated. Since they agree on one product (are in the same strategic group), then, accordingly, they differ only in parameters: a and δ : a_1, δ_1; a_2, δ_2; a_3, δ_3.

Let us consider that there is a flow of risks of intensity λ_0, depends only on the profit margins: a_1, a_2, a_3, that is $\lambda_0 = f(a_1, a_2, a_3)$. This flow is divided into three parts, so the first company receives a risk flow with an intensity of λ_1, the second - a flow with an intensity of λ_2, and the third - a flow with an intensity of λ_3:

$$\lambda_3 = \lambda_0 - \lambda_1 - \lambda_2 \tag{13}$$

We consider that the intensity of these streams is connected with the amounts of inverse proportionality revenues.

$$\frac{\lambda_1}{\lambda_2} = \frac{a_2}{a_1}; \frac{\lambda_1}{\lambda_3} = \frac{a_3}{a_1} \tag{14}$$

The type of dependence $f(a_1, a_2, a_3)$ is quite complicated and depends on some parameter p, which determines some "average" prices of the influence of distinctive characteristics-indexes on the company's profitability in the market. The p parameter equals:

$$p = \frac{a_1 a_2 a_3}{(a_1^\nu + a_2^\nu + a_3^\nu \delta_1)^{\frac{1}{\nu}}} \tag{15}$$

where $\nu > 0$. Let $\nu = 2$.

And then the capital growth of K_1, K_2, K_3 for the first, second and third companies according to the root mean square "a" will be:

$$\begin{cases} K_1 = \dfrac{a_2 a_3 (a_1 + \delta_1) p}{a_1 a_2 + a_2 a_3 + a_3 a_1} \\ K_2 = \dfrac{a_1 a_3 (a_2 + \delta_2) p}{a_1 a_2 + a_2 a_3 + a_3 a_1} \\ K_3 = \dfrac{a_1 a_3 (a_3 + \delta_3) p}{a_1 a_2 + a_2 a_3 + a_3 a_1} \end{cases} \tag{16}$$

Each pair of values (a_1, δ_1), (a_2, δ_2), (a_3, δ_3) in space K_1, K_2, K_3 corresponds to a point, the aggregate of which in spaces forms the K area. Formula (16) define surfaces in space.

5 Experiment

After entering the data, it is mandatory to check the data for correlation because the resulting assessment is levelled out in this case. The critical correlation value for the data arrays is 0.8. This value is the maximum allowable value for calculations in this mathematical package because if the correlation value between the variables is higher than 0.8, the functional relationship is established (multicollinearity). The correlation research data is entered into the matrix with a limit of 600 elements. Therefore, some matrices are split into submatrices, and the correlation relationships are investigated by splitting large matrices. The formula calculates the correlation between the specified indicators:

$$\rho(k_1, k_{i+1}) = \frac{cov(k_1, k_{i+1})}{\sigma_x \cdot \sigma_{x+1}} \tag{17}$$

Investigation of the critical rp^S (subject) correlation correspondence is carried out using a matrix of the following type:

$$rp^S = \begin{pmatrix} \rho(k_1, k_1) & \cdots & \rho(k_1, k_n) \\ \rho(k_2, k_1) & \cdots & \rho(k_2, k_n) \\ \vdots & \ddots & \vdots \\ \rho(k_n, k_1) & \cdots & \rho(k_n, k_n) \end{pmatrix} \tag{18}$$

For determining the weighting factor for each indicator in the matrix form, the necessary statistical data are input:

$$Xp_i = \begin{pmatrix} 1 & \cdots & 1 \\ \phi_i, k_i & \cdots & \phi_i, k_n \\ \vdots & \ddots & \vdots \\ \phi_m, k_1 & \cdots & \phi_m, k_n \end{pmatrix} \tag{19}$$

In order to determine the significance that a certain indicator takes in the subjective assessment, it is necessary to determine a multiple regression of this data set.

First of all, we check the value of optimization functions for correlation. The PCC correlation value shows us how strongly two variables (lists) are correlated. It is important to note that the PCC value ranges from -1 to 1. A value between 0 and 1 indicates a positive correlation. A value of $0 =$ most variation (without correlation). A value between -1 and 0 indicates a negative correlation.

Using the package of libraries Python pandas.DataFrame.corr:

```
import pandas as pd
a = [[1, 2, 3], [5, 6, 9], [5, 6, 11], [5, 6, 13], [5, 3, 13]]
df = pd. DataFrame(data=a) df.corr()
```

In the case of increasing the number of variables using the NumPy library:

```
def corr(data1, data2):
"data1 & data2 should be numpy arrays."
mean1 = data1.mean()
mean2 = data2.mean()
std1 = data1.std()
std2 = data2.std()
# corr = ((data1-mean1)*(data2-mean2)).mean()/(std1*std2)
corr = ((data1*data2).mean()-mean1*mean2)/(std1*std2)
return corr
```

The ratio of the type max (K_1, K_2, K_3) at the set (a_i, δ_i), according to i, allows building a negotiating set with some basic company. At the same time, the coefficients of influence of its indicators apply to other companies, as an agreement in competitive relations.

For the first time, we proposed the use of the method of building Negotiation Sets to evaluate the competitive position of the enterprise.

At construction of negotiating set, one of the enterprises is accepted as the basic competitor and set of its parameters as a basic object of the set. The basic player is accepted the one with the value of $p(x)$ takes the highest values compared to others because according to the nature of this dependence, it reflects the negotiation process most clearly. In our case, we will take the business owner as the basic player, according to (8).

As we can see in Fig. 3, the three-dimensional graph reflects the space of the negotiation set as a combination of points, which reflect the characteristics of the sensitivity parameters of each of their target functions of the players, on which the negotiations $\rho(f(x))$ are held. Building a Negotiation Set is the reproduction of a game with a non-zero sum. This means that there is a benefit to each participant in the business from the compromise formed. These functions have intersection points in a wide variety of options, and the optimal variants are highlighted in yellow. This also corresponds to real market conditions, which makes this approach relevant to this study.

A Negotiation Set is built in accordance with this. To do that, we need:

```
import numpy as np
from numpy import sin, cos, pi
from skimage import measure
import matplotlib.pyplot as plt
from mpl_toolkits.mplot3d import Axes3D

def fun(x, y, z):
return cos(x) + cos(y) + cos(z)

x, y, z = pi*np.mgrid[-1:1:31j, -1:1:31j, -1:1:31j]
vol = fun(x, y, z)
verts, faces = measure.marching_cubes
(vol, 0, spacing=(0.1, 0.1, 0.1))

fig = plt.figure()
ax = fig.add_subplot(111, projection='3d')
ax.plot_trisurf(verts[:, 0], verts[:,1], faces, verts[:, 2],
cmap ='Spectral', lw=1)
plt.show()
```

These points are crossing points built by (15), if we take two functions in pairs, fixing conditional pairs of values a and δ from the third enterprise. On the graph axes the values of conditional variables of the negotiation process from each participant, on which "negotiations" are built, are deferred. The constructed two-dimensional area is a projection of the space on the plane of values, which are further used to establish the formula dependences for calculating the weight coefficients of the integral convolution of the factors of each subject evaluation.

Thus, we obtained two possible scenarios for the optimal result simultaneously for the three parties of the SaaS business (Table 1).

Overall, the company should increase its total subscriptions and revenues while maintaining the cost of software upgrades in order to achieve optimal benefits.

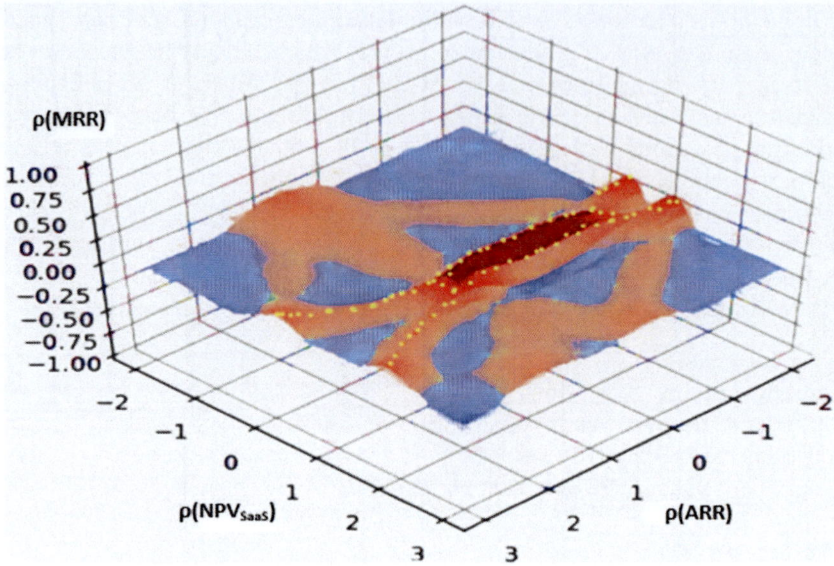

Fig. 3. Presentation of the negotiation plane on the set of model participants.

Table 1. Results of model implementation on Start-up SaaS project data

Indicators	Win scenario 1	Win scenario 2
MRR, *thous. euro*	81(+21)	78(+18)
T_{ec}, *euro*	120	138
UP_S, *euro*	78	83
$LARR_{LTV}$, *euro*	781	450
$LACS_{LTV}$, *euro*	690	453
$LCAC$, *euro*	2921	2520

6 Results and Discussion

The proposed models were used to develop the web and mobile application of the french start-up Antsways. This company provides logistics companies with its SaaS product for route optimisation. Thus, with the help of a web application, the supply chain manager can choose the best route based on the traffic situation (road repairs, strikes, weather conditions, etc.), the preferences of the client (location, day, time) and the driver (current location, vehicle load status, fuel reserves), etc. Subscriptions of logistic companies for web application services were taken into account. We are talking about the savings that logistics companies received as a result of the proposed optimised logistics activities. This fact led to the retention of logistics customers more and more time and the number of such customers gradually increased. At the same time, the income of not only the company of the owner of the startup, but also of two other stakeholders increased. The results of the conducted research during the previous two years are shown in Fig. 4.

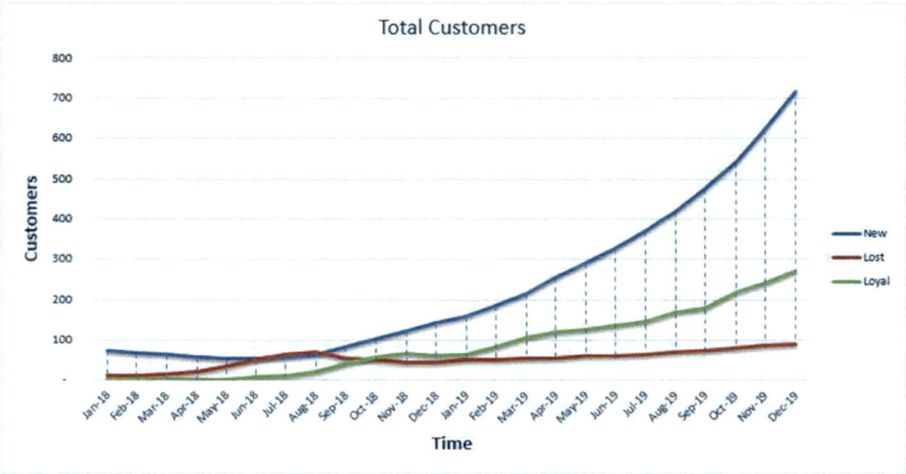

Fig. 4. Status of the client database of the start-up owner.

As we can see, the analysis of customers during interaction with the system was conducted monthly. Three types of such clients were identified: a) passive customers b) unhappy customers who do not meet the service offered to their business and who have logged out c) loyal customers. They were those who were able to offer improvements and signed long-term contracts - they bought a subscription to the product. The number of such loyal clients has been gradually growing, reaching more than 250 in two years, despite the significant growth of clients (700), interacting with the system from 1 to 10 times.

Using the game approach and the models considered, the start-up owner managed to establish an acceptable pricing policy and dynamically change it depending on the growth of its customers.

7 Conclusions

The proposed approach of using game theory to analyse the mutual influences of the three stakeholders considered (customer, owner and developer) allowed to flexibly define interaction strategies based on price policy, the volume and cost of further IT product software development and requirements of loyal customers. Thus, the expanded functions allowed to attract a significant number of one-time customers, which in turn contributed to the development of the start-up, its revenue stream.

The revealed structural components and inter-connections between them formed the basis for the proposed conceptual model "Growing cone" for projects to create and develop complex IT products. The qualitative characteristic of this model is a platform for further improvement of the start-up based on the synergy of a larger number of stakeholders.

Based on the analysis of the conceptual model and identification of three key stakeholders of the game model, we managed to build three groups of mathematical dependencies. Each of these groups describes the interests and values of a certain group of participants.

Using the constructed mathematical descriptions of three key stakeholders, we were able to develop a model of their engagement and mutual influence based on game situations. A simplified company model was used to describe such a cooperative game. Thus, balanced value growth indicators of the game participants were used, which determine some "average" value indicators.

We conducted experimental modelling to investigate the proposed models, for which we developed program codes using the Python libraries. Thus, after entering the data, the correlation check was obligatory. The critical value of correlation by the data sets was 0.8 (maximum allowed for calculations in the used mathematical package). In the correlation research, the data input into the matrix with a limit of 600 elements.

The models and simulation methods considered were used in decision-making processes to develop an international SaaS product for route optimisation. These techniques helped the company to create a product that has value for logistics companies (easy and efficient planning, management and execution of deliveries), to attract more new customers and to retain existing ones. At the same time, the income of not only the company of the owner of the start-up but also of two other stakeholders (customers and developers) increased.

Overall, the approach used has proven to be effective and practical. The authors see its development in further research with an increased number of stakeholders, for example, in digital marketing, finance, R&D etc.

References

1. Blank, S., Dorf, B.: The Startup Owner's Manual: The Step By Step Guide for Building a Great Company. K&S Ranch, Pescadero (2018)
2. Boringer, J., Grehan, B., Kiewell, D., Soenke, L., Mose, P.: Four pathways to digital growth that work for B2B companies (2019). https://www.mckinsey.com
3. Campbell, P.: The SaaS business model: stages, pros and cons, and tools to get ahead (2019). https://www.profitwell.com/blog/saas-business-model
4. CBInsights: The top 20 reasons startups fail (2019). https://www.cbinsights.com/research/startup-failure-reasons-top/
5. Conforto, E.C., Amaral, D.C.: Evaluating an agile method for planning and controlling innovative projects. Project Manag. J. **41**(2), 73–80 (2010). https://doi.org/10.1002/pmj.20089
6. Dermendzhiyska, E.: How basecamp built a $100 billion business by doing less on purpose (2019). https://medium.com

7. Easterbrook, S.: Resolving requirements conflicts with computer-supported negotiation. In: Jirotka, M., Goguen, J. (eds.) Requirements Engineering: Social and Technical Issues, pp. 41–65. Academic Press, Cambridge (1994)
8. Edelman, D., Singer, M.: Competing on customer journeys you have to create new value at every step (2015). https://hbr.org
9. Fundera: Small business statistics: 19 essential numbers to know (2020). https://www.fundera.com/blog/small-business-statistics/
10. Galvao, T., Neto, F., Bonates, M., Campos, M.: A serious game for supporting training in risk management through project-based learning. In: Springer Series Virtual and Networked Organizations, Emergent Technologies and Tools, vol. 248, pp. 52–61. Springer (2012). https://doi.org/10.1007/978-3-642-31800-9_6
11. Gogunskii, V., Kolesnikov, O., Oborska, G., Harelik, S., Lukianov, D.: Representation of project systems using the Markov chain. East.-Eur. J. Enterp. Technol. **1/3**(85), 25–32 (2017). https://doi.org/10.15587/1729-4061.2017.97883
12. Harold, W., Kuhn, E., Nasar, S.: The Essential John Nash. Princeton University Press, Princeton (2007)
13. Harsanyi, J., Reinhard, S.: A general theory of equilibrium selection in games. Economica **18**(5), 63–68 (2001). https://doi.org/10.2307/2554329
14. Hu, J., Wellman, M.: Research of the impact of changes based on external influences in complex IT projects. In: Proceedings of the Fifteenth International Conference on Machine Learning, ICML 1998, pp. 242–250 (1998). https://doi.org/10.5555/645527.657296
15. Li, H., Karray, F., Basir, O.: A framework for coordinated control of multi-agent systems. In: Innovations in Multi-Agent Systems and Applications, vol. 310, pp. 43–67. Springer (2010). https://doi.org/10.1007/978-3-642-14435-6_3
16. Linowski, J.: Agile A/B testing: using stop rules to minimize losses and time wasted (2016). https://goodui.org
17. Loukis, E., Janssen, M., Mintchev, I.: Determinants of software-as-a-service benefits and impact on firm performance. Decis. Supp. Syst. **117**, 38–47 (2019). https://doi.org/10.1016/j.dss.2018.12.005
18. McDuling, J.: Atlassian: the $30 billion tech giant nobody understands (2018). https://www.smh.com.au
19. Morozov, V., Kalnichenko, O., Kolomiiets, A.: Research of the impact of changes based on external influences in complex it projects. In: Proceedings of the IEEE International Conference on Advanced Trends in Information Theory, ATIT 2019, pp. 481–488. IEEE (2019). https://doi.org/10.1109/ATIT49449.2019.9030441
20. Morozov, V., Kalnichenko, O., Mezentseva, O.: The method of interaction modeling on basis of deep learning of neural networks in complex IT-projects. Int. J. Comput. **19**(1), 88–96 (2020)
21. Morozov, V., Kalnichenko, O., Proskurin, M., Mezentseva, O.: Investigation of forecasting methods the state of complex IT-projects with using deep learning neural networks. In: Springer Series Advances in intelligent systems and computing, vol. 1020, pp. 261–280. Springer (2020). https://doi.org/10.1007/978-3-030-26474-1_19
22. Nash, J.: Equalibrium points in n-person games. Econometrica **18**(2), 155–162 (1950)
23. Nash, J.: Non-cooperative games. Ann. Math. **54**(2), 286–295 (1951)
24. Neumann, J., Morgenstern, O.: Theory of Games and Economic Behavior: 60th Anniversary Commemorative Edition. Princeton Classic Editions (2007)
25. Ries, E.: The Lean Startup: How Today's Entrepreneurs Use Continuous Innovation to Create Radically Successful Businesses. Currency, Redfern (2011)

26. Sherstyuk, O., Olekh, T., Kolesnikova, K.: The research on role differentiation as a method of forming the project team. East.-Eur. J. Enterp. Technol. **2/3**(80), 63–68 (2016). https://doi.org/10.15587/1729-4061.2016.65681
27. Smallbiztrends: Startup statistics – the numbers you need to know (2020). https://smallbiztrends.com/2019/03/startup-statistics-small-business.html
28. Swani, K., Brown, P.B., Mudambi, M.S.: The untapped potential of B2B advertising: a literature review and future agenda. Ind. Mark. Manag. (2019). https://doi.org/10.1016/j.indmarman.2019.05.010
29. Tesauro, G.: Extending Q-learning to general adaptive multi agent reinforcement learning: theoretical framework and an algorithm. In: Advances in Neural Information Processing Systems 16 Neural Information Processing Systems, NIPS, pp. 871–878 (2003). https://doi.org/10.5555/2981345.2981454
30. Timinsky, A., Voitenko, O., Achkasov, I.: Competence-based knowledge management in project oriented organisations in bi-adaptive context. In: Proceedings of the IEEE 14th International Scientific and Technical Conference on Computer Sciences and Information Technologies, CSIT-2019, vol. 1, pp. 17–20. IEEE (2019). https://doi.org/10.1109/STC-CSIT.2019.8929806
31. USBLS: Survival of private sector establishments by opening year (2020). https://www.bls.gov/bdm/us_age_naics_00_table7.txt

Modeling and Forecasting Economic and Financial Processes Using Combined Adaptive Models

Peter Bidyuk[1], Aleksandr Gozhyj[3](✉), Yoshio Matsuki[1,2], Natalia Kuznetsova[1], and Irina Kalinina[3]

[1] National Technical University of Ukraine "Igor Sikorsky Kyiv Polytechnic Institute", Kyiv, Ukraine
pbidyke@gmail.com
[2] Kyoto University, Kyoto, Japan
matsuki@wdc.org.ua
[3] Petro Mohyla Black Sea National University, Nikolaev, Ukraine
alex.gozhyj@gmail.com, irina.kalinina1612@gmail.com

Abstract. The article proposes a methodology for modeling and forecasting economic, financial and environmental processes, based on the principles of system analysis and combined models, including linear and non-linear regression models, as well as probabilistic models in the form of Bayesian networks. The basis of the methodology is tracking the computational processes of processing statistical data at all stages using the appropriate sets of statistical quality criteria. The high quality of the intermediate and final results is achieved through the application of statistical quality criteria to the data, the adequacy of the model and the quality of forecasts. The methodology is used in a specialized decision support system aimed at analyzing nonlinear non-stationary processes in various fields.

Keywords: Nonlinear nonstationary processes · Uncertainties · Combined forecasting model · Linear regression · Nonlinear regression · Bayesian model

1 Introduction

Modeling and quality forecasting dynamics of modern processes in economy, finances, ecology, and social systems requires constructing appropriate mathematical models suitable for adequate formal description of complicated nonlinear nonstationary (NN) processes. The process under study can be integrated (processes with trends), cointegrated (when two or more integrated processes are analyzed together and appropriate models are constructed), or heteroscedastic, the variance of which is changing in time on the interval of studying. In this case the variance dynamics should be modeled and forecasted for estimating financial or economic risks and other practical applications. Many processes exhibit

S. Babichev et al. (Eds.): ISDMCI 2020, AISC 1246, pp. 395–408, 2021.
https://doi.org/10.1007/978-3-030-54215-3_25

nonlinearities that appear due to availability of periodical economic crisis, local military conflicts, natural disasters, influence of pandemic processes etc. Usually we consider two types of nonlinearities regarding the processes mentioned: nonlinearities with respect to variables and nonlinearities with respect to parameters [6]. The first type is analyzed and estimated easier, and the second type often requires substantial efforts to perform correct modeling and forecasting.

Such problems could be solved successfully with computer systems available on market, say SAS, SPSS, STATISTICA and many others [9]. However, sometimes such systems cannot be used due to high cost or they do not contain necessary functionality needed in specific applications. Such systems can be used successfully for analyzing banking data, transactions and perform data analysis at large enterprises that possess enough material resources for purchasing them and supporting. Due to existing specific problems of data analysis very often it is reasonable to develop and use specialized decision support systems (DSS) supporting all necessary functions [2,3,8,13]. The main advantages of such systems are functional flexibility, relatively low cost, easy extension of a set of available functions, and adaptation to the user. DSS architecture also allows prototyping that helps in many cases to reduce the terms of system development, refine its functional capabilities, and satisfy customer requirements.

It is reasonable to construct such DSS using systemic approach that supposes application of such useful principles as hierarchical system architecture, application of adaptation and optimization procedures where possible, functional completeness, identification and taking into consideration possible uncertainties characteristic for data processing procedures, model constructing and forecast estimating [1]. The uncertainties in modeling procedures based on statistical data are usually met in the statistical, structural and parametric form. Statistical uncertainties are caused by missing observations, system noise (external random disturbances), measurement errors and outliers. There are multiple numerical procedures for identifying and fighting such uncertainties, such as digital and optimal filters, data imputation procedures, state and parameter estimation methods, outlier processing procedures etc. The structural uncertainties appear due to the use of approximate estimates of model structure parameters, such as model order, delay time (lag), parameters of data probability distribution functions, type of nonlinearity and nonstationarity. The first two types of uncertainties mentioned usually result in parametric uncertainty, i.e. incorrect estimates of model parameters: they may contain bias, as well as be inconsistent or inefficient. All these problems can be taken into account at the stage of DSS development by making use of appropriate numerical methods directed towards decreasing of negative effects produced by the uncertainties mentioned.

In this study we will consider some elements (functions) of a specialized DSS constructed for modeling and forecasting nonlinear nonstationary processes in economy, finances and some other spheres of human activities. The experience available shows that the use of a specialized appropriately constructed DSS provides for a substantial help in analyzing (modeling and forecasting) complex modern processes met in various spheres of human activity and to improve substantially intermediate and final results of computations.

Problem Statement. The purpose of the study is as follows: – to consider some uncertainties met in model constructing procedures; – to propose instruments for coping with uncertainties when processing statistical data; – to consider some possibilities for constructing specialized DSS for short term forecasting of nonlinear nonstationary processes; – to develop efficient combined forecasting model including Bayesian network and regression models; – to show examples of application of the approach developed.

2 Materials and Methods

2.1 Some Processes in Economy, Finances and Ecology

The stochastic processes met in various spheres of human activities are characterized by a wide variety of properties; some types of the processes that are divided into two wide classes are presented in Table 1.

Table 1. Some types of process in economy, finances and ecology

No.	Linear	Nonlinear nonstationary
1	Nonstationary (integrated) with linear 1st order trend	Generally nonlinear but can be piecewise linear
2	Linear, no trend	Integrated of order two and higher
3	Processes with linear periodic components	Cointegrated of order two and higher
4	Linear with noisy components	Heteroscedastic processes
5	–	Bilinear processes
6	–	Nonlinear in parameters
7	–	Processes with outliers
8	–	Levy processes
9	–	Processes with nonlinear periodic components
10	–	Processes with breaks and saturation

We will consider a process as a linear if it can be adequately described by a linear model otherwise the process will be classified as a nonlinear one.

A linear process can be nonstationary, i.e. contain trend of the first order. If the trend is of order two or higher the process goes to the class of nonlinear ones. The wide subclass of nonlinear processes create heteroscedastic processes the variance of which is changing within the interval of study, formally: $Var[\mathbf{x}(k)] \neq const, k \in [1, N]$. Analysis of such processes requires constructing two models: (1) model for the process itself (an amplitude model), and the model describing dynamics the process variance: $Var[x(k)] = F[x^2(k), y(k), y^2(k), \theta, \nu(k)]$, where $y(k)$ is possible extra variable influencing the variance; θ are model parameters; $\nu(k)$ is stochastic process included in the model of variance; k is discrete time that is related to continuous time t via data discretization interval T_s for continuous variables as follows: $t = kT_s$. Today there exist possibilities for selecting and modifying models of various structures to describe correctly conditional variance for the process under study. The models of conditional variance are very popular due to the fact they are required in various situations, for example, in stock trading and diagnostic systems (engineering, medical, financial).

Bilinear processes are widely known in mechanics and economy; they are described by polynomial of order two. The models nonlinear in parameters create an important subclass of models that find their application in the most various areas of human activities. They are used successfully for solving classification as well as forecasting problems, possibly in combination with appropriate linear models. The outliers are correct measurements that exceed in amplitude the main bulk of measurements for some variable. They deteriorate process stationarity (and linearity) and require application of special modeling technics like special forms of distributions. The Levy processes are stochastic processes containing large jumps and transitions between process modes of functioning. This short list of the processes to be studied shows that there is a wide variety of characteristics that need constructing specific models for adequate description, forecasting and control.

Generally all the processes in economy, finances and ecology are considered as stochastic ones due to the fact they contain stochastic components generated (at least) by the random external influences and measurement errors. The theory of time series analysis takes into account this fact in model constructing procedures.

There is a wide variety of statistical tests that help to distinguish between linear and nonlinear process, and then to select appropriate type of models for their formal description. Some tests for frequency domain are as follows: bispectrum analysis; Subba Rao-Gabr tests (for Gaussianity and linearity); Hinich tests (for Gaussianity and linearity); bootstrap based tests, and goodness-of-fit tests [7]. Also there are tests for application in time domain: Lagrange multiplier tests; likelihood ratio tests; Wald test; tests based on second order Volterra expansion; tests based on regression analysis etc. Thus, there is a possibility for choice of appropriate tests to be used in DSS.

2.2 Coping with Some Uncertainties Met in Modeling

Processing Stochastic Uncertainties. Uncertainties are considered as the factors of negative influence on the data analysis procedures that lead to lower quality of intermediate and final results of the analysis. The data processing system proposed includes the functions of data imputation, digital and optimal filtering of data at the stage of preliminary data processing. The optimal filtering subsystem is especially useful due to the following features:

- optimal estimation of states (values of output variables);
- short term forecasting and imputation of missing measurements;
- taking into consideration uncertainty in the form of the two stochastic influences: state and measurement noise (the state noise and measurement errors are always available);
- estimating non-measurable states and some model parameters.

These functions can be accomplished successfully with optimal Kalman filter that has various configurations dependently on specific application [10,12]. The filter requires for its application a model of the process being studied in state space (SS), but usually the model is constructed in the form of autoregression (AR), autoregression with moving average (ARMA) and other types of regression models, say as following:

$$y(k) = \sum_{i=1}^{p} a_i y(k-i) + \sum_{j=1}^{q} b_j x(k-j) + c_1 k + ... + c_m k^m + \varepsilon(k),$$

where $x(k)$, represents input measurements (independent or exogeneous variable); $y(k)$ is output (independent) variable; $\theta = [a_1...a_p \; b_1..b_q \; c_1..c_m]^T$ is vector of model parameters to be estimated with statistical data; p and q are orders of autoregression and moving average parts, respectively; m is order of trend described by the polynomial of time in this case; $\varepsilon(k)$ is random process reflecting influence of random external disturbances (state noise), measurement errors (noise), errors of model structure estimation, and parameter estimation errors. This model can be nonlinear if order of trend m is two or higher: $m \geqslant 2$. It can be transformed into SS form with subsequent application of optimal filtering algorithm.

A linear form of the complete SS model is as follows:

$$\mathbf{x}(k) = \mathbf{A}(k, k-1)\mathbf{x}(k-1) + \mathbf{B}(k, k-d)\mathbf{u}(k-d) + \mathbf{w}(k), \quad (1)$$

$$\mathbf{z}(k) = \mathbf{H}\mathbf{x}(k) + \mathbf{v}(k), \quad (2)$$

where $\mathbf{x}(k)$ is n-vector of states, the elements of which are selected by a researcher; $\mathbf{u}(k)$ is m-dimensional vector of control inputs (if such are available); $\mathbf{w}(k)$ is n-vector of stochastic disturbances; $\mathbf{A}(k, k-1)$ is $(n \times n)$-matrix of dynamics that includes parameters of AR model; $\mathbf{B}(k)$ is $(n \times n)$-matrix of control coefficients consisting of MA part of model shown above; d is input delay time; $\mathbf{z}(k)$ is vector of observations with r elements; $\mathbf{H}(k)$ is $(r \times n)$-matrix of measurement parameters that often has diagonal form; $\mathbf{v}(k)$ is a vector of measurement errors. The double time argument $(k, k-1)$ means that variable with this argument is used at time, k, though it is based on former measurements up to time moment $(k-1)$. To avoid symbols overload in mathematical expressions such argument is rarely used in text.

An advantage of the SS model (1), (2) is that it explicitly takes into account two random process, $\mathbf{w}(k)$, and, $\mathbf{v}(k)$. Thus, formally it is more adequate to the processes being modeled than say regression. The principal task of optimal filter is to compute optimal (in the sense of selected functional) estimates of state vector with taking into account covariance of the two stochastic processes mentioned above. This approach to data processing provides a possibility for improving estimates of state vector $\mathbf{x}$ and computing estimates of its nonmeasurable components if such are available. The basic equation of filtering is as follows:

$$\hat{\mathbf{x}}(k) = \mathbf{A}(k)\hat{\mathbf{x}}(k-1) + \mathbf{B}(k)\mathbf{u}(k-1) + \mathbf{K}(k)[\mathbf{z}(k) - \mathbf{H}(k)\mathbf{A}(k)\hat{\mathbf{x}}(k-1)], \quad (3)$$

where $\mathbf{K}(k)$ is optimal matrix coefficient of the filter. This coefficient can be found as a result of minimizing mathematical expectation of the sum of squared state vector estimate errors:

$$J = \min_K E\{[\hat{\mathbf{x}}(k) - \mathbf{x}(k)]^T [\hat{\mathbf{x}}(k) - \mathbf{x}(k)]\}.$$

where $\mathbf{x}(k)$ is exact state vector value that is computed via deterministic part of the model (1). The matrix coefficient $\mathbf{K}(k)$ is rather easily computed for discrete models by solving known Riccati equation [10].

Generally, the optimal filter considered provides a possibility for fighting uncertainties in the form of two random processes, $\mathbf{w}(k)$, and, $\mathbf{v}(k)$, as well as estimation of nonmeasurable state vector elements when respective components of covariance matrix are known, i.e., covariance matrix for state vector estimation errors: $\mathbf{P}(k) = E\{[\hat{\mathbf{x}}(k) - \mathbf{x}(k)][\hat{\mathbf{x}}(k) - \mathbf{x}(k)]^T\}$. Adaptive versions of the filter are especially useful for practical applications in both on-line and off-line modes of operation. They are applied for repeated estimation of parameters for matrices $\mathbf{A}(k)$ and $\mathbf{B}(k)$ as well as covariance of the two random processes mentioned above [18].

Equation (3) takes simpler form if controls are not taken into account:

$$\hat{\mathbf{x}}(k) = \mathbf{A}(k)\hat{\mathbf{x}}(k-1) + \mathbf{K}(k)[\mathbf{z}(k) - \mathbf{H}(k)\mathbf{A}(k)\hat{\mathbf{x}}(k-1)]. \quad (4)$$

One step ahead forecast can be computed using matrix of dynamics $\mathbf{A}(k)$ from this SS model as follows:

$$\hat{\mathbf{x}}(k+1|k) = E_k[\mathbf{x}(k+1)] = \mathbf{A}\mathbf{x}(k), \quad (5)$$

where, $\hat{\mathbf{x}}(k+1)$, is one step forecast based on the measurements up to the moment k (including). The function (5) can be used for computing forecasts for two and arbitrary number of steps s ahead:

$$\hat{\mathbf{x}}(k+2) = \mathbf{A}\hat{\mathbf{x}}(k+1) = \mathbf{A} \cdot \mathbf{A}\mathbf{x}(k) = \mathbf{A}^2\mathbf{x}(k),$$
$$\hat{\mathbf{x}}(k+s) = \mathbf{A}^s\mathbf{x}(k).$$

The forecast error for two steps forecast is as follows:

$$\mathbf{e}_f(1) = \mathbf{x}(k+1) - \hat{\mathbf{x}}(k+1) = \mathbf{F}\mathbf{x}(k) + \mathbf{w}(k) - \mathbf{F}\mathbf{x}(k) = \mathbf{w}(k),$$
$$\mathbf{e}_f(2) = \mathbf{x}(k+2) - \hat{\mathbf{x}}(k+2) = \mathbf{F}\mathbf{x}(k+1) + \mathbf{w}(k+1) - \mathbf{F}[\mathbf{F}\mathbf{x}(k)]$$
$$= \mathbf{F}\mathbf{x}(k+1) + \mathbf{w}(k+1) - \mathbf{F}[\mathbf{x}(k+1) - \mathbf{w}(k)] = \mathbf{w}(k+1) + \mathbf{w}(k),$$

And the error variance for s steps forecasts is:

$$Var[\mathbf{e}_f(s)] = E\{[\mathbf{w}(k+s-1) + \mathbf{w}\ (k+s-2) + ... + \mathbf{w}(k)] \times [\mathbf{w}^T(k+s-1) \\ + \mathbf{w}^T(k+s-2)\ + ... + \mathbf{w}^T(k)]\} = s\mathbf{Q}_w.$$

Thus, errors of forecasting grow proportionally to the state noise covariance, $\mathbf{Q}_w$.

Estimation of nonmeasurable components of a state vector is accomplished as follows. In this case the dimensionality of measurement and state vectors are different, i.e. $dim[\mathbf{z}] < dim[\mathbf{x}]$. The optimal matrix coefficient of the filter is computed via the expression:

$$\mathbf{K}(k) = \mathbf{P}'(k)\mathbf{H}^T[\mathbf{H}\mathbf{P}'(k)\mathbf{H}^T + \mathbf{R}]^{-1},$$

where dimensions: $dim[\mathbf{P}'(k)] = [n \times n]$; $dim[\mathbf{H}^T] = [n \times r]$ by definition; and $dim[\mathbf{H}\mathbf{P}'(k)\mathbf{H}^T + \mathbf{R}]^{-1} = [r \times r]$. Thus, $dim[\mathbf{K}(k)] = [n \times r]$. The innovation vector $\gamma(k)$ in Eq. (4) is defined as follows:

$$\gamma(k) = \mathbf{z}(k) - \mathbf{H}\mathbf{F}\hat{\mathbf{x}}(k),$$

and has dimensionality $[r \times 1]$; dimension of the product, $dim[\mathbf{K}(k)\gamma(k)] = [n \times 1]$. Thus, if, $dim[\mathbf{x}] = [3 \times 1]$, and $dim[\mathbf{z}] = [2 \times 1]$, then the product $\mathbf{K}(k)\gamma(k)$ takes the form:

$$\mathbf{K}(k)\gamma(k) = \begin{bmatrix} K_{11} & K_{12} \\ K_{21} & K_{22} \\ K_{31} & K_{32} \end{bmatrix} \begin{bmatrix} \gamma_1 \\ \gamma_2 \end{bmatrix}.$$

Now, components of the state vector are defined by the equations:

$$\hat{x}_1(k) = \hat{x}_1(k, k-1) + K_{11}\gamma_1 + K_{12}\gamma_2,$$
$$\hat{x}_2(k) = \hat{x}_2(k, k-1) + K_{21}\gamma_1 + K_{22}\gamma_2,$$
$$\hat{x}_3(k) = \hat{x}_3(k, k-1) + K_{31}\gamma_1 + K_{32}\gamma_2.$$

It can be seen that nonmeasurable component $x_3(k)$ of the state vector can be estimated if matrices, $\mathbf{P}'(k), \mathbf{P}(k)$ and, $\mathbf{K}(k)$, have nonzero respective elements. Thus, the filter generates state estimates that are optimal in the sense of minimum of mathematical expectation of the sum of squared state vector estimate errors, and they are much less influenced by external stochastic disturbances and measurement noise.

As far the filter is used at the stage of preliminary data processing, i.e. preparing the data to model constructing, at this stage still there is no SS model to be used by the filtering algorithm. To overcome the problem, in the DSS constructed the simplified SS model is used of the following form:

$$\mathbf{x}(k) = \mathbf{A}(k)\mathbf{x}(k-1) + \mathbf{w}(k), \tag{6}$$

$$\mathbf{z}(k) = \mathbf{H}\mathbf{x}(k) + \mathbf{v}(k), \tag{7}$$

where, $\mathbf{A}(k) = \mathbf{I}$, i.e. the matrix of dynamics is replaced the unity matrix $\mathbf{I}$. In spite of the simplified approach the model (6), (7) still takes into consideration both stochastic process, $\mathbf{w}(k)$, and, $\mathbf{v}(k)$, that introduce uncertainty into respective computational procedures. An advantage of this approach is that we do not need to construct special model for optimal filtering what is rather problematic at this stage of data processing. Actually, the state model (6) describes random walk processes that are very often met in economy and finances. In spite of their simplicity the models exhibit quite acceptable behavior regarding description of integrated nonstationary process at short time intervals. The computational experiments with a number of nonstationary financial processes like price forming, income and exchange rate forecasting completely justified such simplified approach to preliminary data processing.

Taking Into Account Possible Uncertainties of Model Parameters Estimates. The uncertainties relevant to the model parameter estimates (such as bias and inconsistency) come from the low informative data, or data that do not show normal distribution. This is a requirement for application of least squares (LS) method for parameter estimation. The problem may also arise in a case of multicollinearity of exogeneous variables and influence of nonlinearity that has not been taken into account when the model was constructed. When the data power (size of data sample) is not satisfactory for model construction it could be expanded to some extent by applying special techniques, or computer simulation is applied, or special modeling techniques, such as the group method for data handling (GMDH), are hired. The method very often produces models of acceptable quality even with processing short samples. If data available deviate from the normal distribution, then maximum likelihood (ML) technique can be used, or appropriate Markov Chain Monte Carlo (MCMC) procedures applied [5,11]. The MCMC techniques can be used with quite acceptable computational expenses when number of model parameters is not large.

Model Structure Uncertainties. In the frames of DSS a model structure is estimated in automatic mode of the system operation. If a model is developed for forecasting purpose we construct several candidates and select the best one of them using appropriate adequacy statistics. It is possible to define the following methods to account for structural uncertainties: gradual search for an acceptable model order (AR(p) or ARMA(p, q)) using adaptive modeling and automatic search for the "best" model structure with combined adequacy criteria; adaptive estimation of lag and data distribution parameters; detecting process nonlinearities and describing them with appropriate (alternative) analytical members. As an example of the combined model adequacy and forecast quality criterion can be considered the following:

$$J = \left|1 - R^2\right| + \alpha \ln\left[\sum_{k=1}^{N} e^2(k)\right] + |2 - DW| + \beta \ln(MAPE) + U \underset{\hat{\theta}_i}{\longrightarrow \min},$$

where R^2 is model determination coefficient; $\sum_{k=1}^{N} e^2(k) = \sum_{k=1}^{N} [y(k) - \hat{y}(k)]^2$ is a sum of squared errors; DW is Durbin-Watson statistic determining residual autocorrelations; $MAPE$ is the mean absolute percentage error for analyzing forecasts quality; U is Theil coefficient; α, β are appropriately selected weights; $\hat{\theta}_i$ is a vector of parameters for i-th candidate model. Such criterion is used for automatic (or semi-automatic) selection of the best process model. The criterion is also used for setting adaptive mode of operation for DSS. Also, other forms of the combined criteria were tested. It is important not to overweigh separate members in the right hand side of the expression for criterion.

2.3 Forecasting Methods Which Are Used in the System

The system provides a possibility for using the following forecasting techniques: application of forecasting models based on regression analysis; GMDH and fuzzy GMDH; backpropagation neural nets; support vector regression; the method based upon nearest neighbor analysis; static and dynamic Bayesian networks (BN); generalized linear models (GLM). Bayesian networks are helpful for establishing actual causal relations between the variables selected for model construction [4,14–17]. Besides, the model generates probabilistic forecasts that are a good alternative for regression models.

The set of methods hired for the system construction is suitable for modeling linear stationary and nonlinear nonstationary processes. The regression models may include the trend component and further applied for forecasting integrated nonstationary processes. To compute long-term forecasts the nearest neighbor technique is suitable, but the method requires rather long samples for processing. The DSS allows for performing structural and parametric model adaptation what makes it more functionally flexible and leads to development of high adequacy models in conditions of processes nonstationarity. The principal component analysis is applied to statistical data when necessary what allows to process another uncertainty in the form of highly correlated independent variables in the RHS of a model.

At the stage of model structure estimation the DSS provides for enough information regarding the type of the process being modeled and select correctly a method for parameter estimation. This way it is possible to generate the estimates that satisfy general requirements to the estimates, such as absence of bias, consistency and effectiveness. At each stage of system functioning, i.e. preliminary data processing, model structure and parameter estimation, and generating of forecasts a separate set of statistical quality statistics is used. As a result each stage of data processing is controlled and quality of intermediate and final results is guaranteed.

Statement. The modeling and forecasting information technology proposed on the bases of the set of statistical and probabilistic methods in the frames of decision support system provides for acceptable level of preliminary statistical data processing, adaptive estimation of model structure and parameters with taking into account available uncertainties of stochastic, structural and parametric types, computing forecast estimates and guarantees high quality of final results in the form of possible managerial actions.

3 An Application Example of a Combined Forecasting Model

Now consider application of a combined model for short-term forecasting of a NN financial process at a stock exchange. The model structure is given in Fig. 1. First, the data is prepared to model constructing using optimal filtering and data imputation procedures. As SS model for implementing the filtering procedure the simplified model (6), (7) is used. The missing data imputation is possible by the following methods: simple averaging (provided that the number of missing values does not exceed three), optimization procedure (EM-algorithm) and calculation of predictive values based on regression models.

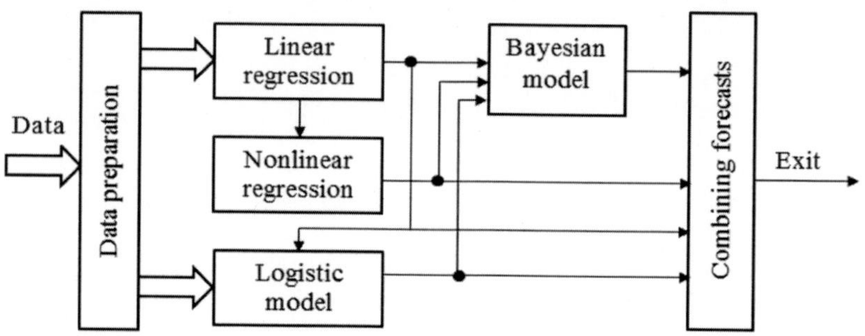

Fig. 1. Combined model for forecasting nonlinear nonstationary processes

Then the data is analyzed with statistical tests for nonlinearity and integration. If the test shows availability of nonlinearity it means that resulting model will consist of linear and nonlinear parts. The linear part is constructed in the form of linear regression. Further on output of the linear regression will be used as one of inputs to logistic regression. These two models together (linear and logistic regression) provide a possibility for formal description of nonlinear processes containing linear component. Thus, to forecast the pricing process the linear regression is used and nonlinear regression together with logistic function provide forecasting of nonlinear component with its further use in the forecast combination procedure. The Bayesian model in the form of Bayesian network generates probabilistic inference regarding the process forecast. This is alternative value of the process forecast that shows the probability of assigning forecasted value to some interval that is used in the block of forecasts combining. Thus, all in all we have for combining four alternative values of forecasts for the process under consideration. The combined forecast is computed in the following way:

$$\hat{y}(k+s) = \sum_{i=1}^{4} w_i \hat{y}_i(k+s), \; \sum_{i=1}^{4} w_i = 1,$$

where, w_i, are weighting coefficients for each alternative forecast; s is a number of forecasting steps. The weights, w_i, can accept equal values if, $\hat{y}(k+s)$, is computed as a simple mean; and alternative calculation of the coefficients is performed as follows:

$$w_i = \frac{1/sse_i}{\sum_{i=1}^{4}(1/sse_i)},$$

where $sse_i = \sum_{k=1}^{N} e^2(k)$ is a sum of forecasting errors on a training sample. For the sake of correctness of the computations in all cases the sum of weighting coefficients should be equal to unity. It is known that combined forecast has a property that its forecasting errors exhibit the variance that is smaller than the lowest variance of the alternative forecasting procedures being used.

The combined model considered also provides a possibility for generating forecasts in the form of probability distribution according to the expression:

$$p(\hat{\mathbf{y}}, \theta|\mathbf{y}) = p(\hat{\mathbf{y}}|\theta, \mathbf{y})p(\theta|\mathbf{y}),$$

where, $p(\hat{\mathbf{y}}, \theta|\mathbf{y})$, is joint probability distribution for the forecasts, $\hat{\mathbf{y}}$, and distribution parameter θ; $p(\hat{\mathbf{y}}|\theta, \mathbf{y})$ is conditional probability for the forecasts; $p(\theta|\mathbf{y})$ is prior probability density function for parameter θ. To compute necessary forecasts in the form of distribution, $p(\hat{\mathbf{y}}|\mathbf{y})$, it is necessary to perform the following integration:

$$p(\hat{\mathbf{y}}|\mathbf{y}) = \int_{R_\theta} p(\hat{y}, \theta|y)d\theta = \int_{R_\theta} p(\hat{y}|\theta, y)p(\theta|y)d\theta.$$

The data sample contains 320 values, 300 of which were used to train the model and 20 were used to validate the forecasting result. The results of one-step-ahead forecasting of the above pricing process are presented in Table 2 for the case of autoregression (AR), Bayesian network (BN), and combined model. The quality of forecasting the process volatility is shown in Table 3. The notations (1) and (2) refer to the training and validation samples respectively.

The best single-step-ahead forecasting result was achieved with the mean absolute percentage error (MAPE) (MAPE = 5.38 in the training sample without data filtering and MAPE = 5.04 with data filtering) was obtained using the model consisting of fifth order linear AR-model and the trend component of the same order. As a result of combining the estimates of the forecasts (the four methods mentioned), the MAPE = 5.18 on the training sample, and the MAPE = 5.19 on the test sample were achieved.

The statistical parameters characterizing quality of short-term volatility forecasting for training (1) and test (2) samples are presented in Table 3. The autoregressive model with conditional heteroskedasticity (ARCH), generalized autoregression with conditional heteroskedasticity (GARCH), exponential GARCH (EGARCH), and stochastic volatility model (SVM) were investigated. The values of MAPE obtained in the training sample for the GARCH (1), EGARCH (1), and SVM (1) models are slightly smaller than for the estimates calculated on the test sample: GARCH (2), EGARCH (2), and SVM (2). Such result of

Table 2. Results of forecasting company stock price

Model type	MAPE without filtering the data	MAPE with filtering	Combined forecast: AR(5)+t5, logistic+ linear regression, BN
AR(5)(1)	8.47	7.95	–
AR(5)+t5(1)	5.38	5.04	–
Logistic + linear regression(1)	6.73	6.11	–
BN(1)	7.49	7.32	5.18
AR(5)(2)	9.26	8.87	–
AR(5)+t5(2)	5.76	5.43	–
Logistic + linear regression(2)	7.15	6.64	–
BN(2)	7.73	7.41	5.19

forecasting was expected. The simplest ARCH model showed very large forecasting errors in both samples. This result was expected; it is explained by very simple model structure that does not reflect the actual behavior of the process variance.

Table 3. Forecasting volatility of earnings for company shares

Model Type	MAPE without filtering the data	MAPE with filtering	Combined forecast for EGARCH and SVM
ARCH(1)	9454.4	9188.7	–
GARCH(1)	45.960	36.270	–
EGARCH(1)	4.9400	3.7530	–
SVM(1)	7.5000	4.9023	3.2475
ARCH(2)	5123.5	2494.4	–
GARCH(2)	51.993	28.396	–
EGARCH(2)	5.93	4.607	–
SVM(2)	10.90	6.9590	3.4583

4 Discussion

The results of modeling and forecasting achieved showed that the systemic approach proposed is useful for developing modeling and forecasting methodology and the DSS based upon it that exhibits the feature of directed search for the best forecasting model in the sets of admissible spaces of structures and model parameters. This feature helps to improve substantially the procedure of data analysis and enhance constructed model adequacy as well as forecast quality based upon the model. The computational experiments carried out with hiring actual data for nonlinear nonstationary processes showed usefulness of the approach for its practical applications to modeling and forecasting as well as necessity for it further refinement in the future studies. It is especially important to improve descriptions for the uncertainties mentioned and to use them for reducing degree of uncertainty in the process of model building and forecast estimation.

Application of optimal filtering procedure has always been helpful regarding further improvement of forecasts quality in the range of about 0,2–0,5%. The same touches upon combination of the forecasts generated by ideologically different techniques. Thus, the combined models turned out to be useful and promising for the future data analysis procedures directed on forecasting NN processes.

5 Conclusions

The methodology is proposed for modeling and forecasting economic, financial, and ecological processes that is based on the system analysis principles and combined models including linear and nonlinear regression models as well as probabilistic models in the form of Bayesian networks. The methodology supposes tracking of computational processes of statistical data processing at all stages with appropriate sets of statistical quality criteria. High quality of intermediate and final results is reached thanks to application of statistical quality criteria to data, model adequacy and quality of forecasts. The methodology was implemented in the frames of specialized decision support system directed to analysis of nonlinear nonstationary processes in the areas mentioned.

Hiring of ideologically different techniques creates a good base for combination of various model types to achieve the best results. An example of the methodology application illustrates that it can be used successfully for solving practical problems of modeling and forecasting NN processes in various areas of human activity. The DSS constructed turned out to be useful instrument for a decision maker to perform quality processing of data using ideologically different techniques, generate alternative forecasts, and combine them to reach high quality results. The system implements tracking procedures of the computational processes using three sets of statistical quality criteria at each stage of modeling and forecasting. The system ideology is flexible regarding adding and modifying existing functions as well as adaptation to a user. It can be applied for supporting decision making in various areas of human activities including banking system, industrial enterprises, investment companies, ecology etc.

Extension of the system possibilities is planned with hiring new forecasting techniques based on probabilistic methods, neuro-fuzzy models and other artificial intelligence techniques. Appropriate attention will also be devoted to constructing user friendly adaptive interface based on the principles of human factors.

References

1. Anfilatov, V.S., Emelyanov, A.A., Kukushkin, A.A.: System analysis in control engineering. Finansy i Statistika, Moscow (2002)
2. Babichev, S., Gozhyj, A., Kornelyuk, A., Lytvynenko, V.: Objective clustering inductive technology of gene expression profiles based on SOTA clustering algorithm. Biopolym. Cell **33**(5), 379–392 (2017). https://doi.org/10.7124//bc.000961
3. Babichev, S., Lytvynenko, V., Gozhyj, A., et al.: A fuzzy model for gene expression profiles reducing based on the complex use of statistical criteria and Shannon entropy. In: Advances in Intelligent Systems and Computing, vol. 754, pp. 545–554. Springer (2018)
4. Bidyuk, P., Gozhyj, A., Kalinina, I.: Modeling military conflicts using Bayesian networks. In: Proceedings of the 2018 IEEE 1st International Conference on System Analysis and Intelligent Computing, pp. 145–151 (2018). https://doi.org/10.1109/SAIC2018.8516861
5. Bidyuk, P., Gozhyj, A., Kalinina, I., Gozhyj, V.: Analysis of uncertainty types for model building and forecasting dynamic processes. In: Advances in Intelligent Systems and Computing, vol. 689, pp. 66–78. Springer (2018). https://doi.org/10.1007/978-3-319-70581-1
6. Bidyuk, P.I., Romanenko, V.D., Timoshchuk, O.L.: Time Series Analysis. Polytechnika NTU Igor Sikorsky KPI, Kyiv (2013)
7. De Gooijer Jan, G.: Elements of Nonlinear Time Series Analysis and Forecasting. Springer, Cham (2017). https://doi.org/10.1007/978-3-319-43252-6
8. Dovgyj, S.O., Bidyuk, P.I., Trofymchuk, O.M.: DSS on the basis of statistical and probabilistic methods. Logos, Kyiv (2014)
9. Fernandez, G.: Data Mining Using SAS Applications. CRC Press LLC, New York (2003). https://doi.org/10.1080/02664763.2012.682451
10. Gibbs, B.P.: Advanced Kalman Filtering, Least Squares and Modeling. Wiley, Hoboken (2011). https://doi.org/10.1002/9780470890042
11. Gilks, W.R., Richardson, S., Spiegelhalter, D.J.: Markov Chain Monte Carlo in Practice. CRC Press LLC, New York (2000). https://doi.org/10.1002/(SICI)1097-0258(19980615)17:11⟨1301::AID-SIM882⟩3.0.CO;2-9
12. Haykin, S.: Adaptive Filter Theory. Prentice Hall, Upper Saddle River (2002)
13. Jao, C.S.: Efficient Decision Support Systems – Practice and Challenges from Current to Future. Intech, Rijeka (2011). https://doi.org/10.5772/682
14. Koski, T., Noble, J.M.: Bayesian Networks. Wiley, New York (2009). https://doi.org/10.1002/9780470684023.ch1
15. Lytvynenko, V., Savina, N., Voronenko, M., et al.: Development of the dynamic Bayesian network to evaluate the national law enforcement agencies. In: The Crossing Point of Intelligent Data Acquisition & Advanced Computing Systems and East & West Scientists, pp. 418–424 (2019)
16. Lytvynenko, V., Savina, N., Voronenko, M., et al.: Development, validation and testing of the Bayesian network of educational institutions financing. In: The Crossing Point of Intelligent Data Acquisition & Advanced Computing Systems and East & West Scientists, pp. 412–418 (2019)
17. Zgurovskii, M.Z., Bidyuk, P.I., Terentyev, A.M.: Method of constructing Bayesian networks based on scoring functions. Cybern. Syst. Anal. **44**(2), 219–224 (2008)
18. Zgurovskii, M.Z., Podladchikov, V.N.: Analytical Procedures of Kalman Filtering. Naukova Dumka, Kyiv (1995). https://doi.org/10.20535/1810-0546.2015.2.91066

Isolation of Informative Features for the Analysis of QRS Complex in ECG Signals

Anatolii Pashko[1(✉)], Iurii Krak[1,2], Oleg Stelia[1], and Oleg Khorozov[2]

[1] Taras Shevchenko National University of Kyiv, Kyiv, Ukraine
aap2011@ukr.net, yuri.krak@gmail.com, oleg.srelya@gmail.com
[2] Glushkov Cybernetics Institute, Kyiv, Ukraine
oleh753@hotmail.com

Abstract. Spatial and frequency analysis are the main approaches to signal classification. The paper investigates statistical algorithms for ECG signal processing, such as correlation function estimation, spectral density (Fourier transform), estimation of R-R intervals distribution, and various variants of wavelet transform. The dynamics of the R-R intervals lengths dispersion variation were also investigated. To distinguish R-R intervals, an ECG approximation was used with the help of a piecewise polynomial continuous and differentiated curve developed by the authors, which has a continuous of first derivative. The algorithm for estimating the first derivative based on proposed splines is developed. The detection of the spline thresholds value for the effective isolation of R peaks in the ECG signal is investigated. Using the spline-approximation derivative allowed us to increase the accuracy of R-R intervals selection up to 6% compared to the known numerical methods. Wavelet transforms were used for the spatial-temporal analysis of the ECG signal, which make it possible to use parts of signals in the systems of operational signal investigation. Also, on the basis of the inverse wavelet transform, the ECG signal was cleared of noise.

These algorithms formed the basis for the creation of a smart system for monitoring and prompt evaluation of cardiac regimens. Combining different approaches to ECG analysis has increased the reliability of the recognition system. Experiments show that the use of a smart system allows us to enough accurately diagnose abnormalities in the heart.

Keywords: Machine learning · Classification · Signal analysis · Expert systems · Wavelet transformation

1 Introduction

Electrocardiography (ECG) as a non-invasive method for the diagnosis of the cardiovascular system including identification of signs characterizing the types of heart failure. The data obtained from patient monitoring consist of a multidimensional series observations of physiological parameters. This data potentially

S. Babichev et al. (Eds.): ISDMCI 2020, AISC 1246, pp. 409–422, 2021.
https://doi.org/10.1007/978-3-030-54215-3_26

contains a lot of information that is used inefficiently due to a heterogeneous data structure that is difficult to interpret. One solution is choice of cardiac cycle features for disease classification.

QRS complex detection is an important step for the analysis of heart rate variability. Algorithms based on the differentiated ECG are computationally efficient for real-time analysis of datasets. For performance, a combination of the squaring function and Hilbert transform-based algorithms can be applied such that differences in detection will point to abnormalities in the signal that can be further analysed. A frequently used mathematical tool in various algorithms for QRS complex selection are derivatives [1,10]. One of the most popular QRS complex detection algorithms - the Pan-Tompkins algorithm [5]. It reliably recognizes QRS complexes based on digital analysis of slope, amplitude, and width. The algorithm automatically adjusts threshold values and parameters periodically to adapt to ECG changes such as QRS morphology and heart rate. The ECG data is a function that responds to variability in "time" and derivatives of which are calculated by differentiating the interpolating functions. As such functions, piecewise polynomial curves are chosen. An analysis of various interpolation methods is contained in [3,4]. So in [4], some interpolation methods as linear interpolation, Lagrange interpolation, Hermits interpolation, and cubic spline interpolation are studied to find the optimal method for selecting the R-R interval in the analysis of heart rate variability. As a result of the experiments, it was shown that the third-order Lagrange interpolation polynomial is the most suitable algorithm for fitting the R-R interval to estimate the autoregressive spectrum since it requires short processing time and shows the lowest error rates when calculating the heart rate variability parameter. Interpolation of the heart rhythm with a cubic spline was studied in [6], where the cubic spline coefficients were used to evaluate and separate different heart diseases. It is shown that, based on the results of the applied methodology, it is possible to categorize cardiac disorders without an actual clinical examination. Despite of the large number of works on this topic, new ones are constantly being created and existing mathematical algorithms for processing and analysing ECG data are being developed. In [12], to increase the accuracy of detecting QRS complexes, a combination of wavelet transform with cubic interpolation by splines is used. Additional requirements for mathematical algorithms are presented in the case of processing the ECG online.

In this study, to approximate the ECG signals, we propose to use the local version of the piecewise polynomial C1 smoothness curve developed by the authors in [9]. The advantages of this spline are the possibilities of fitting its shape to the unsteady behavior of real ECG signals. Moreover, the proposed curve allows us to approximate the cardiac signal and calculate the first derivatives at arbitrary points. The paper proposes also the methods of statistical analysis which allow recursively assessing the characteristics of the ECG signal, which is especially important for real-time processing [13]. The data for the study were taken from [7,11]. In [8,14], various algorithms for processing ECG signals were studied.

2 ECG Signal Spline Approximation for QRS Complex Analysis

The formal problem statement is the following: let the sequence of points be given on time interval $[a, b] : a = \tau_1 < \tau_2 < ... < \tau_N = b$, and at points τ_i, the values of the cardiac signal F_i are given.

Along with the grid Δ_τ, a grid $\Delta_x : \tau_1 < x_2 < ... < x_N < \tau_N$ is introduced, where $x = \alpha\tau + (1-\alpha)\tau$, $0 < \alpha_i < 1$, $i = 3, ...N-1$, $0 < \alpha_2 \leq 1$, $x = \beta\tau + (1-\beta)\tau$, $0 < \beta_i < 1$, $\alpha_{i+1} = 1 - \beta_i$, $i = 2, ...N-2$, $0 < \beta_{N-1} \leq 1$. For segments $[x_i, x_{i+1}]$, $i = 2, ...N-1$ the interpolation polynomial is written as follows:

$$s_i(x) = f_{i+1}\frac{x - x_i}{x_{i+1} - x_i} + f_i\frac{x - x_{i+1}}{x_i - x_{i+1}} + (x - x_i)(x - x_{i+1})(Ax + B), \quad (1)$$

where

$$A = \frac{f_i^{\prime} + f_{i+1}^{\prime}}{(h_i + h_{i+1})^2} - \frac{2(f_{i+1} - f_i)}{(h_i + h_{i+1})^3},$$

$$B = \frac{f_{i+1}^{\prime} - f_i^{\prime}}{2(h_i + h_{i+1})} - \frac{(f_{i+1}^{\prime} + f_i^{\prime})(x_i + x_{i+1})}{2(h_i + h_{i+1})^2} + \frac{(f_{i+1} - f_i)(x_i + x_{i+1})}{(h_i + h_{i+1})^3},$$

$$f_i = \alpha_i F_{i-1} + (1 - \alpha_i)F_i, f_{i+1} = \beta_i F_{i+1} + (1 - \beta_i)F_i,$$

$$f_i^{\prime} = \frac{F_i - F_{i-1}}{\tau_i - \tau_{i-1}}, f_{i+1}^{\prime} = \frac{F_{i+1} - F_i}{\tau_{i+1} - \tau_i},$$

$$h_i = \alpha_i(\tau_i - \tau_{i-1}), h_{i+1} = \beta_i(\tau_{i+1} - \tau_i).$$

Figure 1 shows the approximation graph of the proposed spline of a real ECG signal. Note that the relative error of the proposed spline does not exceed 6%, which allows you to save all the topological characteristics of the original signal and find the first derivatives at arbitrary points, as well as a clear selection of QRS complexes in the signal.

2.1 QRS Complex Detection and R Peak Determination

The use of the piecewise polynomial curve (1), which ensures smoothness in C^1, made it possible to clearly distinguish QRS complexes in the approximated ECG signal, as is shown in Fig. 2.

The next step consists of the use of polynomials (1) to determine the R peak. In determining the position of the R peaks, the values of the first derivatives are used. The search for the peak was carried out among a group of points, the

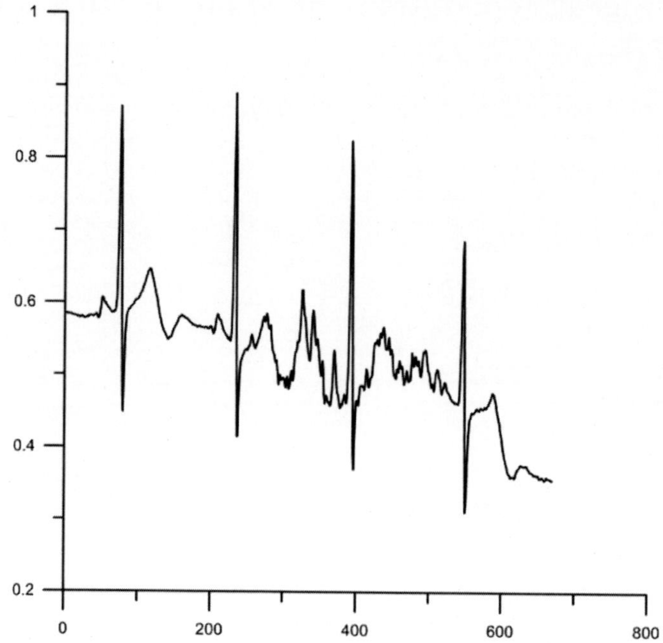

Fig. 1. Spline approximation of the ECG signal

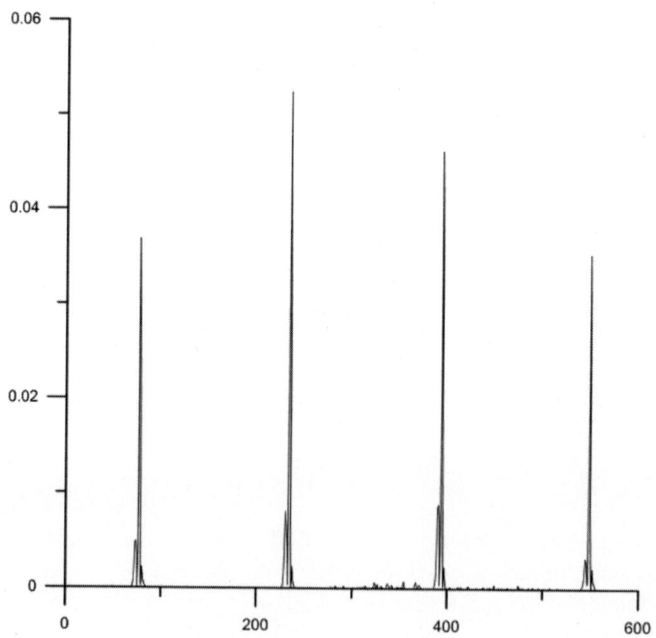

Fig. 2. The QRS-complex isolation

values of the first derivatives at which exceeded a certain threshold. In this case, the values of the derivatives were calculated not only at the starting points, but also at the intermediate ones. In the presented experiments, nine intermediate points were used. As it was shown by computational experiments, this approach is justified, since the values of the derivative exceeding a given threshold can be located between the nodes of the original mesh. In Fig. 3, Fig. 4 and Fig. 5 solid lines show graphs of cardiosignals on which the detected peaks of R are indicated by dots. It was assumed that $\alpha = \beta = \frac{1}{2}$.

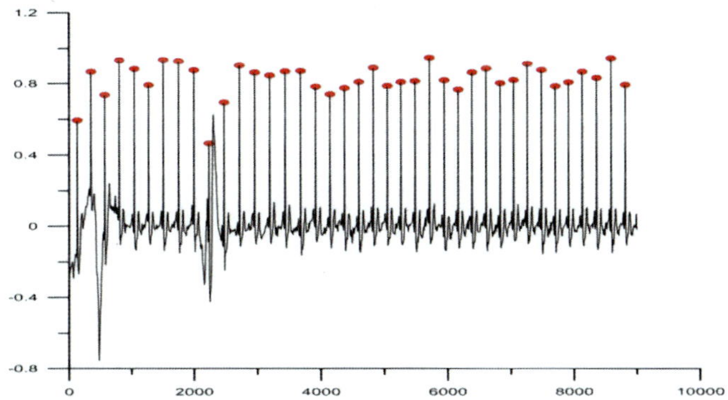

Fig. 3. R peaks for data set 1

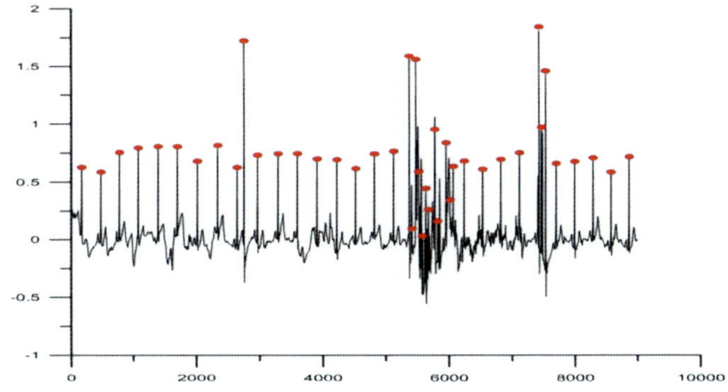

Fig. 4. R peaks for data set 2

The results presented in Fig. 3, Fig. 4 and Fig. 5 indicate a fairly good determination of R peaks using the first derivative. This publication demonstrates and analyses the use of polynomials (1) in the algorithm for detecting R peaks. The percentage of peak detection for various data sets will be studied later.

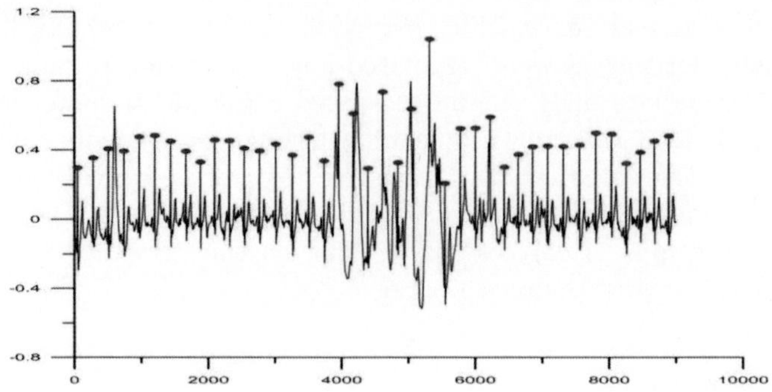

Fig. 5. R peaks for data set 3

It should be noted a significant dependence of peak detection on the selected numerical threshold value. Consider this in the following example. We use cardiac signal data with significant variability of numerical values. The squares of the first derivatives are shown in Fig. 6, and the graph of the initial signal itself is shown in Fig. 7 and Fig. 8.

Note that it is the squares of the first derivatives that play a key role in the algorithm for detecting R peaks. After separating the intervals with the values of the square of the derivative exceeding the specified threshold, the maximum value on the interval and the corresponding abscissa value are extracted from the signal. In Fig. 4 the intervals with maximum values of the square of the derivative are clearly visible. To automate the peak extraction process, a certain threshold value is set.

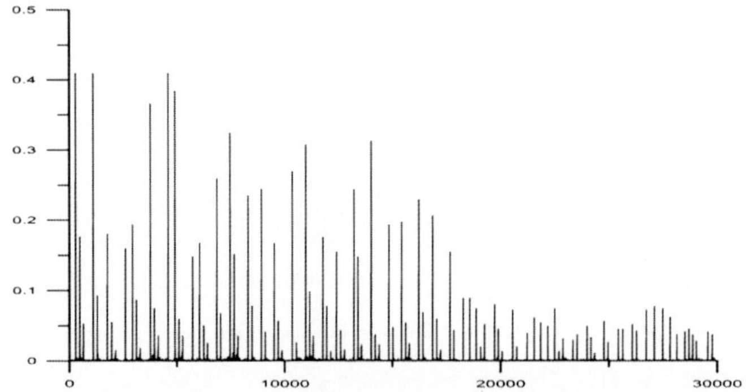

Fig. 6. The square of the first derivative

Figure 7 and Fig. 8 show the graphs of the initial data with the plotted points R for various threshold values. With a threshold value of 0.025, R points were found over the entire interval (see Fig. 7). But "false" points of R were found on

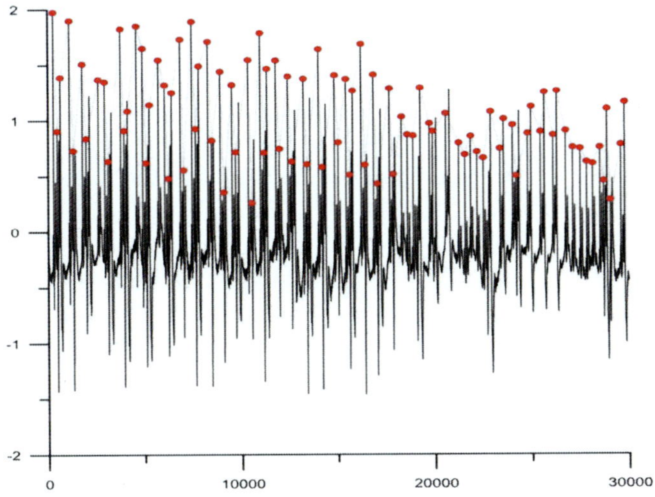

Fig. 7. The threshold value is 0.025

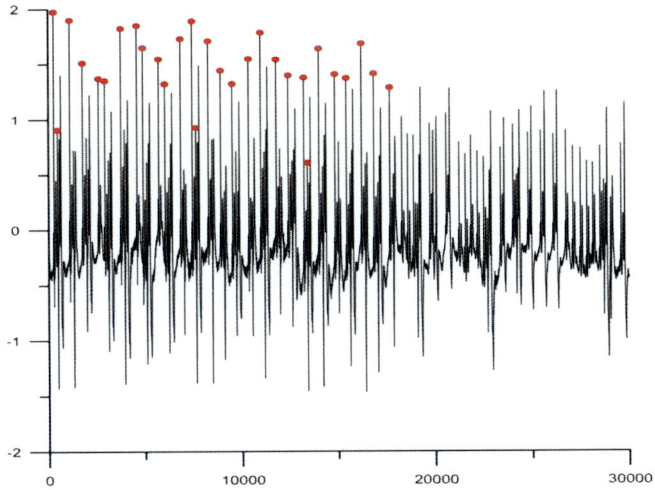

Fig. 8. The threshold value is 0.1

part of the interval. As the threshold value increases, most of the false points disappear, but R points are not found on the other part of the interval as demonstrated in Fig. 8.

3 ECG Signal Processing Based on Discrete Wavelet Transform

The on-line analysis of the ECG signal is based on the correlation-spectral analysis and wavelet analysis. It was noted that the correlation function of the ECG signal during normal heart function clearly indicates the presence of periodic components (Fig. 9).

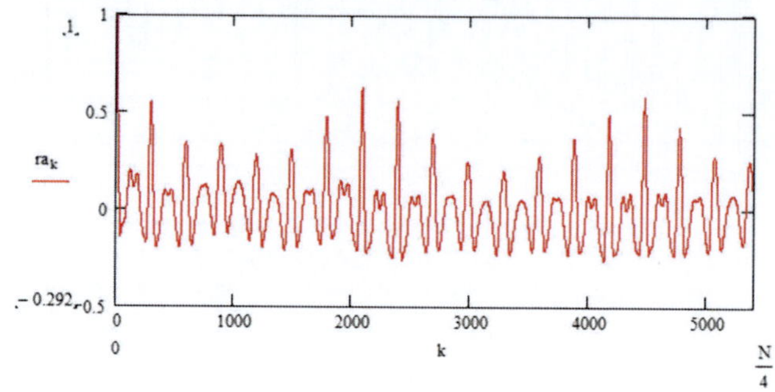

Fig. 9. The correlation function of the ECG signal of a healthy person

In this case, the ECG signal can be considered as a polyharmonic process, i.e. stationary. With abnormalities in the heart, the ECG signal becomes unsteady. In this study, we propose, by assessing the correlation function, to identify the moment of failure of the periodic components, i.e. the moment of violation of the heart. This is important for building smart systems for the rapid assessment of the heart's mode of operation and timely response to changes in conditions that can lead to irreversible consequences. The purpose of the smart system is to ensure the timely detection of abnormal values and timely warning of the patient's condition and predicting possible consequences.

In this case, the ECG signal can be considered as a polyharmonic process. With abnormalities in the heart, the ECG signal becomes unsteady. In this study, we propose, by assessing the correlation function, to identify the moment of failure of the periodic components, i.e. the moment of violation of the heart. This is important for building smart systems for the rapid assessment of the heart's mode of operation and timely response to changes in conditions that can lead to irreversible consequences. The purpose of the smart system is to ensure the timely detection of abnormal values and timely warning of the patient's condition and predicting possible consequences. To evaluate the correlation function, you can use the recurrence formula

$$R_j(N+1) = \frac{N-j-2}{N-j-1} R_j(N) + \frac{1}{N-j-1} x_{N-j} x_N, j = 0, 1, ..., m, \qquad (2)$$

where $R_j(N)$, $j = 0, 1, ..., m$ - it correlation function at a step at a point N; m - the number of points of the correlation function, $\{x_i\}$ - it array of measurements.

Smart systems for monitoring and rapid assessment of the heart's mode of operation should collect data on-line, process and analyze. The considered algorithms for spline approximation and correlation analysis allow us to process data with referential algorithms. For this purpose we can use difference samples of ECG (see Fig. 10).

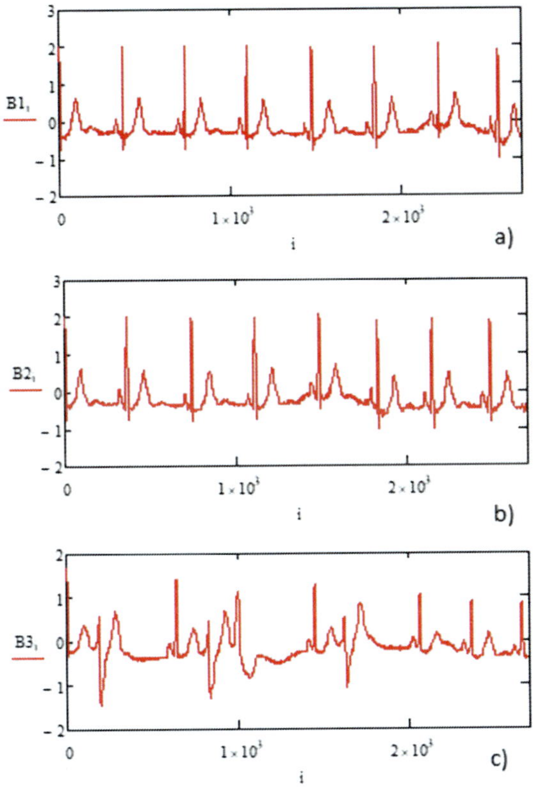

Fig. 10. ECG samples

Algorithm of data collection and processing is the following:

1. The ECG signal data is recorded in the device memory for 5–7 s with a resolution of 300–500 data per second.
2. The resulting implementation is processed, calculated: the average value, the variance of the signal, the assessment of the correlation function, the number of R-R intervals, the average value of the length of the intervals and variance.
3. The received data is saved with compared with the available.
4. If the variance of the length of the R-R intervals changes, then this means a malfunction in the work of the heart. (Variation of variance is checked by statistical methods for testing the hypothesis of equality of variances)

5. If the variance does not change, then go to the first item.

Together with checking the equality of variances, we can test the hypothesis of equality of correlation functions. The presence of differences corresponds to the presence of deviations in the work of the heart. What caused these anomalies is a topic for further research. As a measure of the coincidence of the characteristics of the ECG signal, wavelet transform can be used.

In cardiograms analysis, a special place is occupied study of heart rate variability. The study includes the steps of the R-R intervals measuring and analysis. To study heart rate variability, short-term (several minutes) and long-term (several hours) ECG recordings are used. For the calculations, a cardiogram of 4.5 min was used; data were taken with discreteness $\Delta t = 0.0025$ s. In Fig. 10 the first 8 s of ECG (a) are presented, from 5 to 12 s (b) and from 120 to 128 s (c). Figure 11 shows estimates of the correlation function and estimates of the spectral density of these fragments.

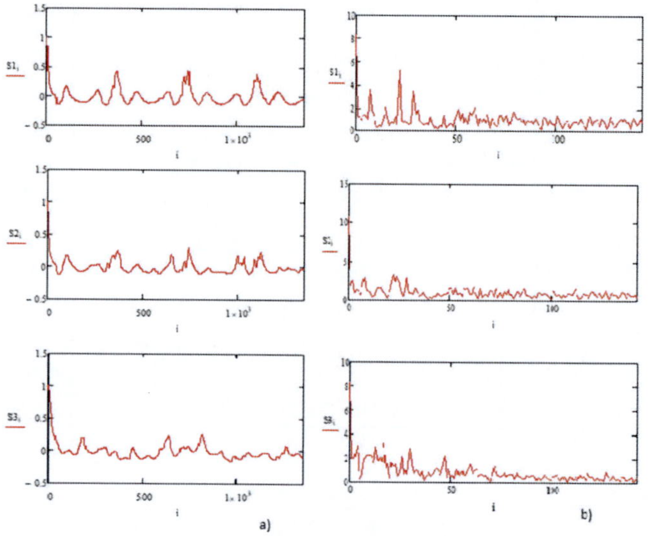

Fig. 11. The correlation function (a) and Spectral Density (Fourier Transform) (b)

The figures show both the periodic component disappears in the estimation of the correlation function over time and the spectral density takes the form of a broad-band process. The Spectral density analysis provides information on the distribution of energy over frequencies. Spectra provide information on the ratio of the energies of different components of heart rate variability. For the entire ECG signal using spline analysis, the distribution density of the length of the R-R intervals was obtained. For the entire ECG signal, the distribution density of the length of the R-R interval was obtained by using spline analysis. In Fig. 12,

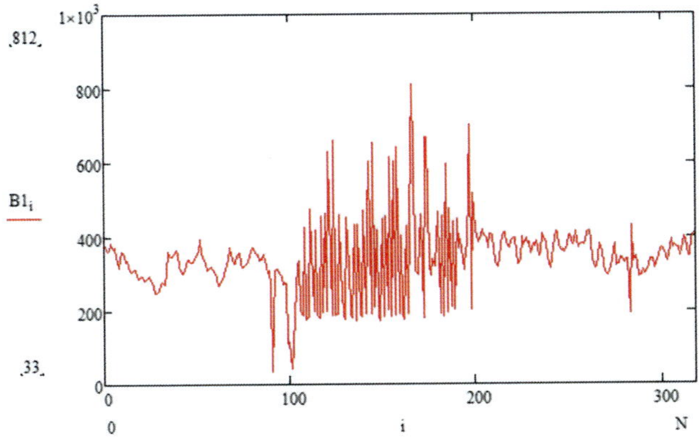

Fig. 12. The lengths of R-R intervals

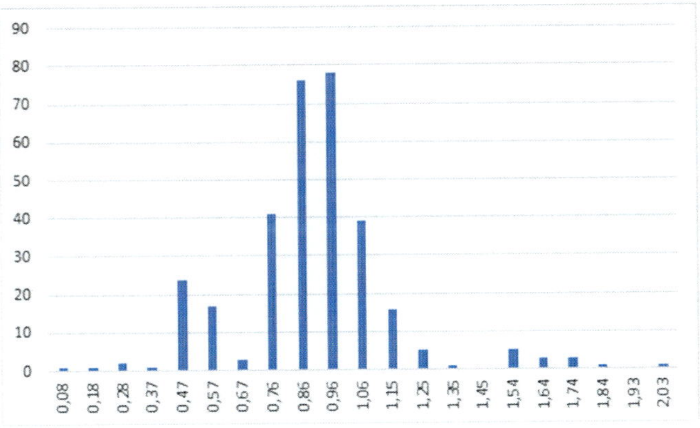

Fig. 13. Histogram of the distribution of R-R intervals

a graph of the lengths of R-R intervals is presented. In Fig. 13, a histogram of the distribution of lengths of R-R intervals is shown.

The average value of the R-R intervals is 338.9, the variance is 10239.8, and the standard deviation is 101.2. The length of R-R intervals in a healthy person has a normal distribution law, and violation indicates a malfunction in the heart. From the graph of the R-R intervals, it can be seen that in the first minute the heart rate variability corresponded to the norm in the third minute the heart rate variability has features. For further analysis, we apply the wavelet transform. Theoretical aspects of the wavelet transform and applied problems were considered in [2]. In Fig. 14, levels of approximating wavelet transform coefficients are presented for three ECG fragments used Haar wavelet:

The wavelet analysis of successive sections of the cardiogram also changes, and this allows to draw conclusions about abnormalities in the hearts. With the most advanced computer-based signal processing, you can quickly get a great

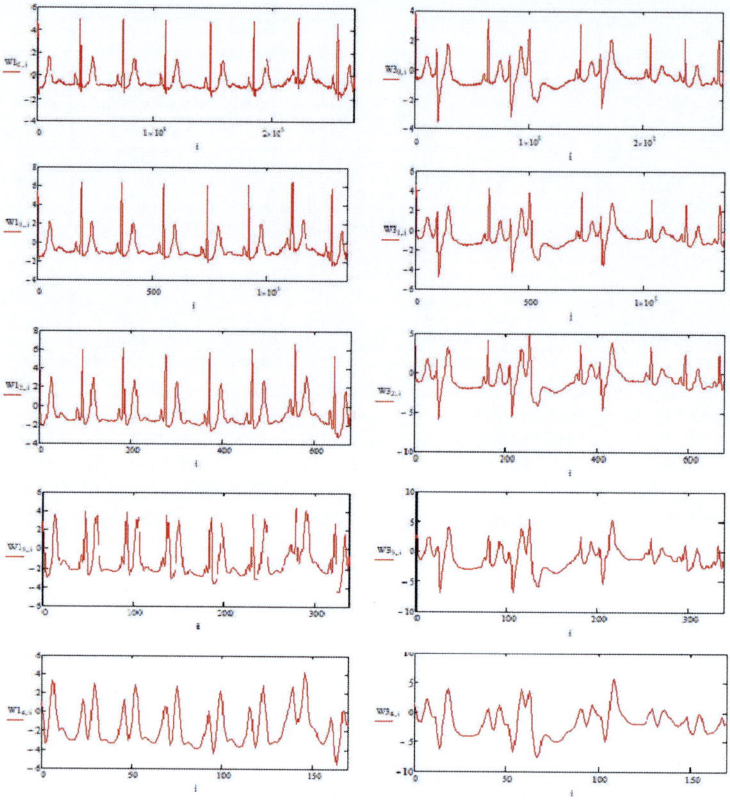

Fig. 14. Wavelet transform levels

look at the big data. By repeating these abilities and traditional methods of ECG analysis, you can do the calculation of the ancillary devices, you can select the power for the automatic analysis of the hourly and frequency parameters of the electrocardiogram, and the digital card will be canceled.

4 Conclusions

This article proposes an approach to processing and analyzing data in cardiology based on statistical analysis and wavelet conversion.

Data analysis and processing were performed on real data. A comparative analysis of the obtained results with the results of the Pan-Thompson algorithm is performed. The analysis showed that the algorithm presented in the article shows better results than the Pan - Thompson algorithm. At the same time, the developed algorithm has the advantage of being able to analyze not only the QRS complex but also other segments of the electrocardiogram.

The use of spline approximation of the R-R intervals and the calculation of the spline approximation derivative made it possible to increase the reliability of the QRS complex isolation. Threshold selection studies have been conducted to efficiently locate QRS complexes. The use of spline-approximation derivatives has allowed up to 6% to increase the accuracy of R-R interval allocation compared to numerical methods. A wavelet transform was used for the spatial-temporal analysis of the ECG signal. Previously, the major problem with complex QRS analysis algorithms was the high complexity of computing. Currently, computing power allows for real-time analysis.

Thus, the approach developed in the article to the processing and analysis of electrophysiological instrumental diagnostic data can be successfully applied in practice of cardiology.

References

1. Ary, L., Valmik, B.: Computerized measurement of the first derivative of the QRS complex. Theor. Pract. Consid. Comput. Biomed. Res. **14**(5), 464–471 (1981)
2. Daubechies, I.: The wavelet transform, time-frequency localization and signal analysis. In: Transactions on Information Theory, vol. 36. IEEE (1990)
3. Ellis, R., Zhu, B., Koenig, J., Thayer, J., Wang, Y.: A careful look at ECG sampling frequency and R-peak interpolation on short-term measures of heart rate variability. Physiol. Meas. **36**, 1827–1852 (2015). https://doi.org/10.1088/0967-3334/36/9/1827
4. Jang, D., Hahn, M., Jang, J., Farooq, U., Park, S.: A comparison of interpolation techniques for RR interval fitting in AR spectrum estimation. In: IEEE Biomedical Circuits and Systems Conference (BioCAS), Hsinchu, pp. 352–355 (2012). https://doi.org/10.1109/BioCAS.2012.6418424
5. Jiapu, P., Willis, J.T.: A real-time QRS detection algorithm. IEEE Trans. Biomed. Eng. **BME-32**(3), 230–236 (1985). https://doi.org/10.1109/TBME.1985.325532
6. Kalovrektis, K., Ganetsos, T., Shammas, N., Taylor, I., Andonopoulos, J.: Development of a computerized ECG analysis model using the cubic spline interpolation method. In: CSS 2011 Proceedings of the 5th WSEAS International Conference on Circuits, Systems and Signals, Corfu Island, Greece, 14–16 July 2011, pp. 186–189 (2011)
7. Khorozov, O., Krak, I., Kulias, A., Kasianiuk, V., Wojcik, W., Tergeusizova, A.: Vital signs monitoring using fuzzy logic rules. In: Information technology in Medical Diagnostics II. Proceedings of the International Scientific Internet Conference on Computer Graphics and Image Processing and 48th International Scientific and Practical Conference on Application of Lasers in Medicine and Biology, pp. 237–244 (2018). https://doi.org/10.1117/12.2501585
8. Nagendra, H.E.A.: Application of wavelet techniques in ECG signal processing: an overview. Int. J. Eng. Sci. Technol. **3**(10), 7432–7443 (2011). https://doi.org/10.7439/Fijbar
9. Olszewski, S., Komada, P., Smolarz, A., et al.: Some features of the numerical deconvolution of mixed molecular spectra. In: Lecture Notes in Computational Intelligence and Decision Making, Series Advances in Intelligent Systems and Computing, vol. 1020, pp. 215–224. Springer (2020). https://doi.org/10.1007/978-3-030-26474-1_2
10. Paul, H., Langner, J., David, B.: Geselowitz, first derivative of the electrocardiogram, circulation research. Circ. Res. **10**(2), 220–226 (1962). https://doi.org/10.1161/01.RES.10.2.220

11. PhysioNet: PhysioNet ECG Data (2020). https://github.com/mathworks/physionet_ECG_data
12. Rodrigues, L., Marengoni, M.: Detecting QRS complex in ECG using wavelets and cubic spline interpolation. In: Biomedical Engineering, BioMed 2012, Innsbruck, Austria, 15–17 February 2012, p. 6 (2012). https://doi.org/10.2316/P.2012.764-135
13. Stelia, O., Krak, I., Potapenko, L.: Controlled spline of third degree: approximation properties and practical application. In: Lecture Notes in Computational Intelligence and Decision Making, Series Advances in Intelligent Systems and Computing, vol. 1020, pp. 20–34. Springer (2020). https://doi.org/10.1007/978-3-030-26474-1_16
14. Tan, R., Perkowski, M.: Toward improving electrocardiogram (ECG) biometric verification using mobile sensors: a two-stage classifier approach. Sensors (Basel) **17**(2) (2017). Art. no. E410. https://doi.org/10.3390/s17020410

Methodological Basis of Causal Forecasting of the Economic Systems Development Management Processes Under the Uncertainty

Marharyta Sharko[1(✉)], Nestor Shpak[2], Olga Gonchar[3], Kateryna Vorobyova[4], Olena Lepokhina[5], and Juliya Burenko[6]

[1] Kherson National Technical University, Kherson, Ukraine
mvsharko@gmail.com
[2] Lviv Polytechnic National University, Lviv, Ukraine
dida_05@ukr.net
[3] Khmelnytsky National University, Khmelnytsky, Ukraine
o.i.gonchar@i.ua
[4] Limkokwing University of Creative Technology, Kuala Lumpur, Malaysia
katrin.vorobyova@gmail.com
[5] Private Higher Education Institution International University of Business and Law, Kherson, Ukraine
Law1569@i.ua
[6] International University of Business and Law, Kherson, Ukraine
burenko-05@mail.ru

Abstract. The methodological bases of information support of the contribution of economic development indicators for variables input are proposed. Structural diagrams, directions and quantitative estimates of the causal relationships of the factors affecting the output dependent variable are developed. It is suggested that the difference between apriori and aposteriori entropy of economic system indicators is to be taken as a measure of removing information uncertainty. To improve the accuracy of quantitative estimates of the likelihood of linkages between economic indicators and input variables, we propose to use the Bayes formula. The algorithm of causal prediction is presented. It is recommended to model the forecasting situations using Bayesian networks, which allow us to encode knowledge of causal and associative relationships to make management decisions for the development of complex economic systems.

Keywords: Causal forecasting · Causality · Uncertainty · Economic development · Information processing · Computational focus

1 Introduction

Forecasting is a science-based prediction based on a system of regularities, identifying the state of diagnostic objects, and identifying ways to change them

S. Babichev et al. (Eds.): ISDMCI 2020, AISC 1246, pp. 423–436, 2021.
https://doi.org/10.1007/978-3-030-54215-3_27

necessary to make adequate management decisions. The economic processes forecasting is at the heart of decision-making efficiency in situations of extreme complexity using models and quantitative methods for predicting possible changes [13,14]. Under uncertainty, the implementation of each economic development strategy and its management is associated with many possible outcomes, the magnitude of potential losses and risks. Most organizations change the purpose of their operations in accordance with changes in external conditions or states of the elements of the system.

The statistical analysis of time series vulnerabilities based on regression and correlation is implemented. It was shown that the obtained results do not allow us to judge causality, but only indicate coincidence of events and phenomena [4,20]. Extrapolation of the development processes based on the analysis of changes in one factor is not correct, as it does not consider the changes in other significant factors that determine the parameters of managerial decision-making [3,16]. The economic systems complexity leads to the search for means of simultaneous modeling of the effects on the predicted characteristics of several independent variables with their development indicators.

In forecasting causal models, called causal or dependent, the predicted value is a function of a large number of input variables. The causal prediction proceeds from the fact that the behavior of the dependent variable in the future is not related to past trends and cannot be predicted on their basis. In other words, each of the variables must have its own set of parameters that determines the probabilistic relationship. This behavior is determined by a complex of cause and effect relationships between the dependent variable and the independent factors and aims to distinguish the most alternative paths of development from the set of possible states. The hereinbefore presented facts indicate the relevance of the current problem.

2 Literature Review

The methods of causal forecasting are based on correlation analysis on the basis of establishing the relationship between economic indicators of input variables and regression analysis based on the derivation of equations linking input and output variables. Causal forecasting has previously been used to study social problems [8], signal isolation against interference [9,15], energy efficiency [6], electrical loads [10], machine learning and product transactions [12], neuroscience [7], associative personalization [19].

Such models are used when independent variables are known in advance or easier to predict than the dependent variable. The difficulty of using them depends on whether the value to be predicted really depends on the independent variable. Existing forecasting models always have concomitant effects that affect the input parameters, which must be varied within well-defined limits. With regard to forecasting the economic systems development in the long run under uncertainty conditions the causal forecasting has not previously been studied.

Obtaining the future performance estimates is an important step in managerial decision-making, which provides the competition advantage [2,17]. A distinctive feature of the economic systems development forecasting is the managerial nature of the decisions made. The role of the economic forecasting increases with the acceleration of the scientific and technological progress, the sophistication of management tasks and the intensification of uncertainty [11]. It is said to use different forecasting methods in order to minimize uncertainty in management [18]. Economic and mathematical models of forecasting give successful results in management decisions only in the presence of stable dynamics of indicators in the short and medium term [21]. The results of their use in the long run, as a rule, are incorrect.

The unsolved parts of the general problem of the development of the methodology for forecasting the processes of managing the development of economic systems under uncertainty include the establishment of causal relationships that affect the output parameters of forecasting objects.

The aim of the research is development of methodological bases for analysis and processing of causal forecasting information under uncertainty which allow us to increase the adequacy of forecasts when making management decisions for the development of economic systems not only in the short but also in the long run.

3 Materials and Methods

The causal link between economic development factors is an information channel in which information coming from the cause must be more than its own statistical uncertainty of the factors. Economic development is influenced by a large number of conditions and factors of different nature: technological, resource, intellectual, organizational, etc. Causal forecasting is possible only for a theoretically grounded model after accumulating a considerable amount of knowledge about the simulated reality.

As a tool for causal forecasting, a system of methods is used to analyze, cause and effect relationships and shape their changes in the future. The analysis and synthesis of the information support methods, deepening of theoretical understanding and modeling of the economic systems development processes are the essence of algorithm of operations sequence of causal forecasting under uncertainty. In the general case, as a rule, it is not possible to distinguish the only property that characterizes economic development. Therefore, we define some sets of properties of the input variables, each of which characterizes a certain local quality of the output predicted quantity $\hat{y}$, and together they characterize this quantity as a whole quite sufficiently.

In set theory, it is possible to combine elements of different groups and levels that form a qualitatively new state of the system with their many characteristics, however, this approach only takes into account the variability of the process, but not the causality and its purpose. To address this shortcoming, it is necessary to develop a methodology for providing information on the specific contribution of economic development indicators to input variables. In causal forecasting, the analyzed processes of economic systems development can be deterministic, in which independent variables uniquely determine the state of the object under consideration and stochastic, if the characteristics of the state are randomly dependent on time. A common requirement for predictive input is that input variables x_i must be mutually exclusive with a collectively exhaustive set of states determined by economic development indicators. The specificity of causal forecasting models is to establish causal relationships between dependent and independent variables and to construct multidimensional regression and simulation models.

The methodological basis of economic development management is a set of factors of production, information systems and forecasting methods in the long run. The causal forecasting process is as follows:

$$\hat{y} = f(x_1, x_2, \ldots, x_n),$$

where x_i is the independent variables, $\hat{y}$ is the response variable.

Causality implies the fulfillment of two postulates:

- independent variables x_i is a necessary condition for the existence of a dependent variable y,
- independent variables x_i exist over time.

4 Methodology and Technique of Casual Forecasting

The result of causal forecasting is determined in the form of verbal statements, the structural analysis of which is related to the separation of the integral characteristics of the development of economic systems y. The central concept of causal prediction is the concept of a system, by which is understood the set of M elements x_i connected by the set of bonds Z, ordering elements into a structure possessing a set of properties V.

$$C = (M \times Z)$$

Therefore, the entire prediction system can be defined as an ordered set of S.

$$S = <(M \times Z) \times V>$$

We denote by C_{ij} the contribution of the j economic development indicator of the i input variable to the value of the output predicted characteristic $\hat{y}$ at time $t = 0$. The information support of the contribution of this economic development indicator to the input variable x_i can be represented as a linear matrix:

$$\| C_{ij} \| = \| C_{i1}, C_{i2}, \ldots, C_{im} \|$$

The process of the causal forecasting system operation is closely related to the determination of changes in the properties of its individual elements over time.

We denote by C'_{ij} the contribution of the j economic development indicator of the i input variable to the value of the output predicted characteristic $\hat{y}$ at time t.

$$\alpha_{ij}(C_{ij})' = f(C_{ij})$$

where α_{ij} is the magnitude of the orientation of the movement of economic indicators in the formation x_i from zero to t.

The values of C_{ij} and C'_{ij} can have different dimensions, so their normalization is necessary.

The contribution to the total value of the predicted output characteristic $\hat{y}$ from the element i is determined by equality.

$$A = \| C_{ij} \| \cdot \| \alpha_{ij} \|^T$$

where A is the column matrix of economic indicators i^{th} input variable.

According to the law of large numbers, as the number of observations in the causal forecast increases, the average value of the discrete predicted value $\hat{y}$ will be equal to its mathematical expectation.

$$M[Y] = \frac{x_1p_1 + x_2p_2 + \cdots + x_np_n}{p_1 + p_2 + \cdots + p_n} = \frac{\sum_{i=1}^{n} x_ip_i}{\sum_{i=1}^{n} p_i}$$

The law of large numbers provides for the effective repayment of random deviations from the regular nature. Structural blocks of information support for the contribution of economic development indicators to input variables are presented in Fig. 1.

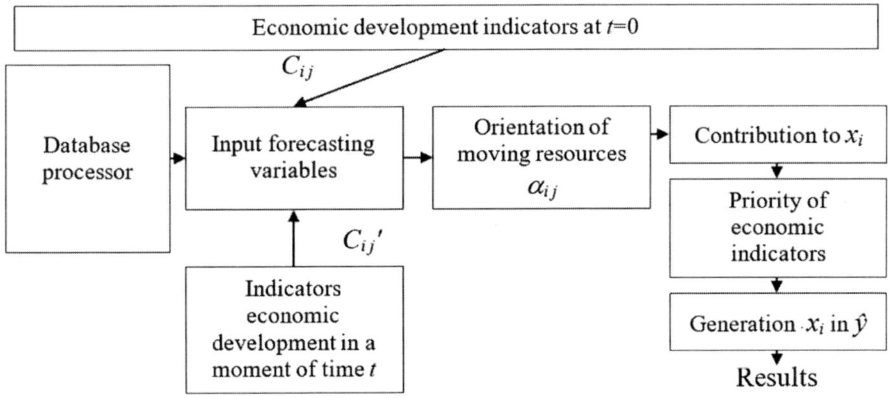

Fig. 1. Structural blocks of information support for the contribution of economic development indicators to the input variables of causal forecasting

The determination of the directionality of causal relationships between the studied factors of influence on the variable $\hat{y}$ can be made by the values of entropy. Entropy is a measure of uncertainty in the choice of random events. With more information, the uncertainty decreases; entropy is a measure of insufficient information about an observed object. The binary cause-and-effect theory of information is based on Shannon's assumptions that the entropy of the output variable y is a real and negative value of H related to the total probability of the prediction system W by the relation.

$$H = k \ln W$$

where k is Boltzmann constant.

Entropy increases with the chaotic distribution of information resources and decreases with their ordering. This creates requirements for input information. Reducing entropy requires large amounts of input. The increase in entropy is due to insufficient and reduced volumes of input information. In this formulation, the entropy of the output variable y is a theoretical-informational degree of uncertainty of a random variable in which the predictor has a probability estimate of Pi.

$$H = \frac{I}{n} = -\sum_{i=1}^{m} P_i \log_2 P_i$$

where $I = -n \sum_{i=1}^{n} P_i \log_2 P_i$.

The total amount of information I depends on the number of elements of the causal prediction system n and the number of their states m. The structural blocks of the technique for calculating the input entropy of causal prediction are presented in Fig. 2.

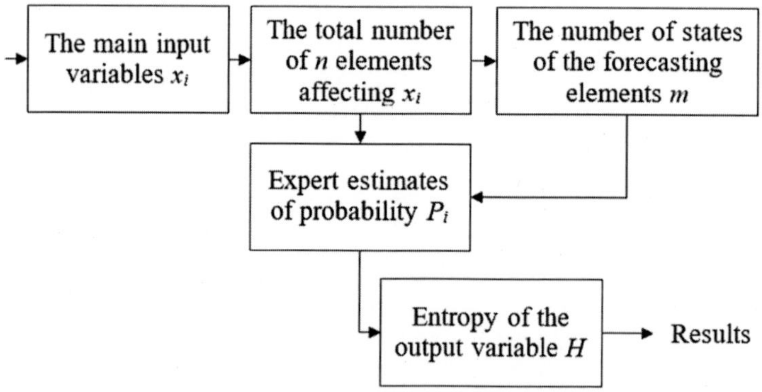

Fig. 2. Structural blocks of the technique of input entropy calculation of causal prediction

The information structure of the a priori data required for the entropy H_0 calculations, which have equal probability, is equal to [1]:

$$H_0 = -\log_2 1 + \log_2 n$$

In this state, the predicted value $\hat{y}$, characterized by the input entropy H_0, will have the maximum amount of information. A posterior entropy H_1, representing the residual uncertainty of information about the state of a prediction object, characterized by variables xi and the probabilities of their manifestation P_i, is equal to:

$$H_1 = P_1(\log_2 N(x_1) - \log_2 N) + P_2(\log_2 N(x_2) - \log_2 N) + \cdots + \\ + P_n(\log_2 N(x_n) - \log_2 N) \quad (1)$$

Then the residual uncertainty of information I is:

$$I = \Delta H = H_0 - H_1$$

The measure of uncertainty is the amount of the missing information. As follows from Formulas (1) and (2), to increase ΔH it is necessary that the quantity H_1, which is taken with the opposite sign in Formula (2), be as large as possible, i.e. so that $\Delta H \to 0$ is the quantity $H_1 \to H_0$.

To improve the accuracy of quantitative estimates of the likelihood of the manifestation of the relationship between x_i and $\hat{y}$, it is proposed to use Bayes' formula that allows one to determine the probability of one event based on knowledge of the probabilities of others [5]. Bayes' formula shows how the probability in the field of obtaining new data should change. It allows the results of experiments to be used to re-evaluate event probabilities and improve decision-making efficiency using statistical data.

If event A can only occur at incompatible positions $B_1, B_2, \ldots, B_n$, then the full probability of event $P(A)$ is:

$$P(A) = P(A \cap B) = P(B_1)P(A|B_1) + P(B_2)P(A|B_2) + \\ + P(B_3)P(A|B_3) + \cdots + P(B_n)P(A|B_n) \quad (2)$$

where the symbol $\cap$ denotes the intersection of varieties.

The probability of event A changes the probability of events $B_1, B_2, \ldots, B_n$:

$$P(B_i|A) = \frac{P(B_i)P(A|B_i)}{P(A)} \quad (3)$$

The conditional probability of the selected specific economic indicators. $P(A|B_j)$ is defined as the partial division of these indicators at the end of the dynamic series by the sum of all previous values. The total probability of events occurrence $P(B_j|A)$, according to Formula (3), is defined as the sum of the products of a priori probabilities $P(B_j)$ for the conditional probabilities $P(A|B_j)$. A posteriori probabilities $P(B_j|A)$ are defined as partial from the division of the corresponding component of the full probability formula by the total value of the total probability $P(A)$.

To implement causal forecasting, it is necessary to have access to the information contained in the elements of the input variable x_i to track the dynamics of the diagnosis process. On this basis, a database is formed, which is a set of inter related indicators that directly or indirectly affect $\hat{y}$. When establishing causal relationships of indicators affecting output forecasting variables, the quality of primary information is an important issue. The manifestation of this information can be done either directly or through indirect connections and interactions of diagnostic parameters. The closeness of these bonds is determined in addition to the interaction of the nearest neighbors, i.e. factors whose participation in the influence of the analyzed dynamic process is indisputable, as well as interaction with more distant neighbors, which, when combined, can have cumulative effects on the prediction results. The structural block chart of technique of calculating the output entropy of causal prediction are presented in Fig. 3.

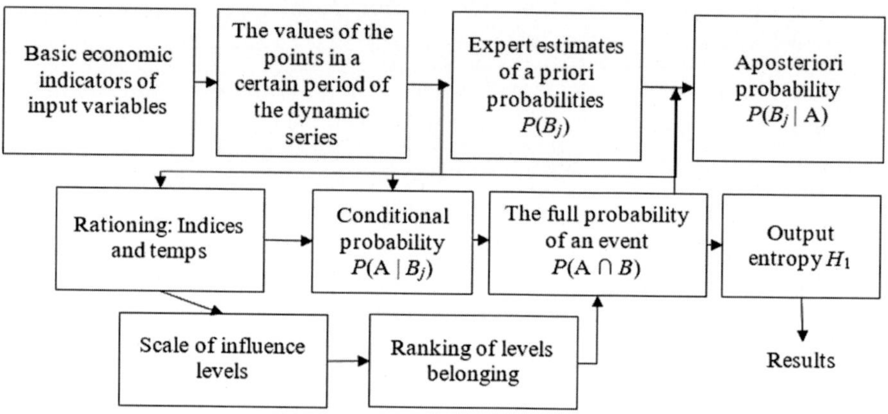

Fig. 3. Structural blocks of the technique of input entropy calculation of causal prediction

5 Results and Discussion

An example of the effect of economic indicators on the input variables x_i of forecasting economic development is presented in Table 1. The values of the probabilities of such indicators are determined by the expert way on the basis of the experience, intuition, analogies and qualifications of the experts, which bears elements of subjectivism. The disadvantages of peer review are unreliable prerequisites information limitations, poor use of results, excessive costs. Building a probability network requires that experts' knowledge be represented by a formalism of probabilities.

The causal forecasting is a technology for developing models of the development of a managed object, taking into account the causal relationships between the indicators that affect the input variables and, through them, the output forecasting variable. Based on a well-founded forecast, appropriate management decisions, economic development are made, program activities and development priorities are specified. Using the Bayes formula leads to a change in the posterior

Table 1. Economic indicators of causal forecasting

Economic indicators	Input variables x_i
Equipment. Technology. Working conditions. Staff qualification	Productivity
Technical and technological resources. Volumes of production. Number of staff. Specialization	Quantitative growth
Innovation. Staff qualification. Competitive advantages. Consumer preferences	Quality of production
Adaptation to market changes. Service. Development of new market niches. Price competition	Marketing

probability, which significantly improves the quality of forecasts by eliminating the subjectivity of experts. Information about the quantitative values of $\hat{y}$ at the end of the dynamic series makes it possible to clarify the values of their posterior probabilities and translate them into the a priori category. Thus, joint statistics of a priori and a posteriori information are organized.

An example of the methodology of such calculations is presented in Table 2. As input, statistical data on the rate of change in the volume of gross agricultural products from 2013–2019, published in the annual statistical collection [5], were used. Based on the obtained refined values of the posterior probability $P(B_j|A)$ of the indices of the economic indices of the input variables x_i, using the formula (1) it is possible to calculate the output entropy.

Table 2. Economic indicators of causal forecasting

Observation intervals	B_1	B_2	B_3	B_4	B_5	B_6	B_7
Growth rate, %	100	102,7	106,4	104,5	112	117,4	113,3
Apriori probability, $P(B_j)$	0,14	0.28	0,42	0,56	0,71	0,84	0,98
Conditional probability, $P(A\|B_j)$	0,149	0,152	0,172	0,204	0,253	0,33	0,491
The full probability of events, $P(A \cap B)$	0,02	0,42	0,071	0,114	0,177	0,277	0,481
Aposteriori probabilities, $P(B_j\|A)$	0,009	0,018	0,031	0,051	0,076	0,119	0,215

The novelty of the proposed forecasting methodology for managing economic processes is the proposal to consider B_j not as a time series date, but as specific economic indicators, the probability of the influence of which $P(B_j)$ on the input variables x_i is estimated expertly by the scale of relations according to graduation: the most important, moderately average, weak, little meaningful. Since economic indicators have different dimensions, they should be normalized by dividing each indicator by the initial indicator of the analyzed time series. Then the components of the time series will express the indices or rates of change of the input values x_i.

A hierarchical structure that reflects cause and effect relationships between elements of the forecasting system is a goal tree whose task is to relate distant

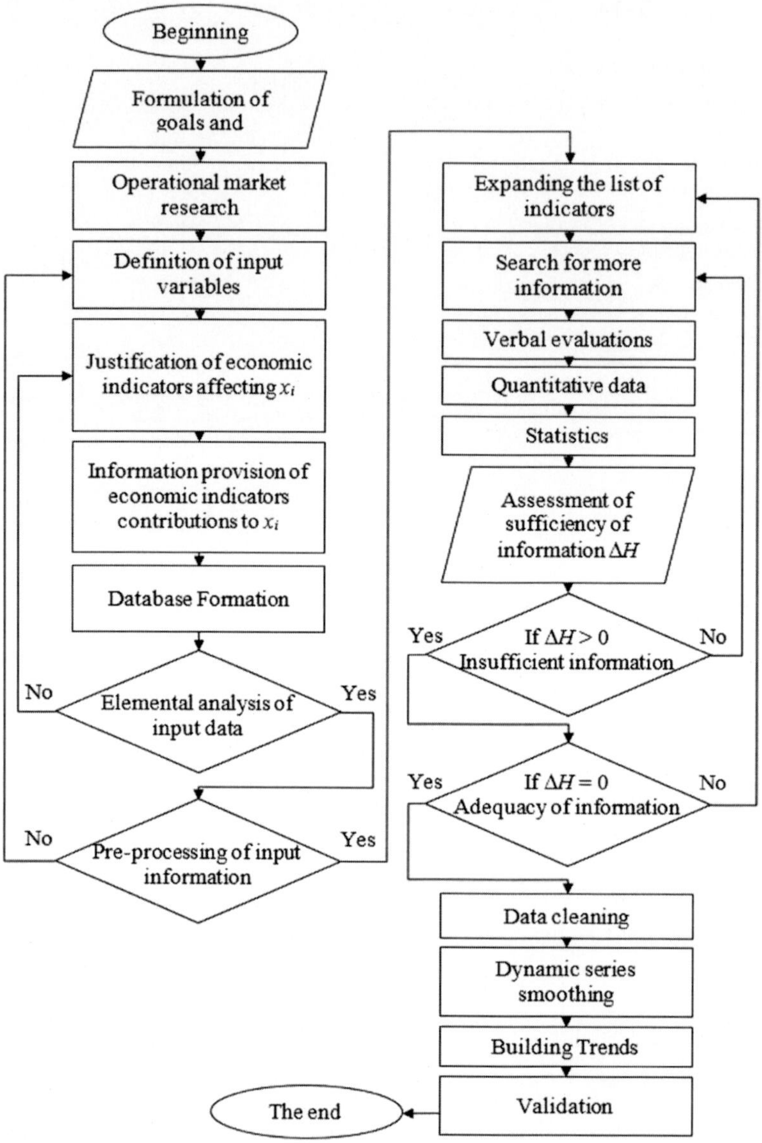

Fig. 4. Casual prediction algorithm

goals to management decisions. This structure allows us to compare possible alternatives and levels of influence on the output variable $\hat{y}$ economic indicators. The causal prediction algorithm is presented in Fig. 4. The basis of causal forecasting of development processes of economic systems in the conditions of uncertainty is the argumentation of the operative study of the market. Market segments and needs characteristics can remain constant over time. Need characteristics can also be predictable. Prediction is based on the laws of statistics using retrospective data. For markets with unpredictable changes in needs characteristics, traditional forecasting methods: cross-industry balance, expert estimates, econometrics, extrapolations are not acceptable here, as they require the study of the causality of predicted changes.

Economic development processes take place under the conditions of constant movement of financial and material resources. With the termination of the flow of resources, such changes are dying out. A necessary condition for the existence of production is constant innovation and change. Therefore, analysis of development processes and trends is a necessary element of market study in causal forecasting. When forming the database, goals and tasks of causal forecasting, it is necessary to pre-process the input information:

- the input data should be observable, not derived from the calculations;
- it is necessary to exclude duplicate and very different data;
- the volume of statistics should be large.

Each state of the external environment that influences a dynamic variable can be matched by a number of variants of control implementations, since the more implementation methods, the greater the information uncertainty. Establishing economic indicators and the probabilities of their realizations affecting the input variables x_i is one of the important steps in the methodology of causal forecasting of the processes of development of economic systems under uncertainty. Grading the indicators by the significance levels of the relationship scale is a necessary programming operation and is explained by the structure of the Bayesian network construction in the direction of obtaining the final control result, which binds the most significant network parameters to obtain a reliable value $\hat{y}$. Using the Bayes formula leads to a change in posterior probability. The implementation of the proposed sequence of actions in the compilation of forecasts consists in computer processing of data and is a set of files stored in the database:

- file where the a priori entropy of the analyzed economic indicator is calculated;
- file that calculates the probability of an indicator occurring in successive periods of a dynamic series;
- file where the specified value of posterior entropy is determined;
- file that calculates the difference between a priori and a posterior entropy as a quantitative measure of residual uncertainty;
- file where the degree of influence of input variables on the predicted output variable is determined.

After establishing the sufficiency of the input information through the difference between a priori and aposterior entropy, it is necessary to carry out a preliminary analysis of the dynamic series of economic indicators, which consists in identifying and eliminating anomalous values of the dynamic series and determining the presence of a trend. If we do not perform preliminary analysis of the input information with the identification of anomalous values, the specific anomalous values that remain in the link of the dynamic series can significantly distort the influence of the main economic indicators influence on the trend model. To clear information when constructing causal forecasting trends, it is recommended to use an exponential smoothing method to obtain an estimate of the trend parameters that characterize not the average process level but the trend that has emerged since the last observation. Later observations are given greater weight than earlier observations, and the weight decreases by exponential.

The construction of the Bayesian network allows at the stage of predictive modeling to perform scenario analysis to clearly present the task, update probabilities when new data arrives, and evaluate the effectiveness of the proposed algorithms in substantiating ideas and hypotheses. In this case, it is possible to determine the informational significance of indicators affecting xi. The time series extrapolation methods used in practice are recommended for short-term prediction, while causal modeling based on a deeper understanding of the complex of cause and effect relationships is recommended for long-term strategic planning.

The proposed methods of informational provision of the specific contribution of economic development indicators to the input variables and quantitative estimates of the causal relationships of the factors affecting the output dependent variable allow, in their practical implementation, to significantly improve the reliability and adequacy of forecasts when managing their economic systems development.

6 Conclusions

The causal prediction of the processes of managing the development of economic systems explains the mechanisms of generation of input variables, taking into account the causal relationships of their occurrence, reveals the uncertainty in the economic development system and substantiates the factors under which the set goals are achieved. Conducted at the lowest level of the causal forecasting management of the development of economic systems statistical analysis and consistent refinement of the probabilities of the events occurrence, the calculations of the priori and posterior entropy can regulate the process information accumulation.

The structural blocks development support for the contributions of economic development indicators to the input variables of causal forecasting, structural blocks of the technique of calculating the input entropy of the causal forecasting, structural blocks of the technique of calculating the output entropy of the causal forecasting served as the basis for the development of algorithms for the Casual Forecasting. The use of Bayesian networks in causal forecasting allows us to prioritize the impact of economic indicators not only on the input variables x_i but also on the output variable $\hat{y}$ and thus quantify the causal relationships of the factors influencing the results of the causal forecasting.

References

1. State Statistics Service of Ukraine (2020). http://www.ukrstat.gov.ua
2. Babichev, S., Škvor, J., Fišer, J., Lytvynenko, V.: Technology of gene expression profiles filtering based on wavelet analysis. Int. J. Intell. Syst. Appl. **10**(4), 1–7 (2018). https://doi.org/10.5815/ijisa.2018.04.01
3. Buleev, I., Bryukhovetsky, Y., Ivanenko, L.: Modelling of increase in intellectualization level of employees labour. Econ. Ind. **78**(2), 80–96 (2017). https://doi.org/10.15407/econindustry2017.02.080
4. Chudý, M., Karmakar, S., Wu, W.: Long-term prediction intervals of economic time series. Empir. Econ. **58**, 191–222 (2020). https://doi.org/10.1007/s00181-019-01689-2
5. Dokuchaev, N.: Near-ideal causal smoothing filters for the real sequences. Sig. Process. **118**, 285–293 (2016). https://doi.org/10.1016/j.sigpro.2015.07.002
6. Elgazzar, M., Hemayed, E.: Electrical load forecasting using Hijri causal events. In: 18th International Middle-East Power Systems Conference (MEPCON), pp. 902–906 (2016). Art. No. 7837003. https://doi.org/10.1109/MEPCON.2016.7837003
7. Lei, W., Qing, F., Zhou, J.: Improved personalized recommendation based on causal association rule and collaborative filtering. Int. J. Distance Educ. Technol. **14**(3), 21–33 (2016). https://doi.org/10.4018/IJDET.2016070102
8. Li, M., Liu, K.: Application of intelligent dynamic Bayesian network with wavelet analysis for probabilistic prediction of storm track intensity index. Atmosphere **9**(16) (2018). Art. No. 224. https://doi.org/10.3390/atmos9060224
9. Li, M.J., Tao, W.Q., Song, C.X., He, Y.L.: Forecasting and evaluation on energy efficiency of China by a hybrid forecast method. Energy Proc. **75**, 2724–2730 (2015). https://doi.org/10.1016/j.egypro.2015.07.703
10. Lu, X., Mamiya, H., Vybihal, J., Ma, Y., Buckeridge, D.: Application of machine learning and grocery transaction data to forecast effectiveness of beverage taxation. Stud. Health Technol. Inform. **264**, 248–252 (2019). https://doi.org/10.3233/SHTI190221
11. Makridakis, S., Hogarth, R.: Forecasting and uncertainty in the economic and business world. Int. J. Forecast. **25**(4), 794–812 (2019). https://doi.org/10.1016/j.ijforecast.2009.05.012
12. Marinescu, I., Lawlor, P., Kording, K.: Quasi-experimental causality in neuroscience and behavioural research. Nat. Hum. Behav. **2**(19), 891–898 (2018). https://doi.org/10.1038/s41562-018-0466-5
13. Mireles-Flores, L.: Recent trends in economic methodology. A literature review. Res. Hist. Econ. Thought Methodol. **36A**, 93–126 (2018). https://doi.org/10.1108/S0743-41542018000036A008
14. Mitra, S.: Is tourism-led growth hypothesis still valid? Int. J. Tour. Res. **21**(5) (2019). Art. No. 107320. https://doi.org/10.1002/jtr.2285
15. Roy, A., Laskar, R.: Non-casual linear prediction based adaptive filter for removal of high density impulse noise from color images AEU. Int. J. Electron. Commun. **72**, 114–124 (2017). https://doi.org/10.1016/j.aeue.2016.12.006
16. Sharko, M., Gusarina, N., Petrushenko, N.: Information model of making management decisions in the economic development of the enterprises. In: Advances in Intelligent Systems and Computing, pp. 304–314 (2019). https://doi.org/10.1007/978-3-030-26474-1
17. Sharko, M.: Management of the Development of Innovation in Industrial Production. Oldie-Plus, Kherson (2010)

18. Sharko, M., Panchenko, Y.: Formation of policy of increasing of intellectual potential. Actual Probl. Econ. **6**(156), 30–40 (2014)
19. Sharko, M., Zaitceva, O., Gusarina, N.: Providing of innovative activity and economic development of enterprise the condition of external environment dynamic change. Naukovyy visnyk Polissya **3**(11), 65–71 (2017). https://doi.org/10.25140/2410-9576-2017-2-3(11)-57-60
20. Shpak, N., Odrekhivskyi, M., Doroshkevych, K., Sroka, W.: Simulation of innovative systems under industry 4.0 conditions. Soc. Sci. **8**, 202 (2019). https://doi.org/10.3390/socsci8070202
21. Zolotin, A., Malchevskaia, E., Kharitonov, N., Tulupyev, A.: Local and global probabilistic-logic inference in algebraic Bayesian networks: a matrix-vector description and issues of sensitivity. Comput. Sci. 133–150 (2018). https://doi.org/10.26456/fssc29

Transition of Combustion to Explosion and Decision Support Systems for Explosion Protection

Victor Volkov[1(✉)] and Yuri Kryvchenko[2]

[1] Odessa I.I. Mechnikov National University, Odesa, Ukraine
viktor@te.net.ua
[2] Odessa National Academy of Food Technologies, Odesa, Ukraine
volkovkryvchenko@gmail.com

Abstract. A simple mathematical model of the combustion-to-explosion transition is constructed. This model is based on solving mathematical problems of the hydrodynamic flame stability, which are reduced to solving eigenvalue problems for linearized differential equations of gas dynamics. The influence of viscosity and compressibility on the development of perturbations is analyzed. The turbulence scale (the average size of the flame cells) in the case of instability of the laminar combustion process is estimated. The possibilities of the transition of slow combustion to both a deflagration explosion and a detonation wave are considered. Theoretical estimates of the explosive induction distance and the time of the combustion-to-explosion transition are obtained. These estimates are expressed by algebraic formulas, the use of which saves computer resources and does not require significant computer time. The latter fact is extremely important for the online control of potentially explosive objects. The calculations of the explosive induction distance and the time of the combustion-to-explosion transition agree with experimental data and the numerical results of other authors. Estimation of the time of a possible combustion-to-explosion transition allows to make timely and correct decision on measures to prevent an explosion or minimize its consequences. The constructed model is universal: it is applicable to the combustion of both homogeneous gas mixtures and heterogeneous media. The application of fuzzy logic makes it possible to use the proposed mathematical model of the transition from slow combustion to explosion for the software of decision support systems on the explosion safety and explosion protection for real potentially explosive objects in industry and transport. The advantage of this decision support system is that it allows the decision makers to do without experts.

Keywords: Combustion · Flame · Deflagration · Explosion · Detonation · Shock wave · Potentially explosive object · Instability · Turbulence · Mathematical model · Explosive induction distance · Decision support system · Control

S. Babichev et al. (Eds.): ISDMCI 2020, AISC 1246, pp. 437–447, 2021.
https://doi.org/10.1007/978-3-030-54215-3_28

1 Introduction and Literature Review

The prevention of accidental explosions in industrial enterprises, everyday life and transport is still an urgent problem. One of the reasons for this state of affairs (along with the complication of technological processes, the emergence of new combustible materials and explosives, and the chemicalization of industry, etc.) is the insufficient efficiency of automatic and automated systems for preventing and suppressing explosions.

The main idea of the modern organization of explosion protection is to prevent the occurrence of accidental fires [14,16,24]. Indeed, if a fire does not occur, then an explosion is impossible. Therefore, the solution to the fire safety problem simultaneously solves the problem of explosion safety. Thus, the problem of explosion safety is not solved as a separate problem, but only within the fire safety problem. Thereby automated control systems for explosive objects are aimed mainly at the timely prevention or suppression of accidental fires and spontaneous combustion [12,14,16]. The idea is sound and generally true, but it has two weaknesses.

The first weakness is that with a relatively low probability of ignition, the possibility of an explosion in the event of a fire may be great [9,22,24]. This is especially true for enterprises where explosive dust-air mixtures are formed during the production process [6,9], as well as for coal mines [4,12,17]. These cases need to be specifically diagnosed, because the damage and loss of life from explosions is much greater than from fires.

The second weakness is that it is not always possible to detect and suppress a fire on time [6,13,14]. For lots of enterprises and for many kinds of equipment the fire safety problem sometimes can't be solved properly, i.e. to guarantee almost complete absence of fires. This is critical if there is a danger of explosion.

In such cases it's necessary to have additional safety "mechanism" to prevent explosions. The main part of such mechanism should be decision support system (DSS) for the decision making on the explosion safety problems. This DSS must help a decision maker to answer these questions:

1. Is the transition of combustion to explosion possible if there is a fire (as a result of accidental ignition or spontaneous combustion)?
2. What is the time for transition of combustion to explosion if such transition is possible?
3. What measures can be taken to prevent an explosion timely or to minimize the possible consequences of an explosion?

The answer to the third question (inhibition, use of shut-off devices, shutdown of equipment, evacuation of personnel, etc.) is determined by the answers to the first two questions. An attempt to project such DSS was done in 2008 [22]. It was shown that the main theoretical problems for decision making are problems of the flame stability, the definition of the explosion induction distance and the time of such induction.

Instability of flames is investigated by different authors and in different formulations of the problem [10]. It is researched also in connection with deflagration-to-detonation transition [11], but these scientific studies are based on numerical simulations of premixed gas combustion. Such numerical simulations are always connected with finite perturbations, while stability of flames should be researched in relation to small perturbations (Darrieus-Landau instability). Hydrodynamic instability of flame may invoke explosion, but there are two kinds of explosions: deflagration explosions and detonations. Most of researches of combustion-to-explosion transition are focused on deflagration-to-detonation transition [5,7,8,11,13,15]. And deflagration explosions are more frequent than detonations, though detonations are more dangerous and destructive.

Numerical simulations of the flame stability [3] and deflagration-to-detonation transition [7,8,11,15] require significant computer resources and time. Therefore they cannot be used for DSS in automated control systems for explosive objects, because the time for decision making is limited.

The aim of the present research is development of a mathematical model of combustion-to-explosion transition that makes it possible to create an adequate mathematical support for DSS of automated control systems for explosive objects. This model should provide opportunities for making simple analytical estimates for the explosive induction distance (a distance between the point of ignition and the shock wave front) and the time of the shock wave formation without working out phenomena in detail.

2 Mathematical Modelling of the Combustion-to-Explosion Transition

The starting point for constructing mathematical model of the combustion-to-explosion transition is the assumption that the main reason for such a transition is the hydrodynamic instability of the laminar flame. Physical model for the combustion-to-explosion transition may be divided into the following stages:

1. There is always possibility of the laminar flame instability.
2. The result of multidimensional flame instability is distortion of the flame front and space-time pulsations. These factors lead to turbulent combustion.
3. Flame accelerates because the combustion surface area increases [3].
4. There are two possibilities for the accelerating flame:
   (a) the accelerating flame transits to the turbulent combustion mode with a constant average speed that is greater than the propagation velocity of the laminar flame (this mode can be called quasi-stationary): in this case, the explosion does not occur;
   (b) the turbulent flame continues to accelerate continuously.
5. If the flame accelerates continuously it accelerates till it generates shock wave. There are also two different possibilities:
   (a) the shock wave is intensive enough to ignite gas behind it: in this case there is detonation;
   (b) the shock wave is not intensive enough to ignite gas behind it: in this case there is no detonation, but there is deflagration explosion, i.e. explosion without detonation.

The flame stability problem is considered by us in various settings (for a viscous incompressible medium and for an ideal incompressible medium, for open space, flat channels and round cylindrical tubes). In the case of a viscous incompressible medium, the flow is described by the incompressibility equation and the Navier-Stokes equations. In the case of an ideal compressible medium, the flow is described by the equations of continuity, Euler, and energy balance. The last equation takes into account the heat release at the flame front.

The problem of flame stability in an open space is considered as a plane (two-dimensional) problem for symmetry reasons. In the two-dimensional case perturbations are set in form $\sim \exp(ihy - i\omega t)$, where $h = 2\pi/\lambda$, λ is wave length, i is unit imaginary number ($i^2 = -1$), ω is complex number, y is spatial coordinate, t is time.

In the case of a flat channel, the boundary condition of the impermeability of its walls imposes an additional condition on the perturbations, namely: an integer number of half-wavelengths must fit along the diameter of the channel.

In the case of flame propagation in a round cylindrical tube, perturbations are specified in the form $\sim \exp(-i\omega t + in\varphi)J_n(\zeta_{nk} r r_0^{-1})$, where r_0 is tube radius, r and φ are cylindrical coordinates, n is the azimuthal wave number ($n = 0, 1, 2, ...$), $J_n(\zeta)$ is the cylindrical functions of the 1-st kind of order n, ζ_{nk} is the k th root of the equation $dJ_n(\zeta)/d\zeta = 0$, $y_{jk}(k = 1, ..., 4)$ is the dimensionless function of z. Thus, the boundary conditions of the impermeability of the tube walls are also satisfied.

Assessment of the stability of combustion is carried out as a result of solving the problem of eigenvalues. Such a problem arises as a result of substituting expressions for perturbations into boundary conditions, which are the fundamental conservation laws at the flame front and the conditions for the finiteness of perturbations at infinity. The value of z is considered as a dimensionless eigenvalue:

$$z = -\frac{i\omega}{hu_1}, \tag{1}$$

where u_1 is normal flame speed.

The value of z for long-wave perturbations ($\xi \equiv hL \equiv \frac{2\pi L}{\lambda} \ll 1$, L is the laminar flame t hickness) can be represented by a series expansion

$$z = z_0 + z_1\xi + ..., \tag{2}$$

where only the first two terms in the expansion play a significant role.

The main formulas for evaluating the stability of combustion of a viscous gas are the formulas [1]:

$$z_0 = \frac{\delta_2}{\delta_2 + 1}\left(-1 + \sqrt{\delta_2 + 1 - \frac{1}{\delta_2}}\right), \tag{3}$$

$$
\begin{aligned}
&-\left(1+\frac{\delta_2}{\delta_2+1}z_0\right)z_1 = \\
&= z_0\left[\frac{1}{2}\left(1+\frac{z_0}{2}\right)(2\delta_2+1)+(\frac{\delta_2}{\delta_3}-1)+\delta_2^m\left(\frac{\delta_2+1}{\delta_2}z_0+\delta_3+3\right)\right]+ \\
&\qquad +\frac{1}{2}(\delta_2-1)\left(1+2z_0+\frac{\delta_0^2}{\delta_3}\right), \quad (4)
\end{aligned}
$$

where $\delta_2 = \dfrac{\rho_1}{\rho_2}$, ρ_1 is the initial inflammable mixture density, ρ_2 is density of combustion products, $\delta_3 = \delta_2 - \dfrac{\delta_2 - 1}{e}$, $m = const$ is exponent in dependence of dynamic-viscosity coefficient on temperature. It is obvious that $z_0 > 0$ (unstable expansion term of Landau) and $z_1 < 0$ (stabilizing expansion term, which is a result of viscosity stabilizing effect).

Solving the inequality

$$z_0 + z_1\xi > 0, \quad (5)$$

allows us to find the wavelength range of unstable perturbations. The wavelength λ_m of the perturbations that grow most rapidly with time is also determined as

$$\lambda_m = -\frac{4\pi L z_1}{z_0} \quad (6)$$

Parameter λ_m can be considered an estimate of the average size of the flame cells, if the flame has a cellular structure as a result of the development of instability [1].

The wavelength λ_m allows to estimate [23] the acceleration of the flame g_f as

$$g_f = \frac{4\pi z u_1^2(1-M_1)}{\lambda_m M_1}, \quad (7)$$

or, in accordance to (8),

$$g_f = -\frac{z_0^2 u_1^2(1-M_1)}{2z_1 L M_1}, \quad (8)$$

where $M_1 = \dfrac{u_1}{a_1}$ is Mach number for the laminar flame propagation ($M_1 \ll 1$).

Solutions for the detonation induction distance X_s and the deflagration-to-detonation transition time τ_s are obtained by H. Jones and M. Nettleton [13] for different types of the flame acceleration g_f. Estimating the detonation induction distance X_s, M. Nettleton determines this distance as a distance between the point of ignition and the shock wave front, whether the shock wave initiates detonation or not. That is in fact the value of X_s is the explosive induction distance.

Under the assumption that $g_f = const$, that is the flame acceleration is constant, distance X_s between the point of ignition and the point, where the flame is situated during the shock wave formation, can be estimated as [10]

$$X_s = \frac{2a_1^2}{g_f \beta_f (\gamma_1 + 1)}, \tag{9}$$

where $\beta_f = const(0 \ll \beta_f \leq 1, \beta_f = 0,9$ in most real cases), a_1 is sonic speed for combustible gas mixture, γ_1 is the ratio of specific heats for this mixture.

Substitution of (8) into (9) leads to

$$X_s = -\frac{4z_1 L}{(1 - M_1)M_1 z_0^2 \beta_f (\gamma_1 + 1)} \tag{10}$$

Formula (10) can be taken as an analytical estimation for the explosive induction distance. For the time of the shock wave formation τ_s such approximate estimate takes place:

$$\tau_s = \frac{X_s}{u_1} \tag{11}$$

Thus, analytical estimates for the explosive induction distance (10) and the time of the shock wave formation (11) are obtained. Calculating by these algebraic formulae is very simple and needs a minimal computer time, but such calculating does not describe the deflagration-to-explosion transition phenomena and its nature in detail as it is done in [2,7,8,11,15,21]. But a detailed analysis of the transition of slow combustion to explosion is just not required for creating of DSS on the explosion safety problems. And simple algebraic formulas (10) and (11) are just fit for this purpose.

It should be noted that eigenvalue z is estimated approximately. To determine the exact value of z, it is necessary to carry out a numerical analysis of the characteristic (secular) equation for the eigen-value $F(z, \xi) = 0$, where F is the multiparameter function of quasipolynomial type [1]. In some models, to study the stability of combustion, characteristic equation $F(z, \xi) = 0$ (polynomial or quasipolynomial for z as usual) needs to be solved numerically, but it also does not require large expenditures of computer time compared to numerical solving differential equations in partial derivatives as in [7,11,15,21].

Viscosity is the main factor stabilizing the normal combustion process. However, the compressibility of the gas medium also exerts a stabilizing effect on the flame propagation. A joint consideration of the effects of viscosity and compressibility on the small perturbations of a stationary flame is not possible due to the extreme mathematical complexity of the problem. Therefore, the effect of compressibility is considered separately. The result of such consideration is characteristic equation $F(z, M_1^2) = 0$, which solution is represented by a series expansion

$$z = z_0 + z_1 M_1^2 + z_2 M_1^4 ..., \tag{12}$$

moreover, only the first two components of this expansion are taken into account, z_0 is unstable positive value from (3), and it is proved that z_1 is negative.

As a criterion for instability, the relation

$$M_1^2 < \left| \frac{z_0}{z_1} \right| \tag{13}$$

can be taken. The value of

$$M_{1cr} = \sqrt{\left| \frac{z_0}{z_1} \right|} \tag{14}$$

may be called the critical Mach number, the excess of which means the possibility of stabilization of the flame.

For example, calculating the critical Mach number for a fast-burning mixture of acetylene with oxygen (37.5% C_2H_2 + 62.5% O_2 by volume) leads to $M_{1cr} = 0,295$; similar calculations for a slow-burning mixture of propane with air (5% C_3H_8 by volume) define $M_{1cr} = 0,335$; it means that stabilizing effect of compressibility at the initial stages of the development of flame instability is very weak (actually $M_1 < 0,03$) either for fast-burning or for slow-burning gas mixtures. At such high critical Mach numbers the flame creates a sufficiently powerful shock wave in front of it, which ignites the mixture.

But sometimes stabilization of the quasi-laminar mode (when the turbulence is small-scale, i.e. the turbulence scale is much smaller than the laminar flame thickness) is really possible before deflagration-to-explosion transition. It corresponds to $M_{1cr} \leq 0,1$. To research fully the problem of the quasi-laminar (or even turbulent) flame stabilization it may be useful to apply fractal theory [18, 19] and theory of strange attractors [20], but it is not the aim of present investigation.

All the conclusions made above can only be attributed not only directly to the combustion of homogeneous gas mixtures, but also for dust-air mixtures. As follows from the dynamics of a two-phase medium, this seems quite possible if the dust-air mixture is a monodisperse system, since the general form of the equations for combustion and shock-detonation processes at a low concentration of the solid phase differs little from those for a homogeneous mixture. The fine dispersion of industrial dust testifies to the fact that a solid substance (flour, etc.) is "smeared" in the gas phase, which allows to consider the medium as quasi-homogeneous, replacing the air parameters with some effective averaged mixture parameters. The finer the dust particles are, the closer the dust-air mixture to the dust-air mixture according to the kinetics of chemical reactions is.

Since the length of the flame zone in the gas-air mixture is much greater than the corresponding length of the combustion zone in a homogeneous gas mixture, the ratio of the perturbation wavelength to the width of the flame zone is much smaller. Correspondingly, the size of the flame front inhomogeneity, which is also an indicator of large-scale turbulence, is smaller (at least in a relative sense). Thus, in the dust-air mixtures instability and large-scale flame turbulence are achieved much more often and faster than in homogeneous mixtures.

3 Experiment, Results and Discussion

To verify the proposed mathematical model calculations of the explosion induction distance X_s and of the time of the shock wave formation τ_s were done for different combustible media. The results are given in Table 1.

Table 1. Explosion induction distance X_s and the time of the shock wave formation τ_s for different combustible media

Combustible media	X_s	τ_s
Gas mixtures near stoichiometry	10 mm–50 cm	0,001 s–5 s
Gas mixtures near concentration combustion limits	50 cm–5 m	5 s–15 s
Fine-dispersed aerosols and dust suspensions	1 m–10 m	5 s – 1 min
Aerosols and dust suspensions	5 m–15 m	15 s–2 min
Aerosols and dust suspensions near concentration combustion limits	More than 15 m	More than 2 min

Those results are in good agreement with experimental data and numerical researches for open spaces, wide channels and wide tubes with smooth walls [5,7, 8,10,11,13,15]. These results can also serve as the basis for designing explosion-proof objects, as well as for making decisions that ensure explosion safety in the operational control mode.

It is well known [13,21] that wall roughness and obstacles in channels and tubes significantly reduce explosion induction distance X_s and the time of the shock wave formation τ_s (in some cases in several times, up to 20–50 times). But, of course, it is impossible to take all these factors into account analytically, and the numerical solutions of such problems require a lot of computer time (while the accuracy of the calculations leaves much to be desired). For this reason, in most real cases, fuzzy logic must be used to make decisions.

The estimates for explosion induction distance X_s and the time of the shock wave formation τ_s made using classical mathematics methods form the basis of fuzzy estimates of the possibility of an explosion. The main idea and principles of such fuzzification is demonstrated in [22]. For example, the ratio $\frac{X_s}{H_{max}}$ of the estimated explosion induction distance X_s (reduced in accordance with the presence of wall roughness and obstacles) to the maximum length of the object H_{max} (silo height, length of the gas pipe, etc.) can be represented as a fuzzy variable $\tilde{H}$. The value of this variable is defined as follows

$$\tilde{H} = \begin{cases} \frac{X_s}{H_{max}}, \ 0 \leq X_s \leq H_{max} \\ 1, \ X_s \geq H_{max} \end{cases} \tag{15}$$

The closer the value of $\tilde{H}$ is to 1, the more explosive the object is. There are of course other fuzzy variables (expressing availability of the detonation concentration limits or of the combustion concentration limits, expressing exceeding of the initial pressure over the critical one and so on). These variables are connected with $\tilde{H}$ by logical conjunction operations. The value of such a fuzzy logical function is expressed by the value of a linguistic variable that provides information for decision making.

4 Conclusions

Thus, a simple mathematical model of the transition of combustion to explosion is constructed. This model is universal: it is applicable to the combustion of both homogeneous gas mixtures and heterogeneous media (primarily dust-air mixtures, but also aerosols and sprays).

Based on the theoretical solutions of the problem of hydrodynamic flame stability, the processes of autoturbulization and flame acceleration, as well as the possibilities of the transition of slow combustion to a deflagration explosion or to detonation, are analyzed. Algebraic formulae for estimations of the explosive induction distance and the time of the combustion-to-explosion transition are obtained. The calculations made using these formulas are in good agreement with experimental data and the results of other studies. The comparative simplicity of the formulas obtained makes it possible to evaluate the possibilities and time of the transition from combustion to explosion without significant expenditure of computer time and computer resources. This is important for on-line control of potentially explosive objects and makes such control less expansive.

The application of fuzzy logic allows us to use the proposed mathematical model of the transition from combustion to explosion for the mathematical support and software of decision support systems on the issues of explosion safety and explosion protection of the real industrial and transport facilities. The advantage of the decision support system of this type is that it allows the decision makers to do without experts. This is especially important, since it is well known that working with experts has always been and remains a weak link in decision support systems.

The development of this study is associated with the possible application of the theory of fractal combustion and strange attractors for the analysis of autoturbulization of the laminar flame and the process of a possible transition of combustion to explosion. In this case, the basis for the research should still be taken from the results of the analysis of the hydrodynamic stability of the flame in the classical sense of Darrieus-Landau. It would also be interesting to try to extend this theory to polydisperse dust-air mixtures, which would significantly expand the capabilities of the corresponding decision support system.

References

1. Aslanov, S., Volkov, V.: The integral method of studying the hydrodynamic stability of laminar flame. Combust. Explos. Shock Waves **27**, 553–558 (1991). https://doi.org/10.1007/BF00784941
2. Bradley, D., Cresswell, T.: Flame acceleration due to flame-induced instabilities in large-scale explosions. Combust. Flame **124**, 551–559 (2001). https://doi.org/10.1016/S0010-2180(00)00208-X
3. Buckmaster, J.: The structure and stability of laminar flames. Ann. Rev. Fluid Mech. **25**, 21–53 (1993). https://doi.org/10.1146/annurev.fl.25.010193.000321
4. Burgherr, P., Hirschberg, S.: Assessment of severe accident risks in the Chinese coal chain. Int. J. Risk Assess. Manag. **7**, 1157–1175 (2007). https://doi.org/10.1504/IJRAM.2007.015299
5. Ciccarelli, G., Dorofeev, S.: Flame acceleration and transition to detonation in ducts. Prog. Energy Combust. Sci. **34**, 499–550 (2008). https://doi.org/10.1016/J.PECS.2007.11.002
6. Eckhoff, R.: Dust explosion prevention and mitigation, status and developments in basic knowledge and in practical application. Int. J. Chem. Eng. **2009**, 1–12 (2009). https://doi.org/10.1155/2009/569825
7. Khokhlov, A.M., Oran, E.S., Thomas, G.O.: Numerical simulation of deflagration-to-detonation transition: the role of shock-flame interactions in turbulent flames. Combust. Flame **117**, 323–339 (1999). https://doi.org/10.1016/S0010-2180(98)00076-5
8. Koksharov, A., Bykov, V., Kagan, L., Sivashinsky, G.: Deflagration-to-detonation transition in an unconfined space. Combust. Flame **195**, 163–169 (2018). https://doi.org/10.1016/j.combustflame.2018.03.006
9. Lemkowitz, S., Pasman, H.: A review of the fire and explosion hazards of particulates. Powder Part. J. **31**, 52–81 (2014). https://doi.org/10.14356/kona.2014010
10. Liberman, M.: Introduction to Physics and Chemistry of Combustion. Springer, Heidelberg (2008). https://doi.org/10.1007/978-3-540-78759-4
11. Liberman, M., Sivashinsky, G.I., Valiev, D., Eriksson, L.E.: Numerical simulation of deflagration-to-detonation transition: the role of hydrodynamic instability. Int. J. Transp. Phenomena **8**(3), 253–277 (2006)
12. Lu, W., Cao, Y.J., Tien, J.: Method for prevention and control of spontaneous combustion of coal seam and its application in mining field. Int. J. Min. Sci. Technol. **27**(5), 839–846 (2017). https://doi.org/10.1016/j.ijmst.2017.07.018
13. Nettleton, M.A.: Gaseous Detonations: Their Nature and Control. Springer, Dordrecht (2013). https://doi.org/10.1007/978-94-009-3149-7
14. Nolan, D.P.: Handbook of Fire and Explosion Protection Engineering Principles for Oil, Gas, Chemical and Related Facilities. William Andrew: Elsevier, Norwich (2011). https://doi.org/10.1016/B978-1-4377-7857-1.00033-1
15. Oran, E.S., Gamezo, V.N.: Origins of the deflagration-to-detonation transition in gas-phase combustion. Combust. Flame **148**, 4–47 (2007). https://doi.org/10.1016/j.combustflame.2006.07.010
16. Pekalski, A., Zevenbergen, J., Lemkowitz, S., Pasman, H.: Review of explosion prevention and protection systems suitable as ultimate layer of protection in chemical process installations. Process Saf. Environ. Protect. **83**(1), 1–17 (2005). https://doi.org/10.1205/psep.04023
17. du Plessis, J.: Active explosion barrier performance against methane and coal dust explosions. Int. J. Coal Sci. Technol. **2**(4), 261–268 (2015). https://doi.org/10.1007/s40789-015-0097-7

18. Sabdenov, K.: Fractal theory of transition of slow burning to detonation in gases. Combust. Explos. Shock Waves **31**, 705–710 (1995). https://doi.org/10.1007/BF00744977
19. Sabdenov, K., Min'rov, L.: On the fractal theory of the slow deflagration-to-detonation transition in gases. Combust. Explos. Shock Waves **34**, 63–71 (1998). https://doi.org/10.1007/BF02671820
20. Swinney, H., Gollub, J.: Hydrodynamic Instabilities and the Transition to Turbulence. Springer, Heidelberg (2014). https://doi.org/10.1007/3-540-13319-4
21. Valiev, D., Bychkov, V., Akkerman, V., Law, C.K., Erikkson, L.E.: Flame acceleration in channels with obstacles in the deflagration-to-detonation transition. Combust. Flame **157**(5), 1012–1021 (2010). https://doi.org/10.1016/j.combustflame.2009.12.021
22. Volkov, V.: Decision support systems on hazards of industrial explosions. In: Seventh International Symposium on Hazards, Prevention and Mitigation of Industrial Explosions: Thirteenth International Colloquium on Dust Explosions and Eighth Colloquium on Gas, Vapor, Liquid, and Hybrid Explosions, St. Petersburg, Russia, 7–11 July 2008, vol. 3, pp. 343–347 (2008)
23. Volkov, V.: Deflagration-to-detonation transition and the detonation induction distance estimation. Proc. Odessa Politechnic Univ. **1**(43), 120–126 (2014). https://doi.org/10.15276/opu.1.43.2014.21
24. Zarko, V., Weiser, V., Eisenreich, N., Vasil'ev, A.: Prevention of Hazardous Fires and Explosions. The Transfer to Civil Applications of Military Experiences. Kluwer Academic Publishers, Springer, Netherlands (1999). https://doi.org/10.1007/978-94-011-4712-5

Information Technology for Decision Making Support and Monitoring in Man-Machine Systems for Managing Complex Technical Objects of Critical Application

Viktor Perederyi[1], Eugene Borchik[2], and Oksana Ohnieva[1(✉)]

[1] Kherson National Technical University, Kherson, Ukraine
viperkms1@gmail.com, oksana_ognieva@meta.ua
[2] Mykolaiv National Agrarian University, Mykolaiv, Ukraine
borchikev@gmail.com

Abstract. The information technology of decision support and monitoring in distributed man-machine systems for managing complex technical objects of critical application is proposed, which complements the theory and methods for solving the problems of increasing the functional sustainability and efficiency of the operational management of the working modes, due to introducing the complex of modern monitoring, evaluation, and analysis means and making relevant decisions to ensure trouble-free operation. The development of this information technology on the example of management of a multilevel regional electrical power network is considered. In calculating the criteria characterizing the degree of workability and technical condition of the electrical power network equipment, in addition to the data obtained from control devices of the monitoring system and diagnostic modules, modern mathematical probabilistic methods and fuzzy models based on the theory of expert evaluation were used. To evaluate the modes of operation, and maintain the full functional sustainability of the regional power network, a Bayesian confidence network was built, with the help of which, based on the experts' knowledge, the probability of the state of both individual components and a complex evaluation of its functioning were performed. For practical substantiation of the obtained results, an experiment was conducted, the results of which confirmed the practical value of the proposed information technology, which can be used in distributed man-machine systems for managing complex technical objects of critical application.

Keywords: Distributed man-machine systems of critical application · Complex organizational and technical objects · Monitoring system · Expert systems · Functional sustainability · Human factor · Bayesian confidence network

S. Babichev et al. (Eds.): ISDMCI 2020, AISC 1246, pp. 448–466, 2021.
https://doi.org/10.1007/978-3-030-54215-3_29

1 Introduction and Literature Review

At present, when creating man-machine systems (MMS) for managing complex technical objects of critical application CTOCA working in real time, the solution to the problems of dispatch control and decision support, as well as monitoring processes and phenomena of anthropogenic and natural nature come to the fore. Despite the fact that the human factor has a significant influence on the decision-making process, the effectiveness and quality of management also depend on how quickly and in a timely manner the required and reliable information on objects and processes is received by the operational and dispatching personnel, who perform the decision-making functions in emergency situations.

Modern MMS of critical application have their own particularities, induced by the sphere of their use, in particular control systems in electrical engineering, transport, petrochemical industries, have a branched hierarchical structure, and are characterized by human involvement in management and decision-making processes. In general, these functions are reduced to the perception of information in a verbal and visual form, its analysis, comparison of alternatives, and, ultimately, to the timely adoption of relevant decisions in conditions of uncertainty under the influence of negative external and production factors, under an uneven information load, limited time for decision-making and significant liability for the consequences of wrong decisions.

In particular, based on the example of electrical power systems, an analysis of data for the results of State Power Supply Inspectorate annual activity reports of accident data in distribution electrical power networks shows that the proportion of accidents that occurred due to the fault of the ODP (operating and dispatching personnel) is up to 40% [9]. This determines the relevance of the research aimed at improving information technologies (IT) for monitoring the state of a critical object in real time, and making relevant decisions to normalize it, under the influence of negative external and production factors on the ODP's cognitive state.

The rapid development of information technology in recent years has made it possible to develop a range of information systems, methods and diagnostic tools in critical application systems. In [7,13] the issues of the IT creation for decision-making systems by ODP are considered. Herewith, the main focus is placed on the development of alternatives to decision-making on the basis of available knowledge bases, providing ODP with necessary information, formation of the optimal sequence of actions to eliminate emergencies.

In [1] the requirements of MMS was focused on the analysis of large amounts of data using special simulation models and tools for solving optimization problems are considered. Such systems are called OLAP system (On Line Analytical Processing), must provide DM or experts with automated means: mathematical modeling of improvement processes of function mechanisms for system of critical application (SCA); formation of databases (DB) and knowledge bases (KB); data warehouse modeling; synthesis of new development mechanisms for SCA.

According to experts in this field, when building a SCA, the most promising approach is based on the technology called Big Data Technology & Big Data

Analysis. A more detailed analysis of the big data technologies application for the construction of such systems in energy is given in [13,16]. In [10] the information technology of control and support for functional sustainability of distributed MMS of critical application was considered, a set of sustainability indicators and system safety, characterizing its fault tolerance and survivability is defined; the set of indicators of human factor, in the management and decision-making, at each hierarchical level of the system was determined. To date, the main trend in the development of man-machine systems (MMS) in particular, in the world electric power industry is the widespread introduction of smart grid technologies (Smart Grid) [11], by means of which high reliability of a critical object operation is ensured. The basis for the implementation of these features is the implementation of systems for monitoring the processes occurring at the object, and creation on their basis of intellectual management systems of their work modes [2,5–7,11,16,17].

As shown in the literature, one of the key tasks in the study of modern science-based approaches to improve the efficiency of distributed MMS critical applications in the field of electricity today is mainly the development of methods and techniques for monitoring and technical diagnostics of expert SCA.

However, the **unsolved part of the problem** there are questions of decision making support and monitoring both in the operational management of operation modes and control for functional sustainability and efficiency of expert SCA, through the introduction of a complex of modern monitoring, evaluation, analysis and decision making relevant decisions, taking into account the indicators of external and cognitive factors on DM at each hierarchical level. In connection with this, for the above analysis of literary sources, in this work, for the development of the theory of efficient operation of distributed MMS, on the example, an IT of monitoring and decision support (DS) in MMS for managing regional electrical power network is proposed, taking into account the influence of external and cognitive factors on DM at each hierarchical level.

The Aim of the Research. For solution of this problem, the following tasks can be formulated:

- to investigate the main causes and factors that most often lead to electrical power network failures to perform a comprehensive assessment of the causes of the emergency condition and the factors that lead to an appropriate cause of failure of the electrical equipment by the coefficient of severity of the critical object (CO) emergency condition,
- to determine the set of indicators of external and cognitive factors and to conduct an expert evaluation by the degree of importance of their influence that directly affect the relevance of DM's decision-making,
- to develop an IT's scheme of the interaction and distribution of individual monitoring modules of basic indicators of the electrical power network, to evaluate the modes of the their workable condition, to build the Bayesian confidence network (BCN), based on the knowledge of experts-specialists of the CO,

– to conduct an experiment, using the proposed IT of managing the regional electrical power network, based on the obtained results.

2 Materials and Methods

In accordance with the research in this subject area, it can be argued that the creation of a dispatching control system for electrical power objects is based largely on the theory of similarity and modeling. When upgrading or creating a new control system, it should be borne in mind that the most important feature is that any model can be executed and function only based on the measurement of its mode parameters. So it is advisable to develop a system that would monitor and control the network parameters and transmit information to the control point for DM's operational decision-making. To solve these issues, the proposed IT scheme is presented in Fig. 1.

The IT for monitoring provides measurement of the set parameters and, on their basis, determination of the established indicators and characteristics or detection of their inconsistencies with the established standards or agreements with automation of obtaining of the generalized data on the emergency of all controlled elements, as well as detection and decision making on elimination of those equipment faults which during short periods of time can lead to the failure of the complex part of the equipment and the possible failure of the electrical power network as a whole.

In accordance with Fig. 1, the operation of the IT is performed according to the following algorithm: the administrator, after logging in, gets access to the MS interface and all system resources. Via the monitoring interface, he is also able to access the state of the monitoring modules, manage the WEB system for the generation and updating of the DS expert knowledge in the Knowledge Base (KB), monitor and control, in real time, the information load and the functional sustainability of the DM. The monitoring system also monitors and controls, in real time, the complex functional sustainability of the object of critical application, interacts with the WEB environment for formulating alternatives to DS and updating the knowledge base. In calculating the criteria characterizing the degree of workability and technical condition of the electrical power network equipment, in addition to the data obtained from control devices of the monitoring system and diagnostic modules, modern mathematical probabilistic methods and fuzzy models based on the theory of expert evaluation were used.

In evaluating the performance of such systems, a qualimetric approach has been well established that involves the following actions [12]:

1. Building a hierarchical structure of indicators.
2. Determining absolute indicators.
3. Determining reference absolute indicators.
4. Determining weighting coefficients.
5. Determining a comprehensive assessment.

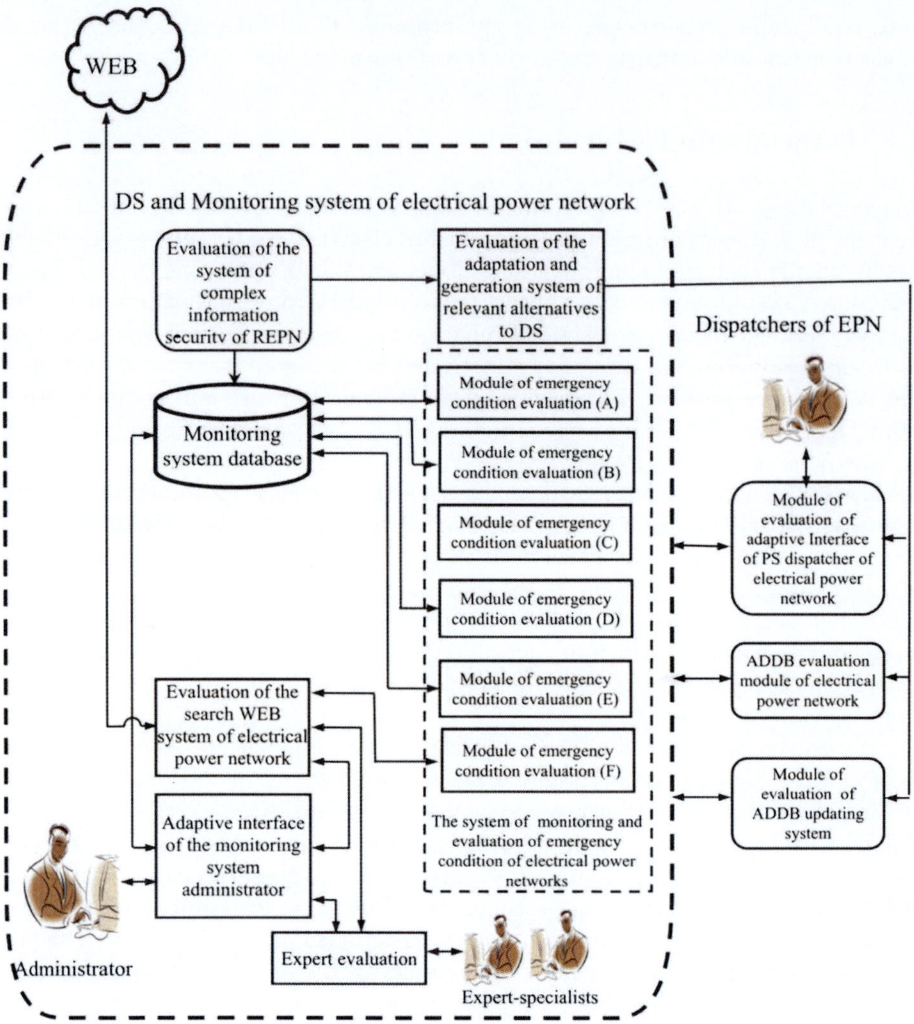

Fig. 1. Scheme of the IT for monitoring, determining and evaluating the state of operating modes of the electrical power network

The absolute indicators of the formulated criteria of the electrical power network operating modes were determined by the experts, expressing their evaluation by each criterion in the following way: normal, workable, and critical.

The estimated indicator of the effectiveness of the emergency causes is calculated as the sum of the products of the importance and significance of the criteria:

$$\Phi_i = \sum K_i^z K_i^\mu \tag{1}$$

where Φ_i – estimated indicators of the emergency causes (A, B, C, D, E, F), influencing the electrical power network performance. K_i^z and K_i^μ are the coefficients of importance and significance respectively of the correspondent i-th factor $(\mathrm{a}_i, \mathrm{b}_i, \mathrm{c}_i, \mathrm{d}_i, \mathrm{e}_i, \mathrm{f}_i)$ leading to a relevant system's emergency cause.

Six groups of the main causes that lead to emergency modes of operation and 24 factors that cause them were identified as the most common and important for all sections and services of the electrical power network. The results of the expert evaluation by the degree of importance of causes and factors are shown in Table 1.

The mathematical apparatus of the Bayesian Confidence Networks (BCN) was expedient used to evaluate the modes of operation of the electrical power network, to date, one of the most relevant models designed to work with incomplete, inaccurate and inconsistent information [8]. To determine and evaluate the state of operation of the CO on the example of the electrical network according to the results of expert evaluation (Table 1), a BCN graph (Fig. 2) was developed, consisting of vertices and arcs connecting these vertices. Vertices are random variables that can be discrete or continuous. The arcs reflect cause and effect relationships between variables, which is why BCN are sometimes referred to as cause-and-effect networks. In cause-and-effect networks, the parent vertices represent the causes, and the child vertices represent the consequences. These dependencies are modeled using appropriate arcs.

The vertices a_1, a_2, a_3, a_4, b_1, b_2, b_3, b_4, c_1, c_2, c_3, c_4, d_1, d_2, d_3, d_4, f_1, f_2, f_3, f_4 are "Decision" type vertices and characterize the degree of operability of the electrical power network according to the corresponding factors of the same name $a_1, a_2, \ldots, f_4$ and can take two states: "sufficient" (the degree of operability of the electrical power network by the corresponding factor is sufficient) and "insufficient" (the degree of operability of the electrical power network by the corresponding factor is insufficient).

The modules of the system scan the measuring-controlling devices, components of each of the factors a_i, b_i, c_i, d_i, f_i $\left(i = \overline{1,4}\right)$, that are attached to them. The symbols a_{ij}, b_{ij}, c_{ij}, d_{ij}, f_{ij} indicate the components with the number j of the corresponding factor with the number i. For example, symbol a_{12} denotes a component with the number 2 of a factor with the number 1. If the modules of the system recorded a deviation of the readings from the normal mode of operation of the electrical power network by the component of one of the factors, then this component shall be referred to as negative. Otherwise, it shall be referred to as positive. For example, let the readings deviate from the normal mode of operation of the electrical power network caused by the absence of voltage by the reason A. Then the component of the factor a_{13} is referred to as negative.

Considering for simplicity that the components of any factor are equivalent, the degree of operability of the electrical power network by the factor a_i is evaluated according to the formula proposed in [3]:

$$Q_{a_i} = \frac{m_{a_i}}{n_{a_i}} \tag{2}$$

where m_{a_i} is the number of positive components of the factor a_i, n_{a_i} is the total number of the components of the factor a_i $\left(i = \overline{1,4}\right)$.

Table 1. Causes and factors that most often lead to emergencies in electrical power networks

	Factors influencing the causes of emergency in electrical power network	Importance
(A) FORCED DEVIATIONS FROM THE LOAD CURVE		
$a1$	Frequency deviation or exceeding the limit values of power flows in sections along separate power supply lines	0,1
$a2$	Malfunction of relay protection and automation devices, or switching off power lines, limiting the power output of the power plant	0,2
$a3$	Exceeding the established contractual magnitudes of power flows on interstate electrical connections	0,2
$a4$	Emergency failure of power units, which can lead to significant power shortages	0,5
(B) FAILURE OF ELECTRICAL POWER EQUIPMENT		
$b1$	Fire outbreak	0,5
$b2$	Breach of switching modes	0,15
$b3$	Breakdown of automatic devices	0,25
$b4$	Failure in the operation of electrical substations	0,1
(C) ELECTRICAL POWER NETWORK OVERLOAD		
$c1$	Failure to observe normal operating modes of equipment, systems, devices	0,3
$c2$	Untimely detection of the threat of technological disruption	0,1
$c3$	Slow renewal of energy supply to consumers and normal parameters of energy supplied to consumers	0,45
$c4$	Failure to identify the condition of equipment that has been shut down due to technological disruption and the ability to put it into operation	0,15
(F) TECHNOLOGICAL DISRUPTIONS (FAILURES, ACCIDENTS)		
$f1$	Disrupted organizational interaction of operating and dispatching personnel	0,6
$f2$	Under-voltage in control circuit	0,1
$f3$	Decrease in frequency	0,1
$f4$	Violation of the permissible flow modes and overload of network elements	0,2
(E) NON-RELEVANT DECISIONS MADE BY OPERATING AND DISPATCH PERSONNEL		
$e1$	Influence of external production factors on DM	0,4
$e2$	DM's cognitive component	0,25
$e3$	DM's information overload	0,35
(D) FAILURE OF THE OPERATION OF SUBSTATION HARDWARE COMPONENTS SET		
$d1$	The use of non-dry contacts of relay protection and automation devices, block contacts of switching equipment, and control keys	0,2
$d2$	The use of industrial converters of E series with a uniform output of 5 mA and a speed of at least 0.5 s	0,15
$d3$	Availability of a built-in auto-diagnostic system that provides constant monitoring of the operability of the components of the electromechanical complex and communication lines	0,3
$d4$	Availability of automatic restart system of hardware components set	0,35

It is obvious that $0 \leqslant Q_{a_i} \leqslant 1$ for any value of $i = \overline{1, 4}$. Herein, the higher is the value of Q_{a_i}, the higher operability has the electrical power network by the

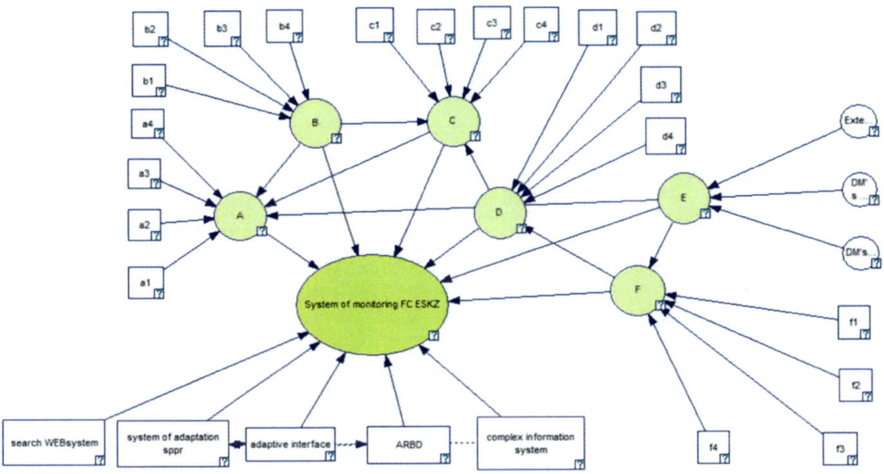

Fig. 2. General BCN graph of definition and assessment of state of critical object

factor a_i. Evaluation of the degree of operability of electrical power network by factors b_i, c_i, d_i, f_i $\left(i = \overline{1,4}\right)$ is carried out similarly.

The modern mathematical apparatus of the theory of expert evaluation [12] was used to evaluate the modes of operation of the electrical power network, where experts were invited to carry out a comprehensive assessment of the degree of the operability of the electrical power network by the relevant factor. As a result of the expert evaluation, the correspondence between the interval of values of the degree of operability of the electrical power network by any factor and the value ("sufficient" or "insufficient") of the corresponding vertix of the BCN was established (Table 2).

Table 2. The results of expert assessment of the degree of operability of the object

A			B			C		
	insufficient	sufficient		insufficient	sufficient		insufficient	sufficient
$a1$	0–0.5	0.5–1	$b1$	0–0.35	0.35–1	$c1$	0–0.4	0.4–1
$a2$	0–0.45	0.45–1	$b2$	0–0.55	0.55–1	$c2$	0–0.55	0.55–1
$a3$	0–0.7	0.7–1	$b3$	0–0.6	0.6–1	$c3$	0–0.6	0.6–1
$a4$	0–0.45	0.45–1	$b4$	0–0.5	0.5–1	$c4$	0–0.5	0.5–1
D			F					
	insufficient	sufficient		insufficient	sufficient			
$d1$	0–0.4	0.4–1	$f1$	0–0.6	0.6–1			
$d2$	0–0.4	0.4–1	$f2$	0–0.7	0.7–1			
$d3$	0–0.4	0.4–1	$f3$	0–0.5	0.5–1			
$d4$	0–0.4	0.4–1	$f4$	0–0.55	0.55–1			

Thus, according to the calculated values by the Formula (2) of the degree of operability of the electrical power network, by some factor, by means of Table 2,

Table 3. The results of expert assessment of DSS components of electrical power network

DSS components of electrical power network	States of operation		
	Normal	Workable	Critical
System of actualization of ADDB	0–0.45	0.45–0.6	0.6–1
System of adaptation and generation of relevant alternatives to DSS	0–0.6	0.6–0.75	0.75–1
Search WEB system	0–0.35	0.35–0.6	0.6–1
Adaptive interface of DM	0–0.5	0.5–0.75	0.75–1
System of complex information security	0–0.5	0.5–0.65	0.65–1

the BCN vertex value corresponding to this factor is easily determined. To characterize the operation of DSS components, the vertices "search WEB system", "system of DSS adaptation", "adaptive interface", "ARBD", "complex information system" are introduced to a Bayesian Confidence Network (Fig. 2). They are vertices of the Decision type and can take three states: "normal", "workable" and "critical". The modern mathematical apparatus of the theory of expert evaluation [12] was also used to evaluate the modes of operation of the components of the DSS in the electrical power network. The results of the expert evaluation are summarized in Table 3.

The vertices A, B, C, D, F correspond to the aforementioned and similarly identified causes for the appearance of faults and breakdowns in the electrical power network. They are binary and take the values "occurs" if the corresponding reason for the occurrence of faults and breakdowns in electrical power network exists and "does not occur" otherwise. The vertex E characterizes the functional sustainability of the DM, it is binary and can take the values "sufficient" if the functional sustainability of the DM is sufficient and "insufficient" otherwise. Finally, the vertex "System of monitoring FC ESKZ" characterizes the state of the system, it takes only two values: "workable", if the state of the system is acceptable, and "critical", if the state of the system is unacceptable, and operational decision making is required to eliminate the corresponding causes of emergency situations.

It should be noted that the vertices under consideration are of the Noisy MAX type: firstly, these vertices have a large number of parent connections, which makes it very difficult to fill in the conditional probability table; secondly, it is much easier for experts to evaluate the value of unconditional probability than that of conditional one. For the noisy vertices A, B, C, D, F, E, E, "System of monitoring FC ESKZ" to be described by experts-specialists, and for the vertix "System of monitoring FC ESKZ" to be described by experts-administrators, it was suggested to evaluate the conditional probabilities of the possible states of the indicated vertices. The results of the evaluation are presented in Table 4, 5, 6, 7, 8, 9, 10.

The Chance – General type vertices "External Environment Influence on the DM", "DM's Cognitive state" and "DM's Psychological state" are parent ones for the vertex E. The vertex "External Environment Influence on the DM" characterizes the degree of influence of the following factors on DM [18]: electric field

Table 4. Description of the node A

Parent	a_1		a_2		a_3		a_4	
State	insufficient	sufficient	insufficient	sufficient	insufficient	sufficient	insufficient	sufficient
occurs	0.25	0	0.3	0	0.25	0	0.25	0
does not occur	0.75	1	0.7	1	0.75	1	0.75	1

Parent	B		C		E	
State	occurs	does not occur	occurs	does not occur	occurs	does not occur
occurs	0.1	0	0.15	0	0.1	0
does not occur	0.9	1	0.85	1	0.9	1

Table 5. Description of the node B

Parent	b_1		b_2		b_3		b_4	
State	insufficient	sufficient	insufficient	sufficient	insufficient	sufficient	insufficient	sufficient
occurs	0.3	0	0.3	0	0.4	0	0.25	0
does not occur	0.7	1	0.7	1	0.6	1	0.75	1

intensity E; magnetic field intensity H; production noise intensity I_N; temperature T. The normative values of these factors are given in Table 11. Assumed that the vertex in question takes two values: "negative" if environmental conditions negatively affect the state of DM, and "positive" otherwise.

To estimate the probability that the random variable "External Environment Influence on the DM" takes the value "negative", it created a system for predicting probability values being based on a fuzzy logical inference according to the Mamdani algorithm on a fuzzy knowledge base [15], in which the input and output variable values are given by fuzzy sets. Given that according to [18], environmental factors, by the increase in the degree of influence on DM, are arranged in the following order: temperature; production noise intensity; magnetic field intensity; electric field intensity, the following fuzzy knowledge base is proposed:

RULE 1: IF u_1 is "above the norm" THEN ν is "high"
RULE 2: IF u_1 is "norm" AND u_2 is "above the norm" AND u_3 is "above the norm" THEN ν is "high"
RULE 3: IF u_1 is "norm" AND u_2 is "above the norm" AND u_3 is "norm" THEN ν is "above the mean"

Table 6. Description of the node C

Parent	c_1		c_2		c_3		c_4	
State	insufficient	sufficient	insufficient	sufficient	insufficient	sufficient	insufficient	sufficient
occurs	0.3	0	0.25	0	0.2	0	0.25	0
does not occur	0.7	1	0.75	1	0.8	1	0.75	1

Parent	B		D	
State	occurs	does not occur	occurs	does not occur
occurs	0.1	0	0.1	0
does not occur	0.9	1	0.9	1

Table 7. Description of the node *D*

Parent	d_1		d_2		d_3		d_4	
State	insufficient	sufficient	insufficient	sufficient	insufficient	sufficient	insufficient	sufficient
occurs	0.2	0	0.35	0	0.25	0	0.2	0
does not occur	0.8	1	0.65	1	0.75	1	0.8	1
Parent	F							
State	occurs	does not occur						
occurs	0.1	0						
does not occur	0.9	1						

Table 8. Description of the node *E*

Parent	External Environment Influence on the DM		DM's Cognitive State		DM's Psychological State	
State	insufficient	sufficient	insufficient	sufficient	insufficient	sufficient
occurs	0.3	0	0.2	0	0.15	0
does not occur	0.7	1	0.8	1	0.85	1

Table 9. Description of the node *F*

Parent	f_1		f_2		f_3		f_4	
State	insufficient	sufficient	insufficient	sufficient	insufficient	sufficient	insufficient	sufficient
occurs	0.1	0	0.05	0	0.2	0	0.15	0
does not occur	0.9	1	0.95	1	0.8	1	0.85	1
Parent	E							
State	occurs	does not occur						
occurs	0.1	0						
does not occur	0.9	1						

Table 10. Description of the node System of monitoring FC ESKZ

Parent	A		B		C		D	
State	occurs	does not occur	occurs	does not occur	occurs	sufficient	occurs	does not occur
critical	0.3	0	0.35	0	0.25	0	0.2	0
workable	0.7	1	0.65	1	0.75	1	0.8	1
Parent	E		F		search WEB system			
State	occurs	does not occur	occurs	does not occur	critical	workable	normal	
critical	0.2	0	0.2	0	0.05	0.01	0	
workable	0.8	1	0.8	1	0.95	0.99	1	
Parent	system of adaptation sppr			adaptive interface				
State	critical	workable	normal	critical	workable	normal		
critical	0.05	0.01	0	0.05	0.01	0		
workable	0.95	0.99	1	0.95	0.99	1		
Parent	ARBD			complex information system				
State	critical	workable	normal	critical	workable	normal		
critical	0.05	0.01	0	0.05	0.01	0		
workable	0.95	0.99	1	0.95	0.99	1		

RULE 4: IF u_1 is "norm" AND u_2 is "norm" AND u_3 is "above the norm" THEN ν is "mean"
RULE 5: IF u_1 is "norm" AND u_2 is "norm" AND u_3 is "norm" AND u_4 is "above the norm" THEN ν is "below the mean"
RULE 6: IF u_1 is "norm" AND u_2 is "norm" AND u_3 is "norm" AND u_4 is "below the norm" THEN ν is "below the mean"
RULE 7: IF u_1 is "norm" AND u_2 is "norm" AND u_3 is "norm" AND u_4 is "norm" THEN ν is "low"

Here, the linguistic variables "electric field intensity", "magnetic field intensity", "production noise intensity", and "temperature" are denoted through u_1, u_2, u_3, u_4 respectively, and ν denotes the linguistic variable "the probability that the random variable (External Environment Influence on the DM) takes value "negative".

Table 11. Normative values of environmental factors

Factors	E, kV/m	H, kA/m	I_N, dB	T, °C
norm	0–0.5	0–8	$\leqslant$50	22–24

To describe linguistic variables u_i $\left(i=\overline{1,3}\right)$, are used the term set {"norm", "above the norm"}, for the variable u_4 - term set { "below the norm", "norm", "above the norm"}, and for the variable ν - term set {"low", "below the mean", "mean", "above the mean", "high"}. The membership functions $\mu_i^1(x_i)$ and $\mu_i^2(x_i)$ of the terms "norm" and "above the norm", respectively, of the linguistic variables u_i $\left(i=\overline{1,4}\right)$, the membership function $\mu_4^3(x_4)$ of the term "below the norm" of the linguistic variable u_4 will be defined as two-sided Gaussian functions:

$$\mu_i^j(x_i) = gauss2fm\left(x_i, \left[\sigma_{i1}^j, a_{i1}^j, \sigma_{i2}^j, a_{i2}^j\right]\right) = \begin{cases} e^{\frac{-\left(x_i - a_{i1}^j\right)^2}{2\left(\sigma_{i1}^j\right)^2}}, & if\ 0 \leqslant x_i \leqslant a_{i1}^j; \\ 1, & if\ a_{i1}^j < x_i < a_{i2}^j; \\ e^{\frac{-\left(x_i - a_{i2}^j\right)^2}{2\left(\sigma_{i2}^j\right)^2}}, & if\ a_{i2}^j \leqslant x_i \end{cases} \tag{3}$$

where the parameters $\sigma_{i1}^j, \sigma_{i2}^j > 0$; $a_{i1}^j, a_{i2}^j \geqslant 0$; $a_{i1}^j \leqslant a_{i2}^j$; $\left(j=\overline{1,2};\ i=\overline{1,4};\right.$ if $j=3$ then $i=4$); x_i – are the elements of the universal sets X_i, $\left(i=\overline{1,4}\right)$, on which the terms "below the norm", "norm" and "above the norm" are defined. Taking into account possible values of environmental factors, universal sets are represented as:

$$\begin{aligned} X_1 &= \{x_1 | x_1 \in [0, 1]\} \\ X_2 &= \{x_2 | x_2 \in [0, 16]\} \\ X_3 &= \{x_3 | x_3 \in [0, 150]\} \\ X_4 &= \{x_4 | x_4 \in [10, 40]\} \end{aligned} \tag{4}$$

The membership functions of the terms {"low", "below the mean", "mean", "above the mean", "high"} of the linguistic variable ν are denoted through $\theta_i(y)$ $\left(i = \overline{1,5}\right)$, where y are elements of the universal set $Y = \{0 \leqslant y \leqslant 1\}$, on which these terms are defined. In this case, the elements of the set are the values of the probability that environmental factors negatively affect the state of the DM. The function $\theta_i(y)$ is defined in the form of a symmetric Gaussian function:

$$\theta_i(y) = gaussfm(y, [\sigma_i, c_i]) = e^{-(y-c_i)^2/2\sigma_i^2} \tag{5}$$

where the parameters are $\sigma_i > 0$; $c_i \geqslant 0$, $\left(i = \overline{1,5}\right)$.

The graphs of the some above functions are given below.

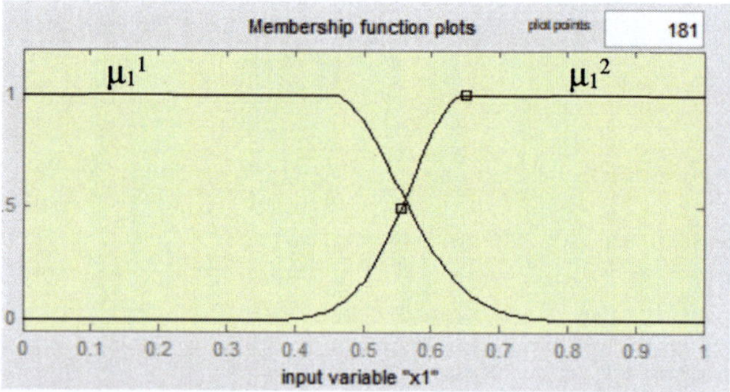

Fig. 3. Membership functions of $\mu_1^1(x_1)$ and $\mu_1^2(x_1)$

When building membership function $\mu_1^1(x_1)$ of the terms "norm" of the linguistic variable u_1 the interval of normative values of the environmental factor "electric field intensity" was taken into account (Table 11): $0 \le E \le 0.5$ (kV/m). Therefore $\mu_1^1(x_1) = 1$ at $0 \le x_1 \le a_{12}^1$ (Fig. 3). In this case, in Formula (2) $a_{11}^1 = 0$ should be considered. The value of the parameter a_{12}^1 can be equal to or even be slightly less than 0.5, which is due to the fact that the set of normative values is a fuzzy set. At $x_1 \ge a_{12}^1$ function graph $\mu_1^1(x_1)$ decreases to zero according to Gaussian law. The speed of this fall is determined by the parameter σ_{12}^1. Membership functions $\mu_1^2(x_1)$, $\mu_i^j(x_i)$ $\left(j = \overline{1,2}; i = \overline{2,3}\right)$ and $\mu_4^3(x_4)$ are built similarly (Fig. 4).

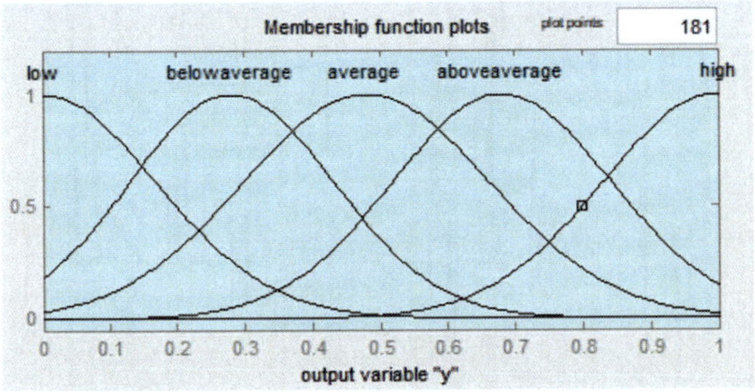

Fig. 4. Membership functions of $\theta_i(y)$ $\left(i = \overline{1,5}\right)$

The fuzzy model F setup (i.e. the determination of the coefficients of the model a_{i1}^j, a_{i2}^j, σ_{i1}^j, σ_{i2}^j, c_l, σ_l $\left(j = \overline{1,2};\ i = \overline{1,4};\ \text{if}\ j = 3\,\text{then}\ i = 4;\ l = \overline{1,5}\right)$) is carried out by solving the minimization problem [15] of the mean-square residual R:

$$R = \frac{1}{n}\sum_{k=1}^{n}\left(y_k - \mathrm{F}\left(K, Z_k\right)\right)^2 \rightarrow min \tag{6}$$

Here n is the sample size of the experimental data linking the inputs $Z = (x_1, x_2, x_3, x_4)$ with the output y of the investigated dependence:

$$\left(Z_k, y_k\right), k = \overline{1,n},$$

where $Z_k = (x_{k,1}, x_{k,2}, x_{k,3}, x_{k,4})$ is input vector, and y_k is the output in k pair.

Herein, $\mathrm{F} = (K, Z_k)$ is the value of the output of the fuzzy model with the value of the inputs specified by the vector Z_k; $K = \left(a_{i1}^j, a_{i2}^j, \sigma_{i1}^j, \sigma_{i2}^j, c_l, \sigma_l\right)$ is the vector of the coefficients of the membership functions of terms of a fuzzy model input and output variables. Taking into consideration the experimental data and the knowledge of experts about the influence of environmental factors on DM, is solved the mathematical programming problem (5) using the Fuzzy Logic Toolbox and Optimization Toolbox packages and thus setup the fuzzy model.

As an example of the fuzzy model considered above, Fig. 5 shows a graph of the dependence of the probability $P(E, H)$ of the negative environmental influence on DM of the main factors: E - electric field intensity and H - magnetic field intensity, provided that the secondary environmental factors, the production noise intensity I_N and temperature T, take normative values: $I_N = 0\,\text{dB}$, $T = 23\,^\circ\text{C}$.

The DM's psychological and cognitive states are determined using the Lusher [14] and Eysenck [4] tests. Through p_1 is denoted the probability that the random variable "DM's Psychological State" takes the value "negative", and through p_2

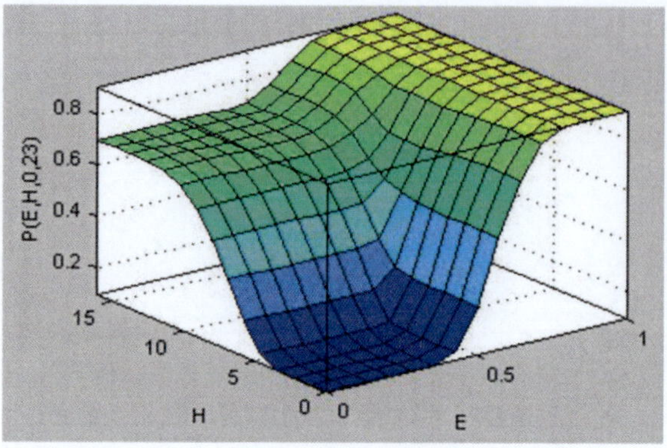

Fig. 5. The dependence of the probability $P(E, H, 0.23)$ of the negative environmental influence on the DM of the factors E and H, provided that the factors $I_N = 0\,\text{dB}$, $T = 23\,^{\circ}\text{C}$

is denoted the probability that the random variable "DM's Cognitive State" takes the value "negative". In assessing these probabilities, is assumed that $p_1 = \frac{q_1}{Q}$, $p_2 = \frac{q_2}{Q}$, where q_1 is the number of tests that characterize the DM's Psychological State negatively; q_2 - the number of tests that characterize the DM's Cognitive State negatively; Q - the total number of tests of the DM's Cognitive State.

3 Experiment, Results and Discussions

The following experiment was carried out to test the proposed information technology. At some point in time, the result of scanning of the measuring and control devices showed that there are deviations from normal values of the factors a_4, b_1, b_2, b_3. This means that the vertices of the BCN a_i, b_i, c_i, d_i, f_i $\left(i = \overline{1,4}\right)$ take values "sufficient", except for the vertices a_4, b_1, b_2, b_3 which take the value "insufficient". The performance indicators of the DMS components are as follows: the "ADDB" vertex is in critical condition, the "complex information system" vertex is in normal one, and the remaining vertices are in workable condition. Finally, the probability of a negative impact of each of the factors ("External Environment Influence on the DM", "DM's Cognitive State" and "DM's Psychological state") on the functional sustainability of the DM was 10%.

The calculation results by the proposed BCN are shown in Fig. 6. The probability that the System of monitoring FC ESKZ is in workable condition is 61%. The study of literature sources for rules of operation of critical application systems shows that a value of not less then 70% can be considered as an acceptable probability value of the operability of the System of monitoring FC ESKZ. This means that in this case, the operation of the System of monitoring FC ESKZ cannot be considered satisfactory.

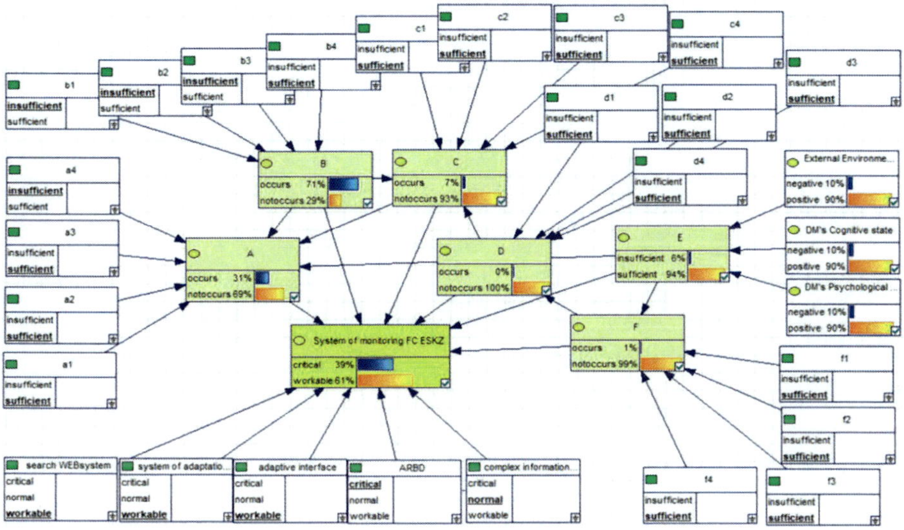

Fig. 6. Results of the calculation of the probability of the operability of the electrical power network under deviation of B factors from normal condition

Let us take into consideration (Fig. 6) the fact that the causes A, B, C, D, F of the faults and breakdowns in the electrical power network in exist with probabilities of 31%, 71%, 7%, 0%, 1%, respectively. Since the conditional probabilities of the influence of causes A, B, C, D, F on the operability of the System of monitoring FC ESKZ set by experts differ slightly (Table 10), the cause with the highest level of probability has the greatest impact on the operation of the system. In this case, this is reason B, which is determined by factors b_1, b_2, b_3, b_4. Among these factors, the factors b_1, b_2, b_3 lead to an insufficient degree of operability of the electrical power network, since the BCN variables $b_i \left(i = \overline{1,4}\right)$ take the value "insufficient". Hence is concluded that in the first instance it is necessary to make a decision to eliminate the causes for the decrease in operability of electrical power network, determined by the above factors. After eliminating these causes, the variables b_1, b_2, b_3 take the value "sufficient".

A repeated calculation by BCN at $b_i = sufficient \left(i = \overline{1,4}\right)$ showed that System of monitoring FC ESKZ is in workable condition with a probability of 84%, which is quite acceptable. In this case, a deviation from the normative values of the factor a_4, negative factors affecting the functional stability of the DM, as well as deficiencies in the operation of the components of the DSS, were not even eliminated.

Thus, according to the results of the experiment, it is possible to make a conclusion that the proposed IT allows to promptly determine the most significant causes of reducing the probability value of the operability of System of monitoring FC ESKZ and to recommend DM to take relevant decisions in real time to eliminate the causes of emergency situations when operating an electrical power network. On the basis of the information technology scheme for the evaluation of the operability degree for a critical object and its full functional sustainability, a Bayesian confidence network was proposed, taking into account expert estimates of the indicators of the object's state presented in the form of tables of BCN conditional probabilities.

It was proposed that noisy MAX type be used for BCN vertices with a large number of parent vertices. First, it allowed to reduce the complexity of filling the conditional probability tables, since in this case, an expert evaluates the impact of the value of the parent vertex on the node under study regardless of what values its other parent vertices take. Second, the conditional probability tables of the Noisy MAX vertices are greatly simplified. For the node "External Environment Influence on the DM", which probability of states are depending on many environmental factors, it was proposed a method of estimating the state probability based on a fuzzy inference by the Mamdani algorithm on a fuzzy knowledge base in which the values of input and output variables are given by fuzzy sets. This made it possible to estimate the probability of the influence on DM of such negative environmental factors as intensity of electric and magnetic fields, intensity of production noise and temperature when deviating from the normative values. On the example of one of the critical objects, a regional multilevel electrical power network, an experiment was conducted on the developed IT. In the first case, during the operation of the critical object, based on the results of scanning of measuring and control devices it was determined that there were deviations from the normative values of the factors a_4, b_1, b_2, b_3 that affect the object's performance. In addition, violations were identified in the DS system. At the same time, the probability of negative influence of each of the factors on DM's functional sustainability [10] was 10%. The calculation on the basis of BCN showed (Fig. 6) that the probability of the System of monitoring FC ESKZ (object of critical application) being in workable state is equal to 61%. Based on the normative requirements for the Rules of Operation of Critical Application Systems, a value of at least 70% is considered as a permissible value of the workable state of System of monitoring FC ESKZ. Therefore, the System of monitoring FC ESKZ cannot be considered workable in this case.

The analysis of the BCN solution shows that the reason B, which is determined by the factors b_1, b_2, b_3 has the greatest influence on the operability of the system. It follows that, first of all, it is necessary to make a decision to eliminate the reasons of the decrease in the operability of the electrical power network, determined by the above factors. Re-calculating by the BCN, provided that the factors b_1, b_2, b_3 took the normative values, and the state of the DS and the DM's functional sustainability did not change, shows that now the System of monitoring FC ESKZ is in operability state with a probability of 84%, which is quite acceptable.

4 Conclusions

The paper considers DS and monitoring information technology in distributed man-machine systems when managing complex technical objects of critical application. The proposed IT complements the theory and methods for solving the problems of improving the operation and functional sustainability of complex distributed man-machine systems of critical application, through the introduction of a complex of modern monitoring, evaluation, analysis and decision making means, taking into account the indicators of human factor, aimed at the assurance of trouble-free operation at each hierarchical level of the system. The construction of this information technology on the example of operational management of multilevel regional electrical power network operation modes is considered.

To achieve these results the work provides a study and a comprehensive assessment of the causes of the emergency condition and the factors that lead to an appropriate cause of failure of the electrical equipment by the coefficient of importance of the object's emergency condition. The expert evaluation of the degree of importance of a set of indicators characterizing one or another cause and factors of an emergency condition corresponding to the hierarchical level of the object is carried out by experts-specialists; a set of indicators of external production and cognitive factors that directly influence the DM's relevant decisions while managing at each hierarchical level of the object is investigated and determined; the functional scheme of information technology of interaction and operation of separate monitoring modules of the basic indicators set of a electrical power network in general hierarchical structure of a complex organizational and technical object is developed; based on the information technology scheme, to monitor the functional sustainability and to evaluate the modes of the operability state of electrical power network a BCN was developed using the table of conditional probabilities obtained by expert methods based on the knowledge of experts at each level of the object.

For practical substantiation of the obtained results, an experiment was conducted, the results of which confirmed the practical value of the proposed information technology, which can be used in distributed man-machine systems when managing complex technical objects of critical application. The proposed IT allows us to promptly identify, in real time, the most significant reasons of the decrease in operability and full functional sustainability of the critical object and to recommend DM making relevant decisions to eliminate the causes leading to emergencies. The proposed DS and Monitoring Information Technology may be recommended in MMS when managing complex organizational and technical objects of critical application.

References

1. Borukaiev, Z.: Mathematical and computer models of processes for perfecting the functioning and development of organizational management systems. Thesis for the degree of Doctor of Technical Sciences in specialty 01.05.02 – mathematical modeling and computing methods. Pukhov Institute for Modeling in Energy Engineering. National Academy of Sciences of Ukraine, Ukraine, Kyiv (2018)
2. Diamantoulakis, P., Kapinas, V., Karagiannidis, G.: Big data analytics for dynamic energy management in smart grids. Big Data Res. **2**, 94–101 (2015). https://doi.org/10.1016/j.bdr.2015.03.003
3. DSTU2389-94: Technical Diagnostics and Control of Technical Condition. Terms and Definitions. State Standard of Ukraine, Kyiv (1994)
4. Eisenk, H.: Test Your Intellectual Abilities. Grif, Chisinau (1992)
5. Huben, D.: On accident rate in 2010 and reliability indicators in electric networks of 6–150 kv of power supply companies that are part of NAC "ECU". Electr. Netw. Syst. **2**, 32–34 (2011)
6. Kreikebaum, F., Das, D., Yang, Y., Lambert, F., Divan, D.: Smart wires – a distributed, low-cost solution for controlling power flows and monitoring transmission lines. In: 2010 IEEE PES Innovative Smart Grid Technologies, pp. 1–8 (2010)
7. Miroshnik, A.: Analysis of the problem of the electrical power quality of in rural networks. In: Bulletin of NTU "KhPI", vol. 41, pp. 100–104 (2011)
8. Murphy, K.: A brief introduction to graphical models and Bayesian networks. Technical report 2001-5-10, Department of Computer Science. Department of Computer Science, University of British Columbia (2001)
9. Paziy, V., Miroshnyk, O.: Analysis of problems of diagnosis and monitoring of technical modes electrical distribution network. Sci. Bull. Tavria Agrotechnol. State Univ. **7**(1), 187–195 (2017)
10. Perederiy, V., Borchik, E., Ohnieva, O.: Information technology of control and support for functional sustainability of distributed man-machine systems of critical application. Lecture Notes in Computational Intelligence and Decision Making, pp. 461–477 (2019). https://doi.org/10.1007/978-3-030-26474-1_33
11. Portal: Smart power grids - talking about a revolution (2009). https://www.ieee.org/about/technologies/emerging/emerging_tech_smart_grids.pdf
12. Qualimetry: Wikipedia Material. https://en.wikipedia.org/wiki/Qualimetry
13. Quitzau, A.: Transforming energy and utilities through big data & analytics (2014). www.slideshare.net/AndersQuitzauIbm/big-data-analyticsin-energy--utilities
14. Semikin, V.: Luscher color test in solving problems of diagnosing the functional state and working capacity of human operator. In: Methods of Research and Diagnostics of the Functional State and Working Capacity of a Human Operator in Extreme Conditions, Moscow, pp. 61–71 (1989)
15. Shtovba, S.: Introduction to the theory of fuzzy sets and fuzzy logic. http://matlab.exponenta.ru/fuzzylogic/book1
16. Stimmel, C.: Big Data Analytics Strategies for the Smart Grid. CRC Print (2015). https://doi.org/10.1201/b17228
17. Yu, N., Johnson, R., Hong, M., Sherick, R., Loparo, K.: Big data analytics in power distribution systems. In: Innovative Smart Grid Technologies Conference, ISGT 2015 (2015). https://doi.org/10.1109/ISGT.2015.7131868
18. Zabolotnyi, I., Sazonov, V.: Expert system for the analysis of emergencies in electrical systems. In: Collection of scientific works of Donetsk National Technical University. Series: Electrical and Power Engineering, vol. 50, pp. 165–171 (2002)

Forecasting Assessment of Printing Process Quality

Vsevolod Senkivskyy(✉), Iryna Pikh, Nataliia Senkivska, Ivan Hileta, Oleh Lytovchenko, and Yuriy Petyak

Ukrainian Academy of Printing, Lviv, Ukraine
senk.vm@gmail.com, pikhirena@gmail.com, senkivskanata@gmail.com, hileta@gmail.com, lytowchenko@gmail.com, yuriy.petyak@gmail.com

Abstract. The printing process is a final stage of both designing and pre-press processing of publications. Implementation of this process materializes the author's design in the form of final-formed printing products. On the one hand, it accumulates the quality metrics (or shortcoming) of the original, on the other hand, it reflects the level and quality of the print process. Existing techniques, models and methods of printing process quality evaluation ensure the proper quality of products. However, implementation of the existing techniques is very expensive. In this paper we present the results of research aimed to the development of a priori evaluation of the quality of the printing process that determines the quality of the original product.

Within the framework of our research, this task is solved at the informational level, since forecasting process is based on the techniques, the main paradigm of which is the difference between factors (linguistic variables) in their pairwise comparison and the priority of the influence of the appropriate factor to investigated process. This approach is more universal because it provides a possibility to apply a general methodology not only for the factors given by numerical characteristics but also covers poorly formalized requirements described using natural language. In this research, we have designed the knowledge matrices and fuzzy logic equations for linguistic variables (factors) of the printing process. Moreover, we have proposed both the calculation algorithm and the technique of defuzzification process implementation in order to obtain the crisp numerical value of the integral index of the quality of the printing process implementation.

Keywords: Printing process quality · Knowledge matrix · Knowledge base · Fuzzy logic equation · Linguistic variable · Membership function · Term set · Fuzzy set · Defuzzification · Forecasting

1 Introduction

The questions of the products quality of any purpose has always been actually not only for direct manufacturers, but also for scientists which research and

S. Babichev et al. (Eds.): ISDMCI 2020, AISC 1246, pp. 467–479, 2021.
https://doi.org/10.1007/978-3-030-54215-3_30

develop the technologies that apply follows for appropriate technological processes implementation. The process of created earlier outputs quality improve or the attempt to obtain a fundamentally new, better product, is inherent to humanity during the all history of its existence. Similar problem is the problem of forming and providing quality printing products, which does not lose its relevance from the time when the first printed book was created. Despite of outward simplicity of both the structural and content representation of the book edition essence, it combines the meaningful and artistic display of the author's intention through compositional and artistic design with regard to textual content, illustrations, drawings, colors and fonts, paper and systems bonding and post-printing processing of book blocks.

Non-ordinality and versatility of the printed products determine the necessity of ensuring the quality of the printing process not only in terms of a technical solution of this problem. So, current solutions concerning evaluation of the qualitative component of the printing process using new technical equipment which is based on microprocess or programmable modules and computerized sections, does not guarantee the expected result. Taking into account presented hereinbefore, we suppose that a theoretically weighted information component that uses a proven practice of foreign firms (not a printing company) within the framework of a new direction - the theory of fuzzy control, could serve as a logical complement to existing systems and means.

Confirmation of this decision is due to the absence of both the logical interrelation between the categories of printing process parameters and the extent of their combined impact to the predicted quality completion of one of the crucial stages of the general printing technology. Equating these parameters to linguistic variables with simultaneous definition of term-sets of corresponding technological characteristics, we obtain the initial database, which is the basis of development of information tools for the formation and forecasting evaluation of the printing process quality on the basis of both the fuzzy set theory and the forecasting techniques based on system approach, which are reasonable solution to obtain the products of the proper quality.

1.1 Formal Problem Statement

The study of the printing process qualitative indicators in the most cases is based on a fuzzy inference model [10,18], which reflects the hierarchy of accumulation of qualitative expressions in a bottom-up scheme, from the lowest level factors through models of linguistic variables, grouped into functional groups of relevant categories, to the top-level models which represent the integral component of the printing output. The use of fuzzy set technique to obtain the expected result dictates the necessity of forming a fuzzy knowledge matrix which is a basis of the initial database for developing a system of fuzzy logical equations. It is obvious that the components of such a knowledge base should take into account expert conclusions concerning the combined influence of a variety of factors (linguistic variables) on the printing process considering the combination of linguistic terms. In order to logically structured the direction of the printing process research, we apply the conditional division of factors (linguistic variables) into categories: technical factors of printing; materials; technological (functional) procedures [12,20]. Consideration and further development of the principle of multilevel hierarchical structuring of the printing process and its top-down outputs on the one hand, and the formation of qualitative indicators in the opposite direction on the other hand, are very important stage of this approach implementation. In this case, the dependence of the printing process quality can be expressed by a function:

$$V = F_V(T, R, L) \tag{1}$$

where T, R, L are the sets of factors of hereinbefore presented categories. The highest level functional, expressed by the linguistic variable V in accordance with equation (1), form an integral index of the typographic process predicted quality. Let a set of linguistic terms will be follows: $X(V)$ $<low, medium, high>$. The level of categories in this case can be divided into the following sets of linguistic terms:

$X(T) = <low, medium, high>$ - quality of technical factors;
$X(R) = <low, medium, high>$ - quality of materials;
$X(L) = <low, medium, high>$ - quality of technological procedures.

As a result, the expert knowledge base which corresponds to the printing process for the highest level can be presented as follows:

$$\begin{aligned} &if\ (T = low)\ or\ (T = medium)\ or\ (T = high)\\ &and\ (R = low)\ or\ (R = medium)\ or\ (R = high)\\ &and\ (L = low)\ or\ (L = medium)\ or\ (L = high)\\ &then\ (V = low)\ or\ (V = medium)\ or\ (V = high) \end{aligned} \tag{2}$$

The main problem within the framework of this research consists of develop of expert knowledge base logical structure based on (2) in the matrix form and apply it for the printing process quality forecasting.

1.2 Reviews of Literature

A lot of works are devoted to evaluation of the printing process quality in various areas of scientific research nowadays. So, in [18] the authors consider the effect of various 3D printing parameters on a transdermal drug delivery system with milliprojections printed using poly lastic acid. The studied parameters assumed printing temperature, layer thickness, extrusion width, infill width, and nozzle orifice diameter. Their effects on the final print quality were evaluated based on the surface finish and dimensional accuracy of the milliprojections. Paper [19] considers issues of laser printing of silver (Ag) nanoparticle lines on PEN substrate at high repetition rates (10–40 kHz) is introduced and an investigation on the factors that affect the printing results, including the experimental setup's parameters and inks' rheological characteristics, is conducted. The obtained by authors indicate that, for high repetition rates, best quality lines derive from high viscosity inks when the distance between the donor and the receiver is relatively smal. A novel hybrid multi-objective optimization method is proposed to optimize the overall printing quality of AJP based on a determined operating window is proposed in [22]. In the proposed by the authors approach, the response surface methodology is combined with the CCD to investigate and quantify the correlations between the printed line features and the key process parameters. In [7] the authors presents the research results concerning improving the quality of generated droplets in direct-write inkjet printing. They have investigated the optimal waveform voltage that influences the size and the velocity of ejected droplets to increase the performance of inkjet printing process. The issues concerning lightfastness properties of printed foil samples due to long-time exposure are considered in [13]. A blister foilsprinted in the gravure printing process have been taken as the sample as they have extensive usage in food and medicinepackaging. A kinetic model to predictthe fading rate of the printed foil has been proposed by the authors in this work. In [1], the effects of ink consumption during the printing process on printing quality parameters were examined. The ink consumption was evaluated in terms of recycling, sustainability, environmental and economic effects, and recommendations were made to reduce the use of ink. The paper [11] presents the results of research concerning improve the roll-to-roll printing quality. The authors have proposed the system in which both ends of the impression roll are driven individually, unlike the conventional printing pressure control system, for accurate printing pressure measurement. To improve the printing process quality, the authors perform a frequency analysis of the printing pressure after compensating for the measurement delay and then apply adaptive feedforward control to uniformize the printing pressure additionally. The issues about simulation of electrohydrodynamic inkjet printing parameters by using regression analysis, a feed-forward neural network trained with backpropagation and a neural network trained with a genetic algorithm separately are presented in [5]. The authors have shown that all models gave good prediction accuracy with less than 10% error in the prediction of the droplet diameter. Furthermore, it had been observed that the performance of genetic algorithm surpasses both the regression- and BPNN-based approaches in most of the test cases.

Studies concerning formation and improving the printing output quality can be carried out based on the use of fuzzy sets mathematical theory [21] using the current tools, techniques and approaches which are successfully applied in various areas of scientific research nowadays [2–4,8,9,16,17]. However, the authors' research are mainly focused on small dimension data processing based on the use of fuzzy logic inference system for control the appropriate object operating. Applying the fuzzy logic inference system for evaluation of the printing process quality create the conditions for complex estimation of various factors influence to integral index, which determines the quality of the printing process. Implementation of the fuzzy inference system to evaluate the printing process quality assumes the following stages: formation of a set of linguistic variables which determine the quality of the printing process; formalization of the input variables into linguistic estimates, i.e. identification of recommended ranges of values of universal term sets and linguistic terms of printing process quality; creation a model to form the integral quality index which include both the membership functions of linguistic variables and the set of fuzzy rules agreed between input variables and output parameter.

The aim of the paper is development of forecasting technique of printing process quality based on the complex use of both a fuzzy knowledge base and a fuzzy inference system which are the basis to determine an integral index of the printing process quality.

2 Materials and Methods

Expert knowledge matrix to evaluate the integral index of printing process quality within the framework of fuzzy inference system is presented in Table 1 [6,15].

Table 1. Knowledge matrix of linguistic variable V - quality of printing process

Quality of technical factors T	Quality of materials R	Quality of technological procedures L	Quality of printing process V
low	medium	low	low
low	low	low	
medium	low	medium	medium
high	high	medium	
high	medium	high	high
high	high	high	

The fuzzy logic equations for the linguistic variable V which contain the data of Table 1 can be presented as follows:

$$\mu_{low}(V) = \mu_{low}(T) \wedge \mu_{medium}(R) \wedge \mu_{low}(L) \vee \mu_{low}(T) \wedge \mu_{low}(R) \wedge \mu_{low}(L);$$

$$\mu_{medium}(V) = \mu_{medium}(T) \wedge \mu_{low}(R) \wedge \mu_{medium}(L)$$
$$\vee \mu_{high}(T) \wedge \mu_{high}(R) \wedge \mu_{medium}(L);$$

$$\mu_{high}(V) = \mu_{high}(T) \wedge \mu_{medium}(R) \wedge \mu_{high}(L) \vee \mu_{high}(T) \wedge \mu_{high}(R) \wedge \mu_{high}(L).$$

The next step involves applying the same procedure to linguistic variables which determine the factors at lower hierarchical level:

$$T = F_T(t_1, t_2, t_3); \; R = F_R(r_1, r_2, r_3); \; L = F_L(l_1, l_2, l_3) \tag{3}$$

where: t_1 is a type of printing process; t_2 is characteristics of bruzer tympan; t_3 is a type of printing plate; r_1 is characteristics of paper; r_2 is characteristics of ink; r_3 is moisturizing solution; l_1 is deformation of a bruzer tympan in the contact with a printing plate; l_2 is deformation of a bruzer tympan in the printing contact zone; l_3 is printing speed.

To confirm presented hereinbefore, let's obtain the fuzzy logic equations for linguistic variable T(quality of technical factors), an expert logical statement for which is the following:

$$\begin{aligned} &if\ (t_1 = sheet)\ or\ (t_1 = roll)\\ &and\ (t_2 = soft)\ or\ (t_2 = medium)\ or\ (t_2 = hard)\\ &and\ (t_3 = low)\ or\ (t_3 = medium)\ or\ (t_3 = high)\\ &then\ (T = low)\ or\ T = medium)\ or\ (T = high) \end{aligned} \quad (4)$$

Table 2 presents the knowledge matrix created based on the statements (4).

Table 2. Knowledge matrix of linguistic variable T - quality of technical factors

Type of printing process t_1	Characteristics of bruzer tympan t_2	Type of printing plate t_3	Quality of technical factors T
sheet	soft	low	low
roll	soft	medium	
sheet	medium	medium	medium
roll	medium	high	
roll	hard	high	high
roll	hard	medium	

Fuzzy logic equations to calculate the membership functions of logic variable T can be presented as follows:

$$\mu_{low}(T) = \mu_{sheet}(t_1) \wedge \mu_{soft}(t_2) \wedge \mu_{low}(t_3) \vee \mu_{roll}(t_1) \wedge \mu_{soft}(t_2) \wedge \mu_{medium}(t_3);$$
$$\mu_{medium}(T) = \mu_{sheet}(t_1) \wedge \mu_{medium}(t_2) \wedge \mu_{medium}(t_3)$$
$$\vee \mu_{roll}(t_1) \wedge \mu_{medium}(t_2) \wedge \mu_{high}(t_3);$$
$$\mu_{high}(T) = \mu_{roll}(t_1) \wedge \mu_{hard}(t_2) \wedge \mu_{high}(t_3) \vee \mu_{roll}(t_1) \wedge \mu_{hard}(t_2) \wedge \mu_{medium}(t_3).$$

Similar systems of fuzzy logical equations can be obtained for the linguistic variables R - quality of materials (Table 3) and L - quality of technological procedures (Table 4). Fuzzy logical equations based on the data of Table 3 are presented below. In this case it is necessary to take into account some peculiarities of the linguistic variables characteristics regarding their influence on the final quality of materials. Thus, during the expert survey it was found that the high quality paper will be provided by medium thickness paper (0.07 mm–0.1 mm), medium viscosity paint and moist acid solution ($4.8 < pH < 5.5$ for offset printing).

Table 3. Knowledge matrix of linguistic variable R - quality of materials

Characteristics of paper (thickness) r_1	Characteristics of ink (viscosity) r_2	moisturizing solution (acidity) r_3	Quality of materials R
little	high	strong	low
large	high	strong	
little	medium	neutral	medium
medium	high	weak	
medium	medium	weak	high
medium	medium	neutral	

$$\begin{aligned}
\mu_{low}(R) &= \mu_{little}(r_1) \wedge \mu_{high}(r_2) \wedge \mu_{strong}(r_3) \\
&\vee \mu_{large}(r_1) \wedge \mu_{high}(r_2) \wedge \mu_{strong}(r_3); \\
\mu_{medium}(R) &= \mu_{little}(r_1) \wedge \mu_{medium}(r_2) \wedge \mu_{neutral}(r_3) \\
&\vee \mu_{medium}(r_1) \wedge \mu_{high}(r_2) \wedge \mu_{weak}(r_3); \\
\mu_{high}(R) &= \mu_{medium}(r_1) \wedge \mu_{medium}(r_2) \wedge \mu_{weak}(r_3) \\
&\vee \mu_{medium}(r_1) \wedge \mu_{medium}(r_2) \wedge \mu_{neutral}(r_3).
\end{aligned}$$

Fuzzy logical equations based on the data of Table 4 can be presented as follows:

$$\begin{aligned}
\mu_{low}(L) &= \mu_{large}(l_1) \wedge \mu_{medium}(l_2) \wedge \mu_{low}(l_3) \\
&\vee \mu_{medium}(l_1) \wedge \mu_{large}(l_2) \wedge \mu_{medium}(l_3); \\
\mu_{medium}(L) &= \mu_{little}(l_1) \wedge \mu_{medium}(l_2) \wedge \mu_{medium}(l_3) \\
&\vee \mu_{medium}(l_1) \wedge \mu_{little}(l_2) \wedge \mu_{medium}(l_3); \\
\mu_{high}(L) &= \mu_{little}(l_1) \wedge \mu_{little}(l_2) \wedge \mu_{high}(l_3) \\
&\vee \mu_{medium}(l_1) \wedge \mu_{medium}(l_2) \wedge \mu_{medium}(l_3).
\end{aligned}$$

Table 4. Knowledge matrix of linguistic variable L - quality of technological procedure

Deformation of a bruzer tympan in the contact with a printing plate l_1	Deformation of a bruzer tympan in the printing contact zone l_2	Printing speed l_3	Quality of technological procedure L
large	medium	low	low
medium	large	medium	
little	medium	medium	medium
medium	little	medium	
little	little	high	high
medium	medium	medium	

The system of fuzzy logical equations reproduces the real tendency of the printing process, the essence of which is that high quality of technological procedures is achieved in the case of small deformations of the bruzer tympan in the fields of both the printing form and the printing contact at medium or high speeds of printing. At this stage of the study, the statements are formed at the level of linguistic terms *small*, *medium*, *high*, which, together with other descriptive characteristics, will be the starting point for establishing a numerical prognostic level of printing process quality.

3 Experiment

The quality of the printing process is determined by a linguistic variable V whose membership functions contain set of the linguistic terms *low, medium, high*. Let's introduce a universal term set $Q = \{q_1, q_2, q_3\}$, in which the values of membership functions of the linguistic variable V are calculated. Taking into account these conditions, we can express the quality index by a fuzzy set defined in the specified points of the interval:

$$V(T, R, L) = \{\frac{\mu_{low}(V)}{q_1}, \frac{\mu_{medium}(V)}{q_2}, \frac{\mu_{high}(V)}{q_3}\} \quad (5)$$

The calculation of the printing process quality index is performed using a matrix of knowledge presented in Table 1, fuzzy logical equations created based on Table 1 data, and the values of the membership functions of the linguistic variables of the distinguished categories. The output database is formed in the form of Table 5. The ranges of the printing process technological parameters are divided into four sections, obtaining five control points at which we calculate the membership functions.

Table 6, Table 7 and Table 8 contain the membership functions values for term-set of the linguistic variable T - *type of a printing process*. In the same way, we have formed the tables with membership functions values for other linguistic variables.

4 Results and Discussion

Taking into account the values of the linguistic variables parameters presemted in Table 5 which can be used to calculate the index of the printing process quality, we can define the values of all linguistic variables membership functions considering the data presented in Table 6, Table 7 and Table 8 in the following way:

- T - quality of technical factors:
  $\mu_{sheet}(t_1) = 0.22$; $\mu_{roll}(t_1) = 0.90$; $\mu_{soft}(t_2) = 0.20$; $\mu_{medium}(t_1) = 0.45$; $\mu_{hard}(t_2) = 0.78$; $\mu_{low}(t_3) = 0.44$; $\mu_{medium}(t_1) = 0.90$; $\mu_{high}(t_1) = 0.30$.

Table 5. Output database of printing process

Variable	Linguistic interpretation	Technological range of variation	Value of the parameter
t_1	Type of printing process	1–5	4
t_2	Characteristics of bruzer tympan	30–90	75
t_3	Type of printing plate	100–500	200
r_1	Characteristics of paper	0.03 mm–0.25 mm	0.15 mm
r_2	Characteristics of ink	15–20 $Pa \cdot c$	17.5 $Pa \cdot c$
r_3	Moisturizing solution	1–7 pH	5 pH
l_1	Deformation of a bruzer tympan in the contact with a printing plate	(1–4)%	2%
l_2	Deformation of a bruzer tympan in the printing contact zone	(5–10)%	2%
l_3	Printing speed	(2000–18000) rpm	6000 rpm

Table 6. Membership functions for term-set $X(t_1)$ - Type of printing process

d_1	1	2	3	4	5
$\mu_{sheet}(d_1)$	1	0.85	0.6	0.22	0.10
$\mu_{roll}(d_1)$	0.06	0.20	0.78	0.90	1

Table 7. Membership functions for term-set $X(t_2)$ - characteristics of bruzer tympan

d_1	30	45	60	75	90
$\mu_{soft}(d_1)$	0.11	0.60	0.45	0.20	0.10
$\mu_{medium}(d_1)$	0.11	0.70	1	0.45	0.11
$\mu_{hard}(d_1)$	0.11	0.22	0.44	0.78	1

Table 8. Membership functions for term-set $X(t_3)$ - type of printing plate

d_1	100	200	300	400	500
$\mu_{low}(d_1)$	1	0.44	0.20	0.11	0.06
$\mu_{medium}(d_1)$	0.25	0.90	1	0.50	0.22
$\mu_{high}(d_1)$	0.11	0.30	0.60	0.85	1

- R - quality of materials:
  $\mu_{little}(r_1) = 0.67$; $\mu_{medium}(r_1) = 1$; $\mu_{large}(r_1) = 0.67$;
  $\mu_{low}(r_2) = 0.20$; $\mu_{medium}(r_2) = 1$; $\mu_{high}(r_2) = 0.50$;
  $\mu_{strong}(r_3) = 0.20$; $\mu_{weak}(r_3) = 1$ $\mu_{neutral}(r_3) = 0.12$.
- L - quality of technological procedure:
  $\mu_{small}(l_1) = 0.45$; $\mu_{medium}(l_1) = 0.80$; $\mu_{large}(l_1) = 0.22$;
  $\mu_{small}(l_2) = 0.55$; $\mu_{medium}(l_2) = 0.75$; $\mu_{large}(r_2) = 0.28$;
  $\mu_{low}(l_3) = 0.44$; $\mu_{medium}(l_3) = 0.90$ $\mu_{high}(l_3) = 0.30$.

Then, we determine the membership functions values of the technical factors, technical materials and technological procedures considering the presented hereinbefore membership functions of the lower level linguistic variables as follows:

- Technical factors:

$$\mu_{low}(T) = 0.22 \wedge 0.20 \wedge 0.44 \vee 0.90 \wedge 0.20 \wedge 0.90 = 0.20;$$
$$\mu_{medium}(T) = 0.22 \wedge 0.45 \wedge 0.90 \vee 0.90 \wedge 0.45 \wedge 0.30 = 0.30;$$
$$\mu_{high}(T) = 0.90 \wedge 0.78 \wedge 0.30 \vee 0.90 \wedge 0.78 \wedge 0.90 = 0.78.$$

- Quality of materials:

$$\mu_{low}(R) = 0.67 \wedge 0.50 \wedge 0.20 \vee 0.67 \wedge 0.50 \wedge 0.20 = 0.20;$$
$$\mu_{medium}(R) = 0.67 \wedge 1 \wedge 0.20 \vee 1 \wedge 0.50 \wedge 1 = 0.50;$$
$$\mu_{high}(R) = 1 \wedge 1 \wedge 1 \vee 1 \wedge 1 \wedge 1 = 1.$$

- Quality of technological procedures:

$$\mu_{low}(L) = 0.22 \wedge 0.75 \wedge 0.44 \vee 0.80 \wedge 0.28 \wedge 0.90 = 0.28;$$
$$\mu_{medium}(L) = 0.45 \wedge 0.75 \wedge 0.90 \vee 0.80 \wedge 0.55 \wedge 0.90 = 0.55;$$
$$\mu_{high}(L) = 1 \wedge 0.55 \wedge 0.30 \vee 0.80 \wedge 0.75 \wedge 0.90 = 0.75.$$

At the next step, we calculate the membership functions values for linguistic variable V - quality of the printing process:

$$\mu_{low}(V) = 0.20 \wedge 0.50 \wedge 0.28 \vee 0.20 \wedge 0.20 \wedge 0.28 = 0.20;$$
$$\mu_{medium}(V) = 0.30 \wedge 0.20 \wedge 0.55 \vee 0.78 \wedge 1 \wedge 0.55 = 0.55;$$
$$\mu_{high}(V) = 0.78 \wedge 0.50 \wedge 0.75 \vee 0.78 \wedge 1 \wedge 0.75 = 0.75.$$

Obtained value of the membership functions for used factors allow us to perform defuzzification procedure of fuzzy set (6) using well known equation [14]:

$$V = \frac{\sum_{i=1}^{m}[\underline{V} + (i-1)\dfrac{\overline{V} - \underline{V}}{m-1}]\mu_i(V)}{\sum_{i=1}^{m}\mu_i(V)} \tag{6}$$

where $\underline{V}$ and $\overline{V}$ are the minimal and maximal values of printing process quality respectively; m is the number of fuzzy terms for linguistic variable V. Let the output values be the following: $m = 3$; $\mu_1(V) = \mu_{low}(V)$; $\mu_2(V) = \mu_{medium}(V)$. Let low and high boundary values of the linguistic variable V will be $\underline{V} = 1\%$ and $\overline{V} = 100\%$ respectively. The points of the range division are follows: 1, 50, 100. As a result of substitution of the initial data in the expression (6) we get:

$$V = \frac{1 \cdot 0.20 + 50 \cdot 0.55 + 100 \cdot 0.75}{0.20 + 0.55 + 0.75} = 68.47\%$$

The integral quality index of the book printing process has been calculated using output parameters which correspond to the real technological characteristics of flat offset printing. Thus, we obtain the possibilities to implement the forecasted management of the printing process quality with the availability of appropriate means, based on the imitation model (software component) of a fuzzy system for forecasting of the printing process quality developed at the Department of Computer Science and Information Technologies of the Ukrainian Academy of Printing.

5 Conclusions

An important aspect for solving the problem of printing process quality forecasting is the application of both the information approach with appropriate theoretical substantiation and development of applications and simulation models for practical application. Practical steps in this case consist in systematic analysis of both the publishing and printing processes. As a result of this step implementation is expert selection of factors and their ranking by their influence to the level of implementation of the technological stages of preparation and production of printed products. Effective and practically valuable results are achieved due the use of general conceptual foundations, applied methods and models of fuzzy set theory, which is based on a linguistic description of technological process factors and a fuzzy logic of creation of the relationships between them using logical inference technique and linguistic variables membership functions. One of the ways of this problem solve is presented in this paper. As a result of the simulation process which is implemented with the use of fuzzy logic methods, we have proposed the knowledge matrix and fuzzy logic equations to calculate the linguistic variables membership functions which determine the various factors of influence to the printing process quality. We have performed also the defuzzification procedure of the obtained fuzzy set in order to determine the crisp value of the integral index of the printing process quality implementation.

References

1. Aydemir, C., Yenidogan, S., Ozsoy, S.A.: Effects of ink consumption on print quality on coated cellulose-based paper surfaces. Cellul. Chem. Technol. **54**(1–2), 89–94 (2020). https://doi.org/10.35812/CelluloseChemTechnol.2020.54.10

2. Babichev, S., Korobchynskyi, M., Lahodynskyi, O., et al.: Development of a technique for the reconstruction and validation of gene network models based on gene expression profiles. East.-Eur. J. Enterp. Technol. **1**(4–91), 19–32 (2018). https://doi.org/10.15587/1729-4061.2018.123634
3. Babichev, S., Škvor, J., Fišer, J., Lytvynenko, V.: Technology of gene expression profiles filtering based on wavelet analysis. Int. J. Intell. Syst. Appl. **310**(4), 1–7. https://doi.org/10.5815/ijisa.2018.04.01
4. Babichev, S., Lytvynenko, V., Korobchynskyi, M., Taif, M.A.: Objective clustering inductive technology of gene expression sequences features. Commun. Comput. Inf. Sci. **716**, 359–372 (2017). https://doi.org/10.1007/978-3-319-58274-0_29
5. Ball, A.K., Das, R., Roy, S.S., Kisku, D.R., Murmu, N.C.: Modeling of EHD inkjet printing performance using soft computing-based approaches. Soft. Comput. **24**(1), 571–589 (2020). https://doi.org/10.1007/s00500-019-04202-0
6. Durnyak, B.V., Senkivskyy, V., Pikh, V.: Information technology of prognostication and providing the quality of publishing and printing processes (methodology of problem solution). Technol. Complexes **1**(9), 21–24 (2014)
7. Hamad, A.H., Salman, M.I., Mian, A.: Effect of driving waveform on size and velocity of generated droplets of nanosilver ink (smartink). Manuf. Lett. **24**, 14–18 (2020). https://doi.org/10.1016/j.mfglet.2020.03.001
8. Jablonski, K., Grychowski, T.: Fuzzy inference system for the assessment of indoor environmental quality in a room. Indoor Built Environ. **27**(10), 1415–1430 (2018). https://doi.org/10.1177/1420326X17728097
9. Jana, D.K., Roy, K., Dey, S.: Comparative assessment on lead removal using micellar-enhanced ultrafiltration (MEUF) based on a type-2 fuzzy logic and response surface methodology. Sep. Purif. Technol. **207**, 28–41 (2018). https://doi.org/10.1016/j.seppur.2018.06.028
10. Kamath, H.N., Rodrigues, L.L.: Influence of overall equipment effectiveness on print quality, delivery and cost: a system dynamics approach. Int. J. Appl. Eng. Res. **11**(8), 5889–5898 (2016)
11. Kim, J., Kim, Y., Kim, T., Lee, B., Park, J., Oh, D.: Printing pressure uniformization through adaptive feedforward control in roll-to-roll printing process. Microsyst. Technol. **26**(1), 265–273 (2020). https://doi.org/10.1007/s00542-019-04640-8
12. Lundstrom, J., Verikas, A.: Assessing print quality by machine in offset colour printing. Knowl.-Based Syst. **37**, 70–79 (2013). https://doi.org/10.1016/j.knosys.2012.07.022
13. Mandal, M., Bandyopadhyay, S.: Study of the lightfastness properties of prints on blister foils by spectral reflectance. Color Res. Appl. **45**(2), 336–344 (2020). https://doi.org/10.1002/col.22449
14. Pikh, I.V., Durnyak, B.V., Senkivskyy, V.M., Holubnyk, T.S.: Information technology of formation of book edition quality. Monograph, Ukrainian Academy of Printing (2017)
15. Senkivska, N.: Synthesis of a model of factors for prognostication of the quality of printing process (on the example of flatbed offset printing technique). Book Qualil. **1**(19), 46–52 (2011)
16. Senkivskyy, V., Pikh, I., Havenko, S., Babichev, S.: A model of logical inference and membership functions of factors for the printing process quality formation. Adv. Intell. Syst. Comput. **1020**, 609–621 (2020). https://doi.org/10.1007/978-3-030-26474-1_42

17. Sichevska, O., Senkivskyy, V., Babichev, S., Khamula, O.: Information technology of forming the quality of art and technical design of books. In: CEUR Workshop Proceedings, vol. 2533, pp. 45–57 (2019)
18. Tang, T.O., Holmes, S., Dean, K., Simon, G.P.: Design and fabrication of transdermal drug delivery patch with milliprojections using material extrusion 3d printing. J. Appl. Polym. Sci. **137**(23), art. no. 48777 (2020). https://doi.org/10.1002/app.48777
19. Tsakona, D., Theodorakos, I., Kalaitzis, A., Zergioti, I.: Investigation on high speed laser printing of silver nanoparticle inks on flexible substrates. Appl. Surf. Sci. **513**, art. no. 145912 (2020). https://doi.org/10.1016/j.apsusc.2020.145912
20. Varepo, L.G., Brazhnikov, A.Y., Volinsky, A.A., et al.: Control of the offset printing image quality indices. J. Phys.: Conf. Seri. **858**(1), art. no. 012038 (2017). https://doi.org/10.1088/1742-6596/858/1/012038
21. Zadeh, L.A.: Fuzzy logic - a personal perspective. Fuzzy Sets Syst. **281**, 4–20 (2015). https://doi.org/10.1016/j.fss.2015.05.009
22. Zhang, H., Choi, J.P., Moon, S.K., Ngo, T.H.: A hybrid multi-objective optimization of aerosol jet printing process via response surface methodology. Addit. Manuf. **33**, art. no. 101096 (2020). https://doi.org/10.1016/j.addma.2020.101096

Organization-Managerial Mechanism of the Counteraction to Non-Observable Economic Processes in the Agricultural Sphere of Ukraine

Olena Moroz[1], Vladimir Semtsov[1], Nataliia Savina[2], Irina Shvarts[1](✉), Oksana Bezsmertna[1], Tetiana Bilokon[1], and Tetiana Piliavoz[1]

[1] Vinnytsia National Technical University, Vinnytsia, Ukraine
moros@vntu.edu.ua, semtsov_vm@ukr.net, s.irinach502@gmail.com, bezsmertnaoksana@gmail.com, tm.bilikon@gmail.com, vitan1975.75@gmail.com

[2] The National University of Water and Environmental Engineering, Rivne, Ukraine
n.b.savina@nuwm.edu.ua

Abstract. The paper substantiates the author's approach to the problem of the counteraction to non-observable economic processes in the agricultural sphere of Ukraine. The approach is based on the monitoring of the actions of the risk groups on the market. It is demonstrated the essence of the problem, characteristic features and possible consequences of the non-observable economic processes in the agricultural sphere. The attention to universal principles-guidelines as the basis of the system of counteraction to non-observable economic processes is focused. It is established that the modern system of state regulation of the national economy does not provide the sufficient risk management. Main risks of non-observable economic processes occurrence in Ukrainian economy (information, political, social, financial economic) are defined, lack of the possibility to form the stable social economic system without the efficient risk management is established. Special attention is paid to the necessity of the improvement of the institutional environment in the context of the counteraction to non-observable economic processes context.

Keywords: Non-observable economy · Non-observable economic processes · Counteraction to non-observable economic processes · Agricultural economy

1 Introduction

Rapid progress of the Ukrainian society, its motion in the direction of the development of the market relations in modern internal and international conditions is impossible without the solution of the problems, emerging as a result of non-observable economic processes.

S. Babichev et al. (Eds.): ISDMCI 2020, AISC 1246, pp. 480–495, 2021.
https://doi.org/10.1007/978-3-030-54215-3_31

Under the definition "non-observable economy" we mean economic activity which is not completely or partially controlled (or monitored) by the organs of power, thus, economic relations of the market agents, participating in the economic activity are not regulated in economic-legal context and, as a result, are not statistically taken into account due to the miscalculations of organization-managerial mechanism. Also, using the expression "non-observable economic processes" we imply the actions, based on deliberate and/or unconscious understatement and/or hiding of the results of own activity.

Modern state-of-art of the national economy shows that nowadays it is necessary to improve the organization-managerial mechanism, aimed at the counter action to the non-observable economic processes. Improvement of the system of interaction between the organs of power, local self-government and the economic entities – is the prerequisite for the obtaining the additional resources and maximization of the total cost.

The scientific novelty of the work is what is offered model of the dependence of the resultant index on the percentage of tillage used by agricultural enterprises, it was concluded that in order to optimize the process of land distribution it is necessary to strengthen the control and accounting of tillage in private households.

2 Review of Literature

Separate aspects of the counteraction to the non-observable economic processes are considered in the works of such modern scholars [7–10, 17, 18, 21, 23] and others. However, in spite of the depth of the research, carried out, still the tools of the organization-managerial mechanism, aimed at the counteraction to the non-observable economic processes in the agricultural sphere are not sufficiently developed. In practical terms all this requires the development and modeling of the institutional technologies, aimed at the obstruction of the non-observable economic processes in the agricultural sphere of Ukraine. To reach the target goal, the given research solves the complex of the following mutually connected tasks: explore the essence of the problem, characteristic features and possible consequences of the non-observable economic processes in the agricultural sphere; determine the methods of the institutional stimulation of the recipient groups of the economic agents regarding the constructive type of the organizational behavior, directed at the prevention of the non-observable economic processes in the course of market modernization. All this stipulated the choice of the subject of the research.

3 Problem Statement

It is obvious that the problem of the realization of the institutional modernization can be solved only on the condition when scientific proposals and recommendations will be closely connected with the real life. Scholastic theoretical assumptions, regarding non-observable economic processes slow down the development of our state. That is why, practical studies, concerning the identification of the mechanisms of the counteracting the non-observable processes in the context of the institutional modernization of the agricultural sphere require special attention.

4 Materials and Methods

We have to take into account the extreme complexity of the task of counter action to non-observable economic processes, as it has been demonstrated globally. These phenomena are observed both in capitalist and communist states. For instance, in the family of the former Minister of Public Security of the communist China Zhau Yun Can in 2019 law enforcement authorities seized $14.5 billion assets [15,26].

Proceeding from the fact that Ukraine is an agrarian country, special attention is paid to the non-observable economy in the agricultural sphere, taking into account its specific character. All the more, namely the Ukrainian businessmen establish the largest agricultural companies of Europe [25].

To illustrate the example of the non-observable economic processes we propose the data regarding the land use in Vinnytsia region in 2017–2018 (Fig. 1) [20,22]. The figure contains data, regarding the land area, used by the personal households (region average – a little more than 23.0%) and agricultural enterprises of the different organization forms (on the average approximately 69.0%). At the same time mathematical calculations show that certain area of the lands of agricultural designation (up to 7.0%) is used by the unknown economic agents. Practically, in every administrative district of the region there are farming lands which are used but the community does not know the official owner (proprietor). Only in one district out of 27 this index is 0.3% but it is rather the exception from the rule, whereas the average index in the districts of region was 6.9%. The fluctuation range of this index was from 0.3% to 20%; in the six districts this value exceeded 10%; the highest index was observed in Shargorod district – 19.8%. Thus, certain part of business is in the "non-observable zone", with the corresponding consequences concerning tax payment, inappropriate accounting of expenses, depreciation, products, etc. At the same time the community knows that certain part of business (probably, close to the authorities) enjoy favorable conditions, inaccessible for the others. In our opinion, mathematical index in the given case can serve as the indirect indicator of the corruption and the reflection of the losses as a result of the imperfection of the system of the monitoring and control.

In our view, the basis of the system of counteraction to non-observable economic processes must be not all-embracing norms (institutions), but the universal principles-guidelines. Naturally, due to various reasons it is impossible to eliminate completely the corruption and non-observable economy. In this connection we must agree with D. Zubrytskyi, who, studying the problem of the Province Committees in the context of the serfdom wrote as far back as in 1858

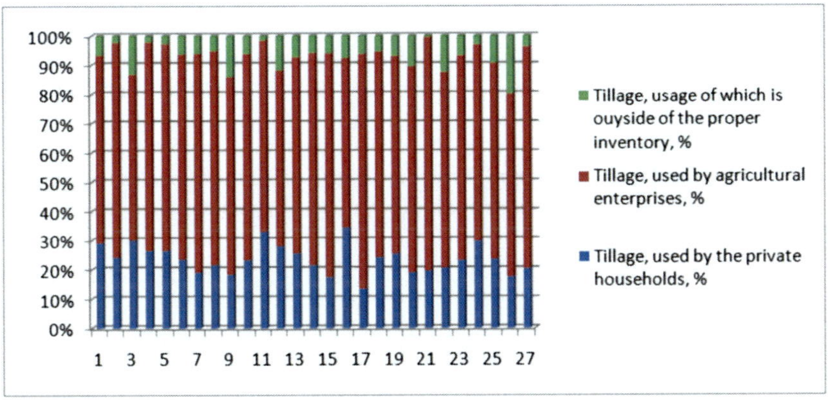

Fig. 1. Share of the non-observable sector of the economy in the context of the land use agricultural branch of Vinnytsia region in 2017–2018 (Source: authors' studies)

"bribes were, are and will be always and everywhere. The Savior has chosen only 12 apostles and one of them was bribe taker" [5]. Then, we do not claim for the exhaustive solution of the problem, aimed at the counteraction to the non-observable economy, but we must concentrate the attention on these targeted priorities/principles of social economic development which are able to provide rapid effect and considerable return. We consider that such basic principles should be: strong system of the law enforcement organs; fair justice; scientifically validated system of the monitoring of the social-economic state. We are convinced that as a result of the realization of these principles the level of the non-observable economy will decrease, however this requires clear scientific substantiation of the mechanism for the counteraction to the non- observable processes as well as empirical analysis of the real economic relations. The simplified attitude to the non- observable economy as to the extremely negative phenomenon stipulates the cost approach of the authorities in the process of the introduction of the counteraction measures. In our opinion the non-observable economy roots both in the attempts to organize the dysfunctional schemes and in superimposed phenomena institutions, world outlook, and ideology. These are interconnected phenomena.

The problem of the non-observable processes is analyzed in the society with great caution [6,16]. The impression is that the representatives of the authority do not want to notice the non-observable economy, meanwhile the society suffers considerable losses because of it. For instance, the scaling down of the customs duties for the imported agricultural products is widespread in Ukraine. For instance, in September of 2012 Zhytomyr customs procured 95 tons of the Swedish margarine (price 15.7 hrs/kg), whereas Kyiv customs procured the margarine of the same company in the volume of 120 tons (price 21.4 hrs/kg). There are many such examples. The result of this doubtful customs cost – loss of the state financial resources.

Unfortunately, the plundering of the national resources, groundless usage of privileges, shady transactions with the taxes is widespread phenomenon in national agricultural sphere. In order to exclude the dysfunction models of the economic behavior, attention should be paid to the relations between the agents of the market. The necessity to improve the organization – managerial mechanism and control over the operations with VAT reimbursement to the business entities becomes very important. But the control measures must not depend on the sum of the reimbursement claim. Nowadays the situation when one business entity (or the agents) creates the system of the enterprises in order to realize the policy of the tax minimization and obtaining the financial resources from the state budget. For instance, (Fig. 2) in 2013 the top management (founder) of the enterprises PE "Repard XXI" PE "Company Rovoks", PIE "New Wave" declared VAT reimbursement for the sum of almost 340 thousand UAH. Although nominally everything was in accordance with the law however it is worth mentioning the fact that the enterprises with the statutory capitals of 100–300 UAH, carry out rather successful export operations, avoiding minor expenses (director and accountant general at the enterprises is a joint position; in August 2013 37 tons of the horse-radish was bought from the self-employed person, who, by the way, was a single tax payer by the interest rate of 5 %). Thus, one of the drawbacks of the acting legislation is that it does not take into account the economic essence of the transactions (agreements) but only provides the carrying out of the verification of certain documents (tax invoices, bank statements, etc.) [4].

The problem of the correct analytical interpretation of the essence of the operations is of primary importance. In our opinion the authorities do much harm in the struggle with the non-observable economy. This becomes apparent in the random policy of the state regulation of the economic activity. Permanent influxes and refluxes in the actions of the authorities block the economic development. For instance, in 2013 according to the Decree of the President of Ukraine "On the Ministry of Revenue and Duties" of 18.03.2013 191 /2013 [3]. Ministry of Revenue and Duties of Ukraine was founded. It should be noted that this organ was one of the basic in the system of the struggle with the non-observable economy. However, Ministry of Revenue and Duties of Ukraine functioned about a year because as a result of the aggravation of the political situation in Ukraine it was liquidated in accordance with the Decree of the Cabinet of Ministers of Ukraine "On the liquidation of the Ministry of Revenue and Duties" of 01.03.2014 67 [2].

Thus, as a result of political changes "Strategic plan of the development of the Ministry of Revenue and Duties for the period of 2013–2018", approved by the Order of the Cabinet of Ministers of Ukraine of 23.10.2013 869-p [1], which was developed with the participation of the international advisers, for instance, company "Price water house Coopers" and approved by the World Customs Organization (WCO) became out-of-date. It is worth mentioning that one of the aims of this plan was the enhancement of the efficiency of the methods of the counteraction to the minimization of the tax payment [1], which, in its turn,

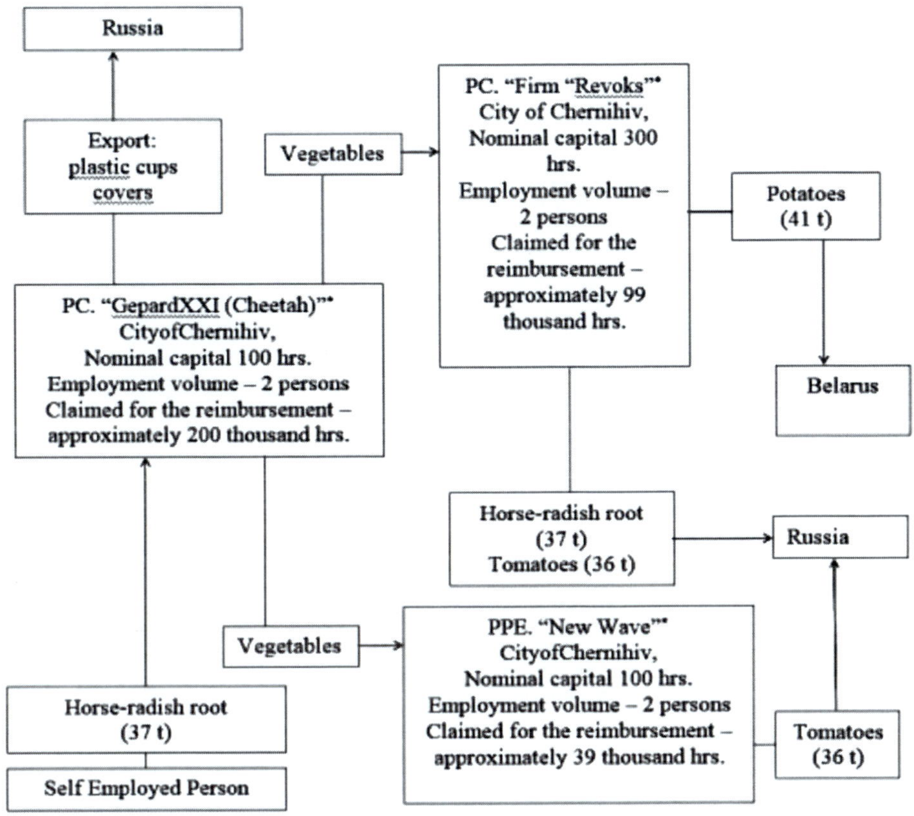

Note: * - connected persons.

Fig. 2. Model of the dysfunctional economic behavior in the agricultural sphere (on the example of VAT reimbursement) (Source: author's research)

provided the following measures: control over the cash payment, introduction of the electronic methods of the verification, remote control over the production of the tobacco products and alcoholic drinks, etc. We are convinced that such ill-considered decisions are not only erroneous but destructive as these decisions put the state on the road of neoapologetics. It is also obvious that the reformation of the top staff of the organs of power does not lay any profound foundation for the qualitative institutional changes in the society and achievement of the efficient results.

The approaches to the study of the real relations, interests of the main agents of the market should be elaborated. As it was already mentioned the authorities actively realize measures, aimed at the decentralization [11]. At the same time we must get rid of the abstract, weakly determined in the economic plan assessments that the local self-government is able to solve economic problems. It is obvious that economic problems are not possible to solve autonomous without coordination with the political, moral and social imperatives [13]. We cannot help seeing the potential danger, hidden in the random process of the self-government development. The essence of the danger may be in the possibility of the opposition of the idea of the self-government to the interests of the state. The decentralization of the power creates obstacles for the concentration of the resources (financial, material, labor, etc.) for the solution of the specific problems, in particular, rapid economic growth. Moreover, decentralization of power without well-considered system of control may fuel corruption and contribute to the separatism, etc. We are convinced that the obligatory condition for the decentralization of power must be the control (so -called "esting"), proceeding from the criteria and principles of the economic safety. There is no doubt that the development of the self-government creates the advantages over the centralized mechanisms of government in the context of the adequate usage of local information and decreases resources, needed for the decision-making. But in practice the self-government is complex system that does not guarantee either efficient managerial decisions or sustainable social-economic development of the region. Motion in the direction of the decentralization development provides the enhancement of management quality in the communities. If the quality and control in the system of management does not improve, then the self-government will inevitably cause the aggravation of social-economic situation.

We should not console ourselves with the illusions that the dismantling of the bureaucratic pyramid enables to provide economic progress of the country. The success of the sustainable social-economic development, in our opinion, greatly depends not on the development of the system of power decentralization, but on the adequate and rapid reaction on those challenges which create hardly foreseeable opportunistic actions of the agents of the market. Thus, decentralization of the power in Ukraine without distinct and subordinated in advance goals dooms to failure. Collective decision making which can be considered as the foundation of the local self-government does not make corruption component impossible. That is why, it is not necessary to overestimate the role of the self-government. It must not be as an end in itself. We must fight the injustice and corruption in the administrative staff but not the state directives. In other case, decentralization of power will inevitably lead to the creation of new opportunistic, by its content, system of stimuli counter balance.

One of the problems of Ukrainian economy is that the determining role is played by the politicized power-business relations. Thus, the success of business is formed under the impact of political institutions and depends, as a rule, not on knowledge, skills, construction of the system of the efficient management at the enterprises but on the political orientations and tastes, prevailing at a certain period of time. The realities of the present time show that the self-government sometimes may distort and enhance the self-will in the sphere of management of the financial resources. Self-government can develop in the context of the imitation democracy where the problem of the electivity is formal. Decentralization is able to mask real personal interests of the separate representatives of the local self-government. We are sure that only the activity of the business combined with the struggle against non-observable economy and corruption are the key symptoms of the social-economic development in Ukraine.

Low level of the salaries of the public servants in combination with the power is nutrition medium for the corruption. Efficient struggle against the corruption is impossible without the solution of the problem of stimulation. The motive of the public servant and/or representative of the organs of local self-government must play main role in the counteraction both to the corruption and non-observable economic processes. In other case the corruption will erode economic-political organism of the state. What is more the elements of the "social decay" as the practice shows often penetrate the higher levels of power. Numerous facts of the attempts of the representatives of the authorities to use the power for the personal enrichment are shown the mass media [19].

5 Experiment and Results

Taking into account the data of the Fig. 1 and using the multivariate correlation-regression analysis technique, two multiple regression models were developed [12,24]. They determine the dependence of the percentage of tillage usage of which is outside of the proper inventory on, first, the percentage of tillage used by the private households, and, second, the percentage of tillage used by agricultural enterprises (Table 1).

A graph of remains (Fig. 3), a graph of normal distribution (Fig. 5) and a graph of selection (Fig. 4)were drawn on the basis of the Fig. 1 data.

The Fig. 3 shows that the models look like the homoscedastic one. Indeed, it is difficult to find clear signs of variability in variance. But there are minor dependencies.

The graphs in Fig. 5 show that the points are maximally concentrated within the range of 5% to 7%. This focus can be used for further forecasts.

The selection graph is a field of correlation of the actual and theoretical (calculated) values of the resultant index [14]. The analysis of selection graphs enables to make speculations that the quality of the developed models can be improved by taking into account only the dynamics of the percentage of tillage used by agricultural enterprises, as the analyzed resultant index of the percentage of tillage usage of which is outside of the proper inventory is much more sensitive to changes of this factor compared to changes of the percentage of tillage used by the private households. As for the private sector, it is necessary to strengthen controls and improve the accounting of tillage used by these households.

On the base of the empirical analysis the authors' transparent approach regarding the problem of the counteraction to the non-observable economic processes in the agricultural sphere is formulated. The given approach is based on the monitoring of the opportunistic actions of the hazardous agents of the market (Fig. 6).

Non observable economic processes are often stipulated by the same factors as the corruption. According to the results of our research the following groups

Table 1. Multiple regression models' data

Index	Model that determines the dependence of the percentage of tillage usage of which is outside of the proper inventory on the percentage of tillage used by the private households	Model that determines the dependence of the percentage of tillage usage of which is outside of the proper inventory on the percentage of tillage used by agricultural enterprises
Model data		
Multiple R	0,219461	0,555738
R-square	0,048163	0,308845
Standardized R-square	0,01009	0,281199
Mean-square error	4,361805	3,716831
Observation count	27	27
Index Y-intersection, a	11,42621	35,56083
Index b_1	−0,19188	−0,41296
Multiple regression model	$\mathbf{y = 11,426 - 0,192 - x_1}$	$\mathbf{y = 35,56 - 0,413 - x_2}$

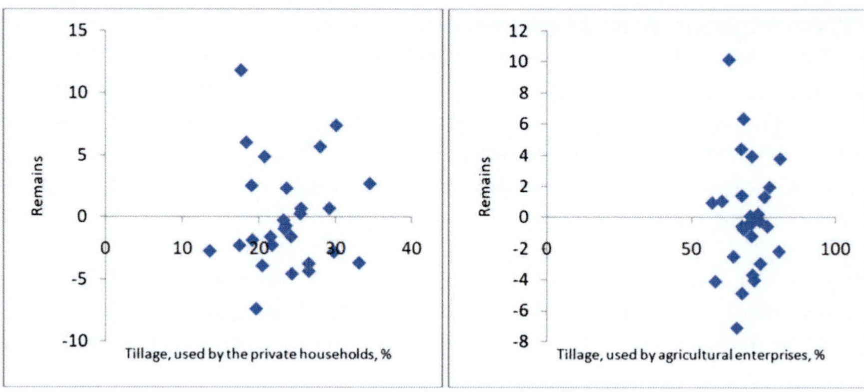

Fig. 3. Graphs of remains for the multiple regression models

of factors can be distinguished: Financial-economic; Cultural; Legal; Mental-psychological.

The accumulated data base regarding the factors of the impact on the non-observable economic processes and the results of the work with the group of the expert (method of expert assessment of the factors weight was applied) enabled to create the dependence model of the studied index (level of the non-observable economic processes in Ukraine) on the selected factors:

$$Y = 0.38 \cdot Y_1 + 0.21 \cdot Y_2 + 0.32 \cdot Y_3 + 0.09 \cdot Y_4, \quad (1)$$

where Y – integrated index of the level of the non-observable economic processes in Ukraine; Y_1 – financial-economic limitations of business (high tax rates for doing business, inappropriate material support of the representatives of power, information asymmetry, considerable transaction costs, etc.); Y_2 – cultural limitations (ethnic, religious peculiarities, lack of ideological fundamentals regarding the formation of the moral and economic relations, etc.); Y_3 – legal limitations

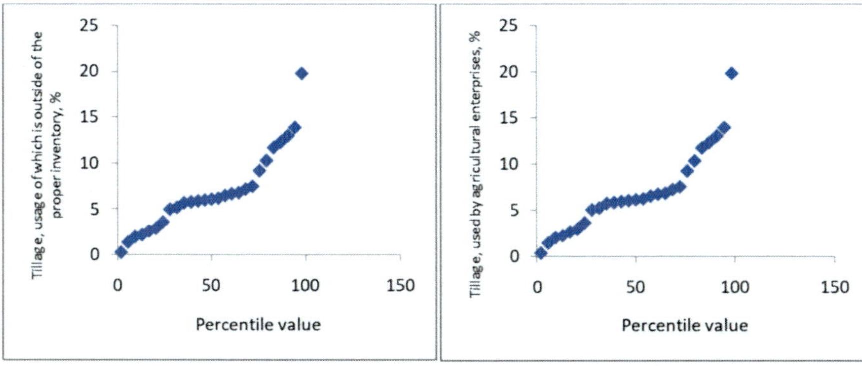

a) for the model that determines the dependence of the percentage of tillage usage of which is outside of the proper inventory on the percentage of tillage used by the private households

б) for the model that determines the dependence of the percentage of tillage usage of which is outside of the proper inventory on the percentage of tillage used by agricultural enterprises

Fig. 4. Graphs of selection for the multiple regression models

Fig. 5. Graphs of normal distribution for the multiple regression models

(permanent random modernization of the organs of power, excessive regulation of the economy, improper transparence regarding the functioning of the organs of power, unjustified liberalization regarding the responsibility in the economic sphere, etc.); Y_4 – mental-psychologic limitations (inertia of the economic behavior, cognitive rigidity of the economic relations, etc.).

6 Discussion

Analysis of the non-observable economic processes in the national agricultural sphere showed that modern system of the state regulation provides the control over the risk factors insufficiently. In our opinion, the main risks are the following:

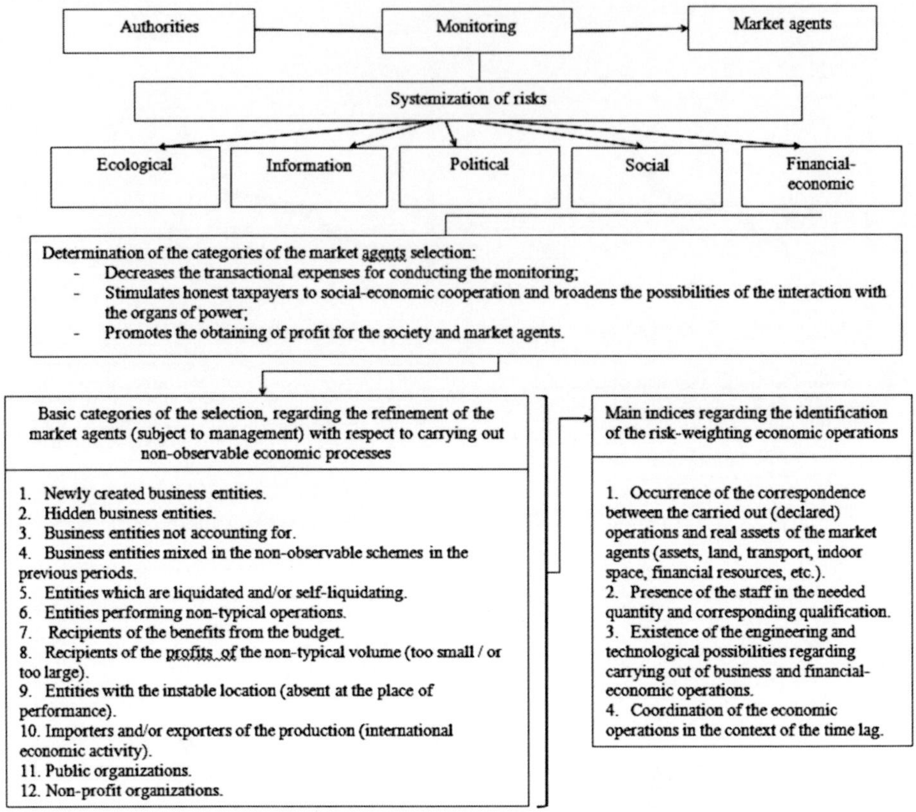

Fig. 6. Scheme of the universal monitoring, regarding the identification of the risky market agents in the national economy (Source: author's research)

1. ecological (degradation of the natural ecologic environment);
2. informational (creation of the information asymmetry cyber crimes in the sphere of information technologies, etc.);
3. political (manifestations of the separatism, threats to the institutions of the private property, loss of the sovereignty, etc.);
4. social (destruction of the infrastructure, formation of the social differentiation, etc.);
5. financial-economic (tax losses, bankruptcy, law rates of development, etc.).

It is obvious that the formation of the stable social-economic system is impossible without efficient risk management. In the conditions of the institutional modernization the improvement of the agricultural sphere management becomes very important. Efficient tool for the solution of the complex problems in the sphere of the counteraction to the non-observable economic processes is the program-target management method, based on the determination of the final aim vector in the conditions of the determined resources and executives. In our view, the efficient counteraction to the non-observable economic processes is possible only on the conditions of the simultaneous introduction of the anticorruption measures (Fig. 7).

There is no doubt that not always the emergence of the non-observable economic processes is caused by the corruption but at the same time it is obvious that stable and large scale development of the dysfunctional models of the economic agents behavior is possible only on the conditions of the existence of the corrupted environment.

In its turn the corruption is always accompanied by the non-observable processes. It is quite clear that problems of moral and ethics play an important role in the emergence of the non-observable processes and corruption. That is why, there exists the critical need regarding the formation in the society the consciousness, characterized by the intolerance to the non-observable economic processes.

In this connection the problems of the coordination of the actions of the power organs and the mass media are of great importance. Due to such permanent communication between the authorities and society measure aimed at the counteraction to the non-observable economic processes can be enriched with high moral content. This, in its turn, will minimize the losses of the resources, spent for the elimination of the consequences, emerging as a result of the economic dysfunctions and will render assistance in the formation of the "social maturity" of the market agents.

It should be noted that for the prevention of the corruption and non-observable economic processes the establishment of the coordination with the international organizations, including law enforcing agencies plays an important role. For instance, only in 2018 our state established contact [16] with the International Anticorruption Coordination Center (London, Great Britain). That promoted rapid obtaining of the information concerning the counteraction to the corruption and economic crimes. Meantime, the Ukrainian authorities must intensify the cooperation with the international organization in this direction, since up till now our law enforcing organs for years wait for the necessary information from foreign countries (for instance, the USA, Great Britain, etc.).

It is worth mentioning the fact, that as a rule, the corruption and the non-observable economic processes emerge in the countries where there exist administrative barriers, for instance, the concession of the permissions or the formation of the prohibitions. Thus, lack of rigid punishment for the non-observable economic transactions, combined with the material incentive of separate representatives of the power promotes the development of the economic dysfunctions in the society.

Practical analysis of the activity of the national market agents gives grounds to state that the existing system of the motivation of the representatives of the state power is a powerful factor for the non-observable economy development. In our opinion, for the construction of the stimulation system theoretically two possible models may be useful.

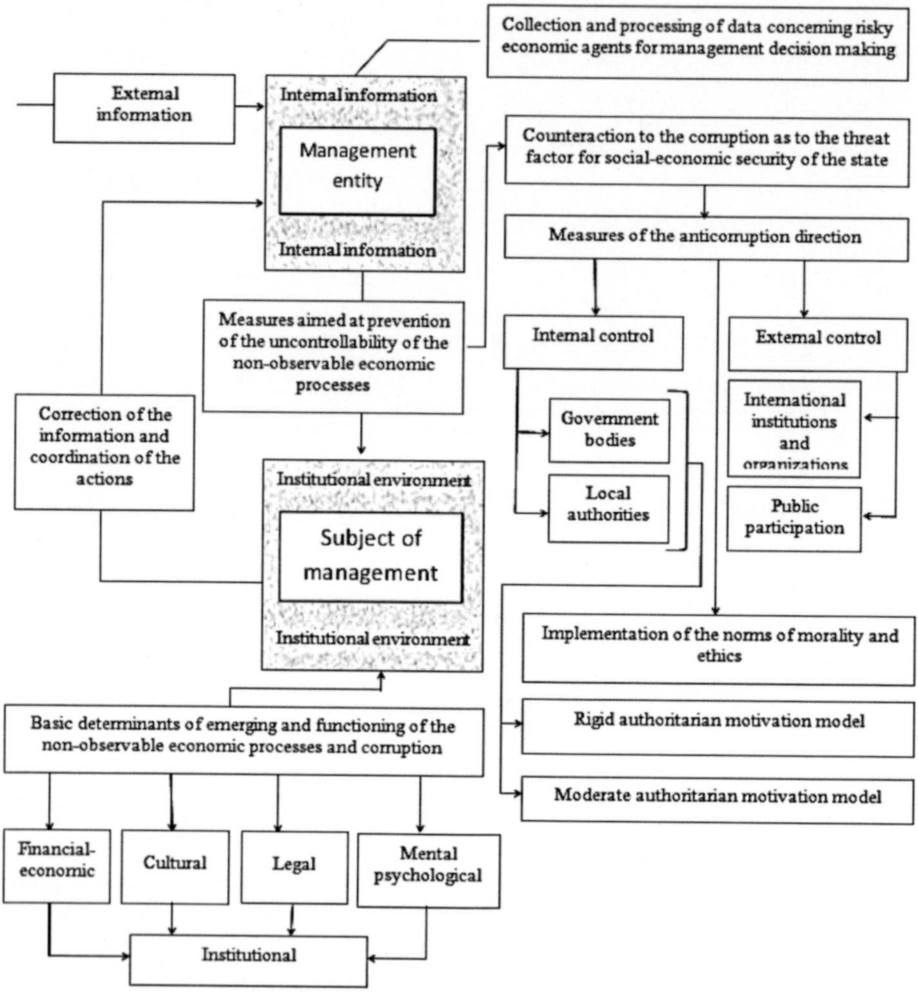

Fig. 7. Organization scheme of the impact on the market agents taking into consideration the institutional component

1. Rigid authoritarian model of motivation. Punishment for the economic and corruption crimes must be extremely severe – long terms of imprisonment with the confiscation of the property (for the extremely dangerous crimes, which caused the considerable losses of the financial resources).
2. Moderate authoritarian motivation model. It is necessary to create for the representatives of the power and members of their families proper social-economic conditions (amenities) with the simultaneous increase of the responsibility, including criminal, for the inobservance of their duties.

As the experience shows it is impossible to reach positive changes in the sphere of counteraction to the non-observable economic processes, only on the base of the convictions regarding the necessity to follow the existing norms without corresponding sanctions in the form of punishments.

It is important that applying punishments for the economic crimes, it is necessary to identify the extent and degree of the guilt of the market agent, the reasons of its economic behavior, etc.

7 Conclusions

Nowadays the imitation of the democratic model of motivation acts in Ukraine. The given model, in conditions of the lack of the morally stable law enforcement and legal system showed its helplessness.

The lack of the efficient stimuli distorted the democratic model of motivation, demonstrated its weakness and aggravated the problems of the non-observable economy and corruption.

If we speak about the categories of the selection, regarding the verification of the market agents on the subject of carrying out the non-observable economic processes, then special attention must be paid to the most profitable spheres of national economy of Ukraine, (for instance, in 2017–2019 the most profitable spheres were agriculture, public health and trade).

It is quite obvious, that the suggested approach, regarding the counteraction to the non-observable economic processes in the context of the anticorruption measures, requires the detailing in the context of the operation and functional principles of the specific organ of power, responsible for this problem. However, tactical correction of the scheme must not contradict to the strategic tasks.

Thus, taking into account present-day situation and new tasks, stipulated by the functioning of the non-observable economy, it is necessary to realize immediately the complex of measures, aimed at the increase of the responsibility of the market agents. Modeling and forecast of the multivariate scenario of the development of the non-observable processes in the country, branches of national economy, on the organization level and on the level of the identification of the recipient groups of the agents of the society.

According to the developed model of the dependence of the percentage of tillage usage of which is outside of the proper inventory on the percentage of tillage used by the private households and the model of the dependence of the resultant index on the percentage of tillage used by agricultural enterprises, it was concluded that in order to optimize the process of land distribution it is necessary to strengthen the control and accounting of tillage in private households.

References

1. Decentralization of power will be fixed in the Constitution. http://www.kmu.gov.ua/control/uk/publish/article?art_id=247209100
2. Order of the Cabinet of Ministers of Ukraine of 23.10.2013, 869-p (amended and augmented). https://zakon.rada.gov.ua/laws/show/869-2013-%D1%80

3. Resolution of the Cabinet of Ministers of Ukraine "On the Liquidation of the Ministry of Revenue and Duties" of 01.03.2014, p. 67 (amended and augmented). http://zakon1.rada.gov.ua/laws/show/67-2014-%D0%BF
4. Tax Code of Ukraine of 02.12.2010 2755-VI (amended and augmented). http://zakon2.rada.gov.ua/laws/show/2755-17/page
5. On the dawn of the peasants freedom (materials for the characteristic of the society). Russ. Antiq. (3), 465–490 (1998)
6. Measuring the Non-Observed Economy: A Handbook. OCED, Paris (2002)
7. Ausan, A. (ed.): New Institutional Economic Theory: Manual. Institutional Economy ed. Infra-M, Moscow (2011)
8. Avdiyskyi, V., Dadalko, V.: Shadow Economy and Economic Security of the State. Alpha-M: Infra-M, Moscow (2010)
9. Bekzhanova, T.: Determination of the boundaries of the shadow economy in the system of national accounts of Kazakhstan. Bull. RUDN Univ. Econ. Libr. **1**, 29–35 (2010)
10. Brendeleva, E.: Neoinstitutional Economic Theory: Study Guide. Deal and Service, Moscow (2006)
11. Brennan, G., Buchanan, J.: The Reason of Rules - Constitutional Political Economy. Cambridge University Press, Cambridge (2008)
12. Darlington, R.B., Andrew, F.H.: Regression Analysis and Linear Models: Concepts, Applications, and Implementation. Guilford Publications, New York (2016)
13. Dovbush, O.: Inconclusive choice. Vesti. Report. **13**(31), 28–31 (2014)
14. Enin, E.: Ukraine will be able to obtain the information from the USA, Great Britain and Interpol in the sphere of the counteraction to the corruption. http://ua.interfax.com.ua/news/general/484092.html
15. Kang, L.B., Blanchard, B.: Exclusive: China seizes $14.5 billion assets from family, associates of ex-security chief: sources. Thomson Reuters is the world's largest international multimedia news agency. http://www.reuters.com/article/2014/03/30/us-china-corruption-zhou-idUSBRE-A2T02S20140330
16. Khorozovn, O., Krak, I., Kulias, A., Wójcik, W., Tergeusizova, A.: Monitoring vital signs using fuzzy logic rules. In: Information Technology in Medical Diagnostics, pp. 237–244 (2018). https://doi.org/10.1201/9780429057618
17. Latov, Y., Kovalev, S.: Shadow Economy: Study Guide for Higher Education Establishments. Norm, Moscow (336)
18. Mandybura, V.: "Shadow" Economy of Ukraine and the Directions of Legislative Strategy of its Elimitation. Parliament Edition, Kyiv (1998)
19. Meyers, L.S., Gamst, G., Guarino, A.J.: Applied Multivariate Research: Design and Interpretation. Sage Publications, Thousand Oaks (2016)
20. Nebava, M., Savina, N., Zayukov, I., Vasazhenko, N., Lesko, O., Smolarz, A., Ormanbekova, A.: Model of electronic public health management on the example of the territorial community of Vinnytsia region. In: Photonics Applications in Astronomy, Communications, Industry, and High-Energy Physics Experiments, vol, 1117634, Proceedings of SPIE 11176 (2019). https://doi.org/10.1117/12.2537421
21. Oleynik, A.: Institutional Economy: Study Guide. INFRA-M, Moscow (2009)
22. Savina, N., Romanko, O., Gromaszek, K., Smailova, S.: Information technology for evaluation of innovation indicators influence and investment activity on competitiveness of the region. In: Photonics Applications in Astronomy, Communications, Industry, and High-Energy Physics Experiment, vol. 1117625, Proceedings of SPIE 11176 (2019). https://doi.org/10.1117/12.2537073

23. Schneider, F., Enste, D.: The Shadow Economy: An International Survey. Cambridge University Press, Cambridge (2003)
24. Shaun, M.: The purposes of multivariate data analysis methods: an applied commentary. J. Afr. Bus. **19**(1), 124–142 (2018). https://doi.org/10.1080/15228916.2017.1374816
25. Shkarpova, E.: Epoch of changes. Focus **16**, 8–37 (2014)
26. Wójcik, W., Pavlov, S., Kalimoldayev, M.: Information Technology in Medical Diagnostics. Taylor & Francis Group, CRC Press, London (2019). https://doi.org/10.1201/9780429057618. Balkema book

Computational Intelligence and Inductive Modeling

Representational Capabilities and Learning of Bithreshold Neural Networks

Vladyslav Kotsovsky[1(✉)] and Anatoliy Batyuk[2]

[1] Uzhhorod National University, Uzhhorod, Ukraine
kotsavlad@gmail.com
[2] Lviv Polytechnic National University, Lviv, Ukraine
abatyuk@gmail.com

Abstract. The paper deals with questions related to the ability of real-weighted bithreshold neurons and neural networks to solve the classification tasks. We study how many partitions of a finite set in n-dimensional vector space can be computed by using bithreshold neurons. The upper and lower bounds on the number of bithreshold dichotomies are proved. Our main theoretical result states that there exists $\Theta\left(m^{n+1}\right)$ bithreshold dichotomies of the set of m points in general position in n-dimensional space. We also estimate the Vapnik-Chervonenkis dimension of the class of bithreshold neurons. The problem of the synthesis of 2-layer bithreshold neural network for binary classification is considered. We demonstrate that an arbitrary dichotomy of m points in general position in n-dimensional space can be realized by 2-layer neural network containing at most $\lceil m/(2n)\rceil$ bithreshold neurons in hidden layer and a single output threshold neuron. The learning algorithm with computational complexity $O\left(mn^2 + m^2\right)$ is proposed for such networks. We give simulation results of the network performance on real data sets and discuss the influence of algorithm parameters on the quality of learning and the representational capabilities of bithreshold networks.

Keywords: Bithreshold neuron · Neural network · Bithreshold network · Classification · Learning algorithm

1 Introduction

Multithreshold neural units were introduced in the 1960s [10]. The main inspiration for dealing with multithreshold devices is their better recognizing ability in comparison with ordinary single threshold units (for example, the single bithreshold unit can solve the famous XOR problem). But unlike the threshold units and networks their multithreshold counterparts are not widely used. So far, we know no examples of modern effective system of computational intelligence developed on the base of multithreshold devices similar to the ones proposed in [2,8]. The hybrid AI systems like that in [18,26] developed on the base of multithreshold paradigm are still missing [23]. So, we had reason to claim in [16]

S. Babichev et al. (Eds.): ISDMCI 2020, AISC 1246, pp. 499–514, 2021.
https://doi.org/10.1007/978-3-030-54215-3_32

that multithreshold units remain a "blank spot" in the theory of artificial neural networks. The main cause of this statement was the lack of appropriate learning techniques for such units and network.

But recently the situation has slightly changed. Izonin et al. announced the use of bithreshold devices in recovery of incomplete data [12] and extended-input neural networks [11]. Lupei et al. stated that the application of bithreshold neurons can improve the quality of authorship identification network [17]. The authors of the present paper proposed bithreshold-like elements for solving the problems in the complex domain [14] and neural networks with smoothed bithreshold activation functions, thereby enabling for them gradient-based learning techniques [16]. This networks can be applied in the design of different managing systems such as the ones developed in [20,22].

But despite the last advances we do not know the exhaustive theoretical research devoted to the ability of multithreshold units and networks. The problem of estimating the capacity of multithreshold elements has been considered in [21] by Takiyama. Olafsson and Abu-Mostafa corrected Takiyama's results and gave the valid upper bound on the number of dichotomies of m-set separable by multithreshold elements in n-dimensional space in [19]. It follows from [19] that this number is $O\left(m^{2n-1}\right)$. We improve considerably this bound in Proposition 1. The unsolved problem was the estimation of the Vapnik-Chervonenkis dimension of the class of bithreshold units with n inputs. We do this in the corollary to Proposition 2.

The another important question concerning the network consisting of multithreshold neurons. Nilsson proved that 2-layer threshold network containing $m-1$ threshold units in hidden layer was capable of computing an arbitrary dichotomy on a set of m points in general position. Then Baum significantly improved this result and showed that $\lceil m/n \rceil$ hidden neurons are sufficient in n-dimensional space [3]. Deolalikar studied bithreshold neurons and 2-layer neural networks, whose hidden layer consists of threshold units, and the single output neuron is bithreshold [6]. But his result concerning such networks was founded on the wrong assumption as was shown in [16], where several sufficient conditions of bithreshold separability were given. We show (see Proposition 3) that the Baum's bound on the number of hidden neurons can be halved in the case of bithreshold networks.

We mentioned in [16] that the effective learning algorithms are unknown for bithreshold neuron and used the fundamental result of Blum and Rivest [4] to explain this. Namely, we proved that the problem of verifying the bithreshold separability is NP-complete even in the case of a single bithreshold unit. But these discouraging result is true only for classical interpretation of learning with the fixed network architecture. In Subsect. 2.2 we describe the synthesis procedure for constructing 2-layer neural network which is capable to compute the given dichotomy of the finite set of points in general position.

The aim of the research is the study of the expressive power of bithreshold neurons and the development of learning techniques for 2-layer neural networks consisting of bithreshold neurons. The paper has the following structure.

First, we give necessary definitions and models. Then, in Subsect. 2.1, we state our theoretical results concerning the number of separation of m-set in n-dimensional space realizable by bithreshold neurons. Next, in Subsect. 2.2, the learning algorithm for 2-layer bithreshold network is proposed. Finally, we give the results of computer simulation on several data sets and discuss them.

2 Materials and Methods

2.1 Model of a Bithreshold Neuron

Let $\mathbf{w} = (w_1, ..., w_n)$ be the vector in a n-dimensional Euclidean space $\mathbb{R}^n$. For an arbitrary vector $\mathbf{x} \in (x_1, \ldots, x_n) \in \mathbb{R}^n$ the value of the inner product $\mathbf{w} \cdot \mathbf{x} = w_1 x_1 + ... + w_n x_n$ is called the *weighted sum* corresponding to $\mathbf{x}$.

A computation unit with n inputs $x_1, \ldots, x_n$ and one output y is said to be a *bithreshold neuron* (BN) with the *weight vector* $\mathbf{w} = (w_1, ..., w_n) \in \mathbb{R}^n$ and *thresholds* t_1, $t_2 \in \mathbb{R}$ $(t_1 < t_2)$ if $y = f_{t_1,t_2}(\mathbf{w} \cdot \mathbf{x})$, where $f_{t_1,t_2}(x)$ is the bithreshold *activation function* given by

$$f_{t_1,t_2}(x) = \begin{cases} -1, \text{ if } t_1 < x < t_2, \\ \ \ 1, \text{ otherwise.} \end{cases}$$

A bithreshold neuron can be conveniently represented graphically as shown in Fig. 1.

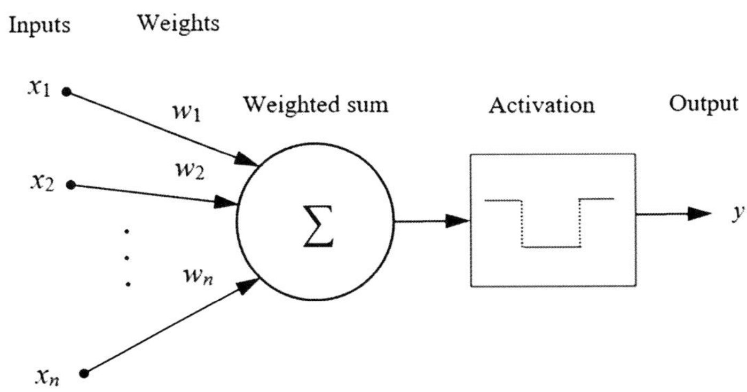

Fig. 1. A bithreshold neuron

The BN is completely defined by the ordered triplet $(\mathbf{w}, t_1, t_2)$. For the sake of brevity we do not distinguish between the BN and its corresponding triplet $(\mathbf{w}, t_1, t_2)$.

B-Separable Sets. We call two sets $A^+ \subset \mathbb{R}^n$ and $A^- \subset \mathbb{R}^n$ *strongly bithreshold separable* (b-separable) if there exists such BN $(\mathbf{w}, t_1, t_2)$ that for all $\mathbf{x} \in A^+ \cup A^-$ $t_1 < \mathbf{w} \cdot \mathbf{x} < t_2 \Leftrightarrow \mathbf{x} \in A^-$. In this case the BN $(\mathbf{w}, t_1, t_2)$ induces the *ordered b-dichotomy* (A^+, A^-) of the set $A = A^+ \cup A^-$ into two disjoint sets A^+ and A^- $(A^+ \cap A^- = \emptyset)$. We call the elements of the sets A^+ and A^- "positive" and "negative" patterns, respectively.

We call two sets $A^+ \subset \mathbb{R}^n$ and $A^- \subset \mathbb{R}^n$ *weakly bithreshold separable* if there exists such BN $(\mathbf{w}, t_1, t_2)$ that either (A^+, A^-) or (A^-, A^+) are ordered b-dichotomies.

The notion of the b-separability is the generalization of the notion of linear separability. The sets B_4 and R_4 in Fig. 2 are not linearly separable, but they are strongly b-separable. This example shows that BN is capable of solving the XOR problem.

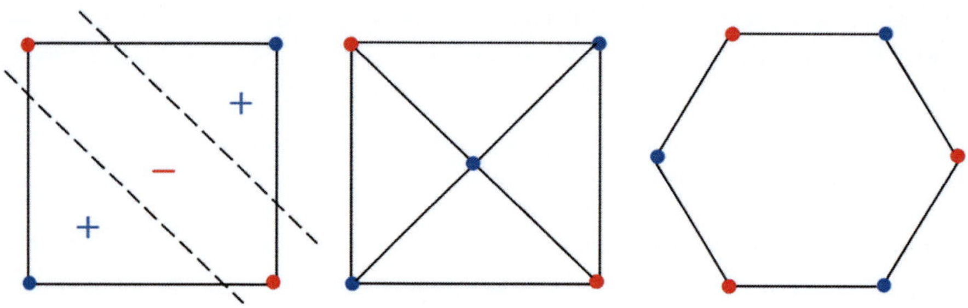

Fig. 2. Illustration of bithreshold separability on the example of sets R_i and B_i containing the red and blue points of respective shapes, $i = 4,\ 5,\ 6$.

It should be noted that the relation of weak separability is symmetrical, and the weakly b-separation of the set A induces the unordered b-dichotomy $\{A^+, A^-\}$. The binary relation of strong separability is not symmetrical. For example, in Fig. 2 the sets B_5 and R_5 are not strongly b-separable, whereas R_5 and B_5 are. So, B_5 and R_5 are only weakly b-separable. And, finally, the sets B_6 and R_6 are neither strongly nor weakly b-separable. We shall henceforth consider only ordered dichotomies separable in the strong sense.

2.2 Bounds on the Number of Bithreshold Dichotomies

Let $D_1(A)$ be the number of linearly separable dichotomies of the finite set A of patterns represented by a set of vector in n-space, and let $|A| = m$, where $|A|$ is the cardinality of set A. Let $D_1(m, n) = \max\{D_1(A) \mid A \subset \mathbb{R}^n,\ |A| = m\}$. The dichotomy-counting theorem state [5] that

$$D_1(m,n) = \begin{cases} 2\sum_{i=0}^{n} \binom{m-1}{i}, & \text{if } m > n+1, \\ 2^m, & \text{if } m \le n+1, \end{cases} \tag{1}$$

where $\binom{m-1}{i}$ is a binomial coefficient, $i = 0, \ldots, n$. Moreover, $D_1(A) = D_1(m, n)$ if and only if all points of the set A are in the general position in n-space, i.e., every subset of n or fewer vectors is linearly independent (or no $n+1$ of them lie in an $(n-1)$-dimensional hyperplane). Let $D_2(A)$ be the number of b-dichotomies (A^+, A^-) of the finite m-set A in the space $\mathbb{R}^n$ and $D_2(m,n) = \max\{D_2(A) \mid A \subset \mathbb{R}^n,\ |A| = m\}$. Let us show that (1) is suitable for establishing the upper bound on $D_2(m,n)$ considerably improving the upper bound from [19].

Proposition 1. *If $n \leq 2m-2$, then the following inequality is true*

$$D_2(m,n) \leq \sum_{i=0}^{n+1} \binom{2m-1}{i}. \tag{2}$$

If all points of the set A are in general position and $m > n+1$, then

$$D_2(A) \geq 2 \sum_{i=0}^{n+1} \binom{m-1}{i}. \tag{3}$$

Proof. Let A be an arbitrary m-set. Every BN $(\mathbf{w}, t_1, t_2)$ performs the b-trichotomy (A_1^+, A^-, A_2^+) of the set A, where $A_1^+ = \{\mathbf{x} \mid \mathbf{w} \cdot \mathbf{w} \leq t_1\}$, $A_2^+ = \{\mathbf{x} \mid \mathbf{w} \cdot \mathbf{w} \geq t_2\}$, $A^- = A \setminus (A_1^+ \cup A_2^+)$. For every point $\mathbf{x} = (x_1, \ldots, x_n) \in A$ we consider two $(n+2)$-dimensional points $\mathbf{x}' = (-x_1, \ldots, -x_n, 1, 0)$ and $\mathbf{x}'' = (x_1, \ldots, x_n, 0, -1)$. It is easy to verify that if BN $(\mathbf{w}, t_1, t_2)$ produces b-trichotomy (A_1^+, A^-, A_2^+) of the set A, then the hyperplane $w_1x_1 + \ldots + w_nx_n + t_1x_{n+1} + t_2x_{n+2} = 0$ performs homogenous linearly separable dichotomy $\left(\tilde{A}^+, \tilde{A}^-\right)$ of the set $\tilde{A} = \{\mathbf{x}' \mid \mathbf{x} \in A\} \cup \{\mathbf{x}'' \mid \mathbf{x} \in A\}$, where

$$\tilde{A}^+ = \{\mathbf{x}' \mid \mathbf{x} \in A_1^+\} \cup \{\mathbf{x}'' \mid \mathbf{x} \in A_2^+\}, \quad \tilde{A}^- = \tilde{A} \setminus \tilde{A}^+.$$

We now show that the mapping $(A_1^+, A^-, A_2^+) \to \left(\tilde{A}^+, \tilde{A}^-\right)$ is injective. Let (B_1^+, B^-, B_2^+) be an arbitrary trichotomy of the set A. If $\left(\tilde{B}^+, \tilde{B}^-\right) = \left(\tilde{A}^+, \tilde{A}^-\right)$, then from the equality $\tilde{A}^+ = \tilde{B}^+$ we can infer that $A_1^+ = B_1^+$ and $A_2^+ = B_2^+$. Thus, $(A_1^+, A^-, A_2^+) = (B_1^+, B^-, B_2^+)$. Therefore, the number of different b-trichotomies of the set A is not greater than the number of homogenous linearly separable dichotomies of the set $\tilde{A}$. If the b-trichotomy (A_1^+, A^-, A_2^+) is induced by BN $(\mathbf{w}, t_1, t_2)$, then the same BN produces corresponding b-dichotomy (A^+, A^-), where $A^+ = A_1^+ \cup A_2^+$. Simultaneously, the BN with the triplet $(-\mathbf{w}, -t_1, -t_2)$ produces the same b-dichotomy but another b-trichotomy (A_2^+, A^-, A_1^+). Hence, in the case $A^+ \neq \emptyset$ at least two different b-trichotomies correspond to given b-dichotomy. So, $D_3(A) \geq 2D_2(A) - 1$, where $D_3(A)$ is the number of all b-trichotomies of the set A. It is known [1], that we can obtain the upper bound on the number of all homogenous linearly separable dichotomies from (1) by decreasing the upper summation limit by 1. Therefore,

$$D_2(A) \leq \frac{1}{2}(D_3(A) + 1) \leq \frac{1}{2}(D_1(2m, n+1) + 1).$$

By applying (1) we obtain (2).

The lower bound (3) follows immediately from the Takiyama's result [21].

Let us estimate the order of growth in the right-hand side of (2) and (3). It is known [25] that $l > n+1$ implies

$$\sum_{i=0}^{n} \binom{l}{i} \leq 1.5 \, \frac{l^n}{n!}$$

Hence, by taking into account the inequality $2^n < (n+1)!$ we have

$$D_2(m,n) \leq \frac{3\,(2m-1)^{n+1}}{2(n+1)!} < 3m^{n+1} = O(m^{n+1}). \tag{4}$$

The estimate (4) is significantly better than the upper bound presented in [19], from where one can conclude that $D_2\,(m,n) = O\left(m^{2n+1}\right)$.

Consider the member in the right-hand side of (3). It is easy to see that if n is fixed, then the sum can be estimated as $\Theta\left(m^{n+1}\right)$. Hence $D_2(m,n) = \Theta\left(m^{n+1}\right)$.

It should be mentioned that $D_1\,(m,n) = \Theta\,(m^n)$ [1]. Therefore, (4) expresses the measure of advantage of bithreshold neurons over threshold ones.

The upper bound (4) can be used for establishing the following.

Corollary 1. *If $n > 5$, then for an arbitrary finite set $A \subset \mathbb{R}^n$ such that $|A| \geq 5n$ there exists a non b-separable dichotomy of the set A.*

Proof. By applying (4) and Stirling's approximation for $(n+1)!$ we find that

$$\begin{aligned} D_2\,(5n,n) &< \frac{3(10n)^{n+1}}{2\,(n+1)!} \\ &= \frac{15 \cdot 10^n n^{n+1}}{(n+1)!} \\ &= \frac{15 \cdot 10^n n^{n+1} e^{n+1}}{\sqrt{2\pi(n+1)}(n+1)^{n+1} e^{\gamma(n+1)}} \\ &< 2.27 \cdot (10e)^n, \end{aligned}$$

where $|\gamma(n)| < (12n)^{-1}$. If

$$n > \frac{\ln 2.27}{4\ln 2 - \ln 5 - 1} \approx 5.02$$

then $D\,(5n,n) < 2^{5n}$. We have obtained that the number of b-dichotomies of the set A is less than the number of all its subset. This implies the existence of non b-separable dichotomies.

Let $\mathcal{BN}_n$ denotes the set of all bithreshold neurons with n inputs. We say that the set A is *shattered* by $\mathcal{BN}_n$ if all its dichotomies are b-separable. The maximum cardinality of a set A shattered by $\mathcal{BN}_n$ is the Vapnik-Chervonenkis dimension, or VCdim, of $\mathcal{BN}_n$ [1,25]. This is one of the most important measure in the learning theory. We shall be interested in the estimating VCdim $\mathcal{BN}_n$.

Proposition 2. *For all $n \in \mathbb{N}$ there exists a $2n$-set A_n in the space $\mathbb{R}^n$ which is shattered by $\mathcal{BN}_n$.*

Proof. We prove this by induction on the dimension of the vector space. Moreover, we show that all possible dichotomies of sets A_n $(n \in \mathbb{N})$ can be computed by BNs satisfying the following condition

$$t_1 = -1, \quad t_2 = 2. \tag{5}$$

If we assume $A_1 = \{-1, 1\}$, then it is easy to verify that all four dichotomies of A_1 can be computed by BN satisfying (5). Suppose that the statement of proposition holds in the case of $(n-1)$-dimensional space and the set A_{n-1} is already constructed. Consider the case of n-dimensional space. Let us define the set $A_n \subset \mathbb{R}^n$ in the following way:

$$A_n = \{(x_1, \ldots, x_{n-1}, 0) \mid (x_1, \ldots, x_{n-1}) \in A_{n-1}\} \cup \{(0, \ldots, 0, -1), (0, \ldots, 0, 1)\}.$$

We shall show that all dichotomies of the set A_n are b-separable. Let (A_n^+, A_n^-) be an dichotomy of A_n and $\left(A_{n-1}^+, A_{n-1}^-\right)$ is the corresponding dichotomy of A_{n-1} such that

$$A_{n-1}^+ = \left\{(x_1, \ldots, x_{n-1}) \mid (x_1, \ldots, x_{n-1}, 0) \in A_n^+\right\}, \; A_{n-1}^- = A_{n-1} \backslash A_{n-1}^+.$$

By the inductive assumption, there exists BN $((w_1, \ldots, w_{n-1}), -1, 2)$ realizing this dichotomy. There are four possibilities, as follows:

1. $\{(0, \ldots, 0, -1), (0, \ldots, 0, 1)\} \subseteq A_n^+$,
2. $(0, \ldots, 0, -1) \in A_n^+, \quad (0, \ldots, 0, 1) \in A_n^-$,
3. $(0, \ldots, 0, -1) \in A_n^-, \quad (0, \ldots, 0, 1) \in A_n^+$,
4. $\{(0, \ldots, 0, -1), (0, \ldots, 0, 1)\} \subseteq A_n^-$.

Let us set $w_n = 3$ in the first case, $w_n = 1.5$ in the second case, $w_n = -1.5$ in the third case, and $w_n = 0$ in the last case. It is easy to verify that in each case the BN $((w_1, \ldots, w_{n-1}, w_n), -1, 2)$ computes b-dichotomy of the set A_n. The proof is complete.

Corollary 2. *If $\mathcal{BN}_n$ is the set all BN with n inputs and $n > 5$, then*

$$2n \leq \text{VCdim}(\mathcal{BN}_n, \mathbb{R}^n) < 5n.$$

For comparison, $\text{VCdim}(\mathcal{T}_n) = n + 1$, where $\mathcal{T}_n$ is the set of all threshold neurons with n inputs [1].

2.3 Learning of Bithreshold Neural Networks

Let us consider a 2-layer feedforward fully-connected *bithreshold neural network* (see Fig. 3). It has n nodes in input layer, several bithreshold neurons in hidden layer and a single output neuron. This network computes a mapping $\mathbb{R}^n \to \{-1, 1\}$ and we are interested in finding a computationally efficient algorithm

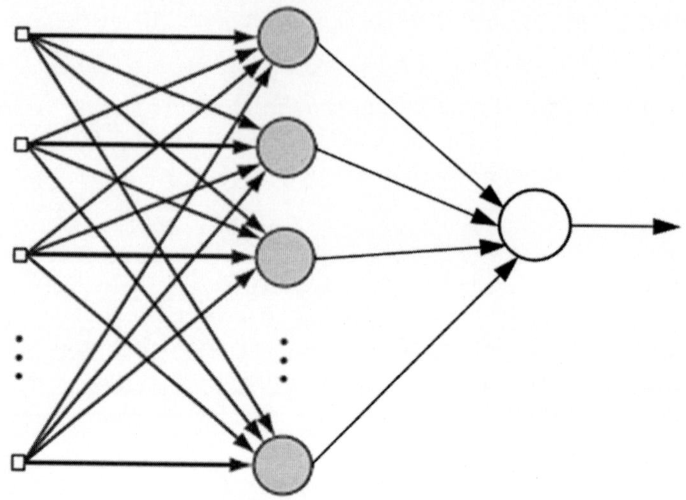

Fig. 3. Architectural graph of a 2-layer bithreshold network with a single output

for training network to compute an arbitrary dichotomy (A^+, A^-) of the finite set A. We adopt Baum's idea [3] in the algorithm design.

Suppose that all pattern of the finite set A are in the general position in n-dimensional space and $|A| = m$. Without loss of generality we can assume that $A^- \neq \emptyset$, $A^+ \neq \emptyset$. Consider the following algorithm:

Stage 1. Initialization

1.1. $k \leftarrow 0$.
1.2. if $|A^-| < |A^+|$ then
$$B \leftarrow A^-,\ C \leftarrow A^+,\ v \leftarrow 1$$
else
$$B \leftarrow A^+,\ C \leftarrow A^-,\ v \leftarrow -1$$

Stage 2. Construction of the hidden layer

2.1. $k \leftarrow k + 1$
2.2. $r \leftarrow \min\{n, |B|\}$
2.3. Extract r patterns from B to X
2.4. Find a solution $\mathbf{w}$ of the linear system

$$\mathbf{x}^i \cdot \mathbf{w} = 1,\ \ \mathbf{x}^i \in X,\ i = \overline{1, r} \tag{6}$$

2.5. $\varepsilon \leftarrow \alpha \min\{\,|\mathbf{w} \cdot \mathbf{x} - 1|\,|\ \mathbf{x} \in C\}$.
2.6. $B \leftarrow B \backslash \{\mathbf{x} \in B\,|\,|\mathbf{w} \cdot \mathbf{x} - 1| < \varepsilon\}$
2.7. Add BN with the triplet $(\mathbf{w}, 1 - \varepsilon, 1 + \varepsilon)$ in the hidden layer

2.8. If $B \neq \emptyset$, then go back to step 2.1

Stage 3. Determination of the output neuron

3.1. Add to the network the output threshold unit with k-dimensional weight vector $(v, \ldots, v)$ and threshold $(k - 0.5)\,v$.

In our algorithm k denotes the index of the hidden neuron, which will be added in the network, B is the current set of unprocessed patterns. Note that during the learning the condition $r < n$ is true at most once.

A two-dimensional example of the performance of the above algorithm is shown in Fig. 4. Blue points are positive patterns and red ones are negative. The hyperplanes corresponding to the solution of (6) are shown in dashed lines. Each BN recognizes at least two negative patterns lying in the hyperplane and, perhaps, some more points inside the "slice" whose width depends on ε.

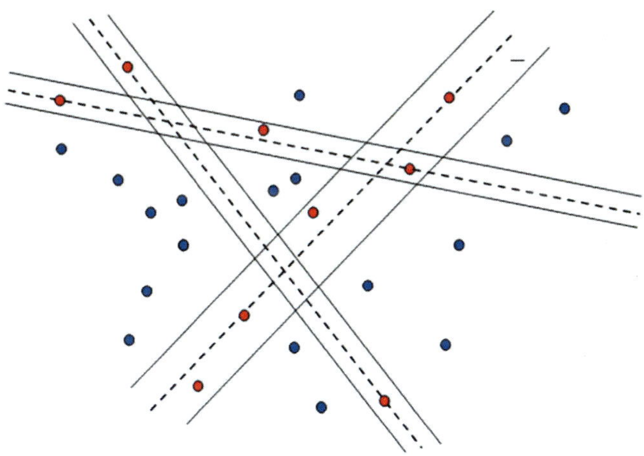

Fig. 4. An example of the network performance

The steps 2.4–2.5 are crucial. The system of linear Eq. (6) always has the solution $\mathbf{w}$, because the set A consists of the points in general position. The value of ε is always positive, because at most n points of the set A lie in the same hyperplane. Thus, each neuron makes valid slice of corresponding patterns belonging to B. Step 2.6 serves to discard point that are already classified correctly. The solving of the linear system in step 2.4 requires $O\left(n^3\right)$ operation. The computational complexity of the next two steps is $O\left(mn\right)$. Since $|B| \leq \min\left\{|A^+|, |A^-|\right\} \leq m/2$ and we extract n pattern in step 2.3, the stage 2 of our algorithm is repeated at most $\left\lceil \frac{m}{2n} \right\rceil$ times.

Let us justify the correctness of the choice of the output neuron weights. If $|A^-| < |A^+|$ and $\alpha < 1$, then the neurons of hidden layer catch only negative patterns, emit -1 for them and 1 for patterns of A^+. In this case the hidden layer neurons are connected with the output neuron with weight 1. Thus, the weighted sum for the output neuron is equal to k for positive input patterns and is at most $k - 1$ for negative ones. So, the threshold value $k - 0.5$ ensures the valid separation of patterns. If $|A^-| \geq |A^+|$, then the hidden layer catches only positive patterns and emits at least one -1 for them and all 1 for negative ones. In this case the weights of output neuron are equal to -1, and the corresponding weighted sum is equal to $-k$ for negative patterns and is at least $1-k$ for positive ones. Hence, the threshold $0.5 - k$ ensures the valid separation of patterns.

The parameter α can be considered as a tolerance measure of our bithreshold network. If $0 < \alpha < 1$, then the network performs well on training set (training error is near 0) but can be "overfitted" and less effective outside the training set. If α is slightly greater than 1, then the network can have the better generalization ability. Moreover, the using of $\alpha > 1$ can result in the reduction of the number of neurons in the hidden layer.

By summarizing observations given above we can state the following.

Proposition 3. *An arbitrary dichotomy of m-set of n-dimensional points in general position can be computed by a 2-layer bithreshold neural network with at most $\lceil \frac{m}{2n} \rceil$ neurons in the hidden layer, for which there exists $O(mn^2 + m^2)$-time learning algorithm.*

Note that our learning algorithm is not valid in the degenerate case. Numerous modifications of the stage 2 are possible. For example, we can add some noise to our data for the purpose of avoiding the degeneracy. The other approach consists of finding on the step 2.4 a number r and linearly independent patterns $\mathbf{x}^i$ $(i = 1, \ldots, r)$ in B such that aff $\langle \mathbf{x}^1, \ldots, \mathbf{x}^r \rangle \cap C = \varnothing$, where aff $\langle \mathbf{x}^1, \ldots, \mathbf{x}^r \rangle$ is the affine hull of the points $\mathbf{x}^i$, $i = 1, \ldots, r$. Notice that the output neuron weights are integer. This is important for hardware implementation (see [7] for further details).

3 Experiment, Results and Discussion

In computer simulation we trained 2-layer bithreshold neural network with single output threshold neuron to solve recognition problems on small- and medium-sized data sets. We were interested in the study of the mean value of k (the number of neurons in the hidden layer) and the influence of the parameter α on the network performance.

In the first simulation we generated training set of m uniformly distributed points in n-dimensional hypercube and applied our learning algorithm, where $m = 5,000$ and $n = 100$. The experiment was repeated 10,000 times. We observed that average, maximum and minimum values of k were 24.98, 25 and 24, respectively, for all $\alpha \in \{0.5, 0.7, 0.9\}$. This result shows that the bound $m/(2n)$ on the size of hidden layer stated in Proposition 3 is tight in the case when patterns are in general position and dichotomies are random. Moreover, the average number of patterns sifted in step 2.6 was near 10.3 (less than 0.42 per one iteration in the stage 2). Therefore, step 2.6 can be omitted without the loss of effectiveness in the case of random data.

In the second simulation we used the well-known iris data set that contains the sepal and petal length and width of 150 iris flowers of three different species: Iris setosa, Iris versicolor, and Iris virginica [9]. We trained a separate networks for recognize every species of flowers. The training set contained 120 randomly chosen four-dimensional patterns, 40 of which were put into the set B and used for determining weights vectors of hidden neurons. The testing set consisted of the rest of the data set. Notice that the training patterns were degenerate, so we needed to reduce the number of equations in step 2.4 of our learning algorithm. We repeated learning 1,000 times to obtain average characteristics of performance. Table 1 presents simulation results for $\alpha \in \{0.5, 0.7, 0.9\}$ (the network accuracy is expressed in percent). We see that Iris setosa was very well recognized by bithreshold networks. Moreover, it sufficed two neurons in hidden layer. In the case $\alpha = 0.9$ it was likely that one BN was sufficient, so the second layer was unnecessary. Furthermore, on average, almost 34 patterns were discarded in step 2.6 of the learning algorithm.

Table 1. Learning results for iris data set

α	Iris setosa			Iris versicolor			Iris virginica		
	k	Training accuracy	Test accuracy	k	Training accuracy	Test accuracy	k	Training accuracy	Test accuracy
0.5	1.82	99.99	99.41	7.20	99.68	83.03	7.72	99.77	79.06
0.7	1.57	99.99	99.48	6.49	99.82	85.64	7.15	99.81	81.73
0.9	1.42	99.99	99.93	5.97	99.77	86.94	6.55	99.80	86.31

The network performance in the cases of two other iris species was worse. Despite the high accuracy on the training set, the mean accuracy on the testing set did not exceed 87%. Table 1 shows that, on average, the use of $\alpha = 0.9$ led to smaller networks with better performance. The obtained values $k \approx 7$ of the number of hidden neurons are considerably less than upper bounds $m/(2n) = 120/(2 \cdot 4) = 15$ or $|B|/n = 40/4 = 10$.

The last simulation concerned the problem of optical recognition on the base of digits data set containing 1797 normalized 8×8 images that were obtained by preprocessing of original 32×32 images of handwritten digits available by NIST [9]. We made numerous experiments in which 80% of patterns were used as a training set and 20%—as a test set. Thus,

$$n = 64, \; m = \lfloor 1797 \cdot 0.8 \rfloor = 1437, \; \lceil m/(2n) \rceil = 12.$$

The results of two experiments are presented in Table 2. The behavior of the learning process is illustrated in Fig. 5 and Fig. 6, respectively.

The first experiment consisted in the identification of a single digits, for example zero. The data set contains 182 instances, so $\lceil |B|/n \rceil = \lceil 182 \cdot 0.8/64 \rceil = 3$. The training patterns were degenerate, so the hidden neurons recognized less

Table 2. Learning results for digits data set

	Single digit recognition			Recognition of even numbers		
α	k	Training error (%)	Test error (%)	k	Training error (%)	Test error (%)
0.5	4	0	3.57	13	0	13.89
0.7	4	0	3.77	13	0	14.28
0.9	4	0	3.51	12	0	13.87
1.1	4	0.37	3.37	12	0.87	15.87

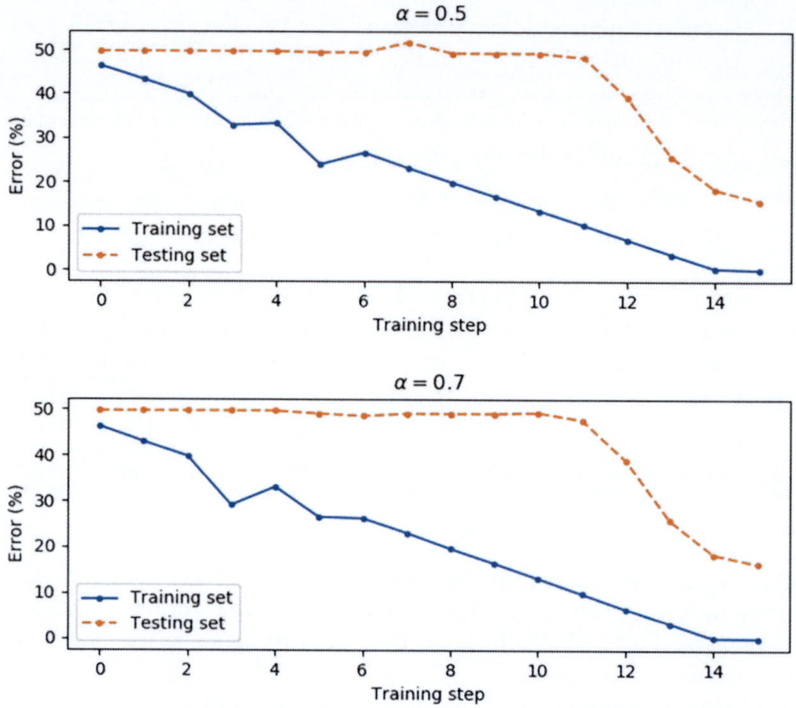

Fig. 5. Learning curves for the recognition of even digits in the case $\alpha \in \{0.5, 0.7\}$

than 64 patterns (on average, the attainable value of recognized patterns was near 55). For all values of α from the set $\{0.5, 0.7, 0.9, 1.1\}$ the hidden layer consisted of $k = 4$ neurons that is almost equal to the theoretical bound $k = 3$ in non-degenerate case, and the network did not make mistakes on the training set. The error on the test set was also acceptable. Note that the value 1.1 of the α did not provide better generalization than $\alpha \in (0, 1)$.

In the second experiment we trained the network to recognize even numbers. For $\alpha \in \{0.9, 1.1\}$ we reached the boundary value $k = 12$. The training reported no errors, except $\alpha = 1.1$. But the test errors were significant enough.

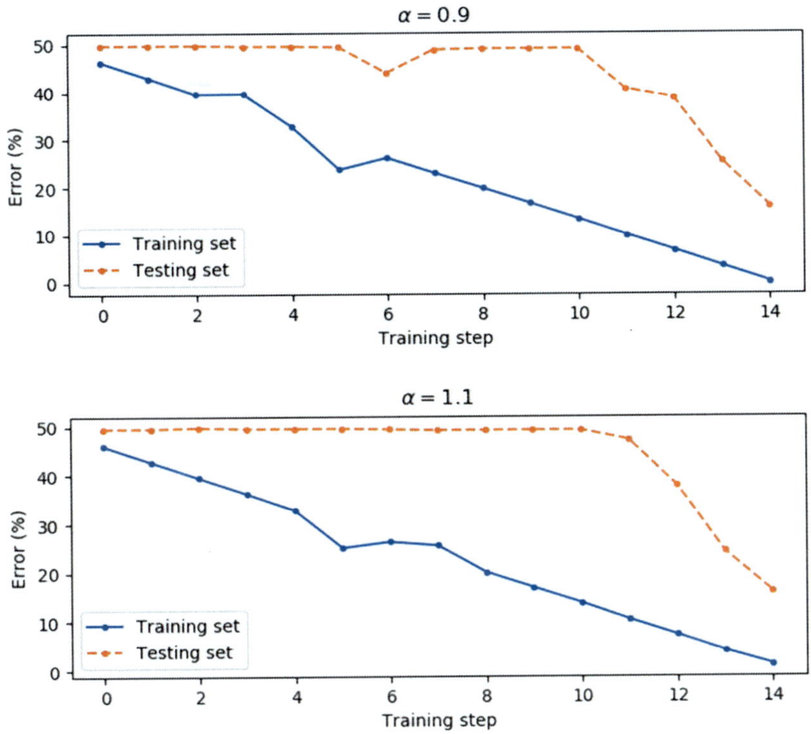

Fig. 6. Learning curves for the recognition of even digits in the case $\alpha \in \{0.9,\ 1.1\}$

In Fig. 5 and Fig. 6 we can see the learning curves depicting the behavior of the training error and the test error. The main part of the training error curve is likely linear with small fluctuations. The test error was very high and almost constant from the start to the 10th learning step and decreased subsequently.

By analyzing data in Table 1, Table 2, Fig. 5, and Fig. 6 we can conclude that $\alpha = 0.9$ seems to be preferable in solving real classification and recognition problems. To improve the generalization capabilities of classifier, we can use the additional hidden layer whose neurons are responsible for the recognition of the single corresponding digit.

Note that our algorithm appears to be more like synthesis than learning, because we change the structure of the network on each iteration of the stage 2. But we can predefine $\lceil m/(2n) \rceil$ nodes in the hidden layer and use the parallel implementation of the algorithm. It is important that the main stage 2 is suitable for parallel computations. Namely, crucial for the performance steps 2.4-2.5 can be parallelized, meanwhile, step 2.3 as well as step 2.6 must be executed in critical sections. This approach allows us to implement effectively our learning algorithm in intelligent software tools like that in [24].

The theoretical results of the paper can be enhanced using the spectral approach in study of generalized neurons which was developed in [15]. The bithreshold network binary classifier admits a generalizations for solving a multiclass classification task. We considered only 2-layer bithreshold networks. The open question remains how bithreshold-like units can be applied in deep neural networks.

4 Conclusions

The new theoretical as well as practical estimates of the representational capabilities of bithreshold neurons and networks have been obtained in the paper. We proved the upper and lower bounds on the number of bithreshold separable dichotomies. Our lower bound is the same as in [21], whereas the upper bound is considerably better than the one stated in [19]. It is very important that for the first time these bounds have the same growth order.

For all $n \in \mathbb{N}$ we constructed the set of $2n$ points in n-dimensional space that is shattered by the set $\mathcal{BN}_n$ of all bithreshold neurons with n inputs. Therefore, we showed that $\mathrm{VCdim}\,(\mathcal{BN}_n) \geq 2n$. Since $\mathrm{VCdim}\,(\mathcal{T}_n) = n + 1$, where $\mathcal{T}_n$ is the set of single-threshold neurons, we can conclude that the expressive power of bithreshold neurons is considerably better. Our upper bound $5n$ on $\mathrm{VCdim}\,(\mathcal{BN}_n)$ does not seem to be tight. We hope shortly to improve this.

The capability of 2-layer bithreshold neural networks was also considered. We proved that an arbitrary binary classification of m points in general position in n-dimensional space can be computed by the bithreshold network with at most $\lceil m/(2n) \rceil$ hidden neurons. Thereby, the bithreshold network needs around half as many hidden neurons as the threshold one. So, bithreshold networks can provide more effective solutions in pattern recognition. We proposed $O\left(mn^2 + m^2\right)$ learning algorithm and justified its correctness. The simulation results on random samples and real data sets show that our algorithm assures high accuracy on the training set. The values of algorithm parameters providing the best performance during the simulation were also found. Consequently, our networks can be useful in pattern recognition units of complex systems such as were proposed in [13]. However, further research is desirable to improve the generalization ability of bithreshold networks.

References

1. Anthony, M.: Discrete Mathematics of Neural Networks: Selected Topics. SIAM, Philadelphia (2001). https://doi.org/10.1137/1.9780898718539
2. Babichev, S., Taif, M.A., Lytvynenko, V.: Inductive model of data clustering based on the agglomerative hierarchical algorithm. In: Proceedings of the 2016 IEEE 1st International Conference on Data Stream Mining and Processing, pp. 19–22 (2016)
3. Baum, E.B.: On the capabilities of multilayer perceptrons. J. Complex. **4**(3), 193–215 (1988). https://doi.org/10.1016/0885-064X(88)90020-9
4. Blum, A., Rives, R.: Training a 3-node neural network is NP-complete. Neural Netw. **5**(1), 117–127 (1992). https://doi.org/10.1016/S0893-6080(05)80010-3

5. Cover, T.: Geometrical and statistical properties of systems of linear inequalities with applications in pattern recognition. IEEE Trans. Electron. Comput. **EC–14**, 326–334 (1965)
6. Deolalikar, V.: A two-layer paradigm capable of forming arbitrary decision regions in input space. IEEE Trans. Neural Netw. **13**(1), 15–21 (2002). https://doi.org/10.1109/72.977261
7. Geche, F., Kotsovsky, V., Batyuk, A.: Synthesis of the integer neural elements. In: 10th International Conference on Computer Sciences and Information Technologies, CSIT 2015, Lviv, pp. 63–66 (2015). https://doi.org/10.1109/STC-CSIT.2015.7325432
8. Geche, F., Kotsovsky, V., Batyuk, A., Geche, S., Vashkeba, M.: Synthesis of time series forecasting scheme based on forecasting models system. In: CEUR Workshop Proceedings, vol. 1356, pp. 121–136 (2015)
9. Géron, A.: Hands-On Machine Learning with Scikit-Learn, Keras, and TensorFlow: Concepts, Tools, and Techniques to Build Intelligent Systems, 2nd edn. O'Reilly Media Inc., Sebastopol (2019)
10. Haring, D.: Multiple threshold perceptron. IEEE Trans. Electron. Comput. **EC–15**, 45–65 (1965)
11. Izonin, I., Kryvinska, N., Tkachenko, R., Zub, K., Vitynskyi, P.: An extended-input GRNN and its application. Procedia Comput. Sci. **160**, 578–583 (2019)
12. Izonin, I., Tkachenko, R., Kryvinska, N., Zub, K., Mishchuk, O., Lisovych, T.: Recovery of incomplete IoT sensed data using high-performance extended-input neural-like structure. Procedia Comput. Sci. **160**, 521–526 (2019). https://doi.org/10.1016/j.procs.2019.11.054
13. Kazarian, A., Teslyuk, V., Tsmots, I., Tykhan, M.: Implementation of the face recognition module for the "smart" home using remote server. In: Shakhovska, N., Medykovskyy, M. (eds.) Advances in Intelligent Systems and Computing III. CSIT 2018. Advances in Intelligent Systems and Computing, vol. 871, pp. 17–27. Springer, Cham (2019). https://doi.org/10.1007/978-3-030-01069-0_2
14. Kotsovsky, V., Geche, F., Batyuk, A.: Artificial complex neurons with half-plane-like and angle-like activation function. In: 10th International Conference on Computer Sciences and Information Technologies, CSIT 2015, Lviv, pp. 57–59 (2015). https://doi.org/10.1109/STC-CSIT.2015.7325430
15. Kotsovsky, V., Geche, F., Batyuk, A.: Finite generalization of the offline spectral learning. In: 2018 IEEE 2nd International Conference on Data Stream Mining and Processing, DSMP 2018, Lviv, pp. 356–360 (2018). https://doi.org/10.1109/DSMP.2018.8478584
16. Kotsovsky, V., Geche, F., Batyuk, A.: On the computational complexity of learning bithreshold neural units and networks. In: Lytvynenko, V., et al. (eds.) Lecture Notes in Computational Intelligence and Decision Making. ISDMCI 2019. Advances in Intelligent Systems and Computing, vol. 1020, pp. 189–202. Springer, Cham (2020). https://doi.org/10.1007/978-3-030-26474-1_14
17. Lupei, M., Mitsa, A., Repariuk, V., Sharkan, V.: Identification of authorship of Ukrainian-language texts of journalistic style using neural networks. East.-Eur. J. Enterp. Technol. **1**(2(103)), 30–36 (2020). https://doi.org/10.15587/1729-4061.2020.195041
18. Okrenets, S., Fefelov, A., Lytvynenko, V., Osypenko, V., Korobchynskyi, M., Voronenko, M.: Synthesis and learning of fuzzy neural networks for solving forecasting problems. In: 2017 IEEE 9th International Conference on Intelligent Data Acquisition and Advanced Computing Systems: Technology and Applications, IDAACS 2017, vol. 2, pp. 1088–1093 (2017). https://doi.org/10.1109/IDAACS.2017.8095253

19. Olafsson, S., Abu-Mostafa, Y.S.: The capacity of multilevel threshold function. IEEE Trans. Pattern Anal. Mach. Intell. **10**(2), 277–281 (1988)
20. Polishchuk, V.: Technology to improve the safety of choosing alternatives by groups of goals. J. Autom. Inf. Sci. **51**(9), 66–76 (2019). https://doi.org/10.1615/JAutomatInfScien.v51.i9.60
21. Takiyama, R.: The separating capacity of multithreshold threshold element. IEEE Trans. Pattern Anal. Mach. Intell. **PAMI–7**(1), 112–116 (1985)
22. Teslyuk, T., Tsmots, I., Teslyuk, V., Medykovskyy, M., Opotyak, Y.: Architecture and models for system-level computer-aided design of the management system of energy efficiency of technological processes at the enterprise. In: Shakhovska, N., Stepashko, V. (eds.) Advances in Intelligent Systems and Computing II. CSIT 2017. Advances in Intelligent Systems and Computing, vol. 689, pp. 538–557. Springer, Cham (2018). https://doi.org/10.1007/978-3-319-70581-1_38
23. Tkachenko, R., Izonin, I.: Model and principles for the implementation of neural-like structures based on geometric data transformations. In: Hu, Z., Petoukhov, S., Dychka, I., He, M. (eds.) Advances in Intelligent Systems and Computing, vol. 754, pp. 578–587. Springer, Cham (2019)
24. Tsmots, I., Skorokhoda, O., Tesliuk, T., Rabyk, V.: Designing features of hardware and software tools for intelligent processing of intensive data streams. In: 2016 IEEE 1st International Conference on Data Stream Mining and Processing, DSMP 2016, Lviv, pp. 332–335 (2016)
25. Vapnik, V.: Statistical Learning Theory. Wiley, New York (1998)
26. Vynokurova, O., Peleshko, D., Borzov, Y., Oskerko, S., Voloshyn, V.: Hybrid multidimensional wavelet-neuro-system and its learning using cross entropy cost function in pattern recognition. In: 2018 IEEE 2nd International Conference on Data Stream Mining and Processing, DSMP 2018, Lviv, pp. 305–309 (2018). https://doi.org/10.1109/DSMP.2018.8478608

On Flexible Co-registration of Optical and SAR Satellite Images

Volodymyr Hnatushenko[1,2(✉)], Peter Kogut[2,3], and Mykola Uvarov[1]

[1] Dnipro University of Technology, Dnipro, Ukraine
vvgnat@ukr.net
[2] EOS Data Analytics Ukraine, Dnipro, Ukraine
{vladimir.gnatushenko,peter.kogut,nikolay.uvarov}@eosda.com
[3] Oles Honchar Dnipro National University, Dnipro, Ukraine
p.kogut@i.ua

Abstract. We consider the problem of Synthetic Aperture Radar (SAR) and optical satellite images co-registration in its special statement. Because of the distinct natures of SAR and optical images, there exist huge radiometric and geometric differences between such images. Their structure, intensity of pixels, and texture are drastically different. As a result, the traditional registration approaches are no longer applicable in this case and it makes the registration process challenging. Mostly motivated by the crop field monitoring problem, we propose a new statement for the co-registration of SAR and optical images. We discuss the consistency of the proposed statement of this problem, propose the scheme for its regularization, and derive the corresponding optimality system.

Keywords: Co-registration · Synthetic aperture radar image · Optical image · Empirical mode decomposition · Speckle suppression · Edge detection · Sustainable contour extraction

1 Introduction

In contrast to the optical observation, the radar images do not depend on reflected sunlight. Instead, a radar satellite sends out its own energy pulse and measures the return signal ('backscatter') after interaction with the surface. In the vegetation case, instead of giving an indication on biophysical processes in the plant, the radar backscatter rather contains information on the structure and moisture content of vegetation and the underlying soil. Therefore, the fusion of SAR and optical images is very important for classification of land cover [15] and estimation of soil moisture to remove vegetation cover effects from radar backscattering coefficient [7,8,19]. At the same time, because of the distinct natures of SAR and optical images, there exist huge radiometric and geometric differences between optical and synthetic aperture of radar images. As a result, their structure and texture are drastically different. That's why the traditional

S. Babichev et al. (Eds.): ISDMCI 2020, AISC 1246, pp. 515–534, 2021.
https://doi.org/10.1007/978-3-030-54215-3_33

registration approaches are no longer applicable in this case and that makes the registration process challenging [11,18,20].

However, even though optical and radar sensors record completely different things, their measurements are nevertheless correlated, as both hold information on the vegetation status. It means that the cloud-induced gaps in optical measurements can be filled in by information from SAR images provided there is 'a feasible' correlation between optical and SAR observations. So, the point is to pave the way for co-registration of optical ($I_{opt} = I_{opt}(x, y)$) and SAR ($I_{SAR} = I_{SAR}(x, y)$) images of the same scenes and propose a rigorous mathematical setting of this problem.

Following Brown [3], the co-registration of two images can be defined as 'the process of overlaying images of the same scene taken at different times, from different viewpoints, and/or by different sensors'. With that in mind, we consider $I_{opt} = \left[I^R_{opt}, I^G_{opt}, I^B_{opt}\right]^t$ and I_{SAR} as two scenes of the same geographical region that were made, probably, at different dates and from different sensors—from a passive optical sensor and an active SAR sensor. We admit that these images can have different resolution, relative translation, rotation, scale, and other geometrical transformations between them and, as a consequence, their supports can also differ from each other, i.e., $I_{opt} : \Omega^{grid}_{opt} \to \mathbb{R}^3$, $I_{SAR} : \Omega^{grid}_{SAR} \to \mathbb{R}$. So, the goal of co-registration is to establish the correspondence between two images $I_{opt} : \Omega^{grid}_{opt} \to \mathbb{R}^3$ and $I_{SAR} : \Omega^{grid}_{SAR} \to \mathbb{R}$, and determine the geometric transformation that 'aligns' the SAR image with the optical one. From the formal point of view, it can be described as follows.

Definition 1. *We say that two images $I_{opt} : \Omega^{grid}_{opt} \to \mathbb{R}^3$ and $I_{SAR} : \Omega^{grid}_{SAR} \to \mathbb{R}$ are co-registered if there exist positive values $\alpha, \beta \in \mathbb{R}$ and a transformation $\mathcal{F} : \mathbb{R}^2 \to \mathbb{R}^2$ such that $\Omega^{grid}_{SAR} = \mathcal{G}\left(\Omega^{grid}_{opt}\right)$, where*

$$\mathcal{G}(x, y) := \begin{bmatrix} \alpha x \\ \beta y \end{bmatrix}, \quad \Lambda := \mathcal{F}\left(\mathcal{G}\left(\Omega^{grid}_{opt}\right)\right) \cap \Omega^{grid}_{SAR} \neq \emptyset, \quad \textit{and}$$
$$\textit{the pixels } (x, y) \in \Omega^{grid}_{SAR} \textit{ and } \mathcal{G}^{-1}\left[\mathcal{F}^{-1}(x, y)\right] \in \Omega^{grid}_{opt}$$
$$\textit{correspond to the same points on the ground, } \forall (x, y) \in \Lambda. \tag{1}$$

As follows from this definition, the core option of any method, which is aimed at the co-registration of images without external geo-data, is to propose a way how condition (1) can be either alleviated or relaxed. From this point of view all existing image co-registration techniques broadly fall into a few categories: the area-based and the feature-based methods, as well as combination of these two.

However, the most methods lose their efficiency in automatic SAR and optical image co-registration. The SAR data are the result of complex interaction of the reflected signals, topographic effects, radar shadow, and effect coming from surface roughness. Because of this SAR images are always contaminated by a multiplicative noise known as speckle and this fact makes the processing and interpretation of SAR images rather difficult. Moreover, SAR images are typically have a large geometrical distortion with optical images. Therefore, the shapes of contours and feature points location are highly erroneous and can change considerably. In most cases, the authors propose to remove the speckle noise in SAR images by using different filters. At the same time, both qualitative and quantitative characteristics show that, after filtering procedure, the SAR images with high-level speckle noise can lose a significant portion of information which is crucial for the monitoring technology.

To counter this problem, a new variational approach is proposed in this paper. The key idea we push forward in this paper is related to the following options. We perform the co-registration of SAR and optical images following, in general, the feature-based method where by distinctive features of such images we mean some stable and sustainable parts of the so-called skeletons for the previously locally smoothed SAR image and the optical master image that we chose in a special way. As for the relaxation of condition (1), we reduce it to a constrained optimization problem on the set of special displacement fields for which the cost functional is the L^p-cross-correlation between two fattened skeletons of the locally smoothed SAR image and the master one, respectively.

The proposed algorithm is quite efficient and can be implemented for automatic SAR-to-optical image co-registration. In contrast to the elastic contour matching scheme (see [14]) which is based on the active contour model, we do not use the optical contours as initial guides for contour detection in the SAR images. We also do not detect the salient segments such as corners of open contours (because of the high-level of speckle noise they can be detected erroneously) in order to use them in the matching process. Instead of this, we carry out a special pre-processing of SAR and optical images. After that we define the so-called sustainable contours for SAR and the luma component of an optical image. Since such contours have zero $\mathcal{L}^2$-Lebesgue measure, their co-registration is a non-trivial problem. Taking this point into account, we propose a new statement for SAR/optical co-registration problem—the so-called flexible co-registration of fattened contours. With that in mind, we give a precise variational setting for this problem in the form of some constrained minimization problem and discuss its sustainability, existence of minimizers, and optimality conditions. In order to evaluate the performance of the proposed approach, the main steps are illustrated by some applications to the real satellite images.

The aim of the research is the development of technique of SAR-to-optical image co-registration based on the co-registration of their fattened sustainable contours.

2 Feature Description of Optical and SAR Images

Let $I_{opt} = \left[I_{opt}^R, I_{opt}^G, I_{opt}^B\right]^t : \Omega_{opt} \to \mathbb{R}^3$ and $I_{SAR} : \Omega_{SAR} \to \mathbb{R}$ be given optical and SAR images, where it is assumed that $\Omega_{SAR} \neq \Omega_{opt}$ and

$$\Omega_{opt} = (0, b_o) \times (0, d_o) \subset \mathbb{R}^2, \quad \Omega_{SAR} = (0, b_s) \times (0, d_s) \subset \mathbb{R}^2.$$

As it was mentioned in Introduction, the speckle noise, blur, and other artifacts may strongly affect the accuracy of the contour detection in the original

Fig. 1. SAR in VH-polarization

Fig. 2. SAR in VV-polarization

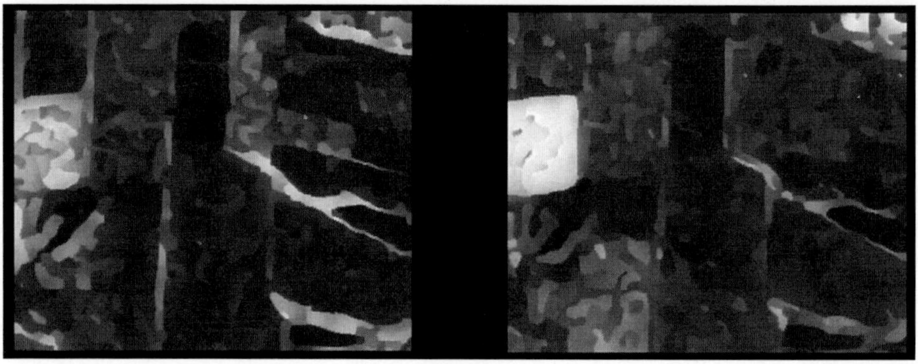

Fig. 3. VH and VV components after DLS-transformation

SAR and optical images. Hence, a specific locally smoothing procedure should be implemented for their pre-processing. With that in mind, we set

$$\widehat{I}_{opt} = \left[\widehat{I}^R_{opt}, \widehat{I}^G_{opt}, \widehat{I}^B_{opt}\right]^t := \left[\mathrm{DLS}\left(I^R_{opt}\right), \mathrm{DLS}\left(I^G_{opt}\right), \mathrm{DLS}\left(I^B_{opt}\right)\right]^t,$$
$$\widehat{I}_{SAR} := \mathrm{DLS}\left(I_{SAR}\right),$$

where DLS $(\cdot)$ stands for the deblurring and local smoothing operator defined on the set of gray-level images. As an particular example of such operator, we can make use of the adopted Alvarez, Lions and Morel model which is based on the following nonlinear parabolic equation:

$$\frac{\partial V(t)}{\partial t} = g\left(|\nabla G_\sigma * V(t)|\right)|\nabla V(t)| \,\mathrm{div}\left(\frac{\nabla V(t)}{|\nabla V(t)|}\right) - \lambda\left(1 - g\left(|\nabla G_\sigma * V(t)|\right)\right)\left(V(t) - I\right), \tag{2}$$

$$\partial_\nu V(t) = 0 \quad \text{on} \quad \text{in} \ (0,\infty)\times\partial\Omega, \tag{3}$$

$$V(0) = \left(G_\sigma * \text{Canny contour of } I\right) \quad \text{in} \ \Omega, \tag{4}$$

where $I = I(x,y)$ represent the initial image (scalar band of optical image or SAR one), $V = V(t,x,y)$ is its deblurred and smoothed version on scale t, $\lambda > 0$ is a weighting parameter, $g(\cdot)$ is the edge stopping function which we define basing on the ideas of Perona and Malik [16] and Alvarez et al. [1]

$$g(z) = \frac{1}{1+\mu z^2}, \quad \mu = \mu(\sigma) \text{ is the gradient threshold parameter,} \tag{5}$$

and $G_\sigma * V$ determines the convolution of function V with the Gaussian kernel G_σ:

$$G_\sigma(\mathbf{z}) = \frac{1}{2\pi\sigma s} e^{-\frac{|\mathbf{z}|^2}{2\sigma s}}, \quad \mathbf{z} = (x,y) \in \mathbb{R}^2, \tag{6}$$

$$(\nabla G_\sigma * V)(\mathbf{z}) := \int_\Omega \nabla G_\sigma(\mathbf{z}-\mathbf{w}) V(\mathbf{w})\, d\mathbf{w}. \tag{7}$$

As a result, we set $\text{DLS}(I) = V_\infty$, where $U_\infty(x,y) = \lim_{t\to\infty} U(t,x,y)$ is a steady state solution to the boundary value problem (2)–(4).

One of the characteristic feature of the model (2)–(4) is the usage of Canny edge detector in the initial condition (4). We utilize this option because of the following well-known performance properties of the Canny edge detector: good edge detection, good their localization, and single response [4]. The second characteristic feature of the model (2)–(4) we would like to emphasize, is an exploitation of extra term

$$\lambda \left(1 - g\left(|\nabla G_\sigma * V(t)|\right)\right)\left(V(t) - I\right),$$

where $(V(t) - I)$ is the forcing term suggested by Nordström while $(1-g)$ is the moderation selector allowing to identify the different homogeneous regions $(g \approx 1)$ on the image that should be separately and more smoothed in relation to the edge regions $(g \approx 0)$ (see [2] for the details).

Remark 1. As it follows from Figs. 1, 2 and 3, the main advantage of the DLS-procedure is the fact that this type of the pre-processing not only reduces the speckle noise, blur, and other artifacts in the original images, but what is the most important issue, the speckle noise is transformed into a set of 'visible spots' with sufficiently small compact supports. As a result, the small size of speckle supports gives rise to suppose that the corresponding perimeters (or length of speckle contours) will be small enough as well. This is a crucial option of our approach and we will essentially utilize this circumstance in determining of the feature characteristics for SAR and optical images.

Since the aim of this section is to propose a sustainable feature description of SAR and optical images, the crucial step consists in the special choice of master image. We define it in the following way: we say that $I^*_{opt} : \Omega_{SAR} \to \mathbb{R}$ is the master image for co-registration of the SAR and optical images $I_{SAR} : \Omega_{SAR} \to \mathbb{R}$, $I_{opt} : \Omega_{opt} \to \mathbb{R}^3$ if its domain is Ω_{SAR} and

$$I^*_{opt} = 0.299\, \mathcal{S}\left(\widehat{I}^R_{opt}\right) + 0.587\, \mathcal{S}\left(\widehat{I}^G_{opt}\right) + 0.114\, \mathcal{S}\left(\widehat{I}^B_{opt}\right), \tag{8}$$

where $\mathcal{S} : [I : \Omega_{opt} \to \mathbb{R}] \to [J : \Omega_{SAR} \to \mathbb{R}]$ stands for a $2D$-interpolation operator with the following conditions of compatibility

$$\mathcal{S}(I)(\alpha x, \beta y) = I(x, y), \ \forall (x, y) \in \Omega_{opt}^{grid},$$

where $\Omega_{opt}^{grid} = \{(i, j) : i = 1, \ldots, N_o, j = 1, \ldots, M_o\} \subset \Omega_{opt}$ and the constants $\alpha > 0$ and $\beta > 0$ are defined as follows $\alpha := b_s/b_o$, $\beta := d_s/d_o$.

Mainly inspired by Remark 1, we also define the cut-off operator $\mathcal{W}_N$ such that Image $\overset{\mathcal{W}_N}{\mapsto}$ Image and $\mathcal{W}_N(J)$ removes all connected components (objects) that have fewer than N pixels from the image $J : \Omega_{SAR} \to \mathbb{R}$, producing other image $\mathcal{W}_N(J) : \Omega_{SAR} \to \mathbb{R}$.

We are now in a position to determine the feature descriptors for SAR and optical images.

Definition 2. *We say that, for a given threshold $N \in \mathbb{N}$, a binary image $\mathfrak{C}_{SAR} : \Omega_{SAR}^{grid} \to \{0, 1\}$ is the sustainable contour for a SAR image $I_{SAR} : \Omega_{SAR}^{grid} \to \mathbb{R}$ if*

$$\mathfrak{C}_{SAR}(x, y) = \mathcal{W}_N \left[\text{Sobel Detector} \left(\widehat{I}_{SAR} \right) \right] (x, y), \tag{9}$$

for all $(x, y) \in \Omega_{SAR}^{grid}$, where $\widehat{I}_{SAR} = \mathrm{DLS}\,(I_{SAR})$.

Definition 3. *We say that, for a given threshold $M \in \mathbb{N}$, a binary image $\mathfrak{C}_{opt} : \Omega_{SAR}^{grid} \to \{0, 1\}$ is the sustainable contour for an optical image $I_{opt} : \Omega_{opt}^{grid} \to \mathbb{R}^3$ if*

$$\mathfrak{C}_{opt}(x, y) = \mathcal{W}_M \left[\text{Canny Detector} \left(I_{opt}^* \right) \right] (x, y), \tag{10}$$

for all $(x, y) \in \Omega_{SAR}^{grid}$, where $I_{opt}^* : \Omega_{SAR} \to \mathbb{R}$ is the luma component of deblurred optical image and defined by (8).

In order to evaluate the concept of sustainable contours in the sense of Definitions 2–3, we refer to Figs. 4, 5, 6, 7, 8 and 9, where we present some results obtained with the application of the proposed technique in several natural SAR and optical images of the same agricultural region. As evidently follows from Fig. 6, a direct contour detection for the original SAR and optical images suffers from many undesired artifacts like double edges, false contours, and others unpleasant things. At the same time, it would be erroneous to suppose that sustainable contour for the SAR image is a sub-graph of the sustainable contour for the corresponding optical image and vise versa. Moreover, even in the case the above mentioned contours, their co-registration is a non-trivial problem.

3 Setting of the Flexible Co-registration Problem

Let $I_{opt} = \left[I_{opt}^R, I_{opt}^G, I_{opt}^B \right]^t : \Omega_{opt} \to \mathbb{R}^3$ and $I_{SAR} : \Omega_{SAR} \to \mathbb{R}$ be given optical and SAR images. Let $\mathfrak{C}_{opt} : \Omega_{SAR} \to \{0, 1\}$ and $\mathfrak{C}_{SAR} : \Omega_{SAR} \to \{0, 1\}$ be the corresponding their sustainable contours. Setting $\Omega := \Omega_{SAR}$, we suppose that

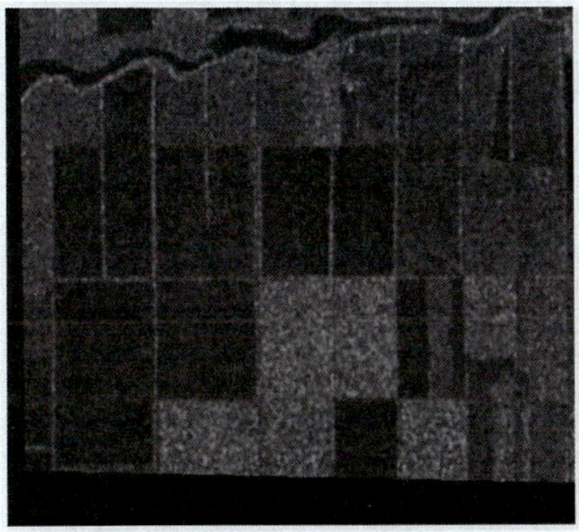

Fig. 4. Original SAR image

Fig. 5. Original optical image

Fig. 6. Contours of the original images detected by the Sobel operator for SAR image (left) and by the Canny method for optical image (right)

each of the contours $\mathfrak{C}_{opt}$ and $\mathfrak{C}_{SAR}$ can be interpreted as a planar and, as a rule, disconnected graph with zero thickness of its edges, i.e.

$$\mathcal{L}^2\left(\text{supp}\left[\mathfrak{C}_{SAR}\right]\right) = \mathcal{L}^2\left(\text{supp}\left[\mathfrak{C}_{opt}\right]\right) = 0, \tag{11}$$

where $\mathcal{L}^2(A)$ stands for the $2D$-Lebesgue measure of a set $A \subset \mathbb{R}^2$, and

$$\begin{aligned} \text{supp}\left[\mathfrak{C}_{SAR}\right] &:= \text{closure}\left\{x \in \Omega \ : \ \mathfrak{C}_{SAR}(x) \neq 0\right\}, \\ \text{supp}\left[\mathfrak{C}_{opt}\right] &:= \text{closure}\left\{x \in \Omega \ : \ \mathfrak{C}_{opt}(x) \neq 0\right\}. \end{aligned}$$

In general, it would be erroneous to suppose that the exists an affine transformations $\mathcal{F} : \mathbb{R}^2 \to \mathbb{R}^2$ of the form

$$\mathcal{F}(x) = Bx + a, \quad \forall x \in \mathbb{R}^2, \tag{12}$$

Fig. 7. SAR and master images after DSIMF-Processing

Fig. 8. Sobel contour of the SAR image after DSIMF-Processing (left); Sustainable contour for the SAR image with the threshold $N = 40$ (right)

Fig. 9. Canny contour of the optical master image after DSIMF-Processing (left); Sustainable contour for the optical image with the threshold $M = 50$ (right)

where $a = \begin{bmatrix} a_1 \\ a_2 \end{bmatrix}$ and $B = \begin{bmatrix} b_{11} & b_{12} \\ b_{21} & b_{22} \end{bmatrix}$ such that the transformed contour $\mathfrak{C}_{SAR}^{coreg} : \Omega_{SAR} \to \{0,1\}$ of the SAR image

$$\mathfrak{C}_{SAR}^{coreg}(x) = \left\{ \begin{matrix} \mathfrak{C}_{SAR}\left(\mathcal{F}_0^{-1}(x)\right), & \text{if } x \in \Lambda, \\ 0, & \text{if } x \in \Omega \setminus \Lambda, \end{matrix} \right\} \tag{13}$$

would be a subset of the optical contour $\mathfrak{C}_{opt} : \Omega_{SAR} \to \{0,1\}$ or wise versa. Here,

$$\mathcal{F}_0^{-1}(x) = B_0^{-1}(x - a_0) \qquad \text{for all } x \in \mathbb{R}^2.$$

The main idea we are going to discuss in this section is to reduce the SAR-to-optical image co-registration problem to the matching problem of contours $\mathfrak{C}_{opt}$ and $\mathfrak{C}_{SAR}$ through their appropriate deformation. To do so, we propose to recover a dense displacement field $u = \begin{bmatrix} u_1 \\ u_2 \end{bmatrix} : \Omega \to \mathbb{R}^2$ in order to maximize the so-called 'similarity' measure between the references optical contour $\mathfrak{C}_{opt} : \Omega_{SAR} \to \{0,1\}$ and a warped SAR-contour over a suitable space of $\mathbb{R}^2$-deformations. As a result, the contour matching problem between optical and SAR images can be stated as follow: Find a displacement field $u^0 : \mathbb{R}^2 \to \mathbb{R}^2$ such that

$$J(u^0) = \sup_{u \in W} J(u), \tag{14}$$

where the energy functional $J(u)$, which is a measure of 'concordance' between two contours, we define as follows

$$J(u) := \mathcal{L}^1\left(\text{supp}\left[\mathfrak{C}_{opt}\right] \cap \left(\text{supp}\left[\mathfrak{C}_{SAR}\right] + u(\cdot)\right)\right).$$

Here, W stands for the class of admissible displacements.

Since it is unknown whether the constrained maximization problem (14) is solvable, it is reasonably to provide some its regularization. A common approach in such cases is to replace the variational problem (14) by the other one, which would be a well-posed problem such that its solutions could be interpreted as suboptimal solutions to the original problem.

For given $\sigma \geq 3$, we associate with the constrained optimization problem (14) the following its regularized version

$$J(u) = \int_{\Omega} \left|F_s(x - u(x)) - F_o(x)\right|^2 dx + \mathcal{R}(u) \to \inf_{u \in W}, \tag{15}$$

where $\mathcal{R} : W \to \mathbb{R}$ is a term which penalizes the undesired properties of the displacement field $u : \mathbb{R}^2 \to \mathbb{R}^2$ like a very fast variations of the function $u \in W$, and

$$F_s(x) = \int_{\Omega} G_\sigma(x - z)\mathfrak{C}_{SAR}(z)\, dz, \quad F_o(x) = \int_{\Omega} G_\sigma(x - z)\mathfrak{C}_{opt}(z)\, dz. \tag{16}$$

Here, the objects $F_o(x)$ and $F_s(x)$ can be interpreted as the fattened contours for the SAR and optical images.

In order to specify the class of admissible displacements W and regularization term $\mathcal{R}(u)$, we set (see [5] for comparison):

$$W = H_0^1(\Omega) \times H_0^1(\Omega), \quad \mathcal{R}(u) = \alpha \int_\Omega \mathrm{Tr}\left[Du\; M(x) Du^t\right] dx. \tag{17}$$

Here, $\alpha = \text{const} > 0$ is a tuning parameter, $H_0^1(\Omega)$ stands for the Sobolev space which is defined as the closure of the set $C_0^\infty(\Omega)$-functions with respect to the norm $\|y\| = \|\nabla y\|_{L^2(\Omega;\mathbb{R}^2)} = \left(\int_\Omega |\nabla y(x)|^2_{\mathbb{R}^2}\, dx\right)^{\frac{1}{2}}$, $Du(x) = \begin{bmatrix} \partial_{x_1} u_1 & \partial_{x_2} u_1 \\ \partial_{x_1} u_2 & \partial_{x_2} u_2 \end{bmatrix}$ is the Jacobian of u at x, and $M(x)$ is a 2×2 symmetric matrix defined at every point x of Ω by the following expression:

$$M(x) := \frac{\left(\lambda + |\nabla F_o(x)|^2\right) Id - \nabla F_o(x) \nabla F_o^t(x)}{|\nabla F_o(x)|^2 + 2\lambda}, \tag{18}$$

where $\lambda > 0$ is a threshold parameter, and Id stands for the unit 2×2 matrix.

As for the tensor $M(x)$, it has been firstly proposed in [5]. Our interest in the specification of this matrix in the form (18), is to prevent the fast variations of the displacement field $u(x)$ across the contours of the fattened graph $F_o(x)$. It is worth to emphasize that the matrix $M(x)$ has one eigenvector equal to $\nabla F_o(x)$, while the second one identifies the direction perpendicular to $\nabla F_o(x)$. The corresponding eigenvalues take the form

$$\kappa_2(x) = \frac{\lambda}{|\nabla F_o(x)|^2 + 2\lambda} > 0, \quad \kappa_1(x) = \frac{\lambda + |\nabla F_o(x)|^2}{|\nabla F_o(x)|^2 + 2\lambda} > 0. \tag{19}$$

We see that the cost functional (15) is the energy functional which, in contrast to the problem (14), measures the disparity between two fattened contours. However, the main benefit to reduce the SAR-to-optical image co-registration problem to consideration of constrained minimization problem (15) is the fact that the intensity of pixels for the objects $F_o(x)$ and $F_s(x)$ can be compared directly. So, in some sense, regularized problem (15) can be interpreted as a particular case of a so-called image registration problem for the monomodal images.

4 Existence of Minimizers

Our main goal in this section is to show that the energy functional (15) has a minimum over the entire set of admissible displacement W and there exists at least one element $u^o \in W$ such that

$$J(u^0) = \inf_{u \in W} J(u). \tag{20}$$

To this end, we make use of some preliminaries.

Lemma 1. *The functional $\mathcal{R} : W \to \mathbb{R}^2$ is strictly convex.*

Proof. Let u and v be arbitrary elements of W such that $u \neq v$. Let $\tau \in (0,1)$ be a given value. Since

$$D(u+\tau(v-u))\ M(x)D(u+\tau(v-u))^t = 2\tau(1-\tau)Dv\ M(x)Du^t + (1-\tau)^2 Du\ M(x)Du^t + \tau^2 Dv\ M(x)Dv^t,$$

it follows that

$$\begin{aligned} D(u+\tau(v-u))\ M(x)D(u+\tau(v-u))^t &- \tau Dv\ M(x)Dv^t \\ - (1-\tau)Du\ M(x)Du^t &= (1-\tau)(-\tau)Du\ M(x)Du^t \\ + \tau(\tau-1)Dv\ M(x)Dv^t &+ 2\tau(1-\tau)Du\ M(x)Dv^t \\ &= -\tau(1-\tau)\Big(Du\ M(x)Du^t + Dv\ M(x)Dv^t - 2Du\ M(x)Dv^t\Big) \\ &= -\tau(1-\tau)D(u-v)\ M(x)D(u-v)^t. \end{aligned} \tag{21}$$

Taking into account that

$$\begin{aligned} \operatorname{Tr}\left[Du\ M(x)Du^t\right] &= \nabla u_1^t M(x)\nabla u_1 + \nabla u_2^t M(x)\nabla u_2 \\ &\overset{\text{by (19)}}{\geq} \kappa_2(x)\left[\sum_{i=1}^{2} |\nabla u_i|^2\right], \end{aligned} \tag{22}$$

it follows from (19) and (21) that

$$\begin{aligned} &\mathcal{R}(\tau u + (1-\tau)v) - (1-\tau)\mathcal{R}(u) - \tau\mathcal{R}(v) \\ &\quad = -\alpha\tau(1-\tau)\int_\Omega \operatorname{Tr}\left[D(u-v)\ M(x)D(u-v)^t\right]\,dx \\ &\quad = -\alpha\tau(1-\tau)\int_\Omega \kappa_2(x)\left[\sum_{i=1}^{2}|\nabla u_i|^2\right]\,dx < 0. \end{aligned}$$

From this the required property follows.

Lemma 2. *The functional $\mathcal{R} : W \to \mathbb{R}^2$ is coercive, namely*

$$\mathcal{R}(u) \geq \alpha\int_\Omega \kappa_2(x)\left[\sum_{i=1}^{2}|\nabla u_i|^2\right]\,dx \geq \alpha\frac{\lambda}{\|\nabla F_o\|^2_{L^\infty(\Omega)} + 2\lambda}\|u\|^2_W, \quad \forall\, u \in W.$$

Proof. This property immediately follows from relation (22) and formula (19). Indeed, following the definition of a fattened graph, we see that $|\nabla F_o|^2 \in L^\infty(\Omega)$. As a result, we have

$$\kappa_2(x) := \frac{\lambda}{|\nabla F_o(x)|^2 + 2\lambda} \geq \frac{\lambda}{\|\nabla F_o\|^2_{L^\infty(\Omega)} + 2\lambda} > 0, \quad \text{for a.a } x \in \Omega.$$

The proof of the following result can be found in Theorem 7.3-2 in [6] (see also [12]).

Lemma 3. *Let $\varphi : \Omega \times \mathbb{R}^2 \to [\beta, \infty)$ be a function with some $\beta > 0$, satisfying the following properties:*

(j) the mapping $\mathbb{R}^2 \ni v \mapsto \varphi(x, v)$ is convex and continuous for almost all $x \in \Omega$;
(jj) the mapping $\Omega \ni x \mapsto \varphi(x, v)$ is measurable for all $v \in \mathbb{R}^2$.

Then the functional $\mathcal{P}(u) := \int_\Omega \varphi(x, u(x))\, dx : L^2(\Omega; \mathbb{R}^2) \longrightarrow \mathbb{R}$ is lower semi-continuous with respect to the weal topology of $L^2(\Omega; \mathbb{R}^2)$, i.e.,

$$\begin{gathered} u_y \rightharpoonup u \quad in \quad L^2(\Omega; \mathbb{R}^2) \quad \Rightarrow \\ \liminf_{k \to \infty} \int_\Omega \varphi(x, u_k(x))\, dx \geq \int_\Omega \varphi(x, u(x))\, dx. \end{gathered} \tag{23}$$

Taking into account Lemmas 1–3, we are in position to prove the following existence result for variational problem (15).

Theorem 1. *There exists a displacement field $u^0 = \begin{bmatrix} u_1^0 \\ u_2^0 \end{bmatrix} \in W := H_0^1(\Omega) \times H_0^1(\Omega)$ such that u^0 is a solution of the contour matching problem* (15).

Proof. As follows from the definition of energy functional $J : W \to \mathbb{R}$ and Lemma 2, this functional is non-negative on W. Moreover, by Lemma 1, we have

$$J(u) \geq \mathcal{R}(u) \overset{\text{by Lemma 1}}{\geq} \frac{\alpha}{2} \frac{\lambda}{\|\nabla F_o\|^2_{L^\infty(\Omega)} + 2\lambda} \|u\|^2_W, \ \forall u \in W. \tag{24}$$

Hence, the infimum in (20) is finite. Let $\{u_k\}_{k \in \mathbb{N}}$ be a minimizing sequence for (20), i.e.

$$\lim_{k \to \infty} J(u_k) = \inf_{u \in W} J(u) = \mu \geq 0. \tag{25}$$

Then, without lost of generality, there exists an integer $k_0 \in \mathbb{N}$ such that (see (24)–(25))

$$\|u_k\|^2_W \leq \frac{2}{\alpha} \frac{\|\nabla F_o\|^2_{L^\infty(\Omega)} + 2\lambda}{\lambda} (\mu + 1), \quad \forall k \geq k_0,$$

that is, the sequence $\{u_k\}_{k \in \mathbb{N}}$ is bounded in $H_0^1(\Omega) \times H_0^1(\Omega)$. In view of reflexivity of the Sobolev space $H_0^1(\Omega)$, there exists a function $u^0 = \begin{bmatrix} u_1^0 \\ u_2^0 \end{bmatrix} \in W$ such that (up to subsequences) $u_{1,k} \rightharpoonup u_1^0$ and $u_{2,k} \rightharpoonup u_2^0$ in $H_0^1(\Omega)$ as $k \to \infty$. By Sobolev embedding theorem, the injection $H_0^1(\Omega) \hookrightarrow L^2(\Omega)$ is compact. Hence, the weak convergence $u_{i,k} \rightharpoonup u_i^0$ in $H_0^1(\Omega)$ implies the strong convergence $u_{i,k} \to u_i^0$ in $L^2(\Omega)$. Setting $\varphi(u) := \operatorname{Tr}[Du\, MDu^t]$, we see that properties (j)–(jj) of Lemma 3 hold true. Hence, the lower semi-continuity property (23) implies that

$$\liminf_{k \to \infty} \mathcal{R}(u_k) \geq \mathcal{R}(u^0). \tag{26}$$

Since the strong convergence in $L^2(\Omega)$ implies (up to a subsequence) the point-wise convergence, we can suppose that $u_k(x) \to u^0(x)$ in $\mathbb{R}^2$ as $k \to \infty$ for almost all $x \in \Omega$, it follows from the initial assumptions that

$$\lim_{k \to \infty} \int_\Omega |F_s(x - u_k(x)) - F_o(x)|^2\, dx = \int_\Omega \left|F_s(x - u^0(x)) - F_o(x)\right|^2\, dx.$$

Utilizing this property together with (26), we finally arrive at the following relations

$$\mu = \inf_{u\in W} J(u) \stackrel{\text{by (25)}}{=} \lim_{k\to\infty} J(u_k) = \liminf_{k\to\infty} J(u_k) \geq J(u^0).$$

Thus, $u^0 \in W$ is a minimizer to the variational problem (15).

5 Optimality System

Following the standard approach [13], we associate with an optimal displacement field $u^0 \in W$ the first variation $\delta J(u^0;h)$ of the energy functional $J : W \to \mathbb{R}$ at u^0 in the direction $h \in W$, where

$$\delta J(u^0;h) := \lim_{\varepsilon\to 0} \frac{J(u^0+\varepsilon h) - J(u^0)}{\varepsilon}.$$

Since $u^0 \in W$ is a minimizer of J, it follows that the equality $\delta J(u^0;h) = 0$ must hold for every $h \in W$. In literature, the equation $\delta J(u^0;h) = 0$ is called the Euler-Lagrange equation associated with the energy functional J.

Let us show that the structure of the first variation $\delta J(u^0;h)$ of the energy functional can be recovered in an explicit form. With that in mind, we make use of the following notation for the duality pairing between W and its dual space W^*,

$$\langle g, h\rangle_{W^*;W} := \sum_{i=1}^{2} \langle g_i, h_i\rangle_{H^{-1}(\Omega);H_0^1(\Omega)}, \quad \forall\, h \in W \ \text{ and } \ \forall\, g \in W^*,$$

where it is assumed that $\langle f, z\rangle_{H^{-1}(\Omega);H_0^1(\Omega)} = \int_\Omega f z\, dx$, $\forall\, z \in H_0^1(\Omega)$ provided $f \in L^2(\Omega)$.

Proposition 1. *The following representation for first variation $\delta J(u^0;h)$ of energy functional J*

$$\delta J(u^0;h) = \left\langle \nabla_G J(u^0), h\right\rangle_{W^*;W}, \quad \forall\, h \in W \tag{27}$$

holds true, where $\nabla_G J(u^0) \in W^*$ and

$$\nabla_G J(u^0) = -2\begin{bmatrix} \left(F_s(x-u^0) - F_o(x)\right)\partial_{x_1} F_s(x-u^0) + \alpha \operatorname{div}\left(M(x)\nabla u_1^0\right), \\ \left(F_s(x-u^0) - F_o(x)\right)\partial_{x_2} F_s(x-u^0) + \alpha \operatorname{div}\left(M(x)\nabla u_2^0\right) \end{bmatrix} \tag{28}$$

is the Gâteaux gradient of J at the point $u = u^0$.

Proof. Following the standard procedure and taking into account the representations (16)–(17), we obtain

$$\lim_{\varepsilon\to 0} \frac{J(u^0+\varepsilon h) - J(u^0)}{\varepsilon} = \Lambda_1(u^0,h) + \Lambda_2(u^0,h),$$

where

$$\begin{aligned}\Lambda_1(u^0,h) &= -2\int_\Omega \Big(F_s(x-u^0(x)) - F_o(x)\Big)\Big(\nabla F_s(x-u^0(x)),h\Big)_{\mathbb{R}^2}\,dx,\\ &= -2\left\langle \left(F_s(\cdot - u^0) - F_o(\cdot)\right)\partial_{x_1}F_s(\cdot - u^0), h_1\right\rangle_{H^{-1}(\Omega);H_0^1(\Omega)}\\ &\quad - 2\left\langle \left(F_s(\cdot - u^0) - F_o(\cdot)\right)\partial_{x_2}F_s(\cdot - u^0), h_2\right\rangle_{H^{-1}(\Omega);H_0^1(\Omega)}, \end{aligned} \tag{29}$$

$$\begin{aligned}\Lambda_2(u^0,h) &= \lim_{\varepsilon\to 0}\frac{\mathcal{R}(u^0+\varepsilon h)-\mathcal{R}(u^0)}{\varepsilon} \overset{\text{by }(22)}{=} 2\alpha\int_\Omega \left[\nabla h_1^t M\nabla u_1^0 + \nabla h_2^t M(x)\nabla u_2^0\right]\\ &= -2\alpha\left\langle \operatorname{div}(M(x)\nabla u_1^0), h_1\right\rangle_{H^{-1}(\Omega);H_0^1(\Omega)}\\ &\quad - 2\alpha\left\langle \operatorname{div}(M(x)\nabla u_2^0), h_2\right\rangle_{H^{-1}(\Omega);H_0^1(\Omega)}. \end{aligned} \tag{30}$$

Since $u_i^0 \in H_0^1(\Omega)$, $i=1,2$, it follows that

$$\left(F_s(x-u^0(x)) - F_o(x)\right)\partial_{x_i}F_s(x-u^0(x)) + \alpha\operatorname{div}(M(x)\nabla u_i^0(x)),\quad i=1,2$$

are elements of the dual space $H^{-1}(\Omega)$. Hence, the energy functional J is Gâteaux differentiable at $u = u^0 \in W$ and the representation (27)–(28) for its Gâteaux gradient $\nabla_G J(u^0)$ is a direct consequence of relations (29)–(30).

In view of Proposition 1, the equality $\delta J(u^0;h) = 0$ leads to the following optimality system for optimal displacements in the contour matching problem (15)

Find $u^0 = (u_1^0, u_2^0) \in H_0^1(\Omega)\times H_0^1(\Omega)$ such that

$$-\operatorname{div}(M(x)\nabla u_1^0) = \alpha^{-1}\left(F_s(x-u^0) - F_o(x)\right)\partial_{x_1}F_s(x-u^0) \quad \text{in } \Omega, \tag{31}$$

$$-\operatorname{div}(M(x)\nabla u_2^0) = \alpha^{-1}\left(F_s(x-u^0) - F_o(x)\right)\partial_{x_2}F_s(x-u^0) \quad \text{in } \Omega, \tag{32}$$

where the symmetric 2×2 matrix $M(x)$ is given by relation (18).

One of the characteristic features of the coupled system (31)–(32) is the fact that we look for optimal displacement field $u(x)$ in $\mathbb{R}^2$ as an element of the space $W = H_0^1(\Omega)\times H_0^1(\Omega)$. Hence, it is assumed that this displacement is zero at the boundary $\partial\Omega$, whereas the homogeneous Neumann boundary conditions

$$(M(x)\nabla u_i, \nu)_{\mathbb{R}^2} = 0 \quad \text{on } \partial\Omega.$$

would be more natural in the image processing. Indeed, we look for a suitable registration transformation $\psi(x) = x - u(x)$ which makes a template fattened graph being similar to a reference graph, it is reasonable to admit a non-zero offset for every pixel in moving template image: consider the rigid rotation as a typical example in image matching. However, such possibility is lost with the homogeneous Dirichlet boundary conditions, which are applied in the case of the space $W = H_0^1(\Omega)\times H_0^1(\Omega)$.

At the same time, if we set $W = H^1(\Omega) \times H^1(\Omega)$, then some technical problems related with the solvability, uniqueness, and coerciveness of the corresponding variational problems can appear. In order to circumstance this obstacle, we assume that the original the fattened contours for SAR and optical images are extended by zero to the entire set $\mathbb{R}^2$ so that instead of Ω we can provide our analysis in a domain $\widehat{\Omega}$ such that $\Omega \subset \widehat{\Omega}$. As a result, the set $W = H_0^1(\widehat{\Omega}) \times H_0^1(\widehat{\Omega})$ can be considered as a plausible choice for a class of admissible displacements.

6 On Non-Uniqueness of Solutions to Optimality System

Since an optimal displacement field $u^0 \in W$ is as a solution of optimality system (31)–(32), we proceed with the study of this coupled boundary value problem. To begin with, we recall some basic properties of nonlinear operators acting from a separable reflexive Banach space X into its dual X^* (for the details we refer to [10, 17]).

The operator $A : X \to X^*$ is called coercive if $\liminf_{\|y\|_X \to \infty} \frac{\langle A(y), y\rangle_{X^*;X}}{\|y\|_X} = \infty$.

The operator $A : X \to X^*$ is called bounded if for any $l_1 > 0$ there exists an $l_2 > 0$ such that $\|A(y)\|_{X^*} \leq l_2$ for all $\|y\|_X \leq l_1$.

The operator $A : X \to X^*$ is called radially-continuous if for any fixed elements $y, \xi \in X$ there exists $\varepsilon > 0$ such that the real-valued function $[0, \varepsilon] \ni t \mapsto \langle A(y + t\xi), \xi\rangle_{X^*;X}$ is continuous.

We say that the operator $A : X \to X^*$ has semi-bounded variation if, for any $\rho > 0$ and any $y, z \in X$ such that $\|y\|_X \leq \rho$ and $\|z\|_X \leq \rho$, the inequality

$$\langle A(y) - A(z), y - z\rangle_{X^*;X} \geq -\gamma\left(\rho, \|y - z\|'\right) \tag{33}$$

holds, where the function $\gamma(\rho, \sigma)$ is nonnegative, continuous, and such that $\lim_{\sigma \to 0} \frac{\gamma(\rho,\sigma)}{\sigma} = 0$, $\forall\, \rho > 0$, and $\|\cdot\|'$ is a norm compact with respect to $\|\cdot\|_X$.

We say that the operator $A : X \to X^*$ possesses the (M)-property if the following implication holds true

$$\left[\begin{array}{c} y_k \rightharpoonup y \ \text{ in } X, \\ A(y_k) \rightharpoonup \eta \ \text{ in } X^*, \\ \limsup\limits_{k\to\infty} \langle A(y_k), y_k\rangle_{X^*;X} \leq \langle \eta, y\rangle_{X^*;X} \end{array}\right] \Longrightarrow \left[A(y) = \eta\right].$$

The main result, we are going to associate to the system (31)–(32), can be stated as follows

Theorem 2. *Let $\mathcal{A} : X \to X^*$ be a bounded coercive operator possessing the property (M). Let $\mathcal{B} : X \to X^*$ be a radially-continuous operator having a semi-bounded variation and satisfying the property*

$$\liminf_{\|y\|_X \to \infty} \frac{\langle \mathcal{B}(y), y\rangle_{X^*;X}}{\|y\|_X} > -\infty. \tag{34}$$

Then the mapping $\mathcal{A} + \mathcal{B} : X \mapsto X^*$ is surjective.

For the proof of this result, we refer to Theorem I.4.20 in [10].

In order to proceed further, we set $X = \left[H_0^1(\Omega)\right]^2$. Then $X^* = \left[H^{-1}(\Omega)\right]$. We also define the operators $\mathcal{A}: X \to X^*$ and $\mathcal{B}: X \to X^*$ as follows

$$\mathcal{A}(u) := \begin{bmatrix} -\operatorname{div}\left(M(x)\nabla u_1\right) \\ -\operatorname{div}\left(M(x)\nabla u_2\right) \end{bmatrix}, \tag{35}$$

$$\mathcal{B}(u) := \begin{bmatrix} -\alpha^{-1}\left(F_s(x-u) - F_o(x)\right)\partial_{x_1} F_s(x-u) \\ -\alpha^{-1}\left(F_s(x-u) - F_o(x)\right)\partial_{x_2} F_s(x-u) \end{bmatrix}. \tag{36}$$

Let us show that these operators satisfy all conditions indicated in Theorem 2.

Lemma 4. *$\mathcal{A}: X \to X^*$ is a bounded operator.*

Proof. Since $\mathcal{A}: X \to X^*$ is a linear operator, this property is a direct consequence of the following chain of inequalities

$$\begin{aligned}
\|\mathcal{A}\| &= \sup_{\|u\|_X \le 1} \|\mathcal{A}(u)\|_{X^*} = \sup_{\|u\|_X \le 1} \sup_{\|v\|_X \le 1} \langle \mathcal{A}(u), v\rangle_{X^*;X} \\
&= \sup_{\|u\|_X \le 1} \sup_{\|v\|_X \le 1} \left[\int_\Omega (\nabla v_1, M(x)\nabla u_1)_{\mathbb{R}^2}\, dx + \int_\Omega (\nabla v_2, M(x)\nabla u_2)_{\mathbb{R}^2}\, dx\right] \\
&\le \left(\sup_{x\in\Omega} \kappa_1(x)\right) \sup_{\|u\|_X \le 1} \sup_{\|v\|_X \le 1} \left[\int_\Omega |\nabla v_1||\nabla u_1|\, dx + \int_\Omega |\nabla v_2||\nabla u_2|\, dx\right] \\
&\le \frac{\lambda + \|\nabla F_o(x)\|^2_{L^\infty(\Omega)}}{2\lambda} \sup_{\|u\|_X \le 1} \|u\|_X \sup_{\|v\|_X \le 1} \|v\|_X < +\infty.
\end{aligned}$$

Lemma 5. *$\mathcal{A}: X \to X^*$ is a coercive operator.*

Proof. To prove the coerciveness, it is enough to apply the following estimate from below

$$\begin{aligned}
\langle \mathcal{A}(u), u\rangle_{X^*;X} &= \int_\Omega (\nabla u_1, M(x)\nabla u_1)_{\mathbb{R}^2}\, dx + \int_\Omega (\nabla u_2, M(x)\nabla u_2)_{\mathbb{R}^2}\, dx \\
&\overset{\text{by Lemma 2}}{\ge} \int_\Omega \kappa_2(x)\left[\sum_{i=1}^2 |\nabla u_i|^2\right] dx \ge \frac{\lambda}{\|\nabla F_o\|^2_{L^\infty(\Omega)} + 2\lambda}\|u\|^2_X.
\end{aligned} \tag{37}$$

Lemma 6. *The operator $\mathcal{A}: X \to X^*$ possesses the property (M).*

Proof. Let $\{u_k\}_{k\in\mathbb{N}}$ be a sequence in X such that

$$\begin{aligned}
u_k \rightharpoonup u \text{ in } X, \quad A(u_k) \rightharpoonup \eta \text{ in } X^* \quad \text{as } k \to \infty, \\
\text{and} \quad \limsup_{k\to\infty} \langle A(u_k), u_k\rangle_{X^*;X} \le \langle \eta, u\rangle_{X^*;X}.
\end{aligned}$$

Then the weak convergence $A(u_k) \rightharpoonup \eta$ in X^* implies that

$$\lim_{k\to\infty} \langle A(u_k), \xi\rangle_{X^*;X} = \langle \eta, \xi\rangle_{X^*;X}, \quad \forall \xi \in X.$$

Combining this relation with the previous inequality, we obtain

$$\limsup_{k\to\infty} \langle A(u_k), u_k - \xi\rangle_{X^*;X} \leq \langle \eta, u - \xi\rangle_{X^*;X}, \quad \forall\, \xi \in X.$$

Since

$$\begin{aligned}\limsup_{k\to\infty} \langle A(u_k), u_k - \xi\rangle_{X^*;X} &\geq \liminf_{k\to\infty} \langle A(u_k) - A(\xi), u_k - \xi\rangle_{X^*;X} \\ &\quad + \lim_{k\to\infty} \langle A(\xi), u_k - \xi\rangle_{X^*;X} \\ &\overset{\text{by (37)}}{\geq} \frac{\lambda}{\|\nabla F_o\|^2_{L^\infty(\Omega)} + 2\lambda}\|u-\xi\|^2_X + \lim_{k\to\infty} \langle A(\xi), u_k - \xi\rangle_{X^*;X} \\ &\geq \langle A(\xi), u - \xi\rangle_{X^*;X},\end{aligned}$$

it follows that $\langle A(\xi), u - \xi\rangle_{X^*;X} \leq \langle \eta, u - \xi\rangle_{X^*;X}$, $\forall\, \xi \in X$. From this and (37) we deduce that this inequality remains valid if only $A(u) = \eta$.

Lemma 7. *The operator* $\mathcal{B} : X \to X^*$ *is radially-continuous, has a semi-bounded variation, and satisfies the property* (34).

Proof. In view of definition of operator $\mathcal{B}$, we have

$$\langle \mathcal{B}(u + t\xi), \xi\rangle_{X^*;X} = -\alpha^{-1} \int_\Omega (F_s(x - u - t\xi) - F_o(x)) \left(\nabla F_s(x - u - t\xi), \xi\right)_{\mathbb{R}^2} dx$$

for each $\xi \in X$ and $t \in \mathbb{R}$. Since $F_s : \mathbb{R}^2 \to \mathbb{R}$ and $F_0 : \mathbb{R}^2 \to \mathbb{R}$ are smooth functions, the radially-continuity property follows. Moreover, due to the smoothness of the indicated functions, we see that

$$\left.\begin{aligned} &0 \leq F_s(x) \leq C_0 = \max\left\{\|F_s\|_{L^\infty(\Omega)}, \|F_o\|_{L^\infty(\Omega)}\right\}, \\ &0 \leq F_o(x) \leq C_0, \\ &|\nabla F_s(x)|_{\mathbb{R}^2} \leq C_3, \quad \forall\, x \in \mathbb{R}^2, \\ &|F_s(x) - F_s(y)|, |F_o(x) - F_o(y)| \leq C_1 |x - y|_{\mathbb{R}^2}, \\ &|\nabla F_s(x) - \nabla F_s(y)|_{\mathbb{R}^2}, \\ &|\nabla F_o(x) - \nabla F_o(y)|_{\mathbb{R}^2} \leq C_2 |x - y|_{\mathbb{R}^2}, \end{aligned}\right\} \tag{38}$$

and therefore, for any $u, v \in X$,

$$\begin{aligned} &\left|\langle \mathcal{B}(u) - \mathcal{B}(v), u - v\rangle_{X^*;X}\right| \\ &\qquad = \alpha^{-1}\Big| \int_\Omega (F_s(x-u) - F_o(x)) \left(\nabla F_s(x-u), u - v\right)_{\mathbb{R}^2} dx \\ &\qquad\quad \pm \int_\Omega (F_s(x-u) - F_o(x)) \left(\nabla F_s(x-v), u - v\right)_{\mathbb{R}^2} dx \\ &\qquad\quad - \int_\Omega (F_s(x-v) - F_o(x)) \left(\nabla F_s(x-v), u - v\right)_{\mathbb{R}^2} dx\Big| \\ &\qquad \leq 2\alpha^{-1} C_0 C_2 \int_\Omega |u - v|^2_{\mathbb{R}^2} dx + 2\alpha^{-1} C_2 C_1 \int_\Omega |u - v|^2_{\mathbb{R}^2} dx \\ &\qquad\qquad = 2\alpha^{-1} C_2 (C_0 + C_1) \|u - v\|^2_{L^2(\Omega;\mathbb{R}^2)}. \end{aligned}$$

Hence, the inequality (33) for operator $\mathcal{B}$ holds with $\gamma(\rho,\sigma) = 2\alpha^{-1}C_2\,(C_0+C_1)\,\sigma^2$, and, therefore, $\mathcal{B} : X \to X^*$ is an operator with semi-bounded variation.

It remains to show that this operator satisfies the property (34). Indeed, due to estimates (38), we get

$$\left|\langle \mathcal{B}(u), u\rangle_{X^*;X}\right| = \alpha^{-1}\left|\int_\Omega (F_s(x-u) - F_o(x))\,(\nabla F_s(x-u), u)_{\mathbb{R}^2}\,dx\right|$$
$$\leq 2\alpha^{-1}C_0C_3\int_\Omega |u|_{\mathbb{R}^2}\,dx \leq 2\alpha^{-1}C_0C_3|\Omega|^{\frac{1}{2}}\,\|u\|_{L^2(\Omega;\mathbb{R}^2)}.$$

Hence, $\liminf_{\|y\|_X\to\infty}\frac{\langle\mathcal{B}(y),y\rangle_{X^*;X}}{\|y\|_X} \geq -2\alpha^{-1}C_0C_3|\Omega|^{\frac{1}{2}} > -\infty$, and we arrive at the announced property.

Utilizing Lemmas 4–7 and Theorem 2, we can give the following conclusion: for a particular choice (35)–(36) of operators $\mathcal{A} : X \to X^*$ and $\mathcal{B} : X \to X^*$, the mapping $\mathcal{A}+\mathcal{B} : X \mapsto X^*$ is surjective. Since the optimality system (31)–(32) can be represented in an operator form as: Find $u^0 = (u_1^0, u_2^0) \in X = \left[H_0^1(\Omega)\right]^2$ such that $\mathcal{A}(u^0) + \mathcal{B}(u^0) = 0$, it follows that the set of its solution is not a singleton, in general. In practical implementations, this circumstance should be taken into account. On the other hand, a direct solving of the coupled Dirichlet boundary value problem (31)–(32) is usually impossible. Hence, it is reasonable to define an optimal displacement u^0 using a 'gradient descent' strategy. Following the standard procedure, we can start from some initial field of displacements $u^* = \begin{bmatrix} u_1^* \\ u_2^* \end{bmatrix} \in W$ and after to solve the following initial value problem for coupled quasi-linear parabolic equations with an anisotropic elliptic operators in their principle part

$$\frac{\partial u_1(t)}{\partial t} - \operatorname{div}\left(M(x)\nabla u_1\right) = \alpha^{-1}\left(F_s(x-u) - F_o(x)\right)\partial_{x_1}F_s(x-u),$$
$$\frac{\partial u_2(t)}{\partial t} - \operatorname{div}\left(M(x)\nabla u_2\right) = \alpha^{-1}\left(F_s(x-u) - F_o(x)\right)\partial_{x_2}F_s(x-u),$$
$$u_1(0) = u_1^*, \quad u_2(0) = u_2^*.$$

The solution of the contour matching problem is then taken as the a steady-state solution $u^0(x) = \lim_{t\to\infty}\begin{bmatrix} u_1(t,x) \\ u_2(t,x) \end{bmatrix}$ of the parabolic system given above. We also notice that in homogeneous region of the fattened graph F_o, as $|\nabla F_o| \ll \lambda$, both eigenvalues κ_i, $i = 1, 2$, tend to $1/2$. Hence, the diffusion in (31)–(32) is nearly isotropic. At the same time, along the contours of F_o, when $|\nabla F_o| \gg \lambda$, we have $\kappa_2 \to 0$ and $\kappa_1 \to 1/2$. As a result, the diffusion becomes anisotropic in this case, taking place mainly along the contours. This circumstance plays a crucial role in correct determination of an optimal displacement field for the contour matching problem.

The other generalizations of the energy functional J for the contour matching problem can based on utilization of the semi-norm of the space of functions with bounded variations instead of the regularized term $\mathcal{R}$ in the form (17) (see, for instance, [9]).

7 Conclusions

In this paper, we have proposed a new technique for the co-registration of satellite images coming from different types of sensors at arguably different times in order to put the similar structures of two different images into correspondence and integrate then the information containing in such images. We mainly focus on the mathematical description of the proposed approach. In particular, we give a precise definition what should be understood by the solution to the co-registration problem for SAR and optical images and announce the formal statement of this problem. We also show that, in general, this problem is ill-posed and propose the scheme for its regularization. The obtained results (see, for instance, Fig. 10) confirmed the efficiency of the proposed algorithm.

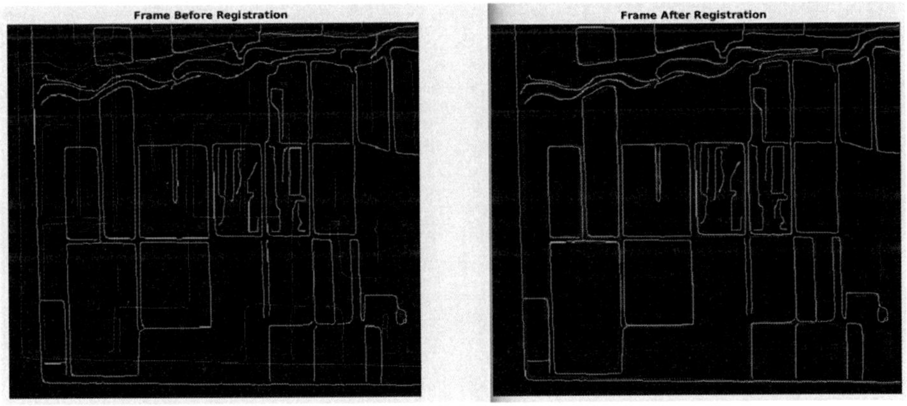

Fig. 10. Result of the contour matching following the solution of variational problem (15).

Acknowledgments. This work has been partially supported by the EOS company (https://eos.com/). EOS—Spatial Data Analytics, GIS Software, Satellite Imagery—is a cloud-based platform to derive remote sensing data and analyze satellite imagery for business and science purposes. We also thank the European Union Copernicus program providing Sentinel data accessible via the Copernicus Open Access Hub.

References

1. Alvarez, L., Lions, P., Morel, J.M.: Image selective smoothing and edge detection by nonlinear diffusion. SIAM J. Numer. Anal. **29**(3), 845–866 (1992). https://doi.org/10.1137/0729012
2. Barcelos, C., Boaventura, M., Silva Jr., E.: A well-balanced flow equation for noise removal and edge detection. IEEE Trans. Image Process. **12**(7), 751–763 (2003). https://doi.org/10.1109/TIP.2003.814242
3. Brown, G.: A survey of image co-registration techniques. ACM Comput. Surv. **24**, 325–376 (1992). https://doi.org/10.1145/146370.146374

4. Canny, J.: A computational approach to edge detection. IEEE Trans. Pattern Anal. Mach. Intell. **8**(6), 679–698 (1986). https://doi.org/10.1109/TPAMI.1986.4767851
5. Chefd'hotel, C., Hermosillo, G., Faugeras, O.: Variational methods for multimodal image matching. Int. J. Comput. Vision **50**(3), 329–343 (2002). https://doi.org/10.1023/A:1020830525823
6. Ciarlet, P.: Mathematical Elasticity, Vol. I: Three-Dimensional Elasticity. North Holland, Amsterdam etc. (1988). https://doi.org/10.1002/zamm.19890691105
7. Garkusha, I., Hnatushenko, V., Vasyliev, V.: Research of influence of atmosphere and humidity on the data of radar imaging by sentinel-1. IEEE 37th International Conference on Electronics and Nanotechnology (ELNANO) (2017). https://doi.org/10.1109/elnano.2017.7939787
8. Garkusha, I., Hnatushenko, V., Vasyliev, V.: Using sentinel-1 data for monitoring of soil moisture. In: IEEE International Geoscience and Remote Sensing Symposium (IGARSS) (2017). https://doi.org/10.1109/igarss.2017.8127291
9. Hnatushenko, V., Kogut, P., Uvarov, M.: On optimal 2-d domain segmentation problem via piecewise smooth approximation of selective target mappings. J. Optim. Differ. Eqn. Appl. (JODEA) **27**(2), 39–74 (2019). https://doi.org/10.15421/141908
10. Ivanenko, V., Mel'nik, V.: Variational Methods in Control Problems for Systems with Distributed Parameters. Naukova Dumka, Kyiv (1988)
11. Kern, J., Pattichis, M.: Robust multispectral image registration using mutual-information models. IEEE Trans. Geosci. Remote Sens. **45**(5), 1494–1505 (2007). https://doi.org/10.1109/TGRS.2007.892599
12. Kogut, P., Kupenko, O.: Approximation Methods in Optimization of Nonlinear Systems. Walter de Gruyter GmbH, Berlin (2010). https://doi.org/10.1515/9783110668520
13. Kogut, P., Leugering, G.: Optimal Control Problems for Partial Differential Equations on Reticulated Domains: Approximation and Asymptotic Analysis. Birkhäuser, Boston (2011). https://doi.org/10.1007/978-0-8176-8149-4
14. Li, H., Manjunath, B., Mitra, S.: A contour-based approach to multisensor image registration. IEEE Trans. Image Process. **4**(3), 320–334 (1995). https://doi.org/10.1109/83.366480
15. Muller, M., Shepherd, J., Dymond, J.: Support vector machine classification of woody patches in New Zealand from synthetic aperture radar and optical data, with lidar training. J. Appl. Remote Sens. **9**(1), art. no. 095984 (2015). https://doi.org/10.1117/1.JRS.9.095984
16. Perona, P., Malik, J.: Scale-space and edge detection using anisotropic diffusion. IEEE Trans. Pattern Anal. Mach. Intell. **12**(7), 629–639 (1990). https://doi.org/10.1109/34.56205
17. Roubiček, T.: Nonlinear Partial Differential Equations with Applications. Birkhäuser, Basel (2013). https://doi.org/10.1007/978-3-0348-0513-1
18. Salehpour, M., Behrad, A.: Hierarchical approach for synthetic aperture radar and optical image coregistration using local and global geometric relationship of invariant features. J. Appl. Remote Sens. **11**(1), art. no. 015002 (2017). https://doi.org/10.1117/1.JRS.11.015002
19. Saradjian, M., Hosseini, M.: Comparison of optical, radar and hybrid soil moisture estimation models using experimental data. J. Appl. Remote Sens. **5**(1), art. no. 053524 (2011). https://doi.org/10.1117/1.3586794
20. Xiang, Y., Wang, F., Wan, L., Jiao, N., You, H.: Os-flow: a robust algorithm for dense optical and sar image registration. IEEE Trans. Geosci. Remote Sens. **57**(9), 6335–6354 (2019). https://doi.org/10.1109/TGRS.2019.2905585

Solving Fraud Detection Tasks Based on Wavelet-Neuro Autoencoder

Olena Vynokurova[1,2,3], Dmytro Peleshko[1,2], Polina Zhernova[3(✉)], Iryna Perova[3], and Andrii Kovalenko[3]

[1] GeoGuard, Kharkiv, Ukraine
vynokurova@gmail.com, dpeleshko@gmail.com
[2] GeoGuard, Lviv, Ukraine
[3] Kharkiv National University of Radio Electronics, Kharkiv, Ukraine
polina.zhernova@gmail.com, rikywenok@gmail.com, andrey.kovalenko@nure.ua

Abstract. The basis of any business is customer databases, which provide information on customer relations with the company. For example, in the field of banking services, the database stores information about the client, account number, data on financial transactions in online trading, purchased goods, their quantity, time of purchase, etc. A fraud detection is a field of data mining, which includes a set of methods for detecting fraudulent activities in the credit and financial sector, telecommunications, and other areas where illegal manipulations with customer accounts, tariff changes, etc. are possible. Typically, a fraud detection technique is based on the detection of events that do not fit into a specific pattern or behavioral pattern specific to a given business process or client that does not correspond to its patterns and trends. Analytical methods of Data Mining are widely used to build fraud detection systems: neural networks, decision trees, associative rules, sequential patterns, etc.

To solve this problem, the architecture of wavelet-neuro autoencoder and its learning algorithm are proposed. Such wavelet-neuro autoencoder has simple architecture, which is based on wavelet neuron. The tuning parameters of this architecture are not only synaptic weights and also parameters of adaptive wavelet activation functions. The tuning parameters of adaptive wavelet activation function are width, center and shape. Proposed wavelet-neuro autoencoder allows solving the fraud detection problem based on both big dataset and small dataset cases.

Keywords: Fraud detection · Wavelet-neuro autoencoder · Data mining · Machine learning

1 Introduction

The problem of fraud detection banking transactions is quite acute. An effective tool to solve the problem of fraud is the use of machine learning algorithms. But for this, first of all, it is necessary to determine the factors that enable the detection of fraudulent operations. A features of this area is that the use of raw

S. Babichev et al. (Eds.): ISDMCI 2020, AISC 1246, pp. 535–546, 2021.
https://doi.org/10.1007/978-3-030-54215-3_34

data does not give an acceptable result [5]. Bank card fraud is usually divided into two broad categories: application fraud and behavioral fraud. Application fraud may occur upon receipt of a new card at the issuing company [6]. Credit scoring is often used to prevent this type of fraud. As for behavioral fraud, it is divided into 4 types: mail theft (a situation where a fraudster gains access to an envelope with a card before it reaches its rightful owner), lost and stolen cards, forged cards (creating a physical copy of the card) and "Without providing a card". Fraud "Without providing a card", unlike the previous three, does not require the card itself - it is carried out remotely with the help of stolen information about the details of the card [5]. This is a very convenient way of fraud, as it is almost completely anonymous. Anti-fraud includes all actions and measures to ensure that fraud never happens.

The sooner fraud is detected, the less losses will be from it, since the card and all transactions on it will be blocked. There are various machine learning algorithms, as well as different approaches to learning. For an effective and high-quality solution to the problem of detecting fraud, it is necessary to choose an approach and algorithm that best fits into the area in question. To solve this problem, using an specific autoencoder which is based on modified wavelet functions. Autoencoder – a neural network with a special architecture that allows the use of teacherless training using the backpropagation method [2]. The simplest architecture of an autoencoder is a direct distribution network without feedback, most similar to a multilayer perceptron (MLP), containing an input layer, a hidden intermediate layer and an output layer. The output layer contains as many neurons as the input layer. The task of the autoencoder is to get the output vector closest to the input. This allows, first of all, data compression when transmitting the input signal vector to the network output. The operation of the autoencoder resembles the operation of the principal component method (PCA) in the sense that the dimensionality of the input data is reduced. PCA - one of the main ways to reduce the dimensionality of data by losing the least amount of information. Sometimes the PCA is called the Karhunen-Loeve transform or Hotelling's transform [11]. At the same time, the autoencoder implements the lossless data compression and decompression algorithm. The standard autoencoder network is shown in Fig. 1.

A distinctive feature of autoencoders is that the number of neurons at the input and output coincides. Autoencoder consists of two parts:

1. Encoder: responsible for compressing the input to latent-space. Represented by the encoding function
$$h = f(x, e^f).$$
2. Decoder: designed to restore input from latent-space. It is represented by the decoding function
$$\hat{x} = g(h, e^g).$$

Thus, the autocoder is described by the function

$$\hat{x} = g(f(x, e^f), e^g)$$

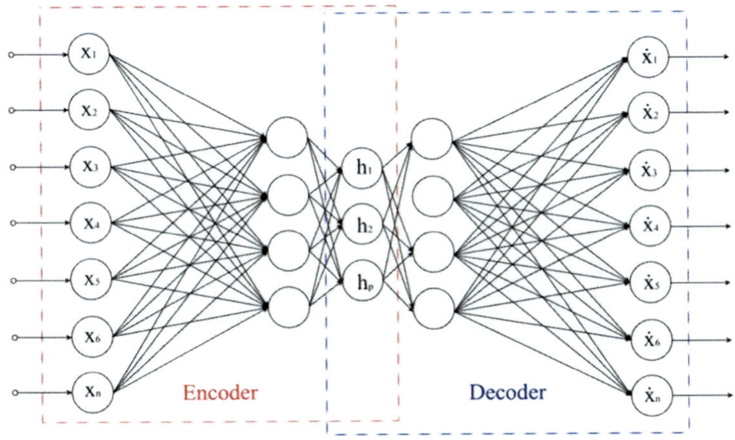

Fig. 1. Autoencoder architecture

where e^f, e^g are the encoder and decoder errors, respectively.

Self-learning process of autoencoders is based on minimizing the difference between the observation vector $x(t)$ and its reconstruction $\hat{x}$, that is, minimizing the cost $\|x - \hat{x}\|^2$ with respect to the parameters e^f and e^g.

There are following autoencoder type:

- Sparse autoencoder is to some extent the antipode of the autoencoder. Instead of training the network to represent blocks of information in a smaller "space", we encode information so that it takes up more space. And instead of forcing the system to converge in the center, and then expand back to its original size, we, on the contrary, increase the middle layers. This type of network can be used to extract many small details from a dataset. If we began to train sparse autoencoder in the same way as autoencoder, we would in most cases get a completely useless network where the output is exactly the same as the input. To avoid this, instead of the input data, we give out the input data plus a penalty for the number of activated neurons in the hidden layer [16].
- Autoencoding variational is the same architecture as that of ordinary ones, but they are taught differently - the approximate probabilistic distribution of input samples. This is to some extent a return to basics, since the autoencoding variational is a bit closer to Boltzmann machines. However, they rely on Bayesian mathematics regarding probabilistic judgments and independence, which are intuitive but require complex calculations. The basic principle can be formulated as follows: take into account the degree of influence of one event on another [10].
- Denoising autoencoder. This is autoencoder by which we input not just data, but data with noise. Nevertheless, we calculate the error using the previous method, comparing the output sample with the original without noise. Thus, the network remembers not small details, but large features, since remem-

bering small details that are constantly changing due to noise often does not lead to anywhere [18].

2 Related Works

Credit card fraud is one of the most common banking problems. With the increase in the number of payment transactions via the Internet, this problem has become much more important and there is a need to find solutions for resolving it. In [8], the use of a sliding window strategy was proposed.

In paper [15] authors have proposed deep auto-encoder model based on restricted Boltzmann machine that can reconstruct normal transactions to find anomalies from normal patterns. In paper [14] authors have applied different supervised methods and ensemble learning methods to detect fraudulent operations using a real dataset.

In paper [17] authors have proposed deep learning topologies with built-in time and memory components such as Long Short-term memory. In [13] author have proposed to focus on data preprocessing, and several algorithms were applied, namely, Local Outlier Factor [7] and Isolation Forest algorithm [12].

In the paper [1] authors have proposed using genetic algorithm to improve classification of imbalanced datasets for credit card fraud detection. The authors in [9] have proposed a deep autoencoder for extracting best features from the data of the transactions and then append a softmax network to define the class labels.

The aim of the research is the developing the wavelet-neuro autoencoder for solving fraud detection tasks. Such wavelet-neuro autoencoder has simple architecture, which is based on wavelet neuron [4]. The tuning parameters of this architecture are not only synaptic weights and also parameters of wavelet activation functions.

3 Wavelet-Neuro Autoencoder Architecture and Its Learning Algorithm

Existing anomaly detection methods are based on autoassociative three-layer perceptron or convolutionary methods. The architectures of such autoencoders do not allow to adjust the appearance and parameters of activation functions. Also this type of autoencoders require large training dataset and may not have time to train on small dataset.

For this, a hybrid two-layer wavelet-neuro autoencoder is proposed for solving the fraud detection problem based on both big dataset and small dataset cases. Also the learning algorithm for tuning synaptic weights and wavelet activation function parameters are proposed.

The proposed architecture of two-layer wavelet-neuro autoencoder is shown in the Fig. 2.

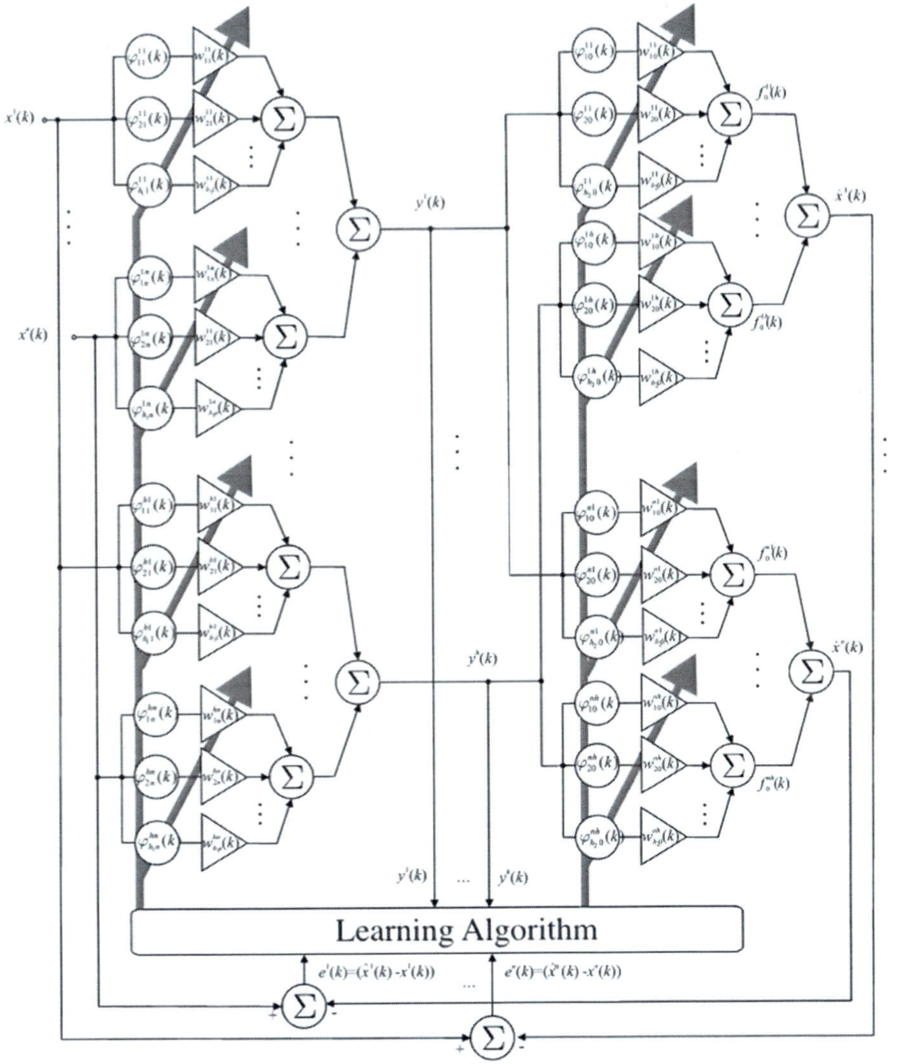

Fig. 2. Wavelet-neuro autoencoder architecture

The building block of this structure is a wavelet neuron with non-linear wavelet synapses [4]. When feeding to the input of the wavelet-neuro autoencoder vector $X = \{x^1(k), x^2(k), ..., x^n(k)\}$ at the output, we obtain signals in the form

$$\hat{x}^i(k) = \sum_{m=1}^{h}\sum_{j=1}^{h_2}\varphi_{j0}^{im}(y^m(k))w_{j0}^{im}(k)$$
$$= \sum_{m=1}^{h}\sum_{j=1}^{h_2}\varphi_{j0}^{im}\left(\sum_{i=1}^{n}\sum_{l=1}^{h_1}\varphi_{li}^{mi}(x^i(k))w_{li}^{mi}\right)w_{j0}^{im}(k) \quad (1)$$

where $\varphi_{li}^{mi}(\bullet)$, $\varphi_{j0}^{im}(\bullet)$ are the wavelet activation functions of the first and second layer, respectively; $w_{li}^{mi}(k)$, $w_{j0}^{im}(k)$ are the synaptic weights of the input and output layers, respectively; $y^m(k)$ is m-th component of the observation vector.

The architecture of the wavelet-neuro autoencoder consists of two layers: a hidden layer in which hn wavelet synapses with h_1 wavelet functions in each and an output layer consisting of hn wavelet synapses with h_2 wavelet functions. In each wavelet synapse, wavelets are implemented that differ in their center, width and shape parameters, which are tuned along with synaptic weights using proposed training methods.

To configure the wavelet-neuro autoencoder, it is advisable to use the adaptive wavelet function, the parameters and shape of which would be adjusted during the learning the autoencoder. In this case, we propose to use the custom one-dimensional wavelet activation function of accessories introduced in [3], which has the form for the first hidden layer

$$\varphi_{li}^{mi}(x^i(x(k))) = (1-\alpha_{mi}^{li}(k)(\tau_{li}^{mi}(x^i(k)))^2)\exp\left(\frac{(\tau_{li}^{mi}(x^i(k)))^2}{2}\right), \quad (2)$$

and for the second layer

$$\varphi_{j0}^{im}(y^m(x(k))) = (1-\alpha_{j0}^{im}(k)(\tau_{j0}^{im}(x^i(k)))^2)\exp\left(\frac{(\tau_{j0}^{im}(x^i(k)))^2}{2}\right), \quad (3)$$

where $\tau_{li}^{mi}(x^i(k)) = (x^i(k) - c_{li}^{mi}(k))(\sigma_{li}^{mi}(k))^{-1}$; $\tau_{j0}^{im}(x^i(k)) = (x^i(k) - c_{j0}^{im}(k))(\sigma_{j0}^{im}(k))^{-1}$; $\alpha_{li}^{mi}(k), \alpha_{j0}^{im}(k)$ are shape parameters $(0 \leq \alpha \leq 1)$.

Based on the learning criteria of the species

$$E^i(k) = \frac{1}{2}\left(x^i(k) - \sum_{i=1}^{n} f_0^{im}(y^m(k))\right)^2 \quad (4)$$

We can write a method for training synaptic weights and all parameters of the wavelet activation function of the first layer in the form

$$w_{li}^{mi}(k+1) = w_{li}^{mi}(k) + \eta^w e^i(k)[f_0^{im}(y^m(k))]'\varphi_{li}^{mi}(x^i(k)), \quad (5)$$

$$\begin{cases} c_{li}^{mi}(k+1) = c_{li}^{mi}(k) + \eta^c e^i(k)[f_0^{im}(y^m(k))]' w_{li}^{mi}(k)\frac{\partial\varphi_{li}^{mi}(x^i(k))}{\partial c_{li}^{mi}(k)}, \\ (\sigma_{li}^{mi})^{-1}(k+1) = (\sigma_{mi}^{li})^{-1}(k) + \eta^\sigma e^i(k)[f_0^{im}(y^m(k))]' w_{li}^{mi}(k)\frac{\partial\varphi_{li}^{mi}(x^i(k)}{\partial(\sigma_{li}^{mi})^{-1}}, \\ \alpha_{li}^{mi}(k+1) = \alpha_{li}^{mi}(k) + \eta^\alpha e^i(k)[f_0^{im}(y^m(k))]' w_{li}^{mi}(k)\frac{\partial\varphi_{li}^{mi}(x^i(k))}{\partial\alpha_{li}^{mi}(k)} \end{cases}$$

where $[f_0^{im}(y^m(k))]' = \sum_{j=1}^{h_2} w_{j0}^{im} \frac{\partial \varphi_{j0}^{im}(y^m(k))}{\partial y^m(k)}$.

The learning process of the second layer is based on the criteria in the form

$$E^i(k) = \frac{1}{2}\left(x^i(k) - \hat{x}^i(k)\right)^2 = \frac{1}{2}e^{i2}(k). \tag{6}$$

Thus, the learning method of synaptic weights and all parameters of the wavelet activation functions of the second layer can be written in the form

$$w_{j0}^{im}(k+1) = w_{j0}^{im}(k) + \eta_0 e^i(k)\varphi_{j0}^{im}(y^m(k)), \tag{7}$$

$$\begin{cases} c_{j0}^{im}(k+1) = c_{j0}^{im}(k) + \eta_0^c e^i(k) w_{j0}^{im}(k) \frac{\partial \varphi_{j0}^{im}(y^m(k))}{\partial c_{j0}^{im}(k)}, \\ (\sigma_{j0}^{im})^{-1}(k+1) = (\sigma_{j0}^{im})^{-1}(k) + \eta_c^\sigma e^i(k) w_{j0}^{im}(k) \frac{\partial \varphi_{j0}^{im}(y^m(k)}{\partial (\sigma_{j0}^{im})^{-1}}, \\ \alpha_{j0}^{im}(k+1) = \alpha_{j0}^{im}(k) + \eta_0^\alpha e^i(k) w_{j0}^{im}(k) \frac{\partial \varphi_{j0}^{im}(y^m(k))}{\partial \alpha_{j0}^{im}(k)}. \end{cases}$$

Thus, the wavelet-neuro autoencoder has simple architecture and allows solving both compressing data and detecting anomalies tasks using both small and big datasets in online mode.

4 Experiment Results and Discussion

To assess the quality of the proposed neural network, a data set from Kaggle "Credit Card Fraud Detection" was taken. The considered data set contains transactions committed in 2013 by cardholders within two days. The data set consists of 284,807 transactions, where 492 transactions are fraud one. As you can see, there are only 0.172% of fraudulent transactions, because of this fact the data set is not balanced. It contains only numeric values that have been converted using the PCA. Features $V_1, V_2 ... V_{28}$ are the main components, which are obtained by PCA. The columns "Time" and "Sum" were not converted. "Time" contains seconds between each of the transactions. The "Class" column contains information about whether the operation is fraudulent or not, i.e. takes the value 1 and 0, respectively.

The Fig. 3 shows a dataset after converting them using PCA.

We divide the dataset into two parts, only normal operations were included in the training dataset, and the test dataset were mixed and both normal and fraud operations were included. Training wavelet-neuro autoencoder will be performed only on the basis of non-fraud (normal) operations. The dataset was normalized and their minimum and maximum values of each features are saved for further use of such system in online mode.

The number of inputs was 29, we have included 28 features with PCA transformation and feature "Sum" with normalization. The number of wavelet activation functions in each synapse was 5. The learning metric is the mean square error (mse).

The training wavelet-neuro authoencoder was performed 10 epochs. The plot of mse change during the training process is shown in Fig. 4. The parameters of

Index	Time	V1	V2	V3	V4	V6	V6	V7	V8	V9	V10	V11	V12
0	0	-1.35981	-0.0727812	2.53635	1.37816	-0.338321	0.462388	0.239599	0.0986979	0.363787	0.0907942	-0.5516	-0.617801
1	0	1.19186	0.266151	0.16648	0.448154	0.0600176	-0.0823608	-0.078803	0.0851017	-0.255425	-0.166974	1.61273	1.06524
2	1	-1.35835	-1.34016	1.77321	0.37978	-0.503198	1.8005	0.791461	0.247676	-1.51465	0.207643	0.624501	0.0660837
3	1	-0.966272	-0.185226	1.79299	-0.863291	-0.0103089	1.2472	0.237609	0.377436	-1.38702	-0.0549519	-0.226487	0.178228
4	2	-1.15823	0.877737	1.54872	0.403034	-0.407193	0.0959215	0.592941	-0.270533	0.817739	0.753074	-0.822843	0.538196
5	2	-0.425966	0.960523	1.14111	-0.168252	0.420987	-0.0297276	0.476201	0.260314	-0.568671	-0.371407	1.34126	0.359894
6	4	1.22966	0.141004	0.0453708	1.20261	0.191881	0.272708	-0.005159	0.0812129	0.46496	-0.0992543	-1.41691	-0.153826
7	7	-0.644269	1.41796	1.07438	-0.492199	0.948934	0.428118	1.12063	-3.80786	0.615375	1.24938	-0.619468	0.291474
8	7	-0.894286	0.286157	-0.113192	-0.271526	2.6696	3.72182	0.370145	0.851084	-0.392048	-0.41043	-0.705117	-0.110452
9	9	-0.338262	1.11959	1.04437	-0.222187	0.499361	-0.246761	0.651583	0.0695386	-0.736727	-0.366846	1.01761	0.83639
10	10	1.44904	-1.17634	0.91386	-1.37567	-1.97138	-0.629152	-1.42324	0.0484559	-1.72041	1.62666	1.19964	-0.67144
11	10	0.384978	0.616109	-0.8743	-0.0940186	2.92458	3.31703	0.470455	0.538247	-0.558895	0.309755	-0.259116	-0.326143
12	10	1.25	-1.22164	0.38393	-1.2349	-1.48542	-0.75323	-0.689405	-0.227487	-2.09401	1.32373	0.227666	-0.242682
13	11	1.06937	0.287722	0.828613	2.71252	-0.178398	0.337544	-0.0967169	0.115982	-0.221083	0.46023	-0.773657	0.323387
14	12	-2.79185	-0.327771	1.64175	1.76747	-0.136588	0.807596	-0.422911	-1.90711	0.755713	1.15109	0.844555	0.792944
15	12	-0.752417	0.345485	2.05732	-1.46864	-1.15839	-0.0778498	-0.608581	0.00360348	-0.436167	0.747731	-0.793981	-0.770407
16	12	1.10322	-0.0402962	1.26733	1.28909	-0.735997	0.288069	-0.586057	0.18938	0.782333	-0.267975	-0.450311	0.936708
17	13	-0.436905	0.918966	0.924591	-0.727219	0.915679	-0.127867	0.707642	0.0879624	-0.665271	-0.73798	0.324098	0.277192
18	14	-5.40126	-5.45015	1.1863	1.73624	3.04911	-1.76341	-1.55974	0.160842	1.23309	0.345173	0.91723	0.970117
19	15	1.49294	-1.02935	0.454795	-1.43803	-1.55543	-0.720961	-1.08066	-0.0531271	-1.97868	1.63808	1.07754	-0.632047
20	16	0.694885	-1.36182	1.02922	0.834159	-1.19121	1.30911	-0.878586	0.44529	-0.446196	0.568521	1.01915	1.29833
21	17	0.962496	0.328461	-0.171479	2.1092	1.12957	1.69604	0.107712	0.521502	-1.19131	0.724396	1.69033	0.406774
22	18	1.16662	0.50212	-0.0673003	2.26157	0.428804	0.0894735	0.241147	0.138082	-0.989162	0.922175	0.744786	-0.531377
23	18	0.247491	0.277666	1.18547	-0.0926025	-1.31439	-0.150116	-0.946365	-1.61794	1.54407	-0.829881	-0.5832	0.524933
24	22	-1.94653	-0.0449005	-0.40557	-1.01306	2.94197	2.95505	-0.0630631	0.855546	0.0499669	0.573743	-0.0812565	-0.215745
25	22	-2.07429	-0.121482	1.32202	0.410008	0.295198	-0.959537	0.543985	-0.104627	0.475664	0.149451	-0.856566	-0.180523

Format Resize Background color Column min/max Save and Close Close

Fig. 3. Fragment of "Credit Card Fraud Detection" dataset

proposed learning algorithm were equal: $\eta^{\bullet} = 0.99$. After 10 epochs, the training process was stopped and then the proposed system was tested.

The idea of detecting fraud operations is that a wavelet-neuro autoencoder trained only in normal operations, and when a fraud operation is submitted, the

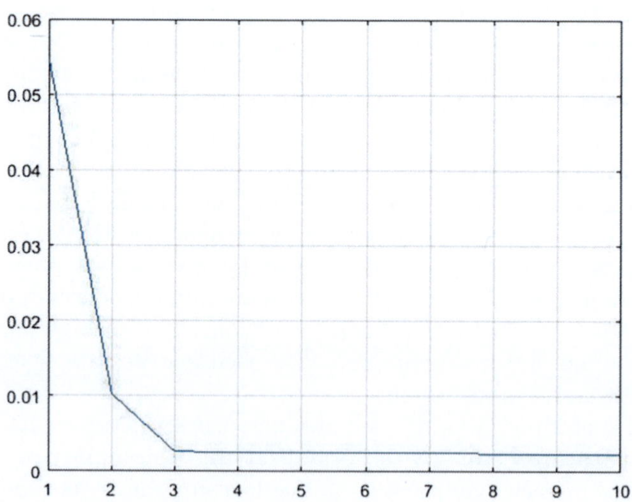

Fig. 4. The plot of mse change during the training process

recovery of the vector which describes such operation will be with a greater mse than non-fraud operation.

The following metrics have been selected to assess the quality of the proposed approach:

$$Accuracy = \frac{TP + TN}{TP + TN + FP + FN} \quad (8)$$

where TP is a true positive decision; TN is a true negative decision; FP is a false positive decision; FN is a false negative decision,

$$Precision = \frac{TP}{TP + FP}, \quad (9)$$

$$Recall = \frac{TP}{TP + FN}. \quad (10)$$

The Table 1 shows the results of experiment with different inputs number in the hidden layer of the wavelet-neuro authoencoder.

Table 1. Classification quality assessment for a dataset Credit Card Fraud Detection

Inputs number of hidden layer	Precision	Recall	Accuracy
2	93%	97%	92%
3	95%	96%	94%
5	96%	96%	95%
10	94%	96%	93%

Figure 5 shows the detection results of fraud transaction (yellow marker is fraud transactions and blue marker is normal transactions). Transaction visualization based on two hidden layer inputs, which are essentially the main components, is performed. For each transaction from test dataset the mse of recovery features vector was evaluation. Based on the analysis of the obtained results, the mse threshold for determining fraud operations was obtained and was equal 0.0111.

Figure 6 shows the ROC curve which allows to evaluate the quality of transactions classification.

Thus, the results show that when the number of hidden layer inputs is equal to 5, we get the best result both in Accuracy and in Precision and Recall balance. In comparison with existing approaches, the proposed auto-encoder has a smaller number of customizable parameters and a similar quality solution, which allows reducing the number of epochs for training and using such system in on-line mode.

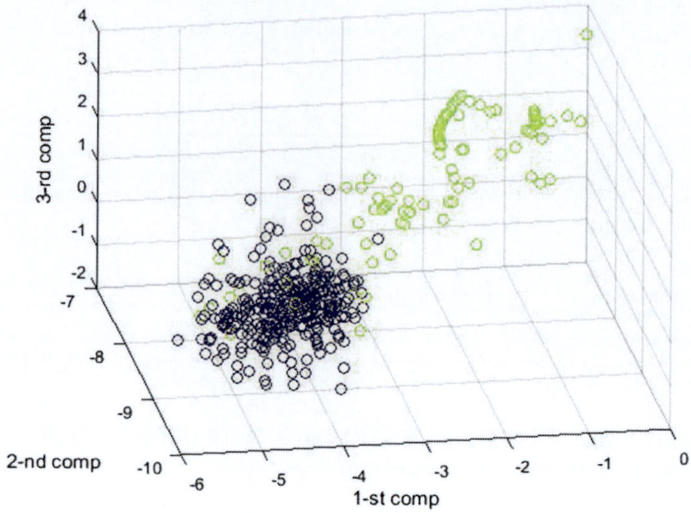

Fig. 5. The results of fraud detection based on wavelet-neuro authoencoder

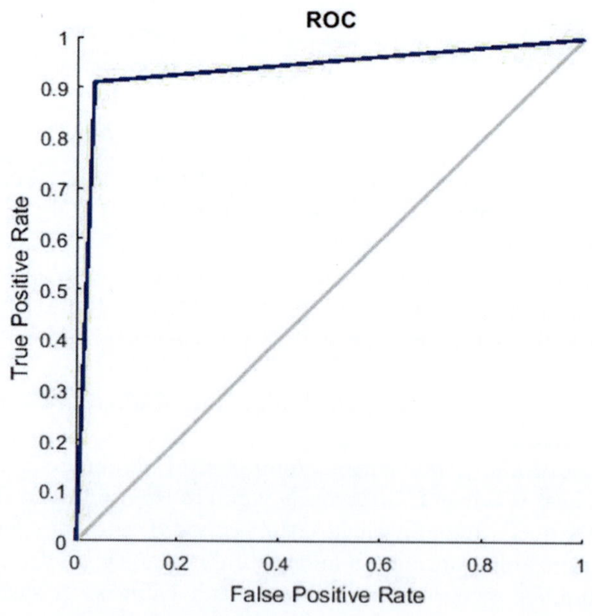

Fig. 6. ROC curve

5 Conclusions

The main goal of data and risk analytics is to delineate hidden dependences and use them to make informed decisions in a variety of tasks. Credit card fraud, operation fraud are escalating significantly with the advancement of the modernized technology and become an easy target for fraudulent. Such cases is a severe problem in the financial service and costs very expensive. Nowadays there are a lot of methods for fraud detection, which is based on different approaches of machine learning algorithms. Not all methods allow solving fraud detection problem based on small and big training dataset simultaneously.

For solving this problem, we have proposed the architecture of wavelet-neuro autoencoder and its learning algorithm. Such wavelet-neuro autoencoder is based on wavelet neuron, which is a modified version of neo-fuzzy neuron. Such neuron has simple arhitecture and can be used as building block for stacked autoencoders. The advantages of wavelet neuron is its activation functions, which can be tuned during learning process all system. So, the tuning parameters of wavelet-neuro autoencoder are not only synaptic weights and also parameters of adaptive wavelet activation functions. The tuning parameters of adaptive wavelet activation function are width, center and shape. Proposed wavelet-neuro autoencoder allows solving the fraud detection problem based on both big dataset and small dataset cases. The effectiveness of proposed approach was examined based on real-world dataset. In comparison with existing approaches, the proposed autoencoder has a smaller number of customizable parameters and a similar quality solution, which allows reducing the number of epochs for training and using such system in on-line mode.

References

1. Benchaji, I., Douzi, S., Ouahidi, B.: Using Genetic Algorithm to Improve Classification of Imbalanced Datasets for Credit Card Fraud Detection: Methods and Protocols (2019)
2. Bengio, Y., Lamblin, P., Popovici, D., Larochelle, H.: Greedy layer-wise training of deep networks. In: Neural Information Processing Systems, Cambridge, MA, USA, pp. 153–160 (2007)
3. Bodyanskiy, Y., Vynokurova, O., Yegorova, E.: Radial-basis-fuzzy-wavelet-neural network with adaptive activation-membership function. Int. J. Artif. Intell. Mach. Learn. **8**, 9–15 (2008)
4. Bodyanskiy, Y., Lamonova, N., Pliss, I., Vynokurova, O.: An adaptive learning algorithm for a wavelet neural network. Expert Syst. **22**, 235–240 (2005). https://doi.org/10.1111/j.1468-0394.2005.00314.x
5. Bolton, R., Hand, D.: Unsupervised profiling methods for fraud detection. In: Conference on Credit Scoring and Credit Control, vol. 7, pp. 5–7 (2001)
6. Bolton, R., Hand, D.: Statistical fraud detection: a review. Stat. Sci. **17**, 235–249 (2002). https://doi.org/10.1214/ss/1042727940
7. Breunig, M.M., Kriegel, H.P., Ng, R.T., Sander, J.: LOF: identifying density-based local outliers. In: Proceedings of the 2000 ACM SIGMOD International Conference on Management of Data, pp. 93–104 (2000). https://doi.org/10.1145/335191.335388

8. Dornadula, V.N., Geetha, S.: Credit card fraud detection using machine learning algorithms. Procedia Comput. Sci. **165**, 631–641 (2019). https://doi.org/10.1016/j.procs.2020.01.057
9. Kazemi, Z., Zarrabi, H.: Using deep networks for fraud detection in the credit card transactions. In: 2017 IEEE 4th International Conference on Knowledge-Based Engineering and Innovation (KBEI), pp. 0630–0633 (2017). https://doi.org/10.1109/KBEI.2017.8324876
10. Kingma, D., Welling, M.: Auto-encoding variational bayes (2014)
11. Levy, A., Lindenbaum, M.: Sequential Karhunen-Loeve basis extraction and its application to images. In: EEE Transactions on Image Processing, pp. 1371–1374 (2000). https://doi.org/10.1109/83.855432
12. Liu, F.T., Ting, K.M., Zhou, Z.H.: Isolation forest. In: 2008 Eighth IEEE International Conference on Data Mining, pp. 413–422 (2008). https://doi.org/10.1109/ICDM.2008.17
13. Maniraj, S., Saini, A., Ahmed, S., Sarkar, S.: Credit card fraud detection using machine learning and data science. Int. J. Eng. Res. **08**, 110–115 (2019). https://doi.org/10.17577/IJERTV8IS090031
14. Dhankhad, S., Mohammed, E., Far, B.: Supervised machine learning algorithms for credit card fraudulent transaction detection: a comparative study. In: IEEE Annals of the History of Computing (2018). https://doi.org/10.1109/IRI.2018.00025
15. Pumsirirat, A., Yan, L.: Credit card fraud detection using deep learning based on auto-encoder and restricted Boltzmann machine. In. J. Adv. Comput. Sci. Appl. **9**, 18–25 (2018). https://doi.org/10.14569/IJACSA.2018.090103
16. Ranzato, M., Poultney, C., Chopra, S., LeCun, Y.: Efficient learning of sparse representations with an energy-based model. In: Proceedings of Neural Information Processing Systems, pp. 1137–1144 (2007)
17. Roy, A., Sun, J., Mahoney, R., Alonzi, L., Adams, S., Beling, P.: Deep learning detecting fraud in credit card transactions. In: Systems and Information Engineering Design Symposium (2018). https://doi.org/10.1109/sieds.2018.8374722
18. Vincent, P., Larochelle, H., Bengio, Y., Manzagol, P.: Extracting and composing robust features with denoising autoencoders. In: Proceedings of the Twenty-Fifth International Conference, pp. 1096–1103 (2008). https://doi.org/10.1145/1390156.1390294

Sound Field Reconstruction from Incomplete Data by Solving Fuzzy Relational Equations

Olexiy Azarov, Leonid Krupelnitskyi, and Hanna Rakytyanska(✉)

Vinnytsia National Technical University, Vinnytsia, Ukraine
{azarov2,rakit}@vntu.edu.ua, krupost@gmail.com

Abstract. The approach to solving inverse problems of source identification in acoustics is proposed based on fuzzy relational calculus. The compositional rule of inference connects the real and observed fuzzy acoustic image using the relationship matrix, which reflects the degree of completeness of the microphone array measurement data. The fuzzy model of the acoustic field is based on 3D membership functions, for which the degree of membership decreases in proportion to the square of the distance to the source. The problem of reconstructing the acoustic field is formulated as the problem of inverse logical inference. The method for reconstructing the acoustic field from incomplete data is proposed based on solving fuzzy relational equations. The problem consists in finding such a number of sound sources, their locations and powers, which minimize the difference between the model and observed fuzzy acoustic image. The solutions of the equation system represent the variants of the acoustic field reconstruction in the form of the main acoustic surface and a set of secondary acoustic surfaces. The main acoustic surface is generated by the least number of sources. The set of secondary acoustic surfaces represents the variants of the sound field reconstruction generated by the upper solutions for the number of sources. Since the sources distribution is completely determined by the properties of the solution set, the proposed approach allows avoiding the generation and selection of candidate sources, that provides simplification of the reconstruction process and reduction of time costs. The genetic and neural algorithm provides accurate and fast reconstruction of the acoustic field for an unknown number of sources and their configuration.

Keywords: Inverse problems in acoustics · Source identification problem · Sound field reconstruction · Solving fuzzy relational equations

1 Introduction

Acoustic vision is a standard technique for localization, tracking and separation of sound sources on the basis of spatial and temporal observations obtained with

S. Babichev et al. (Eds.): ISDMCI 2020, AISC 1246, pp. 547–566, 2021.
https://doi.org/10.1007/978-3-030-54215-3_35

the help of microphone arrays [6,8,28]. The physical principles of constructing microphone arrays with a limited number of microphones cause the problem of incomplete data. For irregular focal point spacing, the theory of inverse problems is adapted for sound field extrapolation with the aim to increase the sound image resolution [10]. The problem of sound field reconstruction, which consists in identifying the coordinates and powers of sound sources, relates to the class of inverse problems with constraints on spatial distribution of the sources. In processing acoustic stream data, the additional constraints are imposed on the reconstruction time. However, according to the basic principles of acoustic vision, the time window of the microphone array is determined by the time of solving the inverse problem of sound field reconstruction [5].

Traditionally, the wave theory models in the form of differential equations are used to describe the sound field [28]. Inverse problems for identifying sound sources are solved on the basis of regularization methods or time reversal methods [28]. If the number of sources and their spatial distribution is unknown, and the measurement data is incomplete, then the application of the traditional methods becomes problematic [14]. At the same time, relying only on available experimental data, experienced acoustic experts can reconstruct events in an acoustic field based on knowledge about the laws of propagation of acoustic energy. In this case, understanding the process requires the use of data mining techniques, which should provide a reliable explanation of the measurement results in the form of the reconstruction of events in time and space [40]. Therefore, it is advisable to use the principles of fuzzy modeling to form the surface of an acoustic image with further use of inverse fuzzy inference for reconstructing the acoustic field from the observed data [15,31].

2 Literature Review

Traditionally, the inverse problem of acoustic field reconstruction is solved with an $l1$ / $l2$ regularization ($l2$ – Tikhonov regularization) [5,10,24]. The inverse problem is defined as an optimization problem, which consists in minimizing the difference between the observed and model acoustic image. For various configurations of acoustic sources, high resolution imaging can only be achieved using sparsity constraints [9,20,24,25]. The sparsity constraint is derived from the sparse distribution of the sources in the acoustic field. The search for the sparse solutions of the optimization problem is carried out by the penalty function method [20]. Sparse regularization methods imply that the number of sparsely distributed sources is known beforehand [25]. To minimize the effects of unavoidable uncertainties arising from the physics of the experiment, a Bayesian solution is used for finding the most probable variant of sound field reconstruction [11,23].

The time reversal method is used to reconstruct the acoustic field of moving sources or sources with transient emission [7,16,21,26]. Inverse problems of sources identification amount to solving the optimization problems; whereas, the time reversal approach relies on the process reversibility admitted by the governing laws of sound energy propagation [16]. To properly refocus the time reversal solutions under uncertainty, an augmentation can be achieved by solving the inverse problem at the selected time moments [21,26]. In practice, a hybrid approach is used to reconstruct the acoustic field of moving or transient sources [21]. At the first stage, the time reversal method determines the approximate location of the sources. At the second stage, the sound field reconstruction is carried out by the sparse regularization methods for the given locations with further interpolated time-domain source identification [21].

Sound field models with interval and fuzzy parameters are used to quantify the effects of limited measurement data [12,13,36]. The direct problem for acoustic pressure identification is modeled by using the discretized convolution integral relationship and the Green's kernel function matrix of the system [13]. To overcome the ill-posedness of the problem in the presence of interval data, the inverse process is restored through the SVD-solutions for an array of evenly distributed point sources [12,13]. Fuzzy variables are associated with the expert opinions aimed to identify the sources with knowing only the bounds of values for the sound field parameters [36]. In this case, events in the sound field are described with the help of both acoustic and visual features [37]. For the models with fuzzy parameters, the inverse process is constructed through the fuzzy decomposition [36]. By using fuzzy level-cut strategy, the original fuzzy parameters are decomposed into the subintervals with the small degree of uncertainty. Transformation functions, which use the subinterval matrix, are approximated by the Taylor series; whereas, the inversion of the subinterval matrix is approximated by the Neumann series [36]. Nevertheless, while representing the formalisms to cope with the problem of incomplete data, the sound field models with interval and fuzzy parameters require the same constraints imposed on the number of sources and their configuration as the classical sparse regularization methods [9,20,24,25].

For the known number of sparsely spaced sources, the gradient descent method can be adapted to jointly reconstruct sources powers and coordinates [18,39]. In this case, the least-squares reconstruction is carried out by adaptively specifying the sound pressure levels at the selected points of a target surface. The coordinates and powers of sound sources are tuned from sparse-view sampling data through the multiple selection of initial conditions for the gradient search [39]. The objective function of the algorithm is modified by adding total variation values, which are considered as controllable variables aimed to select the most probable distribution of sound pressure levels [39]. When the sources are sparsely distributed, a more frequent image update can be obtained that allows real-time imaging. If the sources are closely spaced, the algorithm requires significant time costs to reconstruct the complete solution set of the inverse problem in the form of variants of sources distribution with the equal sound energy.

Under conditions of incomplete measurement data, reconstruction of the acoustic field is complicated if the number of sources and their configuration are unknown [19]. In this case, with an increase in the dimension of the problem, real time reconstruction of the acoustic image becomes problematic [39].

The number of sources is automatically estimated using fuzzy clustering methods based on the analysis of the density of peak values in the sound field [1,17,35]. To reconstruct the sound field generated by many sources, the methods of off-line evolutionary search are used. In order to correctly separate closely spaced sources, a relationship matrix is selected that determines the influence of sources of each class of sound power among the total number of sources whose energy is summed. Tuning the relationship matrix using experimental data provides the optimal number of sound power clusters. Finally, for the found number of clusters, the sources are restored by the $l1/l2$ regularization methods. As a result, reconstruction of the acoustic field requires significant computational costs, since it reduced to the generation and selection of candidate sources, which provide the minimum difference between the observed and model acoustic field in real experiments with a large aperture of the microphone array.

3 Problem Statement

In this paper, we propose an approach to solving inverse problems of source identification in acoustics on the basis of the apparatus of fuzzy relational equations [22,38]. The approach is special due to the resolution of systems of fuzzy relational equations using genetic and neural technology [29–31]. Such an approach, unlike exact analytical methods [22], allows solving systems of equations that describe complex acoustic scenes, where the decision elements are the number, coordinates and powers of sound sources. The use of fuzzy relational equations in telecommunication technologies was considered in [2,3].

A fuzzy model of the acoustic field is based on 3D membership functions of sound sources on a universal set of microphone array nodes. Following the fundamental laws of the sound theory [8,28], the degree of membership decreases in proportion to the square of the distance to the source. The resolution of the microphone array is determined by the system of fuzzy relationship matrices, which reflect the degree of completeness of the measurement data between the rows of the real and observed image [27,32]. The microphone conversion function is modeled by interpreting Zadeh's compositional rule of inference [38], which connects the fuzzy acoustic levels of the real and observed acoustic image using the relational matrices. Then the problem of reconstructing the acoustic field is formulated as an inverse logical inference problem and reduces to solving a system of fuzzy relational equations [29–31].

The problem of solving a system of fuzzy relational equations is formulated as an optimization problem, the solution set of which determines the variants of sources distribution in the sound field. Reconstruction of the acoustic field consists in selecting such a number of sources, their locations and powers, which provide the minimum distance between the model and observed fuzzy acoustic image. Following [22,29–31], the number of sound sources is completely characterized by the unique least solution and a set of upper solutions of fuzzy relational equations. In this case, the problem of reconstructing the sound field has a solution set, which is completely characterized by the main acoustic surface and a set of secondary acoustic surfaces. The main acoustic surface is generated by the least number of sources. The set of secondary acoustic surfaces yields a field reconstruction variants generated by the set of upper solutions for the number of sources. For the least solution, the powers of sound sources are maximal. With the growing number of sources, the powers of sources decrease to provide the equivalent summary sound energy levels at the sparse measurement points [8,28]. Since the distribution of sources in the sound field is completely determined by the properties of the solution set, the proposed approach allows avoiding the generation and selection of candidate sources [33,34]. In this case, the resolution of fuzzy relational equations guarantees a simplification of the reconstruction process, which leads to a reduction in time costs. The optimization problem is solved using the genetic and neural approach [29–31]. Genetic algorithm provides fast but rough reconstruction of the acoustic field. Neuro-fuzzy network provides adaptive correction of the acoustic surface while new experimental data is appearing.

The aim of the research is to develop a method for reconstructing an acoustic field from incomplete data based on the solution of a system of fuzzy relational equations. The method should provide accurate and fast reconstruction of the acoustic field for an unknown number of sources and their configuration. To achieve the goal, it is necessary to develop a fuzzy model of the acoustic field in the form of a system of fuzzy relational equations; to develop a genetic-neural algorithm for solving fuzzy relational equations when reconstructing complex acoustic scenes.

4 Models and Methods of Sound Field Reconstruction

4.1 Fuzzy Model of the Acoustic Field

The conversion function of the microphone array represents the relationship between the acoustic image at the input (the real image to be restored) and the output (the observed image from the measurement results). The acoustic image is built based on a mesh with n rows and m columns. To provide the resolution $n \times m$ in the coordinate system (x, y), the nodes (u_i, v_j) of the acoustic image are formed by discretization the ranges $[\underline{x}, \overline{x}]$ and $[\underline{y}, \overline{y}]$ on n and m levels, respectively:

$$u_i = \underline{x} + (i-1)\Delta_x, \Delta_x = (\overline{x} - \underline{x})/(n-1), i = \overline{1,n},$$
$$v_j = \underline{y} + (j-1)\Delta_y, \Delta_y = (\overline{y} - \underline{y})/(m-1), j = \overline{1,m}.$$

Here $\underline{x}$ ($\overline{x}$) and $\underline{y}$ ($\overline{y}$) are the minimum (maximum) value of the coordinate x (y).

The real and observed fuzzy acoustic image can be represented in the form of fuzzy levels of sound energy in the nodes (u_i, v_j):

$$\boldsymbol{\mu}^A(x,y) = \begin{bmatrix} \mu_{11}^A \cdots \mu_{1m}^A \\ \cdots \; \cdots \; \cdots \\ \mu_{n1}^A \cdots \mu_{nm}^A \end{bmatrix}, \boldsymbol{\mu}^B(x,y) = \begin{bmatrix} \mu_{11}^B \cdots \mu_{1m}^B \\ \cdots \; \cdots \; \cdots \\ \mu_{n1}^B \cdots \mu_{nm}^B \end{bmatrix}.$$

Here $\boldsymbol{\mu}^A(x,y)$ and $\boldsymbol{\mu}^B(x,y)$ are the matrices of fuzzy levels of sound energy for the real and observed image; μ_{ij}^A and μ_{ij}^B are the fuzzy acoustic levels in the point (u_i, v_j), $i = \overline{1,n}$, $j = \overline{1,m}$, of the real and observed acoustic image.

Relations between the real and observed image are defined with the help of the system of matrices $\mathbf{R}_1, ..., \mathbf{R}_n$: $\mathbf{R}_i \subseteq \mu_{ij}^A \times \mu_{ik}^B = [r_{jk}^i, i = \overline{1,n}, j,k = \overline{1,m}]$, where $\mathbf{R}_i$ is the matrix of relations between the acoustic levels μ_{ij}^A and μ_{ik}^B for the i-th row of the real and observed acoustic image. Elements of the matrix $\mathbf{R}_i$ are defined as follows. For the main diagonal elements, $r_{jk}^i = 1$, $j = k$, if the point (u_i, v_j) is the node of the microphone array; $r_{jk}^i = 0$, if the measurement data in the point (u_i, v_j) are absent. If $j \neq k$, then $r_{jk}^i = 0$. In this case, the main diagonal elements of the matrix $\mathbf{R}_i$ are the numbers $r_{jk}^i \in \{0, 1\}$, $j = k$, that characterize the sparsity/completeness of the microphone array measurement data in the point (u_i, v_j) for the i-th row of the acoustic image.

Given matrices $\mathbf{R}_i$, $i = \overline{1,n}$, the microphone array conversion function can be interpreted with the help of Zadeh's compositional rule of inference [38]:

$$\boldsymbol{\mu}_i^A(x,y) \circ \mathbf{R}_i = \boldsymbol{\mu}_i^B(x,y), \tag{1}$$

where $\boldsymbol{\mu}_i^A(x,y) = (\mu_{i1}^A, ..., \mu_{im}^A)$ and $\boldsymbol{\mu}_i^B(x,y) = (\mu_{i1}^B, ..., \mu_{im}^B)$ are the vectors of fuzzy acoustic levels for the i-th row of the real and observed acoustic image; $\circ$ is the operation of *max-min* composition [38].

The following system of fuzzy relational equations

$$\mu_{ik}^B(x,y) = \underset{j=\overline{1,m}}{max}[min(\mu_{ij}^A,\ r_{jk}^i)], \tag{2}$$

is derived from relation (1). Taking into consideration the properties of the matrices $\mathbf{R}_i$, the system (2) can be rewritten in the shortened form:

$$\mu_{ij}^B(x,y) = min(\mu_{ij}^A,\ r_{jj}^i), \tag{3}$$

Fuzzy model of the acoustic field is built based on the fundamental laws of the sound theory [8,28]. Fuzzy acoustic level of the sound strength can be defined as a ratio to the maximum level of the sound source power [28]. Then, fuzzy levels of the observed image $\boldsymbol{\mu}^B(x,y)$ are defined as follows:

$$\mu_{ij}^B = \frac{10}{w_{\max}} \lg \frac{I_\Sigma(u_i, v_j)}{I_0}, \tag{4}$$

where $w_{\max} = 120$ dB is the maximum sound source power; $I_\Sigma(u_i, v_j)$, Wt/m^2, is the measured summary sound energy at the point (u_i, v_j); $I_0 = 10^{-12}$ Wt/m^2 is the audibility threshold intensity.

While constructing the fuzzy model of the acoustic field it is supposed that the powers of the sources are subject to summation, and the degree of membership decreases in proportion to the square of the distance to the source [8,28]. Then, fuzzy levels of the real image $\boldsymbol{\mu}^A(x, y)$ are defined as follows:

$$\mu_{ij}^A(N) = \frac{10}{w_{\max}} \lg \left(\frac{1}{I_0} \cdot \sum_{p=1}^{N} I_p(u_i, v_j) \right), \tag{5}$$

where $I_p(u_i, v_j)$, Wt/m^2, is the strength of the p-th sound source at the point (u_i, v_j); N is the number of sources in the sound field.

Following [8,28], 3D membership function of the p-th sound source on the universal set of nodes (u_i, v_j) can be defined in the form:

$$I_p(u_i, v_j) = \frac{w_p}{4\pi \left[(u_i - x_p)^2 + (v_j - y_p)^2 \right]}, \tag{6}$$

$$\mu_{ij}^A(p) = \frac{10}{w_{\max}} \lg \frac{I_p(u_i, v_j)}{I_0}, \tag{7}$$

where x_p, y_p and w_p are the coordinates and power of the p-th sound source; $\mu_{ij}^A(p)$ is the degree of membership of the node (u_i, v_j) to the location of the sound source with the coordinates (x_p, y_p) and the power w_p.

It is supposed that $\mu_{ij}^A(x = x_p, y = y_p, w = w_{\max}) = 1$.

Correlations (1–7) define the fuzzy model of the acoustic field as follows:

$$\boldsymbol{\mu}^B(x, y) = f_1(\boldsymbol{\mu}^A(x, y), \mathbf{R}_1, ..., \mathbf{R}_n), \tag{8}$$

$$\boldsymbol{\mu}^A(x, y) = f_2(x, y, N, \mathbf{X}, \mathbf{Y}, \mathbf{W}), \tag{9}$$

where $\mathbf{X} = (x_1, ..., x_N)$, $\mathbf{Y} = (y_1, ..., y_N)$ are the vectors of sound sources coordinates; $\mathbf{W} = (w_1, ..., w_N)$ is the vector of sound sources powers; f_1, f_2 are the operators of inputs-output connection, corresponding to formulas (1–7).

4.2 Optimization Problem for Sound Field Reconstruction

The observed acoustic image obtained with the help of the microphone array represents the surface of fuzzy levels $\boldsymbol{\mu}^B = [\mu_{ij}^B]$ of acoustic field energy $I_\Sigma(u_i, v_j)$ at the points (u_i, v_j), $i = \overline{1, n}$, $j = \overline{1, m}$. The real acoustic image, which is subject to restoration, represents the surface of fuzzy levels $\boldsymbol{\mu}^A = [\mu_{ij}^A]$, generated by the summary strength of the sound sources $I_p(u_i, v_j, x_p, y_p, w_p)$, $p = \overline{1, N}$, at the points (u_i, v_j). Given the matrices $\mathbf{R}_i$, $i = \overline{1, n}$, and the observed image $\boldsymbol{\mu}^B$, the problem of sound field reconstruction is reduced to solving the system of fuzzy relational equations (2).

Following [29–31], the problem of solving the system of fuzzy relational equations (2) is formulated as follows. It is necessary to find the number of sources N and the vectors of sound sources coordinates and powers $\mathbf{X}$, $\mathbf{Y}$, $\mathbf{W}$, which provide the minimum distance between the observed and model fuzzy acoustic image:

$$F = \sum_{i=1}^{n}\sum_{j=1}^{m}[\hat{\mu}_{ij}^{B}(I_{\Sigma}(u_i, v_j)) - \mu_{ij}^{B}(\mu_{ij}^{A}(u_i, v_j, N, \mathbf{X}, \mathbf{Y}, \mathbf{W}), R_i)]^2 = \min_{N,\mathbf{X},\mathbf{Y},\mathbf{W}}. \tag{10}$$

In the general case, the system (2) has a solution set $S_1(\mathbf{R}, \boldsymbol{\mu}^B)$ for the number of sources N, which determines a solution set $S_2(\mathbf{R}, \boldsymbol{\mu}^B)$ for the matrix of fuzzy acoustic levels $\boldsymbol{\mu}^A$.

Following [22,29–31], the solution set $S_1(\mathbf{R}, \boldsymbol{\mu}^B)$ is completely characterized by the unique least solution $\underline{N}$ and the set of upper solutions $S_1^* = \{\overline{N}_l,\ l = \overline{1, L}\}$:

$$S_1(\mathbf{R}, \boldsymbol{\mu}^B) = \bigcup_{\overline{N}_l \in S_1^*} [\underline{N}, \overline{N}_l], \tag{11}$$

where the union is taken over all $\overline{N}_l \in S_1^*$.

The set of sound field reconstructions $S_2(\mathbf{R}, \boldsymbol{\mu}^B)$ is completely characterized by the main acoustic surface $\underline{\boldsymbol{\mu}}^A$ and the set of secondary acoustic surfaces $S_2^* = \{\overline{\boldsymbol{\mu}}_l^A,\ l = \overline{1, L}\}$. The main acoustic surface $\underline{\boldsymbol{\mu}}^A$ is generated by the least solution $\underline{N}$, and the set of secondary acoustic surfaces $\overline{\boldsymbol{\mu}}_l^A$ is generated by the set of upper solutions $\overline{N}_l$:

$$S_2(\mathbf{R}, \boldsymbol{\mu}^B) = \bigcup (\underline{\boldsymbol{\mu}}^A(\underline{N}), \bigcup_{\overline{\boldsymbol{\mu}}_l^A \in S_2^*} \overline{\boldsymbol{\mu}}_l^A(\overline{N}_l)). \tag{12}$$

Here $\underline{\boldsymbol{\mu}}^A = [\underline{\mu}_{ij}^A]$ and $\overline{\boldsymbol{\mu}}_l^A = [\overline{\mu}_{ij,l}^A]$ are the matrices of fuzzy acoustic levels μ_{ij}^A for the least and upper reconstructions, where the union is taken over all $\overline{\boldsymbol{\mu}}_l^A \in S_2^*$.

The main acoustic surface $\underline{\boldsymbol{\mu}}^A$ is formed by the least set of sources which is characterized by the vector of sound powers $\mathbf{W}^* = (w_1^*, ..., w_{\underline{N}}^*)$ and the vectors of coordinates $\mathbf{X}^* = (x_1^*, ..., x_{\underline{N}}^*)$, $\mathbf{Y}^* = (y_1^*, ..., y_{\underline{N}}^*)$. The set of secondary acoustic surfaces $\overline{\boldsymbol{\mu}}_l^A$, $l = \overline{1, L}$, is formed by the upper sets of sources which are characterized by the vectors of sound powers $\mathbf{W}_l = (w_1^l, ..., w_{\overline{N}_l}^l)$ and the vectors of coordinates $\mathbf{X}_l = (x_1^l, ..., x_{\overline{N}_l}^l)$, $\mathbf{Y}_l = (y_1^l, ..., y_{\overline{N}_l}^l)$.

Following [29–31], sound field reconstruction is accomplished by way of solving a multiple optimization problem (10) and it begins with the search for its null solution.

Let us impose limitations on the number of sound sources as follows: $\underline{N}_w \leq N \leq \overline{N}_w$, where $\underline{N}_w$ ($\overline{N}_w$) is the minimum (maximum) number of sources the sound power of which is within the range $[\underline{w}, \overline{w}]$; $\underline{w}$ ($\overline{w}$) is the lower (upper) sound power of the sources to be restored.

As the null solution of the optimization problem (10) for the coordinates and powers of sound sources we designate $\mathbf{X}_0 = (x_1^0, ..., x_{N_0}^0)$, $\mathbf{Y}_0 = (y_1^0, ..., y_{N_0}^0)$, $\mathbf{W}_0 = (w_1^0, ..., w_{N_0}^0)$, where N_0 is the number of sound sources for the null solution, $\underline{N}_w \leq N_0 \leq \overline{N}_w$. The null acoustic surface $\boldsymbol{\mu}_0^A = [\mu_{ij,0}^A]$ generated by the null number of sources N_0 corresponds to the null solution $(\mathbf{X}_0, \mathbf{Y}_0, \mathbf{W}_0)$. For the null acoustic surface $\boldsymbol{\mu}_0^A$, the optimization criterion (10) takes the value of $F(\boldsymbol{\mu}_0^A(\mathbf{X}_0, \mathbf{Y}_0, \mathbf{W}_0)) = F_0$.

The lower bound $(\underline{N})$ is found in the range $[\underline{N}_w,\ N_0]$. The upper bound $(\overline{N}_l)$ for $l\ = 1$ is found in the range $[N_0,\ \overline{N}_w]$, and for $l\ > 1$ – in the range $[\underline{N},\ \overline{N}_w]$, where the maximal solutions $\overline{N}_k$, $k < l$, are excluded from the search space.

Let $\mathbf{X}_t\ = (x_1^t, ..., x_{N_t}^t)$, $\mathbf{Y}_t\ = (y_1^t, ..., y_{N_t}^t)$, $\mathbf{W}_t\ = (w_1^t, ..., w_{N_t}^t)$ be some t-th solution of optimization problem (10), where N_t is the number of sound sources for the t-th solution. The t-th acoustic surface $\boldsymbol{\mu}_t^A\ = [\mu_{ij,t}^A]$ generated by the number of sources N_t corresponds to the t-th solution $(\mathbf{X}_t, \mathbf{Y}_t, \mathbf{W}_t)$. Since for all $N \in S_1(\mathbf{R}, \boldsymbol{\mu}^B)$ we have the same value of criterion (10), then $F(\boldsymbol{\mu}_t^A(\mathbf{X}_t, \mathbf{Y}_t, \mathbf{W}_t)) = F_0$. While searching for lower bound $(\underline{N})$ it is suggested that $N_t \leq N_{t-1}$, $w_p^t \geq w_p^{t-1}$; and while searching for upper bounds $(\overline{N}_l)$ it is suggested that $N_t \geq N_{t-1}$, $w_p^t \leq w_p^{t-1}$.

Sound field reconstruction follows the rule [29–31]: if $N_t \neq N_{t-1}$, then $\underline{N}$ $(\overline{N}_l)$=N_t for the number of sources; $\mathbf{X}^*(\mathbf{X}_l)$=$\mathbf{X}_t$, $\mathbf{Y}^*(\mathbf{Y}_l)$=$\mathbf{Y}_t$,$\mathbf{W}^*(\mathbf{W}_l)$=$\mathbf{W}_t$ for the vectors of sound sources parameters; $\boldsymbol{\mu}^A(\overline{\boldsymbol{\mu}}_l^A)\ = \boldsymbol{\mu}_t^A$ for the set of restored acoustic surfaces. If $N_t\ = N_{t-1}$, then the search for the interval solution (11) is stopped. Sound field reconstruction will go on until the condition $\overline{N}_k \neq \overline{N}_l$, $k < l$, has been satisfied.

The genetic and neuro approach is proposed for solving optimization problem (10).

4.3 Genetic and Neuro Method of Sound Field Reconstruction

The chromosome needed in the genetic algorithm for solving the optimization problem (10) is defined as the vector-line of binary codes of the solutions x_p, y_p, w_p, $p\ = \overline{1,N}$, where the length of the chromosome is defined by the current number of sources. The crossover operation is carried out by way of exchanging genes inside each vector $\mathbf{X}$, $\mathbf{Y}$, $\mathbf{W}$. Fitness function is built based on criterion (10). The condition for the completion of the genetic algorithm is the absence of the new upper and lower bounds for the number of sources during a given period of time (number of iterations).

A neuro-fuzzy network isomorphic to the system of fuzzy relational equations (2) is shown in Fig. 1.

The network is designed so that the coordinates and powers of sound sources define the weights of arcs subject to tuning [29,31].

It is shown that the neuro-fuzzy network has four layers. Layer 1 determines coordinates (u_i, v_j), $i\ \overline{1,n}$, $j\ = \overline{1,m}$, of the acoustic field. Layer 2 represents 3D membership functions (6), (7) of the sound sources which generate sound energy I_p, $p\ = \overline{1,N}$. Layer 3 forms 3D fuzzy acoustic surface by finding the summary fuzzy acoustic level (5). Layer 4 represents the fuzzy acoustic level at the output of the microphone array. The arcs between 2th and 3th layers are weighted by the membership functions of the nodes (u_i, v_j) to the location of sound sources with coordinates (x_p, y_p) and powers w_p. The arcs between 3th and 4th layers are weighted by the sparsity/completeness relations r_{jj}^i.

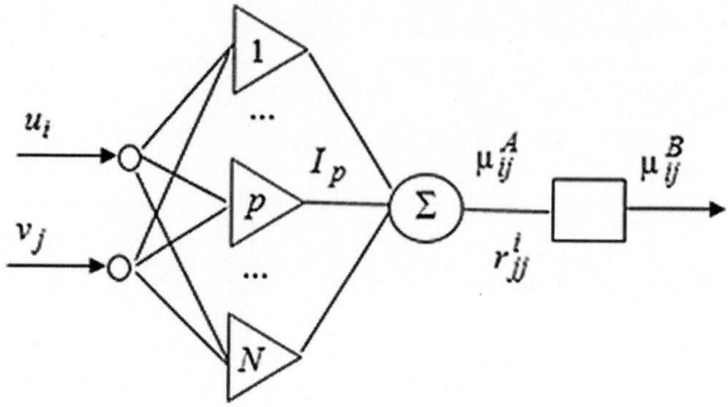

Fig. 1. Neuro-fuzzy model of the system of fuzzy relational equations

Thus, the problem of solving the system of fuzzy relational equations (2) is reduced to the problem of training the neuro-fuzzy network with the use of points $(u_i,\ v_j,\ \mu_{ij}^B)$, $i\ = \overline{1,n}$, $j\ = \overline{1,m}$.

For tuning the parameters of the neuro-fuzzy network, the recurrent relations are used:

$$x_p(t+1) = x_p(t) - \eta \frac{\partial E_t}{\partial x_p(t)}, \quad (13)$$

$$y_p(t+1) = y_p(t) - \eta \frac{\partial E_t}{\partial y_p(t)}, \quad (14)$$

$$w_p(t+1) = w_p(t) - \eta \frac{\partial E_t}{\partial w_p(t)}, \quad (15)$$

which minimize the criterion

$$E_t = \frac{1}{2}(\mu_{ij}^B(t) - \hat{\mu}_{ij}^B(t))^2,$$

where $\mu_{ij}^B(t)$ and $\hat{\mu}_{ij}^B(t)$ are the model and experimental output fuzzy acoustic levels at the t-th training step; $x_p(t)$, $y_p(t)$ and $w_p(t)$ are the sound sources coordinates and powers at the t-th training step; η is a parameter of training, which can be selected according to the results from [29].

Since the coordinate space of the microphone array is discrete, the partial derivatives included in (13)–(15) are obtained using finite differences [29,31].

5 Experiment

An open area of 250 × 250 m, where heavy vehicles are expected to pass, is controlled by acoustic vision. For the controlled area, the speed of vehicles does not exceed 30 km/h. The acoustic scene can be supplemented with short-term signals at the microphone level (shots, explosions, car signals, siren sounds).

Acoustic levels of the sources are in the range of 80–100 dB, i.e., the sound power range is $I_p \in [10^{-4},\ 10^{-2}]$ Wt/m^2. Since the terrain is open, the effect of multiple reflection of acoustic signals is not taken into account. It is assumed that the signal-to-noise ratio is 10:1.

The coverage of the controlled area with the microphone array is shown in Fig. 2.

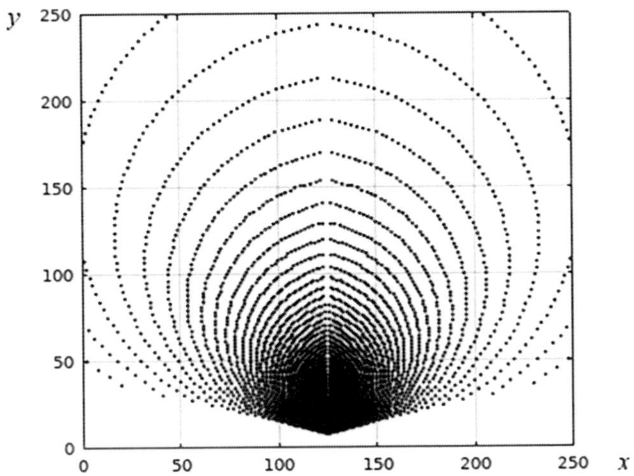

Fig. 2. Coverage of the controlled area with the microphone array

The microphone array is formed by a 32 × 32 matrix of microphones at a distance of 25 m [4]. Nodes of the microphone array are formed at the points of intersection of the rays with the scanning step of 1^0. The resolution of the microphone array is characterized by an irregular distribution of the nodes. The array resolution, which is 3615 points, decreases with the growth of the distance from the center of the array. The image is controlled at a distance of 50 m from the center in the far field of the microphone array, where the measurement data are incomplete.

Let us consider the direct problem to obtain the image $\boldsymbol{\mu}^B$ at the output of the microphone array with the help of the equation system (2). The microphone conversion matrix was formed on the basis of the matrix of coordinates of the rays intersection points (Fig. 2). Let the number of sources be $N = 15$. The acoustic surface was formed by a set of sources defined in the form of triples "coordinates $(\hat{x}_p,\ \hat{y}_p)$ - source power $(\hat{w}_p)$", $p = \overline{1, N}$:

{(22, 207, 6e-3), (58, 163, 2e-2), (65, 203, 5e-3), (81, 189, 7e-3), (94, 211, 2e-2), (32, 78, 3e-4), (43, 91, 1e-2), (102, 87, 2e-3), (114, 115, 7e-4), (146, 138, 2e-3), (171, 116, 7e-4), (180, 127, 3e-4), (205, 156, 4e-3), (212, 140, 3e-3), (218, 61, 5e-4)}

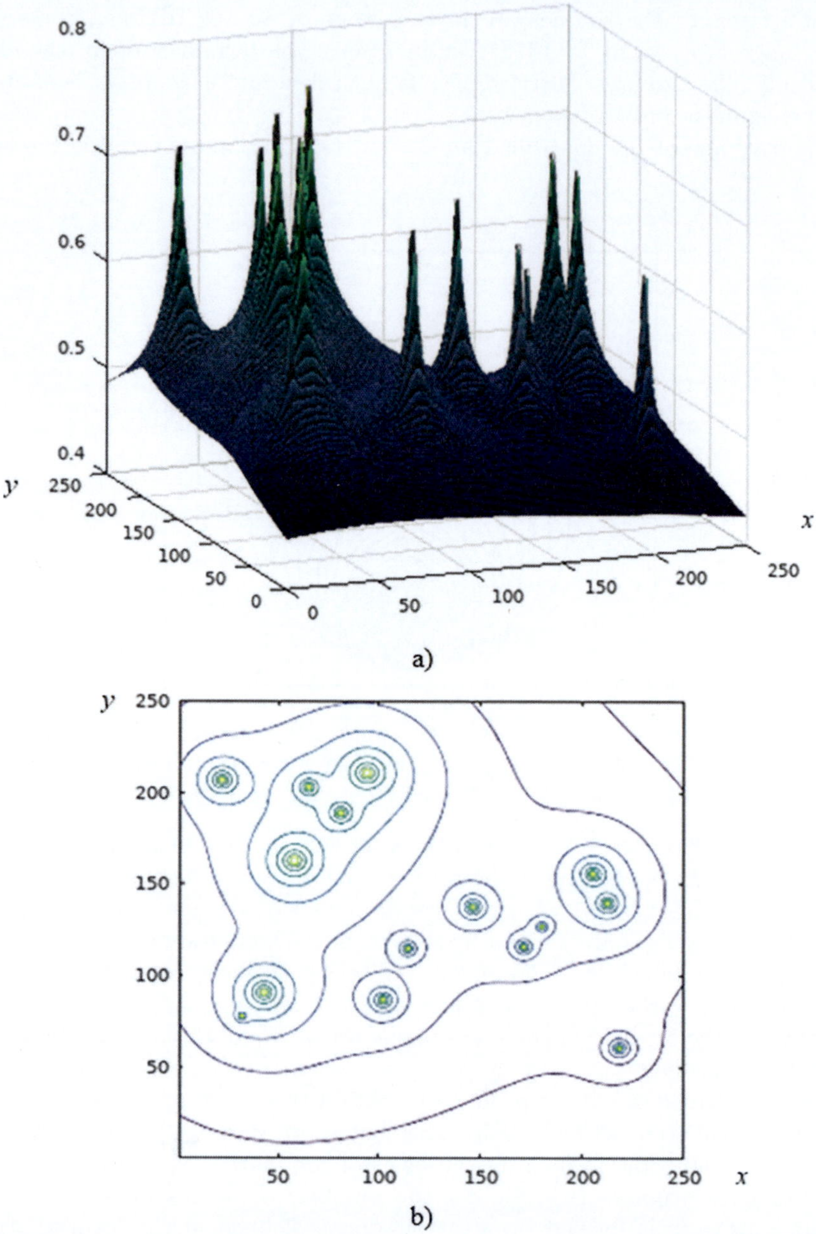

Fig. 3. The real acoustic image for $N = 15$ (a); location of the sources (b)

The real fuzzy acoustic image $\boldsymbol{\mu}^A$ at the input of the microphone array is shown in Fig. 3.

The observed fuzzy image $\boldsymbol{\mu}^B$ obtained as a result of the composition of the real fuzzy image $\boldsymbol{\mu}^A$ and the transformation matrix $\mathbf{R}$ at the output of the microphone array is shown in Fig. 4.

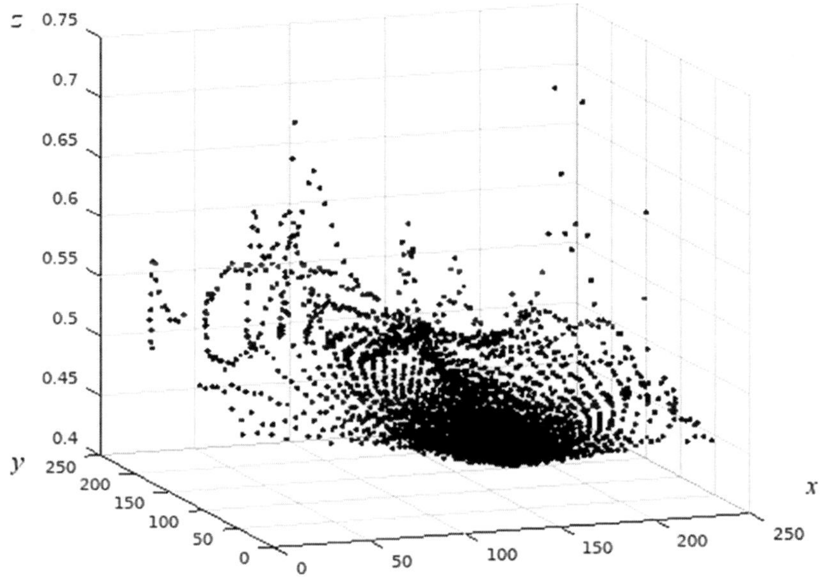

Fig. 4. Observed fuzzy acoustic image at the output of the microphone array

In measuring sound pressure at a point, the permissible error is ±1dB [4]. While the sound image $\boldsymbol{\mu}^B$ can be obtained by field measurements, in the experiment these data are synthesized by solving corresponding direct problem to provide a comparison of the real image $\boldsymbol{\mu}^A$ (Fig. 3) with the reconstructed solution set of the inverse problem (10).

Let us consider the problem of inverse fuzzy inference for reconstructing the acoustic field. Given the observed fuzzy image $\boldsymbol{\mu}^B$ and the transformation matrix $\mathbf{R}$, the real fuzzy image $\boldsymbol{\mu}^A$ should be reconstructed at the input of the microphone array to increase the image resolution to $250 \times 250 = 62500$ points.

Solution set for the number of sources is presented in Table 1. Solution set for the coordinates and powers of the restored sources is presented in Table 2.

For the observed image (Fig. 4), the system of fuzzy relational equations (2) has the least solution and a set of upper solutions, which define the variants of sound sources distribution in 12 areas of the sound field. Areas A1-A12 was formed based on the ranges of change in the coordinates of sound sources in the solutions of the equation system (2). For the areas A1-A12, the sound energy range was restored based on the ranges of change in the powers of sound sources in the solution set. Variants of sound sources configurations are defined by the lower and upper solutions for the number of sources in the solution set.

Table 1. Solution set for the number of sound sources

Area	Coordinates	Sound energy range	Number of sources
A1	[15, 25], [198, 211]	[4e-3, 7e-3]	[1, 3]
A2	[52, 60], [161, 167]	[3e-3, 2e-2]	[1, 4]
A3	[58, 87], [181, 210]	[1e-3, 8e-3]	[2, 7] OR [2, 9]
A4	[91, 96], [208, 217]	[3e-3, 2e-2]	[1, 3]
A5	[29, 45], [72, 93]	[1e-4, 1e-2]	[2, 12]
A6	[98, 107], [83, 92]	[6e-4, 2e-3]	[1, 3]
A7	[109, 123], [112, 118]	[1e-4, 7e-4]	[1, 5]
A8	[145, 150], [133, 143]	[7e-4, 2e-3]	[1, 4]
A9	[165, 184], [110, 131]	[1e-4, 8e-4]	[2, 5] OR [2, 8]
A10	[201, 209], [152, 159]	[1e-3, 4e-3]	[1, 2]
A11	[207, 216], [135, 145]	[1e-3, 3e-3]	[1, 4]
A12	[217, 221], [59, 65]	[1e-4, 6e-4]	[1, 5]

The unique least solution defines the location and powers of $\underline{N} = 15$ sources, that form the main acoustic surface. The four upper solutions define the location and powers of $\overline{N}_1 = 57$, $\overline{N}_2 = 59$, $\overline{N}_3 = 60$ and $\overline{N}_4 = 62$ sources, that form four secondary acoustic surfaces, respectively. The reconstructed acoustic image in the form of the main and the set of secondary acoustic surfaces is shown in Fig. 5.

The root mean square error of reconstruction for the main and secondary surfaces of fuzzy acoustic levels is $RMSE_w = 0.0080$, which corresponds to ± 0.96132 dB and does not exceed the permissible error of sound pressure measurement [4]. The proximity of the level lines of the reconstructed and real surface in Fig. 5 indicates sufficient reconstruction accuracy. Coordinates of the restored sources (x_p, y_p) for the least solution, that corresponds to the main surface, were compared with the experimental values $(\hat{x}_p, \hat{y}_p)$. For the main sources location, reconstruction accuracy was estimated with the help of the root mean square errors $RMSE_x$ and $RMSE_y$:

$$RMSE_x = \sqrt{\tfrac{1}{15}\sum_{p=1}^{15}\left[x_p - \hat{x}_p\right]^2} = 2.4631,$$
$$RMSE_y = \sqrt{\tfrac{1}{15}\sum_{p=1}^{15}\left[y_p - \hat{y}_p\right]^2} = 2.3805,$$

which do not exceed 1% of the lattice aperture at 250 m.

6 Results and Discussion

The papers [1,13,17,35,36,39] propose methods for sound field reconstruction from incomplete data by generating and selecting the appropriate set of sources in the sound field. In [13,36], the events in the sound field are described by the models with fuzzy parameters. In this case, the tuned parameters are the

Table 2. Solution set for the coordinates and powers of sound sources

Area	The least solution	The set of upper solutions
A1	(25, 209, 7e-3)	(15, 198, 5e-3), (24, 209, 5e-3), (20, 204, 4e-3)
A2	(60, 161, 2e-2)	(60, 161, 3e-3), (55, 162, 5e-3), (53, 165, 7e-3), (52, 167, 6e-3)
A3	(61, 205, 6e-3), (85, 192, 8e-3)	(61, 200, 1e-3), (75, 196, 2e-3), (60, 205, 1e-3), (58, 197, 1e-3), (71, 201, 2e-3), (72, 210, 1e-3), (59, 188, 1e-3), (59, 192, 1e-3), (70, 188, 2e-3) OR (75, 190, 1e-3), (83, 182, 2e-3), (78, 186, 2e-3), (80, 194, 1e-3), (86, 181, 2e-3), (70, 195, 1e-3), (71, 190, 2e-3)
A4	(95, 208, 2e-2)	(95, 214, 4e-3), (91, 208, 4e-3), (96, 217, 3e-3)
A5	(30, 81, 5e-4), (42, 90, 1e-2)	(29, 73, 3e-4), (37, 81, 2e-4), (41, 85, 1e-4), (35, 74, 1e-4), (30, 78, 2e-4), (39, 87, 1e-4), (40, 75, 2e-4), (30, 89, 3e-4), (32, 85, 2e-4), (31, 72, 3e-4), (40, 93, 6e-3), (45, 88, 5e-3)
A6	(101, 89, 2e-3)	(98, 92, 7e-4), (105, 83, 6e-4), (107, 86, 7e-4)
A7	(112, 117, 7e-4)	(111, 116, 2e-4), (123, 112, 1e-4), (118, 114, 1e-4), (109, 112, 1e-4), (116, 118, 2e-4)
A8	(149, 139, 2e-3)	(145, 136, 7e-4), (146, 143, 8e-4), (145, 139, 7e-4), (148, 133, 9e-4)
A9	(168, 112, 8e-4), (177, 131, 5e-4)	(169, 115, 2e-4), (181, 116, 2e-4), (172, 119, 3e-4), (179, 123, 3e-4), (175, 127, 2e-4) OR (171, 118, 2e-4), (180, 129, 3e-4), (175, 126, 3e-4), (184, 120, 1e-4), (170, 115, 2e-4), (173, 127, 2e-4), (181, 125, 1e-4),(179, 127, 1e-4)
A10	(201, 155, 4e-3)	(203, 159, 1e-3), (209, 152, 1e-3)
A11	(210, 137, 3e-3)	(209, 145, 1e-3), (215, 138, 1e-3), (216, 135, 1e-3), (207, 139, 2e-3)
A12	(220, 60, 7e-4)	(217, 59, 2e-4), (221, 63, 2e-4), (218, 60, 1e-4), (220, 59, 2e-4), (221, 65, 1e-4)

lower and upper solutions for the sources coordinates x_p, y_p, and powers w_p, $p = \overline{1, N}$. The optimal values of the interval parameters are found using the genetic and gradient algorithms [1,17,35,39]. Multiple selection, that is, finding a set of sources configurations with the equal summary sound energy at the nodes of the microphone array, provides the complete solution set of the inverse problem. The given method reconstructs the sound field in the form of the solution set of fuzzy relational equations. The principal difference of this method is the ability to simultaneously search for the upper and lower solutions separated by a zero

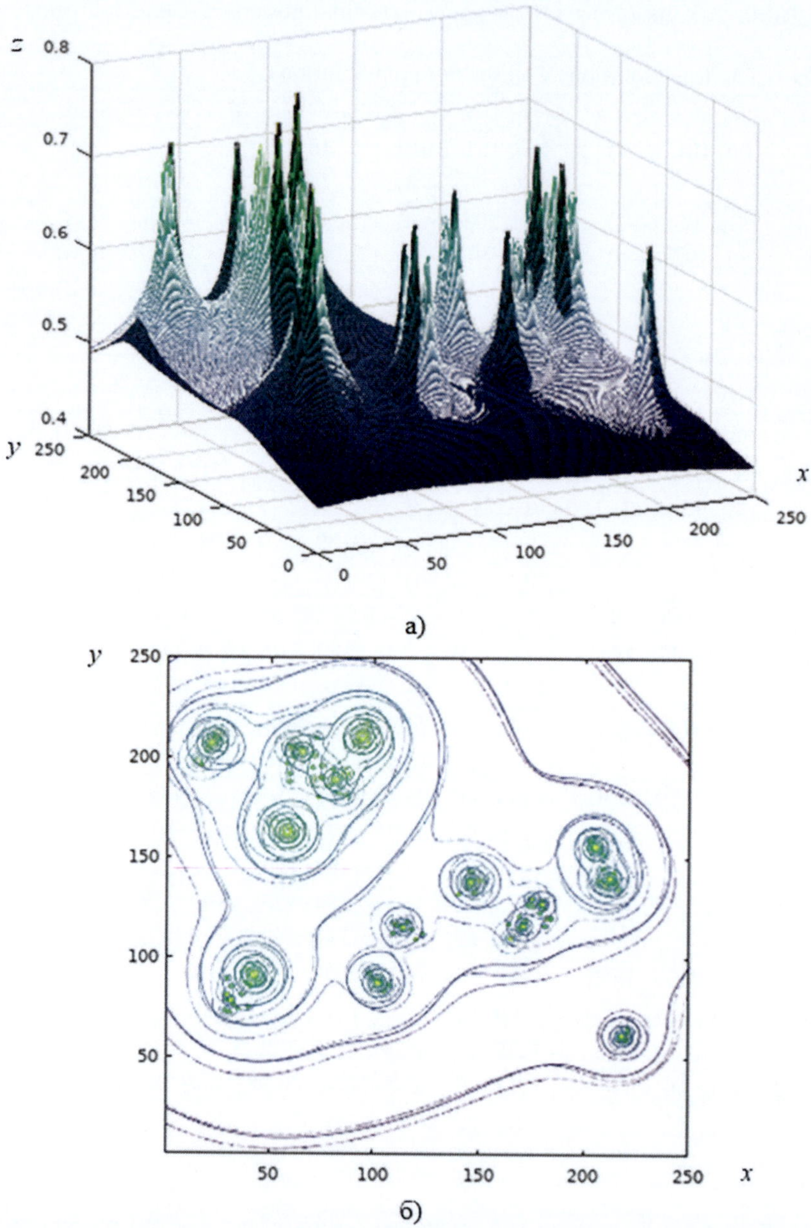

Fig. 5. The reconstructed acoustic image (a); location of the real (–––) and restored (- - -) sources (b)

solution that allows reducing the computational complexity. The effectiveness estimates of the given method are given below.

Sound field reconstruction in the form of the solution set selected from the candidate sources requires solving Z optimization problems with $N(2q+1)$ variables for two-parameter membership functions and binary weights of sources, where Z is the number of reconstructions, $q = 3$ is the number of fuzzy parameters of the sound source (the coordinates and power) [1,17,35]. Sound field reconstruction in the form of the solution set of fuzzy relational equations allows reducing the number of tuning parameters by solving $2Z$ optimization problems with Nq variables for upper and lower bounds of the coordinates and powers of sound sources.

The complexity reduction in solving the optimization problem allows us to obtain the following time estimates. The time window of the microphone array is determined by the reconstruction time and depends on the maximum number of sources of the given power $\overline{N}_w$. In the experiment, the number of sources of the given power was limited to $\overline{N}_w = 100$. Given the speed of vehicles, the time window of the microphone array can not exceed 30 s, since vehicles cover a distance of 250 m of the controlled area during this time. Reconstruction of the acoustic image by methods [1,17,35] requires 56 s, which allows reconstructing every second frame of the data stream. The use of parallel computing to search for the complete solution set to the system of equations made it possible to obtain a single minimum solution in 9 s and the set of maximum solutions in 27 s, which allows reconstructing each frame of the acoustic data stream (Intel Core i5-7400 3.0 Ghz processor). Thus, the proposed method allows to halve the time window of the microphone array, i.e. double the frequency of image updates, that in turn increases the reliability of terrain monitoring.

7 Conclusions

The approach to solving inverse problems of source identification in acoustics is proposed based on fuzzy relational calculus. The compositional rule of inference connects the real and observed fuzzy acoustic image using the relationship matrix. The microphone conversion matrix, which is formed on the basis of the matrix of coordinates of the rays intersection points, reflects the degree of completeness of the observed acoustic image. The fuzzy model of the acoustic field is based on 3D membership functions, for which the degree of membership decreases in proportion to the square of the distance to the source. The problem of reconstructing the acoustic field is formulated as the problem of inverse logical inference and reduces to solving a system of fuzzy relational equations.

The method for reconstructing the acoustic field from incomplete data is proposed based on solving the system of fuzzy relational equations. The problem of solving a system of equations is formulated as an optimization problem. It is necessary to find such a number of sound sources, their locations and powers, which minimize the difference between the model and observed fuzzy acoustic image. The solutions of the equation system represent the variants of the acoustic field reconstruction in the form of the main acoustic surface and a set of secondary acoustic surfaces. The main acoustic surface is generated by the least number of sources. The set of secondary acoustic surfaces represents the variants of the sound field reconstruction generated by the upper solutions for the number of sources. Since the sources distribution is completely determined by the properties of the solution set, the proposed approach allows avoiding the generation and selection of candidate sources, that in turn provides simplification of the reconstruction process and reduction of time costs. The genetic and neural algorithm provides accurate and fast reconstruction of the acoustic field for an unknown number of sources and their configuration.

A further area of research is the development of a method for optimizing the design of a microphone array according to the criteria "accuracy - reconstruction time". The controlled variables are the scanning step, the number and distance between the microphones. The solution to the problem provides for the optimal choice of a relationship matrix, which models the transformation function of the microphone array with different degrees of data completeness.

Acknowledgment. The paper was prepared within the 58–D–393 "Specialized AD systems for audio location and identification of objects on terrain" project.

References

1. Agounad, S., Aassif, E., Khandouch, Y., Maze, G., Decultot, D.: Characterization and prediction of the backscattered form function of an immersed cylindrical shell using hybrid fuzzy clustering and bio-inspired algorithms. Ultrasonics **83**, 222–235 (2018). https://doi.org/10.1016/j.ultras.2017.06.015
2. Azarov, O., Krupelnitsky, L., Rakytyanska, H.: A fuzzy model of television rating control with trend rules tuning based on monitoring results. In: IEEE 2nd International Conference on Data Stream Mining and Processing (DSMP), pp. 369–375. IEEE Publ., Lviv (2018). https://doi.org/10.1109/DSMP.2018.8478583
3. Azarov, O., Krupelnitsky, L., Rakytyanska, H.: Television rating control in the multichannel environment using trend fuzzy knowledge bases and monitoring results. Data **3**(4), 57 (2018). https://doi.org/10.3390/data3040057
4. Azarov, O., Krupelnitsky, L., Steiskal, V., Bilokon, O.: Specialized and measuring equipment of own design and production for TV and radio broadcasting. Catalog of the Scientific and Technical Center "Analog-Digital Systems". VNTU, Vinnitsya (2015)
5. Bai, M., Lin, J.-H.: Source identification system based on the time-domain nearfield equivalence source imaging: fundamental theory and implementation. J. Sound Vib. **307**(1–2), 202–225 (2007). https://doi.org/10.1016/j.jsv.2007.06.025
6. Benesty, J., Chen, J.: Study and Design of Differential Microphone Arrays. Springer, Heidelberg (2013). https://doi.org/10.1007/978-3-642-33753-6
7. Bi, C.X., Li, Y.C., Zhang, Y.B., Zhang, X.Z., Xu, L.: Signal reconstruction of moving sound sources with a fixed microphone array. Mech. Syst. Sig. Process. **118**, 277–289 (2019). https://doi.org/10.1016/j.ymssp.2018.08.052
8. Brandstein, M., Ward, D.: Microphone Arrays: Signal Processing Techniques and Applications. Springer, Heidelberg (2001). https://doi.org/10.1007/978-3-662-04619-7
9. Chu, N., Picheral, J., Mohammad-Djafari, A., Gac, N.: A robust super-resolution approach with sparsity constraint in acoustic imaging. Appl. Acoust. **76**, 197–208 (2014). https://doi.org/10.1016/j.apacoust.2013.08.007

10. Gauthier, P.A., Camier, C., Pasco, Y., Berry, A., Delalay, M.A.: Beamforming regularization matrix and inverse problems applied to sound field measurement and extrapolation using microphone array. J. Sound Vib. **330**(24), 5852–5877 (2011). https://doi.org/10.1016/j.jsv.2011.07.022
11. Gilquin, L., Bouley, S., Antoni, J., Le Magueresse, T., Marteau, C.: Sensitivity analysis of two inverse methods: conventional beamforming and Bayesian focusing. J. Sound Vib. **455**, 88–202 (2019). https://doi.org/10.1016/j.jsv.2019.05.002
12. Hansen, T., Marengo, E.: SVD solutions to inverse source problems in the time domain: application to complex point sources. Wave Motion **89**, 93–103 (2019). https://doi.org/10.1016/j.wavemoti.2019.03.008
13. He, Z., Lin, X., Li, E.: A non-contact acoustic pressure-based method for load identification in acoustic-structural interaction system with non-probabilistic uncertainty. Appl. Acoust. **148**, 223–237 (2019). https://doi.org/10.1016/j.apacoust.2018.12.034
14. Holland, K., Nelson, P.: The application of inverse methods to spatially-distributed acoustic sources. J. Sound Vib. **332**(22), 5727–5747 (2013). https://doi.org/10.1016/j.jsv.2013.06.009
15. Hsu, C.: Advanced Signal Processing Technology by Soft Computing. World Scientific, New York (2000). https://doi.org/10.1142/4555
16. Koo, S., Karve, P., Kallivokas, L.: A comparison of time-reversal and inverse-source methods for the optimal delivery of wave energy to subsurface targets. Wave Motion **67**, 121–140 (2016). https://doi.org/10.1016/j.wavemoti.2016.07.011
17. Kuhne, M., Togneri, R., Nordholm, S.: Robust source localization in reverberant environments based on weighted fuzzy clustering. IEEE Sig. Process. Lett. **16**(2), 85–88 (2009). https://doi.org/10.1109/LSP.2008.2009833
18. Kyle, T., Woods, D., Bolton, J., Rhoads, J.: Least-squares reconstruction of low-frequency inhomogeneous plane waves. J. Sound Vib. **430**, 134–149 (2018). https://doi.org/10.1016/j.jsv.2018.05.009
19. Lu, J., Cheng, W., He, D., Zi, Y.: A novel underdetermined blind source separation method with noise and unknown source number. J. Sound Vib. **457**, 67–91 (2019). https://doi.org/10.1016/j.jsv.2019.05.037
20. Padois, T., Doutres, O., Sgard, F., Berry, A.: Time domain localization technique with sparsity constraint for imaging acoustic sources. Mech. Syst. Sig. Process. **94**, 85–93 (2017). https://doi.org/10.1016/j.ymssp.2017.02.035
21. Pan, S., Jiang, W.: A hybrid approach to reconstruct transient sound field based on the free-field time reversal method and interpolated time-domain equivalent source method. J. Sound Vib. **333**(16), 3625–3638 (2014). https://doi.org/10.1016/j.jsv.2014.03.029
22. Peeva, K., Kyosev, Y.: Fuzzy Relational Calculus. Theory, Applications and Software. World Scientific, New York (2004)
23. Pereira, A., Antoni, J., Leclere, Q.: Empirical Bayesian regularization of the inverse acoustic problem. Appl. Acoust. **97**, 11–29 (2015). https://doi.org/10.1016/j.apacoust.2015.03.008
24. Pham, M., Oudompheng, B., Mars, J., Nicolas, B.: A noise-robust method with smoothed l1/l2 regularization for sparse moving-source mapping. Sig. Process. **135**, 96–106 (2017). https://doi.org/10.1016/j.sigpro.2016.12.022
25. Qiao, B., Zhang, X., Gao, J., Chen, X.: Impact-force sparse reconstruction from highly incomplete and inaccurate measurements. J. Sound Vib. **376**, 72–94 (2016). https://doi.org/10.1016/j.jsv.2016.04.040

26. Rabinovich, D., Turkel, E., Givoli, D.: An augmented time reversal method for source and scatterer identification. J. Comput. Phys. **375**, 99–119 (2018). https://doi.org/10.1016/j.jcp.2018.08.026
27. Rakytyanska, H.: Optimization of knowledge bases on the basis of fuzzy relations by the criteria "accuracy-complexity". East.-Eur. J. Enterp. Technol. **2**(4), 24–31 (2017). https://doi.org/10.15587/1729-4061.2017.95870
28. Rienstra, S., Hirschberg, A.: An Introduction to Acoustics. Eindhoven University of Technology, Eindhoven (2009)
29. Rotshtein, A., Rakytyanska, H.: Adaptive diagnostic system based on fuzzy relations. Cybern. Syst. Anal. **45**(4), 623–637 (2009). https://doi.org/10.1007/s10559-009-9130-4
30. Rotshtein, A., Rakytyanska, H.: Fuzzy logic and the least squares method in diagnosis problem solving. In: Sarma, R. (ed.) Genetic Diagnoses, pp. 53–97. Nova Science Publishers, New York (2011)
31. Rotshtein, A., Rakytyanska, H.: Fuzzy Evidence in Identification, Forecasting and Diagnosis. In: Studies in Fuzziness and Soft Computing, vol. 275. Springer, Heidelberg (2012). https://doi.org/10.1007/978-3-642-25786-5
32. Rotshtein, A., Rakytyanska, H.: Fuzzy genetic object identification: multiple-inputs multiple-outputs case. In: Hippe, Z., Kulikowski, J., Mroczek, T. (eds.) Human-Computer Systems Interaction. Part II. Advances in Intelligent and Soft Computing, vol. 99, pp. 375–394. Springer, Heidelberg (2012). https://doi.org/10.1007/978-3-642-23172-8_25
33. Rotshtein, A., Rakytyanska, H.: Expert rules refinement by solving fuzzy relational equations. In: 6th IEEE Conference on Human System Interaction, pp. 257–264. IEEE Publ., Sopot (2013). https://doi.org/10.1109/HSI.2013.6577833
34. Rotshtein, A., Rakytyanska, H.: Optimal design of rule-based systems by solving fuzzy relational equations. In: Hippe, Z., Kulikowski, L., Mroczek, T., Wtorek, J. (eds.) Issues and Challenges in Artificial Intelligence. Studies in Computational Intelligence, vol. 559, pp. 167–178. Springer, Cham (2014). https://doi.org/10.1007/978-3-319-06883-1_14
35. Toth, B., Vad, J.: A fuzzy clustering method for periodic data, applied for processing turbomachinery beamforming maps. J. Sound Vib. **434**, 298–313 (2018). https://doi.org/10.1016/j.jsv.2018.08.002
36. Wang, C., Qiu, Z., Li, Y.: Hybrid uncertainty propagation of coupled structural-acoustic system with large fuzzy and interval parameters. Appl. Acoust. **102**, 62–70 (2016). https://doi.org/10.1016/j.apacoust.2015.09.006
37. Xie, J., Zhu, M.: Investigation of acoustic and visual features for acoustic scene classification. Expert Syst. Appl. **126**, 20–29 (2019). https://doi.org/10.1016/j.eswa.2019.01.085
38. Yager, R., Filev, D.: Essentials of Fuzzy Modeling and Control. Wiley, New York (1994)
39. Zhang, Y., Wang, Y., Zhang, C.: Total variation based gradient descent algorithm for sparse-view photoacoustic image reconstruction. Ultrasonics **52**(8), 1046–1055 (2012). https://doi.org/10.1016/j.ultras.2012.08.012
40. Zhu, J., Ge, Z., Song, Z., Gao, F.: Review and big data perspectives on robust data mining approaches for industrial process modeling with outliers and missing data. Ann. Rev. Control **46**, 107–133 (2018). https://doi.org/10.1016/j.arcontrol.2018.09.003

Application of Inductive Bayesian Hierarchical Clustering Algorithm to Identify Brain Tumors

Iryna Lurie[1], Volodymyr Lytvynenko[1(✉)], Serge Olszewski[2], Mariia Voronenko[1], Waldemar Woicik[3], Oleg Boskin[1], Ulzhalgas Zhunissova[4], and Michael Sherstiuk[1]

[1] Kherson National Technical University, Kherson, Ukraine
lurieira@gmail.com, immun56@gmail.com, mary_voronenko@i.ua, aandre.lenoge@gmail.com

[2] Taras Shevchenko National University, Kyiv, Ukraine
olszewski.serge@gmail.com

[3] Lublin University of Technology, Lublin, Poland
waldemar.wojcik@pollub.pl

[4] Astana Medical University, Astana, Kazakhstan
Ulzhalgaszhunisova@gmail.com

Abstract. The article presents the results of research concerning development of inductive algorithm for hierarchical Bayesian clustering of gene expression of patients with two types of brain tumors and healthy individuals. The study carried out comparative studies of the clustering quality of inductive and classical methods of Bayesian hierarchical clustering algorithm (BHC). It is proposed to apply the moving average and FFT filtering methods for data dimension reducing. The basic principles of creating an inductive model of objective clustering are formed, the results of clustering are shown at various levels of data dimension reducing, the advantages of objective clustering BHC in comparison with the canonical BHC algorithm are determined.

Keywords: Glioblastoma · Brain tumor · Oligodendroglial tumors · Glioblastoma multiforme profiling of gene expression · Inductive modeling · Bayesian hierarchical algorithm · External balance criterion · Data dimension reducing · Normalizing · Exploratory analysis criterion · Gene expression profiles

1 Introduction

Oligodendroglial tumors make up about 4–7% of all gliomas, but in some cases, the incidence reaches 10–20% due to an improvement in histological diagnosis and the presence of its molecular signature that characterizes these tumors. Oligodendroglial tumors are classified as low-level oligodendroglial tumors (the so-called oligodendroglioma) and high-level anaplastic oligodendroglial tumors

S. Babichev et al. (Eds.): ISDMCI 2020, AISC 1246, pp. 567–584, 2021.
https://doi.org/10.1007/978-3-030-54215-3_36

(anaplastic oligodendroglioma - AO) as well as mixed malignant oligoastrocytic tumors (mixedmalignant glioma).

In addition, mixed tumors (mixedmalignant glioma) can also be low-level (oligoastrocytoma OA) or high-level (anaplastic oligoastrocytoma - AOA) [9].

Glioblastoma multiform (GBM) is one of the most common and most aggressive types of brain tumor [3] and the leading cause of death in these types of adult brain disease. GBM accounts for 52% of all cases of brain tumors. It has a very poor prognosis, despite the existence of many therapeutic methods, including surgical resection of a larger tumor volume followed by concomitant or subsequent chemoradiotherapy. Despite advances in the genomics and classification of glioma subtypes [11,22,30], glioblastoma has a worse prognosis than any other malignant central nervous system tumor with an average life of 14 months. Currently, only two techniques are relevant for generating an array of gene expression: DNA microarray sequencing method [8] and RNA molecules [18].

Identifying types of brain tumor is crucial when choosing the right treatment for patients because different types of brain tumor can respond well to different treatments. Currently, a greater number of computational methods have been developed to detect types of brain tumor. However, existing methods rarely use information from gene regulation networks to facilitate the identification of subtypes [31]. One of the computational methods that allows us to solve this problem is clustering. Clustering methods are divided into hierarchical and iterative [23].

Thus, early diagnosis of brain tumor, their differential diagnosis is an urgent task. In accordance with this, the development of new and improvement of existing clustering methods for microarray and RNASeq data will improve and create information technology for the diagnosis of complex and socially significant diseases.

2 Review of Literature

Hierarchical algorithms are associated with the construction of dendrograms. In agglomerative algorithms, before the start of clustering, all objects are considered separate clusters that are combined during the algorithm. However, the hierarchical cluster analysis procedure works well for a small number of objects and is not suitable for large data due to the complexity of the agglomerative algorithm and too large dendrogram sizes. In iterative algorithms, the data is immediately divided into several clusters, the number of which is estimated based on the conditions. Further, the elements are moved between clusters so that some criterion is optimized, for example, the variability within the clusters is minimized [7]. However, iterative clustering algorithms have several disadvantages:

- it is not guaranteed to achieve the global minimum of the total quadratic deviation, but only one of the local minimums;
- the result depends on the choice of the initial centres of the clusters; their optimal choice is unknown;
- the number of clusters must be known in advance.

Inductive clustering algorithms allow you to minimize these disadvantages significantly. The basic concepts for creating an inductive method of clustering objects are described in [20]. Further evolution of this theory is reflected in the work [29]. In [24,26], the authors show the ways to implement an objective cluster analysis, and advantages of using inductive models compared to traditional methods of data clustering. In [1] the authors proposed an improved inductive model of clustering based on the k-average clustering algorithm. An algorithm is proposed and practically implemented for dividing the source data into two equal-sized subsets. The paper presents studies on the assessment of the stability of the model to the noise component using data "Seeds". However, the application of inductive clustering algorithms, despite the successful results achieved in this area, objective cluster algorithms such as the inductive k-means algorithm, the inductive hierarchical clustering algorithm, the inductive density algorithm DBSCAN, and the inductive SOTA algorithm do not completely solve the problems of data processing in bioinformatics when learning of the expression of genes with using DNA microarrays.

The Bayesian hierarchical clustering algorithm uses limit probabilities to decide which clusters should be joined and to avoid overflow. Basically, he asks about the likelihood that all the data in the potential merge was generated from the same component of the mixture, and compares this with exponentially many hypotheses at lower levels of the tree. The traditional algorithm does not determine the probabilistic data model, so it is difficult to ask how "good" clustering is, compare with other models, make predictions, and group new data into an existing hierarchy. To overcome these limitations, we use statistical inference. Studies on the use of a modified BHC algorithm for clustering gene expression are described in [10,27]. The authors showed that the BHC algorithm leads to a significant increase in speed using minimal loss in the quality of clustering.

In this paper, we proposed a new inductive Bayesian hierarchical clustering algorithm that uses limit probabilities to decide which clusters should be combined to avoid overflow. Basically, he asks about the likelihood that all the data in the potential merge was generated from the same component of the mixture, and compares this with exponentially many hypotheses at lower levels of the tree. One of the unresolved parts of the general problem is the lack of the integrated use of effective models for clustering arrogant noisy data and inductive modelling methods for complex systems, so the second novelty of this work is that it used the combined method of moving average and FFT filtering for data dimension reducing the first time. The practical implementation of the proposed model is based on integrated use as *R*.

Thus, the development of models and methods of objects clustering based on inductive modelling methods to solve the problem of identifying types of brain tumor is an urgent task.

3 Problem Statement

The clustering problem can be formulated as follows. Let $X = \{x_1, x_2, \ldots, x_n\}$ be a set of objects, and $Y = \{y_1, y_2, \ldots, y_n\}$ be a set of cluster labels. The distance function between the objects is specified $\rho(x, x')$. $X^m = \{x_1, \ldots, x_m\} \subset X$ is a final training set of objects. It is necessary to break the sample into disjoint subsets called clusters so that each cluster consists of objects that are close in metric ρ, and the objects of different clusters are significantly different. At the same time, a cluster number y_i is assigned to each object $x_i \in X^m$. The clustering algorithm is a function $a : X \to Y$ that associates a cluster number $y \in Y$ with any object $x \in X$. Multitude Y are known in advance, but more often the task is to determine the optimal number of clusters, from the point of view of one or another criterion for the quality of clustering. Clustering (learning without a teacher) differs from classification (learning with a teacher) in that the labels of the source objects y_i are not initially set, and even the multitude Y may not be known. A block diagram of the identification of experimental data obtained by RNA sequencing with glioblastoma is presented in Fig. 1.

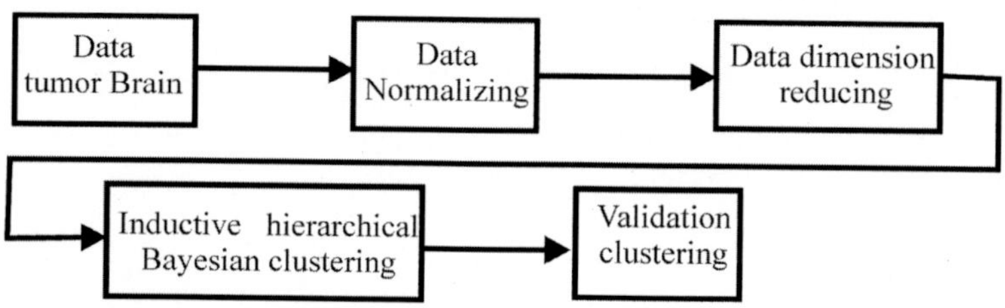

Fig. 1. Procedure for the identification of experimental data obtained by RNA sequencing with glioblastoma

The paper implements the principles of inductive modelling in the framework of inductive clustering, which suggest the following steps [1]:

- normalization of the studied objects signs, i.e. their reducion to the same range;
- dividing the original data set into two equal in power subsets;
- determination of an external criterion or group of relevance criteria for choosing the optimal clustering for two subsets of the same power;
- selection or development of a basic clustering algorithm used as a component of an inductive model of objective clustering.

The solution to the problem of identifying tumor typically consists of three main stages:

1. Data pre-processing and selection of characteristics.
2. Brain tumor type identification methods.
3. Verification of results and visualization.

Evaluation of results is performed using index Dunn, index Calinski-Harabasz, index Entropy and index Silhouette.

The aim of the work is to improve the quality of clustering, to study the proposed method and data dimension reducing method, to develop an inductive Bayesian clustering algorithm for types of brain tumor based on microarray data and to assess the quality of the results.

4 Materials and Methods

4.1 Data

This study used a 50 × 1739 data matrix consisting of 50 human gliomas [4] containing 1739 genes. This dataset consists of three classes: 31 glioblastomas, GBM, 14 tumors with enrichment for oligodendroglial morphology, and 5 normal gliomas.

4.2 Normalization

Normalization of the data was carried out on the basis of the formula:

$$x'_{ij} = \frac{x_{ij} - \text{med}_j}{\max\left(|x_{ij} - \text{med}_j|\right)} \tag{1}$$

where x_{ij} is the meaning of feature i in column j x'_{ij} is normalized value for this parameter, med_j is the median of column j. When using this normalization method, many data features in all columns have the equal median with a maximum range of variation of features from −1 to 1. The data volume for each column falling within the interquartile distance (50%) is the largest compared to other normalization methods.

4.3 Splitting into Equidistant Sets

The algorithm for dividing the original set Ω of objects Ω into 2 equally powerful disjoint subsets Ω^A and Ω^B consists of the following steps [26]:

1. Calculation of $n \cdot (n-1)2$ of paired distances between two objects in the initial data set;
2. Selection of a couple of X_s, X_p, objects, the distance between them is minimal:
$$d(X_s, X_p) = \min_{i,j} d\left(X_i, X_j\right)$$
3. Distribution of the object X_s into the subset Ω^A, and the object X_p into the subset Ω^B;
4. Distribution of object X_s in the subset Ω^A, and the object X_p in the subset Ω^B;
5. Repeat steps 2–3 for the remaining objects until the original set is divided into two subsets. If the number of objects is odd, the last object is added to both subsets.

4.4 Data Dimension Reducing

For reduce the dimension of the studied data array, it is proposed to use the consistent application of the moving average [12,21,25] to reduce the contribution of the noise component and FFT filtering [5,13] to reduce the signal sampling frequency. The scheme of the data dimension reduction method proposed for this work is shown in Fig. 2. The scheme of the algorithm for preliminary data processing by the method of moving average and FFT filtering. At the first stage, the data is smoothed using the moving average method. In this case, the total size of the data array does not change. For smoothed data, the FFT filtering method reduces the sampling rate of the smoothed curve, reducing the size of the processed data array by the number of times corresponding to the size of the moving average window.

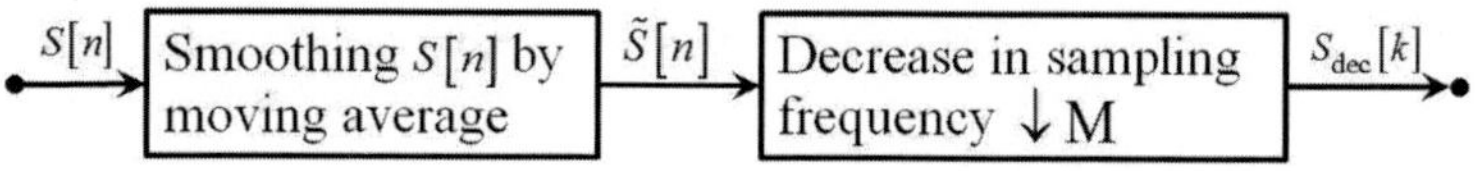

Fig. 2. Scheme of the data dimension reduction method

This allows reducing the dimension of the descriptors without losing the possibility of reliable identification. The essence of the method lies in the fact that many arrays of experimental data that are used to identify semi-empirical dependencies between the heterogeneous characteristics of the studied phenomena are of a fundamental nature. If the category number $S[n]$ is represented as an ordered set of numbers, then the dependence of the quantitative measure of conformity in the category number can be considered as a random process with an uncertain probability distribution. A large number of undefined parameters for the description of such arrays gives rise to a number of problems in the construction of continuous forecasting models. If we consider such ordered arrays as a generalized signal with noise, then this can be expressed by the formula:

$$S[n] = x[n] + \xi[n] \tag{2}$$

where $x[n]$ are the discrete values of a smooth deterministic function that describes the waveform, and $\xi[n]$ are the discrete values of a symmetric random process, then the application of standard discrete methods of signal processing to them will make it possible to isolate the deterministic component and build a predictive model on the array $x[n]$. One of the commonly used methods for processing signals with noise is its conversion using the moving average calculation algorithm:

$$x[n] = \frac{1}{M}\sum_{k=0}^{M-1} S[n-k] \tag{3}$$

where M is the width of the averaging window. For symmetrical noise with a sufficient window width, this operation significantly reduces the noise component of the signal.

To reduce the dimensionality of the data array without losing information about the shape of the generalized signal, the Fourier filtering algorithm was used. To do this, using a discrete Fourier transform, an array of spectral components of the signal processed by the moving average was obtained:

$$X[k] = \sum_{n=0}^{N-1} x[n] e^{-j\frac{2\pi}{N}kn} \tag{4}$$

Array elements $X[k]$ for all $k > \frac{N}{M}$ were discarded creating an array $\widehat{X}[k]$ of lower dimension $\widehat{N}$. For the resulting array, the inverse discrete Fourier transform was performed:

$$\widetilde{x}[n] = \frac{1}{\widehat{N}} \sum_{k=0}^{\widehat{N}-1} \widehat{X}[k] e^{j\frac{2\pi}{N}kn} \tag{5}$$

The choice of just such an algorithm for reducing the dimension of an array is dictated by the presence of its hardware implementation on modern controllers.

4.5 Bayesian Hierarchical Clustering Algorithm

Bayesian hierarchical algorithm was proposed in [14]. Unlike hierarchical clustering algorithms, where the Euclidean distance, Manhattan distance, correlation, and so on, the probability measure of distance is used in the BHC. This means that instead of a fixed distance that determines clustering, the distance in this algorithm is the probability that data elements are present in a particular cluster. This becomes important with iterative cluster joining to form new clusters in a hierarchical structure [19].

BHC is a hierarchical agglomerative clustering algorithm that uses a Bayesian probability measure of distance. This algorithm performs clustering from the bottom up in which clusters are formed in a hierarchy. Initially, all data elements are initialized within individual clusters, after which the clusters are iteratively combined until the clusters merge. The merger is based on a preliminary calculation of probability, which uses Bayes' Theorem. Ultimately, a dendrogram will be obtained that shows the hierarchical structure obtained from the input data set. The BHC algorithm described in [14] consists of the following steps:

Step 1. Initialize all data elements into separate clusters (represented by tree nodes) with probability arrays. Select the model to be used.
Step 2. Calculation of probability arrays for each cluster. The first element of the array is the likelihood that the cluster will remain independent. All other elements of the array are union probabilities with all other clusters.
Step 3. Cluster union based on the highest probability within each array of cluster probabilities.

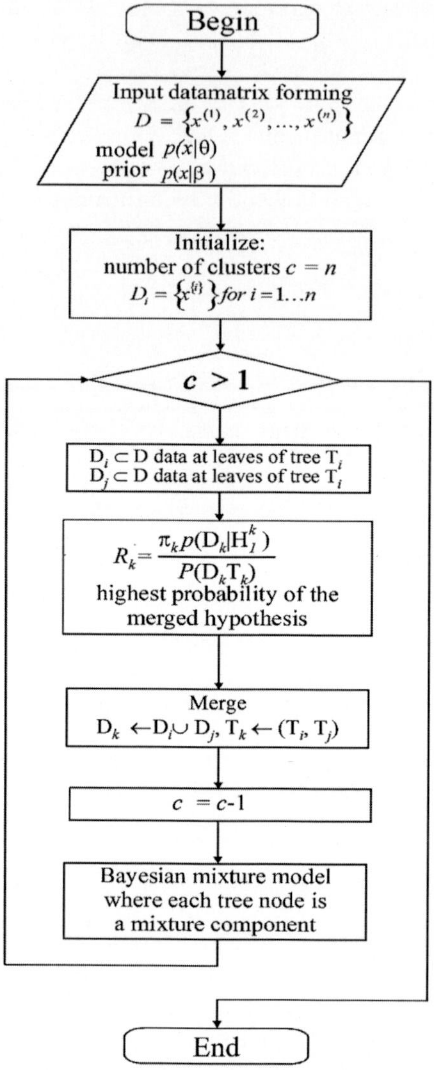

Fig. 3. Scheme of Bayesian Hierarchical Clustering algorithm

Step 4. Create a merge cluster and designate its parent two child clusters. At this step, a probability array is calculated for the parent cluster resulting from the merger.
Step 5. Termination of the algorithm after cluster termination is terminated.

As a result of the algorithm, a dendrogram of a hierarchical structure will be obtained, which will allow you to visually analyze the result of clustering (Fig. 3).

4.6 Inductive Bayesian Hierarchical Algorithm

Currently, much attention is paid to the problems of clustering complex high-dimensional data in conditions of high noise levels. In this paper, by high-dimensional data, we mean data whose dimension of the feature space is equal to or much larger than the number of objects under study. Such data are expression profiles of DNA nucleotide genes, encephalograms of organs of a biological organism, chromatograms of narcotic substances, and the like. In addition to the high dimensionality, the features of the studied data are the level and specificity of the noise component, which is caused by biological processes occurring in the studied object, and the imperfection of the system for creating and generating data for their further processing.

This is primarily due to an increase in the requirements for detection accuracy and the operation of identification systems in various conditions for obtaining information. There are many clustering algorithms these days. Each of them has its advantages and disadvantages and focuses on a specific data type. A high percentage of subjectivity is one of the key shortcomings of existing algorithms, i.e. the high quality of clustering on one data set does not guarantee the same results on another similar data set. Improving the objectivity of clustering is possible using inductive methods of complex modelling systems based on the group method of data processing [16], where data processing is carried out by two equal force subsets. The final decision regarding the nature of the partitioning of objects into clusters is based on the integrated use of external criteria for relevance and internal criteria for assessing the quality of clustering. Thus, the development of hybrid models and methods of clustering objects based on inductive modelling of complex systems is an urgent problem both theoretically and practically.

The construction of inductive technology for clustering high-dimensional data of complex biological nature using the inductive modelling methodology of complex GMDH systems is based on the following principles described in [15]:

1. The principle of heuristic self-organization of the model. In accordance with this principle, enumeration of candidate models in the direction of the increasing complexity of the value of the external quality criteria of the model first decreases, and then increases. The first local minimum of criteria corresponds to the model of optimal complexity.
2. The principle of external addition. To ensure objective verification of the model, it is necessary to specify an external complement, i.e. external criterion of optimality. This criterion evaluates the quality of the model based on a sample of data that was not used to create the model.
3. The principle of consistency. According to this principle, the most optimal model should be selected according to the criterion of minimum displacement, according to which the simulation results on the training and control samples should coincide within the margin of permissible error.

The implementation of the model includes the following steps:

- Formation of a matrix of studied data. The matrix contains n rows or objects under study and m columns or attributes characterizing these objects.
- Dividing the original data set into two equal-sized subsets. These subsets include the same number of paired similar objects.
- In the case of large-scale data, the distance matrix between the objects under study for both subsets is calculated using the correlation metric of proximity. This distance matrix is the input matrix for the next stage of the algorithm.
- Setting and fixing parameters of the BHC algorithm.
- Sequential data clustering on two subsets A and B using the BHC algorithm
- Internal and external quality criteria for clustering at each stage.
- Analysis of the values of the external balance criterion.
- Fixation of clustering at each stage.
- Repeat steps 5.1 and 5.3 of this algorithm if the balance criterion does not reach the specified minimum
- Cluster fixing, validation, and visualization of results.

In more detail, the algorithm is as follows:

Step 1. Start

Step 2. Data preprocessing to reduce the dimension of the feature space using the Shannon entropy:

$$H = -K \cdot \sum_{i=1}^{n} p_i \ln(p_i)$$

Step 3. Formation of the initial set Ω of studied objects.

Step 4. Dividing Ω into two equally powerful subsets in accordance with the above algorithm. The resulting subsets Ω^A and Ω^B can be formally represented as follows:

$$\Omega^A = \{x_{ij}^A\}; \Omega^B = \{x_{ij}^B\},$$

$$i = \overline{1, n_A} = n_B;\ n_A + n_B = n;\ j = \overline{1, m}$$

Step 5. Configuring the BHC clustering algorithm.

Step 6. For the number of clusters $k \in [k_{min}, k_{max}]$:

Step 6.1. Sequential clustering and cluster fixing for $\Omega^A = \{x_{ij}^A\}$; $\Omega^B = \{x_{ij}^B\}$

Step 6.2. Calculation of internal clustering quality criteria for $\Omega^A = \{x_{ij}^A\}$; $\Omega^B = \{x_{ij}^B\}$:

Index Silhouette	$SWC = \frac{1}{K}\sum_{1=1}^{K} S_{x_j}$
Index Dunn	$DI(k) = \min_{i \in K}$
Index Calinski - Harabasz	$QC_{CH} = \frac{QCB \cdot (N-K)}{QCW \cdot (K-1)} \to \max$
Index Entropy	$\left[PE = \frac{\sum_{q=1}^{Q}\sum_{k=1}^{K} \ln(u_{qk})}{Q},\ PE \in [0, \ln Kk]\right]$

Step 6.3. Calculation of external balance criterion

$$ECB = \sqrt{\frac{\sum_{i=1}^{p}\left(IC_A^i - IC_B^i\right)^2}{\sum_{i=1}^{p}\left(IC_A^i + IC_B^i\right)^2}} \to \min,\ p.$$

Step 6.4. If the value of the balance criterion is not minimum, then Step 6.1 is repeated. - 6.3.

otherwise

Step 7. Fixing the minimum value of the external balance criterion.

Step 8. Determining the optimal number of clusters k_{opt}

Step 9. Clustering data (the set Ω of objects under study), fixing the clusters.

Step 10. The End

4.7 Clustering Quality Assessment

As criteria for the quality of clustering were used:

Index Silhouette [17]

$$SWC = \frac{1}{K}\sum_{1=1}^{K} S_{x_j}$$

where K is the number of clusters, S_{x_i} is the "best" membership of the element x_j in the cluster p.

Silhouette refers to a method for interpreting and checking consistency within data clusters. The silhouette value is a measure of how similar the object is in its own cluster (cohesion) compared to other clusters (separation). The silhouette ranges from -1 to $+1$, where a high value indicates that the object is in good agreement with its own cluster and is poorly aligned with neighbouring clusters. If most objects are of high importance, then a clustering configuration is appropriate. If many points have a low or negative value, then the clustering configuration may be too many or too few clusters. The best partition is characterized by the maximum SWC, which is achieved when the distance inside the cluster is small and the distance between the elements of neighbouring clusters is large.

Index Dunn [2]
It is a metric for evaluating clustering algorithms. Compares intercluster dissolution with cluster diameter. The higher the index value, the better the clustering. The purpose of this index is to identify clusters that are compact, with a small difference between cluster members and well-separated, where the objects of different clusters are quite far apart from each other than the dispersion within the cluster. For this purpose of clusters, a higher Dunn index indicates better clustering. One of the disadvantages of using this index is the high computational cost, as the number of clusters and the dimension of the data increase (Fig. 4).

$$DI(k) = \min_{i \in K}$$

Calinski-Harabasz Index [6]:

$$QC_{CH} = \frac{QCB \cdot (N - K)}{QCW \cdot (K - 1)} \to \max$$

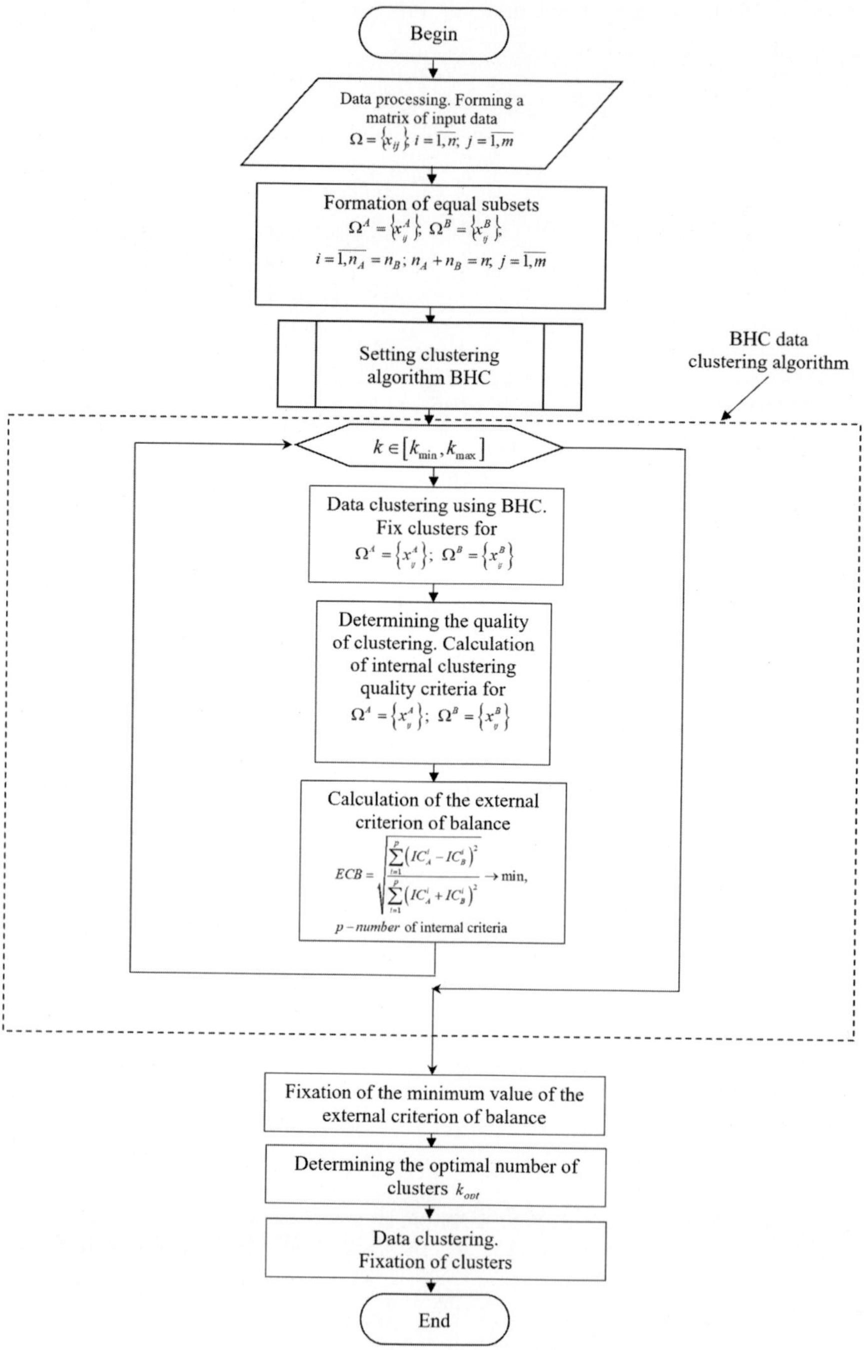

Fig. 4. Scheme of the inductive cluster analysis model based on the Bayesian Hierarchical Clustering algorithm

where N is the number of objects, K is the number of clusters.

The Calinsky-Harabash Index, also known as the Variance Ratio Criterion, is the ratio of the sum of the dispersion between clusters and the inter-cluster dispersion for all clusters. The maximum index value corresponds to the optimal cluster structure.

Entropy [28]

$$\left[PE = \frac{\sum_{q=1}^{Q}\sum_{k=1}^{K}\ln\left(u_{qk}\right)}{Q},\ PE \in [0, \ln Kk]\right]$$

Entropy is known as a numerical expression of the ordering of a system. The entropy of the partition reaches a minimum at the highest ordering in the system (in the case of a clear partition, the entropy is zero). That is, the greater the degree of belonging of an element to one cluster (and the less the degree of belonging to all other clusters), the lower the value of entropy, and the more qualitatively the clustering is performed.

5 Experiment, Results and Discussion

5.1 Pre-processing Results

As a result of applying the moving average FFT filtering algorithms, which are described in Sect. 4.4, three data sets were obtained with 60, 44, and 30 features, respectively, as shown in Fig. 5.

The result of applying the moving average algorithm with the window width is less than the initial number of features of the array of the studied data. As the research results showed, the trend obtained as a result of averaging has a much smaller dispersion. The results of the further Fourier transform of the averaged signal are shown in Fig. 3 b), c) and d). They represent a smooth approximation of the output signal by a finite Fourier series. Moreover, the number of signs obtained is significantly reduced according to the Kotelnikov theorem, in this case 60, 44 and 30, respectively. Despite the result, a visual analysis of the shape of the averaged signal, as well as its Fourier approximation, allows us to say that only two harmonics, the period of which differs by 6 times, and the aperiodic envelope of the low-frequency component are responsible for the shape of the processed signal. The analysis allows us to put forward a hypothesis about the possibility of further reducing in the number of signs (Fig. 6).

5.2 Clustering Results

After normalizing the data and reducing the data, the combined moving average and FFT methods were used in the hierarchical inductive clustering algorithm for three 60×50, 44×50, and 30×50 data sets. The results are presented in Table 1.

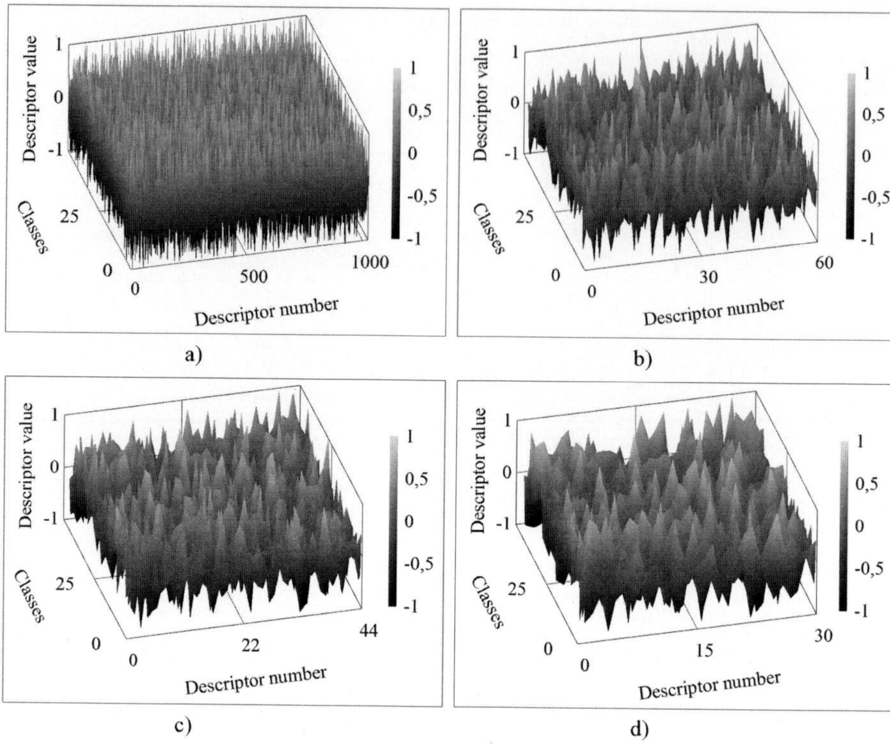

Fig. 5. Data surface before and after pre-processing using the moving average method. Graph a) represents the initial set of descriptor vectors, graph b) represents the set of vectors processed with a moving average with a window of 1.7% of the total vector size, graph c) corresponds to the result of processing with a window of 2.2% and graph d) represents the result of processing with a window of 3.4%.

With an a priori known number of clusters, the correct definition of external criteria, as well as the number of clusters, indicates the quality of the inductive BHC algorithm. The numerical values of the external criterion are given in Table 2.

An analysis of the results allows us to conclude that the data pre-processing, process, normalization and size reducing using the combined moving average and FFT filtering methods in combination with the inductive Bayesian hierarchical clustering algorithm improves the quality of clustering from the point of view of internal and external criteria for the quality of clustering (Table 1).

Analyzing the Silhouette criterion, it should first be noted that their values are a positive value in both variants. The magnitude of the inductive BHC indices in all three data sets significantly exceeds the indices of this criterion obtained using the classical BHC. Which says that the proposed algorithm allows a better data splitting.

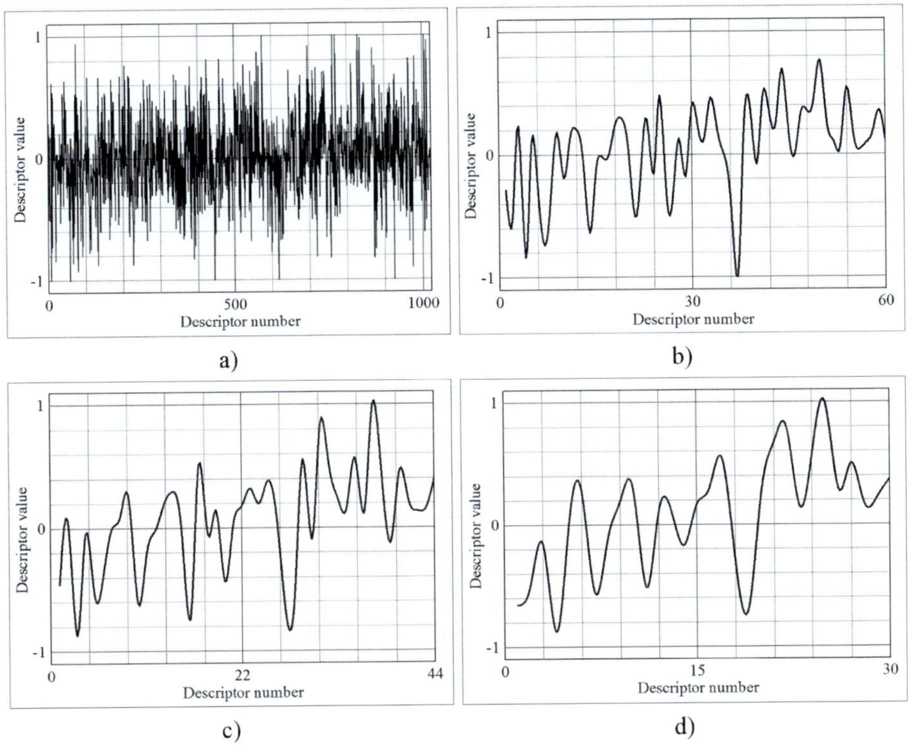

Fig. 6. Descriptor trend before and after pre-processing using the moving average method. Graph a) represents the initial trend of the descriptors, graph b) represents the trend processed by the moving average with a window of 1.7% of the total number of descriptors, graph c) corresponds to the result of processing with a window of 2.2% and graph d) corresponds to the result of processing with a window of 3.4%.

Comparing the value of Index Dunn in pairs, it should be noted that in all cases its value with inductive BHC exceeds the value of classical BHC, which indicates a better clustering of inductive BHC.

The high value of the Calinski - Harabasz Index in inductive BHC confirms the advantage of the proposed algorithm over the classical BHC algorithm.

According to the Entropy Index, the minimum value of entropy indicates a high quality of clustering; therefore, a pairwise comparison of the inductive and classical BHC algorithms in the three studied samples confirms that the resulting clusters are more optimal, therefore, better clustering takes place.

In the inductive BHC algorithm, the use of an external criterion allows you to automatically determine the number of clusters. Table 2 shows that in all three data sets the proposed algorithm accurately determined the number of clusters.

Table 1. Internal criteria for assessing the quality of classical and developed BHC algorithms

Data dimensionality	60/50		44/50		30/50	
Internal quality criterion for clustering	Inductive BHC	Classic BHC	Inductive BHC	Classic BHC	Inductive BHC	Classic BHC
	Number of clusters					
	3	3	3	4	3	3
Index Silhouette	0,19	0,18	0,18	0,03	0,19	0,18
Index Dunn	0,55	0,53	0,43	0,34	0,55	0,53
Index Calinski-Harabasz	10,14	9,15	6,29	5,48	10,14	9,15
Entropy	0,53	0,56	0,62	1,12	0,54	0,56

Table 2. Values of external quality criteria for inductive BHC algorithms for various data dimensions

Data dimensionality	Number of clusters					
	2	3	4	5	6	7
60/50	0,568	0,355	0,453	0,561	0,388	0,366
44/50	0,139	0,124	0,253	0,296	0,278	0,853
30/50	0,888	0,851	0,867	0,856	0,853	0,853

6 Conclusions

This paper presents the results of a study to identify the validation of two types of gliomas and a group of healthy patients based on arrays of genomic data. The results of the simulation have shown high efficiency of the developed inductive Bayesian hierarchical clustering algorithm based on the inductive method of complex systems modelling.

In this work, the Bayesian hierarchical clustering algorithm has been used as the basic one, and the influence of four internal criteria (Silhouette, Dunn Index, Index Calinski - Harabasz, Entropy) on the quality of clustering, as well as the influence of dimension, has been investigated. For the first time, it has been proposed to apply a combination of moving average and FFT filtering algorithms for data dimension reducing. In general, the combined use of the proposed method of data dimension reducing and the inductive approach can significantly improve the quality of the clustering of complex objects. The advantage of the proposed model lies in its stability, due to the use of an external balance criterion for two equally powerful samples.

References

1. Babichev, S., Taif, M., Lytvynenko, V.: Estimation of the inductive model of objects clustering stability based on the k-means algorithm for different levels of data noise. Radio Electron. Comput. Sci. Control **4**, 54–60 (2016). https://doi.org/10.15588/1607-3274-2016-4-7

2. Bezdek, J., Dunn, J.: Optimal fuzzy partitions: a heuristic for estimating the parameters in a mixture of normal distributions. IEEE Trans. Comput. **100**, 835–838 (1975). https://doi.org/10.1109/T-C.1975.224317
3. Bleeker, F., Molenaar, R., Leenstra, S.: Recent advances in the molecular understanding of glioblastoma. J. Neurooncol. **108**(1), 11–27 (2012). https://doi.org/10.1007/s11060-011-0793-0
4. Bredel, M., Bredel, C., Juric, D., et al.: Functional network analysis reveals extended gliomagenesis pathway maps and three novel MYC-interacting genes in human gliomas. Cancer Res. **65**(19), 8679–8689 (2005). https://doi.org/10.1158/0008-5472.can-05-1204
5. Brigham, O.: The Fast Fourier Transform and its applications. In: Prentice-Hall Signal Processing Series, Englewood Cliffs, New Jersey (1998)
6. Calinski, T., Harabasz, J.: A dendrite method for cluster analysis. Commun. Stat. **3**, 1–27 (1974). https://doi.org/10.1080/03610927408827101
7. Celebi, M., Kingravi, H., Vela, P.: A comparative study of efficient initialization methods for the k-means clustering algorithm. Expert Syst. Appl. **40**(1), 200–210 (2013). https://doi.org/10.1016/j.eswa.2012.07.021
8. Cha, Y., Park, S., You, R., Kim, H., Yoon, D.: Microstructure arrays of dna using topographic control. Nat. Commun. **10**(1), 2512 (2019). https://doi.org/10.1038/s41467-019-10540-2
9. Chowdhary, S., Chamberlain, M.: Oligodendroglial tumors. Expert Rev. Neurother. **6**(4), 519–532 (2006). https://doi.org/10.1586/14737175.6.4.519
10. Darkins, R., Cooke, E., Ghahramani, Z., et al.: Accelerating Bayesian hierarchical clustering of time series data with a randomised algorithm. PLoS ONE **8**(4), e59795 (2013). https://doi.org/10.1371/journal.pone.0059795
11. Frattini, V., Trifonov, V., Chan, J.: The integrated landscape of driver genomic alterations in glioblastoma. Nat. Genet. **45**(10), 1141–1149 (2013)
12. Garza-Ulloa, J.: Methods to develop mathematical models: traditional statistical analysis. In: Applied Biomechatronics using Mathematical Models, pp. 239–371 (2018). https://doi.org/10.1016/B978-0-12-812594-6.00005-6
13. Gopinathan, S., Kokila, R., Thangavel, P.: Wavelet and FFT based image denoising using non-linear filters. Int. J. Electr. Comput. Eng. (IJECE) **5**(5), 1018–1026 (2015). https://doi.org/10.11591/ijece.v5i5
14. Heller, K., Ghahramani, Z.: Bayesian hierarchical clustering. In: Proceedings of the 22nd International Conference on Machine Learning, pp. 297–304 (2005). https://doi.org/10.1145/1102351.1102389
15. Ivakhnenko, A.: The group method of data handling - a rival method of stochastic approximation. Soviet Autom. Control **1**, 43–55 (1968)
16. Ivakhnenko, A.: Group method of data handling as competitor to the method of stochastic approximation. Soviet Autom. Control **3**, 64–78 (1968)
17. Kaufman, L., Rousseeuw, P.: Finding Groups in Data. An Introduction to Cluster Analysis. Wiley, Hoboken (2005). https://doi.org/10.1002/9780470316801
18. Lian, B., Hu, X., Shao, Z.M.: Unveiling novel targets of paclitaxel resistance by single molecule long-read RNA sequencing in breast cancer. Sci. Rep. **9**(1), 6032 (2019). https://doi.org/10.1038/s41598-019-42184-z
19. Lowing, N., Bomalaski, R., Mitra, D.: Bayesian Hierarchical Clustering. Nicholas Lowing & Ryan Bomalaski Group 3 CSE 5290 Dr. (2017)
20. Madala, H., Ivakhnenko, A.: Inductive Learning Algorithms for Complex Systems Modeling. CRC Press, Boco Raton (1994). https://doi.org/10.1201/9781351073493

21. Molugaram, K., Rao, S.: Statistical Techniques for Transportation Engineering. Butterworth-Heinemann, Oxford (2017). https://doi.org/10.3846/20294913.2016.1216906
22. Network, T.: Erratum: corrigendum: comprehensive genomic characterization defines human glioblastoma genes and core pathways. Nature **494**(7438), 506–506 (2013). https://doi.org/10.1038/nature11903
23. Omran, M., Engelbrecht, A., Salman, A.: An overview of clustering methods. Intell. Data Anal. **11**(6), 583–605 (2007)
24. Osypenko, V., Reshetjuk, V.: The methodology of inductive system analysis as a tool of engineering researches analytical planning. Ann. Warsaw Univ. Life Sci.-SGGW **58**, 67–71 (2011)
25. Raudys, A., Pabarškaitė, Z.: Optimising the smoothness and accuracy of moving average for stock price data. Technol. Econ. Dev. Econ. **24**(3), 984–1003 (2018). https://doi.org/10.1016/B978-0-12-811555-8.00012-X
26. Sarycheva, L.: Objective cluster analysis of the data on the basis of the group method of data handling. Prob. Manag. Inform. **2**, 86–104 (2008)
27. Savage, R., Heller, K., Xu, Y., et al.: R/BHC: fast Bayesian hierarchical clustering for microarray data. BMC Bioinform. **10**(1), 242 (2009). https://doi.org/10.1186/1471-2105-10-242
28. Sripada, S., Sreenivasa-Rao, M.: Comparison of purity and entropy of k-means clustering and fuzzy c means clustering. Indian J. Comput. Sci. Eng. **2**(3), 343–346 (2011)
29. Stepashko, V.: Elements of Inductive Modeling Theory - State and Prospects of Informatics Development in Ukraine: Monographic arts. K.: Scientific Thought (2010). https://doi.org/10.15407/kvt194.04.041
30. Verhaak, R., Hoadley, K., Purdom, E.: An integrated genomic analysis identifies clinically relevant subtypes of glioblastoma characterized by abnormalities in PDGFRA, IDH1, EGFR and NF1. Cancer Cell **17**(1), 98–110 (2010). https://doi.org/10.1016/j.ccr.2009.12.020
31. Xu, T., Le, T., Liu, L.: Identifying cancer subtypes from mirna-tf-mrna regulatory networks and expression data. PLoS ONE **11**(4), e015279 (2016)

Fingerspelling Alphabet Recognition Using CNNs with 3D Convolutions for Cross Platform Applications

Serhii Kondratiuk[1], Iurii Krak[1,2](✉), Anatolii Kylias[2], and Veda Kasianiuk[1]

[1] Taras Shevchenko National University of Kyiv, Kyiv, Ukraine
sergey.kondrat1990@gmail.com, yuri.krak@gmail.com, veda.kasianiuk@gmail.com

[2] Glushkov Cybernetics Institute, Kyiv, Ukraine
anatoly016@gmail.com

Abstract. The proposed communication technology is developed with cross-platform frameworks and consists of two parts: modeling and recognition of Ukrainian dactyl signs. Modeling is performed using realistic 3d hand model with animations of dynamic gesture and transitions between gesture, developed in Unity3D framework. User is able specify different words in the user interface and adjust number of polygons and step of animation to get satisfying performance. The computations can be done both on the device or in web. Recognition model training and serving is done with Tensorflow, which allows to deploy the model on different devices, including mobile, or to perform the model prediction on server in cloud. The dataset with Ukrainian dactyl signs was collected with 50 persons and 1500 images per each gesture, which allowed to train the model with high enough accuracy and robust in different environment conditions. The model is based on the MobileNetv3 convolutional neural network architecture, and with the optimal configuration of layers and network parameters, also in order to take into account temporal data, 3d convolutions were used. On the collected test dataset, which is 10% of the overall augmented dataset, and is 15000 images, with different light, noise and blurring condition and different personas hands, accuracy of over 98% is achieved.

Keywords: Cross platform · Sing language · Dactyl modeling · Mobilenet · Dactyl recognition · Convolutional neural networks

1 Introduction

There are multiple different ways of communication between human, and one of the widespread option is gesture-based information exchange (alongside with text and speech). Gestures are a common way to transfer information in lots of

S. Babichev et al. (Eds.): ISDMCI 2020, AISC 1246, pp. 585–596, 2021.
https://doi.org/10.1007/978-3-030-54215-3_37

domains, such as Human-Computer Interaction [23], but also is a main way to communicate in communities of people with hearing disabilities and their relatives. Almost every language also has a corresponding sign language, which is used by such communities. Providing a technology for such communities is and actual task, implying that such technology should be able to model and recognize gestures which make up a sign language. Also, such a technology should be available on a wide range of platforms, devices and operating systems (such as Android, iOS, Mac, Windows, Linux). Since most of the modeling and recognition functionality should be the same, this information technology should be also cross-platform [21], in order to utilize the same code-base, hand models for gesture modeling and recognition engine, in order not to reimplement every module under each platform or operating systems from scratch. Different devices vary in computational power and may be energy-efficient but with modest per-formance. Also, as web-based solutions are becoming more common, and web-based platforms are commonly used, allowing to provide the same functionality on different devices, information technology should also be able to run in web, as a web application, also allowing to move computation in the cloud [30].

In this article such a communication technology is proposed to overcome all of possible issues stated beforehand and to take advantage of current cross-platform developments. The proposed technology should be able to run on multiple most common platforms, such as mobile platforms, desktop platforms and web and be adaptable to the hardware it utilizes, meaning that low performance devices should be able to run the computations in cloud, whilst devices with higher performance can run the technology without internet connection on the devices solely. Proposed technology performs both gesture modeling and gesture recognition using cross-platform tools and frameworks, providing the ability to implement it once for all platforms, and also be configurable and apply changes for all platforms at once. The tradeoff tuning in 3d hand modeling and recognition is also proposed as a part of technology, allowing to adjust the step of animation while performing gesture modeling, quality of 3d hand model (adjusting number of polygons) or quality of gesture recognition by using more light-weight model with lower performance. The sing (gesture) modeling and recognition is a part of a single gesture communication technology and this paper is a further development of author's previous works [13,20].

2 Gesture Recognition Approaches Survey

Detection of hand gestures can be considered as a type of task of object detection, which has a set of mature and novel approaches in both classic computer vision and deep learning, with convolutions neural networks specifically. Since release of the convolutional neural network architecture for ImageNet contest, AlexNet [16], this new approach proved to show robust results in different condition of input data. Neural networks show robust recognition quality for object detections, when object have diverse distortions, different scale and various light conditions, noise, blur.

One of the key idea in transferring from static object detection into dynamic object detection is to use multiple subsequent frames from the video input instead of a single image, in order to utilize additional temporal data among with spatial data. As bigger datasets with recorded activities were released (Sports-1M [11], Kinetics [3], Jester [28]), convolutional neural networks with 3-dimensional convolutions because successful. The size of the dataset allowed to train the model without over fitting [5].

Gestures of sign language were detected using different approaches based on classic computer vision with hand-crafted features such as hyperplanes linear classifiers [1,17], geometric characteristics of hand contours defects [15]. Although the state of the art hand gesture recognition architectures are based on CNNs [12], similar to other computer vision tasks [14,18].

Commonly in order to achieve higher performance in terms of accuracy on the gesture dataset, the architecture of the CNN was made more complex [6,9].

However, the proposed technology in this paper aims to work cross-platform, an a various set of platforms and devices, some of them, such as hand-held devices (smartphones and tables) have limited computational resources and capabilities. Thus, research of existing approaches among CNN was accented on lightweight architectures which show satisfying performance on mobile cpus, such as SqueezeNet [10], MobileNet [8], MobileNetV2 [26], ShuffleNet [31] and ShuffleNetV2 [22], MobileNetV3 [7] which aim to reduce computational cost but still keep the accuracy high. In our work, we have used the 2D and 3D versions of MobileNetV3.

3 Problem Statement

The proposed technology should consist of 2 parts: sign language [19] modeling and gesture recognition module. Both modules should be able to run without codebase modification on multiple platforms and should be developed using cross-platform tools. A model to detect and recognize the gesture movement by the user from a camera input should be the in the module for the gesture recognition. The Ukrainian language is limited to a range of gestures but can be further expanded. For the testing of the model a suitable Ukrainian dactyl language dataset should be composed. A gesture specified by a certain set of parameters should be capable of reproducing with the sing language modeling module, using a database and limited to a set of Ukrainian dactylemes signs, but can be extended to other languages.

4 Proposed Approach

Implementing a solution for gestures modeling and recognition for Ukrainian dactyl language without reimplementation under different platforms and operating systems required cross-platform tools to be used. In proposed technology two main modules arc rccognition and hand modeling modules, using cross-platform tools and frameworks for both tasks. Gesture recognition is implemented using framework for deep learning Tensorflow, due to selected approach for this task based on convolutional neural networks with 3d convolutions. Tensorflow provides cross-platform tools to serve a trained mode and deploy it on different types of platforms, either on a server with API access or embedded on a device. MobileNet is used as architecture for a convolutional neural network due to it being optimized for mobile devices, providing ability to run it embedded on low-performance devices. Also, via 3d convolutions it is possible to take into account temporal data from multiple images of input from the camera.

Gesture modeling is implemented using Unity3D - a cross-platform development platform, which allows developing 3d scenes with animations and serving them across different operating systems, platforms and web. The 3d hand model can be adjusted using user interface also implemented in Unity3d. Also, via user interface, the user is able to set current gesture or to specify sequence of gestures, which are then animated. Overall, the novelty of proposed technology is in its cross-platform approach for unified solution for gesture modeling and recognition, also recognition being improved using spatio-temporal data.

5 Gesture Recognition with MobileNet

Gesture recognition, as a part of cross-platform technology for Ukrainian dactyl language modeling and recognition, should be implemented using cross-platform tools. Gesture recognition approach depend on the type of input information they work with. In case of 3D model bases algorithms or skeletal-based algorithms, the approach can use volumetric or skeletal model, or a combination of them [27,29]. Although, these approaches tend to be computationally expensive and require additional hardware from user.

Other type of approaches, appearance-based models derive parameters directly from the image or a sequence of images (in case video is used as an input). As a next step some pattern mining technique or machine learning approach is used to train a recognition model. Due to no need in additional hardware apart from a simple web camera, these type of approaches were selected for the cross-platform technology. Some approaches, for example, Ong et al. [24] propose Sequential Pattern Mining in order to detect signs based on the tree structures.

Convolutional Neural Networks (CNN) is a class of deep neural networks which are regularized versions of multilayer perceptrons, most commonly applied to analyzing images and videos. CNNs are especially good at analyzing images due to ability to take into account locality reference of the data in the image (typically nearby samples at some input data are not related, which is not true in case of an image). Therefore, CNN show state-of-the-art results in image classification and recognition tasks [2,4]. Another benefit of the convolutional neural networks is no need in handcrafted features, unlike conventional pattern matching algorithms. The process of training takes the input data and finds all the features needed for recognition and stores them as weights of the model. CNNs are robust at the task of classification or recognition of the object on an image, independent of input image scale, lightning conditions, occlusions, noise, etc. Although training such a model requires a sufficient dataset. Typically architecture of the CNN consists of a set of convolutional, pooling and ReLU layers. Tensorflow framework provides a cross-platform and performance-efficient implementation of convolutional neural networks.

MobileNetV3 architecture (see Fig. 1) is a new mobile architecture, development of the MobileNet model. MobileNetV3 extends its predecessor with 2 main ideas. Residual blocks connect the beginning and end of a convolutional block with a skip connection. By adding these two states the network has the opportunity of accessing earlier activations that weren't modified in the convolutional block. This approach turned out to be essential in order to build networks of great depth. On the other hand, MobileNetV3 follows a narrow->wide->narrow approach. The first step widens the network using a 1×1 convolution because the following 3×3 depth wise convolution already greatly reduces the number of parameters. Afterwards another 1×1 convolution squeezes the network in order to match the initial number of channels. A factor of 6 opposed to the 4 in our example represents the number of input channels and n how often the block is repeated. Lastly s tells whether the first repetition of a block used a stride of 2 for the down sampling process. This is a common assembly of convolutional blocks.

Input	Operator	exp size	*#out*	SE	NL	*s*
$224^2 \times 3$	conv2d, 3x3	-	16	-	HS	2
$112^2 \times 16$	bneck, 3x3	16	16	✓	RE	2
$56^2 \times 16$	bneck, 3x3	72	24	-	RE	2
$28^2 \times 24$	bneck, 3x3	88	24	-	RE	1
$28^2 \times 24$	bneck, 5x5	96	40	✓	HS	2
$14^2 \times 40$	bneck, 5x5	240	40	✓	HS	1
$14^2 \times 40$	bneck, 5x5	240	40	✓	HS	1
$14^2 \times 40$	bneck, 5x5	120	48	✓	HS	1
$14^2 \times 48$	bneck, 5x5	144	48	✓	HS	1
$14^2 \times 48$	bneck, 5x5	288	96	✓	HS	2
$7^2 \times 96$	bneck, 5x5	576	96	✓	HS	1
$7^2 \times 96$	bneck, 5x5	576	96	✓	HS	1
$7^2 \times 96$	conv2d, 1x1	-	576	✓	HS	1
$7^2 \times 576$	pool, 7x7	-	-	-	-	1
$1^2 \times 576$	conv2d 1x1, NBN	-	1024	-	HS	1
$1^2 \times 1024$	conv2d 1x1, NBN	-	k	-	-	1

Fig. 1. Architecture of the MobileNetv3

Network Improvements have been made in two ways:

1. Layer removal;
2. Swish non-linearity.

In the last block, the 1×1 expansion layer taken from the Inverted Residual Unit from MobileNetV2 is moved past the pooling layer. This means the 1×1 layer works on feature maps of size 1×1 instead of 7×7 making it efficient in terms of computation and latency.

We know that the expansion layer takes a lot of computation. But now that it is moved behind a pooling layer, we don't need to do the compression done by projection layer from the last layer from the previous block. Thus we can remove that projection layer and the filtering layer from the previous bottleneck layer (block).

6 Ukrainian Dactyl Alphabet Dataset Collections for Recognition with MobileNet

Preparing a CNN model relies upon a dataset of significant size with wide representation of gestures in different conditions, including light conditions, scale, skin tone, hand size and proportions, gender, etc. Such a dataset should contain more than 1000 images for each gesture from Ukrainian dactyl alphabet from a significant number of persons. As a part of this paper work, such a dataset was collected with specific distribution on different qualities of images and hands. Specifically, there were in-vited 50 persons to take images of their hands, separated into 70% of male and 30% of female hands. For each gesture 1500 images were collected with various conditions, from poor to high quality of light conditions, with 20% images in poor conditions, 50% images in mediocre and 30% images in high quality light conditions, with similar distribution on other conditions, such as blur and noise. Also different distance from hand to camera were used during collection of the dataset. Different dataset augmentations strategies were used during composition of the dataset, such as rotations, scaling, noising, random sampling, etc. Overall, a dataset of 150000 images was collected. Near 10% of the dataset for used for testing purposes of the final trained model for recognition, as is shown in Fig. 2.

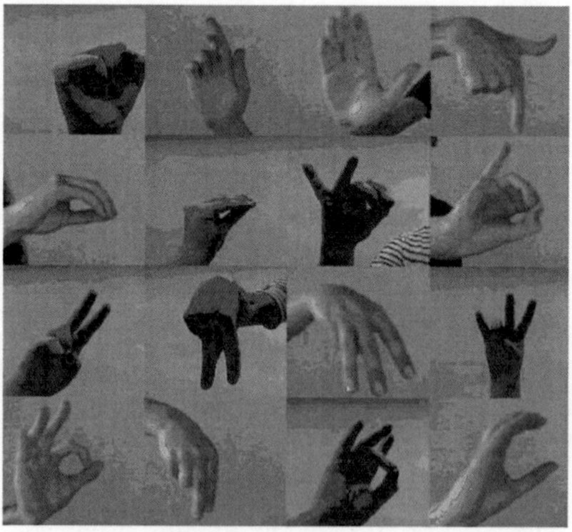

Fig. 2. Sample of collected dataset

For dataset collected a convenient software was developed, using C# and .Net framework for collecting images from camera input and specifying which letter (gesture, dactyl) is currently recorded. This tool shouldn't be considered as a part of information technology, since it's very specific task oriented and non-cross-platform, also it was developed in order to collect Ukrainian dactyl alphabet images dataset, since it wasn't available openly. A typical pipeline for letter recording consisted of such steps:

1. The individual which is being recorded place hand into the specified rect, which is a region of interest.
2. The individual performs needed dactyl sign.
3. The sequence of images is recorded.
4. The individual performing the dactyl slightly moves the hand across various directions.
5. The recording stops and proceeds with the next dactyl sign.

7 MobileNetv3 with 3D Convolutions

Figure 3 shows the pipeline with spatiotemporal modeling approach used for 2D CNN models. We get features of 8 images with 2D CNN and merge them with

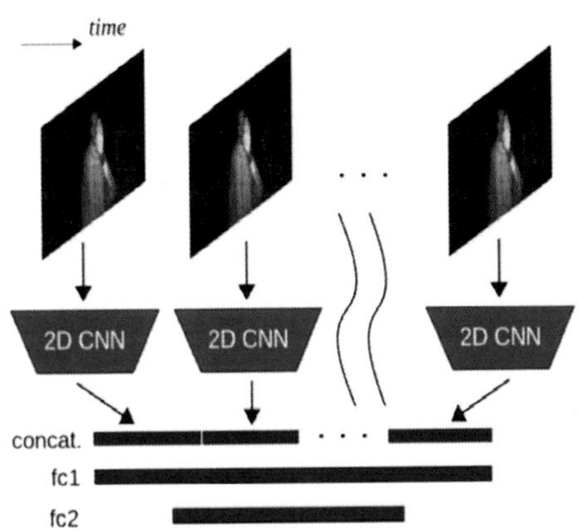

Fig. 3. Pipeline of 3D-convolutions with MobileNetv3

the same sorting. Then, to achieve class-conditional probability values, we use 2 fully connected layers, because they can get the temporal relations, not knowing it's a sequence. Each feature is of size 64 as output form 2D CNN. From first layer, size of feature is 256. With the 2nd layer, size is quantity of classes.

From other side, 3D CNNS have spatio-temporal modeling inherently and don't need additional mechanics. MobileNetV3 is exaggerated so that it receives eight images as input. The information on 3D-MobileNetv3 is presented in Fig. 4.

Layer / Stride	Repeat	Output size
Input clip		c×8×112×112
Conv1(3×3×3)/s(1,2,2)	1	32×8×56×56
Block/s(1,1,1)	1	16×8×56×56
Block/s(1,2,2)	2	24×8×28×28
Block/s(2,2,2)	3	32×4×14×14
Block/s(2,2,2)	4	64×2×7×7
Block/s(1,1,1)	3	96×2×7×7
Block/s(2,2,2)	3	160×1×1×1
Block/s(1,1,1)	1	320×1×1×1
Conv(1×1×1)/s(1,1,1)	1	1280×1×1×1
Linear(1280×*NumCls*)	1	*NumCls*

Fig. 4. Architecture of MobileNetv3 with 3D-convolutions

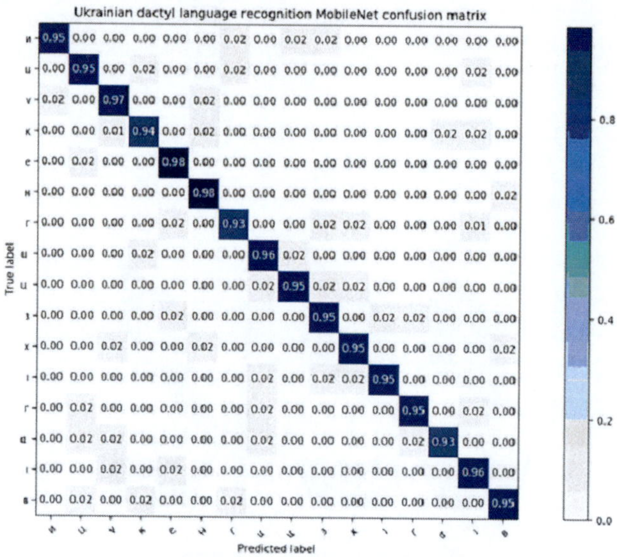

Fig. 5. Confusion matrix of the optimal architecture model

8 Gesture Recognition Experiment

To achieve the best possible result on model training, there were used methods for decreasing overfitting issue of the convolutional network training, such as dropout and early stopping. Multiple variations of MobileNet architecture were considered during training, they were trained and corresponding metrics calculated. Best candidate was chosen based on results which are shown in Fig. 5. Alternative architectures were also considered, with various numbers of convolutional layers, but showed worse accuracy, as are shown in Fig. 6 and Fig. 7.

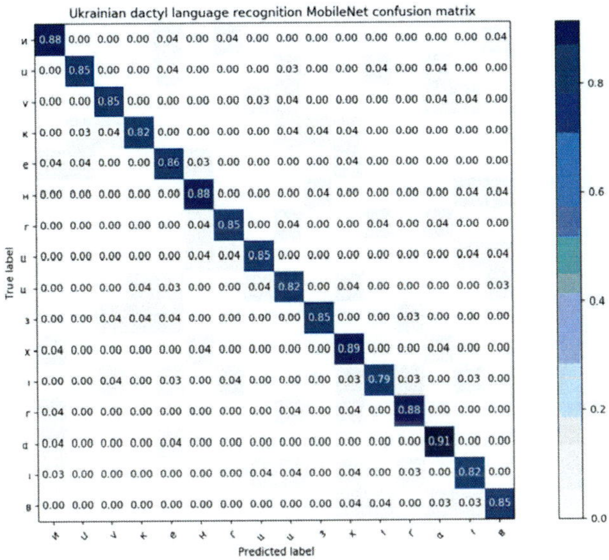

Fig. 6. Confusion matrix of the architecture model with lower number of conv-layers

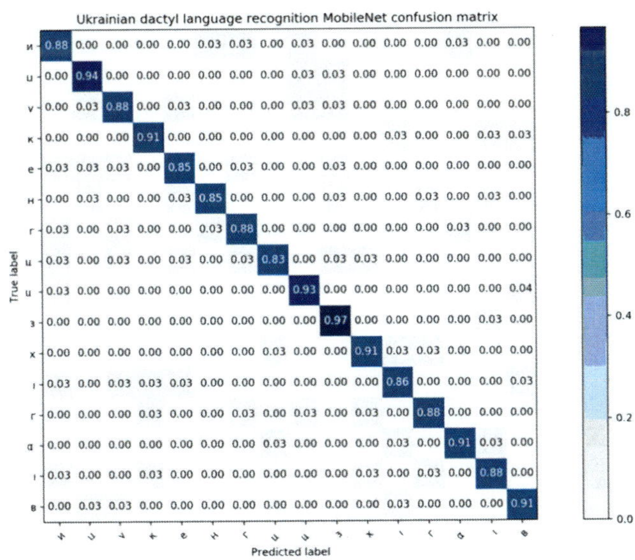

Fig. 7. Confusion matrix of the architecture model with higher number of conv-layers

9 Conclusions

The proposed technology is implemented using cross-platform tools for gesture modeling and gesture recognition modules. Using Tensorflow, Unity3d frameworks and PostgreSQL [25] database allowed to achieve satisfying performance across different platforms, mobile, desktop and web. Technology is implemented with user interface, also built with Unity3d framework, which allows user to select and animate Ukrainian dactyls, with adjustable 3d hand model quality, based on hardware computational capabilities. In order to implement a gesture recognition model with state-of-the-art recognition accuracy on Ukrainian dactyl gestures, a convolutional neural network based on MobileNet architecture was trained. 3d convolutions were utilized, in order to take into account spatio-temporal data, so that model makes predictions on sequence of input images. Trained model achieved state-of-the-art accuracy showed in the paper and served as cross-platform solutions. Light-weight versions of model is also trained in order to deploy on devices. Novelty of the proposed technology is in the collected dataset of Ukrainian dactyl signs images, of size 150000 images overall from more then 50 personas, in different light conditions and distributions of ages, skin tones and gender of the person showing the hand. The proposed technology is cross-platform for both tasks of gesture modeling and recognition, and achieves this using cross-platform frameworks, such as Tensorflow and Unity3D. Novelty of the model lies in using 3d convolutions in the MobileNetv3 architecture in order to take into account temporal information between frames of gesture transition. Satisfying accuracy (more than 98%) on test dataset has been achieved with such model, which shows state-of-the-art results for such class of tasks (sign language recognition) and for Ukrainian dactyl language specifically. The proposed gesture communication technology can be enhanced with other gestures and languages and with other cross-platform modules.

References

1. Barmak, A., Krak, Y., Manziuk, E., Kasianiuk, V.: Information technology of separating hyperplanes synthesis for linear classifiers. J. Autom. Inf. Sci. **51**(5), 54–64 (2019). https://doi.org/10.1615/JAutomatInfScien.v51.i5.50
2. Bobic, V., Tadic, T.: Hand gesture recognition using neural network based techniques. In: 13th Symposium on Neural Networks and Applications (NEUREL), pp. 35–38. IEEE (2016)
3. Carreira, J., Zisserman, A.: Quo vadis, action recognition? A new model and the kinetics dataset. In: Proceedings of the IEEE conference on Computer Vision and Pattern Recognition, pp. 4724–4733. IEEE (2017). https://doi.org/10.1109/CVPR.2017.502
4. Garcia, B.: American Sign language: Real-time American Sign Language Recognition with Convolutional Neural Networks. Stanford University Stanford (2015)
5. Hara, K., Kataoka, H., Satoh, Y.: Can spatiotemporal 3d CNNS retrace the history of 2d CNNS and imagenet? In: Proceedings of the IEEE Conference on Computer Vision and Pattern Recognition, pp. 6546–6555. IEEE (2018)
6. He, K., Zhang, X., Ren, S., Sun, J.: Deep residual learning for image recognition. In: Proceedings of the IEEE conference on Computer Vision and Pattern Recognition, pp. 770–778. IEEE (2016). https://doi.org/10.1109/CVPR.2016.90
7. Howard, A., Sandler, M., Chu, G., et al.: Searching for mobilenetv3. In: arXiv preprint. p. art. no. 5 (2019). arXiv:1905.02244

8. Howard, A., Zhu, M., Chen, B., et al.: Mobilenets: efficient convolutional neural networks for mobile vision applications. In: arXiv preprint. p. art. no. 9 (2017). arXiv:1704.04861
9. Hu, J., Shen, L., Sun, G.: Squeeze-and-excitation networks. In: Proceedings of the IEEE Conference on Computer Vision and Pattern Recognition, pp. 7132–7141. IEEE (2018). https://doi.org/10.1109/CVPR.2018.00745
10. Iandola, F.N., Han, S., Moskewicz, M.W., et al.: Squeezenet: Alexnet-level accuracy with 50x fewer parameters and ¡0.5mb model size. In: arXiv preprint. p. art. no. 13 (2016). arXiv:1602.07360
11. Karpathy, A., Toderici, G., Shetty, S., et al.: Large-scale video classification with convolutional neural networks. In: Proceedings of the IEEE Conference on Computer Vision and Pattern Recognition, pp. 1725–1732. IEEE (2014). https://doi.org/10.1109/CVPR.2014.223
12. Kopuklu, O., Kose, N., Rigoll, G.: Motion fused frames: Data level fusion strategy for hand gesture recognition. In: arXiv preprint. p. art. no. 9 (2018). arXiv:1804.07187v2
13. Krak, I., Kondratiuk, S.: Cross-platform software for the development of sign communication system: Dactyl language modelling. In: Proceedings of the 12th International Scientific and Technical Conference on Computer Sciences and Information Technologies, CSIT 2017, vol. 1, pp. 167–170. IEEE (2017). https://doi.org/10.1109/STC-CSIT.2017.8098760
14. Krak, I., Kudin, G., Kulias, A.: Multidimensional scaling by means of pseudoinverse operation. Cybern. Syst. Anal. **55**(1), 22–29 (2019). https://doi.org/10.1007/s10559-019-00108-9
15. Krak, Y., Golik, A., Kasianiuk, V.: Recognition of dactylemes of ukrainian sign language based on the geometric characteristics of hand contours defects. J. Autom. Inf. Sci. **48**(4), 90–98 (2016). https://doi.org/10.1615/JAutomatInfScien.v48.i4.80
16. Krizhevsky, A., Sutskever, I., Hinton, G.E.: Imagenet classification with deep convolutional neural networks. Adv. Neural Inf. Process. Syst. **25**(2), 1097–1105 (2012). https://doi.org/10.1145/3065386
17. Kryvonos, I., Krak, I.: Modeling human hand movements, facial expressions, and articulation to synthesize and visualize gesture information. Cybern. Syst. Anal. **47**(4), 501–505 (2011). https://doi.org/10.1007/s10559-011-9332-4
18. Kryvonos, I., Krak, I.V., Barmak, O., Kulias, A.: Methods to create systems for the analysis and synthesis of communicative information. Cybern. Syst. Anal. **53**(6), 847–856 (2017). https://doi.org/10.1007/s10559-017-9986-7
19. Kryvonos, I., Krak, I., Barmak, O., Shkilniuk, D.: Construction and identification of elements of sign communication. Cybern. Syst. Anal. **49**(2), 163–172 (2013). https://doi.org/10.1007/s10559-016-9812-7
20. Kryvonos, I., Krak, Y., Barchukova, Y., Trotsenko, B.: Human hand motion parametrization for dactylemes modeling. J. Autom. Inf. Sci. **43**(12), 1–11 (2011). https://doi.org/10.1615/JAutomatInfScien.v43.i12.10
21. LINFO: The linux information project, cross-platform definition. http://www.linfo.org/cross-platform.html
22. Ma, N., Zhang, X., Zheng, H.T., Sun, J.: Shufflenet v2: practical guidelines for efficient CNN architecture design. In: arXiv preprint. p. art. no. 5 (2018). arXiv:1807.11164
23. Mell, P., Grance, T.: The NIST Definition of Cloud Computing, Technical report. National Institute of Standards and Technology: U.S. Department of Commerce. Special publication (2011). https://doi.org/10.6028/NIST.SP.800-145

24. Ong, E.J., Cooper, H., Pugeault, N., Bowden, R.: Sign language recognition using sequential pattern trees. In: Proceedings of the IEEE Conference on Computer Vision and Pattern Recognition, pp. 2200–2207. IEEE (2012)
25. PostgreSQL: The world's most advanced open source relational database (2020). https://www.postgresql.org/
26. Sandler, M., Howard, A., Zhu, M., et al.: Mobilenetv2: inverted residuals and linear bottlenecks. In: Proceedings of the IEEE Conference on Computer Vision and Pattern Recognition, pp. 4510–4520. IEEE (2018). https://doi.org/10.1109/CVPR.2018.00474
27. TensorFlow: Tensorflow framework documentation (2020). https://www.tensorflow.org/api/
28. Twentybn: The 20bn-jester dataset v1 (2019). https://20bn.com/datasets/jester
29. Unity: Unity3d framework (2019). https://unity3d.com/
30. Wang, H., He, W., Wang, F.K.: Enterprise cloud service architectures. Inf. Technol. Manag. **13**(4), 445–454 (2012). https://doi.org/10.1007/s10799-012-0139-4
31. Zhang, X., Zhou, X., Lin, M., Sun, J.: Shufflenet: an extremely efficient convolutional neural network for mobile devices. In: Proceedings of the IEEE Conference on Computer Vision and Pattern Recognition, pp. 6848–6856. IEEE (2018)

Modeling, Analysis and Automation of Wastewater Treatment by Electrocoagulation Method

Andrii Safonyk(✉), Ivan Tarhonii, Andrii Rudyk, and Ivanna Hrytsiuk

National University of Water and Environmental Engineering, Rivne, Ukraine
{a.p.safonyk,a.v.rudyk,i.m.hrytsiuk}@nuwm.edu.ua, targoniy93@ukr.net

Abstract. Theoretical and applied principles of construction of an automated system of sewage treatment on the basis of a generalized spatial model of the process of electrocoagulation purification are proposed. A model of electrocoagulation, which describes the processes taking place in the reactor considering the spatial placement of plates is developed. The solution of the corresponding model problem has been obtained by an asymptotic method. A comparative analysis of the results of computer simulation with the results of full-scale experiment has been conducted. The effect of reactor size, current density, arrangement of electrode plates on the efficiency of chromium extraction was investigated. Comprehensive research has shown that a spatial mathematical model that describes the basic laws of wastewater treatment and the algorithm for solving a singularly perturbed problem is effective for carrying out performance-oriented defining (in particular optimization) the parameters of the treatment process. The obtained ratios are effective for optimizing the parameters of the purification process, in particular, the sizes of the electrocoagulation unit, etc., as well for conducting theoretical studies in cases of the predominance of convective and electrochemical components of the respective process over diffusion, that is typical for the vast majority of electrocoagulation units. The basic scheme of automation of the process of sewage treatment from chromium is constructed. A control system with the ability to change the performance of the entire installation in real-time using SCADA - WinCC Flexible system is provided.

Keywords: m model of electrocoagulation process · Singularly perturbed problem · Asymptotic method · Automated system

1 Introduction

The quality of natural waters changes the most as a result of pollution by sewage from industrial enterprises and communal services, as well as from surface runoff from the territories of settlements, industrial sites, transport routes and agricultural lands. Reagent method of purification and disposal of concentrated technological solutions and washing water is widely used as a method of sewage treatment in the production. Reagent methods include neutralization of acids and

S. Babichev et al. (Eds.): ISDMCI 2020, AISC 1246, pp. 597–611, 2021.
https://doi.org/10.1007/978-3-030-54215-3_38

alkalis, transfer of ions to a slightly soluble state. One of the most progressive trends in water treatment technology is the development and implementation of electrochemical methods. Electrochemical methods (electrodialysis, electrochemical oxidation and hydrolysis) are methods related to the action of electric current on aqueous solutions.

2 Review of the Literature

One of the most promising cleaning methods is the electrocoagulation purification method, because this method requires no additional reagents and can be fully automated. The method of electrocoagulation is based on the physicochemical process of sedimentation (coagulation) of colloidal systems when exposed to them by direct electric current. Electrocoagulation sewage treatment is the process of passage of water into the interelectrode space of the cell, while a certain potential difference is applied to the electrodes [14,15].

At present, a large number of papers devoted to studying the electrocoagulation method for wastewater treatment from different types of pollution [1,3–6,12]. In particular, in [6] was investigated a simple, efficient and inexpensive electrocoagulation method for the treatment of oily mud waters using turbidity measurements. This study has shown that this process can be done by the method of greasy dirt purification in heavily contaminated sewage. In the process of electrocoagulation, microbes can be immobilized by direct adsorption on the surface of the anode by electron transfer and physical elimination through floating pathogens with formed aqueous gas and/or deposition with formed flocs. The integration of electrocoagulation with processes that support free radicals (eg, electro-oxidation), magnetic field and/or ultrasonic field remains a promising method of facilitating its full-scale implantation. Membrane processes should be considered as safe barriers to the disinfection of by-products and hydroxyl radicals [5]. In [1], an electrocoagulation-microfiltration method in batch-mode was used to purify oily wastewater. Microfiltration was performed using indigenous type ceramic membranes to remove these particles formed during the electrocoagulation process at different pressures. In [3], the mechanisms of electrocoagulation of sewage in industrial waters were reviewed and the results of usage of electrocoagulation in cleaning wast water from various contaminants were presented. The authors also investigated the contribution of electric current to the efficiency of electrocoagulation, described the mechanisms of physical elimination and chemical deactivation for the procedure of removing microbes by electrocoagulation. In [12], the method of electrocoagulation is presented as an alternative to conventional water conditioning processes. The main focus of the work was to investigate the effect of operating parameters, such as pH, temperature, electrode material, etc., on the efficiency of removal of contaminants such as Escherichia coli and on increasing concentrations of iron, arsenic, manganese, ammonia, chromium, etc.

However, despite the great interest in the process of electrocoagulation, most of the works are experimental and not sufficiently substantiated theoretically [7].

Also, to date, the non-linear "model mechanisms" that take into account the reverse influence of various characteristics of the electrocoagulation purification process on environmental characteristics are poorly developed, non-systematic or generally absent. There are practically no works aimed at developing mathematical support for automation systems by electrocoagulation purification processes. No less important is the problem of mathematical description (analysis) of experimental data and substantiation of the adequacy of constructed models. Also, the considered problems usually are one-dimensional and cannot sufficiently be effective for the tasks aimed at improving the performance (including optimization) of treatment systems.

Based on the above, **the aim of the research** is to summarize the mathematical model of the process of electrocoagulation wastewater treatment in a spatial case, its analysis and development of a system for automation of the process.

3 Materials and Methods

Let us consider the curvilinear parallelepiped (electrocoagulation unit) of $= ABCDA_*B_*C_*D_*$, limited by smooth orthogonal equipotential surfaces between themselves in angular points and ribs $ABB_*A_* = \{z : f_1(x,y,z) = 0\}$, $CDD_*C_* = \{z : f_2(x,y,z) = 0\}$ and also surfaces of flow $ADD_*A_* = \{z : f_3(x,y,z) = 0\}$, $BCC_*B_* = \{z : f_4(x,y,z) = 0\}$, $ABCD = \{z : f_5 (x,y,z) = 0\}$, $A_*B_*C_*D_* = \{z : f_6(x,y,z) = 0\}$. In the paper, the mathematical model developed in [10] for a spatial case. Thus, to find the distribution of pollution concentration and temperature in the electrocoagulation unit, we come to the following model problem:

$$\begin{cases} \dfrac{\partial c}{\partial t} = -v_c \nabla c + \nabla \left(D_c(\bar{T}) \nabla c \right) - \alpha \bar{T} c - \Phi \\ \dfrac{\partial \bar{T}}{\partial t} = v_{\bar{T}} \nabla \bar{T} + D_{\bar{T}} \Delta \bar{T} + \Psi \end{cases} \tag{1}$$

$$\begin{aligned} & c\,|_{ABB_*A_*} = c_*(M,t), \quad \frac{\partial c}{\partial \vec{n}}\,|_{CDD_*C_*} = 0 \\ & \frac{\partial c}{\partial \vec{n}}\,|_{ADD_*A_* \cup BCC_*B_* \cup ABCD \cup A_*B_*C_*D_*} = 0 \\ & \bar{T}\,|_{ABB_*A_*} = \bar{T}_*(M,t), \quad \frac{\partial \bar{T}}{\partial \vec{n}}\,|_{CDD_*C_*} = 0 \\ & \frac{\partial \bar{T}}{\partial \vec{n}}\,|_{ADD_*A_* \cup BCC_*B_* \cup ABCD \cup A_*B_*C_*D_*} = 0 \end{aligned} \tag{2}$$

where c is the concentration of pollution (ppm); $\bar{T}$ is temperature (°K); $D_{\bar{T}}$ is the diffusion coefficient (m^2/s); q_T is the intensity of internal heat sources; v_C is the rate of spread of pollution (m/s), $v_{\bar{T}}$ is the rate of spread of temperature (m/s). The amount of heat $\Psi = \frac{q_T}{c_T \rho}$ thus released is proportional to the strength of the current, the time of its passage and the voltage drop $q_T = I \cdot U \cdot t$, U is applied voltage (B), I is a current strength (A), c_T is a heat capacity (J/kg), ρ is a density (kg/m^3), $0 < x < L$ is a coagulator length (m), $0 < y < H$ - is a coagulator height (m), $0 < z < B$ is a coagulator width (m). During electrocoagulation, the electrolyte solution is heated. The diffusion motion of particles in a fluid can be considered as friction motion, to which the second Einstein relation can be applied: $D_C(T) = U_C kT$. Here k is Boltzmann constant (J/K), U mobility of the diffusing particles, that is, the coefficient of proportionality between the velocity of the particle and the driving force (Pa·s·m). If the particles are spherically symmetric, then $U_C = 1/(6\pi hr)$, where h is the coefficient of viscosity of the liquid, r is the radius of the particle [8], $\Phi = \frac{It}{V_A nF} + k_h S_k t$ is the function responsible for the electrochemical reaction, V_A here is the volume of the apparatus (m^3), n is the number of reaction electrons, F is the Faraday constant (C/mol), k_h is the rate constant of the heterogeneous reaction (mol/m^5·s), S_k is the cathode area (m^2).

Taking into account the relationship between the values of parameters that characterize certain components of the process (in particular the dominance of convective and mass-exchange factors over diffusion) leads to the complication of the mathematical model of the process. One of the effective ways of solving the corresponding problems in the case of the predominance of convective factors of the process over diffusion during filtration of aqueous solutions in the model regions limited by equipotential or quasi-equipotential lines and flow lines is: phased fixation of the characteristics and components of the process and environment; solving filtration problems using conformal or quasi-conformal mappings of a complex potential or quasi-potential area on a physical area; transition in equations of convective diffusion and boundary and initial conditions from physical variables to the coordinates of the region of complex potential or quasi-potential, which greatly simplifies their recording and provides the possibility of autonomous research, parallel computing.

Given that the problem of finding the velocity field (1)–(2) is solvable [2], in particular, the velocity field and a number of other quantities, such as filtration flow, are calculated, and by replacing the variable $x = x(\varphi, \psi, \eta)$, $y = y(\varphi, \psi, \eta)$, $z = z(\varphi, \psi, \eta)$, in system (1) and conditions (2), the corresponding "diffusion problem" is obtained [2,8,9,11,13]:

$$\begin{cases} \dfrac{\partial C}{\partial t} + v\nabla C = \varepsilon\nabla\left(d_C T\nabla C\right) - \varepsilon\bar{\alpha}T\,C - \left(\dfrac{I}{V_A nF} + k_h S_k\right)t \\ \dfrac{\partial T}{\partial t} - v\nabla T = \varepsilon d_T \Delta T + \dfrac{I \cdot U \cdot t}{c_T \rho} \end{cases} \tag{3}$$

$$\begin{gathered} C(\varphi, \psi, \eta, 0) = C_0^0(\varphi, \psi, \eta),\ T(\varphi, \psi, \eta, 0) = T_0^0(\varphi, \psi, \eta) \\ C(\bar{\varphi}_*, \psi, z, t) = C_*(\psi, \eta, t),\ T(\bar{\varphi}_*, \psi, \eta, t) = T_*(\psi, \eta, t) \\ \left.\frac{\partial C}{\partial \varphi}\right|_{\varphi=\bar{\varphi}^*} = 0\ \left.\frac{\partial T}{\partial \varphi}\right|_{\varphi=\bar{\varphi}^*} = 0 \end{gathered} \tag{4}$$

$$\left.\frac{\partial C}{\partial \psi}\right|_{\psi=0} = 0, \left.\frac{\partial C}{\partial \psi}\right|_{\psi=Q_*} = 0$$

$$\left.\frac{\partial T}{\partial \psi}\right|_{\psi=0} = 0, \left.\frac{\partial T}{\partial \psi}\right|_{\psi=Q_*} = 0$$

$$\left.\frac{\partial C}{\partial \eta}\right|_{\eta=0} = 0, \left.\frac{\partial C}{\partial \eta}\right|_{\eta=Q^*} = 0$$

$$\left.\frac{\partial T}{\partial \eta}\right|_{\eta=0} = 0, \left.\frac{\partial T}{\partial \eta}\right|_{\eta=Q^*} = 0$$

the solution of which with precision $O\left(\varepsilon^{n+1}\right)$ where ε is a small parameter that characterize the predominance of the convective and mass exchange component of the process $(D_C(T) = \varepsilon d_C T = \varepsilon U_C kT,\ D_T = \varepsilon d_T,\ \bar{\alpha} = \varepsilon\alpha,\ |v| > v_* \gg \varepsilon)$ was obtained as an asymptotic series [8]:

$$C = C_0 + \sum_{i=1}^{n} \varepsilon^{i} C_i + \sum_{i=0}^{n} \varepsilon^{i} \Pi_i + \sum_{i=0}^{n} \varepsilon^{i} \breve{\Pi}_j + \sum_{i=0}^{n+1} \varepsilon^{i/2}\, \tilde{\Pi}_i$$
$$+ \sum_{i=0}^{n+1} \varepsilon^{i/2}\, \tilde{\tilde{\Pi}}_i + \sum_{i=0}^{n+1} \varepsilon^{i/2}\, \widehat{\Pi}_i + \sum_{i=0}^{n+1} \varepsilon^{i/2}\, \widehat{\widehat{\Pi}}_i + R_C \quad (5)$$

$$T = T_0 + \sum_{i=1}^{n} \varepsilon^{i} T_i + \sum_{i=0}^{n} \varepsilon^{i} \bar{P}_i + \sum_{i=0}^{n} \varepsilon^{i} \breve{P}_i + \sum_{i=0}^{n+1} \varepsilon^{i/2}\, \tilde{P}_i$$
$$+ \sum_{i=0}^{n+1} \varepsilon^{i/2}\, \tilde{\tilde{P}}_i + \sum_{i=0}^{n+1} \varepsilon^{i/2}\, \widehat{P}_i + \sum_{i=0}^{n+1} \varepsilon^{i/2}\, \widehat{\widehat{P}}_i + R_T \quad (6)$$

where $R_C(\varphi, \psi, \eta, t, \varepsilon)$, $R_T(\varphi, \psi, \eta, t, \varepsilon)$ are the remaining members, $C_i(\varphi, \psi, \eta, t)$, $T_i(\varphi, \psi, \eta, t)$ are regular parts of the asymptotic series $(i = \overline{0,\ n})$; $\Pi_i(\xi, \psi, \eta, t)$, $\bar{P}_i(\xi, \psi, \eta, t)$, are regular parts of the asymptotic series $\varphi = \bar{\varphi}^*$ (amendments to the output of the electrocoagulation unit) $\left(i = \overline{0, n+1}\right)$, $\breve{\Pi}_j\left(\breve{\xi}, \psi, \eta, t\right)$, $\breve{P}_j\left(\breve{\xi}, \psi, \eta, t\right)$ are the boundary layer functions in the vicinity $\varphi = \bar{\varphi}_*$ (corrections at the entrance to the electrocoagulation unit) $\left(i = \overline{0, n+1}\right)$, and functions $\tilde{\Pi}_i\left(\varphi, \tilde{\psi}, \eta, t\right)$, $\tilde{\tilde{\Pi}}_i\left(\varphi, \tilde{\tilde{\psi}}, \eta, t\right)$, $\widehat{\Pi}_i(\varphi, \psi, \tilde{\eta}, t)$, $\widehat{\widehat{\Pi}}_i(\varphi, \psi, \tilde{\tilde{\eta}}, t)$, and $\tilde{P}_i\left(\varphi, \tilde{\psi}, \eta, t\right)$, $\tilde{\tilde{P}}_i\left(\varphi, \tilde{\tilde{\psi}}, \eta, t\right)$, $\widehat{P}_i(\varphi, \psi, \tilde{\eta}, t)$, $\widehat{\widehat{P}}_i(\varphi, \psi, \tilde{\tilde{\eta}}, t)$, $\left(i = \overline{0, n+1}\right)$ are the boundary layer functions in the vicinity $\psi = 0,\ \psi = Q_*,\ \eta = 0,\ \eta = Q^*$ (amendments around the side walls of the electrocoagulation unit), respectively; $\xi = (\varphi^* - \varphi)/\varepsilon$, $\breve{\xi} = (\varphi - \varphi_*)/\varepsilon$, $\tilde{\psi} = \psi/\sqrt{\varepsilon}$, $\tilde{\tilde{\psi}} = (Q_* - \psi)/\sqrt{\varepsilon}$, $\tilde{\eta} = \eta/\sqrt{\varepsilon}$, $\tilde{\tilde{\eta}} = (Q^* - \eta)/\sqrt{\varepsilon}$ - "stretches" of the relevant variables. By substituting the relations (5)–(6) in to (3)–(4) and performing the standard procedure of "equating" the coefficients for the same powers of ε, we obtain the following problems for obtaining $C_i(\varphi,\psi,\eta,t)$, $T_i(\varphi,\psi,\eta,t)$ $(i=\overline{0,n})$.

$$\begin{cases} \dfrac{\partial C_0}{\partial t} + v\nabla C_0 = -\varPhi, \quad \dfrac{\partial T_0}{\partial t} - v\nabla T_0 = \varPsi \\ C_0\left(\varphi,\psi,\eta,0\right) = C_0^0(\varphi,\psi,\eta),\ T_0\left(\varphi,\psi,\eta,0\right) = T_0^0(\varphi,\psi,\eta) \\ C_0\left(\bar{\varphi}_*,\psi,z,t\right) = C_*(\psi,\eta,t),\ T_0\left(\bar{\varphi}_*,\psi,\eta,t\right) = T_*(\psi,\eta,t) \end{cases}$$

$$\begin{cases} \dfrac{\partial C_i}{\partial t} + v\nabla C_i = \nabla\left(d_C T_{i-1}\nabla C_{i-1}\right) - \bar{\alpha}T_{i-1}\,C_{i-1} - \varPhi \\ \dfrac{\partial T_i}{\partial t} - v\nabla T_i = d_T \varDelta T_{i-1} + \varPsi \\ C_i\left(\varphi,\psi,\eta,0\right) = 0,\ T_i\left(\varphi,\psi,\eta,0\right) = 0 \\ C_i\left(\bar{\varphi}_*,\psi,z,t\right) = 0,\ T_i\left(\bar{\varphi}_*,\psi,\eta,t\right) = 0 \end{cases}$$

The result of solving these problems is:

$$T_0 = \begin{cases} \displaystyle\int_{\varphi_0}^{\varphi} \frac{\varPsi\left(s,\psi,\eta,f\left(s,\psi,\eta\right)-f+t\right)}{v^2\left(s,\psi,\eta\right)} ds + T_*\left(t-f,\psi,\eta\right), & t \geq f \\ \displaystyle\int_0^t \varPsi\left(f^{-1}\left(s+f-t,\psi,\eta\right),\psi,\eta,s\right) ds + T_0^0\left(f^{-1}\left(f-t,\psi,\eta\right),\psi,\eta\right), & t < f \end{cases}$$

$$C_0 = \begin{cases} \displaystyle\int_{\varphi_0}^{\varphi} \frac{\varPhi\left(s,\psi,\eta,f\left(s,\psi,\eta\right)-f+t\right)}{v^2\left(s,\psi,\eta\right)} ds + C_*\left(t-f,\psi,\eta\right), & t \geq f \\ \displaystyle\int_0^t \varPhi\left(f^{-1}\left(s+f-t,\psi,\eta\right),\psi,\eta,s\right) ds +_0^0 \left(f^{-1}\left(f-t,\psi,\eta\right),\psi,\eta\right), & t < f \end{cases}$$

$$T_i = \begin{cases} \displaystyle\int_{\varphi_0}^{\varphi} \frac{\varPsi_i\left(s,\psi,\eta,f\left(s,\psi,\eta\right)-f+t\right)}{v^2\left(s,\psi,\eta\right)} ds, & t \geq f \\ \displaystyle\int_0^t \varPsi_i\left(f^{-1}\left(s+f-t,\psi,\eta\right),\psi,\eta,s\right) ds, & t < f \end{cases}$$

$$C_i = \begin{cases} e^{-\lambda_i} \displaystyle\int_{\varphi_0}^{\varphi} \frac{\varPhi_i\left(s,\psi,\eta,f\left(s,\psi,\eta\right)-f+t\right)}{v^2\left(s,\psi,\eta\right)} e^{\lambda_i(s,\psi,\eta,t)} ds, & t \geq f \\ e^{-\lambda_i} \displaystyle\int_0^t \varPhi_i\left(f^{-1}\left(s+f-t,\psi,\eta\right),\psi,\eta,s\right) e^{\lambda_i(\varphi,\psi,\eta,s)} ds, & t < f \end{cases}$$

where $\Phi_i(\varphi,\psi,\eta,t) = \nabla(d_C T_{i-1}\nabla C_{i-1}) - \Phi$, $\Psi_i(\varphi,\psi,\eta,t) = d_T \Delta T_{i-1} + \Psi$, $\lambda_i(\varphi,\psi,\eta,t) = \alpha \int\limits_{\varphi_0}^{\varphi} \frac{T_{i-1}}{v^2(s,\psi,\eta)} ds$, $f\left(\varphi,\bar{\psi},\bar{\eta}\right) = \int\limits_{\varphi_0}^{\varphi} \frac{ds}{v^2\left(s,\bar{\psi},\bar{\eta}\right)}$ a function that determines the passage time of the corresponding part of the path from point $(x(\varphi_*,\bar{\psi},\bar{\eta}), y(\varphi_*,\bar{\psi},\bar{\eta}) z(\varphi_*,\bar{\psi},\bar{\eta}))$ to point $(x(\varphi,\bar{\psi},\bar{\eta}), .y(\varphi,\bar{\psi},\bar{\eta}), z(\varphi,\bar{\psi},\bar{\eta}))$ along the corresponding flow line (as the intersection of some two surfaces $\psi(x,y,z) = \bar{\psi}$, $0 \le \bar{\psi} \le Q_*, \eta(x,y,z) = \bar{\eta}$, $0 \le \bar{\eta} \le Q^*$), f^{-1} - function inverted to f with respect to the variable φ (note that such a function exists because $v^2(\varphi,\psi,\eta)$ is a continuously differentiable, bounded, positively defined function. Functions $\Pi_i(\xi,\psi,\eta,t)$, $\bar{P}_i(\xi,\psi,\eta,t)$ $(i = \overline{0, n+1})$, $\breve{\Pi}_i\left(\breve{\xi},\psi,\eta,t\right)$, $\breve{P}_i\left(\breve{\xi},\psi,\eta,t\right)$, $\tilde{\Pi}_i\left(\varphi,\tilde{\mu},\eta,t\right)$, $\tilde{P}_i\left(\varphi,\tilde{\mu},\eta,t\right)$, $\tilde{\tilde{\Pi}}_i\left(\varphi,\tilde{\tilde{\psi}},\eta,t\right)$, $\widehat{\Pi}_i\left(\varphi,\psi,\tilde{\eta},t\right)$, $\widehat{\Pi}_i\left(\varphi,\psi,\tilde{\tilde{\eta}},t\right)$ and $\tilde{\tilde{P}}_i\left(\varphi,\tilde{\tilde{\psi}},\eta,t\right)$, $\tilde{\tilde{P}}_i\left(\varphi,\tilde{\tilde{\psi}},\eta,t\right)$, $\widehat{P}_i\left(\varphi,\psi,\tilde{\tilde{\eta}},t\right)$ functions was obtained similarly to [2]. The evaluation of the residual terms is carried out similarly [2]. The justification for the convergence of solutions is similar to [9].

4 Experiment

The calculation of the change in the concentration of chromium over time at the exit of the electrocoagulation unit was performed to check the adequacy of the generalized model [15]. The following input data were specified [15]: $L = 0.3$ m; $B = 0.3$ m; $H = 1$ m; $\varepsilon = 0.00005$; $D_c = 0.001$ m^2/min; $D_t = 1$ m^2/min; $v_c = 0.4/60$ m/min; $v_t = 0.4/60$ m/min ; $n = 2$; $Rg = 8.314 \cdot 60 \cdot 60$ kg$\cdot$m^2/s^2; $\mu = 0.85$; $F = 96485.3365$ kg $\cdot$ m^2/s^2 ; $As = 0.3 \cdot 1 \cdot 10$ m^2; $P = 101325 \cdot 60 \cdot 6$ kg/m $\cdot$ s^2; $k_h = 2.907$; $U = 2.71$ V; $c_t = 4218 \cdot 60 \cdot 60$ (kg $\cdot$ m^2/s^2)/(kg$\cdot$K); $\rho = 1000$ kg/m^3; $j_{st} = 50$ A/m^2.

The peculiarity of this experiment is the size of the reactor $L = 0.3$ m; $B = 0.3$ m; $H = 1$ m; current density $j_{st1} = 50$ A/m^2, $j_{st1} = 50$ A/m^2. The results of the full-scale experiment and the calculation of the model problem (1)–(2) are presented in Fig. 1 and Fig. 2.

It is obvious that the electrocoagulation technology for cleaning water from chromium can use several parameters to control the quality of treated water. These parameters according to the proposed model include the speed of fluid flow, the time of the electrocoagulation process, the size of the reactor, the amount of current between the plates, the number of cathodes and anodes, the distance between the plates. There are known researches [15] where the influence of the time of the electrocoagulation process, the distance between the plates, the thickness of the cathode and the anode, the current density on the efficiency of chromium extraction from wastewater are determined. That is why the investigation of the effect of geometric dimensions of the reactor at a constant volume of the electrocoagulation unit on the extraction of heavy metal ions in the example of chromium is a relevant and urgent task. The main problem that needs to be solved is to make the decision at the design stage in particular about the shape

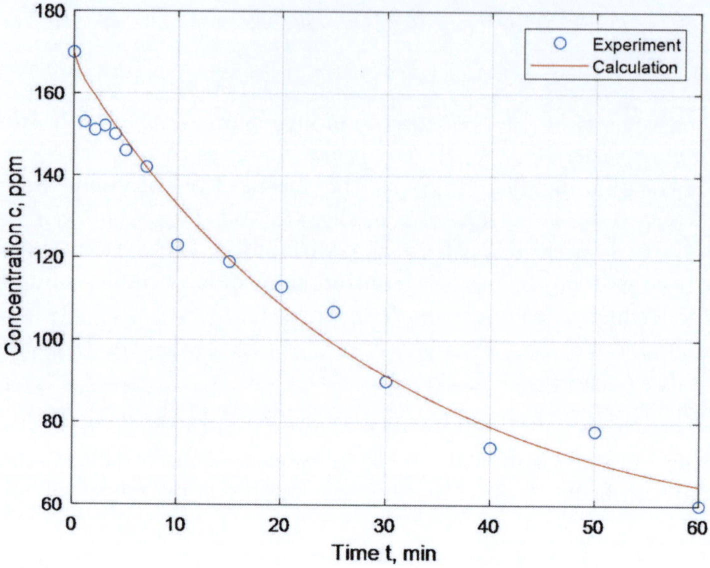

Fig. 1. The distribution of chromium concentration over time at the output of the electrocoagulation unit according to the model problem and experiment at a current density of $50\,A/m^2$ (error = 4.3%)

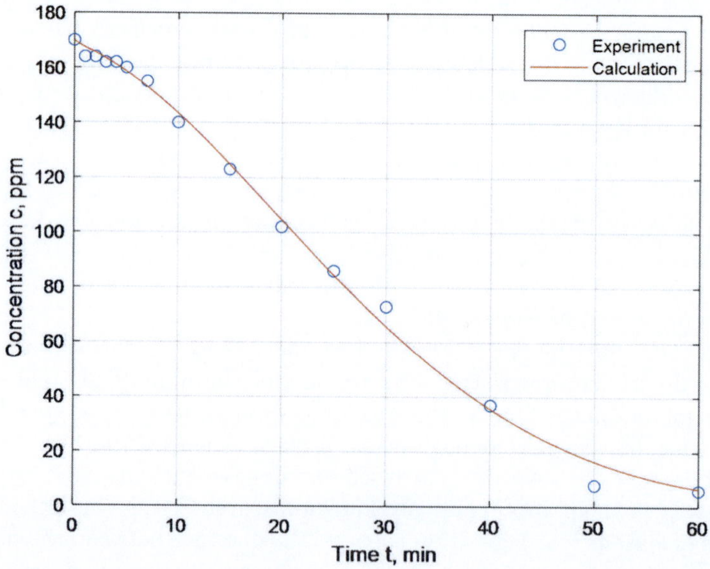

Fig. 2. The distribution of chromium concentration over time at the output of the electrocoagulation unit according to the model problem and experiment at a current density of $100\,A/m^2$ (error = 6.1%)

of the electrocoagulation unit, the number of plates, the design dimensions. The minimization of energy consumption during the electrocoagulation process also is an important factor. To study the effect of reactor sizes on the initial concentration of chromium, a calculation was made for different current densities at variable reactor sizes. The following data were used for the experiment: the sizes of the electrocoagulation unit $L = 0.3$ m, $B = 0.6$ m, $H = 1$ m, as well as all possible combinations, given that in each case the length of the electrocoagulation unit requires a different number of plates with a constant distance between them - 3 cm, current densities, $j_{st1} = 50$ A/m^2, $j_{st1} = 100$ A/m^2. The simulation results are shown in Fig. 3 and Fig. 4, and the main characteristics of the process are presented in Table 1.

Table 1. Variation of chromium concentration at variable current density and size of electrocoagulation unit

		The concentration (ppm) at the output at time t = 60 min at current density:	
		50 A/m^2	100 A/m^2
Dimensions of electrocoagulation unit (m)	L = 1, B = 0.3, H = 0.6	72.0008	9.6835
	L = 1, B = 0.6, H = 0.3	72.0008	9.6835
	L = 0.6, B = 1, H = 0.3	63.1508	5.1835
	L = 0.3, B = 1, H = 0.6	57.3008	0.0835
	L = 0.6, B = 0.3, H = 1	63.1508	5.1835
	L = 0.3, B = 0.6, H = 1	57.3008	0.0835

5 Results and Discussion

In Fig. 1 and Fig. 2 comparison between computer and full-scale experiments is shown. As we can see, the relative error is 4.3% and 6.1% respectively. Increasing the current density at constant reactor sizes leads to a decrease in the initial concentration of chromium. However, as can be seen in Fig. 3, Fig. 4 at a constant length of the electrocoagulation unit and accordingly the number of plates, the geometric dimensions of the cathode and the anode do not affect the efficiency of chromium extraction from wastewater. According to the data in Table 1 and Fig. 3, Fig. 4, at constant reactor volume and current density, more efficient purification will occur in the case of the smallest length of the electrocoagulation unit at the same time. That is, to remove chromium from wastewater for 60 min in two different electrocoagulation units at current density $j_{st} = 50$ A/m^2 the first with length $L = 0.6$ m, 20 plates with dimensions $B = 1$ m, $H = 0.3$ m and the second with length $L = 0.3$ m, 10 plates with dimensions $B - 1$ m, $H - 0.6$ m at the outlet will have a contamination concentration of 63.15 ppm and 57.3 ppm,

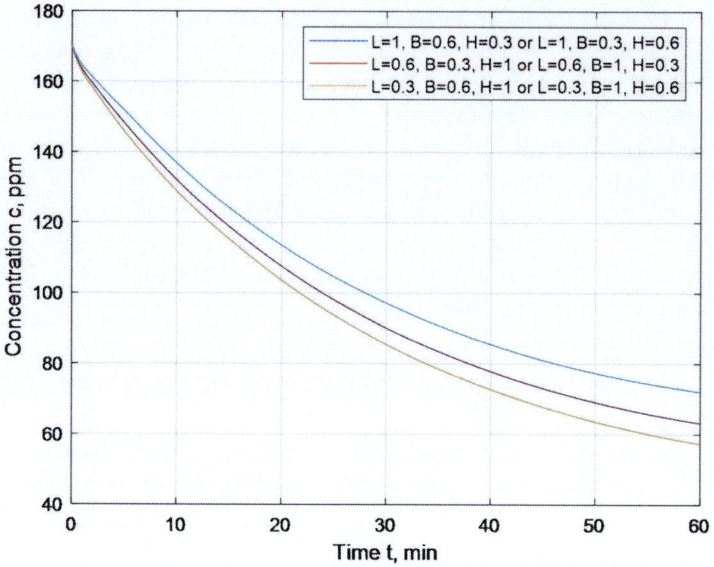

Fig. 3. Distribution of chromium concentration over time at the output of the electro-coagulation unit at current density for various design data

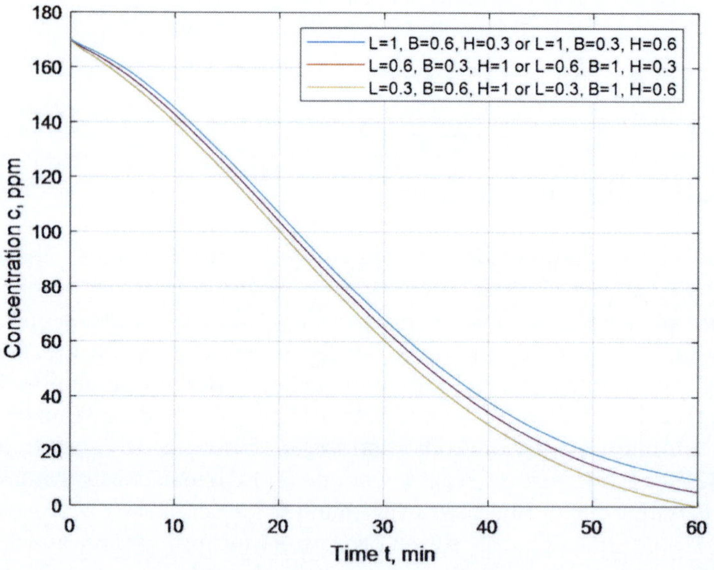

Fig. 4. Distribution of chromium concentration over time at the output of the electro-coagulation unit at current density for various design data

respectively. Thus, the smaller the chromium sewage, the more efficiently it will be removed. This gives prerequisites for the creation of an automated control system for the treatment of chromium from wastewater.

To automate the process of electrocoagulation water purification, a functional automation scheme has been developed (see Fig. 5), according to which control of the system occurs automatically in real-time mode by the controller using the program algorithm and in the manual mode with operational control exercised by a person. However, regardless of the mode, data are collected on the basic control parameters for archiving, and alarming occurs when exceeding acceptable values, if there are problems in the system.

All functions of regulation and control of basic parameters in the projected automation system are performed by the programmable logic controller S7-300 of Siemens enterprise with modules of expansion, analog inputs SM331, analog outputs SM332 and discrete inputs/outputs SM323, which in this configuration has 2 analog inputs, 2 analog outputs, 8 discrete inputs and 8 discrete outputs that are sufficient to control the system.

To ensure acceptable concentration of chromium in process waters, a system consisting of an electrocoagulation unit, sandbox, primary sump, primary clarifier, aeration pool, 3 sludge level sensors, nickel ion concentration sensor in water, programmable logic controller S7-300, 3 regulators is used, a frequency-controlled compressor.

This system has 4 control circuits that provide the following technological parameters at the appropriate level:

1. The concentration of chromium at the outlet of the electrocoagulation unit. (1-analog input, 1-analog output).
2. The level of heavy mineral impurities (sand) in the sand. (1 - discrete input, 1 - discrete output).
3. The level of sediment in the sump. (1 - discrete input, 1 - discrete output).
4. The level of sediment in the electrocoagulation unit. (1 - discrete input, 1 - discrete output).

Also, control the following technological parameters:

1. Control of the rotation of the aeration basin mixer (1 - discrete inlet, 1 - discrete outlet).
2. Emergency water level in sandblasts (1 - discrete inlet, 1 - discrete outlet).
3. Emergency water level in the aerator (1 - discrete inlet, 1 - discrete outlet).
4. Emergency water level in sedimentation tanks (1 - discrete inlet, 1 - discrete outlet).

In order to detect deviations in the technological process and in case of deviation of critical parameters, to notify the operator for his intervention by turning on the siren. In view of the above, a schematic diagram of the control of chromium concentration in wastewater was developed, which is presented in Fig. 5.

1 - electrocoagulation unit; 2 - primary sump; 3 - secondary sump; 4 - aeration pool; 5 - meta tank; Z - the problem of the permissible concentration of

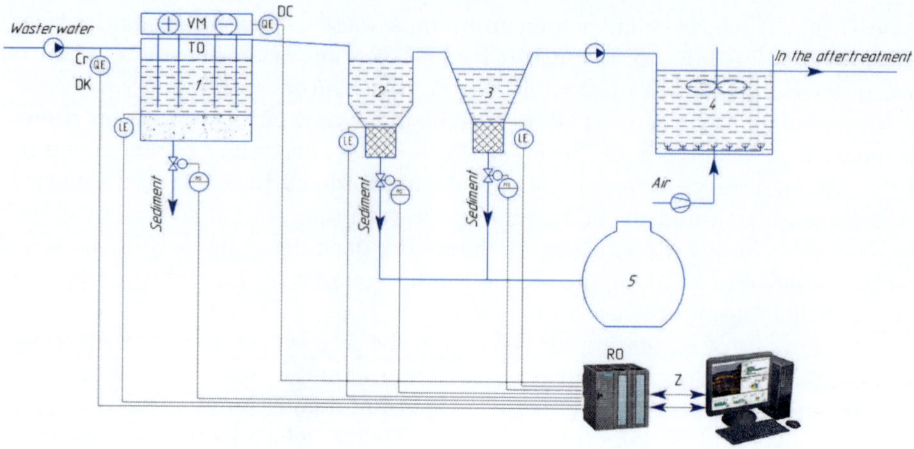

Fig. 5. Schematic diagram of control of permissible concentration of impurities in wastewater

the component to be purified; P1 - standard PI - regulator for calculating the current flowing in the electrocoagulation unit; P2 - system PID - regulator for maintaining the set current strength at the appropriate level when compared to the real current flowing between the walls of the electrocoagulation unit; RO is a regulator that acts as a controller in this system, controlling the magnitude of the applied voltage from the rectifier to the cathode and the anode; VM is an actuator (rectifier) that supplies voltage to the plates of an electrocoagulation unit; TO is a technological object that the reactor stands for; DC is a current encoder that measures the real current flowing between the walls of an electrocoagulation unit; DK is the concentration sensor of the target component at the inlet of the reactor, which shows the change in input concentration for the implementation of the optimal control mode.

According to the block diagram, the system will work as follows: the problem of the permissible concentration of pollution (C) is compared with the value of the input concentration (DK) and if the system enters pollution with a concentration exceeding the set permissible norms, the automation system is put into operation.

The first step is to represent the magnitude of the input concentration of contamination on the PI - regulator (P1), which according to dependence, calculates the amount of current flowing between the cathode and the anode to ensure the allowable the concentration of contamination of the target component at the outlet of the treatment plant. The obtained value is compared with the magnitude of the current flowing in the electrocoagulation unit at the moment. The difference obtained is fed to a standard PID controller (P2), which converts the input current signal to a uniform voltage 0...5 V, which controls the rectifier. All of the described actions occur in the controller by exchanging signals with visualization, sensors and processing of the control program according to the obtained results.

The next step is to transmit control signals from the analog output of the controller (PO) to the rectifier (VM), which is connected to a 220 V AC power supply and produces a DC 24 V high-voltage signal. The analog signal from the controller is fed to the analog input of the rectifier, which produces a proportional signal at the +24 VDC outputs supplied to the cathode and the anode of the electrocoagulation unit. At such a voltage, a current flows in the water, which causes the processes of chromium precipitation to precipitate directly to the electrocoagulation unit (TO).

To maintain the current level, the encoder (DS) is used to feedback to the controller.

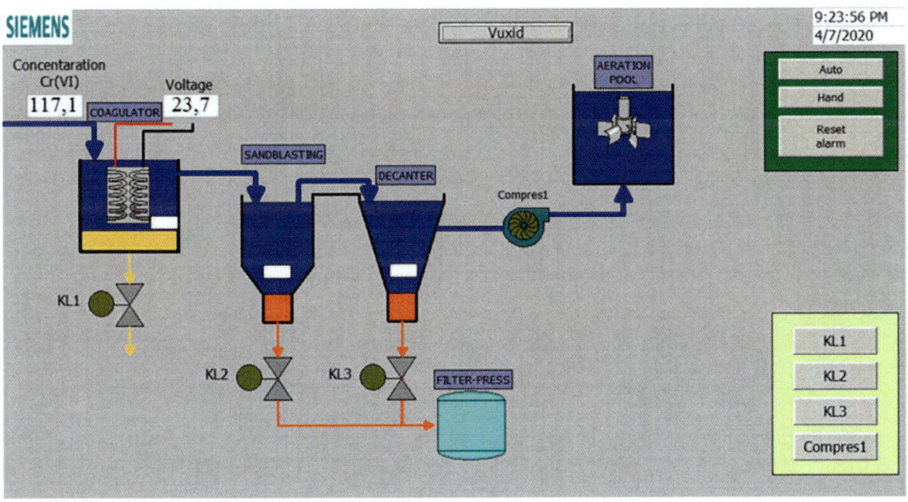

Fig. 6. Graphic window for displaying the process of chromium sewage treatment by electrocoagulation

The hardware implementation of this project involves the use of a programmable logic controller S-7 300 from Siemens with expansion modules. Using the Simatic Step7 Manager, a process control program was written, preconfiguring the controller hardware and creating a character table that describes all the inputs, outputs, labels with their character, hardware names, and comments. The main components of the program are the implementation of the method of definition and maintenance of the required current between the plates of the electrocoagulation unit. To accomplish this task, we use a standard PID controller, which performs the function of adjusting the current according to the task.

To visualize the project, a graphical screen was created on which automatic and manual control of the process is carried out (Fig. 6), display of errors in the operation of the respective devices, as well as display of the main informative parameters of the technological process, such as the direction of the flow of liquid by the glowing green indicator of the respective parts of the valves, filter pumps, also providing information on the applied voltage, current, concentration of chromium at the inlet of the electrocoagulation unit.

6 Conclusions

The model of the process of water treatment from chromium depending on the applied current and geometric size of the electrocoagulation has been developed. A solution to the corresponding model problem has been found. The results of calculations of the distribution of chromium concentration over time at the exit of the coagulator taking into account the change of the applied current and the size of the electrocoagulation unit are presented. The effect of the length of the electrocoagulation unit at a constant volume of the reactor on the efficacy of wastewater treatment has been investigated. The reliability of the obtained results has been confirmed by comparison with the results of the experiment, and the relative error does not exceed 6.1%. A system for automation of wastewater treatment from chromium, which provides established environmental standards for concentrations of pollution in treated wastewater has been developed. A mnemonic scheme for ease and efficiency of process control by an operator has been developed using WinCC Flexible software (Simatic) and a set of automation hardware used in systems for reliable operation with a wide range of applications has been chosen.

Acknowledgments. Research in this work was provided as a part of project "Innovative principles of design and adaptation of resource-saving solutions for optimization of water treatment processes with automated control" (0119U000083).

References

1. AI-Rubaiey, N., Albrazahjy, M.G., Abdulkareem, M.: Electrocoagualation treatment of oil-based mud wastewater. In: IOP Conference Series Materials Science and Engineering (2020). https://doi.org/10.1088/1757-899X/737/1/012191
2. Bomba, A., Safonyk, A., Voloshchuk, V.: Spatial modeling of multicomponent pollution removal for liquid treatment under identification of mass transfer coefficient. Math. Model. Comput. **5**(2), 108–118 (2018). https://doi.org/10.23939/mmc2018.02.108
3. Changmai, M., Pasawan, M., Purkait, M.K.: Treatment of oily wastewater from drilling site using elecrocoagulation followed by microfiltration. Sep. Purif. Technol. **210**, 463–472 (2019)
4. Ghernaout, D., Eloughdiri, N.: Electrocoagulation process in the context of disinfection mechanism. Open Access Libr. J. **7**(2), 1–14 (2020). https://doi.org/10.4236/oalib.1106083
5. Kumar, V., Kumar, R.: Optimization and assessment of electrocoagulation treatment technology for sewage treatment, October 2019

6. Nasrullah, M., Siddique, M.N.I., Abd Wahid, Z.: Effect of high current density in electrocoagulation process for sewage treatment. Asian J. Chem. **26**(14), 4281–4285 (2014). https://doi.org/10.14233/ajchem.2014.16134
7. Posavicic, H., Halkijevic, I., Vukovic, Z.: Applicatoin of electrocoagulation for water conditioning. Environ. Eng. **6**(2), 59–70 (2019). https://doi.org/10.37023/ee.6.2.3
8. Safonyk, A., Bomba, A., Tarhonii, I.: Modeling and automation of the electrocoagulation process in water treatment. In: Advances in Intelligent Systems and Computing, vol. 871, pp. 451–463 (2019). https://doi.org/10.1007/978-3-030-01069-0_32
9. Safonyk, A., Burduk, A., Targoniy, I.: Research of simulation model of electrocoagulation process. In: Modeling, Control and Information Technologies, vol. 3, pp. 117–118 (2019). https://doi.org/10.31713/MCIT.2019.67
10. Safonyk, A., Prysiazhniuk, O.: Modeling and simulation in engineering modeling of the electrocoagulation processes in nonisothermal conditions. Model. Simul. Eng. **2019**, 6 (2019). https://doi.org/10.1155/2019/9629643
11. Safonyk, A., Targoniy, I., Martyniuk, Y., Bomba, A.: Research and automation of the process of wastewater treatment electrocoagulation. In: Computer Science and Information Technologies, CSIT 2018, vol. 1, pp. 84–88 (2018). https://doi.org/10.1109/STC-CSIT.2018.8526661
12. Ungureanu, N., Vladut, V., Paraschiv, G.: Electrocoagulaytion in wastewater treatment. In: The 15th Annual Meeting Durable Agriculture Agriculture of the Future Agriculture of the Future (2019)
13. Vlasyuk, A., Zhukovskii, V.: Mathematical simulation of the migration of radionuclides in a soil medium under nonisothermal conditions with account for catalytic microparticles and nonlinear processes. J. Eng. Phys. Thermophys. **90**(6), 1386–1398 (2017). https://doi.org/10.1007/s10891-017-1697-4
14. Yildiz, S., Oran, E.: Sewage sludge disintegration by electrocoagulation. Int. J. Environ. Health Res. **29**(9), 1–13 (2018). https://doi.org/10.1080/09603123.2018.1554181
15. Zongo, I., Merzouk, B., Palm, K., Wethe, J., Maiga, A.H., Leclerc, J.P., Lapicque, F.: Study of an electrocoagulation (EC) unit for the treatment of industrial effluent of Ouagadougou, Burkina Faso. Adv. Appl. Sci. Res. **3**(1), 572–582 (2012)

Optimization of Ingredients for Two-Layer Epoxy Coating for Protection of Sea and River Vehicles

Andriy Buketov[1], Serhii Yakushchenko[1(✉)], Tetyana Cherniavska[1], Danyl Zhytnyk[1], Natalia Buketova[1], Tetyana Ivchenko[1], Igor Fesenko[2], and Roman Negrutsa[1]

[1] Kherson State Maritime Academy, Kherson, Ukraine
buketov@tntu.edu.ua, yakushchenko.sv@gmail.com, tanya2181@ukr.net, danylzhytnyk@gmail.com, nataliyabuketona@gmail.com, ivchenko_tata@i.ua, roma3d3@gmail.com

[2] V. Bakul Institute for Superhard Materials of the National Academy of Sciences of Ukraine, Kiev, Ukraine
igorfesenko@ukr.net

Abstract. It was proved, that in order to increase the operational characteristics of parts of river and sea transport, including their physical and mechanical properties, it is advisable to use protective polymer composite coatings. It was shown, that in order to increase the performance of the physical and mechanical properties of the epoxy binder, it is necessary to introduce additives: dispersed and fibrous fillers. It is important to use them in complex. The effect of the modifier and the filler on the adhesive strength and impact strength of the epoxy composite coating has been preliminary analyzed. The critical content of the components in the epoxy oligomer was determined by the method of mathematical planning of the experiment: the modifier is 2, 4-diaminoazobenzene-4'-carboxylic acid (DAABCA) and the synthesized powdery titanium-aluminum charge (PTAC). The introduction of such ingredients into the epoxy binder in the complex allows to increase the adhesion strength and impact strength of a two-layer epoxy composite coating. The results obtained make it possible to create a protective coating with improved adhesion and physical-mechanical properties.

Keywords: Composite · Epoxy matrix · Mathematical planning of experiment · Regression equation

1 Introduction

Today, the scientific schools in the leading countries of the world engaged in solving an important scientific and technical problem focused to development of protective polymer anti-corrosion and wear-resistant coatings for sea and river transport. The operational characteristics of polymeric composite materials

S. Babichev et al. (Eds.): ISDMCI 2020, AISC 1246, pp. 612–626, 2021.
https://doi.org/10.1007/978-3-030-54215-3_39

(CM), including epoxy, are enhanced by the pre-treatment of the matrix ingredients by external fields, as well as by the introduction of nano- and microdispersed fillers in the form of powders [10,14,15,21]. Various technologies of processing and forming of epoxy compositions with nano- and microdispersed filler have been developed. The new materials with high performance characteristics for protective coatings have been created. In addition, as a result of directional introduction into the epoxy matrix of dispersed fillers a synergistic effect is realized, which allows to significantly increase the operational characteristics of protective coatings when operating equipment in critical conditions.

2 Review of Literature

In [5,7–9] the authors developed new epoxy composite coatings to protect process equipment from corrosion and wear. It may be interesting for customers that during the development of coatings, the microdispersed fillers, which are industrial wastes, were used. This allows us not only to improve the properties of coatings, but also significantly reduce their cost. The advantage of these developments over foreign and domestic counterparts takes into account the main factors of aggressive environments, where technological equipment is operated, and the development of materials which in the complex have high physical and mechanical, thermophysical properties, corrosion resistance and wear resistance. At the same time, it should be noted, that the problem of improvement of mechanical, anticorrosive properties and wear resistance of CM in the complex has not yet been finally resolved. The field of polymeric materials science, which covers the substantiation of the interaction mechanism of the composition ingredients in their structure formation at the micro-, meso- and macro-levels, needs further basic research. These processes in the structure formation of materials in the future determine their cohesion and, as a consequence, anti-corrosion properties and wear resistance. Taking into account hereinbefore, it is important to introduce the microdispersed fillers into the epoxy binder, that are active in the interaction, including synthesized by high-voltage electric spark discharge. At the same time, it is important to conduct a study to determine the critical content of the PTAC synthesized by high-voltage electric discharge in the formation of complex systems in the form of protective epoxy coatings for sea and river transport.

The aim of the research is development of a technique to optimize the components content of the two-layer coating with high performance to protect vehicles.

3 Materials and Methods

The epoxy diane oligomer ED-20 was selected as the main component for the binding when forming epoxy CM, which is distinguished by high adhesive and cohesive strength. For polymerization of compositions, polyethylene polyamine hardener was used. PEPA is a low molecular weight substance consisting of monomer units with the following form: [- CH2-CH2-NH-]$_n$. The DAABCA was used as a modifier. The modifier's molar mass is Mn = 256,25998 g/mol, molecular formula – $C_{13}H_{12}N_4O_2$. The synthesized powdery titanium-aluminum charge was used as the microdisperse filler. The PTAC filler (d = 10–12 µm) consists of a mixture of powders of the following composition: Ti (75%) + Al_3Ti (15%) + Ti_3AlC_2 (10%).

Epoxy CM's forming technology is following: heating of the resin up to the temperature T = (353 ± 2) K and exposure at this temperature during τ = (20 ± 0.1) min; dosage of the filler and its further loading into epoxy binder; hydrodynamic blending of the oligomer ED–20 and filler during τ = (10 ± 0.1) min; ultrasonic processing of the composition during τ = (1.5 ± 0.1) min; cooling of the composition to the ambient temperature during τ = (60 ± 5) min; introduction of the curing agent PEPA and mechanical blending of the composition during τ = (5 ± 0.1) min. [5,7] The CM was cured during τ = 72.0 ± 0.1 hours at the temperature T = 393 ± 2 K.

The following characteristics of CM, as adhesive strength and impact strength were studied in the work. The adhesive strengt of CM to the metal base was investigated by measuring the destructive stresses (the "fungus method") at a uniform separation of a pair of glued specimens according to ASTM D897-08(2016). The diameter of the working part of the specimens at separation was d = 25 mm. The impact strength was measured in accordance with the Charpy impact test (ASTM D6110) on the pendulum MK-30 at T = 298 ± 2 K and relative humidity φ = 50 ± 5%. Samples with the following parameters were used: (63.5 × 12.7 × 12.7) ± 0.5 mm. The distance between the supports is 40 ± 0.5 mm.

To obtain high performance of CMs, after the preliminary experiments, a method of mathematical planning of the experiment was used, which simultaneously allowed to take into account the ratio of components and technological parameters of obtaining composites with a considerable reduction of experimental works [6]. A model of the full factor experiment 2^2 was used for the design of the experiments. Factor levels and variation intervals were selected based on the previous experiments. The root mean square deviation for the coefficients was determined by the formula:

$$S(a_i) = \sqrt{S^2(a_i)} = \frac{S^2(y)}{\sum_l^n x_i l} \tag{1}$$

$$S^2(y) = \sum_{l=1}^{N} \sum_{i=1}^{n} (Y_l - y_{il})^2 \tag{2}$$

$$S^2(Y) = \frac{S^2(y)}{n} \tag{3}$$

where: $S^2(y)$ is the dispersions of adequacy; $S^2(Y)$ is the average dispersions of adequacy; N is the number of experiments according to the plan; n is the number of parallel experiments; x_i is the coded variable.

The weighted value of the coefficients of the equation was determined using Student's criterion:

$$t_i = |\frac{d_i}{S(a_i)}| \tag{4}$$

where a_i is the coefficient value.

The condition of significance of the equation coefficients has the form $t_i > t_{calc}$. t_{calc} was determined from the table in [1] in 5% significance level. The adequacy of the description of the process of obtaining the CM was tested using the Fisher test:

$$F_i = \frac{S_R^2}{S^2(y)} \tag{5}$$

$$S_R^2 = \frac{\sum_l^N Y_l^2 - \sum_i^N a_i \sum_l^N x_{il} y_l}{f_1} \tag{6}$$

where: S_R^2 is the residual dispersion; f_1 is the freedom degrees.

The number of degrees of freedom f_1 and f_2 was determined by the formulas:

$$f_1 = N - (k + 1) \tag{7}$$

$$f_2 = N(n - 1) \tag{8}$$

where k is the number of polynomial coefficients.

The table value F of the criterion F_{calc}, taking into account f_1 and f_2, was found for the 5% significance level. Given $F_i < F_{calc}$, it was considered, that the polynomial adequately describes the process under study [13].

In order to reduce the error, in the study of properties and search for optimal values, five or six parallel measurements were performed and the arithmetic mean was found:

$$Y = \frac{y_1 + \ldots + y_n}{n} \tag{9}$$

where: $y_1 + \ldots + y_n$ is the results of individual experiments; Y is the arithmetic mean of all parallel experiments; n is the number of parallel experiments.

To eliminate false values, the Student's criterion was used [18]:

$$\frac{y_i - y}{S} \geq t \tag{10}$$

where: y_i is the results of a single experiment; y is the arithmetic mean of all results; S is the quadratic error; t is the tabular value of the Student test.

The quadratic error was determined by the formula:

$$S = \frac{\sqrt{\sum_l^n (y_i - Y)^2}}{n} \tag{11}$$

An experiment was considered false if the experimental value of the criterion is greater than the table value: $|t_{exper}| \geq t_{tab}$.

Among the wide range of mathematical shells available for processing experimental data, spline interpolation was used [2,3,20]. Cubic spline interpolation allows the curve to be formed through a set of points in such a way that the first and second derivatives of the curves will be continuous at each point. Moreover, the specified curve is formed by introducing a series of cubic polynomials passing through a set of three adjacent points. Subsequently, cubic polynomials are combined in Mathcad 11 software to form a single curve. At the same time, using the cubic spline interpolation, the smoothing function was integrated with the processing of experimental data $Supsmootx(x, y)$ [20]. This function returns an n-dimensional vector formed by the local use of a symmetric linear least-squares smoothing procedure by the K-nearest-neighbor rule, where K is chosen adaptively. Here x and y are n-dimensional vectors of the original data. It is important to use this feature to filter experimental data. Most often the purpose of such filtering is to prevent rapid variations in (x_i) caused by noise. The result is a smoothed dependence that is dominated by the low-frequency component.

4 Experiment

On the previous stage, the effect of the amount of modifier and dispersive filler on the adhesive and mechanical properties of epoxy CM [5,7] were investigated. Critical content of the DAABCA modifier and PTAC in the epoxy matrix was determined. It is important to combine the above-mentioned ingredients in the formation of composites for adhesive and surface layers to form coatings of different functional purpose. The method of mathematical planning was used to reduce the number of experiments.

Formation of the Adhesive Layer

The adhesive strength of composites with different content of modifier and dispersant filler was investigated. Each component was coded by units with consideration of the variation step, for standardization and for simplification of calculations (Table 1).

Table 1. Coded and natural values of components

Components	Coded value	Optimum value, q, pts.wt.	Deviation, Δq, pts.wt.	Coded and natural values of components		
				−1	0	+1
DAABCA	x_1	1.00	0.50	0.50	1.00	1.50
PTAC	x_2	0.10	0.05	0.05	0.10	0.15

The results of the conducted experiments are shown in Table 2. In addition, the values of the factor x' were determined by the formula [6]:

$$x_i' = (x_i)^2 - \frac{\sum_{u-1}^{N} x_{iu}^2}{N} \tag{12}$$

Table 2. Scheme of experiment planning

№	x_0	x_1	x_2	$x_3 = x_1^2 - d$	$x_4 = x_2^2 - d$	x_1x_2
1	1	−1	−1	0.33	0.33	+1
2	1	+1	−1	0.33	0.33	−1
3	1	−1	+1	0.33	0.33	−1
4	1	+1	+1	0.33	0.33	+1
5	1	0	0	−0.67	−0.67	0
6	1	+1	0	0.33	−0.67	0
7	1	−1	0	0.33	−0.67	0
8	1	0	+1	−0.67	0.33	0
9	1	0	−1	−0.67	0.33	0
$\sum_{u-1}^{N} x_{iu}^2$	9	6	6	2	2	4

The experimental results are shown in Table 3.

Table 3. Results of study of adhesion strength of CM

№	Content of components, q, pts.wt.		Adhesion strength, σ_a, MPa
	x_1	x_2	y_2
1	0.05	0.05	40.2
2	0.15	0.05	39.3
3	0.05	0.15	37.8
4	0.15	0.15	36.8
5	0.10	0.10	44.2
6	0.15	0.10	43.4
7	0.05	0.10	38.4
8	0.10	0.15	41.0
9	0.10	0.05	43.8

Formation of the Surface Layer
The composition of CM was optimized by indexes of impact strength. The components were encoded with arbitrary units according to the Table 4.

The results of experimental study of impact strength of PCM are shown in Table 5.

5 Results and Discussion

The mathematical model $y = f(x_1, x_2)$ was formed as a regression equation [6]:

$$y = b_0 + b_1x_1 + b_2x_2 + b_{11}x_1^2 + b_{22}x_2^2 + b_{12}x_1x_2 \tag{13}$$

Table 4. Coded and natural values of components

Components	Coded value	Optimum value, q, pts.wt.	Deviation, Δq, pts.wt.	Coded and natural values of components		
				−1	0	+1
DAABCA	x_1	1.00	0.50	0.50	1.00	1.50
PTAC	x_2	0.50	0.25	0.25	0.50	0.75

Table 5. Results of study of impact strength of PCM

№ of experiment	Content of components, q, pts.wt.		Impact strength, W, kJ/m^2
	x_1	x_2	y_2
1	0.50	0.25	9.8
2	1.50	0.25	9.4
3	0.50	0.75	10.8
4	1.50	0.75	11.0
5	1.00	0.50	12.3
6	1.50	0.50	11.4
7	0.50	0.50	11.8
8	1.00	0.75	10.2
9	1.00	0.25	10.4

The regression coefficients were determined in accordance with [19]:

$$b_i = \frac{\sum_{u=1}^{N} x_1 y_1}{\sum_{u=1}^{N} x_{1u}^2} \tag{14}$$

The coefficients of regression equation are given in Table 6.

Table 6. The coefficients of mathematical model

b_0	b_1	b_2	b_{11}	b_{22}	b_{12}
44.46	0.52	−1.28	−3.68	−2.18	−0.02

During the analysis of adhesion strength, the regression equation was obtained:

$$y = 44.46 + 0.52x_1 - 1.28x_2 - 3.68x_1^2 - 2.18x_2^2 - 0.02x_1x_2 \tag{15}$$

According to Cochran test [20], the adequacy of the model was examined:

$$G = \frac{S_{umax}^2}{\sum_{u=1}^{N} S_{ui}^2} \leq G_{(0.05;f_1;f_2)} \tag{16}$$

where: S_{ui}^2 is the dispersion of results deviation; m is the number of experiments; S_{umax}^2 is the maximal dispersion;

Dispersions deviation were obtained as follows:

$$S_{ui}^2 = \frac{\sum_{i=1}^{m}(y_i - \bar{y_i})^2}{m-1} \tag{17}$$

where: y_{im} is the current value; $\bar{y_i}$ is the average value. The current dispersions were obtained by equation:

$$\sigma^2\{y\} = \frac{\sum_{i=1}^{N=9}\sigma^2\{y\}_i}{N(m-1)} \tag{18}$$

where $\sigma^2\{y\}_i = \sum_{i=1}^{m=3}(y_i - \bar{y_i})^2$;

$$\sigma^2\{y_{av}\} = \frac{a^2\{y\}}{N} \tag{19}$$

or

$$S_{b_0}^2 = \frac{S_0^2}{N} \tag{20}$$

Dispersion values are shown in Table 7.

Table 7. Dispersions of deviation (S_{ui}^2) and current dispersions ($\sigma^2\{y\}_i$)

№	The dispersions of deviation		Current dispersions	
	Conditional designation	Value	Conditional designation	Value
1	S_{u1}^2	8.770	$\sigma^2\{y\}_1$	17.540
2	S_{u2}^2	1.710	$\sigma^2\{y\}_2$	3.420
3	S_{u3}^2	8.590	$\sigma^2\{y\}_3$	17.180
4	S_{u4}^2	3.370	$\sigma^2\{y\}_4$	6.740
5	S_{u5}^2	0.840	$\sigma^2\{y\}_5$	1.680
6	S_{u6}^2	0.963	$\sigma^2\{y\}_6$	1.927
7	S_{u7}^2	4.630	$\sigma^2\{y\}_7$	9.260
8	S_{u8}^2	1.560	$\sigma^2\{y\}_8$	3.120
9	S_{u9}^2	1.120	$\sigma^2\{y\}_9$	2.240

Moreover:

$$\sum_{i=1}^{N} S_{ui}^2 = 18.703 \tag{21}$$

$$\sigma^2\{y\} = S_0^2 = 2.078 \tag{22}$$

Cochran criterion was determined by the equation [20]:

$$G_{calc} = \frac{S^2_{u_{max}}}{\sum_{i=1}^{N} S^2_{ui}} \tag{23}$$

$$G_{calc} = \frac{4.63}{18.703} = 0.248 \tag{24}$$

Dispersion of all factors: $S^2_{u_{max}} = 8.77$. According to the tables in [19,20]: $G_{tab} = 0.478$.

That is:

$$G_{calc} = 0.248 \leq G_{tab} = 0.478 \tag{25}$$

The results of the study of the adhesive strength of the materials are shown in Table 8.

Table 8. The adhesion strength of materials

№	Adhesion strength, σ_a, MPa			Average value, σ_a, MPa
	1	2	3	
1	39.9	41.2	39.5	40.2
2	37.8	40.2	39.9	39.3
3	37.5	38.1	41.1	37.8
4	35.5	36.0	38.9	36.8
5	44.0	45.2	43.4	44.2
6	42.8	44.5	42.8	43.4
7	36.2	38.5	40.5	38.4
8	42.4	40.6	40.0	41.0
9	43.0	43.4	45.0	43.8

The dispersions of the coefficients of the model was determined as follows (Table 9):

$$S^2_{b_i} = \frac{S^2_0}{\sum_{u-1}^{N} x^2_{iu}} \tag{26}$$

Weighted values of the coefficients of the model were found by Student's test. According to the work of the authors [18], Student's value is $t_T = 2.1$. Subsequently, the Student's criterion (t_{calc}) were determined: $t_0, t_1, t_2, t_{11}, t_{22}, t_{12} > t_T$.

Moreover:

$$t_i = \frac{|b_i|}{S_{b_i}} \tag{27}$$

Calculated values of Student's criterion t_0, t_2, t_{11}, t_{22} are larger than t_T, so it was considered that b_0, b_2, b_{11}, b_{22} coefficients of the model are significant.

Table 9. Dispersions of the model (S_b^2) and Student's criterion (t_{calc})

№	Dispersion of model		Student's criterion	
	Conditional designation	Value	Conditional designation	Value
1	$S_{b_0}^2$	0.231	t_0	88.68
2	$S_{b_1}^2$	0.346	t_1	0.88
3	$S_{b_2}^2$	0.346	t_2	2.18
4	$S_{b_{11}}^2$	1.039	t_{11}	3.61
5	$S_{b_{22}}^2$	1.039	t_{22}	2.14
6	$S_{b_{12}}^2$	0.520	t_{12}	0.03

Calculated values t_1, t_{12} are lower than t_T, so it was considered that b_1, b_{12} indexes are insignificant. The final equation of the model was:

$$y = 44.46 - 1.28x_2 - 3.68x_1^2 - 2.18x_2^2 \tag{28}$$

The equation was examined by Fisher test [4]:

$$F_{calc} = \frac{S_{umax}^2}{S_y^2} \leq F_{(0.05;f_{ad};f_y)} \tag{29}$$

where $S_{umax}^2 = 4.63$ is the current dispersion:

$$S_y^2 = \frac{\sum_{i=1}^{N} S_{ui}^2}{N} \tag{30}$$

$S_y^2 = 2.078$ is the dispersion of deviation. Then: $F_c = 2.228$.
in accordance with tables in [16]: $F_t = 2.93$.
It was proved, that $F_{calc} < F_t$. That's why the model is adequate.
At the next stage, the encoded values were converted to natural:

$$x_i = \frac{q_i - q_{i0}}{\Delta q_i} \tag{31}$$

where: q_i is the test results, q_{i0} is the initial value, Δq_i is the deviation.
As a result of transformations the model with natural values was obtained:

$$\sigma_a = 0.24 + 1.49q_2 - 0.15x_1^2 - 8.73x_2^2 \tag{32}$$

Based on experimental studies it was found, that both fillers are significant. It should be noted that the square dependence of the main and additional fillers on the adhesive strength values has a greater influence. Analyzing the calculated response surface, it was determined that the optimum adhesion strength at tear ($\sigma_a = 43.8$–44.2 MPa) has developed epoxy composite, that contains the DAABCA modifier in the amount of $q = 0.5$–1.0 pts.wt. and disperse filler -

Table 10. The coefficients of regression equation for impact strength

b_0	b_1	b_2	b_{11}	b_{22}	b_{12}
12.01	−0.10	0.40	−0.27	−1.57	0.15

synthesized powder titanium-aluminum charge with a content of $q = 0.05$–0.10 pts.wt. in the epoxy binder.

The coefficients of mathematical model are shown in Table 10. During the analysis of impact strength values, the mathematical model was obtained:

$$y = 12.01 - 0.10x_1 + 0.40x_2 - 0.27x_1^2 - 1.57x_2^2 + 0.15x_1x_2 \quad (33)$$

The adequacy of the model were examined by Cochran test [17]. Dispersions are shown in Table 11.

Table 11. Dispersions of deviation (S_{ui}^2) and current dispersions ($\sigma^2\{y\}_i$)

№	The dispersions of deviation		The current dispersions	
	Conditional designation	Value	Conditional designation	Value
1	S_{u1}^2	0.03	$\sigma^2\{y\}_1$	0.06
2	S_{u2}^2	0.03	$\sigma^2\{y\}_2$	0.06
3	S_{u3}^2	0.03	$\sigma^2\{y\}_3$	0.06
4	S_{u4}^2	0.04	$\sigma^2\{y\}_4$	0.08
5	S_{u5}^2	0.03	$\sigma^2\{y\}_5$	0.06
6	S_{u6}^2	0.03	$\sigma^2\{y\}_6$	0.06
7	S_{u7}^2	0.01	$\sigma^2\{y\}_7$	0.02
8	S_{u8}^2	0.03	$\sigma^2\{y\}_8$	0.06
9	S_{u9}^2	0.03	$\sigma^2\{y\}_9$	0.06

In this case:

$$\sum_{i=1}^{N} S_{ui}^2 = 0.26 \quad (34)$$

$$\sigma^2\{y\} = S_0^2 = 0.029 \quad (35)$$

The Cochran criteria is (23):

$$G_{calc} = \frac{0.04}{0.26} = 0.154 \quad (36)$$

The dispersion for all factors is: $S_{u_{max}}^2 = 0.04$. According to table in [17,20]: $G_{tab} = 0.478$.

Then:

$$G_{calc} = 0.154 \leq G_{tab} = 0.478 \tag{37}$$

The results of the study of the impact strength of CM are shown in Table 12.

Table 12. Impact strength of CM

№	Impact strength, W, kJ/m^2			Average value, W, kJ/m^2
	1	2	3	
1	9.7	9.7	10.0	9.8
2	9.5	9.5	9.2	9.4
3	10.6	10.9	10.9	10.8
4	11.2	10.8	11.0	11.0
5	12.4	12.4	12.1	12.3
6	11.5	11.2	11.5	11.4
7	11.7	11.9	11.8	11.8
8	10.3	10.3	10.0	10.2
9	10.5	10.5	10.2	10.4

Dispersion coefficients of the model coefficients were found by Eqs. (18–20). Student's criterion are shown in Table 13.

Table 13. Dispersion of model (S_b^2) and student's criterion (t_{calc})

№	Dispersion of model		Student's criterion	
	Conditional designation	Value	Conditional designation	Value
1	$S_{b_0}^2$	0.003	t_0	201.83
2	$S_{b_1}^2$	0.005	t_1	1.44
3	$S_{b_2}^2$	0.005	t_2	5.76
4	$S_{b_{11}}^2$	0.014	t_{11}	2.22
5	$S_{b_{22}}^2$	0.014	t_{22}	13.04
6	$S_{b_{12}}^2$	0.007	t_{12}	1.8

Calculated indexes of Student's criterion t_0, t_2, t_{11}, t_{22} are larger than t_T, so it was considered that b_0, b_2, b_{11}, b_{22} coefficients of the model are significant. Calculated values t_1, t_{12} are lower than t_T, so it was considered that b_1, b_{12} indexes are insignificant. The final equation of the model is:

$$y = 12.01 + 0.40x_2 - 0.27x_1^2 - 1.57x_2^2 \tag{38}$$

The Fisher index was used to verify the equation [11, 12].

Carrent dispersins: $S_{umax}^2 = 0.04$.

The dispersions of deviation: $S_y^2 = 0.029$.
So: $F_{calc} = 1.385$.
According to [4,16] the table criteria is $F_t = 2.93$.
It was proved, that $F_{calc} < F_t$, that's why the model is adequate.

At the next stage, the encoded values were converted to natural and the following model was obtained:

$$W = 3.882 + 2.13q_1 + 26.67q_2 - 1.07x_1^2 - 25.07x_2^2 \tag{39}$$

During analysis, it was determined that the impact strenght show maximum values ($W = 11.8$–$12.3\ \mathrm{kJ/m^2}$) with the modifier's contents: DAABCA – 0.5–1.0 pts.wt., PTAC – 0.50–0.75 pts.wt. With further increase the content of ingredients the decreasing of impact strength was observed. Therefore, it is advisable to modifier and dispersed filler with the aforementioned content into epoxy binder to improve operational characteristics of functional layer of protective coating.

6 Conclusions

The influence of modifier and filler on adhesive strength and impact strength of epoxy composite coating is analyzed in the paper. The critical content of the components in the epoxy oligomer was determined by the method of mathematical planning of the experiment: the modifier DAABCA and the disperse filler PTAC. Introduction of such ingredients with the optimum content into the epoxy binder in the complex allows to increase the adhesive strength and impact strength of the two-layer epoxy composite coating for the protection of sea and river transport.

It was found, that during the formation of the adhesive layer of the protective coating, it is necessary to use material, which contains DAABCA ($q = 0.5$–1.0 pts.wt) and PTAC ($q = 0.05$–0.10 pts.wt.) in the epoxy binder. The composite is characterized by improved adhesive strength - $\sigma_a = 43.8$–44.2 MPa.

It was proved, that during the formation of the surface layer of the protective coating, it is necessary to use material, which contains DAABCA ($q = 0.5$–1.0 pts.wt) and PTAC ($q = 0.50$–0.75 pts.wt.) in the epoxy binder. The composite is characterized by improved impact strength - $W = 11.8$–$12.3\ \mathrm{kJ/m^2}$.

Protective coatings with improved adhesive strength (when forming the adhesive layer) and cohesive strength (when forming the surface layer) was created. It is advisable to use the composites in the form of protective coatings with high performance characteristics for the restoration of details of various functional purpose.

References

1. Asghar, A., Raman, A., Aziz, A., et al.: A comparison of central composite design and Taguchi method for optimizing Fenton process. Sci. World J. **2014**, 1–14 (2014). https://doi.org/10.1155/2014/869120

2. Babichev, S., Lytvynenko, V., Korobchynskyi, M., Taif, M.A.: Objective clustering inductive technology of gene expression sequences features. In: Communications in Computer and Information Science, vol. 716, pp. 359–372 (2017). https://doi.org/10.1007/978-3-319-58274-0_29
3. Babichev, S., Taif, M.A., Lytvynenko, V.: Inductive model of data clustering based on the agglomerative hierarchical algorithm. In: Proceedings of the 2016 IEEE 1st International Conference on Data Stream Mining and Processing, pp. 19–22 (2016)
4. Basu, D.: Randomization analysis of experimental data: the fisher randomization test. In: Selected Works of Debabrata Basu, pp. 305–325. Springer (2011). https://doi.org/10.1007/978-1-4419-5825-9_28
5. Brailo, M., Buketov, A., Yakushchenko, S., et al.: The investigation of tribological properties of epoxy-polyether composite materials for using in the friction units of means of sea transport. Mater. Perform. Charact. **7**(1), 275–299 (2018). https://doi.org/10.1520/MPC20170161
6. Brailo, M.V., Bezbakh, O.M., Husiev, V.M., Yakushchenko, S.: Modified epoxy matrix with improved properties for protection of transport vehicles. Bull. Karaganda Univ. Math. Ser. **3**(95), 88–100 (2019). https://doi.org/10.31489/2019M2/88-100
7. Buketov, A.V., Sapronov, A., Buketova, N., et al.: Impact toughness of nanocomposite materials filled with fullerene C 60 particles. Compos. Mech. Comput. Appl. Int. J. **9**(2), 141–161 (2018). https://doi.org/10.1615/CompMechComputApplIntJ.v9.i2.30
8. Buketov, A.V., Stukhlyak, P.D., Dobrotvor, I.G., et al.: Effect of the nature of fillers and ultraviolet irradiation on the mechanical properties of epoxy composite coatings. Strength Mater. **41**(4), 431–435 (2009). https://doi.org/10.1007/s11223-009-9136-1
9. Dobrotvor, I.H., Stukhlyak, P.D., Buketov, A.V.: Determination of the ranges of the optimal content of a dispersed filler in epoxy composites. Mater. Sci. **45**(6), 790 (2009). https://doi.org/10.1007/s11003-010-9244-x
10. Duleba, B., Greškovič, F., Dulebová, L., Jachowicz, T.: Possibility of increasing the mechanical strength of carbon/epoxy composites by addition of carbon nanotubes. In: Materials Science Forum, vol. 818, pp. 299–302. Trans Tech Publications (2015). https://doi.org/10.4028/www.scientific.net/MSF.818.299
11. Forastiere, L., Mealli, F., Miratrix, L., et al.: Posterior predictive p-values with fisher randomization tests in noncompliance settings: test statistics vs discrepancy measures. Bayesian Anal. **13**(3), 681–701 (2018). https://doi.org/10.1214/17-BA1062
12. Guigo, N., Van Berkel, J., de Jong, E., Sbirrazzuoli, N.: Modelling the non-isothermal crystallization of polymers: application to poly (ethylene 2, 5-furandicarboxylate). Thermochim. Acta **650**, 66–75 (2017). https://doi.org/10.1016/j.tca.2017.02.008
13. Kim, H.Y.: Statistical notes for clinical researchers: chi-squared test and Fisher's exact test. Restor. Dent. Endod. **42**(2), 152–155 (2017). https://doi.org/10.5395/rde.2017.42.2.152
14. Liu, D., Zhao, W., Liu, S., Cen, Q., Xue, Q.: Comparative tribological and corrosion resistance properties of epoxy composite coatings reinforced with functionalized fullerene C60 and graphene. Surf. Coat. Technol. **286**, 354–364 (2016). https://doi.org/10.1016/j.surfcoat.2015.12.056
15. Lv, J., Liu, Z., Zhang, J., Huo, J., Yu, Y.: Bio-based episulfide composed of cardanol/cardol for anti-corrosion coating applications. Polymer **121**, 286–296 (2017). https://doi.org/10.1016/j.polymer.2017.06.036

16. McCrum-Gardner, E.: Which is the correct statistical test to use? Br. J. Oral Maxillofac. Surg. **46**(1), 38–41 (2008). https://doi.org/10.1016/j.bjoms.2007.09.002
17. Napitupulu, D.: Cochran Q-test for criteria validation of PeGI framework. In: Journal of Physics: Conference Series, vol. 1430, p. art. no. 012037. IOP Publishing (2020). https://doi.org/10.1088/1742-6596/1430/1/012037
18. Peruzzi, V.V., da Silva, G.A., Renaux, C., et al.: Using statistical student's t-test to qualify the electrical performance of the diamond MOSFETs. In: 2018 33rd Symposium on Microelectronics Technology and Devices (SBMicro), pp. 1–4 (2018). https://doi.org/10.1109/SBMicro.2018.8511495
19. Rajmohan, T., Palanikumar, K.: Application of the central composite design in optimization of machining parameters in drilling hybrid metal matrix composites. Measurement **46**(4), 1470–1481 (2013). https://doi.org/10.1016/j.measurement.2012.11.034
20. Wilrich, P.T.: Critical values of Mandel's h and k, the Grubbs and the Cochran test statistic. AStA Adv. Stat. Anal. **97**(1), 1–10 (2013). https://doi.org/10.1007/s10182-011-0185-y
21. Zhan, Y., Zhang, J., Wan, X., Long, Z., He, S., He, Y.: Epoxy composites coating with Fe_3O_4 decorated graphene oxide: modified bio-inspired surface chemistry, synergistic effect and improved anti-corrosion performance. Appl. Surf. Sci. **436**, 756–767 (2018). https://doi.org/10.1016/j.apsusc.2017.12.095

The Approach for the Definition of Hemodynamic State in Pregnant Women with Extragenital Malformations

Iryna Perova[1(✉)], Oleh Datsok[1], Polina Zhernova[1], Olha Velychko[2], and Svitlana Bahan[3]

[1] Kharkiv National University of Radio Electronics, Kharkiv, Ukraine
rikywenok@gmail.com, oleh.datsok@nure.ua, polina.zhernova@gmail.com
[2] Yaroslav Mudryi National Law University, Kharkiv, Ukraine
umasbul@gmail.com
[3] Institute of Pharmacy Professionals Qualification Improvement, Kharkiv, Ukraine
bagan.v.o@gmail.com

Abstract. One of the problems leading to the maternal mortality and therapeutic abortion is recognition extra genital pathologies on time. The proposed approach bases on the analyzing several hemodynamic parameters such as systolic and diastolic arterial blood pressure, stroke volume, blood volume per minute, the total peripheral resistance of vessels using the means of cluster analysis. The basic block diagram of the system for express diagnostics of EGP's pregnant women for realization of this approach is described. The learning of the proposed system were performed with the results of medical investigations measured in 147 pregnant women with the gistosis sequela.

Keywords: Gestosis · Express diagnostics of extra genital pathology's pregnant women · Gestosis severity · Hemodynamic parameters · Manhattan distance · Cluster's center

1 Introduction

According to the data of the World Health Organization, maternal mortality is emergency high. About 830 women daily dead in the world due to the pregnancy or childbirth sequela. Accordance with rough estimation about 303 thousand women had died during the pregnancy and after childbirth in 2015 [24].

Recent investigations of the health status of pregnant women reveal that 60–70% of gestation course pathology cases associate with hidden or inveterate extra genital pathology (EGP). The extra genital diseases specific gravity among mortality factors is 34,1% last five years. Cardiovascular diseases (40,3%) predominate over the nosological entities, virulent disease (24,7%) rank the second fiddle, neoplasms (14,7%) rank next, and other probable diseases' prevalence is 20,3%. EGP is the plentiful group of miscellaneous and different meaning

S. Babichev et al. (Eds.): ISDMCI 2020, AISC 1246, pp. 627–638, 2021.
https://doi.org/10.1007/978-3-030-54215-3_40

diseases, syndromes and statuses of pregnant women conjoined together as few as that they are not the gynecological disease or obstetric sequela. There are diseases and pathoses with different severity and are amenable to the therapy, correction or require a therapeutic abortion. The specific gravity of mentioned diseases among the maternal mortality factors is given in Fig. 1. The primary hazard of EGP is the possibility of significant impair to the fetus state and stimulate the various diseases's propagation within the perinatal period and be the maternal mortality factor. Recommendations on observation of pregnant women with hemodynamic system malformations are given in [7].

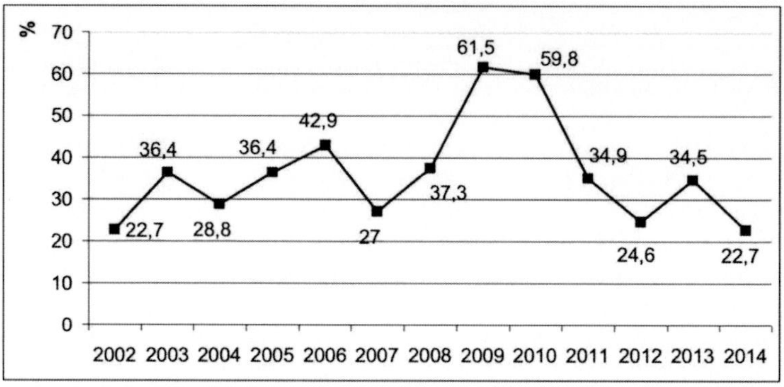

Fig. 1. EGP specific gravity among the maternal mortality factors in Ukraine

It has been proved that precaution with EGP existence should be performed in both directions: specific and non-specific (Fig. 2). The specific precaution generally intends to prevent decompensation or prior disease sequela and depends on the diagnosis of woman. Non-specific precaution is necessary for all EGP's pregnant women despite EGP's type. The hard gestosis of EGP's patients are observed in 85% of cases. The gestosis is the pregnancy sequela with malformation of vital organs, particularly the vascular system and blood flow. The term "pregnancy sequela" refers to entire pathological states originating entirely during the pregnancy and become to pregnancy sequela both mother and fetus. The gestosis is not a standalone disease and is the clinical evidence of the maternal organism adaptive mechanism's failure to adequately provide fetation [5,11].

The timeliness is conditioned with hard diseases's sequela. Chronic kidney malformation and hypertensive disease are observed in women after EGP. Perita mortality due to gestosis is 18–30% and prevalence is 50–80%. Every third infant has sequela of hypoxia and antenatal subpar growth curve, delay in physical and psychomotor development. In the structure of maternal mortality, the late gestosis ranks 2–3 fiddle according to frequency appearance rising 15–25% of common cases. Recently, it is the primary reason for life-threatening mother's states, raising the risk of peritoneal loss in 5 times and results in more than 50 thousand maternal mortality annually in the world [8].

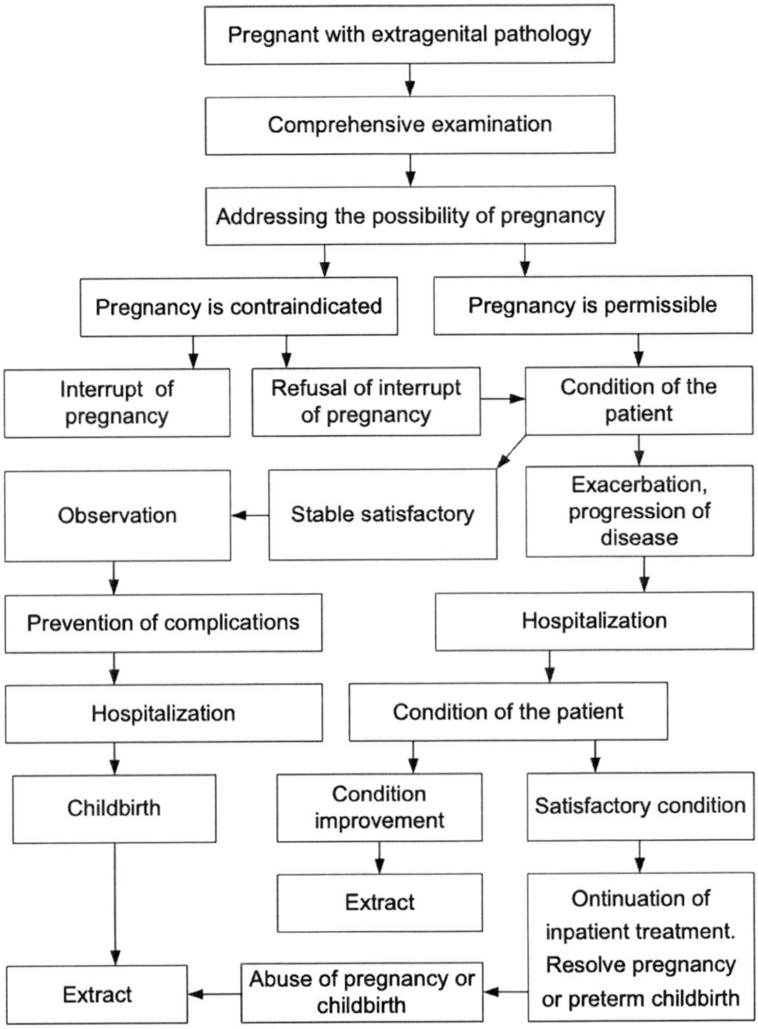

Fig. 2. EGP's pregnant women surveillance

2 Literature Review, Problem Statement and Goal of the Research

Analysis of recent investigation the pregnant women with different etiology complications pointed to important omissions for identification and correction of these diseases which impact on women and fetus health. Features of the pregnancy complicated with infectious diseases, namely HIV/AIDS, tuberculosis, malaria, helminth and virus infections and non-communicable disease, namely mental disability, gestational diabetes, gestoses requires particular attention and reliable diagnosis. Impact of these diseases on the pregnancy had been considered in [21]. Differential diagnostic of vessel crease based on the biomechanical parameters gives a complete picture of the peripheral blood circulation state and allow to monitor the foremost indicators during the physiotherapy, drug treatment and function tests. Changing in hemodynamics and more other physiological factors depending on gestation had been described in [10,22].

The goal of investigation [10] was the arterial blood pressure forecasting during the second trimester at hypertensive disorders complicated with gestoses on early pregnancy. The retrospective single-centre cohort investigation had been determining the high arterial blood pressure indicators with multifactorial regression analysis. Authors proposed the method of the blood protein biomarkers detection for estimation the level of pathological disorders at the early diagnostic's stages. The cardio vessel system load rises due to additional placental blood circle during the pregnancy. The features of pregnancy complications caused by cardio vessel disorders were considered in detail in [17]. Features of primary reasons for ischemic heart disease and attendance methods forecasting, initial estimation and recommendations for primary and secondary prevention of cardiovascular diseases at pregnant women were described in [19]. Importance of early gestoses' reasons determination for reducing of gestoses' severity and pregnancy prolonging was shown in [3,14,15,23]. Authors [15] emphasize the importance and need to observing gestoses' severity during the pregnancy. In [2,6,18] it was shown that hemodynamic system disorder at pregnant women is the leading pathogenetic mechanism of fetal state and growth at various pregnancy complications.

Recently, the gestosis is one of the most pressing challenges of the modern tocology regarding wide prevalence, complicate aetiopathogenesis, missing of premature and confidence diagnostic criteria, effectual precaution and therapy solutions, a high indicator of maternal and peritoneal morbidity and mortality, high economic efforts [9] to intensive care of patients [25].

3 Features of the Information Supply of EGP's Pregnant Women Diagnostic Process

Cardiovascular system (CVS) is the universal indicator of compensating and accommodating reactions, depicting the state of regulatory mechanisms and adaptive capacities of the organism [13]. Hemodynamic system malformations in pregnant women are the prior mechanism of abnormal fetation in different pregnancy sequela. Changing in hemodynamic and other physiological factors in the organism during the pregnancy has gone on permanently and recessively and sometimes incidentally as pregnancy is the so dynamic process.

The monitoring of hemodynamic parameters on any phase within the perinatal period must be full and includes the next parameters: weight, arterial blood pressure (ABP) (systolic, diastolic, average), heart rate (HR), heart rate variability (HRV), electrocardiogram (ECG), stroke volume (SV), blood volume per minute (BPM), total peripheral resistance of vessels (TPRV) and systolic index

(SI) (Fig. 3). Early diagnosis of gestosis signs allows reducing the severity of disease and prolong the pregnancy to a certain period. The gestosis diagnostics is based on the anamnesis and results of clinical research, Additional research such as Ultrasonic investigation, the Doppler sonography are avoided in suspected case.

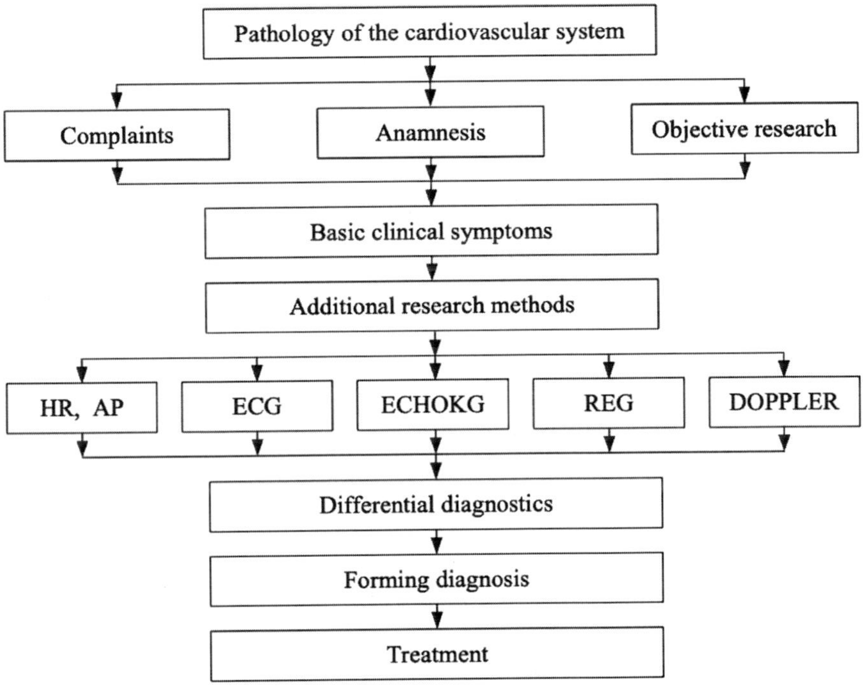

Fig. 3. Monitoring of hemodynamic parameters within the perinatal period

4 Analysis and Data Processing During Pregnant Women Investigations

Follow parameters characterizing the state of cardiovascular system of a pregnant woman [1] had been selected: weight (kg), heart rate (HR, beat per minute), systolic arterial blood pressure (SAP, mm Hg), diastolic arterial blood pressure (DAP, mm Hg), stroke volume (SV, ml), blood volume per minute (BV, l), the total peripheral resistance of vessels (TPRV, $\frac{GPa*s}{m^3}$). Proposed data set and the priory known diagnoses are used to determine the membership functions in indeterminacy conditions for "complex of symptoms – diagnosis".

In reference to indeterminacy of symptoms and complicate diagnostic process the modification of adaptive fuzzy clustering approach based on Manhattan metrics [4,16] has been proposed. As shown in Fig. 4, the rational monitoring and gestosis risk definition system consists of the measuring and information subsystems (hardware and software). The initial data (training dataset) are collected from 147 pregnant women 17–43 years old with various gestosis severity and gestation period equals to 12–38 weeks. The initial data set is denoted as numerical observations [12] (Fig. 5).

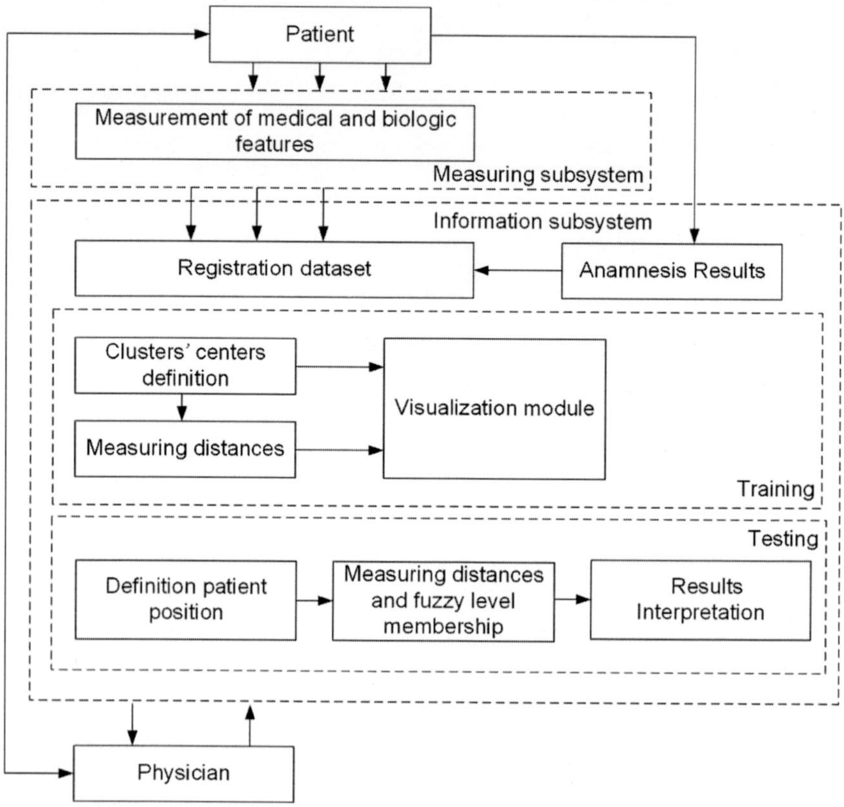

Fig. 4. The basic block diagram of the system for express diagnostics of EGP's pregnant women

The gestosis severity emerged with underlying of EGP had been determined with scale, characterizing the pregnancy state during physician examination [20]. The clinical symptoms, duration of disease, EGP existence and fetal state are taken into account (Table 1).

Patients had been partitioned with 7 groups:

Group 1 – 4 patients with hard gestosis severity selected by "extra-high criteria";

	Weight	HR	SAP	DAP	SV	BV	TPRV	Gestosis
0	67	95	110	70	74.70	4.89	1785	3
1	82	92	110	60	69.23	4.76	1598	5
2	101	113	120	80	71.66	4.84	1435	3
3	100	107	135	80	84.38	5.47	1814	2
4	64	77	120	80	68.86	4.92	1820	5
...	...	...	...	...	...	...	...	...
142	71	115	120	95	82.82	5.97	1923	4
143	70	61	130	80	79.58	5.46	1436	6
144	60	58	105	80	38.60	4.73	1846	7
145	86	86	135	90	78.23	5.34	1359	6
146	79	107	105	70	70.26	4.67	1439	5

Fig. 5. Parameters characterizing the state of cardiovascular system of a pregnant woman

Table 1. Scale of the gestosis severity

Severity	Scores
Light gestosis severity	≤7
Middle gestosis severity	8–11
Hard gestosis severity	≥12

Group 2 – 19 patients with light gestosis severity selected by "extra-low criteria";
Group 3 – 39 patients with middle gestosis severity selected by "middle criteria";
Group 4 – 20 patients with light gestosis severity directed to the middle gestosis severity selected by "low criteria";
Group 5 – 22 patients with middle gestosis severity directed to the hard gestosis severity selected by "greater than middle criteria";
Group 6 – 30 patients with middle gestosis severity directed to the light gestosis severity selected by "less than middle criteria";
Group 7 – 13 patients with hard gestosis severity directed to the middle gestosis severity selected by "high criteria".

Training of information subsystems consist of cluster's centers definition, measuring distances between clusters and visualization module.

The principal component analysis (PCA) were proposed for data visualization. (Fig. 6). Each group of gestosis type was marked by different colors.

The module of cluster's centers definition, based on the Manhattan metrics, define a position of cluster centers. In Fig. 6 position of each patient was marked by circles, cluster's centers were marked with squares. Further, distances between cluster centers according to growing gestosis severity were calculated with formula:

$$dist_{ij} = \sum_{i,j=1}^{n} |c_i - c_j|, \tag{1}$$

where n is a total number of medical features (7 in our case).

All distances between clusters were presented in Fig. 7.

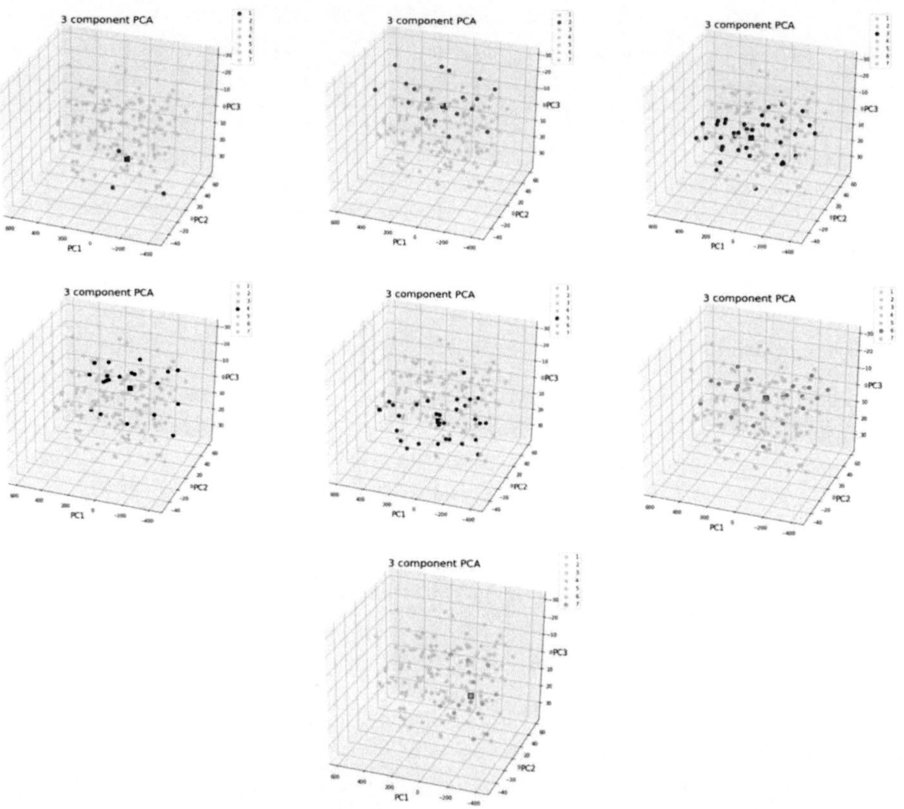

Fig. 6. PCA-visualization of each gestosis type

dist17 = 103.13;
dist75 = 201.69;
dist53 = 94.98;
dist36 = 80.6;
dist64 = 30.1;
dist42 = 91.7.

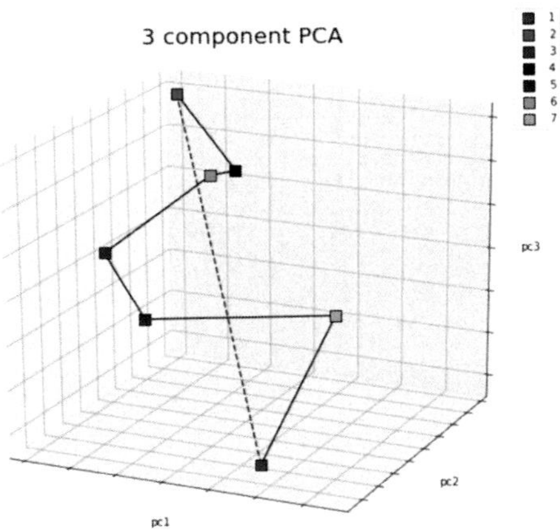

Fig. 7. Centers' clusters location for gestosis severity (from high (red) to low (green))

The distance between patients with hard gestosis severity selected by "extra-high criteria" and patients with light gestosis severity selected by "extra-low criteria" have the most high values:

dist12 = 301.94.

Patient's location relatively the clusters' centers is computed on the testing step with formula:

$$dist(x) = \sum_{i=1}^{n} |x_i - c_j|, \tag{2}$$

Membership to the gestosis type is performed with the formula:

$$mu_i(x) = \frac{\frac{1}{dist_i}}{\sum_{i=1}^{m} \frac{1}{dist_i}}, \tag{3}$$

where m is the total number of the cluster's centers.

Therefore, the membership value to the gestosis type of a pregnant woman can be defined personally. As illustration of proposed approach, the medical indicators of pregnant woman with hard gestosis severity directed to the middle gestosis severity selected by "high criteria" (7 type) are presented in Table 2.

Data visualization is presented in Fig. 8.

Distances between the patient with gestosis and cluster's centers were calculated according to (2):

dist1 = 201.88,
dist2 = 460.95,
dist3 = 387.33,
dist4 – 369.23,
dist5 = 300.17,

Table 2. Medical indicators of a pregnant woman

Weight	59
Heart rate	78
Systolic arterial blood pressure	105
Diastolic arterial blood pressure	70
Stroke volume	70.71
Blood volume per minute	4.87
The total peripheral resistance of vessels	1904

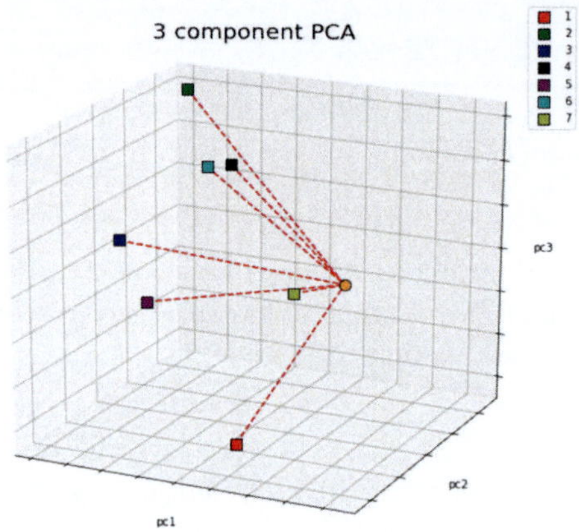

Fig. 8. Data visualization

dist6 = 343.53,
dist7 = 98.75.

Values of the membership functions to the gestosis type are equal to:

mu1 = 0.17
mu2 = 0.07
mu3 = 0.08
mu4 = 0.09
mu5 = 0.11
mu6 = 0.10
mu7 = 0.38

Therefore, the highest membership value points to the hard gestosis severity directed to the middle gestosis severity (type 7).

5 Conclusions

The problem of maternal mortality and therapeutic abortion caused by gestosis had been highlighted. The approach based on the monitoring of hemodynamic parameters set, such as parameters such as systolic and diastolic arterial blood pressure, stroke volume, blood volume per minute, the total peripheral resistance of vessels and weight was proposed. The block diagram of the information medical system realizing the approach of the parameters' evaluation and recognizing the gestosis severity with successive disease trend was proposed. The method of indicators' analysis based on the calculation Manhattan distances and membership value to a certain cluster was developed. The system was trained with using data sets of 147 pregnant women with light, middle and hard gestosis severity and orients to recognition three gestosis severity (light, middle, hard) and four boundary states: light gestosis severity directed to the middle gestosis severity, middle gestosis severity directed to the hard gestosis severity, middle gestosis severity directed to the light gestosis severity, hard gestosis severity directed to the middle gestosis severity.

References

1. The order of ministry of health of Ukraine no 676 of 31.12.2004 "about on approval of clinical protocols on obstetric and gynecological aid". https://zakononline.com.ua/documents/show/38238_485322
2. Abraham, D., Adamovich, D., Turan, O., Blitzer, M., Baschat, A.: Maternal blood pressures during pregnancy and the risk of delivering a small-for-gestational-age neonate. Hypertens. Pregnancy **35**(3), 350–360 (2016). https://doi.org/10.3109/10641955.2016.1150487
3. Allen, R., Aquilina, J., Zamora, J., Thangaratinam, S., Velauthar, L.: External validation of preexisting first trimester preeclampsia prediction models. Eur. J. Obstet. Gynecol. Reprod. Biol. **217**, 119–125 (2017). https://doi.org/10.1159/000490385
4. Bodyanskiy, Y., Perova, I.: Online medical data stream mining based on adaptive neuro-fuzzy approaches. J. Appl. Comput. Sci. **198**(4), 3–25 (2017). https://doi.org/10.15407/kvt198.04.003
5. Bokslag, A., Teunissen, P.W., Franssen, C., van Kesteren, F., Kamp, O., Ganzevoort, W.: Effect of early-onset preeclampsia on cardiovascular risk in the fifth decade of life. Am. J. Obstet. Gynecol. **216**(5), 523.e1–523.e7 (2017). https://doi.org/10.1016/j.ajog.2017.02.015
6. Boyd, H.: Clinical follow-up of women after hypertensive disease in pregnancy. BJOG Int. J. Obstet. Gynaecol. **124**(10), 1614 (2017). https://doi.org/10.1111/1471-0528.14707
7. Foo, F.L., Lees, C., McEniery, C.M., Thilaganathan, B., Wilkinson, I.: Assessment of arterial function in pregnancy: recommendations of the international working group on maternal hemodynamics. Ultrasound Obstet. Gynecol. **50**(3), 324–331 (2017). https://doi.org/10.1002/uog.17565
8. Khmil, S.V., Franchuk, U.Y.: Peculiarities of pregnancy in women at high risk of developing late gestosis on the background of metabolic syndrome (literature review). Bull. Sci. Res. **4**, 16–19 (2017)
9. Li, R., Callaghan, W.M., Tsigas, E.Z.: Health and economic burden of preeclampsia: no time for complacency. Am. J. Obstet. Gynecol. **217**(3), 235–236 (2017). https://doi.org/10.1016/j.ajog.2017.06.011

10. Mabuchi, A., Yamamoto, R., Ishii, K., Muto, H., Mitsuda, N.: Significance of high-normal blood pressure during early second trimester for predicting the onset of hypertensive disorders in pregnancy. Hypertens. Pregnancy **35**(2), 234–241 (2016). https://doi.org/10.3109/10641955.2016.1139719
11. Medved, V.I.: Extragenital pathology of pregnant women as a multidisciplinary problem. Med. Aspects Woman's Health **3**, 15–18 (2016)
12. Mustetsov, N.P., Bahan, S.A.: Neural network of the gestosis diagnosis system. Visnyk NTUU KPI Seriia-Radiotekhnika Radioaparatobuduvannia **72**, 42–46 (2018)
13. O'Gorman, N., Poon, L.C., Rolnik, D.L., Plasencia, W., Singh, M.: Contingent screening for preterm pre-eclampsia. Ultrasound Obstet. Gynecol. **47**(5), 554–559 (2016). https://doi.org/10.1002/uog.15807
14. Olaya-Garay, S., Velásquez-Trujillo, P., Vigil-De Gracia, P.: Blood pressure in adolescent patients with pre-eclampsia and eclampsia. Int. J. Gynaecol. Obstet. **138**(3), 335–339 (2017). https://doi.org/10.1002/ijgo.12237
15. Pauli, J., Repke, J.: Preeclampsia: short-term and long-term implications. Obstet. Gynecol. Clin. North Am. **42**(2), 299–313 (2015). https://doi.org/10.1016/j.ogc.2015.01.007
16. Perova, I., Bodyanskiy, Y.: Adaptive fuzzy clustering based on manhattan metrics in medical and biological applications. Bull. Natl. Univ. "Lviv Polytechnic" **826**, 8–12 (2015)
17. Quinn, M., Cedars, M.: A canary in the coal mine: reproductive health and cardiovascular disease in women. Semin. Reprod. Med. **35**(3), 250–255 (2017). https://doi.org/10.1055/s-0037-1603570
18. Ramadan, H., Rana, S., Nasim, R.: Myocardial performance index in hypertensive disorders of pregnancy: the relationship between blood pressures and angiogenic factors. Hypertens. Pregnancy **36**(2), 161–167 (2017). https://doi.org/10.1080/10641955.2017.1280048
19. Sanghavi, M., Gulati, M.: Cardiovascular disease in women: primary and secondary cardiovascular disease prevention. Obstet. Gynecol. Clin. North Am. **43**(2), 265–285 (2016). https://doi.org/10.1016/j.ogc.2016.01.001
20. Savelyeva, G.M.: Eclampsia in modern obstetrics. Obstet. Gynecol. **6**, 4–9 (2010)
21. Sina, B.: Pregnancy and the global disease burden. Reprod. Health **170**(14) (2017). https://doi.org/10.1186/s12978-017-0420-4
22. van der K, T., C, K., van den P, B.: Blood pressure patterns in women with gestational hypertension or mild preeclampsia at term. Eur. J. Obstet. Gynecol. Reprod. Biol. **210**, 360–365 (2017). https://doi.org/10.1016/j.ejogrb.2017.01.021
23. Wang, Y., Hao, M., Sampson, S., et al.: Elective delivery versus expectant management for pre-eclampsia: a meta-analysis of RCTs. Arch. Gynecol. Obstet. **295**, 607–622 (2017). https://doi.org/10.1007/s00404-016-4281-9
24. World Health Organization: Maternal mortality (2020). https://www.who.int/news-room/fact-sheets/detail/maternal-mortality
25. Wright, D., Gallo, D.M., Gil Pugliese, S., Casanova, C., Nicolaides, K.H.: Contingent screening for preterm pre-eclampsia. Ultrasound Obstet. Gynecol. **47**(5), 554–559 (2016). https://doi.org/10.1002/uog.15807

Complex Internet Data Management System

Viacheslav Zosimov, Oleksandra Bulgakova(✉), and Valeriy Pozdeev

V.O. Sukhomlynsky Mykolaiv National University, Mykolaiv, Ukraine
zosimovvv@gmail.com, sashabulgakova2@gmail.com, pozdeev1405@gmail.com

Abstract. The article presents the system of complex data management on the Internet. The system consists of several modules, each of which solves separate tasks of web data processing, such as extracting web data, methods for automatically integrating semantic markup into HTML code of existing and new web pages, search results ranking. At the heart of the kernel is a developed object-oriented web data operating language that implements a set of basic data processing and conversion functions. This architectural solution allows to select the system core as an abstraction level, which on the one hand provides all the modules of the system with the necessary set of software functions, and on the other - hides the details of their implementation.

Keywords: Information search · Ranking model · Meta-search system · Semantic markup · Ontologies · Web resource structure · Data mining · Domain-specific language · Content management system

1 Introduction

Each year the growth rate of new information is increasing. The World Wide Web is becoming available to a growing number of subscribers, both in terms of distribution geography and age range. Hardware devices are being improved, and new bandwidth standards are being developed that could not be dreamed of 10 years ago. New software is being developed at a rapid rate to create web applications that are more functional, easy to use, accessible to a wide range of users. Programming languages, frameworks, content management systems, site designers, online editors are estimated in the hundreds [10]. Such a variety of web development tools gives network users ample opportunity, but the downside of rapid technology development is the inability to create versatile software to efficiently manage data in terms of the diversity of structures and approaches to their generation [20].

2 Review of Literature

One of the solutions to the effective web data management problem in the presence of a large number of approaches to the web pages creation is to develop

S. Babichev et al. (Eds.): ISDMCI 2020, AISC 1246, pp. 639–652, 2021.
https://doi.org/10.1007/978-3-030-54215-3_41

methods that will be one level of abstraction higher than parsers-programs and grabbers [8]. As such approach, a subject-oriented web data management language is developed that allows performing standard data operations: extraction, storage, processing, display, regardless of the way they are presented on a web page.

Another negative factor in having a large variety of development platforms is the fact that most of them do not support the ability to add semantic markup directly to web page code in basic versions [17]. And those with semantic markup functions implement them at the level of third-party plugins and applications [1, 9, 11]. However, it is clear that further qualitative development of the World Wide Web is not possible without the close integration of semantic markup into the HTML of web pages. The very presence of semantic markup makes it possible to develop smart search using search agents, to effectively apply web data aggregation and analysis.

With all the above, the article is devoted to the description of the complex Internet data operating system (CIDO) within the semantic web concept, focused on solving the problem of improving the quality of intelligent search and data processing on the Internet using software search agents, combining new methods, models and means of operating web data at the semantic level in real time.

3 Abstract Model of CIDO System

In the course of researching information and business processes on the Internet, a number of perspective directions were allocated, the development of which will allow to increase the efficiency of web environment use for all participants from developers to users [2] (Fig. 1).

The World Wide Web is a distributed system that provides access to related documents located on different computers connected to the Internet. The main user tool for retrieving information from the World Wide Web is search engines [3]. A further step in the development of information retrieval methods is meta-search systems and search agents that perform additional processing of search results.

The main tool of the information owner is to develop a web resource for submitting the necessary information in the form of HTML code, which is the standard of representation in the world wide web. Another participant in information processes in the web environment is web development specialists.

Based on the study of the activities of each type of participants in the web environment, the main system modules were identified, on the basis of which a general scheme is presented in Fig. 1. The CIDO system provides web developers, web resource owners, and global network users with extensive web data management capabilities to fulfill their own information needs, and satisfies the following requirements:

- integration of all software solutions into one system with modular architecture;

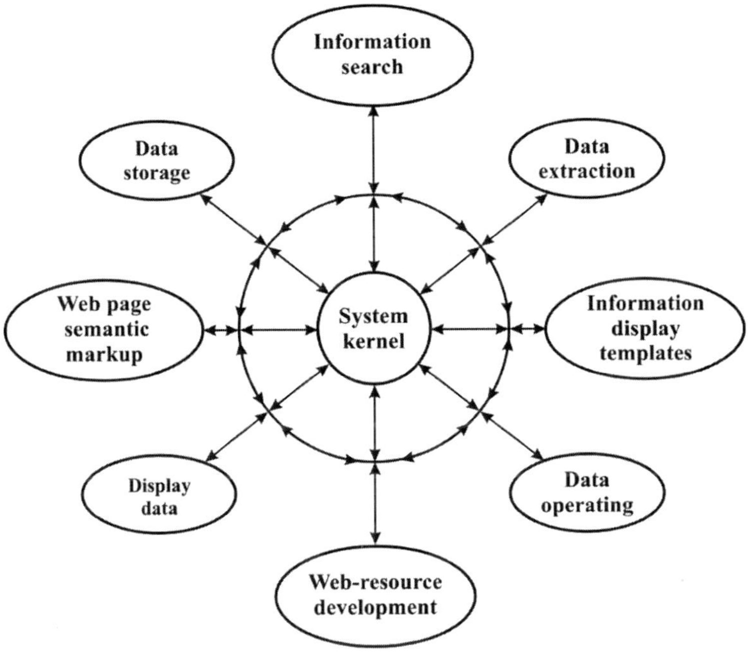

Fig. 1. General scheme of the proposed technology

- providing system users with effective software tools for the full cycle of web data management from web development to information retrieval;
- use of the latest research results and modern software tools in development [19];
- high level of abstraction in the software modules design that allows scaling, and updating the system without the need to interact with the program code of each individual module;
- high speed [16];
- user-friendly, intuitive interface, extensive customization options [19].

3.1 CIDO System Architecture

The CIDO system general scheme, presented in Fig. 1 is an abstract model of the web data operating process. In general, the CIDO system is a set of individual modules that can interact with each other. At the heart of the kernel is a developed domain-specific language for operating web-data, which implements a set of basic data processing and conversion functions.

Such an architectural solution makes it possible to select the system kernel as an abstraction level, which on the one hand provides all the modules of the system with the necessary set of software functions, and on the other - hides the details of their implementation [16]. Similar architectural solutions underpin the implementation of an input/output model based on byte streams for Unix family operating systems [14, 16]. The system modules are logically divided into two groups:

1. Base modules:
   - data extraction;
   - data storage;
   - data processing;
   - display of data.
2. Functional modules:
   - creating web pages with integrated semantic markup;
   - integration of semantic markup into the HTML code of existing web pages;
   - display content based on the semantic profile of a web property (display templates);
   - additional processing of search engine results:
     - meta-search;
     - search results ranking;
     - search agents application.

Figure 2 presents a general diagram of the CIDO system components interaction. The base modules inherit the necessary functions of the system kernel

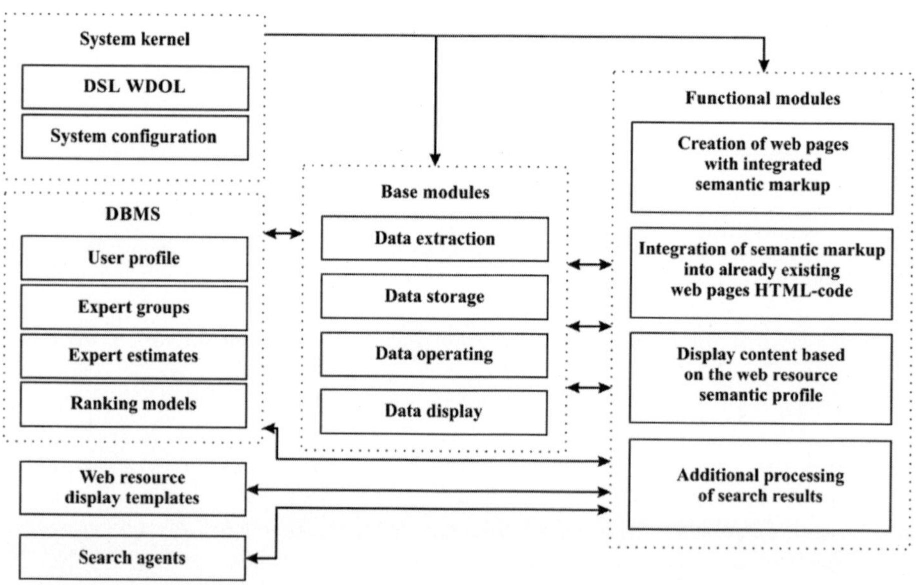

Fig. 2. The general diagram of the CIDO system components interaction

and add the level of abstraction between the kernel and the functional modules. This architectural solution implements a multi-level approach to systems design, where each layer serves its part of the interaction process. With this structure, the collaboration of the base and function modules becomes much easier, more transparent and understandable. An example of such an approach to systems implementation is the Open Systems Interconnection Basic Reference Model (OSI), an abstract network model for communications and network protocol development [13]. The implementation and operation details of the base modules are hidden from the end-user of the system for security and reliability.

Functional modules implement the system's application functions of in the form of web applications to meet the information needs of the end user. A significant advantage of using web applications is the correct operation of the program, regardless of the client's operating system. These modules are provided by services that provide the base modules and the system core. Some functional modules may provide services to others. For example, semantic profiles are used when displaying web resources and search results. Each of the functional module has its own visual and text interfaces for client interaction with the system.

The dominant concept in the creation of distributed network applications is the client-server architecture, which is why it formed the basis for the CIDO system construction. The server part contains the following components:

- base modules;
- domain-specific language interpreter;
- database;
- version control system;
- computing part of functional modules.

The client side is responsible for user interaction and contains functional module interfaces.

3.2 Data Mining Module

The data extraction module is one of the most important in the system, since it cannot be further processed without efficient data extraction. Figure 3 presents the web data extraction module components interaction diagram. The basis of the module is the developed domain-specific web data operating language [18], the interpreter of which is part of the system core.

The module architecture was designed to abstract the searching and extraction data process from the type of platform on which the web resource was developed. To find and extract the necessary code snippets, the browser's built-in capabilities are used, namely the developer console [4], which gives access to the web resource structure. Data extraction can be performed once, or a production template can be created to customize the timetable for specific operations.

3.3 Data Storage Module

The data mining module results can be stored in the system for later use. Data storage made in a separate level of the CIDO system abstraction, which allows

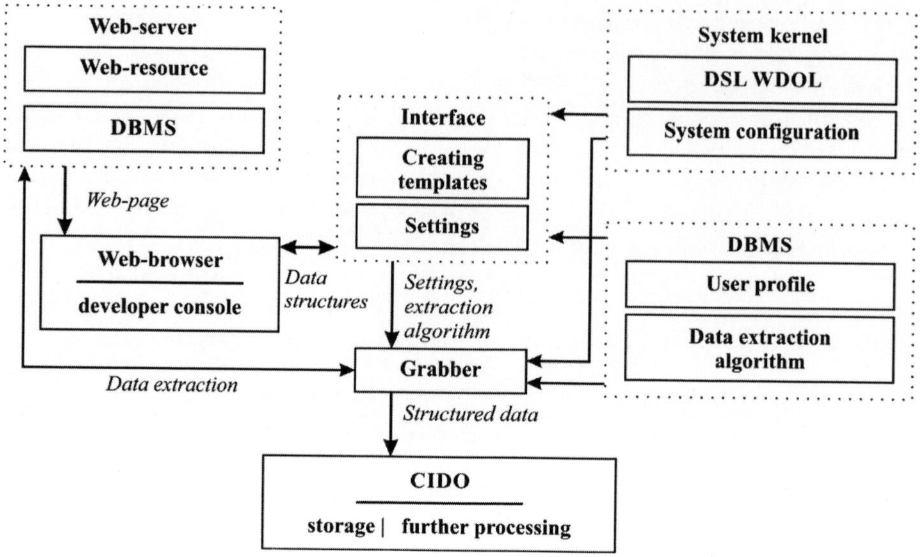

Fig. 3. Web data extraction module components interaction scheme

to make changes to the module operation without interfering with the kernel code and other modules. Till the system development time, this module involves the use of several storage methods:

- database;
- Version Control System;
- delimiter-separated values text.

Also, bringing the module to a system abstraction separate layer makes the process of adding new ways of data storing, or disabling existing ones, faster and more transparent. Figure 4 presents the web storage module components interaction diagram. The module interface accepts system and user settings and sends commands to the storage controller. The interface is also used to the module adjust operations, choose how to store data, access data manually. The controller accepts input from one of the system modules and processes them according to commands and settings received through the interface.

3.4 Data Operating

In a broad sense, data processing is a systematic sequence of operations performed on the data. For the processing of electronic data, the sequence of operations is specified by a computer program. The data processing module provides three applications for the following programs:

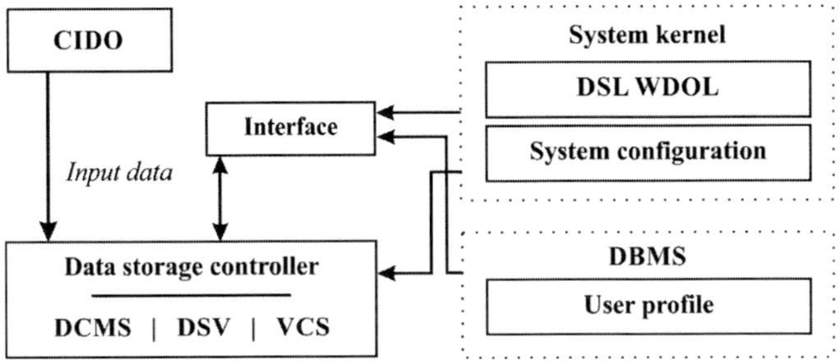

Fig. 4. The web storage module components interaction diagram

1. Application of the system built-in software functions.
2. Application of custom programs added to the system as information processing templates.
3. Application of external software solutions.

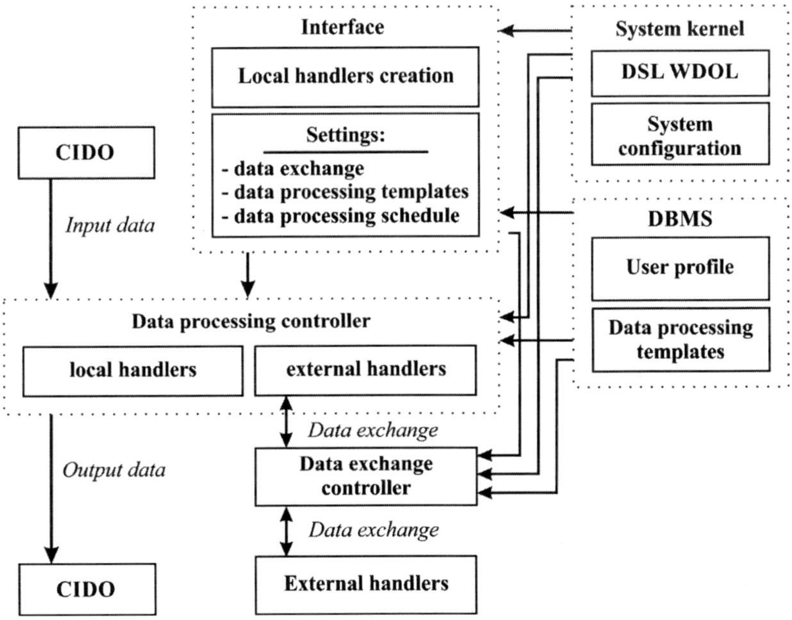

Fig. 5. Web data processing module components interaction scheme

Custom data processing software can be built using the built-in language or one of the general-purpose programming languages such as PHP or Javascript. Figure 5 presents a diagram of the interaction of components of the web data processing module. Using external software solutions to perform data operations requires additional measures to be taken to:

- ensuring the security and integrity of data transmission and retrieval from external software products;
- ensuring the data processing results correctness.

These functions are performed by a special data controller, the data module interaction with external software products.

3.5 Display of Data

Since the use of the CIDO system involves the web data processing, therefore, all the system results are presented to the end user in the form of web pages. The CIDO system uses a template-based approach to display data. Two types of templates are used to display two groups of system results:

1. Templates, based on a web resource semantic profile. This type of template is essential to the system. The architecture of these templates relies heavily on similar templates used in the most common site content management system, WordPress, and called themes.
   According to this approach, the final webpage is based on three components:
   - directly data;
   - HTML template that includes web page structure and style;
   - a semantic profile that integrates semantic markup into the HTML template, for example, for effective and correct use of smart search methods. This type of template is used to display web resources and search results.
2. Templates for displaying aggregation and data processing results such as tables, lists, charts, etc.

Figure 6 presents the data display module components interaction diagram. Obviously, one template cannot take into account all potential users information peculiarities of the perception. This can cause some groups of users to have difficulty finding the information they need. The main difference of the CIDO approach is that the templates are display not on cite-developer side but on the user-side. That is, all web resources created on the basis of the corporate web resources semantic profile, can be displayed equally according to one custom template [18].

The scheme in Fig. 7 demonstrates the differences between the conventional approach to web resource content displaying from the proposed in paper. Conventions on Fig. 7: S1,... Sn - Web Resources, T1,... Tn - Display Templates, U1,... Un - Users. Displaying all web resources of the same type according to a single template with a defined structure of elements and their style design has significant advantages over the existing approach:

1. The user spends no time exploring the structure of the new web resource, searching the navigation menu and the main content block. All typical blocks are always located in the same places.

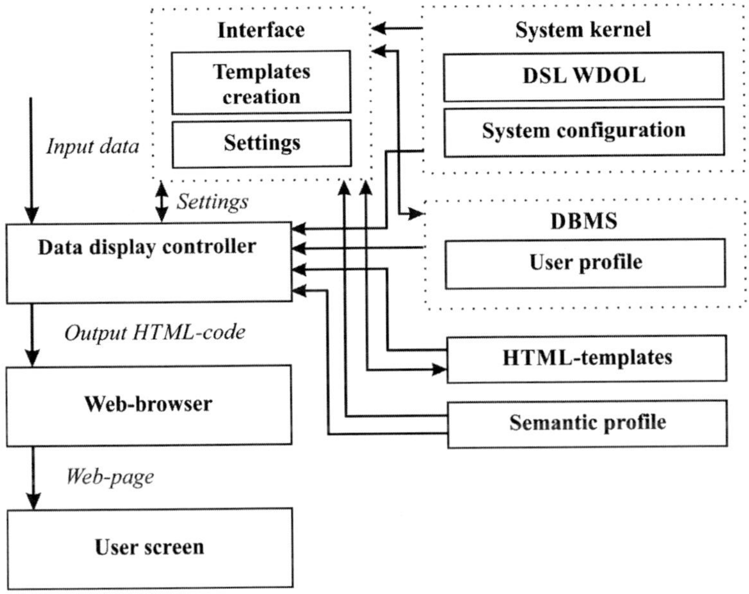

Fig. 6. Data display module components interaction scheme

2. The user has the opportunity to customize the style of the web resource according to their wishes and individual features - text size, colors, block sizes and locations, etc.
3. The first two points are obvious, but the system potential is much wider. The absence of hard-set templates by the web resource developer allows the user to change the whole web resource structure. For example, it is possible to disable the display of any blocks, as well as display information from different pages of the web resource on one screen, which greatly speeds up the information perception and decision making process.

3.6 Development of Web Resources with Integrated Semantic Markup

There are currently no web development software platforms that have semantic markup integration in the web page code in the basic feature set. For some of them, semantic markup functions are implemented as plugins and modules (Drupal, Wordpress, Movable type) [5,12,15]. However, they have limited functionality and do not allow full and qualitative markup of all web resource information content.

As was noted hereinbefore, further World Wide Web qualitative development is not possible without the close integration of semantic markup into web page code. There are two ways to solve this pressing problem:

- integration of semantic markup into existing web resources;

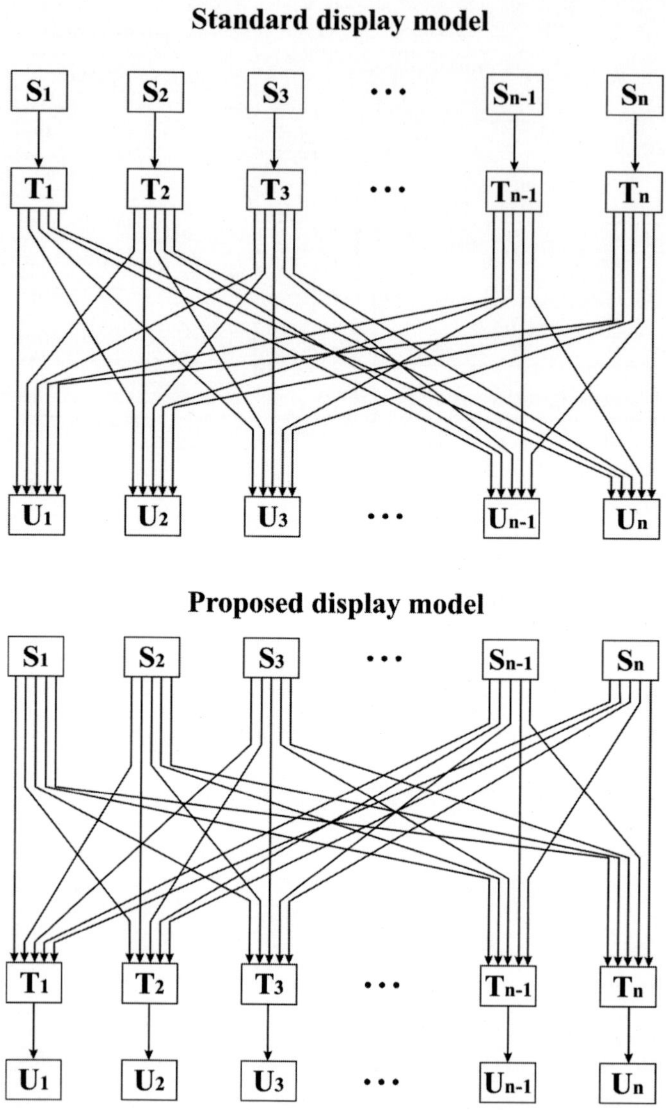

Fig. 7. Scheme for applying web resource displaying templates

- development of software tools for creating new web resources with integrated semantic markup.

This module is the basis for the creation of software with built in ability to integrate semantic markup in the new web pages HTML code. The proposed approach involves creating separate semantic profiles for the most common types of web resources:

- corporate;

- online store;
- news;
- blogs;
- informational
- forums, etc.

3.7 Semantic Markup of Existing Web Pages

In the last decade of rapid development, web-based development systems have been acquired, based on real-time generation of web page end-code at the time an HTTP request from a user is received on the server [6]. Ready-made web pages do not exist at the time of request. Accordingly, it is not possible to add semantic markup to the web-page code manually. HTML code generation of the page is based on a display template that describes the HTML structure of the future page, and the content is stored in the database. Adding a semantic markup integration feature to the web page's final code on such systems is impossible without significant interference with the system source code. As noted earlier, for some of them there are partial solutions, implemented in the form of plugins and modules. They are mainly used to add extended snippets and do not allow web resource full semantic markup.

With all of the above, integrating semantic markup into the existing web pages HTML code is an up-to-date and challenging task.

Solving this problem with external methods without interfering with the web pages generation requires the following solutions:

- effective methods of data extraction;
- web resource semantic profile;
- creation and storage the semantic markup template;
- display a web page with a semantic markup overlay.

Data extraction is implemented in the form of CIDO system corresponding module. The developed corporate web resources semantic profile fully provides the necessary set of classes to describe the structures. The semantic markup template is a meta description of an existing web property. It contains a set of user-defined classes for describing the content-based web resource semantic structure and is stored in a database. If an HTTP request is made to the server, HTML code is generated and the semantic markup template overlaps the existing web page rendering template by including additional tag attributes in the HTML blocks that directly contain the content. This semantic markup layout is called microdata. Since semantic attribute binding is applied to the HTML tags of the template and not to the data itself, updating information on the site does not require re-updating the semantic markup. This is very important because most modern services use JSON-LD format to add semantic markup to existing sites [7]. Its peculiarity is that it generates Javascript code as a "snippet": a "property" that is inserted by the <script> tag at the beginning of the HTML document. This code is static and requires regular re-tagging of the updated data fragments manually.

Table 1. Comparison of the results of the developed CMS with the CMS Wordpress

Number of web-resource/saved time, %	Web resource development time, in hours	
	CMS wordpress	Developed CMS
1/6.38	47	44
2/6.25	32	30
3/17.65	17	14
4/3.45	29	28
5/5.41	37	35
6/12	50	44
7/12.5	48	42
8/25	24	18
9/11.11	27	24
10/10	30	27
11/18.75	32	26
12/20	18	15
13/16.67	20	16
14/4.35	23	22
15/4.76	21	20
16/20	30	24
Avarage 12.14 %		
	Semantic markup, in hours	
	CMS wordpress	Developed CMS
1/63.64	121	44
2/61.54	78	30
3/65.85	41	14
4/66.27	83	28
5/62.37	93	35
6/62.07	116	44
7/62.83	113	42
8/59.09	44	18
9/65.22	69	24
10/64	75	27
11/55.17	58	26
12/53.13	32	15
13/55.56	36	16
14/62.07	58	22
15/62.26	53	20
16/58.62	58	24
Avarage 61.23%		

4 Results and Discussion

One of the most important practical embodiments of the system modules is the content management system (CMS). The developed CMS includes a list of typical pages. For each page the interface of adding specific fields according to the developed structure of corporate web resources is implemented. To solve the displaying content problem, the visual and text interfaces for building display templates are integrated into the basic set of CMS functions. The user has the opportunity to develop a template that meets the requirements for the web pages style, such as colors, text and block size, information blocks location, navigation menu design, etc., and view all corporate web resources in a single style. If there are no display templates on the visitor's side, the web-resource content will be displayed according to the developed default templates.

To determine the developed CMS benefits, several experiments were conducted. During the period 2019–2020 years, 16 corporate web resources were developed using this CMS. In parallel, the same web resources were developed by means of the most actual today CMS Wordpress (Table 1).

5 Conclusions

The article is devoted to developing approaches to solving current research problems such as:

- improvement of web data extraction methods;
- developing new approaches to displaying the web resources content;
- improving the quality of keyword information search:
  - search spam fight;
  - personalization of search results.
- two directions of the semantic web concept development:
  - development of web resources with integrated semantic markup;
  - developing methods for integrating semantic markup into the HTML code of existing web pages;
- development the effective methods of search agents application for realization of intellectual search concept.

To enhance the effectiveness of the solutions offered, they have all been integrated into a single web data management system on the Internet, which gives web developers, web resource owners and WAN users extensive web data management capabilities to meet their own information needs. The system has been developed within the concept of semantic web and is focused on improving methods of information search, creation of web resources and software search agents. The developed CMS gives users the opportunity to create web resources with integrated semantic markup, as well as to choose one display template for all web resources of the same type.

This approach saves the owner of the web resource from the need to spend extra resources on design development and forces to focus on the formation of quality information content. Visitors, in turn, eliminates the need to spend time studying the design structure and navigation of each individual web resource to find the necessary information.

References

1. Amerland, D.: Google Semantic Search: Search Engine Optimization (SEO) Techniques That Get Your Company More Traffic, Increase Brand Impact, and Amplify Your Online Presence. Que Publishing, Indianapolis (2013)
2. Bernsen, N., Dybkjar, L.: Multimodal Usability Human. Computer Interaction Series. Springer, Heidelberg (2009)
3. Castillo, C., Davison, B.: Adversarial Web Search. Now Publishers Inc., Boston (2011)
4. devtools (2018). https://learn.javascript.ru/devtools/
5. Drupal: CMS drupal (2017). https://www.drupal.org/
6. HTML (2017). https://html5book.ru/poleznye-servisy-html/
7. JSONLD: JSON-LD (2017). https://www.w3.org/TR/json-ld/
8. Khenyk, B.: HTML y CSS. Put k sovershenstvu, BKhV-Peterburh (2011)
9. Kosara, T., Tsoglin, S., Mernika, M.: Domain-specific languages: a systematic mapping study. Inf. Softw. Technol. **71**(March), 77–91 (2016)
10. Kuznetsov, M.: PHP. Praktyka sozdanyia Web-saitov. BKhV-Peterburh (2012). in Russian
11. Maass, W., Kowatsch, T.: Semantic Technologies in Content Management Systems: Trends, Applications and Evaluations. Springer, Heidelberg (2012)
12. Movable: CMS movable (2017). https://movabletype.org/
13. OSI (2016). https://uk.wikipedia.org/wiki/OSI/
14. UNIX (2016). https://www.unix.com/
15. Wordpress: CMS wordpress (2017). https://wordpress.com/
16. Zosimov, V.: Usage the operating systems multitasking for organization multi-access to complex systems modeling computing system. Inductive Model. Complex Syst. **5**, 146–150 (2013)
17. Zosimov, V.: Web-based information analysis using inductive modeling algorithms. In: Proceedings of the International Workshop on Inductive Modelling, pp. 81–84. Zhukyn IWIM (2015)
18. Zosimov, V.: An integrated approach to improving the efficiency of web data processing based on semantic markup. Upr. sist. mas **276**(4), 32–45 (2018)
19. Zosimov, V., Khrystodorov, O., Bulgakova, O.: Dynamically changing user interfaces: software solutions based on automatically collected user information. Program. Comput. Soft **44**, 492–498 (2018). https://doi.org/10.1134/S036176881806018X
20. Zosimov, V., Stepashko, V., Bulgakova, O.: Inductive building of search results ranking models to enhance the relevance of text information retrieval. In: 26th International Workshop on Database and Expert Systems Applications, Valencia, pp. 291–295 (2015). https://doi.org/10.1109/DEXA.2015.70

Dynamic Coordination of Strategies for Multi-agent Systems

Petro Kravets[1], Vasyl Lytvyn[1], Victoria Vysotska[1(✉)], Yuriy Ryshkovets[1], Svitlana Vyshemyrska[2], and Saule Smailova[3]

[1] Lviv Polytechnic National University, Lviv, Ukraine
{Petro.O.Kravets,Vasyl.V.Lytvyn,Victoria.A.Vysotska, Yuriy.V.Ryshkovets}@lpnu.ua

[2] Department of Informatics and Computing Technology, Kherson National Technical University, Kherson, Ukraine
printvvs@gmail.com

[3] 5D. Serikbayev East Kazakhstan State Technical University, Ust-Kamenogorsk, Republic of Kazakhstan
smailova@mail.ru

Abstract. Dynamic coordination of multi-agent systems (MAS) strategies under uncertainty based on a stochastic game model is solved. Dynamic coordination is teaching of the system of generating spatially distributed periodic signals. A stochastic game model is built, criteria for dynamic coordination of player strategies are determined, a recurrent method, algorithmic and software for stochastic game solving are developed. The developed model, method and algorithm for stochastic game resolution provide dynamic MAS coordination, which is manifested in locally deter-mined spatial and temporal alignment of players' strategies. Dynamic coordination is ensured by an adaptive search method for resolving stochastic play, taking into account current penalties for relevant spatial coordination and rhythm disturbances. It is established that the effectiveness of training players to perform coordinated rhythmic actions is determined by the balance between these penalties, which is achieved by the influence of white noise. Dynamic coordination of agent strategies is achieved as the real-time stochastic game is unleashed based on gathering current information and its adaptive processing. The considered method belongs to the class of reactive methods and simulates the reflexive behavior of living organisms. The method allows finding stochastic game solutions in pure strategies.

Keywords: Multi-agent system · Stochastic game · Dynamic coordination of strategies

1 Introduction

The increase in productivity of modern computing systems, the development of in-formation network technologies, the tools of parallel and distributed

S. Babichev et al. (Eds.): ISDMCI 2020, AISC 1246, pp. 653–670, 2021.
https://doi.org/10.1007/978-3-030-54215-3_42

programming have given a new impetus to the origins in the mid-twentieth century of research in the fields of collective behavior of automata, "soft" computing technologies, distributed artificial intelligence, decentralization of intelligence, and decentralization [29,53,54]. Because of the knowledge reconstruction and integration in these areas, a new agent-oriented scientific direction is formed [16,17].

An agent is an autonomous system of making and decision-making with elements of artificial intelligence, which can interact with other agents and people using the resources of the information network to solve this task [10,42]. The agent functions according to the goals set by its developer or owner [35,55]. Multi-agent system (MAS) is a system formed by several interacting intellectual agents with properties: autonomy - agents fully or partially independent, specialization and limited knowledge - each agent performs highly specialized functions and does not have a holistic understanding of the system, decentralization-distributed system [20,31,43,44].

2 Related Works

Modern MAS research is related to solving complex problems of distributed artificial intelligence: knowledge management, coordination and cooperation, determining the optimal structural organization of the MAS, planning scenarios of collective behavior of agents, developing methods, languages and means of communication agents, methods and means of computer-aided design MAS, multi-agent training, agent mobility, plasticity, reliability and fault tolerance [29,53,54]. MASs are used for solve problems that cannot solved with a single agent or centralized way [11,46,47,51], such as

- e-commerce,
- military,
- emergency management,
- modeling of social structures and phenomena,
- workforce management,
- distance learning,
- scheduling,
- securities and investment market management,
- transportation management,
- logistics,
- geo-information systems,
- strategic computer games,
- simulators,
- network technologies and distributed computing,
- control ting network flows,
- information search on the Internet,
- mobile communications and more.

The multi-agent system performs a distributed solution of the task. The success of a multi-agent solution to a joint task depends on the agents' actions coordination level [24–26]. Coordination is the assurance of coordinated, orderly operation of all units of the multi-agent system [17,27,34]. Coordination is centralized or decentralized. With centralized coordination, there is a lead agent (or coalition of such agents) that sends the decision to all other agents. Agents can consider the recommended solution when designing their own behavior strategies. Decentralized coordination is based on interaction between agents. Such interaction may be implicit when agents act independently and influence one another through changing states of the common environment, and overt, communicative, when agents additionally exchange information directly with one another. By interacting with the environment, agents collect and process information about the states studied, build a model of the environment in the state-of-action space, for example, in the form of a cognitive map of decision-making rules and share knowledge with each other [15,50]. The information exchange can be global or within locally defined coalitions. Agents can share knowledge about the environment, current and predicted strategy choices, the value of winnings, and more. Agents use special language and established interaction protocols to share knowledge [18,19,28,32,52].

With the increasing size of MASs operating within distributed computer information systems, centralized coordination methods are inferior to the effectiveness of decentralized coordination methods due to locally determined agent interaction properties, specialization and limited knowledge of the managed environment [5,6,22,38]. Taking into account the factors of decentralized decision-making, competition, interaction, cooperation, training, self-organization inherent in the collective of agents, we use the stochastic game model for uncertainty of the MAS [16,17]. Game coordination of agents' strategies in the process of MAS self-organization is an urgent scientific and practical problem that is not well understood to date [3,4,37].

From a game-theoretic point of view, coordination is a concerted choice of strategies that satisfy the conditions imposed on the values of the payoffs or losers obtained during their optimization by a team of agents [12,21,45]. Collective decisions are coordinated if they meet the requirements of profitability, sustainability and fairness for all decision makers. In practice, Nash, Slater, Joffrion, Bayes, correlation equilibrium, Pareto optimality, etc. are used [1,2,30,36,39]. Game coordination can be spatial and dynamic over time. Spatial coordination is the achievement of a steady map of players' strategies as they unleash a stochastic game. A strategy map is a spatially distributed value of player strategies that satisfy the criteria for the performance of the MAS agents. Dynamic coordination is the achievement of a steady rhythmic change in the map of players' strategies during the stochastic game. Dynamic coordination of player strategies and dynamic repetition of stochastic game play should distinguished. Dynamic coordination is considered as the final solution of the stochastic game, its dynamic equilibrium. Dynamic repetition is a way of unleashing a stochastic game under uncertainty [36,40,41,48,49].

The practical application of dynamic coordination is the ability of the MAS to generate complex signals that can used, for example, in robotic systems for spatial orientation and agent movement [7,9,23,33]. Coordination leads to the conditions of self-organization of the system. Self-organization is the process of converting locally coordinated actions of agents into globally coordinated actions within the MAS at the expense of internal factors, without corresponding external influence. It is the ability of a team of agents with locally based relationships and goals to achieve consistent, coordinated behavioral strategies through self-learning. Self-organization and the complex behaviors of the MAS can arise from the collective implementation of the simplest actions of agents. Synergetic, an interdisciplinary science that studies nonlinear open dissipative (energy-dissipating) systems, deals with the study of self-organization of systems of different nature [8,9,13,14].

The purpose of the work is develop a game model, method, algorithm and software to provide conditions for dynamic coordination and self-organization of MAS.

3 Materials and Methods

3.1 Stochastic Game Model

Dynamic coordination of strategies will be performed on the basis of a stochastic game model of agents $\Gamma = \left(D, \left\{U^i\right\}_{i \in D}, \left\{Z\left(\xi^i\right)\right\}_{i \in D}\right)$, where $D \neq \varnothing$ is many players, $U^i = \left(u^i\left(1\right), u^i\left(2\right), ..., u^i\left(N_i\right)\right)$ is vector clean strategies i-player, $Ni \geq 2$ is number of clean strategies, $Z\left(\xi^i\right)$ is random loss distribution ξ^i $\forall i \in D$.

Agents interact with the stochastic environment $E : \left\{U^i\right\}_{i \in D} \to \left\{\xi^i\right\}_{i \in D}$, given inputs $\left\{U^i\right\} \forall i \in D$ and random output values $\left\{\xi^i\right\} \forall i \in D$. Environment inputs are determined by pure player strategies, and current player programs set outputs. Model environment E in the form of random distributions $Z\left(\xi^i\right) \forall i \in D$ not known to a priori agents.

Let dynamic coordination be determined by the repetition of pure player strategies over time $\tau = N_i$. The structure of the MAS determines the locally determined mechanism for the formation of random losses $\xi^i = \xi^i\left(\tau, u^{D_i}, \omega\right)$ agents that are functions of common strategies $u^{D_i} \in U^{D_i} = \underset{j \in D_i}{\times} U^j$ from local subsets $Di \subseteq D$, $D_i \neq \varnothing$ $\forall i \in D$ for a period τ and a chance $\omega \in [0, 1]$.

Repetitive stochastic play unfolds at discrete times $n = 1, 2, ...$ Current losers $\xi_n^i = \xi_n^i\left(\tau, u^{D_i}, \omega\right)$ are defined for values $n \geq \tau$. It is considered to be accidental losers $\left\{\xi_n^i\right\}$ are independent $\forall u_n = u \in U$, $\forall i \in D$, $n = 1, 2, ...,$ have a constant mathematical expectation $M\left\{\xi_n^i\right\} = \nu\left(\tau, u^{D_i}\right) = const$ and limited second moment $\underset{n}{\sup} M\left\{\left[\xi_n^i\right]^2\right\} = \sigma^2\left(\tau, u^{D_i}\right) < \infty$. A priori agents do not know the stochastic characteristics of random losers.

Now n agent average losses take on value:

$$\Xi_n^i\left(\left\{\tau, u_n^{D_i}\right\}\right) = \frac{1}{n}\sum_{t=1}^{n} \xi_t^i \; \forall i \in D. \tag{1}$$

The purpose of the agents is to minimize the own functions of the middle losers

$$\overline{\lim_{n\to\infty}}\, \Xi_n^i \to \min_{p_n^i} \forall i \in D. \tag{2}$$

So, interacting with the environment is based on observing the current losers $\left\{\xi_n^i\right\}$ players must learn how to choose clean strategies $\left\{u_n^i\right\}$ so that over time $n = 1, 2, ...$ ensure that the system of criteria is fulfilled (2).

The solution of the game problem will satisfy one of the conditions of collective equilibrium, for example, Nash, Slater, Pareto, depending on the method of formation of sequences of strategies $\left\{u_n^i\right\} \forall i \in D$. To solve problem (2), you must define a sequence formation rule $\left\{u_n^i\right\}$ variants of decisions in time.

3.2 A Method of Unleashing a Stochastic Game

Taking into account that $\tau = N_i$, formation of a sequence of variants of decisions $\left\{u_n^i\right\}$ with the desired properties, we execute on the basis of probability matrices

$$p_n^i = \begin{bmatrix} p_n^i(1,1), & ..., & p_n^i(1,N_i) \\ ..., & ..., & ... \\ p_n^i(N_i,1), & ..., & p_n^i(N_i,N_i) \end{bmatrix}, \forall i \in D.$$

Elements $p_n^i(j,k)$ vectors of mixed strategies are conditional probabilities of choosing pure strategies depending on the current variant of action u_n^i and the resulting loss ξ_n^i. Matrix rows p_n^i are mixed strategies i-player when in a state of being $u_n^i \in U^i$. Mixed strategies take on a single simplex value

$$S^{N_i} = \left\{ p \left| \sum_{k=1}^{N_i} p(j,k) = 1; p(j,k) \geq 0 \;\; (j = 1..N_i) \right. \right\}.$$

The game starts with unskilled strategies with the same probabilities of their choice $p_0^i(j,k) = \dfrac{1}{N_i}$, where $j,k = 1..N_i$. To provide adaptive properties, the likelihood of choosing strategies with smaller programs should increase over time $n = 1, 2, ...$ As a result, a probability distribution that minimizes the functions of the middle losers (1) of all players should formed. To ensure this, we apply a recurrent method of modifying mixed strategy vectors:

$$p_{n+1}^i(u_n) = \pi_{\varepsilon_{n+1}}\left\{p_n^i(u_n) - \gamma_n \xi_n^i \left[e\left(u_{n+1}^i\right) - p_n^i(u_n)\right]\right\}, \tag{3}$$

where $p_n^i(u_n)$ is mixed strategy i-player in the state $u_n \in U^i$; $\pi_{\varepsilon_{n+1}}$ is single projector ε is simplex [42], which is a subset of a single simplex; $\gamma_n > 0$ is a monotonically descending sequence of nonnegative quantities that governs the step size of a method; $\varepsilon_n > 0$ is a monotonically decreasing sequence of non-negative quantities that governs the rate of expansion ε - simplex; $e\left(u_{n+1}^i\right)$ is determined at time n single state transition vector indicator $u_{n+1}^i \in U^i$.

The value of a pure strategy is determined randomly based on the current probability distribution $p_n^i\left(u_n^i\right)$:

$$u_n^i = \left\{ u^i(l) \,|l = \arg\min_l \sum_{k=1}^{l} p_n^i(j,k) > \omega\,(j,l = 1..N_i) \right\}, \tag{4}$$

where $\omega \in [0,1]$ is random variable with uniform distribution.

Being in the moment of time t_n in the state u_n^i, i-player based on a mixed strategy $p_n^i(u_n)$ chooses an action u_n^i according to (4) and by the time t_{n+1} gets the current loser ξ_n^i , then calculates a mixed strategy $p_{n+1}^i(u_n)$ according to (3) and goes into a new state u_{n+1}^i. Method (3) is obtained on the basis of a stochastic approximation of the complementary rigidity condition that holds for mixed strategies at the Nash equilibrium point [10]. To construct method (3), the condition of complementary rigidity is weighed down by elements of mixed strategies, which makes it possible to find solutions to the game in pure strategies. Due to the dynamic reorganization of mixed strategies based on the processing of current losers, method (3) provides an adaptive choice of agent strategies over time. Options γ_n and ε_n can calculated as:

$$\gamma_n = \gamma n^{-\alpha}, \varepsilon_n = \varepsilon n^{-\beta}, \tag{5}$$

where $\gamma > 0$; $\alpha > 0$, $\varepsilon > 0$; $\beta > 0$. In the general case, the parameters γ_n, ε_n may change depending on your current status u_n^i the pace of time $n\left(u_n^i\right)$.

Convergence $\lim\limits_{n\to\infty} \left\|p_n^i\left(u_n^i\right) - p^{i^*}\left(u_n^i\right)\right\| \to 0$, $\forall i \in D$ method (3) to optimal strategies $p^{i^*}\left(u_n^i\right)$ $\forall u_n^i \in U^i$, which minimize the mean loss (1) is provided by stochastic approximation conditions [35].

4 Experiment

4.1 Reference Example

We dynamically coordinate strategies for stochastic play with a regular structure of communication between agents. The structure of the game is determined by the number of agents $L = m \times m$, $m \geq 2$, subsets of neighboring agents D_i, number of clean strategies $N_i = N = 2$, $i = 1..L$.

Dynamic coordination consists of spatial and temporal coordination of strategies. Spatial coordination is about keeping the player strategies in locally defined areas D_i $\forall i \in D$ as shown in Fig. 1. Temporal coordination is determined by keeping in mind the ratio of players' strategies over a period $\tau = 2$.

Let pure player strategies take on binary values $u_n^i \in \{0,1\}$, simplifying the visual analysis of dynamically coordinated strategies. Spatial coordination is defined as equal values of player strategies (players try to repeat each other's actions), and temporal coordination is defined as the rhythmic change of inverse values of strategies in two consecutive times. As a result, dynamic coordination is about repeating strategy matrices $[0]_{m\times m} - [1]_{m\times m} - [0]_{m\times m} - [1]_{m\times m} - \dots$. To provide dynamically coordinated strategies, we define current player programs using difference features:

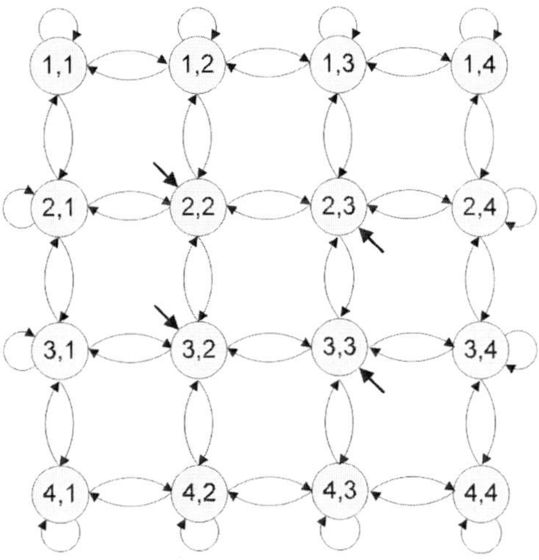

Fig. 1. MAS structure

$$\xi_n^i = \lambda \sum_{s\in D_i} \left| u_n^i - u_n^s \right| / L_i + (1 - \lambda) \left| \overline{u}_n^i - u_{n-1}^i \right| + \mu_n, \quad (6)$$

where $\xi_n^i \in R^1$; $\lambda \in [0, 1]$ is weight factor; D_i is a set of adjacent players corresponding to those shown in Fig. 1 connections; $L_i = |D_i|$ is the number of neighboring players; u_n^i is the importance of pure strategy; $\overline{u}_n^i$ is the inverse value of pure strategy; $\mu_n \sim Normal\,(0, d)$ is additive white Gaussian noise, normally distributed random variable with zero mathematical expectation and variance $d > 0$.

The first component of expression (6) determines the penalty for violation of spatial coordination, the second component determines the penalty for violation of temporal coordination of player strategies, and the third component determines the effect of random interference in the form of white noise.

The current white noise values can calculated by the formula:

$$\mu_n = \sqrt{d} \left(\sum_{j=1}^{12} \omega_{j,n} - 6 \right)$$

where $\omega \in [0, 1]$ is a valid random number with a uniform distribution law.

Dynamic coordination efficiency is evaluated by:

1. The average loss function:

$$\Xi_n = \frac{1}{L}\sum_{i=1}^{L}\Xi_n^i \tag{7}$$

where $L = |D|$ is number of players;

2. The average number of spatial-coordinated player strategies:

$$K_n = \frac{1}{n}\sum_{t=1}^{n}\sum_{i=1}^{L}\chi\left(\sum_{s\in D_i}\left|u_t^i - u_t^s\right| = 0\right) \tag{8}$$

where $\chi\left(\right) \in \{0,1\}$ is event indicator function;

3. The average number of player-coordinated strategies over time that determine the rhythmic change in strategy sequences:

$$M_n = \frac{1}{n}\sum_{t=1}^{n}\sum_{i=1}^{L}\chi\left(\left|\overline{u}_t^i - u_{t-1}^i\right| = 0\right) \tag{9}$$

4. The average deviation of mixed player strategies from optimal values $p^{i^*}\left(u_n^i\right)$:

$$\Delta_n = \frac{1}{Ln}\sum_{t=1}^{n}\sum_{i=1}^{L}\sum_{u_t^i\in U^i}\left\|p_t^i\left(u_t^i\right) - p^{i^*}\left(u_i^t\right)\right\| \tag{10}$$

where $\|.\|$ is Euclidean vector norm.

4.2 Stochastic Game Learning Algorithm

1. Set the initial parameter values:
   $n = 0$ is the starting point of time;
   $L = |D|$ is number of players;
   $D_i\ \forall i \in D$ is many neighboring players who determine the structure of the game;
   $N_i,\ i = 1..L$ is the number of clean player strategies;
   $\tau = N_i$ is a period of dynamic coordination of player strategies;
   $U^i = \left\{u^i(1), u^i(2), ..., u^i(N_i)\right\},\ u^i \in \{1..N_i\},\ i = 1..L$ is vectors of pure player strategies;
   $p_0^i = [1/N_i]_{N_i\times N_i},\ i = 1..L$ is matrix of mixed player strategies;
   $\gamma > 0$ is the step of learning step;
   $\alpha \in (0,1]$ is the order of the learning step;
   ε is parameter of ε-simplex;
   $\beta > 0$ is the speed of expansion of ε-simplex;
   $d > 0$ is the variance of the plants;
   $\mu_n \sim Z(0,d)$ is law of distribution of interferences;
   $\lambda \in [0,1]$ is weight loss ratio;
   n_{max} is the maximum number of method steps.

2. Choose action options $u_n^i \in U^i$, $i = 1..L$ according to (4).
3. Get current loser values ξ_n^i, $i = 1..L$ according to (6).
4. Calculate parameter values γ_n, ε_n according to (5).
5. Compute the elements of vectors of mixed strategies p_n^i, $i = 1..L$ according to (3).
6. Calculate the characteristics of decision quality Ξ_n (7), K_n (8), M_n (9), and Δ_n (10).
7. Set the next point in time $n := n + 1$.
8. If $n < n_{max}$, then go to step 2, otherwise - end.

5 The Results of Computer Simulation

Simulation results $n_{max} = 10$ thousand steps of stochastic game are shown in Fig. 2, 3, 4, 5, 6, 7, 8, 9, 10 and Fig. 11. Figure 2 shows, on a logarithmic scale, graphs of functions that characterize the efficiency of dynamic coordination of a stochastic game $L = 4 \times 4$ agents for parameter values $\gamma = 1$, $\varepsilon = 0.999/2$, $\alpha = 0.01$, $\beta = 2$, $\lambda = 0.5$, $d = 0.36$. Graph 1 shows the function of the average number of spatial-coordinated strategies K_n, graph 2 – the average number of rhythmic changes in players' strategies M_n, graph 3 is the average current losing players Ξ_n.

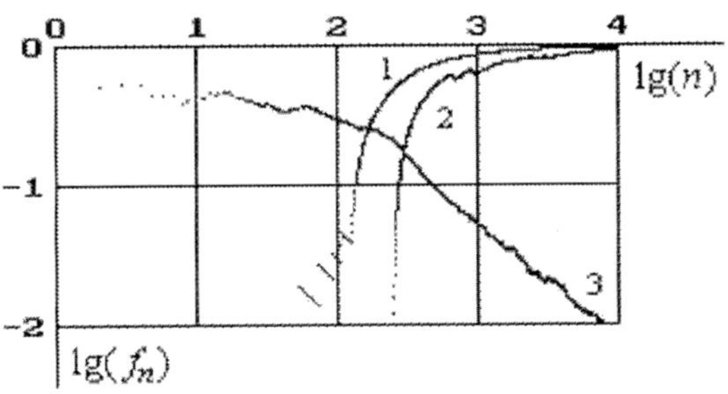

Fig. 2. Characteristics of stochastic game self-organization

The decrease in the function of the average losers over time indicates the convergence of the game method. The number of spatially coordinated strategies illustrates the coordination of players 'strategies, and the number of rhythmic strategies changes the inverse coordination of players' strategies at successive times. The growth of these features over time indicates the self-organization of the stochastic game.

Figure 3–4 shown the dynamics of choosing pure stochastic game strategies. The points on the charts indicate the strategies of the players $u^i = 1$, where $i = 1..L$. The graph period corresponds to two consecutive time points. At the initial interval, there is a chaotic choice of strategies by untrained agents (Fig. 3).

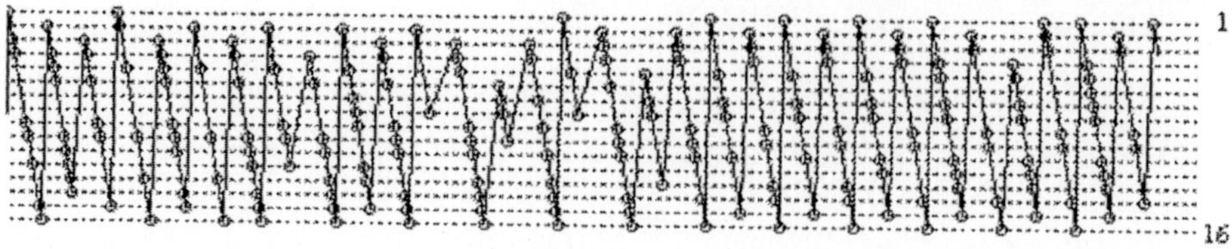

Fig. 3. Uncoordinated choice of strategies for unlearned stochastic play

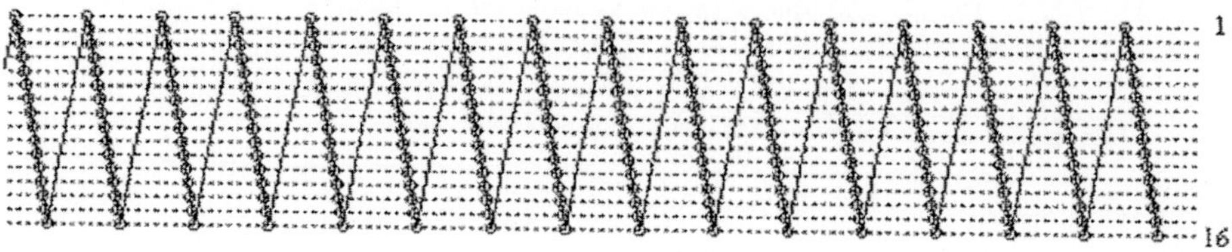

Fig. 4. Dynamically coordinated choice of strategies of the trained stochastic game

After training ($n \sim 10^3$ steps) is a time-coordinated, rhythmic change in binary strategies (Fig. 4).

Method (3) ensures that stochastic games are solved in pure strategies (on top of a single simplex). Shown in Fig. 4 A rhythmic change in strategies is generated by trained transition probability matrices $p_n^{i^*} = \rightarrow \begin{bmatrix} 0,1 \\ 1,0 \end{bmatrix} \forall i \in D$.

The arrow marked the players' states $\forall i \in D$ at the current time n. According to the generated probability, matrices $p_n^{i^*}$ at time $n+1$ players go into inverted states. Weighting factor λ determines the level of impact on the current loser ξ_n^i (6) spatial-coordinated strategies and the rhythmic change of players' strategies in successive moments. A certain balance between these impacts determines dynamic coordination. As you grow λ spatial coordination is provided, but the rhythm of their change over time is disturbed. The results of the effect of the parameter λ the effectiveness of the game method (3) is shown in Fig. 5 in the form of function graphs K_n (8) and M_n (9), obtained at $n = n_{max}$ and $d = 0$. As can be seen in Fig. 5, using the parameter λ it is not possible to achieve a

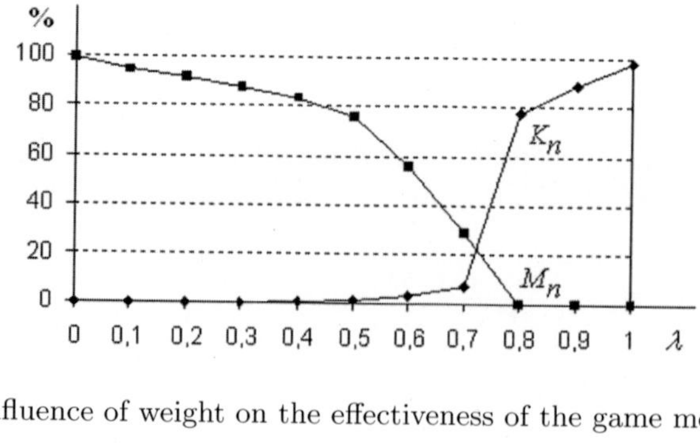

Fig. 5. The influence of weight on the effectiveness of the game method at $d = 0$

proper balance of mutual coordination and rhythmic change in players' strategies. It has experimentally established that such a balance can achieved by the influence of white noise, which provides a random, unbiased fluctuation of the balance between different types of penalties. The corresponding results obtained at $\alpha = 0.01$, $d = 0.36$, filed in Fig. 6.

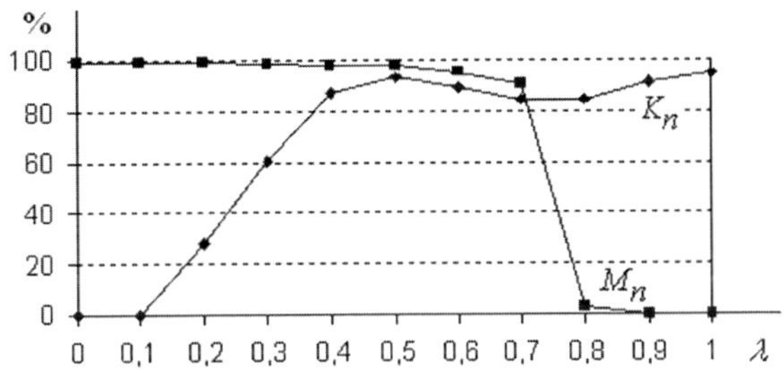

Fig. 6. Effect of weight factor λ on the effectiveness of the game method at $d = 0.36$

Under white noise, the conditions of coordinated rhythmic choice of strategies are fulfilled for parameter values $\lambda \in [0.4; 0.7]$ at a level exceeding 80%.

Figure 7 shown impact of interference dispersion d the effectiveness of the game method (3). The results were obtained at $\alpha = 0.01$, $\lambda = 0.5$. At magnitude $\sqrt{d} \in [0.5; 0.7]$ the performance characteristics of the method exceed the level 80%. Outside this segment, one of the performance parameters is deteriorating.

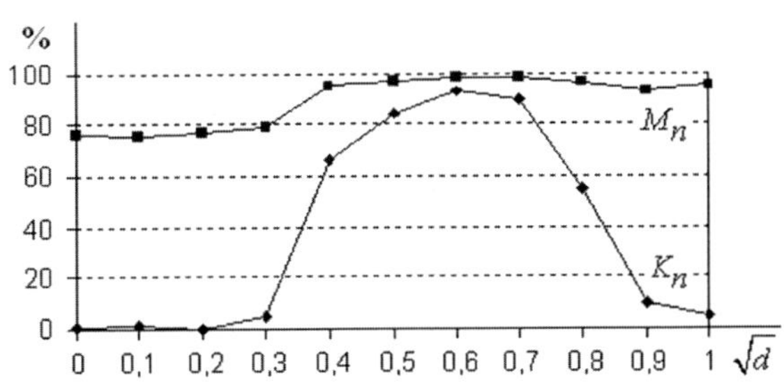

Fig. 7. The effect of variance d on the effectiveness of the game method

The effectiveness of the game method depends largely on the parameters α and β, which determine the speed of convergence of the game method. The values of these parameters must satisfy the basic conditions of stochastic approximation. Figure 8 shown dependence of stochastic game performance parameters on α. Data obtained for parameters $d = 0.36$ and $\lambda = 0.5$.

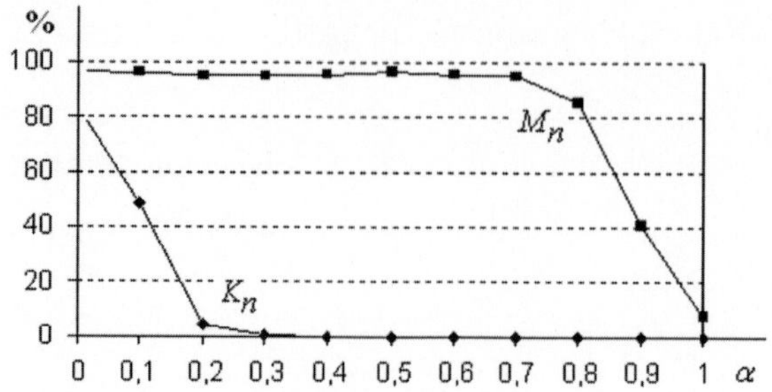

Fig. 8. The effect of the parameter α on the effectiveness of the game method

Growth α leads to poor spatial coordination of players. This maintains a high degree ($\alpha \leq 0.8$) the timing of players' strategies. Coordinated and rhythmic actions of players are achieved at small values $\alpha \leq 0.01$.

Dependence of spatial and temporal coordination coefficients on quantity $L = m \times m$ players are shown in Fig. 9 for parameters $\alpha = 0.01$, $d = 0.36$, $\lambda = 0.5$.

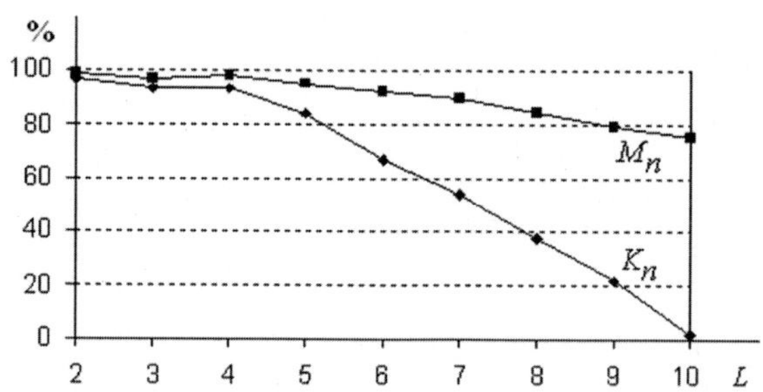

Fig. 9. Dependence of MAS coordination parameters on the number of players

The coefficients are obtained at time $n = 10^4$ game steps. Acceptable (greater than 80%) strategy coordination is the place to play with the maximum number $L = 5 \times 5$ agents, each of which has a $N = 2$ clean strategy.

6 Discussions About the Computer Simulation Results

As the dimension of the stochastic game grows, the average number of steps required to achieve the proper level of coordination of agents' strategies increases. Figure 10 shows the dependence of the average number of steps in the game $\overline{n}$, necessary to achieve a coefficient of coordination $K_n = 80\%$, the number of players $L = m \times m$, each of which has a $N = 2$ clean strategy.

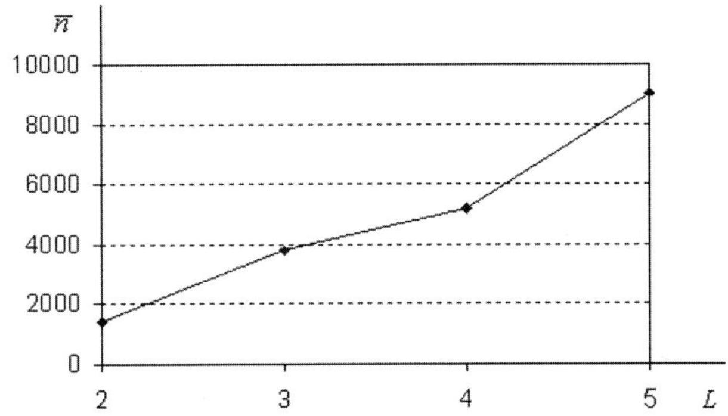

Fig. 10. The dependence of the average number of steps of learning the game on the number of players

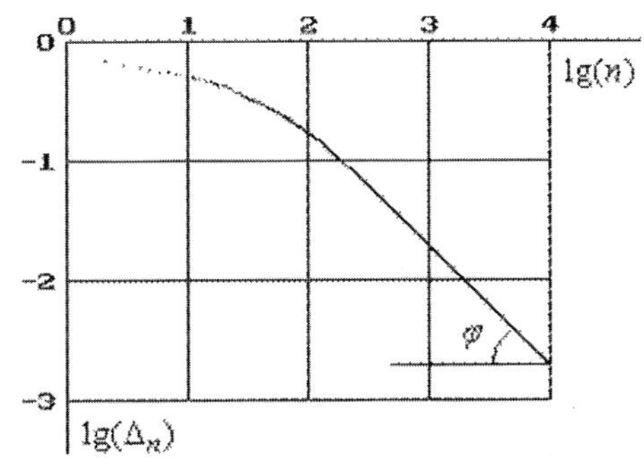

Fig. 11. Dynamics of average stochastic game decoupling error

Proposed method (3) for stopping stochastic agent play belongs to the class of reactive methods and has a relatively low degree of convergence. This is because, at the beginning of the game, agents do not have any information about the environment with which they interact. Information is collected through learning through adaptive rearrangement of vectors of mixed strategies in proportion to the values of current losers. In general, the speed of convergence of the game method depends on all the parameters of the model. In Fig. 8 is a graph of the average rate (10) of deviation of mixed strategies from optimal values for parameters $\gamma = 1$, $\varepsilon = 0.999/2$, $\alpha = 0.01$, $\beta = 2$, $\lambda = 0.5$, $d = 0.36$. The order of convergence rate can be estimated by the angle tangent ϕ, formed between the graph of linear approximation of the norm of deviation of mixed strategies of players from optimal values and the time axis.

With the appropriate selection of parameters, you can achieve close to 1 order of convergence speed of the stochastic game.

7 Conclusions

The developed model, method and algorithm for stochastic game resolution provide dynamic MAS coordination, which is manifested in locally determined spatial and temporal alignment of players' strategies. Dynamic coordination is ensured by an adaptive search method for resolving stochastic play, taking into account current penalties for relevant spatial coordination and rhythm disturbances. It is established that the effectiveness of training players to perform coordinated rhythmic actions is determined by the balance between these penalties, which can be achieved by the influence of white noise. Dynamic co-ordination of agent strategies is achieved as the real-time stochastic game is unleashed based on gathering current information and its adaptive processing. The considered method belongs to the class of reactive methods and simulates the reflexive behavior of living organisms. The method allows finding stochastic game solutions in pure strategies. The effectiveness of the dynamic coordination of system strategies is controlled by the characteristic functions of the mean losers, the coefficients of spatial and temporal coordination, and the Euclidean rate of deviation of the dynamic mixed strategies from optimal values. The decrease in the mean loss function, the increase in the coefficients of coordination and the decrease in the rate of rejection of mixed strategies indicate the convergence of the game method.

The speed of dynamic coordination of pure agent strategies depends on the dimension of the stochastic game, the magnitude of the interference, and the parameters of the game method. As the number of players increases and the intensity of the interference, the speed and efficiency of game coordination of the MAS decreases. The peculiarity of the proposed recurrent stochastic game resolution method is the relatively low degree of convergence, which is due to the a priori uncertainty of the system. This disadvantage is overcome by the high speed of modern computing facilities and the ability to parallelize the problem or use networked distributed computing. The convergence of the method is ensured by observance of the basic conditions of stochastic approximation. The convergence parameters are theoretically determined and refined experimentally. With proper selection of the parameters of the game method, the values of which must meet the conditions of stochastic approximation, it is possible to achieve close to 1 degree order of convergence rate.

The reliability of the obtained results is confirmed by the repetition of the values of the calculated characteristics of the stochastic game for different sequences of random variables. The results obtained can used to generate spatially distributed periodic signals in multi-agent and robotic systems. The development of the proposed game method is possible in the direction of increasing the intellectual level of agents.

References

1. Anisimova, O., Vasylenko, V., Fedushko, S.: Social networks as a tool for a higher education institution image creation. In: CEUR Workshop Proceedings, vol. 2392, pp. 54–65 (2019)
2. Antonyuk, N., Medykovskyy, M., Chyrun, L., Dverii, M., Oborska, O., Krylyshyn, M., Vysotsky, A., Tsiura, N., Naum, O.: Online tourism system development for searching and planning trips with user's requirements. In: Advances in Intelligent Systems and Computing IV. vol. 1080, pp. 831–863. Springer (2020). https://doi.org/10.1007/978-3-030-33695-0_55
3. Babichev, S., Korobchynskyi, M., Lahodynskyi, O., et al.: Development of atechnique for the reconstruction and validation of gene network models basedon gene expression profiles. Eastern-Eur. J. Enterp. Technol. **1**(4–91), 19–32 (2018). https://doi.org/10.15587/1729-4061.2018.123634
4. Babichev, S., Lytvynenko, V., Korobchynskyi, M., Taif, M.A.: Objective clustering inductive technology of gene expression sequences features, vol. 716, pp. 359–372 (2017). https://doi.org/10.1007/978-3-319-58274-0_29
5. Berko, A., Alieksevyeva, K.: Quality evaluation of information resources in web-projects. Actual Prob. Econ. **136**(10), 226–234 (2012)
6. Boreiko, O., Teslyuk, V., Zelinskyy, A., Berezsky, O.: Development of models and means of the server part of the system for passenger traffic registration of public transport in the "smart" city. Eastern-Eur. J. Enterp. Technol. **1**(2–85), 40–47 (2017). https://doi.org/10.15587/1729-4061.2017.92831
7. Borovska, T., Grishin, D., Kolesnik, I., Severilov, V., Stanislavsky, I., Shestakevych, T.: Research and development of models and program for optimal product line control. In: Advances in Intelligent Systems and Computing IV, vol. 1080, pp. 186–201 (2020). https://doi.org/10.1007/978-3-030-33695-0_14
8. Bublyk, M., Rybytska, O., Karpiak, A., Matseliukh, Y.: Structuring the fuzzy knowledge base of the it industry impact factors. In: Computer Sciences and Information Technologies (CSIT), pp. 21–24 (2018). https://doi.org/10.1109/STC-CSIT.2018.8526760
9. Bublyk, M., Rybytska, O.: The model of fuzzy expert system for establishing the pollution impact on the mortality rate in Ukraine. In: Computer Sciences and Information Technologies (CSIT), vol. 1, pp. 253–256 (2017). https://doi.org/10.1109/STC-CSIT.2017.8098781
10. Byrski, A., Kisiel-Dorohinicki, M.: Evolutionary Multi-Agent Systems: From Inspirations to Applications. Springer, Heidelberg (2017)
11. Chen, B.S.: Stochastic Game Strategies and Their Applications. CRC Press, Boca Raton (2019)
12. Cherednichenko, O., Vovk, M., Yanholenko, O., Yakovleva, O.: Towards the technology of employers' requirements collection development. In: Integrated Computer Technologies in Mechanical Engineering, pp. 228–239. Springer (2020). https://doi.org/10.1007/978-3-030-37618-5_21
13. Chukhray, N., Shakhovska, N., Mrykhina, O., Bublyk, M., Lisovska, L.: Consumer aspects in assessing the suitability of technologies for the transfer. In: Computer Sciences and Information Technologies (CSIT), pp. 142–147 (2019). https://doi.org/10.1109/STC-CSIT.2019.8929879
14. Chukhray, N., Shakhovska, N., Mrykhina, O., Bublyk, M., Lisovska, L.: Methodical approach to assessing the readiness level of technologies for the transfer. In: Advances in Intelligent Systems and Computing IV, vol. 1080, pp. 259–282 (2020). https://doi.org/10.1007/978-3-030-33695-0_19

15. Demchuk, A., Lytvyn, V., Vysotska, V., Dilai, M.: Methods and means of web content personalization for commercial information products distribution. In: Advances in Intelligent Systems and Computing, vol. 1020, pp. 332–347 (2020). https://doi.org/10.1007/978-3-030-26474-1_24
16. Dignum, F., Bradshaw, J., Silverman, B., Doesburg, W.: Agent for Games and Simulations: Trends in Techniques. Concepts and Design. Springer, Heidelberg (2009)
17. Fudenberg, D., Levine, D.: The Theory of Learning in Games. MIT Press, Cambridge (1998)
18. Greenwald, A., Hall, K.: Correlated Q-learning. In: Proceedings of the Twentieth International Conference on Machine Learning, pp. 242–249 (2003)
19. Hu, J., Wellman, M.P.: Nash Q-learning for general-sum stochastic games. Mach. Learn. Res. **4**, 1039–1069 (2003)
20. Kaelbling, L., Littman, M., Moore, A.: Reinforcement learning: a survey. J. Artif. Intell. Res. **4**, 237–285 (1996)
21. Kanishcheva, O., Cherednichenko, O., Sharonova, N.: Image tag core generation. In: CEUR Workshop Proceedings, pp. 35–44 (2019)
22. Kazarian, A., Teslyuk, V., Tsmots, I., Tykhan, M.: Implementation of the face recognition module for the "smart" home using remote server. In: Advances in Intelligent Systems and Computing III, AISC, vol. 871, pp. 17–27. Springer (2019). https://doi.org/10.1007/978-3-030-01069-0_2
23. Komar, M., Golovko, V., Sachenko, A., Bezobrazov, S.: Development of neural network immune detectors for computer attacks recognition and classification. In: International Conference on Intelligent Data Acquisition and Advanced Computing Systems, pp. 665–668 (2013). https://doi.org/10.1109/IDAACS.2013.6663008
24. Kravets, P.: The methodology of multi-agent systems: a modern state and future trends. In: Proceedings of the International Conference on Computer Science and Information Technologies, CSIT, pp. 125–127 (2006)
25. Kravets, P.: The control agent with fuzzy logic. In: Perspective Technologies and Methods in MEMS Design, MEMSTECH, pp. 40–41 (2010)
26. Kravets, P.: Game method for coalitions formation in multi-agent systems. In: Proceedings of the IEEE 13th International Scientific and Technical Conference on Computer Sciences and Information Technologies, CSIT 2018, vol. 1, pp. 1–4 (2018). https://doi.org/10.1109/STC-CSIT.2018.8526610
27. Kravets, P., Burov, Y., Lytvyn, V., Vysotska, V.: Gaming method of ontology clusterization. Webology **16**(1), 55–76 (2019)
28. Kushner, H., Yin, G.: Stochastic Approximation and Recursive Algorithms and Applications. Springer, Heidelberg (2013)
29. Lee, Y., Chong, Q.: Multi-agent systems support for community-based learning. Interact. Comput. **15**(1), 33–55 (2003). https://doi.org/10.1016/S0953-5438(02)00057-7
30. Lozynska, O., Savchuk, V., Pasichnyk, V.: Individual sign translator component of tourist information system. In: Advances in Intelligent Systems and Computing IV, vol. 1080, pp. 593–601. Springer, Cham (2020). https://doi.org/10.1007/978-3-030-33695-0_40
31. Nazin, A., Poznyak, A.: Adaptive Choice of Variants: Recurrence Algorithms (in Russian). Science, Moscow (1986)
32. Neogy, S., Bapat, R., Dubey, D.: Mathematical Programming and Game Theory. Springer, Heidelberg (2018)

33. Paliy, I., Turchenko, V., Koval, V., Sachenko, A., Markowsky, G.: Approach to recognition of license plate numbers using neural networks. In: IEEE International Conference on Neural Networks – Conference Proceedings, pp. 2965–2970 (2004). https://doi.org/10.1109/IJCNN.2004.1381137
34. Puterman, M.L.: Markov Decision Processes: Discrete Stochastic Dynamic Programming. Wiley, New York (2005)
35. Radley, N.: Multi-Agent Systems – Modeling, Control, Programming, Simulations and Applications. Scitus Academics LLC (2017)
36. Rusyn, B., Pohreliuk, L., Rzheuskyi, A., Kubik, R., Ryshkovets, Y., Chyrun, L., Chyrun, S., Vysotskyi, A., Fernandes, V.: The mobile application development based on online music library for socializing in the world of bard songs and scouts' bonfires. In: Advances in Intelligent Systems and Computing IV, vol. 1080, pp. 734–756. Springer (2020). https://doi.org/10.1007/978-3-030-33695-0_49
37. Rusyn, I.B., Hamkalo, K.R.: Bioelectricity production in an indoor plant-microbial biotechnological system with Alisma plantago-aquatica. Acta Biol. Szeged. **62**(2), 170–179 (2019). https://doi.org/10.14232/abs.2018.2.170-179
38. Rusyn, I., Valko, B.: Container landscaping with festuca arundinaceae as a minibioelectrical systems in a modern buildings. Int. J. Energy Clean Environ. **20**(3), 211–219 (2019). https://doi.org/10.14232/abs.2018.2.170-179
39. Rzheuskyi, A., Kutyuk, O., Voloshyn, O., Kowalska-Styczen, A., Voloshyn, V., Chyrun, L., Chyrun, S., Peleshko, D., Rak, T.: The intellectual system development of distant competencies analyzing for it recruitment. In: Advances in Intelligent Systems and Computing IV, vol. 1080, pp. 696–720. Springer, Cham (2020). https://doi.org/10.1007/978-3-030-33695-0_47
40. Sachenko, A., Kochan, V., Turchenko, V., Tymchyshyn, V., Vasylkiv, N.: Intelligent nodes for distributed sensor network. In: Conference Record - IEEE Instrumentation and Measurement Technology Conference, pp. 1479–1484 (1999). https://doi.org/10.1109/IMTC.1999.776072
41. Sachenko, S., Pushkar, M., Rippa, S.: Intellectualization of accounting system. In: IEEE International Workshop on Intelligent Data Acquisition and Advanced Computing Systems: Technology and Applications, Dortmund, Germany, 6–8 September 2007, pp. 536–538 (2007). https://doi.org/10.1109/IDAACS.2007.4488477
42. Scerri, P., Vincent, R., Mailler, R.: Coordination of Large-Scale Multiagent Systems. Springer, Heidelberg (2010)
43. Sun, Z.: Cooperative Coordination and Formation Control for Multi-Agent Systems. Springer, Heidelberg (2018)
44. Sutton, R., Barto, A.: Reinforcement Learning: An Introduction. MIT Press (1998). https://doi.org/10.1017/S0263574799271172
45. Tkachenko, V., Cherednichenko, O., Godlevskyi, M.: The concept of device metamodel for real-time communication in the transboundary environment monitoring system. In: International Scientific-Practical Conference,. PIC S&T, pp. 64–70 (2018). https://doi.org/10.1109/INFOCOMMST.2018.8632028
46. Ummels, M.: Stochastic Multiplayer Games: Theory and Algorithms. Amsterdam University Press, Amsterdam (2014)
47. Ungureanu, V.: Pareto-Nash-Stackelberg Game and Control Theory: Intelligent Paradigms and Applications. Springer, Heidelberg (2018)
48. Varetskyy, Y., Rusyn, B., Molga, A., Ignatovych, A.: A new method of fingerprint key protection of grid credential. In: Advances in Intelligent and Soft Computing, pp. 99–103 (2010). https://doi.org/10.1007/978-3-642-16295-4_11

49. Veres, O., Rusyn, B., Sachenko, A., Rishnyak, I.: Choosing the method of finding similar images in the reverse search system. In: CEUR Workshop Proceedings, pp. 99–107 (2018)
50. Vysotska, V., Fernandes, V., Emmerich, M.: Web content support method in electronic business systems. In: CEUR Workshop Proceedings, vol. 2136, pp. 20–41 (2018)
51. Watkins, C., Dayan, P.: Q-learning. In: Machine Learning, vol. 8, pp. 279–292. Kluwer Academic Publishers, Boston (1992)
52. Weinberg, M., Rosenschein, J.: Best-response multiagent learning in non-stationary environments. In: AAMAS 2004, New York, USA (2004)
53. Weiss, G.: Multiagent Systems, 2nd edn. The MIT Press, Cambridge (2013)
54. Wooldridge, M.: An Introduction to Multiagent Systems. Wiley, Hoboken (2002)
55. Yang, S., Xu, J.X., Li, X., Shen, D.: Iterative Learning Control for Multi-Agent Systems Coordination. Wiley-IEEE Press (2017)

Information Technology for Processing the Parameters of the Rolling Spectra at Various Positions of the Human Body

Yurii Bezsmertnyi[1], Sergii Pavlov[2(✉)], Victor Shevchuk[1], Stanislav Iaremyn[1], and Viktoria Dumenko[3]

[1] Scientific Research Institute of Invalid Rehabilitation on the base of Vinnytsia Pirogov National Medical University, Vinnytsia, Ukraine
bezsmertnyiyurii@gmail.com, stanislav.iaremin@gmail.com
[2] Vinnytsia National Technical University, Vinnytsia, Ukraine
psv@vntu.edu.ua
[3] Vinnitsa State Pedagogical University named after M. Kotsyubynsky, Vinnytsia, Ukraine
vk.dumenko@ukr.net

Abstract. A mathematical model of the human body in the form of an inverted three-link pendulum is developed that allows you to calculate the natural frequencies of the system's oscillations, describes the increase in frequencies during the transition from a two-leg to a one-leg stand. As diagnostically significant parameters, the maximum oscillation amplitudes in two directions, the ratio of the oscillation amplitudes for a single-support and two-support strut, and the spectral density of the signal in certain frequency ranges corresponding to the main vibration frequencies of the body were identified. When transferring body weight to one of the feet with a double support, the center of mass is shifted somewhat posteriorly and asymmetrically for the left and right foot. Patterns of swing amplitude variation are not detected. With a single-leg stand, there is a significant (on average 2 times) increase in the swing amplitude, especially in the frontal plane, which is associated with a significant decrease in the area of the supporting surface in this direction. The model of body oscillation in the sagittal plane with a two-support and one-support strut is displayed in the system of second-order differential equations with respect to the angles of deviation of the links from the vertical. Stabilographic studies of body position showed that the type of the corresponding trajectories of the projection of the center of mass of the body, the amplitude of the oscillations in the planes and the average projection time of the study differ significantly among healthy subjects.

Keywords: Mathematical model · Common center of mass · Statography · Body position · Rolling

S. Babichev et al. (Eds.): ISDMCI 2020, AISC 1246, pp. 671–686, 2021.
https://doi.org/10.1007/978-3-030-54215-3_43

1 Introduction

When a person maintains a given posture, individual segments of his body experience slight fluctuations that are not always visible to the naked eye, but can be detected using special biomedical electronic systems. One of the most widely used methods is stabilography (statography). In stabilographic studies, reaction forces are recorded R_j, arising in separate parts of the supporting surface of the subject [17]. The simplest model of the human foot includes two sections - the front (toe) and the back (heel), which for two-legged standing gives $j = 4$ reaction forces. By measured values R_{1-4} it is possible to calculate the coordinates (x, y) of the projection of the center of human body mass on the supporting surface and thus register the functions $(x(t), y(t))$, which have a complex look and represent the result of a superposition of irregular oscillations with different characteristic times [11,17]. The various parameters of these curves and the indices calculated from them are widely used in medicine to diagnose the state of the musculoskeletal and nervous systems, the inner ear and balance organs, the presence of spinal pathologies, dizziness [5,15], in the rehabilitation of patients with sclerosis, in Parkinson's disease, for restoration of locomotor function and speech after a stroke [4,19]. The use of stabilography in sports and sports medicine allows timely assessing the athletes' functional preparedness, adjusting training modes, treating injuries, rehabilitation measures, developing special training exercises and practicing weightlifting techniques, optimizing the choice of stance for shooters and positions for skaters [8,21]. During the improvement of sportsmanship observed a minimization of body rocking amplitudes and increase of stability when performing standard and a number of complicated poses of static and dynamic equilibrium. At the same time, the mechanisms of the emergence of certain patterns of body swing remain unknown, and in some cases, the data of stabilization remain within normal limits and with some pronounced pathological changes [10,11,23].

The scientific novelty of the presented results consists in the developed mathematical model of body vibrations in two-support and one-support poses and in the analysis of the corresponding stabilographic data, first carried out in order to identify patterns and mechanisms of maintaining the vertical posture in normal and pathological conditions.

2 Review of Literature

It is known that the angle fluctuations in the knee and femoral joints are correlated, which is associated with the presence of kinematic synergies that maintain a constant position of the center of mass of the body due to the control of the position of individual segments - the trunk and limbs [7]. Various hypotheses have been proposed on the mechanisms maintaining a constant posture, which are oriented to auditory, visual, tactile and other signals and have different characteristic times [3,12,20]. Rocking dynamics has common features with the processes of correlated random walks [9]. Perhaps maintaining the balance of the human body is focused on a certain pattern of motion of the center of mass of the body [14].

Separation of rational and diagnostically significant information from stabilograms is a difficult and not completely solved problem, despite thousands of studies on the method of registration and analysis of stabilographic data. Particularly difficult is the study of stabilograms obtained during various dynamic tests (jump, descent from the platform, raising the arm, etc.), since there are hundreds of different kinematic schemes for moving individual body segments from one fixed position to another [6,16,22].

By registered values R_j it is possible to calculate the coordinates of the application point of the vertical component of the support resultant reaction force, the center of pressure (CP). If we assume that the horizontal component of the reaction forces is equal to zero, the CP is an instantaneous equilibrium point of the body [12]. By connecting a continuous line (for example, approximating by splines) the positions of the equilibrium point at consecutive moments of time, we can obtain the trajectory of random walks of the central of mass (CM) [10]. Thus, you can expand the curves $x(t)$ and $y(t)$ to chaotic (random walks) and regular (oscillations with a certain frequency) components, the analysis of which allows a deeper understanding of posture maintenance mechanisms [3,13,14].

3 Problem Statement

The most popular method for studying stabilograms is spectral analysis. The calculation of the power spectral density shows that most of it (>90%) falls on the share of low frequencies ($f < 2\,\text{Hz}$) [18]. At the same time, different authors indicate somewhat different ranges corresponding to vibrations with different spectral powers and different mechanisms of origin [1,2,18,24]. Perhaps these differences are associated with the features of the measurements (different poses and dynamic tests and different models of stabilographs). Assessment of the amplitude-frequency spectrum of stabilograms is extremely important and is used in medical diagnostics, sports, sports medicine and in such cases when it is necessary to determine the blood alcohol content or the degree of alcohol dependence [8,11,12,23]. Therefore, the study of the question of what is the range of frequencies of oscillations of the human body in the norm, with pathologies of locomotor system and various conditions, whether there is a dependence of the results on the conditions of the measurements and the type of test, is extremely relevant.

The aim of the research is a comparative analysis of stabilograms at various positions of the human body and the development of an adequate mathematical model for processing recorded data.

The task of the work is to identify those parameters of stabilograms that can be effectively used in medical diagnostics and justified within the framework of a mathematical model that allows the analysis of the norm and various pathologies of the system for maintaining the balance of the posture.

4 Materials and Methods

Stabilographic studies of 12 healthy subjects were performed (age $33,5 \pm 16,5$ years, weight $67,7 \pm 20,8$ kg, height $1,73 \pm 0,17$ m) without pronounced physical and neurological pathologies with the help of the Statograph device. During the measurements, the subject was asked to maintain a conventional two-post stance for 30 s. In this case, support reaction forces were recorded $R_{1,2}^{l,r}$ front (1) and rear (2) departments of the left (l) and right (r) feet (Fig. 1), by which the coordinates of the CM of the human body were automatically calculated. Then, the same measurements were successively carried out with a two-leg stand with the transfer of body weight to one of the feet, as well as when balancing on one leg. Since the oscillation amplitudes in the sagital and frontal planes significantly depend on the test area of the test subject, it should be expected that the parameters of the vibrational spectrum will differ with a single and double support.

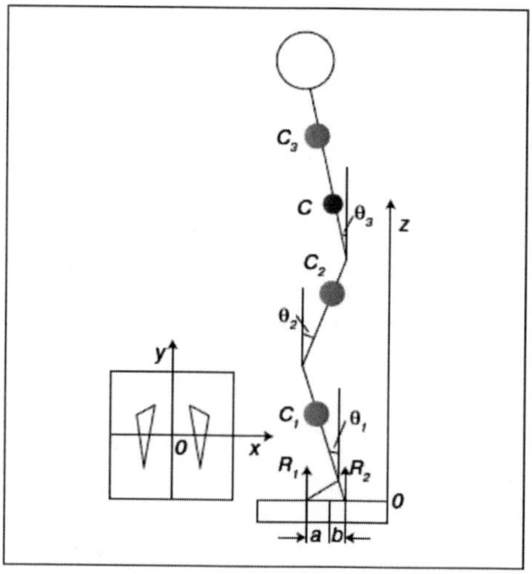

Fig. 1. Scheme of a three-link kinematic chain of a human body for assessing swing parameters in the sagittal plane with a two-leg stand

The recorded time series ($x(t)$, $y(t)$) were then amplified ($\times 100$). Subsequently, low-pass filtered ($f < 0,01$ Hz) and high frequency ($f > 10$ Hz) oscillations using a standard $2n$ order Butterworth filter. To eliminate the trend of the contour, the x, $y(t)$ oscillations were shifted relative to the average value $<x, y(t)>$ so that after shifting get $<x, y> = 0$. The first 2 s from recording rows $x(t)$, $y(t)$ were deleted to eliminate errors associated with filtering and averaging.

For the purpose of the subsequent construction of the corresponding mathematical model of the body as a multilink system, the lengths of segments of the lower extremities between the ankle, knee, and femur joints, as well as the trunk and segments of the upper extremities were measured.

5 Experiment and Results

Measurement Results and Analysis of Stabilograms. The characteristic initial form of the dependences $x(t)$, $y(t)$, as well as the trajectory $y(x)$ of the projection of the CM (point C in Fig. 1) are presented in Fig. 2 and 3, respectively. Most of the subjects showed some asymmetry in the location of the projection of the center of mass relative to the axis OY (Fig. 3), which is associated with the individual features of a convenient two-post rack [11]. With a two-leg stand with weight transfer, the projection of the center of mass (x_C, y_C) shifted in the direction of the supporting foot, however, symmetry in the location of the corresponding trajectories (curves 2 and 3 in Fig. 3) was observed only in one of the subjects.

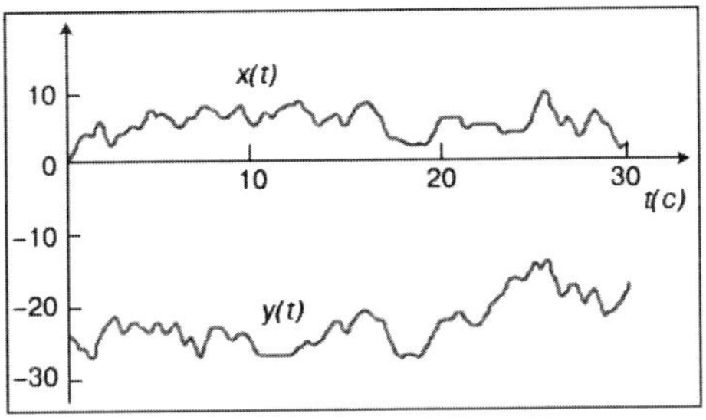

Fig. 2. Oscillations $(x(t), y(t))$ of the projection of the center of mass of the body of one of the subjects with a double support

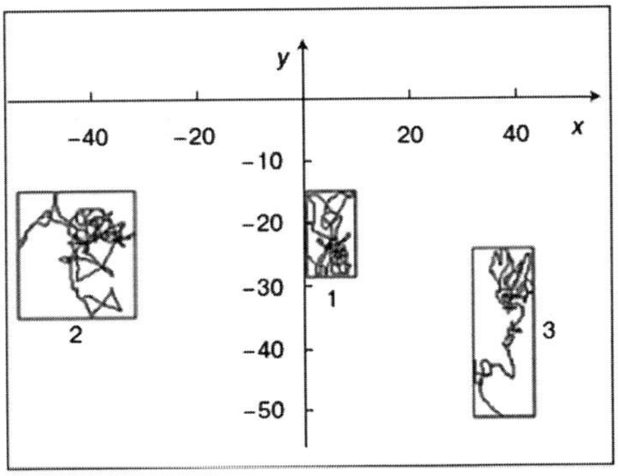

Fig. 3. The trajectories $y(t)$ of the projection of the center of mass with a vertical two-support (1) stand and two-support stand with the transfer of body weight to the left (2) and right (3) foot

Linear dimensions $h'_x h_y$ of the rectangles bounding the area filled with the corresponding trajectories (Fig. 3, 4) determine the maximum swing amplitudes in the sagittal (OX) and frontal (OY) planes. Moreover, in the corresponding trajectories, it is easy to isolate the component corresponding to higher amplitude random walks and high-frequency tremor.

The projections of the CM at a stand with transfer of body weight were accompanied in most cases by a shift of the corresponding point (x_C, y_C) backward. When standing on one foot, the projection of the center of mass also shifted somewhat backward and towards the outer surface of the corresponding foot (Fig. 4). The amplitude of the swing with a single support rack increased significantly both in the sagittal and in the frontal plane. The trajectories of the walks of the CM are characterized by a greater length of the sections of unidirectional displacement of the center of mass and a slightly smaller tremor (Fig. 4). Any regular patterns of swing patterns with a two-support and one-support stance in healthy subjects are not visually determined.

For the entire array of measurements, the distribution of the projections of the CM is uniform and individual features in it are leveled (Fig. 5). In accordance with the results of previous large-scale measurements [20], our results relate to variants of the norm of a two-support rack of healthy individuals.

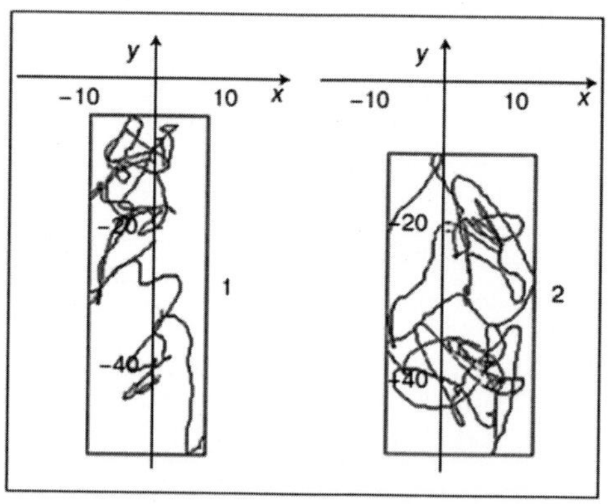

Fig. 4. The trajectories $y(t)$ of the projection of the center of mass with a one-leg stand on the left (1) and right (2) feet

Spectral analysis showed that vibrations in the sagittal plane have a greater power (Fig. 6a), which is most often explained by the large linear size of the support area in the direction of the axis OY. With an artificial decrease in the area of the support, an increase in the power of oscillations in the sagittal plane up to the loss of stability was observed in the experiments [20]. Our estimates showed that several main ranges with different power densities can be distinguished in the spectrum: for oscillations in the direction of the axis OX these are ranges $f \in [0.2; 0.4]$ (I), $f \in [0.4; 1]$ (II), $f \in [1; 1.4]$ (III), $f \in [1.4; 4]$ (IV), and for oscillations in the direction of the axis OY - ranges $f \in [0.2; 0.3]$ (I), $f \in [0.3; 0.9]$ (II), $f \in [0.9; 1.3]$ (III), $f \in [1.3; 3]$ (IV). Although the kind of distributions $W(f)$

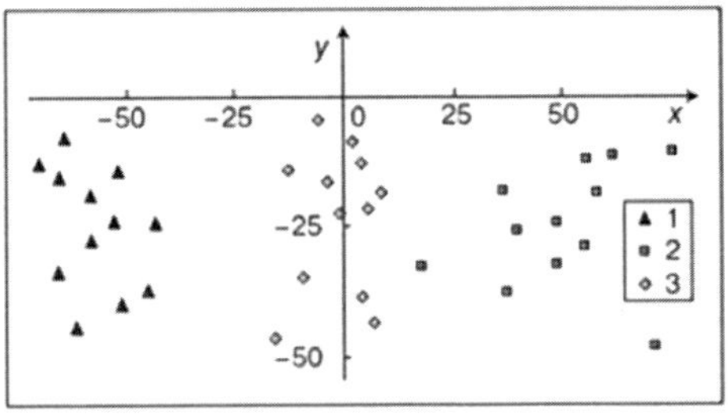

Fig. 5. Averaged $y(x)$ values for all subjects with a two-leg stand (1) and a stand with weight transfer to the left (2) and right (3) feet

different in different subjects, the ratio $W_I : W_{II} : W_{III} : W_{IV}$ weakly changing and close to $0.5 : 0.4 : 0.07 : 0.03$. Thus, the diagnostic value may not be so much the form of dependence $W(f)$ or values of individual peaks, how much is the ratio between the energy density corresponding to certain frequency ranges corresponding to different control systems for maintaining body position.

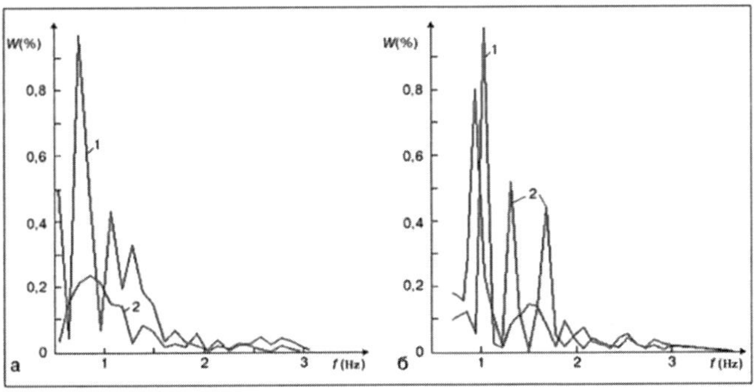

Fig. 6. The dimensionless density of the spectral power of the signal for swinging in the sagittal (curve 1) and frontal (curve 2) planes with two-support (a) and one-support (b) struts of the same subject

A natural decrease in the area of the support can be obtained by taking measurements with a single support stand. In this case, the swing power in the sagittal plane slightly increases in the region W_I and decreases for the remaining frequencies, however, the sway in the frontal plane is significantly enhanced. At the same time, three main peaks corresponding to swings in three frequency ranges are clearly distinguished W_I, W_{II}, W_{III} (Fig. 6b). The frequency values corresponding to the maxima of the spectral plane with a two-support and one-support strut are slightly different. In general, with a single support rack, all frequency values are shifted towards slightly higher frequencies. Similar qualitative regularities are described in the literature for stabilographic studies of the rack of athletes.

The patterns of oscillations of the center of mass also differ with a two-support and one-support strut. In this case, it is possible to identify statistically significant relationships between the amplitudes of oscillations when resting on one foot (A) and with a standard two-post rack (A_0), calculated as $A = \left(h_x^2 + h_y^2\right)^{0.5}$ (Fig. 7). Normally, the average swing amplitudes during support with the transfer of the body coincide and linearly depend on A_0. With a single-leg stand, the dependences are similar, and the swing amplitudes increase almost twice as compared with a double-leg stand. So, the average values across all subjects $A_0 = 12.028$, with a two-post stand with body transfer $A_t = 14.286$ and 14.684 for the right and left legs and with one support stand $A_t = 24.582$ and 24.862 for the right and left legs. Such a change in amplitude cannot be unambiguously explained due to a decrease in the size of the support area alone, since different racks are associated with different participation of support and non-support limbs.

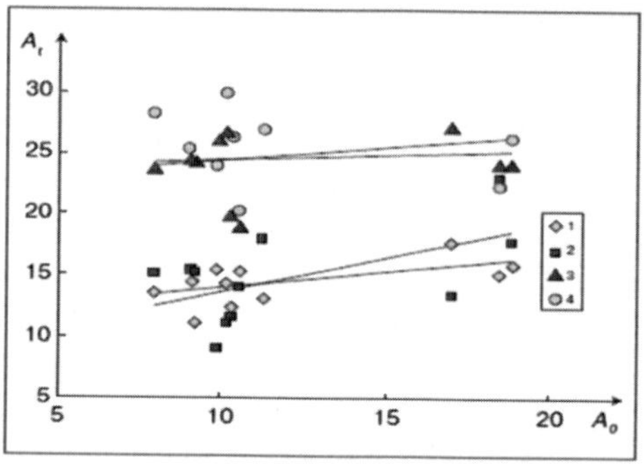

Fig. 7. Swing amplitude A_t during the test as a function of swing amplitude A_0 with a standard two-leg stand for different tests: two-leg stand with transfer of body weight to the right (1) and left (2) foot and with a one-leg stand on the right (3) and left (4) leg

A mathematical model of the oscillations of the human body in the sagittal plane with a double support. Consider the corresponding model of the human body in the form of an inverted three-link pendulum (Fig. 1). Lengths and masses of segments L_{1-3} and m_{1-3}, segment CM C_{1-3} located at a distance d_{1-3} from the beginning of the segment, counting along the direction of the axis OZ. Assuming that the position of the support remains unchanged, the geometry of the pendulum can be completely described by the deflection angles q_{1-3} links from the vertical. The movement of the pendulum is described by the Lagrange equations:

$$\frac{d}{dt}\frac{\partial K}{\partial \dot{\theta}_j} - \frac{\partial K}{\partial \theta_j} + \frac{\partial P}{\partial \theta_j} = 0, \qquad (1)$$

where K and P are the kinetic and potential energy of the system, and the dot above the symbol denotes the time derivative.

The first segment (lower leg) is involved in rotational motion relative to a fixed hinge (ankle joint). The second (thigh) and third (trunk) segments participate in rotational motion relative to their center of mass and in the figurative motion of the center of mass arising from the movement of the previous links. Moreover, the expressions for the energies K_j, P_j individual links can be written as:

$$\begin{aligned}
&K_1 = 0.5\dot{\theta}_1^2\left(J_1 + m_1 d_1^2\right),\\
&K_{2,3} = 0.5J_{2,3}^c\dot{\theta}_{2,3}^2 + 0.5m_1 V_{2,3}^2,\\
&P_1 = m_1 g d_1\left(1 - \cos\theta_1\right),\\
&P_2 = m_1 g L_1\left(1 - \cos\theta_1\right) + m_2 g d_2\left(1 - \cos\theta_2\right),\\
&P_3 = m_1 g L_1\left(1 - \cos\theta_1\right) + m_2 g L_2\left(1 - \cos\theta_2\right) + m_3 g d_3\left(1 - \cos\theta_3\right),
\end{aligned} \tag{2}$$

where J_j - the moment of inertia of the link relative to its center of mass C_j, g - acceleration of gravity, V_j - the speed of movement of the links relative to their CM:

$$\begin{aligned}
V_2^2 &= L_1^2\dot{\theta}_1^2 + d_2^2\dot{\theta}_2^2 + 2L_1 d_2\dot{\theta}_1\dot{\theta}_2\cos\left(\theta_1 - \theta_2\right)\\
V_3^2 &= L_1^2\dot{\theta}_1^2 + L_2^2\dot{\theta}_2^2 + d_3^2\dot{\theta}_3^2 + 2L_1 L_2\dot{\theta}_1\dot{\theta}_2\\
&\times\cos\left(\theta_1 - \theta_2\right) + 2L_1 d_3\dot{\theta}_1\dot{\theta}_3\cos\left(\theta_1 - \theta_3\right) + 2L_2 d_3\dot{\theta}_2\dot{\theta}_3\cos\left(\theta_2 - \theta_3\right).
\end{aligned}$$

Substituting expressions (2) into (1), expanding the expressions for trigonometric functions in series in powers of the corresponding angle and neglecting the terms $\sim \mathrm{o}\left(\theta_i^2\right)$, we get a system of linear differential equations, which in matrix form can

$$M \cdot \frac{d^2}{dt^2}\overrightarrow{\theta} + N \cdot \overrightarrow{\theta} = 0, \tag{3}$$

where $\overrightarrow{\theta}^T = (\theta_1, \theta_2, \theta_3)$, sign T means transposition. The coefficients of the matrix have the following values:

$$\begin{aligned}
&M_{11} = J_1 + m_1 d_1^2 + \left(m_2 + m_3\right)L_1^2 \quad M_{12} = m_2 L_1 d_2 + m_3 L_1 L_2\\
&M_{13} = M_{31} = m_3 L_1 d_3\\
&M_{22} = J_2 + m_2 d_2^2 + m_3 L_2^2 \qquad\qquad M_{21} = m_2 L_1 d_2 + m_3 L_1 L_3\\
&M_{23} = M_{32} = m_3 L_3 d_3\\
&M_{33} = J_3 + m_3 d_3^2
\end{aligned}$$

$$N = \begin{pmatrix} m_1 g\left(d_1 + 2L_2\right) & 0 & 0\\ 0 & m_2 g\left(d_2 + L_2\right) & 0\\ 0 & 0 & m_3 g d_3 \end{pmatrix}$$

We study the system (3). First, we determine the natural oscillations of the pendulum by substituting in (3) a solution of the form

$$\theta_j = \alpha_j \sin\left(\omega t + \psi\right),$$

where α_j, ω, ψ - amplitude, circular frequency and the initial phase of the oscillations. Then we get the following system of algebraic equations:

$$\left(M\omega^2 + N\right)\cdot\overrightarrow{\alpha} = 0, \tag{4}$$

where $\overrightarrow{\alpha}^T = (\alpha_1, \alpha_2, \alpha_3)$.

System solvability condition (4) det $|M\omega^2 + N| = 0$ leads to an equation for determining the natural frequencies of oscillations w, the solution of which can be easily obtained by numerical methods, if all the coefficients of the model are given.

As a mathematical model of a one-support strut, we also consider a three-link pendulum (Fig. 8). We assume that the subject stands on a straight supporting leg, and holds the non-supporting leg straight and uses it to maintain balance, so that the angle of deviation of the straightened leg from the vertical changes q_2, but not the relative position of the segments of the straightened leg (Fig. 8). The subject's hands are located along the torso and remain motionless.

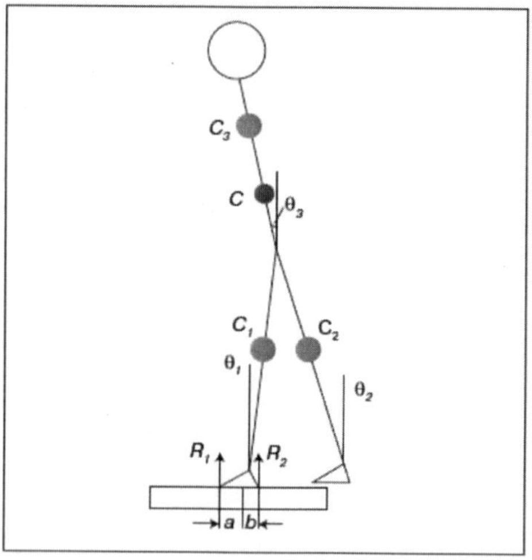

Fig. 8. Diagram of a three-link model of the human body for evaluating the swing parameters in the sagittal plane with a single support stand

We introduce the length, mass, and position of the center of mass of all links of the model similarly to the previous case. We will assume that the left and right limbs are the same, and their length, mass and distance to the center of mass of the link are L_1, m_1, d_1. The movement of the pendulum in this case will be described by the same canonical equations (1), where the expressions for the potential and kinetic energy of the first link will take the same form (2), and for the second and third links we get:

$$\begin{aligned} &K_{2,3} = 0.5J_{2,3}^c\dot{\theta}_{2,3}^2 + 0.5m_1V_{2,3}^2 \\ &P_2 = m_1g\left(d_1 + d_1\cos\theta_2 - L_1\cos\theta_1\right) \\ &P_3 = m_3g\left(L_1 + d_3 - d_3\cos\theta_3 - L_1\cos\theta_1\right) \\ &V_2^2 = L_1^2\dot{\theta}_1^2 + d_1^2\dot{\theta}_2^2 + 2L_1d_1\cos\left(\theta_1+\theta_2\right)\dot{\theta}_1\dot{\theta}_2 \\ &V_3^2 = L_1^2\dot{\theta}_1^2 + d_3^2\dot{\theta}_3^2 + 2L_1d_3\cos\left(\theta_1-\theta_3\right)\dot{\theta}_1\dot{\theta}_3. \end{aligned} \qquad (5)$$

Substituting expressions (5) into (1), expanding the trigonometric functions in series and neglecting small terms, we also obtain a system of differential equations of the form (3), where

$$M_{11} = J_1 + m_1 d_1^2 + (m_1 + m_3) L_1^2 \quad M_{12} = M_{21} = m_1 L_1 d_1$$
$$M_{13} = M_{31} = m_3 L_1 d_3$$
$$M_{22} = J_2 + m_1 d_1^2 \quad M_{33} = J_3 + m_3 d_3^2$$
$$M_{23} = M_{32} = 0$$

$$N = \begin{pmatrix} m_1 g (d_1 + L_1) + m_3 g L_1 & 0 & 0 \\ 0 & m_1 g d_1 & 0 \\ 0 & 0 & m_3 g d_3 \end{pmatrix}.$$

In this case, it also makes sense to first investigate the free vibrations of the system, determine the natural frequencies of vibrations, and compare the cases of a two-support and one-support racks. The procedure for investigating the solution of the problem in this case will be the same as described in the previous section for the two-support rack.

6 Discussion

The study of the eigen oscillations of systems was carried out by numerical methods. The lengths of model segments are known from measurements, and the calculations of masses, moments of inertia, and the positions of the centers of mass of the segments were carried out on the basis of regression ratios obtained from large-scale measurements on dissected corpses and given in the literature [24]. Since in the three-link model, the trunk and upper limbs fixedly held along it were considered as one segment, the calculation of the position of the center of mass and the moment of inertia of the upper link (link 3 in both models) was carried out on the basis of the corresponding formulas of classical mechanics. The mass of the link was determined by summing the masses of the segments included in it, calculated by the corresponding regression dependences for the known body mass and height of the test person and the length of the segment [24]. Similar calculations were carried out for the model of a single-support rack, where the lower extremities were considered as whole links.

The calculation results of the frequencies of free oscillations of the system are given in Table 1. Comparison of calculated values with the results obtained by us in the analysis of recorded stabilograms and similar data presented in the literature [1,2,18] show that the calculated frequencies of natural vibrations correspond to three main ranges of oscillations W_I, W_{II}, W_{III}. At the same time, higher-frequency components ($f_{2,3}$) caused by physiological tremor [2], and the low-frequency component (f_1) - mechanical properties of the system. The somewhat overestimated values of the calculated frequency compared with the data of spectral analysis of stabilograms are due to the fact that the damping effect of friction in the joints, the presence of articular moments, the appearance of which is oriented towards maintaining an unchanged position of the center of mass and

Table 1. Eigen frequencies of oscillations of the human body in two-legged standing

Study number	Height (m)	Body weight (kg)	f_1	f_2	f_3
1	1.82	80.75	1.51	1.88	4.95
2	1.56	52.45	1.35	1.83	4.98
3	1.72	60.95	1.38	1.83	4.98
4	1.80	61.1	1.41	1.88	5.63
5	1.69	55.7	1.35	1.91	5.57
6	1.90	84.75	1.51	1.98	6.65
7	1.74	61.5	1.41	1.89	5.65
Average value	1.75	65.3	1.42	1.89	5.49

is characterized by a certain delay time associated with the mathematical model, were not taken into account reaction of balance organs, nervous and muscular systems.

Numerical calculations according to the model of a one-support strut lead to frequency values that are slightly higher than the corresponding frequencies of body vibrations with a two-support strut for a given individual, but belong to the same ranges highlighted above W_I, W_{II}, W_{III}. Thus, the results of numerical calculations of changes in vibrational frequencies during the transition from a two-support to a one-support rack are consistent with stabilographic data.

Our results are consistent with the hypothesis that any decrease in the area of the supporting surface (both in the longitudinal and in the transverse direction) leads to an increase in postural instability [11]. At certain critical values of the dimensions of the supporting surface, an individual may lose stability of a given test position of the body. In this case, the oscillation amplitude increases both in the sagittal and in the frontal plane, and the character of the component corresponding to low-frequency random walks of the center of mass of the system changes - the average "mean free path" of the center of mass during random walks increases, which leads to a significant increase in the maximum swing amplitude. The contribution of the high-frequency component corresponding to the tremor, which is superimposed on random walks, also changes.

The results of measurements and studies on the proposed mathematical models can clarify the mechanisms involved in maintaining the position of the body, and their possible changes in pathologies. As mentioned above, the complex dynamics of body swings is considered as the result of the balance control system, which is aimed at finding out the degree of instability of a given position by small deviations in the position of individual segments and evaluating the resulting displacements of the center of mass. A feedback system using such a mechanism is very effective for normal patients, since even with large amplitudes of oscillations in both planes with a single support, none of the examined volunteers lost stability. The series of experiments we started on a comparative study of the parameters of a two-support and one-support stance is very promising for diagnosing the condition of patients with pathologies of the nervous and musculoskeletal systems and with injuries. It is known that an increase in the amplitude of swings in the frontal plane in the elderly is a reliable parameter for predicting the incidence and fracture of bone in osteoporosis. The study of the dynamics of oscillations of the center of mass at a two-support and one-support strut will allow developing new criteria for the early diagnosis of such conditions. The fact is that with the usual double-support stance, the physiological mechanisms for finding a stable position are little involved. In order to reveal the degree of readiness of feedback systems to maintain the balance of the body under extreme conditions preceding, for example, a fall, it is necessary to artificially set some uncomfortable stance, and the most natural at the same time is a vertical single-stance, which allows you to activate physiological feedback mechanisms and make them choose optimal conditions for the movement of body segments to maintain stance stability. The more uncomfortable stance the subject is forced to support, the larger the area of possible body movements that has to be "scanned" by the equilibrium maintenance system. An analysis of the patterns of such scanning movements, the trajectories of the CM $y(x)$ and their individual components - random walks and tremors - will allow individual diagnostics of patients. The mathematical models developed above can be effectively used to study forced body vibrations in a particular position (with mechanical muscle stimulation, a sudden jerk). The presence of feedback mechanisms can also be introduced into the model in the form of a returning pair of forces in the joints, taking into account the corresponding delay time of the neuromuscular system of regulation, which will be the further subject of the authors' study.

7 Conclusions

The results of stabilographic studies of several types of a two-leg vertical posture and a stand on one foot showed that the type of corresponding trajectories of the projection of the center of mass of the body on the horizontal plane, the oscillation amplitudes in the sagittal and frontal planes, and the average projection time of the study differ significantly in different healthy subjects. Due to the impossibility of an unambiguous visual assessment of stabilographic data, mathematical methods of signal processing are of great importance. As diagnostically significant parameters, the maximum swing amplitudes in two directions, the ratio of the swing amplitudes for a one-support and two-support strut, and the spectral density of the signal in certain frequency ranges corresponding to the main vibration frequencies of the body were identified. When transferring body weight to one of the feet with a two-leg stand, the center of mass shifts somewhat posteriorly and asymmetrically for the left and right foot. Patterns of changes in the amplitude of the swings are not detected. With a single-leg stand, there is a significant (on average 2 times) increase in the swing amplitude, especially in the frontal plane, which is associated with a significant decrease in the area of the support surface in this direction.

Thc mathematical model of the human body in the form of an inverted three-link pendulum allows you to calculate the natural frequencies of the system's oscillations, describes the increase in frequencies during the transition from a two-leg to a one-leg stand. The model allows further generalization to the case of pathologies associated with limiting joint mobility, and also allows the introduction of a physiological regulation system for maintaining an upright post as control of rotation moments in joints with some delay time.

A comparative analysis of the stabilogram with a standard two-leg stand and balancing on one foot it allows you to activate and evaluate the work of the neuromuscular system in managing the maintenance of the balance of the human body. The scientific novelty of the presented results consists in the developed mathematical model of body vibrations in two-support and one-support poses and in the analysis of the corresponding stabilographic data, first carried out in order to identify patterns and mechanisms of maintaining the vertical posture in normal and pathological conditions.

References

1. Alessandrini, M., Lanciani, R., Bruno, E., Napolitano, B., Di Girolamo, S.: Posturography frequency analysis of soundevoked body sway in normal subjects. Eur. Arch. Otorhinolaryngol. **263**, 248–252 (2006). https://doi.org/10.1007/s00405-005-0965-7
2. Apthorp, D., Nagle, F., Palmisano, S.: Chaos in balance: non-linear measures of postural control predict individual variations in visual illusions of motion. PLoS ONE **9**(12), e113897 (2014). https://doi.org/10.1371/journal.pone.0113897
3. Bottaro, A., Yasutake, Y., Nomura, T., Casadio, M., Morasso, P.: Bounded stability of the quiet standing posture: an intermittent control model. Hum. Mov. Sci. **27**(3), 473–495 (2008). https://doi.org/10.1016/j.humov.2007.11.005
4. Devetak, G., Bohrer, R., Rodacki, A., Manffra, E.: Center of mass in analysis of dynamic stability during gait following stroke: A systematic review. Gait Posture **72**, 154–166 (2019). https://doi.org/10.1016/j.gaitpost.2019.06.006
5. Domergue, H., Rodríguez-Mañas, L., Laosa Zafra, O., Hood, K., Gasq, D., Regueme, S., Sinclair, A., Bourdel-Marchasson, I.: The use of posturography in investigating the risk of falling in frail or prefrail older people with diabetes. J. Frailty Aging **9**(1), 44–50 (2020). https://doi.org/10.14283/jfa.2019.27
6. Ferry, M., Martin, L., Termoz, N., Cote, J., Prince, F.: Balance control during an arm raising movement in bipedal stance: which biomechanical factor is controlled? Biol. Cybern. **91**, 104–114 (2004). https://doi.org/10.1007/s00422-004-0501-7
7. Freitas, S., Duarte, M., Latash, M.: Two kinematic synergies in voluntary whole-body movements during standing. J. Neurophysiol. **95**, 636–645 (2006). https://doi.org/10.1152/jn.00482.2005
8. Kizilova, N., Karpinsky, M., Griskevicius, J., Daunoraviciene, K.: Posturographic study of the human body vibrations for clinical diagnostics of the spine and joint pathology. Mechanika **80**(6), 37–41 (2009). https://doi.org/10.5755/j01.mech.80.6.15500
9. Martinerie, J., Gagey, P.: Chaotic analysis of the stabilometric signal. In: Woollacott, M. (ed.) Posture and Gait, no. 1, pp. 404–407. University of Oregon Books, Portland (1992)

10. Michalak, K., Przekoracka-Krawczyk, A., Naskrecki, R.: Parameters of the crossing points between center of pressure and center of mass signals are potential markers of postural control efficiency. PLoS ONE **14**(7), e0219460 (2019). https://doi.org/10.1371/journal.pone.0219460
11. Mochizuki, L., Duarte, M., Amadio, A., Zatsiorsky, V., Latash, M.: Changes in postural sway and its fractions in conditions of postural instability different postural control mechanisms. J. Appl. Biomech. **22**, 51–60 (2006). https://doi.org/10.1123/jab.22.1.51
12. Okamoto, A.: Biomechanical analysis of the moment about the center of mass during the downswing phase in women's driver shot. In: 37th International Society of Biomechanics in Sport Conference, Oxford, OH, United States, pp. 356–359 (2019)
13. Pavlov, S., Wójcik, W., Smolarz, A.: Information Technology in Medical Diagnostics. CRC Press, Leiden (2017)
14. Rey-Martinez, J., Pérez-Fernández, N.: Open source posturography. Acta Otolaryngol. **136**(12), 1225–1229 (2016). https://doi.org/10.1080/00016489.2016.1204665
15. Ruhe, A., Fejer, R., Walker, B.: Center of pressure excursion as a measure of balance performance in patients with non-specific low back pain compared to healthy controls: a systematic review of the literature. Eur. Spine J. **20**(3), 358–368 (2011). https://doi.org/10.1007/s00586-010-1543-2
16. Serkova, V., Pavlov, S., Romanava, V., et al.: Medical expert system for assessment of coronary heart disease destabilization based on the analysis of the level of soluble vascular adhesion molecules. In: Proceedings of SPIE, Photonics Applications in Astronomy, Communications, Industry, and High Energy Physics Experiments, vol. 10445, p. 104453O (2017). https://doi.org/10.1117/12.2280984
17. Shams, A., Vameghi, R., Shamsipour-Dehkordi, P., Allafan, N., Bayati, M.: The development of postural control among children: repeatability and normative data for computerized dynamic posturography system. Gait Posture **78**, 40–47 (2020). https://doi.org/10.1016/j.gaitpost.2020.03.002
18. Soames, R., Atha, J.: The spectral characteristics of postural sway behaviour. Eur. J. Appl. Physiol. **49**, 169–177 (1982). https://doi.org/10.1007/bf02334065
19. Sologubov, E., Yavorskii, A., Kobrin, V., Nemkova, S., Sinel'nikova, A.: Use of computer stabilography and computer-assisted biomechanical examination of gait for diagnosis of posture and movement disorders in patients with various forms of infantile cerebral paralysis. Biomed. Eng. **34**(3), 138–143 (2000). https://doi.org/10.1007/BF02389845
20. Tyazhelov, O., Fischenko, V., Iaremin, S., Karpinsky, M., Karpinska, O.: Modeling of processes of support of a vertical posture. Orthop. Traumatol. Prosthet. **598**(1), 42–49 (2015)
21. Tyazhelov, O., Karpinsky, M., Karpinskaya, O., Iaremin, S.: Features of dynamic characteristics of statograms at fixation of joints of the lower extremity. Trauma **15**(2), 88–93 (2014). https://doi.org/10.22141/1608-1706.2.15.2014.81375
22. Wójcik, W., Pavlov, S., Kalimoldayev, M.: Information Technology in Medical Diagnostics II: Taylor & Francis Group, CRC Press, Balkema Book, London (2019)

23. Yamamoto, T., Smith, C., Suzuki, Y., Kiyono, K., Tanahashi, T., Sakoda, S., Morasso, P., Nomura, T.: Universal and individual characteristics of postural sway during quiet standing in healthy young adults. Physiol. Rep. **3**(3), e12329 (2015). https://doi.org/10.14814/phy2.12329
24. Zatsiorsky, V., Selujanov, V., Chugunova, L.: Methods of determining mass-inertial characteristics of human body segments. In: Contemporary Problems of Biomechanics, pp. 272–291 (2000)

Classification of Time Realizations Using Machine Learning Recognition of Recurrence Plots

Lyudmyla Kirichenko(✉), Petro Zinchenko, and Tamara Radivilova

Kharkiv National University of Radio Electronics, Kharkiv 61166, Ukraine
lyudmyla.kirichenko@nure.ua

Abstract. In the article, the machine learning classification of time realizations using the recurrence plot visualization is considered. Every time realization is converted into a matrix of recurrence states and it is presented as a black-and-white image. The resulting images of realizations are classified using deep neural networks. A deep residual neural network is used as an image classifier. The binary classification of EEG realizations is carried out. The result of the binary classification is the detection of an epileptic seizure. The data for the experiment are records of brain activity containing 178 values. The results of the study show that the considered method has a high classification accuracy. The proposed classification approach can be readily used in practice.

Keywords: Machine learning classification · Time series classification · Recurrence plot · EEG realizations · Deep residual networks

1 Introduction

One of the most difficult classification tasks is time series classification. A time series can be represented as a time realization of some dynamic system, which reflects the many internal and external relations of this system. Complex dynamic systems include biological, informational, technical, and social ones. The time series of such complex systems are partially or completely random, having an inverse nonlinear relationship and a long-term correlation dependence. A well-known example of biomedical time series is realizations of electroencephalogram (EEG).

The human brain has an electric field, which is characterized by electric surges from several milliamps to several hundred milliamps. Each neuron of the brain generates a change in electrical potential, this process can be measured. Analysis of the electric field of the brain is widely used in the diagnosis of various diseases. Thus, the task of classifying EEG realizations arises.

Currently, machine learning is widely used in time series classification. Various approaches to machine learning methods for time series classifying, including

S. Babichev et al. (Eds.): ISDMCI 2020, AISC 1246, pp. 687–696, 2021.
https://doi.org/10.1007/978-3-030-54215-3_44

algorithms, the selection of features and applications, are considered in re-view papers [2,5,9,11,12,28]. It should be noted a large number of algorithms, metrics and features, which can be used for comparison and further classification of time series. In addition, many of them are highly specialized and rather cumbersome.

There are much fewer methods that allow to represent time series in the form of some structures, for example, images, which are easy to compare with each other. One of these approaches, which allows us to visualize the time series in the form of a recurrence plot, is presented in this work using the classification of EEG realizations as an example.

2 Literature Review

Most often, after the time series preprocessing, a set of some features is extracted from it, which are the input of the classifier. The selection of features depends on the task of classification. Features can characterize various time series attributes, for example, the structure of the time series components [32], fractal [20] or chaotic [34] properties of series.

The simplest features to classify might be the values of the time series themselves. In [3,4,22], it was shown that in the case of strongly expressed fractal properties, it is sufficient to consider the values of the series without calculating the fractal characteristics. Also, various distance measures of the time series can be used as features [5,21]. A popular method of feature extraction is the conversion of the time series into the frequency domain applying the Fourier or the wavelet transform [10].

The above-mentioned approaches are widely used to classify EEG realizations. The review [24] presents the majority of modern machine learning methods used to classify EEG signals. The review [7] focused on the use of deep learning for EEG classification tasks.

At the same time, in recent decades, methods of time series analysis based on chaotic dynamics methods have become widespread. The starting point, in this case, is the fact that a single time realization of a dissipative system is enough to restore its attractor, which contains all the information about the system [31]. In particular, the brain (or part of it) can be considered as a nonlinear dynamic system that is sensitive to initial conditions, and EEG is the trajectory of this dynamic system [1].

One of the methods originating in nonlinear dynamics is recurrence analysis (the method of recurrence plots), originally proposed in [8] and further developed in [25,26]. Currently, recurrence analysis is widespread and used to characterize various systems, including those with fractal properties [17,18]. The recurrence plot contains state repeatability information of a dissipative dynamical system represented by a time series. Recurrent properties can be visualized in the form of geometric structures and the time series dynamics may be clearly demonstrated. Thus, the technique of recurrence plots converts time series into images that can be classified using computer vision methods [14,15,27]. However, such studies are quite new and have not yet become widespread.

The aims of the present work is carried out of the short EEG classification based on the recurrence plots, which are converted into images and further classified using a residual neural network.

3 Time Series Classification Based on Recurrence Plots

The method of recurrence plots is based on the restoration of the system attractor from single time realization (Packard-Tackens procedure) [31]. In this case, the pseudophase state space X of the system can be obtained using realization values u_t:

$$x_i = (u_i, u_{i+\tau}, ..., u_{i+(m-1)\tau}) \quad (1)$$

where x_i is the value of the pseudophase trajectory at the i-th moment of time, u_i is the value of the time realization at the i-th moment of time, m is the dimension of the pseudophase space, τ is the time lag. The authors of [8] proposed a method that allows mapping the m -dimensional pseudophase trajectory x_i, $i = 1, ..., N$ into a binary matrix of size $N \times N$. An element with coordinates (i, j) takes a value equal to 1 if the state $\vec{x}_j$ is a recurrence to the state $\vec{x}_i$, that is, $\vec{x}_j$ falls into some given neighborhood of $\vec{x}_j$, and the element (i, j) is 0 in the opposite case. Such a matrix (recurrence plot) contains information on the recurrence behavior of the time series. The recurrence plot RP is a matrix, where value $RP_{i,j}$ is equal to 1 when the distance between x_i and x_j is less than ε:

$$RP_{i,j} = \Theta(\varepsilon||x_i - x_j||), x_i, x_j \in R^m, i, j = 1, ..., N \quad (2)$$

where ε is the radius of the m-dimensional neighborhood of a point x_i, is the distance between the points, $\Theta(\cdot)$ is the Heaviside function. Recurrent plots are easy to visualize in black and white. Then the recurrent states are displayed in black, and the rest in white. The upper part of Fig. 1 shows a realization of a sequence of independent random variables (a) and a realization of a discrete chaotic system (b). The lower part of Fig. 1(c), (d) demonstrates the recurrence plots corresponding to these time realizations. Chaotic realizations have a specific structure related to the autocorrelation properties of the system, which is reflected in the recurrence plots. It was shown in [17–19,29] that the change in the correlation structure of a system causes the change in the topology of the recurrence plot. Thus, the analysis of recurrence plots makes it possible to classify the observed time series.

4 Input Data

To carry out the classification, popular data of brain activity records were selected. The initial dataset contained records of brain activity for various human conditions: an epileptic seizure, from the tumor zone, from the healthy part of the brain, when a person has closed eyes, and when a person has open eyes. Thus, the data was divided into 5 classes. Each class contained 100 files, where each file corresponded to one object (person). Each file contained records of brain

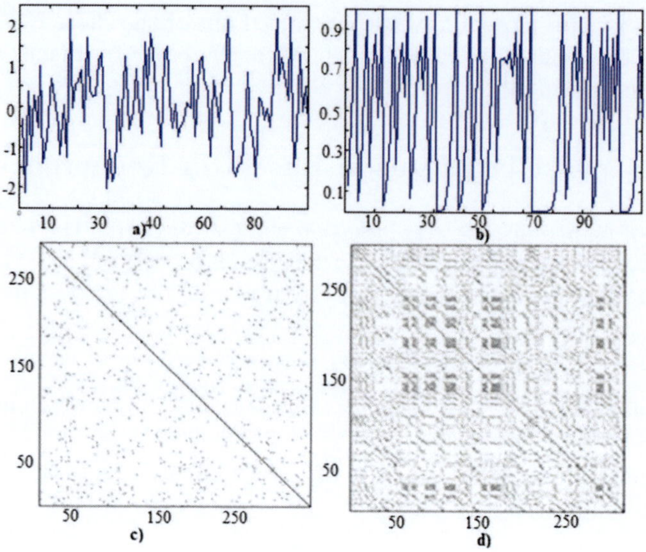

Fig. 1. Time realizations and corresponding recurrence plots

activity for 23.5 s, which corresponded to a time series of length 4097 values. In [1], a detailed description of these brain activity data was given and their nonlinear properties were shown, in particular, the chaotic properties of time series corresponding to an epileptic seizure. Although the dataset contained 5 classes of time series, most studies conducted a binary classification, where the epileptic seizure class was compared with the rest. In [16], epileptic seizures were diagnosed using machine learning methods such as Artificial Neural Networks, Naive Bayesian, k-Nearest Neighbor, Support Vector Machines and k-Means. The experiment showed very good results: for most algorithms, the accuracy of the binary classification of records lasting 23 s was more than 99 A more difficult task is to determine the seizure in the case of significantly shorter data. In the data set [33], each time realization of 4097 values was divided into 23 parts of 178 values, which corresponded to 1 s. In [30], presented on the web platform Kaggle, using machine learning algorithms from [16], a classification of 1 s length records was carried out. In this work, we also used the above mentioned time realizations of brain activity with a length of 178 values. Figure 2 shows the time realizations of EEG from the 1-st class (epileptic seizure) at the top (a) and realizations from other classes at the bottom (b). Figure 3 shows the recurrence plots corresponding to the time realizations shown in Fig. 2. It is worth noting that it is almost impossible to visually find differences between the EEG classes both in realizations and recurrence plots.

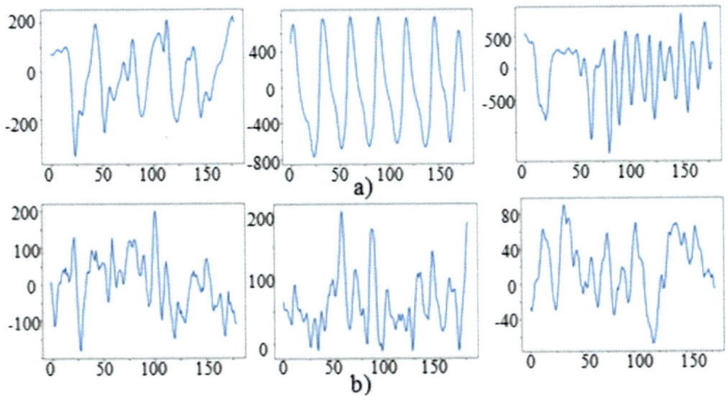

Fig. 2. Realizations of EEG of an epileptic seizure a) and other classes b)

5 Neural Networks for Image Classification

Currently, in connection with the increased computing power and the appearance of database images, computer vision technologies have become very common. It became possible to apply these technologies not only in various fields of scientific and technical activity but also in everyday life. The best results of image recognition have been shown by deep neural networks, in particular the convolutional neural network, which in contrast to the multilayer perceptron, are taken into account two-dimensional image topologies [6,23].

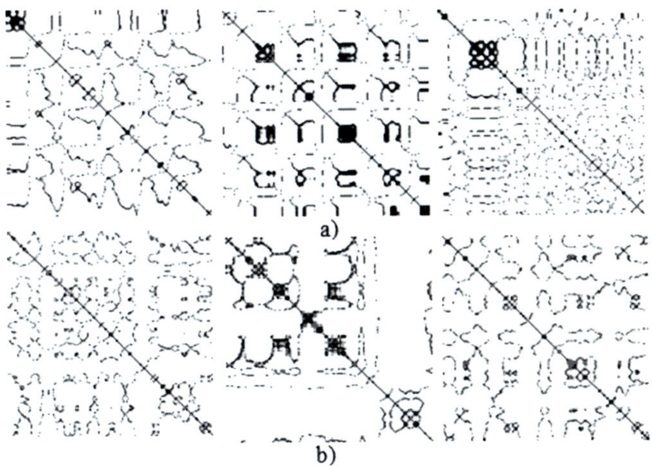

Fig. 3. Recurrence plots of an epileptic seizure a) and other classes b)

Convolutional neural networks receive features in an end-to-end multilayer manner. Studies have shown that with an increase in the number of layers, the recognition accuracy does not increase significantly: at first, the accuracy can grow a little, but the more we increase the depth of the network, the worse it starts to classify. This is due to the fact that the networks are trained by the method of backward propagation of errors, and as a result of differentiation according to the chain rule, the gradient attenuates for sufficiently deep layers. To solve this problem, the structure of a residual neural network was developed [13]. The main feature of the residual network is that the classification result of the extended network will be no worse than the base one. The main idea is to replace the usual sequential layer connection with a shortcut connection. Shortcut connections skip several layers and match identifiers. Their outputs are added to the grouped layers outputs. The main unit of the residual network consists of two layers with weights which are not necessarily convolutional, and a shortcut connection that simply transmits a signal to the output. The structure diagram of the residual network is shown in Fig. 4.

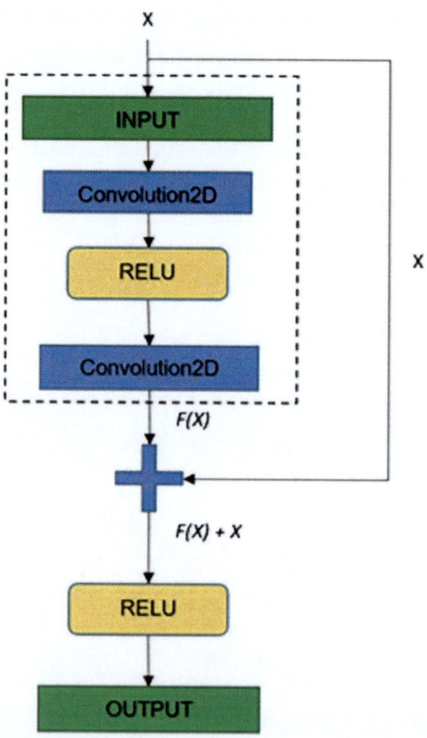

Fig. 4. Layout of the structure of the residual network

6 Experiment, Results and Discussion

In the experiment, a residual neural network which had 131 layers with weights was used. The network consisted of 11 separate and 120 connected in blocks of a shortcut connection. The first 127 layers are intended for revealing characteristics (features), the last 4 used for constructing the classification function. The deep residual neural network contained ten blocks, three convolutional layers, and one fully connected layer. The output of the last fully connected layer is fed to the logistic function, which distributes into two classes. Neurons in a fully connected layer are connected to all neurons in the previous layer. Sub-sample layers follow after the second and third convolutional layers. The ReLU function is used to the output of each convolutional and fully connected layer. For training on the network, the adaptive learning rate optimization algorithm Adam was used.

The experimental data are taken from [30]. The dataset consists of 5 classes that contain records of the brain activity of different human conditions. The experiment was a binary classification of EEG realizations, where the epileptic seizure class was compared with the rest.

Each class consists of 100 files, where each file contains records about a single person. Each file contains records of brain activity for 23.5 s. Each time series contains 4097 values. So, respectively, the dataset contains records of brain activity about 500 individuals for 23.5 s, which consists of 4097 values. As a result, 11500 time series were obtained with a length of 178 values for 1 s interval.

The experiment was first conducted on a sample of 7500 realizations (6000 without a seizure/1500 with a seizure), where 6000 (4800 without a seizure/1200 with a seizure) were used for training and 1500 (1200 without a seizure/300 with a seizure) for the test. Then, to determine the effect of sample size on the quality of classification, a whole dataset was used (11500 realizations), where 8500 realizations were chosen for training (6800 without a seizure/1700 with a seizure) and 3000 for a test (2400 without a seizure/600 with a seizure).

The classification results are presented in the confusion matrix (Table 1), where the symbol P means that the object was classified as belonging to a positive class (seizure), N means belonging negative class (without seizure), the symbol T shows that the class was determined correctly, and F shows incorrect determination.

Table 1. Confusion matrix

Value	Seizure (true)	Without seizure (false)
Positive	282 (TP)	6 (FP)
Negative	18 (FN)	1194 (TN)

Based on the confusion matrix, the classification evaluation metrics were calculated and they are presented in Table 2. It is worth noting that the classification results obtained in the present work based on the method of recurrence plot visualization are not inferior in accuracy to the results obtained in [30] for several different classifiers.

Table 2. Classification evaluation metrics

Measure	Seizure (true)
Accuracy	0.984
Precision	0.9792
Recall	0.94
F1 score	0.9592

7 Conclusions

In the work, a fairly new approach to solving the task of time series classifying by the example of EEG realizations has been considered. The proposed approach, instead of extracting quantitative features of time series, has been based on the series transformation in the images and the further classification of the obtained images by deep neural networks. As conversion from time series to black and white image, the construction of the recurrence plot has been used.

Residual neural network has been selected as an image classifier. Binary classification of EEG realizations, which contained records of an epileptic seizure have been carried out. The data for the experiment were records of brain activity of 178 values, which corresponds to 1 s. The binary classification result was the detection of an epileptic seizure. The results have shown that the considered classification method has a fairly high classification accuracy even with a short length of EEG realizations. Indicators of classification quality the accuracy and F1-score were equal, respectively, 0.984 and 0.959.

The conducted experiment has confirmed the success of the approach to the time series classification by visualizing them. Further quality improvement may be possible in two ways. The first is to improve the architecture of the neural network, and the second is to develop methods for increasing the visual difference between recurrence plots of different time series classes. In our future studies, we intend to focus on study and training the deep residual neural networks designed to recognize recurrence plots of time series from various datasets.

References

1. Andrzejak, R.G., Lehnertz, K., Mormann, F., Rieke, C., David, P., Elger, C.E.: Indications of nonlinear deterministic and finite-dimensional structures in time series of brain electrical activity: dependence on recording region and brain state. Phys. Rev. E **64**(6), 061907 (2001)
2. Bagnall, A., Lines, J., Bostrom, A., Large, J., Keogh, E.: The great time series classification bake off: a review and experimental evaluation of recent algorithmic advances. Data Min. Knowl. Disc. **31**(3), 606–660 (2017)
3. Bulakh, V., Kirichenko, L., Radivilova, T.: Classification of multifractal time series by decision tree methods. In: 14th International Conference ICTERI, pp. 1–4 (2018)

4. Bulakh, V., Kirichenko, L., Radivilova, T.: Time series classification based on fractal properties. In: 2018 IEEE Second International Conference on Data Stream Mining & Processing (DSMP), pp. 198–201 (2018). https://doi.org/10.1109/DSMP.2018.8478532
5. Buza, K.: Time series classification and its applications. In: Proceedings of the 8th International Conference on Web Intelligence, Mining and Semantics, pp. 1–4 (2018)
6. Ciresan, D.C., Meier, U., Masci, J., Gambardella, L.M., Schmidhuber, J.: Flexible, high performance convolutional neural networks for image classification. In: Twenty-Second International Joint Conference on Artificial Intelligence, pp. 1237–1242 (2011)
7. Craik, A., He, Y., Contreras-Vidal, J.L.: Deep learning for electroencephalogram (EEG) classification tasks: a review. J. Neural Eng. **16**(3), 031001 (2019). https://doi.org/10.1088/1741-2552/ab0ab5
8. Eckmann, J.P., Kamphorst, S.O., Ruelle, D.: Recurrence plots of dynamical systems. World Sci. Ser. Nonlinear Sci. Ser. A **16**, 441–446 (1995)
9. Esling, P., Agon, C.: Time-series data mining. ACM Comput. Surv. (CSUR) **45**(1), 1–34 (2012)
10. Faraggi, M., Sayadi, K.: Time series features extraction using Fourier and Wavelet transforms on ECG data (2019). https://blog.octo.com/en/time-series-features-extraction-using-fourier-and-wavelet-transforms-on-ecg-data/
11. Fawaz, H.I., Forestier, G., Weber, J., Idoumghar, L., Muller, P.A.: Deep learning for time series classification: a review. Data Min. Knowl. Disc. **33**(4), 917–963 (2019)
12. Fulcher, B.D.: Feature-based time-series analysis. In: Feature Engineering for Machine Learning and Data Analytics, pp. 87–116 (2018)
13. Fung, V.: An overview of ResNet and its variants. Towards Data Science (2017)
14. Hatami, N., Gavet, Y., Debayle, J.: Classification of time-series images using deep convolutional neural networks. In: Tenth International Conference on Machine Vision (ICMV 2017). International Society for Optics and Photonics (2018)
15. Hatami, N., Gavet, Y., Debayle, J.: Bag of recurrence patterns representation for time-series classification. Pattern Anal. Appl. **22**(3), 877–887 (2019)
16. Karlık, B., Hayta, B.: Comparison machine learning algorithms for recognition of epileptic seizures in EEG. In: Proceedings IWBBIO 2014 (2014)
17. Kirichenko, L., Kobitskaya, Y., Habacheva, A.: Comparative analysis of the complexity of chaotic and stochastic time series. Radioelectronics Inform. Manag. **2**(31), 126–134 (2014)
18. Kirichenko, L., Radivilova, T., Bulakh, V.: Classification of fractal time series using recurrence plots. In: 2018 International Scientific-Practical Conference Problems of Infocommunications. Science and Technology (PIC S&T), pp. 719–724 (2018). https://doi.org/10.1109/INFOCOMMST.2018.8632010
19. Kirichenko, L., Radivilova, T., Bulakh, V.: Binary classification of fractal time series by machine learning methods. In: Advances in Intelligent Systems and Computing, vol. 1020, pp. 701–711 (2019). https://doi.org/10.1007/978-3-030-26474-1_49
20. Kirichenko, L., Radivilova, T., Bulakh, V.: Machine learning in classification time series with fractal properties. Data **4**(1) (2019) art. no. 5. https://doi.org/10.3390/data4010005
21. Kirichenko, L., Radivilova, T., Tkachenko, A.: Comparative analysis of noisy time series clustering. In: Proceedings of the 3rd International Conference on Computational Linguistics and Intelligent Systems (COLINS), pp. 184–196 (2019)

22. Kirichenko, L., Radivilova, T., Zinkevich, I.: Comparative analysis of conversion series forecasting in e-commerce tasks. In: Conference on Computer Science and Information Technologies, pp. 230–242 (2017). https://doi.org/10.1007/978-3-319-70581-1_16
23. LeCun, Y., Bengio, Y.: Convolutional networks for images, speech, and time series. In: The Handbook of Brain Theory and Neural Networks, vol. 3361, no. 10 (1995)
24. Lotte, F., Bougrain, L., Cichocki, A., et al.: A review of classification algorithms for EEG-based brain-computer interfaces: a 10 year update. J. Neural Eng. **15**(3), 031005 (2018)
25. Marwan, N., Romano, M.C., Thiel, M., Kurths, J.: Recurrence plots for the analysis of complex systems. Phys. Rep. **438**(5–6), 237–329 (2007)
26. Marwan, N., Wessel, N., Meyerfeldt, U., Schirdewan, A., Kurths, J.: Recurrence-plot-based measures of complexity and their application to heart-rate-variability data. Phys. Rev. E **66**(2), 026702 (2002)
27. Michael, T., Spiegel, S., Albayrak, S.: Time series classification using compressed recurrence plots. In: Proceedings of ECML-PKDD (2015)
28. Pietrow, D., Matuszewski, J.: Objects detection and recognition system using artificial neural networks and drones. In: 2017 Signal Processing Symposium (SPSympo), pp. 1–5 (2017). https://doi.org/10.1109/SPS.2017.8053689
29. Radivilova, T., Kirichenko, L., Ageiev, D., Bulakh, V.: Classification methods of machine learning to detect DDoS attacks. In: 2019 10th IEEE International Conference on Intelligent Data Acquisition and Advanced Computing Systems: Technology and Applications (IDAACS), vol. 1, pp. 207–210 (2019). https://doi.org/10.1109/IDAACS.2019.8924406
30. Harun-Ur-Rashid, S.: Machine learning algorithms for epileptic seizures (2018). https://www.kaggle.com/harunshimanto/machine-learning-algorithms-for-epileptic-seizures
31. Takens, F.: Detecting strange attractors in turbulence. In: Rand, D., Young, L.S. (eds.) Dynamical Systems and Turbulence, Warwick 1980. Lecture Notes in Mathematics, vol. 898 (1980)
32. Trovero, M.A., Leonard, M.J.: Time series feature extraction. In: SAS, pp. 2020–2018 (2018). https://doi.org/10.1145/3227609.3227690
33. Wu, Q., Fokoue, E.: Epileptic seizure recognition data set (2018). https://archive.ics.uci.edu/ml/datasets/Epileptic+Seizure+Recognition
34. Yerokhin, A., Turuta, O., Babii, A., Nechyporenko, A., Mahdalina, I.: Usage of phase space diagram to finding significant features of rhinomanometric signals. In: 2016 XIth International Scientific and Technical Conference Computer Sciences and Information Technologies (CSIT), pp. 70–72 (2016). https://doi.org/10.1109/STC-CSIT.2016.7589871

SEE-Management by the Force of the Process Functioning System Based on the Output-Input Ratio: The Energy Aspect

Viktor Yarmolenko[1], Nataliia Burennikova[1(✉)], and Roman Akselrod[2]

[1] Vinnytsia National Technical University, Vinnytsia, Ukraine
{01559yarmol,n.burennikova}@ukr.net
[2] Kyiv National University of Construction and Architecture, Kyiv, Ukraine
akselrod.knuba@ukr.net

Abstract. The purpose of the study, the results of which are being introduced, is further improvement of the theoretical and practical procedures (methods) for measuring the energy of products and of OIR of the processes of functioning systems based on author's models for efficiency components and their use to appropriate management by processes. The article specifies procedures of the specified type, which were given our previously published works. It was found that the OIR of the process and its effectiveness in the classical sense are equivalent with viewpoint of the characteristic process (as a qualitative component of its efficiency), but the first of these rates (OIR) describes the qualitative component of the efficiency process mainly from the point of view of benefits (profit), and the second states from the standpoint of losses (costs). It is substantiated that the rate of the quantitative component of OIR of the process is the same as the rate of the quantitative component effectiveness in the classical sense, but their qualitative components are distinguished. On the examples of certain processes we showed the practical realization of the methods measuring of OIR for the processes of functioning systems, invented by the authors and its application (together with methods of measuring other author's rates) to manage by the force of the functioning such systems (precisely that is the scientific novelty of the proposed research results). It is emphasized that the proposed methods can be used for the study of complex systems of different types and levels of hierarchy.

Keywords: Energy approach · Energy of products for the functioning systems processes · Force of the process · Models of the efficiency components of Burennikova (Polischuk) – Yarmolenko · Output-input ratio (OIR) of the process · Scale · Effectiveness and efficiency of the process · SEE-analysis · SEE-management · Total · Net and scale products of the process

S. Babichev et al. (Eds.): ISDMCI 2020, AISC 1246, pp. 697–713, 2021.
https://doi.org/10.1007/978-3-030-54215-3_45

1 Introduction

Periodically in the scientific literature there are publications concerning the analysis of systems (in particular physical and economic) based on the so-called energy approach. At the same time effectiveness of the energy conversion is characterized by the output-input ratio (OIR) of the system, which, as proved, is always less than one hundred percent, that is the energy conversion in principle without loss is impossible. Clarifying the concept of "energy" for the physical value of motion [10] (in contrast to the definition of this concept in any textbook or reference book as a generalized measure of the motion matter) emphasizes that it was necessary to say: "no measure" and the characteristic, since the measure of this value is its unit – joule [10]. Just therefore, in our understanding the energy of the system is such a generalized characteristic of a system that determines its quantitatively-qualitative state and determines the transformation of its structure by changing the space-time arrangement of the elements system [29]. It is believed that energy is the driving force of any changes, including the processes of the functioning and development of systems; it is a generalized characteristic of the motion matter.

From the viewpoint of the energy approach the theory and practice of evaluating of the force processes functioning complex systems on the basis of the concept of OIR as a model of effectiveness for the purpose of management these processes are relevant. This necessitates to development of theoretical and practical approaches to the methodology for evaluating the force of the process functioning complex system on the basis of the concept of OIR with the help of author's efficiency rates [29]. This determined the purpose of the paper. The same technique, in our opinion, can be applied as well in relation to the so-called theory of strings, that was considered [24]. Incorrect understanding by stated authoress of notion OIR leaded her to false conclusions.

The scientific novelty of the work is what is offered methodological approaches to the theory and practice of measuring energy products and of the OIR of subprocesses of the processes of functioning of systems of various types and hierarchical levels require the appropriate algorithms implementation based on modeling.

2 Review of Literature

Regarding the problem of research related to the use of energy approaches, it should be noted that scientists studied the energy of systems from different points of view. In particular [1] investigated the time as an economic space for the development of the economic system, considered the generation of so-called economic energy [23] asserted that all sciences should come to a unified theory; that the notion of energy is universal (it is conditioned), to his idea, by the possibility of application of the energy notion for all without excluding of complex forms of phenomena, and not only to simple substances) [9] systematized and classified the definitions and additions to the concept of "energy"

[13] investigated the mechanisms of development on the basis of energy, entropy and information [17] initiated into the theory of energy [18] studied the problems of managing by complex systems from the point of view of entropy and other system regularities [19] formed the ideas about energy of progress) [11,15] considered in essence the energy of systems from the standpoint of problems of the physical economy [24] proved the rationality of the allocation of economic energy and, along with other issues of economic energy, also considered some issues, related to the OIR [24]; authors [24] offered and justified expediency of using, in their view, economic force as an indicator of the economic efficiency of the national economy, described its structure and algorithm of measurement, on the basis of the conducted studies calculated the economic force of Ukraine in the course of the period chosen by them [7]; Authors [24] devoted his doctoral dissertation to the theoretical and methodological principles of advancing national economic development, formed, in particular, the methodology of research of the national economy on the basis of the energy approach, developed the concept of an advancing national economic development [25]. It should be noted in work [24] along with other issues of economic energy and some issues related to the OIR, did not provide the algorithm for calculating this coefficient, making mistakes in the calculations and conclusions, believing that it is investigating the OIR, rather than the quantitative component of efficiency; she transferred these errors to her doctoral dissertation [24].

By scientific research of the processes based on modeling authors use appropriate to assigned purpose rates, that characterize the obtained results. With regard to the rates of the force of various processes (as the ability of these processes to give a certain result), which is of interest to us at the moment, one can argue that often researchers use the notion of efficiency, considering it to be the concept equivalent to the effect [2,12,14,20]. This sometimes leads, in our opinion, to a fuzzy understanding of both efficiency and effectiveness. Some authors use other approaches, distinguishing these concepts, such as the authors of works [6,8,16,21,22,27,32]. The results of our researches (for more than 20 years) of the force for process functioning systems based on the efficiency category of any process as of such category, which simultaneously is characterized as with quantitative side (in the form of scale product of the process) so and from the qualitative (taking into account the effectiveness of the process) proved that it makes sense to act exactly the same way. Processes were investigated, starting from the labor process (1996) to any process (2012). Studies of various processes in this aspect we continue to this day. The use along with hereby of package of author's rates, proposed by us, as indicators of the process has expanded the innovative approaches to the knowledge of the process functioning system [3,4,26,28,30,31].

3 Problem Statement

Regarding the research problems associated with the use of energy approaches, the unsolved part of this problem is betterment of the design and disclosure of the theory and practice of measuring the energy products and output-input ratios (OIR) of subprocesses of the processes functioning systems for the purpose of management these processes. Author's approaches to the application of the efficiency components, mentioned the above, should contribute to solving this part of the problem. The article, which is considering, is a continuation of articles [29,30], where some results of our studies concerning the measurement of product energies and the measurement of the OIR of the sub processes of the processes functioning systems on the basis of the author's rates for efficiency components are presented, but this part of the problem needs of further researches.

The purpose of the study is further improvement of the theoretical and practical methods for measuring the energy products and the output-input ratios (OIR) of the processes functioning systems based on author's models of efficiency components and their use in process management. The object of research is the process of managing complex systems of different types and levels of the hierarchy. The subject of the research is theoretical and methodological aspects of the effective management by complex systems based on the measurement of energy products and OIR processes of these systems functioning. The practical significance of the work is implementation models of efficiency components for measuring in practice the energy of products and the OIR of subprocesses processes of functioning systems based on the author's rates for efficiency components and their application in management by processes.

4 Materials and Methods

The article uses: method of the generalization to find out the essence, role and basic provisions relative to the components of efficiency, measuring of the energy products and measuring of OIR of subprocesses of the processes of functioning of systems, which are analyzed; situational and systematic approaches, methods of analysis and synthesis, induction and deduction for the purpose of investigate the force of the functioning process of the systems which were being studied; computational, comparative and tabular methods for analyzing the influence of factors on the management of the investigated processes; author's models of efficiency components and author's program for calculating relevant rates; SEE-analysis; SEE-management, etc.

We will use our models of efficiency components of any process and appropriate rates as indicators of the force of the process. The models are based on the following hypothesis about the consequences of any process. The consequences of any process are its products: as benefit (utility); as a cost; the total product in the form of a product as a benefit and a product as a cost; a scale product in the form of a product as a benefit and a part of the product as a cost, proportional to the share product as a benefit in the total product. Rates of the efficiency components we will express using rates of the appropriate products. In the study of processes, as usual, we will use the following equations of change its efficiency:

$$J_R = J_K \cdot J_E = J_K \cdot J_{V/Z} = J_G \cdot J_{1+V/Z} \qquad (1)$$

$$J_R = J_G \cdot J_{1+Z/V} \cdot J_{G/Z} \cdot J_{V/G} = J_G \cdot J_{1+Z/V} \cdot J_{G/Z} \cdot J_{1+Z/G} \tag{2}$$

where the indices J_R, J_K, J_E and others are indexes of the change of certain rates as the ratios of the corresponding rates to the basic. Rates in the last formula are the following: V - rate of the total product of the process; Z - rate of product of the process as a cost (loss); $G = (V - Z)$ – rate of the product of the process as a benefit; $K = G + Z \cdot G/V$ is rate of scale products of the process; $E = V/Z$ – rate of the effectiveness of the process as the ratio of the rates V and Z; $R = K \cdot E = K \cdot V/Z = G(1 + V/Z)$ is rate of the process efficiency [3,4,26,28,30,31].

An information base for calculating of rates for efficiency components in the case, for example, enterprises are their annual financial statements. Rates for calculations should be taken in monetary dimension in actual prices per worker. In other cases certain information base is used with a glance of the fact, that the practical application of the proposed approaches to the study of the particular process on the basis of modeling depends from the specifics of this process and needs special consideration. This is related to the peculiarities of measuring process products [31]. It is known that the energy of the system uniquely depends on the parameters characterizing its state. In any system (for example, social, economic, pedagogical, organizational, environmental, etc.) its elements are combined with various types of ties. Termination or creation of connections requires the consumption of the corresponding energy, the value of which is determined by the type of connections. Some systems require more communication energy, while others require less power [30].

In accordance with our hypothesis of the efficiency reaction existence there is a reaction of system to the appropriate type of communication (social, economic, environmental and organizational, etc.). The estimation of this reaction, which correlates with the expenditure of energy in relation to the process functioning system, requires of simultaneous consideration of both the quantitative component of a certain rate and its qualitative component. Energy is transmitted to the elements of the system and causes its movement (change of states of the system, its structure, behavior, etc.) [30]. For example, the quantitative component of the rate $K = G + Z \cdot G/V = G \cdot (1 + Z/V)$ of the scale product is rate G, and the qualitative component is rate $(1 + Z/V)$; the quantitative component of the rate $E = V/Z = G/Z \cdot V/G$ of the effectiveness of the process is rate G/Z, and the qualitative component is rate V/G; the rate K is a quantitative component of the rate R of the efficiency of the process, and rate E is a qualitative component. The product of the values of quantitative and qualitative components of a particular rate is equal to the value of this rate. We use stated above for improving and disclosing of a methodology for measuring of the OIR of the process of functioning system on the basis of author's efficiency rates for the purpose of management by this process (together with other rates) on examples of processes related to wages of hired worker of the types of economic activity "agriculture, forestry and fishing" (type of activity 1) and "mining and quarrying" (type of activity 2) of Poland and Ukraine in examined year, and also for systems related to the theory of strings (in the sense of the authoress of work [24]). In the introduction we emphasized that energy is a generalized characteristic of the motion matter. It is known that the energy is produced, transmitted, and its quantity is measured. Any process of a certain type and hierarchical level, developing in space and time, acts as a form of implementation in the spatial and temporal arrangement of energy, specific for this type and level. The consequence (result) of the action of the energy of the process functioning system is its products; therefore it is possible to measure the OIR of the specified process for its products. This can be done, for example, based on author's rates for efficiency components [29,30].

K. Yahelska, considering the essence and structure of economic energy, naturally stops in sphere of the effectiveness of its transformation on some issues related with OIR [24], but, as we noted earlier, does not provide the algorithm for its calculation, making mistakes in the calculations and conclusions, believing that she researches OIR, rather than the quantitative component of efficiency. This authoress, remembered on page 100 of this work the article by Veynik A.I., who argued that phenomena at various quantitative levels of the cosmos (universe) should have certain similar features should obey by certain general laws [23], concluded that, according to the paradigm of this scientist, all the sciences have to come to a unified theory. She also gives an example that in physics theories are united already into the general theory of strings. Yagelskaya Ye.Yu., in article [24] presents the structure of economic energy as following:

$$E = mV^2, \tag{3}$$

where E is the economic energy, m is the mass of the economic system; and V^2 is acceleration of economic processes (in the latter case K. Yahelska admitted inaccuracy, because V^2 is not acceleration, but the square of speed).

Therefore, considering in the article [30] the practice of measuring the energy of products of the process functioning system on the basis of the rates efficiency components, we clarified that V is the system speed, and suggested in the calculations to use formula (4), which is essentially analogous to formula (3):

$$En_D = D \cdot S^2, \tag{4}$$

where En_D, in this case, is the rate of the description energy of a particular product of the process functioning of system: effectiveness, scale (scale product), efficiency, costs, etc.; D is the rate for the characteristic of the product of the process, S is rate of the speed of change for this product under the influence of certain factors (in particular, the factor of time, since the products are taken over a specific period of time). Using the fact that the physical content of a partial derivative du/dx (or du/dy) is description of the value of the speed for change of a function $U = f(x, y)$ in a point (x, y) when only x (or y) is changed in the direction of the axis Ox (or Oy), we in accordance with (4) obtained formulas for calculating the energies of products of the process functioning system based on the rates of efficiency components [30].

At a later date we emphasized: the values of the rates of the energy of the total product, of the costs (losses) and of the pure product of the process are equal to the value of the rates of these products [30]). The latter result is of particular importance, because it indicates that investigating any process with use of the rates of efficiency components is investigating of the process in the energy aspect.

We considered the practical use of derived results of the energy measuring of the efficiency components on the examples of the economic and pedagogical systems [30]. For example, in relation to the economic object of the study we chose the process of capital investment in the protection and rehabilitation of Ukraine's soil, underground and surface waters in 2011–2015, and in relation to the pedagogical object study we chose the process of students teaching. In the paper [29] we proposed to use the formula (5) to calculate output-input ratio η of the process in the form of the ratio of the rate G of the product as the utility of the process to the rate V of the total product of the process (since, as is known, the output-input ratio is defined as the ratio of the useful used energy to the total amount of energy received by the system):

$$\eta = G/V. \tag{5}$$

Using the theory of boundaries, in the stated above paper we received 8 formulas [29], which in the extreme cases characterize the fundamental connection between the OIR and the rates of the efficiency components of processes with point of view of their change (increase or decrease).

In the same article we considered the practical use of the obtained results of measuring of the OIR of the process on the examples of economic and educational systems, without considering the system in the aspect of string theory. As an example of the economic object of the study in the same article [29], we chose the process of capital investment in the protection and rehabilitation of Ukraine's soil, underground and surface waters in 2011–2015, and as an example of a non-economic research object, namely, pedagogical are the process of students teaching. The peculiarity of this object consist in the measurement of the valuation rates V, Z, G, K, R in the ball-hours; but not the hryvnias.

5 Experiment and Results

In the proposed article for example in calculating of the OIR of research objects we select initially the types of economic activity "agriculture, forestry and fishing" (type of activity 1) and "mining and quarrying" (type of activity 2) with point of view of processes related to wages for hired workers; then we shall consider the processes, which Yagelskaya Ye. Yu. [24] connects with theory of strings. We shall used the results of our work [26], where on the basis of appropriate modeling approaches to comparative characteristic of force of the employees compensation by types of economic activity at intergovernmental aspect (on the example of designated types of economic activity of Poland and Ukraine) are examined with a glance of the parity of purchasing power (PPP). It should be noted that the use of some "outdated" data (which relates to 2012) in the following Tables 1, 2, 3, 4 and 5 involves with the need to apply the results, which we obtained in last specified article [26], for the purpose of volume lessening of the article, that is presented, at the expense already obtained by us formerly of results, which can be used to obtaining of new research results. The Table 1 marks the macroeconomic rates, related to the process of production of gross domestic product in Poland and Ukraine during examined year (in actual prices, milliards of national currencies, and also milliards of dollars by PPP), and also value of PPP.

Table 1. Macroeconomic rates, related to the process of production of gross domestic product of Poland and Ukraine in examined year****

Rates	Country	
	Poland	Ukraine
1. Gross domestic product:		
a) (milliards of national currencies)*	1615.894	1459.1
b) (milliards of dollars)**	872	387
2. PPP: (date of line 1a)/(date of line 1b)***; national currency/dollar	1.853089	3.770284

Source: Table data is a copy of the data in the table 1 of article [26].

The Table 2 shows macroeconomic rates, related to the types of activity "agriculture, forestry and fishing" (type of activity 1), and "mining and quarrying" (type of activity 2) in Poland and Ukraine during the same year Using the data in Table 2, we found rates of these types of activities in Poland and Ukraine during the examined year, related to compensation of the employees (these rates are marked in a Table 3).

Based on the Table 3 data, we are calculated the rates of efficiency constituents, related with the process of formation of gross profit, mixed income indicated of activity types of Ukraine in examined year in comparing with the corresponding rates of this process of Poland with a glance of PPP (it is marked in a Table 4).

Data of Table 4 shows that in our case the rating assessments of the OIR of the process are the same as the rating evaluations of effective nesses, but it proves that they are not always the same as rating assessments of scale product and of efficiency process (the latter concerns, for example, a scale product of a type of activity 2 in Table 4, and also follows from Table 2 of the article [29].

The rating assessment coincidence of the OIR of the process and rating evaluation of effectiveness (optional coincidence of rating assessments of the OIR the process and, accordingly, rating assessments of scale product and of efficiency) can be explained as follows. Determined the functional relationship between the rates of effectiveness E, the scale product K, the efficiency of R and the rate of output-input ratio η, we have [29]:

$$E = V/Z = V/(V - G) = 1/(1 - G/V) = 1/(1 - \eta), \tag{6}$$

$$\eta = G/V = (V - Z)/V = 1 - Z/V = 1 - 1/(V/Z) = 1 - 1/E, \tag{7}$$

Table 2. Macroeconomic rates, related to the types of economic activity 1 and 2 of Poland and Ukraine in examined year***

"Agriculture, forestry and fishing" (type of activity 1)

Output V_0				Compensation of the employees Z_1			
Poland		Ukraine		Poland		Ukraine	
millions of national currencies*	millions of dollars**	millions of national currencies*	millions of dollars**	millions of national currencies*	millions of dollars**	millions of national currencies*	millions of dollars**
132602	71557.275	269983	71608.133	8362	4512.465	26739	7092.039

Gross profit, mixed income G_0				Number of employees	
Poland		Ukraine		Poland	Ukraine
millions of national currencies*	millions of dollars**	millions of national currencies*	millions of dollars**	millions of persons*	millions of persons*
46806	25258.366	86330	22897.479	0.1429	0.7418

"Mining and quarrying" (type of activity 2)

Output V_0				Compensation of the employees Z_1			
Poland		Ukraine		Poland		Ukraine	
millions of national currencies*	millions of dollars**	millions of national currencies*	millions of dollars**	millions of national currencies*	millions of dollars**	millions of national currencies*	millions of dollars**
62430	33689.693	153036	40590.046	16162	8721.653	44739	11866.215

Gross profit, mixed income G_0				Number of employees	
Poland		Ukraine		Poland	Ukraine
millions of national currencies*	millions of dollars**	millions of national currencies*	millions of dollars**	millions of persons*	millions of persons*
15260	8234.899	54462	14445.066	0.1735	0.4380

*** Source: Table data is a copy of the data in the table 2 of article [26].

$$K = G + Z \cdot G/V = G + (V - G) \cdot G/V = G + G - G/V = G(2 - \eta), \quad (8)$$

$$R = K \cdot E = G(2 - \eta)/(1 - \eta). \quad (9)$$

Since the rate E is the function of only one variable η (formula (6)), and the rates K and R are the functions of two variables G and η (formula (8) and (9) respectively), this explains the stated above. In addition to formulas (6) and (7) it follows that really OIR characterizes the effectiveness of the process (OIR of the process and effectiveness are equivalent with the point of view of the process characteristics. Formulas (8) and (9) indicate that the OIR of the process partly influences on the levels of its scale product and efficiency [29].

Data of Table 4 it follows, that the levels of scale product, effectiveness, efficiency, and OIR of the of the economic activity type 1, related to the compensation of the Ukrainian employees, in examined year are lower in comparing with the corresponding rates of Poland, but the levels of effectiveness, efficiency, and OIR of the type 2 of economic activity are higher, and the level of magnitude is lower.

For SEE-management by the process, that considers, we will use a methodology based on the components of the efficiency proposed by us in [4]. To this effect we compile a matrix (map) of the SEE management based on the results of the SEE-analysis of compensation of the employees as a constituent of costs of the process formation of gross profit, mixed income of types of activity 1 and 2 Ukraine in examined year in comparing with the corresponding rates of this process of Poland (Table 5). The data in this table is based on the data in Table 4 as the difference between the values of indexes of the respective rates.

Table 3. Rates the types of economic activity 1 and 2 of Poland and Ukraine in examined year, related to compensation of the employees ***

Rates	Poland		Ukraine	
	Type 1	Type 2	Type 1	Type 2
The effectiveness of process of formation of gross profit, mixed income $E = V_0/(V_0 - G_0)$**	1.5455499	1.3235107	1.4700713	1.5524986
Part of output, which corresponds to the compensation of the employees $V_1 E \cdot Z_1$ (millions of dollars)**	6974.240	11543.201	10425.803	18422.282
Number of employees (millions of persons)*	0.1429	0.1735	0.7418	0.4380
Compensation of the employees Z_1 (millions of dollars)*	4512.465	8721.653	7092.039	11866.215

***Source: Table data is a copy of the data in the table 3 of article [26].

It should be noted that SEE-management is based naturally on SEE-analysis. Under the author's SEE-analysis of the processes of functioning complex systems, we understand the analysis of the force of these processes with the help of author's models and corresponding rates of the scale product, effectiveness, and the efficiency of the subprocesses of these processes. The letter triplex SEE in the title SEE-analysis is composed by the authors from the first letters of the lexeme scale, effectiveness, efficiency; the order of letters is explained by the order of calculation of the following rates: scale product is K, effectiveness is E, efficiency is R. Costs on functioning of the process, its net and total products; scale product, effectiveness and efficiency etc. are products of the process [4]. The factors that serve to form the products of this process are called F-impulses of the process ($F = R, K, E, \eta, G, 1{+}Z/V, G/Z, V/G, V, Z$); these impulses can be both positive and negative [28]. The efficiency components rates of the functioning of systems, the method of using F-impulses as indicators of the direction of the consequences of subprocesses of the process of functioning of systems, F-impulses as elements of the author's SEE-analysis of the processes of functioning of complex systems are a constituent parts of the mechanism of the SEE-management by complex systems [4].

The values of the rates ΔJ_R, ΔJ_K, ΔJ_E, ΔJ_η, ΔJ_G, $\Delta J_{1+Z/V}$, $\Delta J_{G/Z}$, $\Delta J_{V/G}$, ΔJ_V, ΔJ_Z of Table 5 describes the degree of influence of the appropriate factors on the formation of products of the investigated process. These factors of the studied process respectively are: the efficiency of the process (with the rate of R); scale of the process (with rate of K); effectiveness (with the rate of E); net product (with the rate of G); qualitative component of a scale product (with the rate of $(1 + Z/V)$); quantitative component of effectiveness (with the rate of G/Z); qualitative component of effectiveness (with the rate of V/G); total product (with the rate of V); costs (with the rate of Z).

Matrix of the SEE-management on the results of SEE-analysis of the process (Table 5) clearly illustrates the F impulses and their impact on the investigated process, SEE-reserves and SEE-risks, recommended managerial SEE-actions [4].As examples of research objects, we chose the types of economic activity "agriculture, forestry and fishing" (type of activity 1) and "mining and quarrying" (type of activity 2) with point of view of processes related to wages for hired workers. Similarly you can explore other economic processes. In relation to other economic processes, we emphasize that in [3] we considered the sufficient number of performance of the processes of functioning of complex economic systems. They can be used for practical verification for the hypothesis about the possibility of using the author's rates of the components of efficiency for measuring of the OIR of economic processes.

6 Discussion

We pass on to the application of invented by us of method of calculating of the OIR for processes that are related to the theory of strings, which Yagelskaya Ye. Yu. considered [24].

Cited on Yegerev [5], the authoress of the article [24] for a minimum unit of economic energy takes a string whose vibrations in accord with the theory of strings, which she mentioned before, set the properties of matter. K. Yahelska notes, that the more frequency of vibration correspond to the more energy accumulated in this vibration, and according to the concept of equivalence for mass and energy ($E = m\dot{c}^2$), it is greater the mass of the particle, in the role of which manifests itself a string that vibrates [24].

K. Yahelska believes that the mass of economic systems can be represented by a set of masses of factors (elements) that form the economic system, that is, for each element i am inherent its own energy. Thus, economic energy is a defined kind of sub-energy matrix. Each string (S) can take the value of f (force) or l (loss), that is, at a certain time have force or characterized by dissipation, by predominance of losses [24]. The effectiveness of the energy conversion is characterized by the OIR of the system, which, as was proved in thermodynamics, is always less than one hundred percent, that is, the conversion of energy without loss is impossible in principle [24]. Consequently, each string can take the value of f or l; this means: that each "sub-energy" will depend on the fluctuation of its string, which actually determines the nature of "sub-energy", that is [24]:

$$s_i = \frac{f_i}{l_i} \tag{10}$$

Table 4. Rates description of types of economic activity 1 and 2, related to the of compensation of the employees of Ukraine in examined year, in comparing with the corresponding rates of Poland*

Type 1

Country	Description of total product of process		Description of product of process as costs		Description of net product of process	
	V	J_V	Z	J_Z	G	J_G
Poland	48805.04 (1)	1	31577.78 (1)	1	17227.26 (1)	1
Ukraine	14054.74 (2)	0.2879771	9560.581 (2)	0.3027629	4494.154 (2)	0.2608746

Country	Description of qualitative component of scale product of process		Description of scale product of process		Description of effectiveness of process	
	$1 + Z/V$	$J_{1+Z/V}$	K	J_K	E	J_E
Poland	1.6470188 (2)	1	28373.62 (1)	1	1.54555 (1)	1
Ukraine	1.6802391 (1)	1.02017	7551.254 (2)	0.2661364	1.470071 (2)	0.9511637

Country	Description of efficiency of process		Description of quantitative component of effectiveness	
	R	J_R	G/Z	$J_{G/Z}$
Poland	43852.85(1)	1	0.545550(1)	1
Ukraine	11100.88(2)	0.2531393	0.470071(2)	0.8616464

Country	Description of qualitative component of effectiveness		Description of output-input ratio (OIR) of process	
	V/G	$J_{V/G}$	$\eta = G/V$	J_η
Poland	2.833012(2)	1	0.352981(1)	1
Ukraine	3.127338(1)	1.103891	0.319761(2)	0.905887

Type 2

Country	Description of total product of process		Description of product of process as costs		Description of net product of process	
	V	J_V	Z	J_Z	G	J_G
Poland	66531.42 (1)	1	50268.89 (1)	1	16262.53 (1)	1
Ukraine	42060 (2)	0.6321825	27091.81 (2)	0.5389379	14968.19 (2)	0.9204098

Country	Description of qualitative component of scale product of process		Description of scale product of process		Description of effectiveness of process	
	$1 + Z/V$	$J_{1+Z/V}$	K	J_K	E	J_E
Poland	1.7555662 (1)	1	28549.94 (1)	1	1.323511 (2)	1
Ukraine	1.644123 (2)	0.93652	24609.54 (2)	0.8619823	1.552499 (1)	1.173015

Country	Description of efficiency of process		Description of quantitative component of effectiveness	
	R	J_R	G/Z	$J_{G/Z}$
Poland	37786.16(2)	1	0.323511(2)	1
Ukraine	38206.29(1)	1.011119	0.552499(1)	1.707821

Country	Description of qualitative component of effectiveness		Description of output-input ratio (OIR) of process	
	V/G	$J_{V/G}$	$\eta = G/V$	J_η
Poland	4.091087(1)	1	0.244434(2)	1
Ukraine	2.809959(2)	0.6868491	0.355877(1)	1.455923

*Money rates are presented on the average for a year in dollars on one of the employee; indexes are the coefficients of rates of types of activity 1 and 2 of Ukraine to the corresponding rates of Poland; V is part of producing, that answers of compensation of the employees; Z is compensation of the employees; $G = (V - Z)$ is part of gross profit, that answers of compensation of the employees (rates are expected with a glance PPP).

Source: Table data is derived from the data of Table 4 of work [26] by adding columns of values of rates G/Z, V/G, $\eta = G/N$ and J_η, expected by authors. In parentheses in Table 4 we also provided ratings for the respective rates $(J_K = J_G J_{1+D_Z}, J_E = J_{G/Z} J_{V/G}, J_R = J_K J_E)$.

Table 5. Matrix (map) SEE-management on the results of SEE-analysis of compensation of the employees as a constituent of costs of the process formation of gross profit, mixed income of types of activity 1 and 2 in Ukraine during examined year, in comparing with the corresponding rates of this process of Poland*

Parameters of F-impulses ($F = R, K, E, \eta, G, 1 + Z/V, G/Z, V/G, V, Z$) as increment of indices of corresponding process rates*:									
efficiency ΔJ_R	scale product ΔJ_K	effectiveness ΔJ_E	OIR ΔJ_η	net product ΔJ_G	qualitative component of scale product $\Delta J_{1+Z/V}$	quantitative component of effectiveness $\Delta J_{G/Z}$	qualitative component of effectiveness $\Delta J_{V/G}$	total product ΔJ_V	product as costs ΔJ_Z
Type 1									
−74.69	−73.39	−4.88	−9.41	−3.91	+0.02	−13.84	+10.39	−71.20	−69.72
Influence of F-pulses on the process being investigated									
negative	negative	negative	negative	negative	positive	negative	positive	negative	negative
SEE-reserves (+) and SEE-risks (−)									
−	−	−	−	−	+	−	+	−	−
Recommended managerial SEE-actions									
Ensure growth	Ensure growth	Ensure growth	Ensure growth	Ensure growth	Provide the no less level	Ensure growth	Provide the no less level	Ensure growth	Ensure growth
Type 2									
+1.11	−13.80	+17.30	+45.59	−7.96	−0.06	+70.78	−31.32	−36.78	−46.11
Influence of F-pulses on the process being investigated									
positive	negative	positive	positive	negative	negative	positive	negative	negative	negative
Recommended managerial SEE-actions									
+	−	+	+	−	−	+	−	−	−
Provide the no less Level	Ensure growth	Provide the no less level	Provide the no less level	Ensure growth	Ensure growth	Provide the no less level	Ensure growth	Ensure growth	Ensure growth

*Growth of the indexes of efficiency component forthe employees compensation process as a constituent of costs for the gross profit formation process, mixed income of types of activity 1 and 2 are considered in comparison: defined rates of the process in Ukraine compared with appropriate rates of this process in Poland in a percent. Source: calculated by authors.

In the future, the authoress singles out several basic elements that, in her view, form the mass of the economic system: the technological reserves, accumulated experience, intellectual capital, goodwill, money supply, resource capital, spiritual and cultural component, political stability. Thus, economic energy can be represented by an array of "sub-energies":

$$E = (e_1 e_2 e_3 \cdots e_n) \cdot V^2 \tag{11}$$

She assigns to each f_i and l_i the values that, in her opinion, most correspond to the present state of the element of the mass of the economic system (the national economy of Ukraine), calling it expert, and proceeding from the formula (11) and operating by velocity V, the value of which accepts the maximum – equal to 1. As we noted in the problem statement, K. Yahelska, in this case presented two conditional examples (we saved the terminology of K. Yahelska, although this terminology is false concerning OIR):

1. an example of the calculation (determination) of the expert value of the current economic energy of the country for the case if the energy does not exceed 100% (OIR 100 pct., Table 6 [24]);
2. an example of calculating the value of the country's economic energy for the case if the energy exceeds 100% (OIR $\leq$ 100 pct., Table 7 [24]).

Without going into controversy about the correctness of guiding by K. Yahelska these examples, as those which are related to the theory of strings, we note the following. With our notation we have:

$$E = \frac{V}{Z} = \frac{G}{Z} \cdot \frac{V}{G} = \frac{G}{Z} \cdot \frac{G+Z}{G} = \frac{G}{Z} \cdot \left(1 + \frac{Z}{G}\right), \tag{12}$$

$$\eta = \frac{G}{V} = \frac{G}{Z} \cdot \frac{Z}{V} = \frac{G}{Z} \cdot \frac{Z}{Z+G} = \frac{G}{Z} \cdot \frac{1}{1+G/Z}, \tag{13}$$

where the first multiplier in formulas (12), (13) is a quantitative effectiveness component (with a rate E or η, respectively) of the subprocesses of the process functioning systems as of a qualitative force component, and the second is a qualitative effectiveness component. Formula (12) characterizes effectiveness mainly with the point of view of the cost of the process, and formula (13) – with the point of view of benefit (utility).

Table 6. Finding the expert value of the current economic energy of the country for the case if the energy does not exceed 100 pct

Rate	Technological reserves	Accumulated experience	Intellectual capital	Goodwill	Money supply	Resource capital	Spiritual and cultural component	Political stability
F	0.3	0.4	0.7	0.5	0.4	0.8	0.4	0.3
L	0.7	0.6	0.3	0.5	0.6	0.2	0.6	0.7
E	0.43	0.67	2.33	1.00	0.67	4.00	0.67	0.43

Source: [24], Table 1

In the notation of K. Yahelska formula (12), (13) will have the following form (in our understanding f is the rate of utility (benefit), and l is the rate of losses (costs)):

$$E = \frac{f+l}{l} = \frac{f}{l} \cdot \frac{f+l}{l} = \frac{f}{l}\left(1 + \frac{l}{f}\right), \tag{14}$$

$$\eta = \frac{f}{f+l} = \frac{f}{l} \cdot \frac{l}{l+f} = \frac{f}{l} \cdot \frac{1}{1+f/l}, \tag{15}$$

It is clear that for each of the mentioned by K. Yahelska main elements that form, in her opinion, the mass of the economic system, corresponds a certain economic process, for which the above formulas can be used. Since in Table 6 and in Table 7 value $f + l = 1$, then from formulas (14), (15) it follows that the value $E = 1/l$, and the value of output-input ratio $\eta = f$. Consequently, the values of the OIR of the corresponding processes are contained in the first lines of these tables and the values f/l of the quantitative components of effectiveness – in the third.

Table 7. Finding the value of the economic energy of the country for the case if the energy exceeds 100 pct

Rate	Technological reserves	Accumulated experience	Intellectual capital	Goodwill	Money supply	Resource capital	Spiritual and cultural component	Political stability
F	0.7	0.8	0.9	0.7	0.8	0.9	0.7	0.7
L	0.3	0.2	0.1	0.3	0.2	0.1	0.3	0.3
E	2.33	4.00	9.00	2.33	4.00	9.00	2.33	2.33

Source: [24], Table 2.

The specified authoress affirms that in the second case, if $e > 100$ pct (Table 7), this indicates the future transformation of the system, about the coming of crisis [24]. According to our calculations, inversely, in the second case value of the OIR is greater than the corresponding value of the OIR in the first case, and therefore the corresponding processes in the second case are more effective in comparison with the processes in the first case (the values of e are not the values of the OIR of the systems, but are the values of the quantitative components of effectiveness these systems). Consequently, K. Yahelska mistakenly believes that she is talking about the OIR of the country, and not the quantitative component of effectiveness. She also made false conclusions, indicating the death of the system [24], about the future transformation of the system, about the crisis in the second case, etc. [24]. In this regard, some other judgments by K. Yahelska are incorrect [24]. These mistakes and incorrect judgments K. Yahelska transferred to her doctoral dissertation [25]. In our previously published works we proves that the rate of the quantitative component of the effectiveness can exceed 100 pct., while the system is stable (then $G > Z$; for example [3,29]).

7 Conclusions

Methodological approaches to the theory and practice of measuring energy products and of the OIR of subprocesses of the processes of functioning of systems of various types and hierarchical levels require the appropriate algorithms implementation based on modeling. The measuring approaches of the indicated energies and of the OIR of processes based on the use of certain authors' models are being suggested. These models are based on the use of author's rates of efficiency components for subprocesses of processes of functioning of systems. In particular, the following is emphasized: so far as (as we proved earlier) the values of energy rates for general products, products as costs (losses) and pure products (products in the form of profits, benefits) of the subprocesses of the processes functioning systems are equal to the values of rates the mentioned products, the study of certain processes based on the rates of these products automatically means their scientific consideration in the energy aspect. It was found that the OIR of the process and its effectiveness in the classical sense are equivalent with viewpoint of the characteristic process (as a qualitative component of its efficiency), but the first of these rates (OIR) describes the qualitative component of the efficiency process mainly from the point of view of benefits (profit), and the second states from the standpoint of losses (costs). It is distinguished quantitative and qualitative components in the output-input ratio of the process; it allows to characterize the effectiveness of the process (from the point of view OIR) with quantitative aspect as well as and qualitative aspect accordingly. It is substantiated that the rate of the quantitative component of OIR of the process is the same as the rate of the quantitative component effectiveness in the classical sense, but their qualitative components are distinguished On the examples of certain processeswe showed the practical realization of the methods measuring of OIR for the processes of functioning systems, invented by the authors and its application (together with methods of measuring other author's rates) to manage by the force of the functioning such systems (precisely that is the scientific novelty of the proposed research results). By the proposed method, we amplify and diversify existing aspects in understanding the factors necessary for the study of concepts and processes based on the energy approach. Further research is planned to be directed to SEE-analysis of power efficient systems.

References

1. Biryukov, V.: Time as an economic space for the development of an economic system. Thesis ... doctor of economic sciences: 08.00.01, St. Petersburg (2000)
2. Borisov, A.: Large Economic Dictionary. Book World, Moscow (2007)
3. Buriennikova, N., Yarmolenko, V.: Efficiency of Functioning of Complex Economic Systems of Agrarian Direction. VNAU, Vinnitsa (2017)
4. Buryennikova, N., Yarmolenko, V.: See-controlling on the basis of components of efficiency as a way to improve the force of complex systems functioning: essence. Methodol. Bus. Inform **1**, 145–152 (2016). http://www.irbis-nbuv.gov.ua/cgi-bin/irbis_nbuv/cgiirbis_64.exe?I21DBN=LINK&P21DBN=UJRN&Z21ID=&S21REF=10&S21CNR=20&S21STN=1&S21FMT=ASP_meta&C21COM=S&2_S21P03=FILA=&2_S21STR=binf_2016_1_24
5. Egerev, S.: String: Physical Encyclopedia, vol. 5. Soviet Encyclopedia, Moscow (1998)
6. Fedulova, I.: Management of Organizations. Swan, Kyiv (2004)
7. Kendyuhov, A., Yagelskaya, E.: Economic force and economic energy as new indicators of economic efficiency. Collection of Scientific Memoirs of State Economic and Technologic University of Transport **30**, 213–221 (2014)
8. Klymash, N.: Scientific and theoretical aspects of the essence of concepts of "efficiency" and "effectiveness". Sci. Mem. NUKhT **28**, 124–125 (2009)
9. Kogan, I.: Systematization and classifcation of definitions and additions to the concept of "energy". Autom. IT Energy Sector **2–3**, 56–63 (2009)
10. Kolotylo, D.: Ecology and Economics. KNEU, Kyiv (1999)
11. Larush, I.: Physical Economics. Schiller Institute of Science and Culture. Scientific book, Moscow (1997)

12. Liamets, V., Teviashev, A.: System Analysis. Introductory Course. KhNURE, Kharkiv (2004)
13. Melnikl, G.: Fundamentals of Development. University Book, Sumy (2003)
14. Mochernyi, S.: Economic Theory. Academy (Alma-mater), Kyiv (2003)
15. Naiko, D., Shevchuk, O.: Physical economics and its problems. Collection of Scientific Memoirs of VNAU **1**(48), 265–272 (2011)
16. Oleksiuk, O.: Economics of Efficiency. KNEU, Kyiv (2008)
17. Podolynskyi, S.: Selected Works. KNEU, Kyiv (2000)
18. Prangishvili, I.: Entropy and Other Systemic laws: Issues of Managing Complex systems. Nauka, Moscow (2003)
19. Rudenko, M.: The Energy of Progress. Dzhura, Ternopil (2005)
20. Sheremet, A., Sayfulin, R.: Finance of Enterprises. INFRA, Moscow (1997)
21. Teslenok, I., Mykhailova, O., Bohachenko, O.: Modern approaches to determining the effectiveness of enterprise management. Econ. Herald Donbas **1**(27), 208–212 (2012)
22. Tishchenko, A., Kizim, N., Dogadaylo, Ya.V.: Economic performance of the enterprise. ID "INZhEK", Kharkiv (2003)
23. Veynik, A.: Thermodynamics of real processes. Science and Engineering, Minsk (1991)
24. Yagelskaya, Y.: The essence and structure of economic energy. Prob. Econ. Manag. **8**(24), 98–111 (2013)
25. Yahelska, K.: Theoretical and methodological principles of preventive national development. thesis ... doctor of economic sciences: 08.00.03, Pokrovsk (2016)
26. Yarmolenko, V., Burennikova, N.: Comparative characteristic of force of compensation of the employees by type of economic activity: intergovernmental aspect. SWorld J. Econ. **J21517**(9), 66–73 (2015). http://www.sworldjournal.com/e-journal/j21517.pdf. Downloaded from SWorld. Terms of Use http://www.sworld.education/index.php/ru/e-journal/about-journal/terms-of-use
27. Yarmolenko, V., Burennikova, N.: Measuring the efficiency of the processes of functioning of a component system based on the models of the efficiency constituents: the energy aspect. Bus. Inform **12**, 102–110 (2019). https://doi.org/10.32983/2222-4459-2019-12-102-110
28. Yarmolenko, V., Polishchuk, N.: Application of f-impulses as indicators of directionality of results of processes of functioning of complicated systems in author's see-analysis of the force of processes on the basis of components of the efficiency. SWorld J. Econ. **J11517**, 110–119 (2015). http://www.sworldjournal.com/e-journal/j11517.pdf. Downloaded from SWorld, Terms of Use http://www.sworld.education/index.php/ru/e-journal/about-journal/terms-of-use
29. Yarmolenko, V., Buryennikova, N.: The practice of measuring the efficiency coefficient of the process of system operation based on indicators of efficiency components. Prob. Econ. **3**(37), 260–266 (2018). http://nbuv.gov.ua/UJRN/Pekon_2018_3_35
30. Yarmolenko, V., Buryennikova, N.: The practice of measuring the energies of products of the process of system functioning on the basis of performance components. Bus. Inform **7**, 115–121 (2018). www.sworldjournal.com/e-journal/j11517.pdf
31. Yarmolenko, V., Polishchuk, N.: Components of the efficiency of the functioning of complex systems as objects of modeling. Herald Cherkassy Univ. **33**(246), 86–93 (2012)
32. Zahorna, T.: Economic diagnostics. Centre of Educational literature, Kyiv (2007)

An Algorithm for Constructing Additive and Multiplicative Voronoi Diagrams Under Uncertainty

Elena Kiseleva, Olga Prytomanova(✉), and Vadim Padalko

Oles Honchar Dnipro National University, Dnipro, Ukraine
kiseleva47@ukr.net, olga.prytomanova@gmail.com

Abstract. An algorithm is proposed to construct additive and multiplicative Voronoi diagrams under uncertainty in the initial data of a problem, which may be inaccurate, unreliable or fuzzy. To remove the uncertainty in the initial data having a fuzzy-multiple nature, a method of neurolinguistic identification of complex nonlinear dependences is used. The developed algorithm is based on a synthesis of the theory of optimal set partitioning and neuro-fuzzy technologies. Methods of optimal set partitioning are a universal mathematical apparatus for constructing various types of Voronoi diagrams. This apparatus uses an approach based on the statement of continuous problems of optimal partitioning of sets from n-dimensional Euclidian space into subsets with a partition quality criterion that provides the corresponding types of Voronoi diagrams. The universality of such approach to the construction of Voronoi diagrams makes it possible to generalize the methods of solving the optimal set partitioning problems for the case of fuzzy setting of the initial problem parameters. Along with the problem of constructing a Voronoi diagram with fuzzy parameters, this approach also allows us to state and solve a problem of finding optimal, in a sense, coordinates of the diagram's generator points. The developed algorithm is software implemented. The work of the algorithm is demonstrated through examples of constructing additively weighted and multiplicatively weighted Voronoi diagrams with fuzzy parameters and optimal placement of generator points in a bounded set of n-dimensional Euclidian space.

Keywords: Additive Voronoi diagram · Multiplicative Voronoi diagram · Generator points · Optimal set partitioning · Nondifferentiable optimization · Fuzzy parameters

1 Introduction and Literature Review

A Voronoi diagram of a given finite set M of the plane (space) points called generator points is a mathematical object that is the plane (space) partitioning such that each domain (a Voronoi cell) of this partitioning forms a set of elements

S. Babichev et al. (Eds.): ISDMCI 2020, AISC 1246, pp. 714–727, 2021.
https://doi.org/10.1007/978-3-030-54215-3_46

located closer to one of the points of set M than to any other point of this set [11]. Voronoi diagrams in two- and three-dimensional spaces are used in varios fields of applied sciences: crystallography, physics, astronomy, chemistry, microbiology, computer graphics, and ophthalmology; they also are used to solve artificial intelligent problems and to recognize images. Substantial reviews of applications of Voronoi diagrams in various fields are given in [1,2,10].

Note that many of the known algorithms for constructing Voronoi diagrams in two- and three-dimensional spaces are quite complex. As for the Voronoi diagram for the case of the arbitrary dimensionality space, its construction is associated with considerable algorithmic challenges. Indeed, it is known [11] that for a given number N of the generator points, the number of elements needed to describe the Voronoi diagram, grows exponentially depending on the space dimensionality.

In [6–8], algorithms are proposed for constructing a standard Voronoi diagram and its various generalizations. These algorithms have a number of significant advantages compared to the ones known in the scientific literature. They are based on the application of the theory of optimal set partitioning (OSP) in n-dimensional Euclidian space $E_n (n \geq 2)$ into non-intersecting subsets [6]. Paper [8] contains a brief review of some generalizations of the standard Voronoi diagram, available in the scientific literature, in particular: additively weighted Voronoi diagram and multiplicatively weighted Voronoi diagram. In [8], some other generalizations of the standard Voronoi diagram are also mentioned, such as a Laguerre diagram (a power Voronoi diagram), approximated (fuzzy) Voronoi diagrams, Voronoi diagrams with the constraints on the power of generator points, dynamic Voronoi diagrams etc. In addition, [8] contains a table of correspondence between the concrete type of the Voronoi diagram and the mathematical model of the continuous problem of optimal set partitioning, as a result of which this diagram can be obtained. The development of algorithms for constructing Voronoi diagrams under uncertainty, which are characteristic for practical problems, is one of the promising trends in the development of the subject under study.

Several decades ago, the theory of probability was a unique tool to process uncertainty situations arising in problems of science, technology, and everyday life [14]. Note that the probability based on the principles of traditional binary logics is sufficient to solve the problems under uncertainty associated with randomness. However, situations of uncertainty in objectives, conditions, constraints, initial data, which may be inaccurate, unreliable, and fuzzy, are characteristic for many practical problems of constructing Voronoi diagrams. To remove or reduce uncertainties that have not a probabilistic, but a fuzzy-multiple nature, the apparatus of the theory of fuzzy sets, of fuzzy logics, of neural networks and their combinations (neuro-fuzzy systems and technologies) can be used [3,5,13,15–17].

The purpose of this paper is to develop an algorithm for constructing additive and multiplicative Voronoi diagrams under uncertainty, which is based on a synthesis of methods for the solution of continuous OSP problems in n-dimensional Euclidian space E_n [6] and neuro-fuzzy technologies [4,5,9,13].

2 Materials and Methods

Before formulating a mathematical statement of the problem under consideration, we give a definition of a standard (classical) Voronoi diagram [11].

A standard (classic) Voronoi diagram of a finite set $M = \{\tau_1, ..., \tau_N\} \subset E_n$ of the generator points $\tau_i = (\tau_i^{(1)}, \tau_i^{(2)}, ..., \tau_i^{(n)}), i = 1, 2, ..., N$ in n-dimensional Euclidian space $E_n (n \geq 2)$ is a set of Voronoi polyhedrons

$$Vor(\tau_i) = \{x \in \mathrm{E}_n : c(x, \tau_i) \leq c(x, \tau_j),\ j \neq i,\ j = 1, ..., N\}, i = 1, 2, ..., N \quad (1)$$

of the initial points $\tau_1, \tau_2, ..., \tau_N$, where $c(x, \tau_i)$ is a given metric. This is usually the standard Euclidian metric $c(x, \tau_i) = ||x - \tau_i||_2 = \sqrt{\sum_{j=1}^{n} (x^{(j)} - \tau_i^{(j)})^2}$.

In other words, a Voronoi diagram of a finite set $M = \{\tau_1, ..., \tau_N\}$ from $E_n (n \geq 2)$ is the set

$$Vor(M) = \bigcup_{\tau_i \in M} Vor(\tau_i), \quad (2)$$

where $mes(Vor(\tau_i) \cap Vor(\tau_j)) = 0,\ i, j = 1, 2, ..., N (i \neq j)$; here $mes(\cdot)$ is a Lebesgue measure. Partitioning of space E_n into Voronoi polyhedrons $Vor(\tau_i)$, $i = 1, 2, ..., N$ of the given set of points $M = \{\tau_1, ..., \tau_N\}$ is unique.

We represent a definition of a Voronoi diagram of a finite number of the generator points $\tau_1, \tau_2, ..., \tau_N$, *placed optimally* in a bounded set $\Omega \subset E_n$.

Assume that Ω is some given bounded set of E_n, $\tau_1, \tau_2, ..., \tau_N$ is a finite set of generator points in Ω. In cases when the location of the points $\tau_1, \tau_2, ..., \tau_N$ in Ω is unknown and they should be placed (selected) in Ω, it is possible to introduce another variant of the Voronoi diagram on the set $\Omega \subset E_n$, namely, a Voronoi diagram of a finite number of points placed optimally in a bounded set, which generalizes the standard Voronoi diagram.

By a *Voronoi diagram of a finite number of generator points* $\tau_1, \tau_2, ..., \tau_N$, *located optimally in a bounded set* $\Omega \subset E_n$, we will call a set of Voronoi polyhedrons

$$Vor(\tau_i) = \{x \in \Omega : c(x, \tau_i)/w_i + a_i \leq c(x, \tau_j)/w_j + a_j; i, j = 1, ..., N; j \neq i\}, \quad (3)$$

of the points $\tau_1, \tau_2, ..., \tau_N$, for which the total weighted distance from the points of set Ω to the corresponding generator points of Ω is the minimal one, i.e. a functional

$$J(\{\tau_1, ..., \tau_N\}) = \sum_{i=1}^{N} \int_{Vor(\tau_i)} (c(x,\ \tau_i)/w_i + a_i) dx \quad (4)$$

takes the minimal value. Here, $w_i > 0$, a_i, $i = 1, 2, ..., N$ are given numbers (weights).

Remark: While giving the values of the parameters $a_1, ..., a_N, w_1, ..., w_N$ and the type of the function $c(x, \tau_i)$ in the expression for functional (4), it is possible to obtain various variants of the Voronoi diagram with the optimal placement of the generator points: the additively weighted ones ($w_i = 1, a_i \neq 0, i = 1, 2, ..., N$), the multiplicatively weighted ones ($w_i > 1, a_i = 0, i = 1, 2, ..., N$), the additively-multiplicatively weighted ones ($w_i > 1, a_i \neq 0, i = 1, 2, ..., N$) etc.

Thus, according to the definition of the Voronoi diagram of a finite number of generator points $\tau_1, \tau_2, ..., \tau_N$, located optimally in a bounded set $\Omega \subset E_n$, its construction requires to form a set of Voronoi polyhedrons (3) and to find such coordinates of the corresponding generator points $\tau_1, \tau_2, ..., \tau_N$, which will minimize the functional (4). The formulated problem is the problem under uncertainty.

However, the construction of Voronoi diagram is complicated significantly under uncertainty, in particular, when a number of parameters in (4) may be inaccurate, fuzzy, or the mathematical description of some dependences in (4) is unknown. For instance, Voronoi diagrams with fuzzy parameters appear in case when the parameters $a_1, ..., a_N, w_1, ..., w_N$ are given as the linguistic variables which depend on external factors; moreover, the form of these dependencies is unknown in advance. To remove the fuzziness of the parameters $a_1, ..., a_N$, $w_1, ..., w_N$, we apply the method of neurolinguistic identification of complex nonlinear dependencies from [4,9]. Let us briefly describe its essence.

2.1 A Metod of Neurolinguistic Identification

To simplify the description of the neurolinguistic identification method to restore the values of fuzzy parameters $a_1, ..., a_N, w_1, ..., w_N$, we denote their restored values as y and consider a functional dependence of the output y on inputs $z_1, ..., z_q$ of an identification object in the form

$$y = y(z_1, ..., z_q), \tag{5}$$

where $z_1, ..., z_q$ are factors effecting on y. Assume for the identification problem that the domains of definition of the inputs $z_1, ..., z_q$, the range of changes of the output y for (5) as well as the expert-experimental information about dependence (5) in the form of a sample of data on the inputs and output of the identification object are known.

The problem of identification (restoration) of a complex nonlinear dependence of the form (5) is considered as constructing a model of the object according to the expert-experimental data on the relationship <inputs> - <output>; this problem is usually solved in two stages [9]:

- structural identification: formation of a fuzzy knowledge base about the object and construction on its basis of a fuzzy model of the object with several inputs and one output; this fuzzy model reproduces roughly the dependence of the output on the inputs, with the help of linguistic rules "if-then" being generated from the experimental data;

– parametric identification (setting): search for such parameters of a fuzzy model, which minimize deviations of the model values from the experimental ones.

As a result of applying the method of neurolinguistic identification, we obtain the exact (clear) value of the output variable, which is calculated according to the following formulas:

$$y = \frac{\sum_{k=1}^{L} d_k \cdot \mu *_{D_k}(y)}{\sum_{k=1}^{L} \mu *_{D_k}(y)}, \tag{6}$$

$$\mu *_{D_k}(y) = \begin{cases} \sum_{j=1}^{s_k} p *_j^k (z_1, z_2, ..., z_q), \ if \ \sum_{j=1}^{s_k} p *_j^k (z_1, z_2, ..., z_q) \leq 1, \\ 1 \ otherwise, \end{cases} \tag{7}$$

$$p *_j^k (z_1, z_2, ..., z_q) = v *_j^k \prod_{i=1}^{q} \mu *_{ij}^k (z_i), \tag{8}$$

$$\mu *_{ij}^k (z_i) = \frac{1}{1 + \left(\frac{z_i - t*_{ij}^k}{e*_{ij}^k}\right)^2}, i = 1, ..., q, j = 1, ..., s_k, k = 1, 2, \ldots, L, \tag{9}$$

where $\mu *_{D_k}(y)$ is a function of membership of the output variable y to a class D_k, $k = 1, 2, \ldots, L$; L is a number of classes (linguistic terms) of the output variable y; d_k is a center of the class D_k; $p *_j^k (z_1, z_2, ..., z_q)$ is fuzzy production rules obtained from the expert-experimental information about dependence (5); j is a number of the rule in the k-th class, $j = 1, 2, \ldots, s_k$; s_k is a number of rules in the k-th class; $v*_j^k$ is a weight of the j-th rule in the k-th output class; $\mu *_{ij}^k (z_i)$ is the bell-shaped function of membership of z_i to its linguistic term in the j-th rule of the k-th output class of the output variable y; $t*_{ij}^k$ is a maximum coordinate; $e*_{ij}^k$ is a concentration coefficient of this membership function.

Remark 1. The values $v*_j^k$ of the weights of the rules in (8) and $t*_{ij}^k$, $e*_{ij}^k$ of the parameters of membership function (9), which are marked with asterisk as the optimal ones are those obtained as a result of the parametric identification stage of the neurolinguistic identification method. For these value, the deviation of the experimental data from the model ones obtained after tuning fuzzy model of the object (5), reaches the minimum value. Values $\mu *_{D_k}(y)$, $p *_j^k (z_1, z_2, ..., z_q)$ and $\mu *_{ij}^k (z_i)$ in (6)–(9) are calculated under the optimal values $v*_j^k$, $t*_{ij}^k$, $e*_{ij}^k$.

Remark 2. To tune the fuzzy model, the Shor's r-algorithm is used [12].

Thus, in functional (4), each of the fuzzy parameters $a_1, ..., a_N$, $w_1, ..., w_N$, mentioned before as the output y, which depends on the inputs $z_1, ..., z_q$ in the neurolinguistic identification method, is calculated according to formulas (6)–(9). When the values of fuzzy parameters in (4) are restored with the help of neurolinguistic identification, we obtain their restored values before tuning $\hat{a}_1, ..., \hat{a}_N$, $\hat{w}_1, ..., \hat{w}_N$ and $\bar{a}_1, ..., \bar{a}_N$, $\bar{w}_1, ..., \bar{w}_N$ after tuning.

Let us now formulate a mathematical statement of the problem of constructing the Voronoi diagram with the optimal placement of a finite number N of generator points in a bounded set Ω of n-dimensional Euclidian space $E_n (n \geq 2)$ based on methods for solving the OSP problems and with the restored values of its fuzzy parameters.

Assume that Ω is some given bounded set of E_n, $\tau_1, \tau_2, ..., \tau_N$ is a finite set of the generator points in Ω. For the case when the location of points $\tau_1, \tau_2, ..., \tau_N$ in Ω is unknown and they should be placed (selected) in Ω, we introduced above the Voronoi diagram of a finite number of points placed optimally in a bounded set.

By a *Voronoi diagram of a finite number of generator points* $\tau_1, \tau_2, ..., \tau_N$*, located optimally in a bounded set* $\Omega \subset E_n$ *with fuzzy parameters*, we will call a set of Voronoi polyhedrons:

$$Vor(\tau_i) = \{x \in \Omega : c(x, \tau_i)/\bar{w}_i + \bar{a}_i \leq c(x, \tau_j)/\bar{w}_j + \bar{a}_j; \\ j \neq i; i, j = 1, ..., N\}, \quad (10)$$

of the points $\tau_1, \tau_2, ..., \tau_N$, for which the total weighted distance from the points of set Ω to the corresponding generator points $\tau_1, \tau_2, ..., \tau_N$ is the minimal one, i.e. a functional:

$$J(\{\tau_1, ..., \tau_N\}) = \sum_{i=1}^{N} \int_{Vor(\tau_i)} (c(x, \tau_i)/\bar{w}_i + \bar{a}_i) dx \quad (11)$$

takes the minimal value. Here, $\bar{a}_1, ..., \bar{a}_N$, $\bar{w}_1, ..., \bar{w}_N$ are values of the fuzzy parameters of the Voronoi diagram restored after turning.

2.2 Solution Method

We present an approach to constructing Voronoi diagram (10)–(11) based on the use of the mathematical apparatus of the theory of optimal set partitioning [6] and neuro-fuzzy technologies [9]. To do this, we first present the corresponding continuous linear problem of optimal partitioning of a set of n-dimensional Euclidian space E_n into the subsets with previously unknown coordinates of some points, characteristic for each subset, called the subsets' "centers" [6].

Assume that Ω is a bounded Lebesgue measurable set in n-dimensional Euclidian space E_n. We will refer to a set of Lebesgue measurable subsets $\Omega_1, ..., \Omega_N$ from $\Omega \subset \mathrm{E}_n$ as a possible partitioning of set Ω into its non-intersecting subsets $\Omega_1, ..., \Omega_N$, if

$$\bigcup_{i=1}^{N} \Omega_i = \Omega, mes(\Omega_i \cap \Omega_j) = 0, i, j = 1, ..., N, (i \neq j),$$

where $mes(\cdot)$ is a Lebesgue measure.

By $\sum_{\Omega}^{N}$ we denote the class of all possible partitionings of a set Ω into subsets $\Omega_1, ..., \Omega_N$, i.e.

$$\sum_{\Omega}^{N} = \left\{ (\Omega_1, ..., \Omega_i, ..., \Omega_N) : \bigcup_{i=1}^{N} \Omega_i = \Omega,\ mes(\Omega_i \cap \Omega_j) = 0,\ i, j{=}1, ..., N,\ i \neq j \right\}.$$

We introduce a functional

$$F(\{\Omega_1, ..., \Omega_N\}, \{\tau_1, ..., \tau_N\}) = \sum_{i=1}^{N} \int_{\Omega_i} (c(x,\ \tau_i)/\bar{w}_i + \bar{a}_i) dx, \tag{12}$$

where $c(x, \tau_i)$ is a given real-valued function defined and bounded on $\Omega \times \Omega$, measurable by argument $x = (x^{(1)}, x^{(2)}, ..., x^{(n)}) \in \Omega$ for any fixed $\tau_i = (\tau_i^{(1)}, \tau_i^{(2)}, ..., \tau_i^{(n)}) \in \Omega$, $i = 1, 2, ..., N$; $\bar{a}_1, ..., \bar{a}_N$, $\bar{w}_1, ..., \bar{w}_N$ are values of the fuzzy parameters restored after turning.

Here and in the following, we consider Lebesgue integrals. We assume that the measure of the set of boundary points of subsets Ω_{i}, $i = 1, 2, ..., N$ is equal to zero. To the following problem [6] we will refer to as *a continuous linear problem of the optimal partitioning of a set* Ω from E_n *into its non-intersecting subsets* $\Omega_1, ..., \Omega_N$ *(among which there could be the empty ones) with finding coordinates of centers* $\tau_1, \tau_2, ..., \tau_N$ *of these subsets respectively.*

Problem A. It is required to find a partition of a set Ω into N Lebesgue-measurable subsets $\Omega_1^*, ..., \Omega_{\mathrm{N}}^*$ and coordinates of centers $\tau_1, ..., \tau_N$ of these subsets in the domain Ω , which provide $\min\limits_{\substack{\{\Omega_1, ..., \Omega_{\mathrm{N}}\} \in \sum_{\Omega}^{\mathrm{N}} \\ \{\tau_1, ..., \tau_N\} \in \Omega^{\mathrm{N}}}} F(\{\Omega_1, ..., \Omega_{\mathrm{N}}\},$ $\{\tau_1, ..., \tau_{\mathrm{N}}\})$, where functional $F(\{\Omega_1, ..., \Omega_{\mathrm{N}}\}, \{\tau_1, ..., \tau_{\mathrm{N}}\})$ has the form (12).

We define a pair $(\{\Omega_1^*, ..., \Omega_{\mathrm{N}}^*\}, \{\tau_1^*, ..., \tau_{\mathrm{N}}^*\})$, which minimizes functional (12) on a set $\sum_{\Omega}^{N} \times \Omega^{\mathrm{N}}$ as an *optimal solution* to Problem A. Wherein, in Problem A we define a partition $\{\Omega_1^*, ..., \Omega_{\mathrm{N}}^*\} \in \sum_{\Omega}^{\mathrm{N}}$ as an *optimal partitioning* of set $\Omega \subset \mathrm{E_n}$ into N subsets, and a set $\tau^* = (\tau_1^*, ..., \tau_N^*) \in \underbrace{\Omega \times ... \times \Omega}_{N} = \Omega^{\mathrm{N}}$ of centers $\tau_i^* \in \Omega_{\mathrm{i}}^*$, $i = 1, 2, ..., N$ as *optimal centers* of subsets Ω_{i}^*, $i = 1, 2, ..., N$.

To construct a method for solving Problem A, we introduce characteristic functions Ω_{i}:

$$\lambda_i(x) = \begin{cases} 1, & x \in \Omega_{\mathrm{i}}, \\ 0, & x \in \Omega \backslash \Omega_{\mathrm{i}},\ i = 1, ..., N. \end{cases}$$

of subsets Ω_{i}, $i = 1, 2, ..., N$ respectively. The following Theorems 1 and 2, which will be used later to formulate an algorithm for solving Problem A, are given in [6].

Theorem 1. For a possible partitioning $\{\Omega_1^*, ..., \Omega_{\mathrm{N}}^*\} \in \sum_{\Omega}^{\mathrm{N}}$ of set Ω for each fixed $\tau \in \Omega$ would be an optimal partitioning for Problem A, it is necessary and sufficient that $c(x, \tau_i)/\bar{w}_i + \bar{a}_i \leq c(x, \tau_j)/\bar{w}_j + \bar{a}_j$, $i, j = 1, ..., N, j \neq i$, almost everywhere (a.e) for $x \in \Omega_{\mathrm{i}}^*$. Wherein, at points $x \in \Omega$ belonging to the optimal boundary of subsets Ω_{i}^* and Ω_{j}^*, an equal sign is reached in the above inequality.

Theorem 2. Components of a characteristic vector-function $\lambda_*(x) = (\lambda_{*1}(x), ..., \lambda_{*i}(x), ..., \lambda_{*N}(x))$, corresponding to an optimal partitioning $\{\Omega_1^*, ..., \Omega_N^*\}$ for Problem A are determined as follows:

$$\lambda_{*i}(x) = \begin{cases} 1, & c(x, \tau_{*i})/\bar{w}_i + \bar{a}_i \leq c(x, \tau_{*j})/\bar{w}_j + \bar{a}_j, \\ & j = 1, ..., N, i \neq j \text{ a.e. for } x \in \Omega, \text{ then } x \in \Omega_{*i}, \\ 0 & \text{otherwide,} \end{cases} \quad (13)$$

for $i = 1, 2, ..., N$ and almost all $x \in \Omega$ where $\tau_1^*, ..., \tau_N^* \in \Omega^N$ is an optimal solution to the problem

$$G(\tau) = \int_\Omega \min_{i=1,...,N} [c(x, \tau_i)/\bar{w}_i + \bar{a}_i] dx \rightarrow \min, \tau \in \Omega^N. \quad (14)$$

Next, we present an algorithm for solving Problem A which is based on Theorems 1 and 2 as well as on one variant of the generalized gradient descent method with space expansion in the direction of the difference of two successive generalized antigradients (the Shor's r-algorithm [12]).

The essence of the generalized gradient descent methods with space expansion is based on constructing linear operators changing the metric of space while successive approximations and on choosing the descent direction corresponding to the antigradient in space with a new metric. In the iteration formula of the r-algorithm [12] which has the form

$$\tau^{[k+1]} = \tau^{[k]} - h_k B_{k+1}^\tau \left[B_{k+1}^\tau\right]^T g_G(\tau^{[k]}), k = 0, 1, 2, ..., \quad (15)$$

B_{k+1}^τ is an operator that maps the transformed space into the main space E_n (wherein, $B_0^\tau = I$ is an indentity matrix); h_k is a stepping factor which is chosen from the condition of the minimum of function G in direction $B_{k+1} B_{k+1}^T g_G(\tau^{[k]})$; $g_G(\tau^{[k]})$ is a generalized gradient of function $G(\tau)$ at a point $\tau^{[k]}$.

In this study, the r-algorithm is used in the H-form [12] (H_k is a symmetric matrix such that $H_k = B_k B_k^T$), for which iteration formula (15) becomes as follows

$$\tau^{[k+1]} = \tau^{[k]} - h_k \frac{H_{k+1} g_G(\tau^{[k]})}{\sqrt{(H_{k+1} g_G(\tau^{[k]}), g_G(\tau^{[k]}))}}, k = 0, 1, 2, ...,$$

where

$$H_{k+1} = H_k + (1/\alpha_k^2 - 1) \frac{H_k \Delta_k \Delta_k^T H_k}{(H_k \Delta_k, \Delta_k)}; \Delta_k = g_G(\tau^{[k]}) - g_G(\tau^{[k-1]}).$$

Coefficient α_k of space expansion is taken to be equal to 3. For the stepping factor h_k, the adaptive adjustment method described in [12] is used. We define the i-th component of a vector of generalized gradient $g_G^\tau(\tau) = (g_G^{\tau_1}(\tau), ..., g_G^{\tau_i}(\tau), ..., g_G^{\tau_N}(\tau))$ of function

$$G(\tau_1^{(1)}, ..., \tau_1^{(n)}; ...; \tau_i^{(1)}, ..., \tau_i^{(n)}; ...; \tau_N^{(1)}, ..., \tau_N^{(n)}) = \int_\Omega \min_{i=1,...,N} [c(x^{(1)}, ..., x^{(n)}; \tau_i^{(1)}, ..., \tau_i^{(n)})/w_i + a_i] dx^{(1)} ... dx^{(n)}$$

at a point $\tau = \tau_1^{(1)}, ..., \tau_N^{(n)}$ as follows:

$$g_G^{\tau_i}(\tau) = g_G^{\tau_i}(\tau_1^{(1)}, ..., \tau_N^{(n)}) = \frac{1}{\bar{w}_i} \int_\Omega g_c^{\tau_i}(x^{(1)}, ..., x^{(n)}; \tau_i^{(1)}, ..., \tau_i^{(n)}) \\ \times \lambda_i(x^{(1)}, ..., x^{(n)}) dx^{(1)}...dx^{(n)}, i = 1, ..., N, \tag{16}$$

where $g_c^{\tau_i}(x, \tau)$ is the i-th component of the N-dimensional vector of generalized gradient $g_c^{\tau}(x, \tau)$ of function $c(x, \tau_i)$ at a point $\tau = (\tau_1, ..., \tau_i, ..., \tau_N)$ (here, $\tau_i = (\tau_i^{(1)}, ..., \tau_i^{(n)})$); for a fixed $x \in \Omega$, it has the form

$$g_c^{\tau_i}(x^{(1)}, ..., x^{(n)}; \tau_1^{(1)}, ..., \tau_N^{(n)}) = \begin{pmatrix} g_c^{\tau_i^{(1)}}(x^{(1)}, ..., x^{(n)}; \tau_1^{(1)}, ..., \tau_N^{(n)}) \\ \dots\dots\dots\dots\dots\dots \\ g_c^{\tau_i^{(n)}}(x^{(1)}, ..., x^{(n)}; \tau_1^{(1)}, ..., \tau_N^{(n)}) \end{pmatrix}.$$

Remark. In the case when $c(x, \tau_i)$ is the Euclidian metrics, the i-th component of the N-dimensional vector of generalized gradient $g_c^{\tau}(x, \tau)$ of function $c(x, \tau_i) = \sqrt{\sum_{i=1}^{n}(x^{(k)} - \tau_i^{(k)})^2}$ at a point $\tau = (\tau_1, ..., \tau_i, ..., \tau_N)$, for a fixed $x \in \Omega$, has the following form:

$$g_c^{\tau_i}(x, \tau) = \begin{pmatrix} (x^{(1)} - \tau_i^{(1)}) / \left(\bar{w}_i \sqrt{\sum_{k=1}^{n}(x^{(k)} - \tau_i^{(k)})^2} \right) \\ \dots\dots\dots\dots\dots\dots\dots\dots \\ (x^{(n)} - \tau_i^{(n)}) / \left(\bar{w}_i \sqrt{\sum_{k=1}^{n}(x^{(k)} - \tau_i^{(k)})^2} \right) \end{pmatrix}.$$

2.3 Algorithm

We enclose a domain Ω in an n-dimensional parallelepiped Π, whose sides are parallel to the axes of the Cartesian coordinate system. We cover the parallelepiped Π with a rectangular grid and specify the initial approximation $\tau = \tau^{[0]}$. Then we calculate values $\lambda^{[0]}(x)$ at the grid nodes according to formulas (13) for $\tau = \tau^{[0]}$; we calculate values of $g_G(\tau)$ according to formulas (16) under $\lambda(x) = \lambda^{[0]}(x)$, $\tau = \tau^{[0]}$, and under restored values of fuzzy parameters $\bar{a}_1, ..., \bar{a}_N$, $\bar{w}_1, ..., \bar{w}_N$, which, in their turn, are calculated by formulas (6)–(9). We specify the initial test stepping factor $h_0 > 0$ of the r-algorithm and fine

$$\tau^{[1]} = \tau^{[0]} - h_0 \frac{H_1 g_G(\tau^{[0]})}{\sqrt{(H_1 g_G(\tau^{[0]}), g_G(\tau^{[0]}))}},$$

where P_Π is an operator of projecting onto Π.

We pass to the next step. Suppose that certain values $\tau^{[k]}$, $\lambda^{[k-1]}(x)$ have been obtained at the grid nodes as a result of calculations after k steps of the algorithm ($k = 1, 2, ...$). Describe the $[k+1]$-th step.

1. Calculate values of $\lambda^{[k]}(x)$ at the grid nodes according to formulas (13) for $\tau = \tau^{[k]}$.
2. Find values of $g_G(\tau)$ according to formulas (16), for $\lambda(x) = \lambda^{[k]}(x)$, $\tau = \tau^{[k]}$.

3. Carry out the $[k+1]$-th iteration of the r-algorithm in the H-form [12] by the formula

$$\tau^{[k+1]} = P_{\Pi}\left(\tau^{[k]} - h_k \frac{H_{k+1}g_G(\tau^{[k]})}{\sqrt{(H_{k+1}g_G(\tau^{[k]}), g_G(\tau^{[k]}))}}\right).$$

4. If the conditions

$$\left\|\tau^{[k]} - \tau^{[k+1]}\right\| \leq \varepsilon, \varepsilon > 0, \quad (17)$$

does not hold true, pass to the next, $[k+2]$-th step of the algorithm; otherwise, pass to p. 5.

5. Assume that $\lambda_*(x) = \lambda^{[l]}(x)$, $\tau_* = \tau^{[l]}$, where l is the iteration's number at which condition (17) held true.
6. Calculate an optimal value of objective function $G(\tau_*)$ from (14) using the formula $G(\tau_*) = \int_{\Omega} \min_{i=1,...,N} [c(x, \tau_{*i})/\bar{w}_i + \bar{a}_i]dx$.

Thus, as a result of solving Problem A using the described algorithm based on Theorems 1, 2 and on the neurolinguistic identification method for restoring values of fuzzy parameters, we obtain a set of Voronoi polyhedrons (3) of generator points τ_i, $i = 1, ..., N$:

$$Vor(\tau_i) = \{x \in \Omega \subset E_{\rm n} : c(x, \tau_{\rm i})/\bar{w}_{\rm i} + \bar{a}_{\rm i} \leq c(x, \tau_{\rm j})/\bar{w}_{\rm j} + \bar{a}_{\rm j}, \ \ {\rm i, j} = 1, ..., {\rm N}, {\rm j} \neq {\rm i}\}.$$

But, unlike the standard Voronoi diagram (1) in which points $\tau_1, ..., \tau_N$ are fixed, coordinates of generator points $\tau_1, ..., \tau_N$ placed optimally in $\Omega \subset {\rm E_n}$ are found as a solution to the finite-dimensional optimization problem

$$G(\tau) = \int_{\Omega} \min_{i=1,...,N} [c(x, \tau_i)/\bar{w}_i + \bar{a}_i]dx \rightarrow \min, \tau \in \Omega^{\rm N},$$

with an nondifferentiable objective function $G(\tau)$ and restored values of fuzzy parameters.

3 Results and Discussion

To illustrate the described algorithm, a C#-based software has been developed in the Visual Studio 2019 environment, with the help of which the presented below Voronoi diagrams were constructed. Figures 1 and 2 show additively weighted and multiplicatively weighted Voronoi diagrams are shown for $N = 7$ generator points located optimally in domain $\Omega = \{x = (x^{(1)}, x^{(2)}) \in {\rm R}^2 : 0 \leq x^{(\rm i)} \leq 10, i = 1, 2\}$; these Voronoi diagrams were obtained as a result of solving Problem **A** using the described algorithm for zero initial approximations $\tau_i = (0; 0), i = 1, ..., N, N = 7$.

For the additive-weighted Voronoi diagram with fuzzy parameter values $\hat{a}_i$, $i = 1, ..., N$ restored before turning (Fig. 1a)), we obtained the following results. After 92 iterations of the described algorithm, the minimal value of the objective functional of the direct problem is F=297.23; the optimal coordinates of

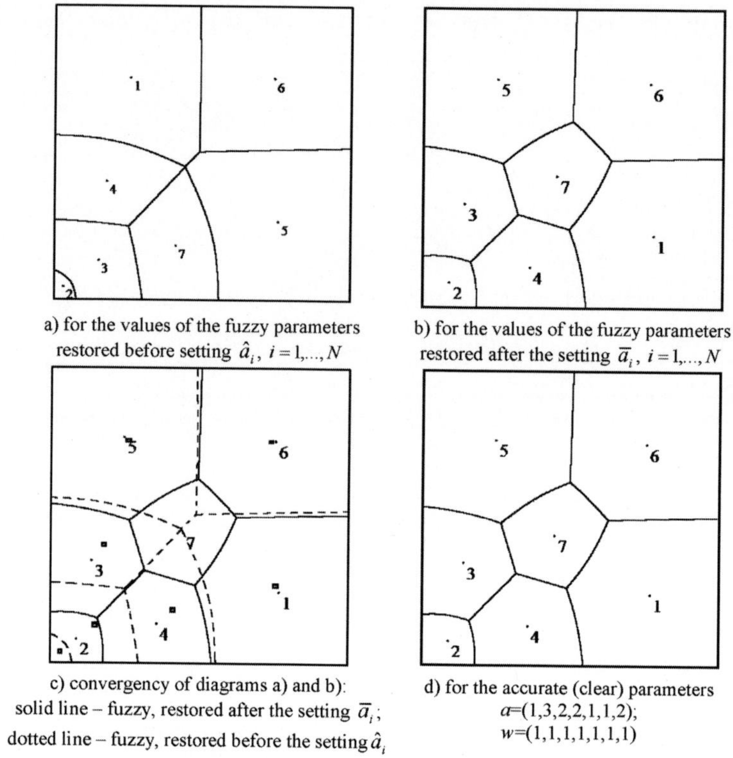

Fig. 1. Additively weighted Voronoi diagrams with the optimal placement of seven generator points: a) for fuzzy parameter values $\hat{a}_i$, $i = 1, ..., N$ restored before tuning; b) for fuzzy parameter values $\bar{a}_i$, $i = 1, ..., N$ restored after tuning; c) superposition of diagrams a) and b): solid line – fuzzy, restored after tuning $\bar{a}_i$, dotted line – fuzzy, restored before tuning $\hat{a}_i$; d) for exact (clear) parameters a = (1, 3, 2, 2, 1, 1, 2); w = (1, 1, 1, 1, 1, 1, 1)

the generator points are $\tau_1 = (2.49; 7.59)$, $\tau_2 = (0.28; 0.42)$, $\tau_3 = (1.46; 1.32)$, $\tau_4 = (1.72; 4.05)$, $\tau_5 = (7.56; 2.69)$, $\tau_6 = (7.41; 7.58)$, $\tau_7 = (4.06; 1.84)$. For the additive-weighted Voronoi diagram with fuzzy parameter values $\bar{a}_i$, $i = 1, ..., N$ restored after turning (Fig. 1b)) we obtained the following results. After 91 iterations of the described algorithm, the minimal values of the objective functional of the direct problem is F=295.01; the optimal coordinates of the generator points are $\tau_1 = (7.69; 2.43)$, $\tau_2 = (0.84; 0.81)$, $\tau_3 = (1.34; 3.48)$, $\tau_4 = (3.54; 1.33)$, $\tau_5 = (2.42; 7.68)$, $\tau_6 = (7.55; 7.56)$, $\tau_7 = (4.42; 4.43)$.

To compare the results, a superposition of the additively weighted Voronoi diagrams is shown in Fig. 1c) for the cases of fuzzy values of parameters restored before tuning and after one; in Fig. 1d) the additively weighted Voronoi diagram is shown with clear parameter values. For clear values of the parameters, the following results were obtained after 91 iterations: the minimal value of the objective functional of the direct problem is F = 295.01; the optimal coordinates

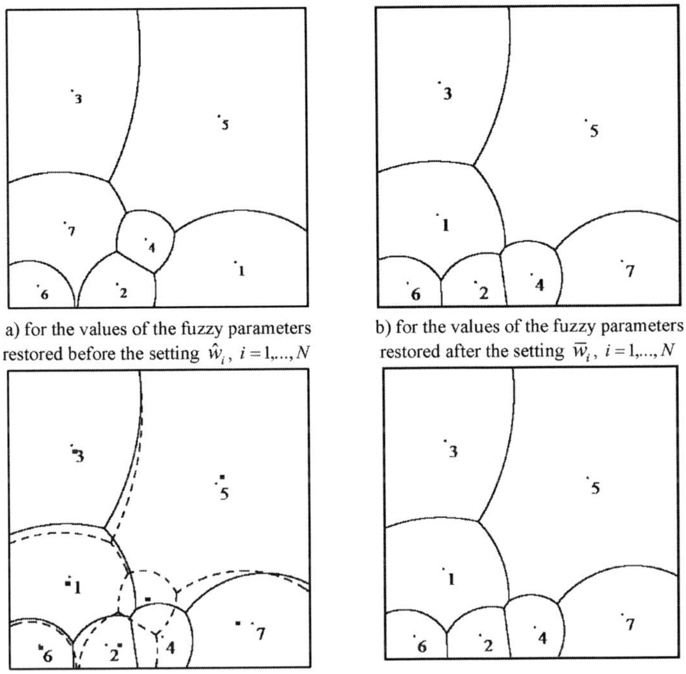

a) for the values of the fuzzy parameters restored before the setting $\hat{w}_i$, $i = 1,...,N$

b) for the values of the fuzzy parameters restored after the setting $\bar{w}_i$, $i = 1,...,N$

c) convergency of diagrams a) and b): solid line - fuzzy, restored after the setting $\bar{w}_i$; dotted line - fuzzy, restored before the setting $\hat{w}_i$

d) for the accurate (clear) parameters $a=(0,0,0,0,0,0,0)$; $w=(2,1,3,1,4,1,2)$

Fig. 2. Multiplicatively weighted Voronoi diagrams with the optimal placement of seven generator points: a) for fuzzy parameter values $\hat{w}_i$, $i = 1, ..., N$ restored before tuning; b) for fuzzy parameter values $\bar{w}_i$, $i = 1, ..., N$ restored after tuning; c) superposition of diagrams a) and b): solid line – fuzzy, restored after tuning $\bar{w}_i$, dotted line – fuzzy, restored before tuning $\hat{w}_i$; d) for exact (clear) parameters a = (0, 0, 0, 0, 0, 0, 0); w = (2, 1, 3, 1, 4, 1, 2)

of the generator points are $\tau_1 = (7.69; 2.43)$, $\tau_2 = (0.84; 0.81)$, $\tau_3 = (1.34; 3.48)$, $\tau_4 = (3.54; 1.33)$, $\tau_5 = (2.42; 7.68)$, $\tau_6 = (7.55; 7.56)$, $\tau_7 = (4.42; 4.43)$. For the multiplicative-weighted Voronoi diagram with fuzzy values of parameters $\hat{w}_i$, $i = 1, ..., N$ restored before turning (Fig. 2a)), the following results were obtained after 84 iterations: the minimal value of the objective functional of the direct problem is F = 65.63; the optimal coordinates of the generator points are $\tau_1 = (7.96; 1.52)$, $\tau_2 = (3.74; 0.77)$, $\tau_3 = (2.23; 7.35)$, $\tau_4 = (4.67; 2.35)$, $\tau_5 = (7.05; 6.34)$, $\tau_6 = (1.04; 0.72)$, $\tau_7 = (2.02; 2.91)$. For the multiplicative-weighted Voronoi diagram with fuzzy values of parameters $\hat{w}_i$, $i = 1, ..., N$ restored after turning (Fig. 2b)), the following results were obtained after 86 iterations: the minimal value of the objective functional of the direct problem is F = 64.55; the optimal coordinates of the generator points are $\tau_1 = (2.07; 3.10)$, $\tau_2 = (3.74; 0.79)$, $\tau_3 = (2.07; 7.52)$, $\tau_4 = (5.09; 1.04)$, $\tau_5 = (6.91; 6.21)$, $\tau_6 = (0.96; 0.77)$, $\tau_7 = (8.11; 1.49)$.

To compare the results, a superposition of the multiplicatively weighted Voronoi diagrams is shown in Fig. 2c) for the cases of fuzzy values of parameters restored before tuning and after one; in Fig. 2d) the multiplicatively weighted Voronoi diagram is shown with clear parameter values. For clear values of the parameters, the following results were obtained after 86 iterations: the minimal value of the objective functional of the direct problem is F=64.55; the optimal coordinates of the generator points are $\tau_1 = (2.07; 3.10)$, $\tau_2 = (3.74; 0.79)$, $\tau_3 = (2.07; 7.52)$, $\tau_4 = (5.09; 1.04)$, $\tau_5 = (6.91; 6.21)$, $\tau_6 = (0.96; 0.77)$, $\tau_7 = (8.11; 1.49)$.

4 Conclusions

The study has resulted in an algorithm developed to construct additive and multiplicative Voronoi diagrams under uncertainty for the case when some parameters may be fuzzy, inaccurate, or linguistically specified. To restore fuzzy parameters, the method of neurolinguistic identification of complex unknown dependencies was used. Further, to construct additively and multiplicatively weighted Voronoi diagrams with optimal placement of a finite number N of generator points in a bounded set Ω of n-dimensional Euclidean space $E_n (n \geq 2)$, we formulated the corresponding continuous problem of optimal partitioning of a set from E_n into its non-intersecting subsets with a partition quality criterion that provides the corresponding type of Voronoi diagram.

To solve this problem, a mathematical and algorithmic apparatus was used for solving problems of optimal partitioning of a set of n-dimensional Euclidian space E_n. The apparatus is based on an unified approach, which consists in reducing the initial infinite-dimensional optimization problem in a certain way (for example, via the Lagrange functional) to non-smooth, as a rule, finite-dimensional optimization problems; for the numerical solution of the last ones modern effective methods of non-differentiable optimization are used – various variants of the Shor's r-algorithm. The same r-algorithm was applied at the stage of turning fuzzy parameters of Voronoi diagrams in the method of neurolinguistic identification of complex nonlinear dependencies.

The C#-based software has been developed to implement the proposed algorithm in the Microsoft Visual Studio 2019 environment. The results of constructing of both multiplicatively and additively weighted Voronoi diagrams with fuzzy parameters obtained using the developed software are presented. While comparing the results of constructing both additively and multiplicatively weighted Voronoi diagrams, we can see (Fig. 1, 2) that the obtained optimal solutions for clear parameter and for fuzzy parameters restored by the method of neurolinguistic identification after turning coincide with a sufficient degree of accuracy.

Thus, it may be concluded that the synthesis of the methods of optimal set partitioning in the n-dimensional Euclidian space and neuro-fuzzy technologies allows us to construct successfully the additively wieghted and multiplicatively weighted Voronoi diagrams with optimally placed generator points under uncertainty.

References

1. Atamturk, A., Nemhauser, G., Savelsbergh, M.: A combined lagrangian, linear programming and implication heuristic for large-scale set partitioning problems. J. Heuristics **1**, 247–259 (1996). https://doi.org/10.1007/BF00127080
2. Aurenhammer, F., Klein, R., Lee, D.T.: Voronoi Diagrams and Delaunay Triangulations. Springer, New York (2013). https://doi.org/10.1007/978-1-4939-2864-4507
3. Bezdek, J., Dubois, D., Prade, H. (eds.): Fuzzy Sets in Approximate Reasoning and Information Systems. Springer, Boston (1999). https://doi.org/10.1007/978-1-4615-5243-7
4. Borisov, V., Kruglov, V., Fedulov, A.: Fuzzy models and networks. Hotline-Telecom, Moscow (2016)
5. Kacprzyk, J., Pedrycz, W. (eds.): Springer Handbook of Computational Intelligence. Springer, Heidelberg (2015). https://doi.org/10.1007/978-3-662-43505-2
6. Kiseleva, E.: The emergence and formation of the theory of optimal set partitioning for sets of the n-dimensional euclidean space. Theory and application. J. Autom. Inf. Sci. **50**(9), 1–24 (2018). https://doi.org/10.1615/JAutomatInfScien.v50.i9.10
7. Kiseleva, E., Koriashkina, L.: Theory of continuous optimal set partitioning problems as a universal mathematical formalism for constructing voronoi diagrams and their generalizations. Algorithms for constructing Voronoi diagrams based on the theory of optimal set partitioning. Cybern. Syst. Anal. **51**(4), 489–499 (2015). https://doi.org/10.1007/s10559-015-9740-y
8. Kiseleva, E., Koriashkina, L.: Theory of continuous optimal set partitioning problems as a universal mathematical formalism for constructing Voronoi diagrams and their generalizations. Theoretical foundations. Cybern. Syst. Anal. **51**(3), 325–335 (2015). https://doi.org/10.1007/s10559-015-9725-x
9. Kiseleva, E., Prytomanova, O., Zhuravel, S.: Valuation of startups investment attractiveness based on neuro-fuzzy technologies. J. Autom. Inf. Sci. **48**(9), 1–22 (2016). https://doi.org/10.1615/JAutomatInfScien.v48.i9.10
10. Okabe, A., Boots, B., et al.: Spatial Tessellations: Concepts and Applications of Voronoi Diagrams. Wiley, West Sussex (2000)
11. Preparata, F., Sheimos, M.: Computational Geometry: An Introduction. Springer, New York (1985). https://doi.org/10.1007/978-1-4612-1098-6
12. Shor, N.: Nondifferentiable Optimization and Polynomial Problems. Springer, Boston (1998). https://doi.org/10.1007/978-1-4757-6015-6
13. Siddique, N.: Fuzzy Sets Theory and Its Applications. Springer, Cham (2014). https://doi.org/10.1007/978-3-319-02135-5
14. Voskoglou, M. (ed.): Fuzzy Sets, Fuzzy Logic and Their Applications. MDPI, Basel (2020). http://www.mdpi.com/journal/mathematics/special_issues/
15. Zadeh, L.: The concept of a linguistic variable and its application to approximate reasoning. Inf. Sci. **8**(3), 199–249 (1975). https://doi.org/10.1016/0020-0255(75)90036-5
16. Zadeh, L., Kacprzyk, J. (eds.): Computing with Words in Information/Intelligent Systems 2. Applications. Physica, Heidelberg (1999). https://doi.org/10.1007/978-3-7908-1872-7
17. Zimmerman, H.J.: Fuzzy Sets Theory and Its Applications. Springer, Dordrecht (2001). https://doi.org/10.1007/978-94-010-0646-0

Author Index

S. Babichev et al. (Eds.): ISDMCI 2020, AISC 1246, pp. 729–731, 2021.
https://doi.org/10.1007/978-3-030-54215-3

MIX
Papier aus verantwortungsvollen Quellen
Paper from responsible sources
FSC® C105338

Printed by Books on Demand, Germany